Lecture Notes in Computer Science 6597

Commenced Publication in 1973
Founding and Former Series Editors:
Gerhard Goos, Juris Hartmanis, and Jan van Leeuwen

Yuval Ishai (Ed.)

Theory of Cryptography

8th Theory of Cryptography Conference, TCC 2011
Providence, RI, USA, March 28-30, 2011
Proceedings

Springer

Volume Editor

Yuval Ishai
Technion
Computer Science Department
Haifa 32000, Israel
E-mail: yuvali@cs.technion.ac.il

ISSN 0302-9743 ISSN 1611-3349 (eBook)
ISBN 978-3-642-19570-9 ISBN 978-3-642-19571-6 (eBook)
DOI 10.1007/978-3-642-19571-6
Springer Heidelberg Dordrecht London New York

Library of Congress Control Number: 2011921972

CR Subject Classification (1998): E.3, C.2, D.4.6, K.6.5, G.2, I.1

LNCS Sublibrary: SL 4 – Security and Cryptology

Typesetting: Camera-ready by author, data conversion by Scientific Publishing Services, Chennai, India

Printed on acid-free paper

Springer is part of Springer Science+Business Media (www.springer.com)

Preface

These are the proceedings of TCC 2011, the 8th Theory of Cryptography Conference, which was held in Brown University, Providence, RI, during March 28–30, 2011. The conference was sponsored by the International Association for Cryptologic Research (IACR). The General Chair was Anna Lysyanskaya.

The Program Committee, consisting of 18 members, selected 35 of 108 submissions for presentation at the conference. Two closely related submissions were presented by a single joint talk, resulting in a technical program with 34 talks. The program also included two invited lectures: by Luca Trevisan, titled "Dense Model Theorems and Their Applications," and by Rafael Pass, titled "Concurrent Security and Non-Malleability." The conference featured a rump session for informal short presentations of new results, chaired by Tal Malkin.

The Best Student Paper Award was given to Stefano Tessaro for his paper "Security Amplification for the Cascade of Arbitrarily Weak PRPs: Tight Bounds via the Interactive Hardcore Lemma." Three additional papers were selected for invitation to the *Journal of Cryptology*: "Input Locality and Hardness Amplification," by Andrej Bogdanov and Alon Rosen, "PCPs and the Hardness of Generating Private Synthetic Data," by Jonathan Ullman and Salil Vadhan, and "Round-Optimal Password-Based Authenticated Key Exchange," by Jonathan Katz and Vinod Vaikuntanathan.

I am grateful to all those who helped make this conference possible. First and foremost I wish to thank all authors who contributed to the excellent pool of submissions. There were more high-quality submissions than we could fit into the program, a clear indication that the TCC conference is going strong. I deeply thank the Program Committee members for their dedication and hard work. The initial review stage was followed by intensive discussions which helped shape the program and resulted in additional feedback to the authors. Many of the reviews were provided by external reviewers whose names are listed in the following page. I thank them all for their time and effort.

Special thanks go to Anna Lysyanskaya, who as a General Chair was in charge of the local arrangements and also doubled as a Program Committee member, to Geetha Jagannathan for administering the conference website, and to Tal Malkin for chairing the rump session. I thank Vipul Goyal, Krzysztof Pietrzak, Mike Rosulek, and Dominique Unruh for their help with shepherding and verifying accepted papers. We had the pleasure of using the submissions and review software developed by Shai Halevi. I thank Shai for providing quick technical assistance whenever it was needed. Finally, I am indebted to Oded Goldreich, the Chair of the TCC Steering Committee, as well as the recent TCC Chairs Tal Rabin, Shai Halevi, Salil Vadhan, Ran Canetti, Omer Reingold, and Daniele Micciancio for their help and advice.

January 2011 Yuval Ishai

TCC 2011

The 8th Theory of Cryptography Conference

Brown University, Providence, RI
March 28–30, 2011

Sponsored by the *International Association for Cryptologic Research (IACR).*

General Chair

Anna Lysyanskaya	Brown University, USA

Program Chair

Yuval Ishai	Technion and UCLA, Israel and USA

Program Committee

Benny Applebaum	Weizmann Institute and Tel-Aviv University, Israel
Boaz Barak	Microsoft Research and Princeton University, USA
Melissa Chase	Microsoft Research Redmond, USA
Ronald Cramer	CWI Amsterdam and Leiden University, The Netherlands
Juan Garay	AT&T Labs – Research, USA
Vipul Goyal	Microsoft Research India
Shai Halevi	IBM Research, USA
Yuval Ishai	Technion and UCLA, Israel and USA
Hugo Krawczyk	IBM Research, USA
Anna Lysyanskaya	Brown University, USA
Vadim Lyubashevsky	INRIA and ENS Paris, France
Mohammad Mahmoody	Cornell University, USA
Chris Peikert	Georgia Institute of Technology, USA
Krzysztof Pietrzak	CWI Amsterdam, The Netherlands
Manoj Prabhakaran	University of Illinois, Urbana-Champaign, USA
Guy Rothblum	Princeton University, USA
Gil Segev	Microsoft Research SVC, USA
Dominique Unruh	Saarland University, Germany

External Reviewers

Foteini Baldimtsi
Amos Beimel
Sasha Berkoff
Bruno Blanchet
Xavier Boyen
Elette Boyle
Zvika Brakerski
Ran Canetti
Ignacio Cascudo
André Chailloux
Nishanth Chandran
Seung Geol Choi
Ashish Choudhary
Kai-Min Chung
Véronique Cortier
Dana Dachman-Soled
Ivan Damgård
Yevgeniy Dodis
Orr Dunkelman
Omid Etesami
Sebastian Faust
Serge Fehr
Dario Fiore
Matthias Fitzi
Georg Fuchsbauer
Sebastian Gajek
Clint Givens
Dov Gordon
Ronen Gradwohl
Jens Groth
Iftach Haitner
Gerhard Hancke
Kristiyan Haralambiev
Moritz Hardt
Danny Harnik
Carmit Hazay
Nadia Heninger
Martin Hirt
Dennis Hofheinz
Susan Hohenberger
Thomas Holenstein
Abhishek Jain
Charanjit Jutla
Yael Tauman Kalai
Seny Kamara
Aniket Kate
Jonathan Katz
Markulf Kohlweiss
Vladimir Kolesnikov
Boris Köpf
Daniel Kraschewski
Sara Krehbiel
Steve Kremer
Ralf Kuesters
Ranjit Kumaresan
Eyal Kushilevitz
Allison Lewko
Huijia Rachel Lin
Yehuda Lindell
Eran Omri
Feng-Hao Liu
Hemanta Maji
Sarah Meiklejohn
Ilya Mironov
Tal Moran
Daisuke Moriyama
Steven A. Myers
Jesper Buus Nielsen
Tatsuaki Okamoto
Claudio Orlandi
Ilan Orlov
Carles Padró
Rafael Pass
Arpita Patra
Tal Rabin
Dominik Raub
Mariana Raykova
Alon Rosen
Mike Rosulek
Arnab Roy
Amit Sahai
Christian Schaffner
Aaron Segal
Adam Smith
Marc Stevens
Madhur Tulsiani
Yevgeniy Vahlis
Vinod Vaikuntanathan
Muthu Venkitasubramaniam
Oriol Farrás Ventura
Damien Vergnaud
Ivan Visconti
Bogdan Warinschi
Brent Waters
Daniel Wichs
Jürg Wullschleger
Arkady Yerukhimovich
Hila Zarosim
Hong-Sheng Zhou

Table of Contents

Hardness Amplification

Invited Talk 1

Leakage Resilience

Tamper Resilience

Encryption

Composable Security

Secure Computation

Privacy

Coin Tossing and Pseudorandomness

Invited Talk 2

Black-Box Constructions and Separations

Black-Box Separations

Input Locality and Hardness Amplification

Andrej Bogdanov[1,⋆] and Alon Rosen[2,⋆⋆]

[1] Dept. of CSE and ITCSC, Chinese Univ. of Hong Kong
andrejb@cse.cuhk.edu.hk
[2] Efi Arazi School of Computer Science, IDC Herzliya
alon.rosen@idc.ac.il

Abstract. We establish new hardness amplification results for one-way functions in which each input bit influences only a small number of output bits (a.k.a. input-local functions). Our transformations differ from previous ones in that they approximately preserve input locality and at the same time retain the input size of the original function.

Let $f\colon \{0,1\}^n \to \{0,1\}^m$ be a one-way function with input locality d, and suppose that f cannot be inverted in time $\exp(\tilde{O}(\sqrt{n} \cdot d))$ on an ε-fraction of inputs. Our main results can be summarized as follows:

- If f is injective then it is equally hard to invert f on a $(1-\varepsilon)$-fraction of inputs.
- If f is regular then there is a function $g\colon \{0,1\}^n \to \{0,1\}^{m+O(n)}$ that is $d + O(\log^3 n)$ input local and is equally hard to invert on a $(1-\varepsilon)$-fraction of inputs.

A natural candidate for a function with small input locality and for which no sub-exponential time attacks are known is Goldreich's one-way function. To make our results applicable to this function, we prove that when its input locality is set to be $d = O(\log n)$ certain variants of the function are (almost) regular with high probability.

In some cases, our techniques are applicable even when the input locality is not small. We demonstrate this by extending our first main result to one-way functions of the "parity with noise" type.

Keywords: one-way function, input locality, hardness amplification, parity with noise.

1 Introduction

In this paper we are interested in amplifying the hardness of inverting a one-way function. Our goal is to do so without significantly deteriorating the function's parallel complexity and/or efficiency. To the best of our knowledge, these objectives are not simultaneously achieved by any of the previous methods for amplifying hardness.

⋆ Research supported by RGC GRF grant 2150617.
⋆⋆ Research supported by BSF grant 2006317 and ISF grant 334/08.

Y. Ishai (Ed.): TCC 2011, LNCS 6597, pp. 1–18, 2011.

Volume Editor

Yuval Ishai
Technion
Computer Science Department
Haifa 32000, Israel
E-mail: yuvali@cs.technion.ac.il

ISSN 0302-9743 ISSN 1611-3349 (eBook)
ISBN 978-3-642-19570-9 ISBN 978-3-642-19571-6 (eBook
DOI 10.1007/978-3-642-19571-6
Springer Heidelberg Dordrecht London New York

Library of Congress Control Number: 2011921972

CR Subject Classification (1998): E.3, C.2, D.4.6, K.6.5, G.2, I.1

LNCS Sublibrary: SL 4 – Security and Cryptology

Typesetting: Camera-ready by author, data conversion by Scientific Publishing Services, Chennai, India

Printed on acid-free paper

Springer is part of Springer Science+Business Media (www.springer.com)

They first present one-way functions with constant *output locality*, meaning that each output bit depends on a most a constant number of input bits [3]. These functions are constructed using *randomized encodings*, a tool that allows them to transform well known candidate one-way functions that have low (but not necessarily constant) parallel complexity into ones with constant output locality. They then go on and show that, in some specific cases, the functions resulting from their randomized encodings also satisfy constant input locality [4].

An alternative source for candidate one-way functions with small input and output locality is given by Goldreich [8]. These candidates are arguably more natural than the ones resulting from the AIK transformations. They also seem to offer a more attractive tradeoff between input length and security (as in many cases randomized encodings necessitate a significant blow up in the input size of the original function). Goldreich's constructions are quite general, and allow flexibility in the choice of the function, both in terms of the way in which inputs are connected to outputs, as well as in the choice of the predicates used to compute the function's output bits. To date, no sub-exponential time inversion algorithm is known for any variant of his functions.

Known hardness amplification methods are not well suited for functions of the above sort. Being inherently sequential, the GILVZ and HHR transformations do not preserve parallelism. Yao's transformation, on the other hand, does not increase parallel time, but it does incur a significant loss in efficiency (cf. Lin et al. [15]). This presents us with the challenge of coming up with efficient hardness amplification methods that are well suited for parallelizable functions. Our approach to the problem will be to utilize properties implied by the highly parallel structure of the function, and specifically small input-locality.

1.3 Main Results

Let $f\colon \{0,1\}^n \to \{0,1\}^m$ be a one-way function with input locality d, and suppose that f cannot be inverted in time $\exp(\tilde{O}(\sqrt{n}\cdot d))$ on an ε-fraction of inputs. Our first main result falls into the category of *self-amplification*, meaning that the hardness amplification does not require modifying the underlying function.

Theorem 1 (Self-amplification for injective functions): *Suppose that f is injective. Then, f cannot be inverted in time $\exp(\tilde{O}(\sqrt{n}\cdot d))$ on a $(1-\varepsilon)$-fraction of inputs.*

Based on the ideas used in the proof Theorem 1, we prove an analogous theorem for functions of the "parity with noise" type. Specifically, consider a family, $\{M_n\}$, of $m(n)\times n$ matrices with entries in $\{0,1\}$ and let $p \in [0,1]$ be a parameter. Define a function family $f_n\colon \{0,1\}^n \to \{0,1\}^m$ as $f_n(x,e) = M_n x + e \pmod 2$, where x is a vector chosen uniformly at random from $\{0,1\}^n$, and $e \in \{0,1\}^m$ is a vector of hamming weight at most $2pm$ chosen from the following distribution: Each entry of e is chosen independently from a p-biased distribution, conditioned on e having hamming weight at most $2pm$.

We assume that $\{f_n\}$ is one-way against randomized time $\exp(\tilde{O}(\sqrt{m}))$ on some ε fraction of inputs. We also require that the functions f_n are 1-1. This happens when M_n is a generator matrix of a code of minimum distance $4pm$. In such a case, the input locality of f_n will be as large as $\Omega(n)$. Nevertheless, we can prove the following analogue of Theorem 1.

Theorem 2 (Self-amplification for parity with noise): *Suppose that $\{f_n\}$ is injective. Then, (under appropriate constraints on parameters) $\{f_n\}$ cannot be inverted in randomized time $\exp(\tilde{O}(\sqrt{m}))$ on a $(1-\varepsilon)$-fraction of inputs.*

To make our results applicable to a wider class of functions, we also consider a generalization of Theorem 1 to the case where the function we wish to amplify is regular (every output has the same number of preimages). As before, we assume that the function $f\colon \{0,1\}^n \to \{0,1\}^m$ has input locality d, and that f cannot be inverted in time $\exp(\tilde{O}(\sqrt{n}\cdot d))$ on an ε-fraction of inputs. This time, however, we are not able to prove self-amplification and settle for some increase in output length and input locality, while still preserving input length.

Theorem 3 (Amplification for regular functions): *Suppose that f is regular. Then, there is a function $g\colon \{0,1\}^n \to \{0,1\}^{m+O(n)}$ that is $d + O(\log^3 n)$ input local and that cannot be inverted in time $\exp(\tilde{O}(\sqrt{n}\cdot d))$ on a $(1-\varepsilon)$-fraction of inputs.*

A natural candidate for a function with small input locality and for which no sub-exponential time attacks are known is Goldreich's one-way function [8]. Given a bipartite graph G with n vertices on the left, m vertices on the right, and regular right-degree d_{out} and given a predicate $P\colon \{0,1\}^{d_{\mathsf{out}}} \to \{0,1\}$, the function $f_{G,P}\colon \{0,1\}^n \to \{0,1\}^m$ is defined by setting the ith bit of $f_{G,P}(x)$ to be equal to $P(x_{\Gamma(i,1)}, \ldots, x_{\Gamma(i,d_{\mathsf{out}})})$, where $\Gamma_{(i,j)}$ is the jth neighbor of right vertex i of G. Goldreich proposed setting $m = n$ and considered d_{out} ranging from a constant to $O(\log n)$. He conjectured that when G is a good expander graph and P is randomly chosen, with high probability $f_{G,P}$ is one-way when n is sufficiently large.

We consider instantiations of Goldreich's functions with a certain class of balanced predicates, which we call *d_{out}-parity-blowup* predicates, and assume that G is chosen at random. Relying on Theorem 3 we can prove the following.

Theorem 5 (Amplification for Goldreich's function): *Suppose that for at least half the graphs G, the function $f_{G,P}$ is hard to invert on an ε-fraction of inputs for circuits of size $\exp(\tilde{O}(\sqrt{n}))$. Then there exists a function $g\colon \{0,1\}^n \to \{0,1\}^{n+O(\log(1/\varepsilon))}$ of circuit size $O(n \log n)$ that is hard to invert on a $(1-\varepsilon)$-fraction of inputs by circuits of size $\exp(\tilde{O}(\sqrt{n}))$.*

By observing that parity-blowup predicates can be represented by constant degree polynomials over $GF(2)$ we can apply the randomized encodings of AIK [3], and obtain a function with constant output locality and slightly longer input and output length.

Finally, we state a result that applies in the setting where d_{out} is constant and $m \geq Dn$, where $D = D(d_{\mathsf{out}})$ is a sufficiently large constant. Invoking a recent result of Bogdanov and Qiao [6], we prove that for any P and with high probability over the choice of G if $f_{G,P}$ is hard to invert on an ε fraction of inputs in time $\exp(\tilde{O}(\sqrt{n}))$, then $f_{G,P}$ is hard to invert on a $1-\varepsilon$ fraction of inputs in time $\exp(\tilde{O}(\sqrt{n}))$.

1.4 Applicability

Generally speaking, our results are not applicable to functions that are obtained via the randomized encodings of AIK. This is because these encodings typically incur at least a quadratic blow up in the input size. Thus, even if the original function is exponentially hard to invert, we cannot hope to prove that the resulting function is more than $\exp(O(\sqrt{n}))$ hard to invert (at least not based on the hardness of the original function).

It is conceivable that in some specific cases the randomized encodings can be performed in a way that does not significantly increase the input length of the original function. However, even if such cases exist, we are currently not aware of any natural candidate one-way function that would potentially satisfy Theorem 1's hypothesis. While AIK give several injective functions with constant output locality, none of these seems to have small input locality, and moreover they are all known to be invertible in time less than $\exp(\tilde{O}(\sqrt{n}))$ (e.g., ones that are based on the hardness of factoring and of finding discrete-logarithms). Other, presumably harder to invert, candidates are not known to be injective (though they may be regular, making Theorem 3 applicable).

Nevertheless, we feel that Theorem 1 is worth stating and proving. First of all, the fact that we could not think of any appropriate example does not mean that such does not exist. Secondly, the proof of the theorem contains the core ideas behind our reductions, and gives us the opportunity to present them without any irrelevant complications. Finally, and most importantly, using the main ideas of the theorem, we are able to prove an analogous result for functions of the "parity with noise" type, which are generally not known to be invertible in less than $\exp(O(n/\log n))$ time [5].

As we mentioned above, there is no known sub-exponential time algorithm that succeeds in inverting Goldreich's function on a non-negligible fraction of inputs. Applebaum, Barak, and Wigderson [2] prove that, when based on d-parity blowup predicates, the output of Goldreich's function is pseudorandom against linear functions, low-degree polynomials, and constant-depth circuits. In light of this, it currently seems reasonable to conjecture that no algorithm can invert such variants of the function on a small $\varepsilon = \varepsilon(n)$ fraction of inputs in time $\exp(\tilde{O}(\sqrt{n} \cdot d))$. Under this assumption, we obtain a function with poly-logarithmic input locality and constant output locality that cannot be inverted by algorithms with comparable running time on a significantly larger, $(1-\varepsilon)$, fraction of inputs.

Even though not stated explicitly in Section 1.3, our reductions offer a concrete tradeoff between the running time of the reduction and the error $\varepsilon = \varepsilon(n)$ we are able to amplify from. The actual overhead incurred by the reduction is $\exp(O(\sqrt{n \cdot \log(1/\varepsilon) \cdot d \cdot \log n}))$. Thus, assuming that the original function is hard to invert in roughly this time, we can amplify starting from errors as small as say $\varepsilon(n) = 2^{-n^{O(1)}}$. Note that previous amplification methods are not applicable for such ranges of parameters, even if we assume sub-exponential hardness. This is because the input lengths of the functions resulting from their transformations grows proportionally to $\tilde{O}(1/\varepsilon)$.

1.5 Ideas and Techniques

Our self-amplification result is based on the following simple idea. Suppose f is a 1-1 function with input locality d and x and x' are two inputs that differ in one coordinate. Suppose we can invert $f(x)$. Then with a little bit more work we can invert $f(x')$: By input locality, $f(x)$ and $f(x')$ can differ in at most d coordinates. We change d coordinates of $f(x')$ until we find $f(x)$, recover x, and change x in one coordinate to recover x'.

By repeating this argument r times, we can invert $f(x')$ where x and x' are within distance r using $O(n^{dr})$ invocations to the original inverter. So if we can invert f at x, we can also invert f at any x' within distance r of x. Therefore, assuming f is easy to invert on some set that covers an ε-fraction of $\{0,1\}^n$, we can also invert f at any input within distance r of this set. By setting $r = O(\sqrt{n})$, we obtain Theorem 1, the self-amplification result for 1-1 functions.

Amplifying regular functions. The assumption that f is 1-1 is important in this argument. If f was not 1-1, the inverter could return some other preimage which is very far from x and therefore also far from x'. In Theorem 3 we show that if the function $f\colon \{0,1\}^n \to \{0,1\}^m$ is not 1-1 but regular (i.e. K-to-1 for some K), then there exists a new function $f'\colon \{0,1\}^n \to \{0,1\}^{m'}$, $m' = m+O(n)$ such that if f is hard on an small fraction of inputs, then f' is hard on almost all of its inputs.

The transformation from f to f' effectively isolates inputs by applying an appropriate hash function. Hashing is a standard way to reduce a regular function to a 1-1 function [13,12]. However, applying a pairwise-independent hash increases input locality by $\Omega(\log K)$ (see Section 5.1) and makes Theorem 1 inapplicable when K is large. In Claim 1 we describe a new construction of a hash function which increases input locality only by $O((\log n)^3)$ and maps most preimages of f to unique values. Combining this hash with Theorem 1, we obtain Theorem 3, our amplification result for regular input-local functions.

Parity with noise. In Section 4 we apply our ideas to show self-amplification for functions of the parity with noise type. Although these functions do not have low-input locality, we are able to apply our techniques. The reason is that these functions consists of two parts: A linear component, which is randomly self reducible, and the noise component, which is input-local. By combining an application of Theorem 1 to the noise component with a random self-reduction on the linear component, we prove Theorem 2.

Goldreich's function. As we explain in Section 6, Goldreich's function is unlikely to be 1-1 (except in special cases which are easy to invert), so Theorem 1 does not apply directly. However, we show that when m/n is a sufficiently large constant, if $f(x_1) = f(x_2)$, then x_1 and x_2 must be substantially correlated. Assuming f can be inverted on an ε-fraction of inputs, using our self-reduction from Theorem 1, for most x' we can invert $f(x)$ at some x that is close to x'. The inverse we obtain may not be equal to x, but it will be substantially correlated to x'. Using a result of Bogdanov and Qiao [6], we then recover an inverse for $f(x')$.

Our second application concerns functions $f\colon \{0,1\}^n \to \{0,1\}^m$ where $m = n$, but the degree is $O(\log n)$ and the predicate that f is based on is a "parity blowup" predicate (see Section 6). First we prove that such functions are likely to be at most K-to-1 for some constant K. Using the hash from Claim 1, we obtain a new function f' that is almost 1-1 and is almost as hard to invert. Finally, using the randomized encodings of Applebaum et al. [3], we can transform f' into a function with constant *output* locality at a polylogarithmic cost in the input and output length.

1.6 Open Questions

We believe it is interesting to investigate if our methods apply to a wider class of candidate one-way functions. In Section 6 we show that our amplification methods apply to variants of Goldreich's function where either (1) the degree is constant but the output to input length ratio is sufficiently large, or (2) the function is length-preserving, but the degree is logarithmic (so the function is not output-local) and the predicate is of a special form.

It would be interesting to investigate the range of parameters where the function is length-preserving and the degree is constant. We conjecture that when the predicate is balanced, such functions are "almost 2^{cn}-to-1" for some constant c, in the sense that for most x, $f(x)$ has $2^{cn \pm o(n)}$ preimages. If this was the case, we could apply Theorem 3 (and Corollary 1) to obtain very hard to invert functions with better locality parameters.

1.7 Paper Organization

Section 2 contains the most basic definitions relating to input locality, output locality, and regularity. The proof of Theorem 1, which holds the key ideas for our results, as well as the proof of Theorem 3, which deals with the regular case and involves the construction of a new input local hash function, are included in the main body of the paper. Other results are stated in corresponding sections. Due to lack of space, their proofs are deferred to the full version.

2 Definitions

Let $f\colon \{0,1\}^n \to \{0,1\}^m$ be a function. We say that the ith output $f(x)_i$ *depends* on the jth input x_j if there exists a setting of the inputs $x_1, \ldots, x_{j-1}, x_{j+1}, \ldots, x_n$ such that $f(x_1, \ldots, x_{j-1}, 0, x_{j+1}, \ldots, x_n)_i \neq f(x_1, \ldots, x_{j-1}, 1, x_{j+1}, \ldots, x_n)_i$. We define the *degree* of the jth input to be the number of outputs that depend on the jth input. We say f has *input locality* d if the degree of every input is at most d. We define the *degree* of an output as the number of inputs it depends on and the *output locality* as the maximum degree of an output.

We say that f is *K-to-1* if for every $x \in \{0,1\}^n$, there exist exactly K inputs $x' \in \{0,1\}^n$ such that $f(x') = f(x)$. We say f is *regular* if it is K-to-1 for some K. We say f is *at most K-to-1* (resp., *at least K-to-1*) if for every x there are at most K (resp., at least K) x' such that $f(x') = f(x)$. We say f is *ε-close to K-to-1* if for at least a $(1-\varepsilon)$ fraction of the inputs $x \in \{0,1\}^n$, there exist exactly K inputs $x' \in \{0,1\}^n$ such that $f(x') = f(x)$.

In this work we consider both uniform and non-uniform constructions of one-way functions. The security of such functions can be defined against deterministic, randomized, and non-uniform inverters. We do not attempt to state our results in the most general setting. Instead, we use the definition that is most natural for the proof, in order to avoid distracting technical issues. For our purposes, it will be sufficient to define non-uniform one-way functions against non-uniform adversaries and uniform one-way functions against uniform (possibly randomized) adversaries.

In the non-uniform setting, we say $f\colon \{0,1\}^n \to \{0,1\}^m$ is *hard against circuits of size s on an α-fraction of inputs* if for every circuit C of size at most s, $f(C(f(x))) = f(x)$ for at most $\alpha \cdot 2^n$ inputs $x \in \{0,1\}^n$.

In the uniform setting, a function family $f = \{f_n\colon \{0,1\}^n \to \{0,1\}^{m(n)}\}$ is *one-way against (randomized) time $t(n)$ on an $\alpha(n)$-fraction of inputs* if (1) f is computable in deterministic polynomial time and (2) for every (randomized) algorithm A that runs in time $t(n)$ and every sufficiently large n, $f_n(A(1^n, f_n(x))) = f_n(x)$ for at most an $\alpha(n) \cdot 2^n$ fraction of inputs $x \in \{0,1\}^n$ (and with probability at most $1/2$ over the coin tosses of A). (To simplify notation, we will omit the length parameter 1^n as an input to the inverter in our proofs.)

3 Self-amplification for 1-1 Functions

Let $f\colon \{0,1\}^n \to \{0,1\}^m$ be any 1-1 function, and let d_j be the degree of the jth input. Set $\Delta = \sum_{j=1}^n d_j^2$.

Theorem 1. *Let $f = \{f_n\colon \{0,1\}^n \to \{0,1\}^{m(n)}\}$ be a 1-1 function family. Suppose f is one-way against time $\exp(O(\sqrt{r\Delta}\log n))$ on a e^{-r}-fraction of inputs ($r = r(n)$). Then f is one-way against time $\exp(O(\sqrt{r\Delta}\log n))$ on a $(1-e^{-r})$-fraction of inputs.*[1]

When f is a 1-1 function with input locality d, we get that if f is one-way against time $\exp(O(\sqrt{rn}\cdot d\log n))$ for a e^{-r}-fraction of inputs, then the same function is also one-way for a $(1-e^{-r})$-fraction of inputs.

---

[1] Usually, hardness amplification results are stated in terms of two parameters, the initial hardness ε and the "derived hardness" $(1-\delta)$. Since the complexity of our inverter is dictated by the minimum of ε and δ, without loss of generality we state our results for the special case $\varepsilon = \delta = e^{-r}$.

The proof is based on the following idea. For simplicity let us consider the case where the degree of every input is at most d. Assume that f can be inverted in time $\exp(O(\sqrt{r\Delta}\log n))$ on an e^{-r}-fraction of inputs and let S' be the set of inputs on which this inversion algorithm succeeds. Let us consider all inputs x that are within hamming distance $\sqrt{2rn}$ from S'. By a standard probabilistic argument (Lemma 1, based on Theorem 7.5.3 in [1]) it follows that at least $1 - e^{-r}$ fraction of inputs x have this property. Now if x and $x' \in S'$ differ in at most $\sqrt{2rn}$ coordinates, then $y = f(x)$ and $y' = f(x')$ will differ in at most $\sqrt{2rnd}$ coordinates. Therefore we can invert f at $y = f(x)$ by flipping the given set of $\sqrt{2rnd}$ coordinates on which y and y' differ, inverting f at y' to obtain x', and then moving back from x' to x by changing at most $\sqrt{2rn}$ coordinates.

We first state and prove the probabilistic inequality which is the technical heart of our argument. We prove the inequality in slightly more general form than is needed to prove Theorem 1 for later applications.

Lemma 1. *Consider the space $\{0,1\}^n$ with the p-biased distribution (i.e., each coordinate takes value 1 independently at random with probability p) for some $p \in [0,1]$. Let $X \subseteq \{0,1\}^n$ be any set of measure e^{-r} and let $d_1, \dots, d_n$ be positive numbers. Let*

$$Z = \left\{z \colon \textstyle\sum_{j\in[n]\colon x_j \neq z_j} d_j \leq \sqrt{2r\Delta} \text{ for some } x \text{ in } X\right\}.$$

where $\Delta = \sum_{i=1}^n d_i^2$. Then Z has measure at least $1 - e^{-r}$.

Proof. Define $d(z) = \min_{x\in S} \sum_{j\in[n]\colon x_j\neq z_j} d_j$. Then any change in z_j changes $d(z)$ by at most d_j. By Azuma's inequality, we have

$$\Pr[d(z) \leq \mathrm{E}[d(z)] - t] < e^{-2t^2/\Delta} \quad \text{and} \quad \Pr[d(z) \geq \mathrm{E}[d(z)] + t] < e^{-2t^2/\Delta}$$

Setting $t = \mathrm{E}[d(z)]$, from the first inequality we get $e^{-2t^2/\Delta} > e^{-r}$, and therefore $t < \sqrt{r\Delta/2}$. From the second one, $\Pr[z \notin Z] = \Pr[d(z) \geq \sqrt{2r\Delta}] < e^{-r}$. □

Alternatively, Lemma 1 follows from a simple application of Talagrand's inequality.

Proof (of Theorem 1). Let $\varepsilon = e^{-r}$. We prove the contrapositive. Assume A inverts f_n on an ε-fraction of inputs in time $\exp(O(\sqrt{r\Delta}\log m))$. We construct an algorithm B that inverts f_n on a $(1-\varepsilon)$-fraction of inputs as follows: On input y, perform the following procedure: For any set of at most $\sqrt{2r\Delta}$ coordinates of $[m]$, flip the value of y in these coordinates to obtain y', compute $x' = A(y')$, then flip any set of $\sqrt{2r\Delta}$ coordinates of x' to obtain x. If $f_n(x) = y$, output x. The running time of B is

$$\binom{m}{\sqrt{2r\Delta}} \cdot (\text{running time of } A) \cdot \binom{n}{\sqrt{2r\Delta}} \cdot (\text{eval. time of } f_n) = \exp(O(\sqrt{r\Delta}\log n)).$$

We now argue that B inverts f on a $(1-\varepsilon)$-fraction of inputs. Let S' be the set of those x' such that $A(f(x')) = x'$. For each $j \in [n]$, let d_j denote the degree of the jth input. Now let

$$S = \left\{x \colon \textstyle\sum_{j\in[n]\colon x_j\neq x'_j} d_j \leq \sqrt{2r\Delta} \text{ for some } x' \text{ in } S'\right\}.$$

If x' is in S' and x is its closest element in S, then $f(x)$ and $f(x')$ differ in at most $\sqrt{2r\Delta}$ coordinates. Moreover, x and x' can also differ in at most this many coordinates. It follows that if x is in S, then B successfully inverts $f(x')$. By Lemma 1, S contains at least a $1 - \varepsilon$ fraction of inputs. □

Remark 1. The proof of Theorem 1 easily generalizes to function families that are $e^{-r}/2$-close to 1-1. A non-uniform version, where "running time" is replaced by "circuit size", is also straightforward. We will use these extensions in our applications in Sections 5 and 6.

Theorem 1 gives a non-trivial result only when the sum of the squares of the input degrees D is at most $o(n^2/\log n)$. This assumption could be violated even if there is a single input of f whose degree is $\Omega(n)$. It is natural to ask if the self-amplification argument could be modified so as to allow for a small number of inputs that have unusually large degree.

We argue that this is unlikely to be the case: In the full version of the paper, we give an example showing that if non-trivial self-amplification can be achieved for functions where all but one of their inputs have degree at most $d + 1$, then every function of input locality d has a non-trivial inversion algorithm.

4 Linear Functions with Noise

We now state a self-amplification result for functions of the "parity with noise" type. We consider the following type of function. Let $\{M_n\}$ be a family of $m(n)$ by n matrices with entries in $\{0, 1\}$ and $p \in [0, 1]$ be a parameter. We define the function family $f_n : \{0, 1\}^n \to \{0, 1\}^{m(n)}$ as follows:

$$f_n(x, e) = M_n x + e$$

where x is a vector chosen uniformly at random from $\{0, 1\}^n$, and $e \in \{0, 1\}^m$ is a vector of hamming weight at most $2pm$ chosen from the following distribution: Each entry of e is chosen independently from a p-biased distribution, conditioned on r having hamming weight at most $2pm$. The matrix multiplication and vector addition are performed modulo two.

We will consider functions f_n that are 1-1. This happens when M_n is a generator matrix of a code of minimum distance $4pm$. In such a case, the input locality of f_n will be as large as $\Omega(n)$. Nevertheless, we can prove an analogue of Theorem 1 in this setting. One difference is that our self-amplification argument here is randomized, so we require that the function family is hard to invert even for randomized adversaries.

Theorem 2. *Suppose the function family* $\{f_n : f_n(x, e) = M_n x + e\}$ *is 1-1 and one-way against randomized time* $\exp(O(\sqrt{rm}\log m))$ *on a* e^{-r} *fraction of inputs. Assume* $r < pm/10$*. Then* $\{f_n\}$ *is one-way against randomized time* $\exp(O(\sqrt{rm}\log m))$ *on a* $1 - e^{-r}$ *fraction of inputs.*

The proof of Theorem 2 is given in the full version of this paper.

5 Hardness Amplification for Regular Functions

Theorem 1 shows how to achieve self-amplification for functions with small input locality that are 1-1. The assumption that the function is 1-1 was crucial in the argument for the following reason. Suppose f is a 1-1 function with input locality d and x and x' are two inputs that differ in exactly one coordinate. Suppose we can invert $f(x)$. Then with a little bit more work we can invert $f(x')$: Since $f(x)$ and $f(x')$ can differ in at most d coordinates, we change d coordinates of $f(x')$ until we find $f(x)$, recover x, and move back from x to x'.

An important point in this argument is that because f is 1-1, the inversion algorithm is guaranteed to return x and not some other preimage for $f(x)$. If f were not 1-1, the inverter could return some other preimage which is very far from x and therefore also far from x'. So in general we do not know how to achieve self-amplification for input-local functions that are not 1-1.

We now argue that if $f\colon \{0,1\}^n \to \{0,1\}^m$ is not 1-1 but regular, then there exists a new function $f'\colon \{0,1\}^n \to \{0,1\}^{m'}$, $m' = m + O(n)$ such that if f is hard on an small fraction of inputs, then f' is hard on almost all of its inputs. Moreover, the input locality of f' is not much larger than the input locality of f.

To simplify notation, let $\alpha(d, r, \log n) = (d + r + (\log n)^3) \cdot (\log n)$.

Theorem 3. *Suppose there exists a K-to-1 function $f\colon \{0,1\}^n \to \{0,1\}^m$ with input locality d which is hard against circuits of size $\exp(O(\sqrt{rn} \cdot \alpha(d, r, \log n))$ on a e^{-r}-fraction of inputs. Then there exists $f'\colon \{0,1\}^n \to \{0,1\}^{m+\log_2 K+O(r)}$, with input locality $d + O(r + (\log n)^3)$ which is hard against circuits of size $\exp(O(\sqrt{rn} \cdot \alpha(d, r, \log n)))$ on a $(1 - e^{-r})$-fraction of inputs. Moreover, if f is computable by a circuit of size s, then f' is computable by a circuit of size $s + O(n(\log n)^3)$.*

The construction of f' from f is non-uniform. In fact, our proof provides a randomized construction but for simplicity we present the argument in the non-uniform setting. We follow the standard approach of turning a general function into an almost 1-1 function via hashing [13,12]. The function f' will have the form $f'(x) = (f(x), h(x))$, where h is a suitably chosen hash function that does not increase input locality by much. If f is regular, then f' will be almost 1-1 in the sense that for most x, $f(x)$ has a unique preimage. Moreover, if f has input locality d, then f' will have input locality $d + O(r + (\log n)^3)$. We then amplify the hardness of f using Theorem 1 (and Remark 1).

Theorem 3 can be combined with the randomized encodings of Applebaum et al. [3,4] to obtain a hardness amplification result that preserves *output locality*, at the expense of increasing the input length by logarithmic factors.

Corollary 1. *Suppose there exists a regular function $f\colon \{0,1\}^n \to \{0,1\}^m$ with input locality d_{in} and output locality $d_{\mathsf{out}} \geq 3$ that is hard against circuits of size $\exp(O(\sqrt{rn} \cdot \alpha(d_{\mathsf{in}}, r, \log n)))$ on a e^{-r}-fraction of inputs. Then there is a function $f'\colon \{0,1\}^{n'} \to \{0,1\}^{m'}$, where $n' = O(n(\log n)^3)$ and $m' = m + O(n(\log n)^3)$ with output locality d_{out} that is hard against circuits of size $\exp(O(\sqrt{rn} \cdot \alpha(d_{\mathsf{in}}, r, \log n)))$ on a $(1 - e^{-r})$-fraction of inputs. If f is computable by a circuit of size s, then f' is computable by a circuit of size $s + O(n(\log n)^3)$.*

5.1 A Hash with Small Input Locality

A standard way to reduce a K-to-1 one-way function to a 1-1 one-way function is by hashing. Namely, we would like to define $f'(x) = (f(x), h(x))$, where $h\colon \{0,1\}^n \to \{0,1\}^{\log_2 K + O(1)}$ is a pairwise independent hash function. However, known constructions of pairwise independent hash functions have input locality as large as $\Omega(\log_2 K)$. This is in fact necessary: Mansour et al [14] showed that pairwise independent hash functions have *average sensitivity* $\Omega(n)$. By averaging, it follows that the input locality of such functions must be $\Omega(\log_2 K)$.

We need to construct a function f' from f which preserves not only the hardness of f but also its small input locality. Our function f' will also have the form $f'(x) = (f(x), h(x))$, where h is a suitably chosen hash function. However, our hash function h will only be approximately pairwise independent, chosen in a manner to have small input locality.

We note that Appelbaum et al. [4] (Appendix C in the journal version) give a different construction of an "almost pairwise-independent" hash function. However, the almost pairwise independence property they establish for their construction, while sufficient for their application, appears too weak to derive Claim 1.

Claim 1. *Suppose $f\colon \{0,1\}^n \to \{0,1\}^m$ is at most K-to-1, where $2^{k-1} \le K < 2^k$. Then there exists a function $h\colon \{0,1\}^n \to \{0,1\}^{k+3r+3}$ such that the function $f'(x) = (f(x), h(x))$ is $e^{-r}/2$-close to 1-1. Moreover, h is a linear function over $\{0,1\}^n$ with input locality $O(r) + \min\{k, O((\log n)^3)\}$.*

We now prove Claim 1. The construction of h will be probabilistic.

Construction of h. Assume that f is at most K-to-1, where $2^{k-1} \le K < 2^k$. The function h has the form $h(x) = (h_a(x), h_b(x))$ where

$$h_a(x) = (a_k \cdot x + a'_k, a_{k-1} \cdot x + a'_{k-1}, \ldots, a_{k_0+1} \cdot x + a'_{k_0+1})$$
$$h_b(x) = (b_1 \cdot x + b'_1, b_2 \cdot x + b'_2, \ldots, b_{3r+k_0+3} \cdot x + b'_{3r+k_0+3}).$$

and $k_0 = \min\{8(\log n)^2, k\}$. (In particular, if $k < 8(\log n)^2$, h only consists of the h_b part.)

To generate a random h, we choose the vectors $a_i, b_i \in \{0,1\}^n$ from the following distributions: Each a_i is chosen independently at random from the p_i-biased distribution over $\{0,1\}^n$, where $p_i = 4(\log n)^2/i < 1/2$. Each b_i is chosen independently at random from the uniform distribution over $\{0,1\}^n$, and a'_i, b'_i are uniformly random bits.

We now argue that if f is regular, then with probability at least $1/2$ over the choice of h, f' is regular over all but an $e^{-r}/2$ fraction of its inputs.

The proof will have two stages. In the first stage, we argue that for all but an $e^{-r}/8$ fraction of inputs x, there are at most 2^{r+k_0} inputs x' such that $(f(x), h_a(x)) = (f(x'), h_a(x'))$. In the second stage, we finish the proof by showing that h_b hashes all but an $e^{-r}/8$ fraction of those xs uniquely.

The first stage. If $k_0 = k$, the conclusion is trivial, so let us assume that $k_0 < k$. Let us fix an input x and let $S = \{x' \colon f(x) = f(x')\}$. Without loss of generality, we may assume that $2^{k-1} \leq |S| < 2^k$. (If $|S|$ is smaller, we disregard the effect of the first few hashes in h_a.) We consider the following sequence of random sets defined recursively: $S_k = S$, $T_i = \{x' \in S_i \colon a_i \cdot x' = a_i \cdot x)\}$ and

$$S_{i-1} = \begin{cases} T_i, & \text{if } (1-1/n)|S_i|/2 \leq |T_{i-1}| \leq (1+1/n)|S_i|/2 \\ S_i, & \text{otherwise.} \end{cases}$$

Here is the intuition for this definition: We want to think of the ith hash a_i as "successful" if it decreases the size of siblings of x by roughly a factor of two (not much more and not much less). If all but r of the hashes are successful, then the size of S_0 can not be much more than 2^r, and so x will not have more than 2^r siblings that map to $(f(x), h_a(x))$. It is sufficient to show that the probability that more than r of the hashes fail to be successful is quite small.

Notice that by definition of the sets S_i, it must be that $S_i \geq \prod_{j=i}^{k}(1-1/n) \cdot 2^{i-1} \geq 2^{i-2}$. So we are left with the following question: Given a set S of size at least 2^{i-2}, how likely is it to be split successfully at stage i?

Lemma 2. *Assume* $|R| \geq 2^{i-2}$. *Let* $a \sim \{0,1\}^n_p, b \sim \{0,1\}_{1/2}$, *where* $p = 4(\log n)^2/i < 1/2$. *Then for* n *sufficiently large and any* $\varepsilon > 0$,

$$\Pr\big[\#\{y \in R \colon a \cdot y + b = 0\} \notin (1 \pm \varepsilon)|R|/2\big] \leq \frac{1}{n^4\varepsilon^2}.$$

Applying this lemma with $R = S_i$ and $\varepsilon = 1/n$, we have that each hash is successful with probability at least $1 - 1/n^2$, and the events are independent of one another. By a union bound, the probability of having more than r unsuccessful splits is at most $\binom{n}{r} \cdot (1/n^2)^r \leq n^{-r} < e^{-r}/8$. So for any $x \in \{0,1\}^n$,

$$\Pr\big[|S_{k_0}| \geq 2^{k_0+r}\big] \leq e^{-r}/8.$$

Proof. Let $X = \sum_{y \in R}(-1)^{a \cdot y + b}$. Then $\mathrm{E}[X] = 0$ and

$$\begin{aligned} \mathrm{E}[X^2] &= \sum_{y,z \in R} \mathrm{E}[(-1)^{a\cdot(y+z)}] \\ &\leq |R| \max_z \sum_{y \in R} \mathrm{E}[(-1)^{a\cdot(y+z)}] \\ &= |R| \max_z \sum_{y \in R_z} \mathrm{E}[(-1)^{a\cdot y}] \\ &= |R| \max_z \sum_{y \in R_z} (1-2p)^{|y|} \end{aligned}$$

where $R_z = \{y + z \colon y \in T\}$, and $|y|$ denotes the hamming weight of y. Notice that the summation is maximized when R_z is a threshold set T – the set of all strings of hamming weight up to $k-1$ and possibly some of hamming weight k. Then we have

$$\mathrm{E}[X^2] \le |R| \cdot \sum_{y\in T}(1-2p)^{|y|} \le |R| \cdot \sum_{w=0}^{k}\binom{n}{w}\cdot(1-2p)^w.$$

We look at two cases: If $i \ge n$, then

$$\mathrm{E}[X^2] \le |R|\sum_{w=0}^{n}\binom{n}{w}\cdot(1-2p)^w = |R|\cdot 2^n\cdot(1-p)^n \le 4|R|^2\cdot e^{-pn} \le |R|^2/n^4,$$

for n sufficiently large. If $i < n$, the ratio of consecutive terms in the summation is

$$\binom{n}{w+1}(1-2p)^{w+1}\Big/\binom{n}{w}(1-2p)^w = (1-2p)\cdot\frac{n-w}{w+1} > 1$$

for every $w \le k$, and so

$$\mathrm{E}[X^2] \le |R|\cdot k\cdot\binom{n}{k}\cdot(1-2p)^k \le |R|^2\cdot n\cdot(1-2p)^k \le |R|^2\cdot n\cdot e^{-2pk}.$$

Since $k \ge i/\log n$, we get that $\mathrm{E}[X^2] \le |R|^2/n^4$ in this case also. By Chebyshev's inequality, it follows that

$$\Pr\big[\#\{y\in R\colon a\cdot y+b=0\}\notin(1\pm\varepsilon)|R|/2\big] = \Pr\big[|X|>\varepsilon|R|\big] \le 1/n^4\varepsilon^2. \qquad \square$$

The second stage and conclusion. Now fix an x such that $|S_{k_0}| < 2^{k_0+r}$. We now argue that by the end of the second stage, x is very likely to have a unique hash:

$$\Pr[\exists x'\in S_{k_0}-\{x\}\colon h(x')=h(x)] \le \sum_{x'\in S_{k_0}-\{x\}}\Pr[h(x')=h(x)] < e^{-r}/8.$$

Putting the analysis of both stages together, it follows that by the end of stage 2, for any specific x,

$$\Pr_h[\exists x'\colon f'(x')=f'(x)] \le e^{-r}/4.$$

Averaging over x and applying Markov's inequality, we get that for at least half the functions h,

$$\Pr_x[\exists x'\colon f'(x')=f'(x)] \le e^{-r}/2.$$

Now let us calculate the locality of a typical function h. For any fixed input bit, say x_1, let Y_a and Y_b be the number of occurrences of x_1 in h_a and h_b respectively. Then $\mathrm{E}[Y_a] = \sum_{i=k_0}^{k}4(\log n)^2/i \le 4(\log n)^3$ and $\mathrm{E}[Y_b] = (3r+k_0+3)/2$, so $\mathrm{E}[Y_a+Y_b] = O((\log n)^3+r)$. By Chernoff bounds and a union bound, we get that with probability at least 3/4, no input bit has more than $O((\log n)^3+r)$ occurrences in h.

Therefore, there exists a hash function h that has input locality $O((\log n)^3+r)$ and such that f' is 1-1 on all but $e^{-r}/2$ fraction of its inputs.

Remark 2. The conclusion of Claim 1 also holds under the weaker assumption that the function h is $e^{-r}/4$-close to at most K-to-1. We will use this generalization in Section 6.

5.2 Proof of Theorem 3

We now prove Theorem 3. To do so, first we show that the transformation from f to f' is hardness-preserving in the following sense: If f is hard to invert on an e^{-r}-fraction of inputs, then f' is hard to invert on an $\Omega(e^{-r})$-fraction of inputs. Since f' is almost 1-1, we can apply self-amplification to conclude that f' is in fact hard on a $1 - e^{-r}$ fraction of inputs.

Claim 2. *Assume $f\colon \{0,1\}^n \to \{0,1\}^m$ is K-to-1 where $2^{k-1} \le K < 2^k$. Let f' and h be as in Claim 1. Assume that f' can be inverted on an $(1 - e^{-r}/400)$-fraction of inputs by a circuit of size s. Then f can be inverted on a $(1 - e^{-r})$-fraction of inputs by a circuit of size $O(s \cdot r \cdot 2^{3r})$.*

The proof of Claim 2 can be found in the full version of this paper.

Proof (of Theorem 3). Suppose $f\colon \{0,1\}^n \to \{0,1\}^m$ is a regular function with input locality d which is hard against circuits of size $\exp(O(\sqrt{rn} \cdot \alpha(d, r, \log n)))$ on a e^{-r}-fraction of inputs. Let $f'(x) = (f(x), h(x))$, where h is chosen as in Claim 1. It is easy to check that f' has the desired input locality and circuit complexity.

Now suppose f' can be inverted by a circuit of size $\exp(O(\sqrt{rn}\cdot\alpha(d, r, \log n)))$ on a e^{-r} fraction of its inputs. By Claim 1, f' is $e^{-r}/2$-close to 1-1. By Theorem 1 and Remark 1, f' can be inverted on a $(1-e^{-r}/400)$-fraction of inputs by a circuit of size $s = \exp(O(\sqrt{rn} \cdot \alpha(d, r, \log n)))$. By Claim 2, f can then be inverted on a $(1 - e^{-r})$ fraction of inputs by circuits of size $\exp(O(\sqrt{rn} \cdot \alpha(d, r, \log n)))$. □

Proof (of Corollary 1). Since $h(x)$ is a linear function, we can apply the randomized encoding of Applebaum et al. to reduce its output locality at the cost of increasing the input and output length of f'. Specifically, we perform the following transformation on f' to obtain a new function f''. Suppose the ith output $h(x)_i$ has the form

$$h(x)_i = x_{i1} + x_{i2} + \cdots + x_{ik_i}.$$

We introduce new inputs $r_{i1}, r_{i2}, \ldots, r_{i(k_i-1)}$ and replace the output $h(x)_i$ by the sequence of outputs:

$$(x_{i1} + r_{i1}, r_{i1} + x_{i2} + r_{i2}, \ldots, r_{i(k_i-1)} + x_{ik_i}).$$

It is easy to check that f'' has the desired input and output length, and its output locality is $\max\{d_{\mathsf{out}}, 3\}$.

Applebaum et al. [3,4] show that if f'' can be inverted on an ε-fraction of inputs by a circuit of size s, then f' can be inverted on a $\Omega(\varepsilon)$-fraction of inputs by a circuit of size $O(s/\varepsilon)$. Plugging in $\varepsilon = e^{-r}$ and $s = \exp(O(\sqrt{rn}\cdot\alpha(d_{\mathsf{in}}, r, \log n)))$, the corollary follows. □

6 Goldreich's Function on a Random Graph

We now consider two applications of our techniques to the candidate one-way function proposed by Goldreich [8]. Given a bipartite graph G with n vertices on the left, m vertices on the right, and regular right-degree d_{out} and a predicate $P\colon \{0,1\}^{d_{\text{out}}} \to \{0,1\}$, the function $f_{G,P}$ from $\{0,1\}^n$ to $\{0,1\}^m$ is defined by

$$f_{G,P}(x)_i = \text{the } i\text{th bit of } f(x) = P(x_{\Gamma(i,1)}, \ldots, x_{\Gamma(i,d_{\text{out}})})$$

where $\Gamma_{(i,j)}$ is the jth neighbor of right vertex i of G.

Goldreich [8] considered such constructions for the setting of parameters $m = n$ and d_{out} ranges from a constant to $O(\log n)$. He conjectured that when G is a good expander graph and P is a randomly chosen predicate, with high probability $f_{G,P}$ is one-way.

Cook et al. [7] showed that when G is random and P is suitably chosen, $f_{G,P}$ is secure against adversaries that implement *myopic algorithms*. Bogdanov and Qiao [6] studied a variant of Goldreich's function in the setting where G is random, d is constant, and $m = Dn$, where $D = D(d_{\text{out}})$ is a sufficiently large constant. They showed that for a large class of predicates P (those that correlate with one or a pair of their inputs) and for most G, $f_{G,P}$ can be inverted on most inputs. It is conceivable that $f_{G,P}$ could be one-way for all predicates P that are not linear and do not belong to the class ruled out by Bogdanov and Qiao.

We establish two results regarding local hardness amplification of Goldreich's function. Informally, we show that

1. In the setting where d is constant and $m \geq Dn$, where $D = D(d_{\text{out}})$ is a sufficiently large constant, for any P and with high probability over the choice of G if $f_{G,P}$ is hard to invert on an e^{-r} fraction of inputs in time $\exp(O(\sqrt{rn} \cdot d_{\text{out}} \cdot \log n))$, then $f_{G,P}$ is hard to invert on a $1 - e^{-r}$ fraction of inputs in time $\exp(O(\sqrt{rn} \cdot d_{\text{out}} \cdot \log n))$.
2. When $d_{\text{out}} = O(\log n)$ and $m = n$, for a certain class of predicates P and with high probability over G, if $f_{G,P}$ is hard to invert on a e^{-r} fraction of inputs, then there exists a function $f'\colon \{0,1\}^{n'} \to \{0,1\}^{m'}$, where $n', m' = n \cdot \text{polylog}n$, of similar complexity to f and constant output locality that is hard to invert on a $1 - e^{-r}$ fraction of inputs.

   Our result applies to all $O(\log n)$-*parity-blowup* predicates, which we define as follows. Let $P_c\colon \{0,1\}^c \to \{0,1\}$ be any balanced predicate, where c is some constant. The d_{out}-parity-blowup of P_c is the predicate $P\colon \{0,1\}^{d_{\text{out}}} \to \{0,1\}$ which is obtained by replacing each of the variables in P_c by a parity of $\lfloor d_{\text{out}}/c \rfloor$ inputs, where all the inputs are distinct. Applebaum, Barak, and Wigderson [2] showed that the output of Goldreich's function based on such predicates is pseudorandom against linear functions, low-degree polynomials, and constant-depth circuits.

The random graph G is chosen from the following distribution: For each of the m right vertices of G, choose all of its d_{out} neighbors independently at random among the n left vertices of G. We will call such graphs (n, m, d_{out}) *random graphs*.

6.1 Self-reducibility for Functions with Long Output

Theorem 4. *Let $D \geq 2^{K d_{\text{out}}}$ where K is a sufficiently large constant, and $P\colon \{0,1\}^{d_{\text{out}}} \to \{0,1\}$ be any predicate. Let G be an (n, m, d_{out}) random graph.*

With probability at least $1-o(1)$ *over the choice of* G*, if* $f_{G,P}$ *is hard for circuits of size* $\exp(O(\sqrt{rn}\cdot Dd_{\text{out}}\log n))$ *on an* e^{-r}*-fraction of inputs, then* $f_{G,P}$ *is hard for circuits of size* $\exp(O(\sqrt{rn}\cdot Dd_{\text{out}}\log n))$ *on a* $1-e^{-r}$*-fraction of inputs.*

We prove Theorem 4 (in the full version of this paper) by an argument similar to the one used in the proof of Theorem 1. The principal obstacle to applying Theorem 1 here is that with high probability, the function $f_{G,P}$ is not 1-1. There are several reasons for this. One reason is that $f_{G,P}$ is likely to contain inputs that do not appear in any output. A more important reason is that unless the predicate P is linear, for most inputs x, it is likely that there is a linear number of coordinates i such that the ith coordinate does appear in the output, but changing the value of x_i does not change the value of $f_{G,P}(x)$.

We show that although $f_{G,P}$ is unlikely 1-1, with high probability every pair of inputs that map to the same output is highly correlated (or anticorrelated), that is they agree (or disagree) in value on most of the coordinates. Using the argument from the proof of Theorem 1, we show that if $f_{G,P}$ can be inverted on an ε-fraction on inputs by a circuit of suitable size, then for a $1-\varepsilon$ fraction of inputs x, it is possible to find an x' such that x and x' are highly correlated. We then use a result of Bogdanov and Qiao [6] which says that for most inputs x, given x' that is correlated with x, then we can invert $f_{G,P}(x)$.

6.2 Amplification for Certain Length-Preserving Functions

Let $P\colon \{0,1\}^c \to \{0,1\}$ be a balanced predicate. The d_{out}*-parity-blowup* of P is the predicate obtained by replacing each variable in P by $\lfloor d_{\text{out}}/c\rfloor$ variables, where all the new variables are disjoint.

Theorem 5. *Let* $c\geq 3$ *and* $d_{\text{out}}=\max\{130c\log n, 4c^2\}$*. Let* P *be the* d_{out}*-parity-blowup of some balanced predicate on* c *bits and let* G *be an* (n,n,d_{out})*-random graph. Suppose that for at least half the graphs* G*,* $f_{G,P}\colon \{0,1\}^n \to \{0,1\}^n$ *is hard to invert on an* e^{-r}*-fraction of inputs against circuits of size* $\exp(O(r^{3/2}\sqrt{n}\log n))$*. Then there exists*

1. *A function* $f'\colon \{0,1\}^n \to \{0,1\}^{n+O(r)}$ *of circuit size* $O(n\log n+r)$ *that is hard on a* $1-e^{-r}$*-fraction of inputs for circuits of size* $\exp(O(r^{3/2}\sqrt{n}\log n))$*.*
2. *A function* $f''\colon \{0,1\}^{n'} \to \{0,1\}^{m'}$*, where* $n',m'=O(n\cdot d(n)^c)$*,* f'' *is hard on a* $1-e^{-r}$*-fraction of inputs for circuits of size* $\exp(O(r^{3/2}\sqrt{n}\log n))$ *and where every output of* f'' *depends on at most* $c+1$ *inputs.*

Theorem 5 is proved in the full version of this paper. To prove part 1, we first show that the function $f_{G,P}$ is likely to be $e^{-r}/4$-close to $O(e^r)$-to-1. Using Claim 1, we then transform $f_{G,P}$ into a function $f'\colon \{0,1\}^n \to \{0,1\}^{n+O(r)}$ which is $e^{-r}/2$-close to 1-1. Using the self-reduction of Theorem 1, we then argue that if f' is easy to invert on an e^{-r} fraction of inputs, then it is also easy to invert on a $1-e^{-r}$-fraction of inputs, and so $f_{G,P}$ is also easy to invert on the same fraction of inputs by circuits of similar size. To prove part 2, we observe that parity-blowup predicates can be represented constant degree polynomials over

$GF(2)$. Applying the randomized encodings of Applebaum et al. [3] to these polynomials, we obtain a function with constant output locality and slightly longer input and output length.

Acknowledgements. We thank the TCC'11 reviewers for useful comments.

References

1. Alon, N., Spencer, J.: The Probabilistic Method. John Wiley, Chichester (1992)
2. Applebaum, B., Barak, B., Wigderson, A.: Public-key cryptography from different assumptions. In: STOC 2010, pp. 171–180 (2010)
3. Applebaum, B., Ishai, Y., Kushilevitz, E.: Cryptography in NC0. In: FOCS 2004, pp. 166–175 (2004)
4. Applebaum, B., Ishai, Y., Kushilevitz, E.: Cryptography with Constant Input Locality. In: Menezes, A. (ed.) CRYPTO 2007. LNCS, vol. 4622, pp. 92–110. Springer, Heidelberg (2007); Journal version in J. Cryptology 22(4), 429-469 (2009)
5. Blum, A., Kalai, A., Wasserman, H.: Noise-tolerant learning, the parity problem, and the statistical query model. In: STOC 2000, pp. 435–440 (2000)
6. Bogdanov, A., Qiao, Y.: On the security of goldreich's one-way function. In: Dinur, I., Jansen, K., Naor, J., Rolim, J. (eds.) APPROX 2009. LNCS, vol. 5687, pp. 392–405. Springer, Heidelberg (2009)
7. Cook, J., Etesami, O., Miller, R., Trevisan, L.: Goldreich's One-Way Function Candidate and Myopic Backtracking Algorithms. In: Reingold, O. (ed.) TCC 2009. LNCS, vol. 5444, pp. 521–538. Springer, Heidelberg (2009)
8. Goldreich, O.: Candidate One-Way Functions Based on Expander Graphs. Electronic Colloquium on Computational Complexity (ECCC) 7(90) (2000)
9. Goldreich, O., Impagliazzo, R., Levin, L.A., Venkatesan, R., Zuckerman, D.: Security Preserving Amplification of Hardness. In: FOCS 1990, pp. 318–326 (1990)
10. Goldreich, O., Krawczyk, H., Luby, M.: On the Existence of Pseudorandom Generators. SIAM J. Comput. 22(6), 1163–1175 (1993)
11. Haitner, I., Harnik, D., Reingold, O.: On the Power of the Randomized Iterate. In: Dwork, C. (ed.) CRYPTO 2006. LNCS, vol. 4117, pp. 22–40. Springer, Heidelberg (2006)
12. Impagliazzo, R., Levin, L., Luby, M.: Pseudo-random Generation from One-way Functions. In: STOC 1989, pp. 12–24 (1989)
13. Impagliazzo, R., Luby, M.: One-way Functions are Essential for Complexity Based Cryptography. In: FOCS 1989, pp. 230–235 (1989)
14. Mansour, Y., Nisan, N., Tiwari, P.: The Computational Complexity of Universal Hashing. Theor. Comput. Sci. 107(1), 121–133 (1993)
15. Lin, H.C., Trevisan, L., Wee, H.: On Hardness Amplification of One-Way Functions. In: Kilian, J. (ed.) TCC 2005. LNCS, vol. 3378, pp. 34–49. Springer, Heidelberg (2005)
16. Yao, A.C.-C.: Theory and Applications of Trapdoor Functions (Extended Abstract). In: FOCS 1982, pp. 80–91 (1982)

General Hardness Amplification of Predicates and Puzzles*
(Extended Abstract)

Thomas Holenstein[1,**] and Grant Schoenebeck[2,***]

[1] Department of Computer Science, ETH Zurich, Switzerland
thomas.holenstein@inf.ethz.ch
http://www.complexity.ethz.ch/people/holthoma

[2] Department of Computer Science, Princeton University, Princeton NJ 08544, USA
gschoene@princeton.edu
http://www.cs.princeton.edu/~gschoene

Abstract. We give new proofs for the hardness amplification of efficiently samplable predicates and of weakly verifiable puzzles which generalize to new settings. More concretely, in the first part of the paper, we give a new proof of Yao's XOR-Lemma that additionally applies to related theorems in the cryptographic setting. Our proof seems simpler than previous ones, yet immediately generalizes to statements similar in spirit such as the extraction lemma used to obtain pseudo-random generators from one-way functions [Håstad, Impagliazzo, Levin, Luby, SIAM J. on Comp. 1999].

In the second part of the paper, we give a new proof of hardness amplification for weakly verifiable puzzles, which is more general than previous ones in that it gives the right bound even for an arbitrary monotone function applied to the checking circuit of the underlying puzzle.

Both our proofs are applicable in many settings of interactive cryptographic protocols because they satisfy a property that we call "non-rewinding". In particular, we show that any weak cryptographic protocol whose security is given by the unpredictability of single bits can be strengthened with a natural information theoretic protocol. As an example, we show how these theorems solve the main open question from [Halevi and Rabin, TCC2008] concerning bit commitment.

1 Introduction

In this paper, we study two scenarios of hardness amplification. In the first scenario, one is given a predicate $P(x)$, which is somewhat hard to compute given x. More concretely: $\Pr[A(x) = P(x)] \leq 1 - \frac{\delta}{2}$ for any A in some given

* A full version of this paper is available [22].

** Work was done while the author was at Microsoft Research, Silicon Valley Campus.

*** Work was partially done while author was a summer intern at Microsoft Research Silicon Valley Campus and partially supported by a National Science Foundation Graduate Fellowship.

Y. Ishai (Ed.): TCC 2011, LNCS 6597, pp. 19–36, 2011.

complexity class, where typically δ is not too close to 1 but at least polynomially big (say, $\frac{1}{\text{poly}(n)} < \delta < 1 - \frac{1}{\text{poly}(n)}$). One then aims to find a predicate which is even harder to compute.

In the second scenario, one is given a computational search problem, specified by some relation $R(x, y)$. One then assumes that no algorithm of a certain complexity satisfies $\Pr[(x, A(x)) \in R] > 1 - \delta$, and again, is interested in finding relations which are even harder to satisfy. It is sometimes the case that R may only be efficiently computable given some side information generated while sampling x. Such problems are called "weakly verifiable puzzles".

Our aim is to give proofs for theorems in both scenarios which are both simple and versatile. In particular, we will see that our proofs are applicable in the interactive setting, where they give stronger results than those previously known.

1.1 Predicates

Overview and Previous Work. Roughly speaking, Yao's XOR-Lemma [39] states that if a predicate $P(x)$ is somewhat hard to compute, then the k-wise XOR $P^{\oplus k}(x_1, \ldots, x_k) := P(x_1) \oplus \cdots \oplus P(x_k)$ will be even harder to compute. While intuitive, such statements are often somewhat difficult to prove. The first proof of the above appears to be by Levin [31] (see also [11]). In some cases, even stronger statements are needed: for example, the extraction lemma states that one can even extract several bits out of the concatenation $P(x_1)P(x_2)\ldots P(x_k)$, which look pseudorandom to a distinguisher given $x_1, \ldots, x_k$. Proving this statement for tight parameters is considered the technically most difficult step in the original proof that one-way functions imply pseudorandom generators [17]. Excluding this work, the easiest proof available seems to be based on Impagliazzo's hard-core set theorem [23], more concretely the uniform version of it [19,1]. A proof along those lines is given in [20,13]. Similar considerations are true for the more efficient proof that one-way functions imply pseudorandom generators given by Haitner et al.[15].

Contributions of this Paper. In this paper, we are concerned with statements of a similar nature as (but which generalize beyond) Yao's XOR-Lemma. We give a new theorem, which is much easier to prove than the hard-core set theorem, and which is still sufficient for all the aforementioned applications.

Our main observation can be described in relatively simple terms. In the known proof based on hard-core sets ([23,19]), the essential statement is that there is a large set S, such that for $x \in S$ it is computationally difficult to predict $P(x)$ with a non-negligible advantage over a random guess. Proving the existence of the set S requires some work (basically, boosting, as shown in [30]). We use the idea that the set S can be made *dependent* on the circuit which attempts to predict P. The existence of a hard set S for a particular circuit is a much easier fact to show (and occurs as a building block in some proofs of the hard-core theorem). For our idea to go through, S has to be made dependent on some of the inputs to C as well as some other fixed choices. This technique of switching quantifiers resembles a statement in [2], where Impagliazzo's hard-core set theorem is used to show that in some definitions of pseudo-entropy it is also possible to switch quantifiers.

Besides being technically simpler, making the set S dependent on C has an additional advantage. For example, consider a proof of the XOR Lemma. To get a contradiction, a circuit C is assumed which does well in predicting the XOR, and a circuit D for a single instance is built from C. On input x, D calls C as a subroutine several times, each time "hiding" x as one of the elements of the input. Using our ideas, we can ensure that x is hidden always in the same place i, and even more, the values of the inputs $x_1, \ldots, x_{i-1}$ are constant and independent of x. This property, which we call non-rewinding, is useful in the case one wants to amplify the hardness of interactive protocols.

We remark that in this paper we are not concerned with efficiency of XOR-Lemmas in the sense of derandomizing them (as in, e.g., [28,24,26]).

1.2 Weakly Verifiable Puzzles

Overview and Previous Work. The notion of weakly verifiable puzzles was introduced by Canetti et al. [4]. A weakly verifiable puzzle consists of a sampling method, which produces an instance x together with a circuit $\Gamma(y)$, checking solutions. The task is, given x but not necessarily Γ, to find a string y for which $\Gamma(y) = 1$. One-way functions are an example: $\Gamma(y)$ just outputs 1 if $f(y) = x$ (since Γ depends on the instance it can contain x). However, weakly verifiable puzzles are more general, since Γ is not given at the time y has to be found.

Canetti et al. show that if no efficient algorithm finds solutions with probability higher than δ, then any efficient algorithm finds k solutions simultaneously with probability at most $\delta^k + \epsilon$, for some negligible ϵ. This result was strengthened by [25], showing that requiring some $\delta' > \delta + 1/\operatorname{poly}(n)$ fraction of correct answers already makes efficient algorithms fail, if k is large enough. Independently of the current work, Jutla [29] improved their bound to make it match the standard Chernoff bound. A different strengthening was given in [16], where it was noted that the algorithm in [4] has an additional property which implies that it can be applied in an interactive cryptographic setting, also they studied how much easier solving a weakly verifiable puzzle becomes if one simply asks for a single correct solution from k given puzzles. Also independently of our work, Chung et al. [6] give a proof for the threshold case (similar to Jutla) which is also applicable in an interactive setting; however, their parameters are somewhat weaker than the ones given by most other papers. Finally, [9] gives yet another strengthening: they allow a weakly verifiable puzzle to have multiple solutions indexed by some element q, and the adversary is allowed to interactively obtain some of them. They then study under what conditions the hardness is amplified in this setting.

Contributions of this Paper. In this work, we present a theorem which unifies and strengthens the results given in [4,16,25,29,6]: assume a monotone function $g : \{0,1\}^k \to \{0,1\}$ specifies which subpuzzles need to be solved in order to solve the resulting puzzle (i.e., if $c_1, \ldots, c_k$ are bits where c_i indicates that a valid solution for puzzle i was found, then $g(c_1, \ldots, c_k) = 1$ iff this is sufficient to give a valid solution for the overall case.) Our theorem gives a tight bound for any such g (in this sense, previous papers considered only threshold functions for g). Furthermore, as we will see our proof is also applicable in an interactive setting (the proof given in [25,29] does not have this property). Our proof is heavily inspired by the one given in [4].

1.3 Strengthening Cryptographic Protocols

Overview and Previous Work. Consider a cryptographic protocol, such as bit commitment. Suppose that a non-perfect implementation of such a protocol is given, which we would like to improve. For example, assume that a cheating receiver can guess the bit committed to with some probability, say $3/5$. Furthermore, suppose that a cheating sender can open the commitment in two ways with some probability, say $1/5$. Can we use this protocol to get a stronger bit commitment protocol?

Such questions have been studied in various forms both in the information theoretic and the computational model [8,7,10,19,21,38,16].

However, all of the previous computational work except [16] focused on the case where the parties participating in the protocol are at least semi-honest, i.e., they follow the protocol correctly (this is a natural assumption in the case for the work on key agreement [10,19,21], as in this case the participating parties can be assumed to be honest). An exception to this trend was the work by Halevi and Rabin [16], where it was shown that for *some* protocols, the information theoretic bounds also apply computationally.

The above are results in case where the protocol is repeated *sequentially*. The case where the protocol is repeated in parallel is more complicated [3,35,34,18,12,5].

Contributions of this Paper. We explicitly define "non-rewinding" (which was, however, pointed to in [16]) which helps to provide a sufficient condition for transforming complexity theoretic results into results for cryptographic protocols. Using, the above results, and specifically that the above results are non-rewindable, we show that we can strengthen any protocol in which the security goal is to make a bit one party has unpredictable to the other party, in the case where an information theoretic analogue can be strengthened. We also study interactive weakly verifiable puzzles (as has been done implicitly in [16]), and show that natural ways to amplify the hardness of these work.

We only remark that our proof is applicable to parallel repetition for non-interactive (two-round) protocols (e.g. CAPTCHAs).

Due to space restrictions, many of the proofs and even some of the formal statements of theorems have been omitted. We encourage the interested reader to look at the full version of this paper [22].

2 Preliminaries

Definition 1. *Consider a circuit C which has a tuple of designated input wires labeled $y_1, \ldots, y_k$. An oracle circuit $D(\cdot)$ with calls to C is* non-rewinding *if there* *is a fixed i and fixed strings y_1^* to y_{i-1}^* such that for any input y to D, all calls to C use inputs $(y_1^*, \ldots, y_{i-1}^*, y)$ on the wires labeled $y_1, \ldots, y_i$.*

Definition 2. *Let C be a circuit which has a block of input wires labeled x. An oracle circuit D which calls C (possibly several times) treats x obliviously if the input x to D is forwarded to C directly, and not used in any other way in D.*

We say that an event happens *almost surely* if it has probability $1 - 2^{-n}\,\mathrm{poly}(n)$.

We denote by $[m]$ the set $\{1, \ldots, m\}$. The density of a set $S \subseteq \{0,1\}^n$ is $\mu(S) = \frac{|S|}{2^n}$. We sometimes identify a set S with its characteristic function $S : \{0,1\}^n \to \{0,1\}$. We often denote a tuple $(x_1, x_2, \ldots, x_k)$ by $x^{(k)}$.

If a distribution μ over some set is given, we write $x \leftarrow \mu$ to denote that x is chosen according to μ. We sometimes identify sets with the uniform distribution over them. We let μ_δ be the Bernoulli distribution over $\{0,1\}$ with parameter δ, i.e., $\Pr_{x \leftarrow \mu_\delta}[x = 1] = \delta$. Furthermore, μ_δ^k is the distribution over $\{0,1\}^k$ where each bit is i.i.d. according to μ_δ.

When two interactive algorithms A and B are given, we will denote by $\langle A, B \rangle_A$ the output A has in an interaction with B, and by $\langle A, B \rangle_B$ the output which B has. We sometimes consider probabilities like $\Pr[\langle A, B \rangle_A = \langle A, B \rangle_B]$, in which case the probability is over random coins of A and B (if any), but they are chosen the same on the left and the right hand side.

3 Efficiently Samplable Predicates

3.1 Single Instance

Informal Discussion. Fix a predicate $P : \{0,1\}^n \to \{0,1\}$ and a circuit $C(x, b, r)$ which takes an arbitrary $x \in \{0,1\}^n$, a bit $b \in \{0,1\}$, and some randomness r as input. We may think of C as a circuit which tries to distinguish the case $b = P(x)$ from the case $b = 1 - P(x)$. Our idea is to identify a set S for which we can show the following:

1. If x is picked randomly from S, then $\Pr[C(x, P(x), r) = 1] \approx \Pr[C(x, 1 - P(x), r) = 1]$.
2. C can be used to predict $P(x)$ for a uniform random x correctly with probability close to $1 - \frac{1}{2}\mu(S)$

On an informal level, one could say that S explains the hardness of computing P from C's point of view: for elements from S the circuit just behaves as a uniform random guess, on the others it computes (or, more accurately, *helps* to compute) P. Readers familiar with Impagliazzo's hardcore lemma will notice the similarity: Impagliazzo finds a set which explains the computational difficulty of a predicate for *any* circuit of a certain size. Thus, in this sense Impagliazzo's theorem is stronger. The advantage of ours is that the proof is technically simpler, and that it can be used in the interactive setting (see Section 3.3) which seemingly comes from the fact that it helps to build non-rewinding proofs.

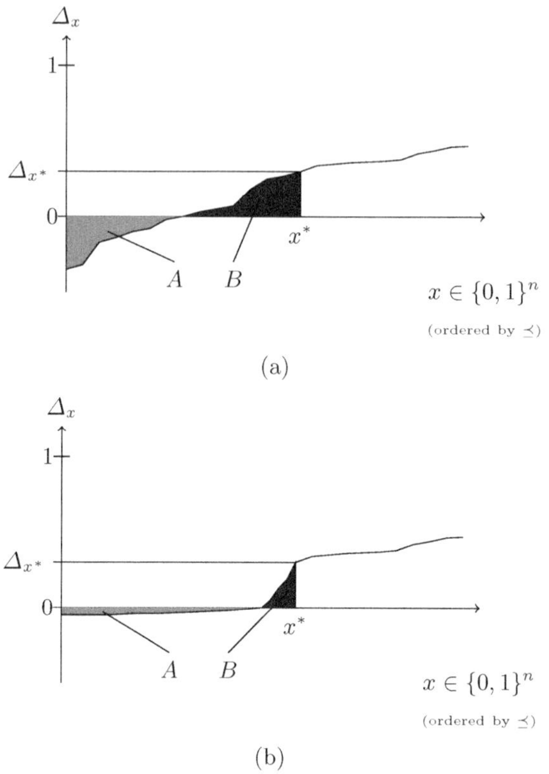

Fig. 1. Intuition for the proof of Theorem 1. In both pictures, on the vertical axis, the advantage of the circuit in guessing right over a random guess is depicted. The elements are then sorted according to this quantity. The point x^* is chosen such that the area of A is slightly smaller than the area of B.

The Theorem. The following theorem formalizes the above discussion. It will find S by producing a circuit which recognizes it, and also produces a circuit Q which uses C in order to predict P.

Theorem 1. *Let $P : \{0,1\}^n \to \{0,1\}$ be a computable predicate. There is an algorithm* Gen *which takes as input a randomized circuit $C(x,b,r)$ and a parameter ϵ, and outputs two deterministic circuits Q and S, both of size* $\text{size}(C)\cdot\text{poly}(n,\frac{1}{\epsilon})$, *as well as $\delta \in [0,1]$, such that almost surely the following holds:*

Large Set: *$S(x,P(x))$ recognizes a set $S^* = \{x|S(x,P(x)) = 1\}$ of density at least $\mu(S^*) \geq \delta$.*

Indistinguishability: *For the above set S^* we have*

$$\Big|\Pr_{x\leftarrow\{0,1\}^n,r}[C(x,P(x),r)=1] - \Pr_{x\leftarrow\{0,1\}^n,r}[C(x,P'(x),r)=1]\Big| \leq \epsilon, \quad (1)$$

where $P'(x) := P(x) \oplus S(x)$, i.e., P' is the predicate which equals P outside S and differs from P within S.

Predictability: *Q predicts P well:* $\Pr_{x \leftarrow \{0,1\}^n}[Q(x) = P(x)] \geq 1 - \frac{\delta}{2}$.

Additionally, these algorithms have the following properties:

1. *Unless $\delta = 1$ algorithm Q predicts slightly better:*[1] $\Pr[Q(x) = P(x)] \geq 1 - \frac{\delta}{2} + \frac{\epsilon}{4}$.
2. *If P is efficiently samplable (i.e., pairs $(x, P(x))$ can be generated in polynomial time),* Gen *runs in time* $\mathrm{poly}(n, \frac{1}{\epsilon})$.
3. Gen, *S, and Q can be implemented with oracle access to C only (i.e., they do not use the description of C).*
4. *When thought as oracle circuits, S and Q use the oracle C at most $\mathcal{O}(\frac{n}{\epsilon^2})$ times. Also, they both treat x obliviously, and their output only depends on the number of 1's obtained from the oracle calls to C and, in case of S, the input $P(x)$.*

The proof uses no new techniques. For example, it is very similar to Lemma 2.4 in [19], which in turn is implicit in [31,11] (see also Lemma 6.6 and Claim 7 on page 121 in [20]). Our main contribution here is to give the statement and to note that it is very powerful. The proof itself is only given in the full version of the paper [22], which we encourage the reader to view. It is only remarkable for how straight-forward it is (given the statement).

Proof Overview. We assume that overall $C(x, P(x), r)$ is more often 1 than $C(x, 1-P(x), r)$. Make S the largest set for which the Indistinguishability property is satisfied as follows: order the elements of $\{0,1\}^n$ according to $\Delta_x := \Pr_r[C(x, P(x), r) = 1] - \Pr_r[C(x, 1-P(x), r) = 1]$, and insert them into S sequentially until both $\Pr_{x \leftarrow S,r}[C(x, P(x), r) = 1] > \Pr_{x \leftarrow S,r}[C(x, 1-P(x), r) = 1]$ and indistinguishability is violated. Then, it only remains to describe Q. For any $x \notin S$ note that $\Pr[C(x, P(x), r) = 1] - \Pr[C(x, 1-P(x), r) = 1] \geq \epsilon$, as otherwise x could be added to S. Thus, for those elements $P(x)$ is the bit b for which $\Pr[C(x, b, r) = 1]$ is bigger. In this overview we assume that $\Pr[C(x, b, r) = 1]$ can be found exactly, so we let $Q(x)$ compute the probabilities for $b = 0$ and $b = 1$, and answer accordingly; we will call this rule the "Majority Rule". Clearly, $Q(x)$ is correct if $x \notin S$, and in order to get "predictability", we only need to argue that Q is not worse than a random guess on S.

Consider now Figure 1 (a), where the elements are ordered according to Δ_x. The areas depicted A and B are roughly equal, which follows by the way we chose S (note that $\Pr_{x \leftarrow S,r}[C(x, P(x), r) = 1] - \Pr_{x \leftarrow S,r}[C(x, 1-P(x), r) = 1] = \mathbf{E}_{x \leftarrow S}[\Delta_x]$).

At this point our problem is that the majority rule will give the incorrect answer for all elements for which $\Delta_x < 0$, and as shown in Figure 1 (b), this can be almost all of S, so that in general the above Q *does* perform worse than a random guess on S. The solution is to note that it is sufficient to follow the majority rule in case the gap is bigger than Δ_{x^*}. In the full proof we will see that if the gap is small so that $-\Delta_{x^*} \leq \Pr[C(x, 0, r) = 1] - \Pr[C(x, 1, r) = 1] \leq \Delta_{x^*}$

[1] This implies that $\delta \geq \frac{\epsilon}{2}$, which can always be guaranteed.

then a randomized decision works: the probability of answering $b = 0$ is 1 if the gap is $-\Delta_{x^*}$, the probability of answering $b = 0$ is 0 if the gap is Δ_{x^*}. When the gap is in between then the probability of answering $b = 0$ is linearly interpolated based on the value of the gap. So for example, if the gap is 0, then $b = 0$ with probability $\frac{1}{2}$.[2] A bit of thought reveals that this is exactly because the areas A and B in Figure 1 are almost equal.

In the full proof, we also show how to sample all quantities accurately enough (which is easy) and how to ensure that S is a set of the right size (which seems to require a small trick because Δ_x as defined above is not computable exactly, and so we actually use a different quantity for Δ_x). We think that the second is not really required for the applications later, but it simplifies the statement of the above theorem and makes it somewhat more intuitive.

3.2 Multiple Instances

Informal Discussion. We explain our idea on an example: suppose we want to prove Yao's XOR-Lemma. Thus, we are given a predicate $P : \{0,1\}^n \to \{0,1\}$ which is somewhat hard to compute, i.e., $\Pr[C^{(1)}(x) = P(x)] < 1 - \frac{\delta}{2}$ for any circuit $C^{(1)}$ coming from some family of circuits (the superscript (1) should indicate that this is a circuit operating on a single instance). We want to show that any circuit $C^{(\oplus k)}$ from a related family predicts $P(x_1) \oplus \cdots \oplus P(x_k)$ from $(x_1, \ldots, x_k)$ correctly with probability very close to $\frac{1}{2}$, and aiming for a contradiction we now assume that a circuit $C^{(\oplus k)}$ exists which does significantly better than this is given.

As a first step, we transform $C^{(\oplus k)}$ into a circuit $C^{(k)}(x_1, b_1, x_2, b_2, \ldots, x_k, b_k)$ as follows: $C^{(k)}$ invokes $C^{(\oplus k)}(x_1, \ldots, x_k)$ and outputs 1 if the result equals $b_1 \oplus \cdots \oplus b_k$, otherwise it outputs 0. We see that we would like to show $\Pr[C^{(k)}(x_1, P(x_1), \ldots, x_k, P(x_k)) = 1] \approx \frac{1}{2}$.

Here is the key idea: we apply Theorem 1 sequentially on every position i of $C^{(k)}$. Done properly, in each position one of the following happens: (a) we can use $C^{(k)}$ to predict $P(x)$ from x with probability at least $1 - \frac{\delta}{2}$, or (b) we find a large set S_i^* such that if $x_i \in S_i^*$, $C^{(k)}$ behaves roughly the same in case b_i equals $P(x_i)$ and in case b_i is a uniform random bit. If (a) happens at any point we get a contradiction and are done, so consider the case that (b) happens k times. Recall now how $C^{(k)}$ was built from $C^{(\oplus k)}$: it compares the output of $C^{(\oplus k)}$ to $b_1 \oplus \cdots \oplus b_k$. If x_i lands in the large set for any i we can assume that b_i is a random bit (and it is very unlikely that this happens for no i). Then, $C^{(k)}$ outputs 1 exactly if $C^{(\oplus k)}$ correctly predicts a uniform random bit which is independent of the input to $C^{(\oplus k)}$. The probability such a prediction is correct is exactly $\frac{1}{2}$, and overall we get that $C^{(\oplus k)}$ is correct with probability close to $\frac{1}{2}$.

The theorem gives the formal statement for $C^{(k)}$, in the full version the transformation to $C^{(\oplus k)}$ is done as an example.

[2] It may be instructive to point out another rule which does not work: if one produces a uniform random bit in case the gap is smaller than Δ_{x^*} then elements in the region marked A with negative gap larger than Δ_{x^*} are problematic.

The Theorem. Fix a predicate $P : \{0,1\}^n \to \{0,1\}$ and a boolean circuit $C^{(k)}(x_1, b_1, \ldots, x_k, b_k)$. We are interested in the probability that the circuit outputs 1 in the following Experiment 1:

Experiment 1:
$\forall i \in \{1, \ldots, k\} : x_i \leftarrow \{0,1\}^n$
$\forall i \in \{1, \ldots, k\} : b_i := P(x_i)$
$r \leftarrow \{0,1\}^*$
output $C^{(k)}(x_1, b_1, \ldots, x_k, b_k, r)$

We will claim that there are large sets $S_1^*, \ldots, S_k^*$ with the property that for any x_i which falls into S_i^*, we can set b_i to a random bit and the probability of the experiment producing a 1 will not change much. However, we will allow the sets S_i^* to depend on the x_j and b_j for $j < i$; we therefore assume that an algorithm GenS is given which produces such a set on input $t_i = (x_1, b_1, \ldots, x_{i-1}, b_{i-1})$.

Experiment 2:
for $i := 1$ to k **do**
  $t_i := (x_1, b_1, \ldots, x_{i-1}, b_{i-1})$
  $S_i^* := \text{GenS}(t_i)$
  $x_i \leftarrow \{0,1\}^n$
  if $x_i \in S_i^*$ **then** $b_i \leftarrow \{0,1\}$ **else** $b_i := P(x_i)$ **fi**
end for
$r \leftarrow \{0,1\}^*$
output $C^{(k)}(x_1, b_1, \ldots, x_k, b_k, r)$

Theorem 2 essentially states the following: assume no small circuit can predict $P(x)$ from x with probability $1 - \frac{\delta}{2}$. For any fixed circuit $C^{(k)}$, any ϵ, and any k there is an algorithm GenS which produces sets S_i^* with $\mu(S_i^*) \geq \delta$ and such that the probability that Experiment 1 outputs 1 differs by at most ϵ from the probability that Experiment 2 outputs 1.

Theorem 2. *Let P be a computable predicate, $k, \frac{1}{\epsilon} \in \text{poly}(n)$ parameters. There are two algorithms* Gen *and* GenS *as follows:* Gen *takes as input a randomized circuit $C^{(k)}$ and a parameter ϵ and outputs a deterministic circuit Q of size* $\text{size}(C^{(k)}) \cdot \text{poly}(n)$ *as well as $\delta \in [0,1]$.* GenS *takes as input a circuit $C^{(k)}$, a tuple t_i, and a parameter ϵ and outputs a deterministic circuit $S_{t_i}(x, b)$ of* $\text{size}(C^{(k)}) \cdot \text{poly}(n)$. *After a run of* Gen, *almost surely the following properties are satisfied:*

Large Sets: *For any value of $t_i := (x_1, b_1, \ldots, x_{i-1}, b_{i-1})$ the circuit $S_{t_i}(x_i, P(x_i))$ recognizes a set $S_i^* := \{x_i | S(t_i, x_i, P(x_i)) = 1\}$. The probability that in an execution of Experiment 2 we have $\mu(S_i^*) < \delta$ for any of the S_i^* which occur is at most ϵ.*

Indistinguishability: *Using sets $S_{t_i}^*$ as above in Experiment 2 gives*

$$\Big|\Pr[\textit{Experiment 1 outputs 1}] - \Pr[\textit{Experiment 2 outputs 1}\Big| \leq \epsilon. \tag{2}$$

Predictability: *Q predicts P well:* $\Pr_{x \leftarrow \{0,1\}^n}[Q(x) = P(x)] \geq 1 - \dfrac{\delta}{2}$.

Additionally, these algorithms have the following properties:

1. *Unless* $\delta = 1$ *algorithm* Q *predicts slightly better:* $\Pr[Q(x) = P(x)] \geq 1 - \frac{\delta}{2} + \frac{\epsilon}{16k}$.
2. *If* P *is efficiently samplable (i.e., pairs* $(x, P(x))$ *can be generated in polynomial time),* Gen *and* GenS *run in time* poly(n).
3. Gen, GenS, S_{t_i}, *and* Q *can be implemented with oracle access to* C *only (i.e., they don't use the description of* C*).*
4. *When thought of as oracle circuits,* S_{t_i} *and* Q *use the oracle* C *at most* $\mathcal{O}(\frac{k^2 n}{\epsilon^2})$ *times. Also, they both treat* x *obliviously and are non-rewinding. Finally, their output only depends on the number of* 1*'s obtained from the oracle calls to* C *and, in case of* S_{t_i}, *the input* $P(x)$.

The proof is given in the full version, but follows the informal discussion above. We encourage the interested reader to see the full version [22].

3.3 Cryptographic Protocols Which Output Single Bits

Again we start with an example: consider a slightly weak bit commitment protocol, where the receiver can guess the bit the sender committed to with probability $1 - \frac{\delta}{2}$. In such a case, we might want to strengthen the scheme. For example, in order to commit to a single bit b, we could ask the sender to first commit to two random bits r_1 and r_2, and then send $b \oplus r_1 \oplus r_2$ to the receiver. The hope is that the receiver has to guess both r_1 and r_2 correctly in order to find b, and so the protocol should be more secure.

In the case where the protocol has some defect that sometimes allows a sender to cheat, we might also want to consider the protocol where the sender commits twice to b, or, alternatively, that he commits to r_1, then to r_2, and sends both $b \oplus r_1$ and $b \oplus r_2$ to the receiver. In this case, one can hope that a cheating receiver still needs to break the protocol at least once, and that the security should not degrade too much.

Just how will the security change? We want to consider a scenario in which the security is information theoretic. We can do this by assuming that instead of the weak protocol, a trusted party distributes a bit X to the sender and some side information Z to the receiver. The guarantee is that for any f, $\Pr[f(Z) = X] \leq 1 - \frac{\delta}{2}$. In such a case, one can easily obtain bounds on the security of the above protocols, and the hope is that the same bounds hold in the computational case. The theorem below states that this is indeed true (for protocols where the security consists of hiding single bits).

We remark that while the two aforementioned examples of protocol composition are already handled in [16] (their result applies to any direct product and any XOR as above), Theorem 3 handles any information theoretic amplification protocol as long as it can be implemented efficiently.

Definition 3. *A pair* (X, Z) *of random variables over* $\{0, 1\} \times \mathcal{Z}$*, where* $\mathcal{Z}$ *is any finite set, is* δ*-hiding if*

$$\max_{f:\mathcal{Z}\to\{0,1\}} \Pr[f(Z) = X] \leq 1 - \frac{\delta}{2}. \tag{3}$$

Theorem 3. *Let a cryptographic protocol (which we think of as "weak") $W = (A_W, B_W)$ be given in which A_W has as input a single bit c. Assume that there is a function δ such that for any polynomial time adversary B^*_W there is a negligible function ν such that*

$$\Pr_{x \leftarrow \{0,1\}}[\langle A_W(x), B^*_W \rangle_B = x] \leq 1 - \frac{\delta}{2} + \nu(n), \tag{4}$$

where the probability is also over the coins of A_W and B^*_W (if any).

Let further an information theoretic protocol $I = (A_I, B_I)$ be given. In I, A_I takes k input bits $(X_1, \ldots, X_k)$ and has a single output bit. Furthermore, assume that I is hiding in the sense that for k independent δ-hiding random variables (X_i, Z_i), any (information theoretic) adversary B^*_I, and for some function $\eta(k)$:

$$\Pr\left[\begin{matrix}\langle A_I(X_1,\ldots,X_k), B^*_I(Z_1,\ldots,Z_k)\rangle_A = \\ \langle A_I(X_1,\ldots,X_k), B^*_I(Z_1,\ldots,Z_k)\rangle_B\end{matrix}\right] < \frac{1}{2} + \eta(k). \tag{5}$$

Let $S = (A_S, B_S)$ be the protocol where A and B first execute $k(n)$ copies of W sequentially, where A uses uniform random bits as input. Then, they run a single execution of protocol I. In the execution to I, A uses his k input bits to the weak protocols as input. The output of A in S is the output of A in the execution of I. We also need that (A_I, B_I) and $k(n)$ are such that I can be run in time poly(n) *for $k = k(n)$.*

Then, for any polynomial time B^*_S there is a negligible function ν' such that

$$\Pr[\langle A_S, B^*_S\rangle_A = \langle A_S, B^*_S\rangle_B] \leq \frac{1}{2} + \eta(k) + \nu'(n)\ . \tag{6}$$

Proof. Let $x \in \{0,1\}^n$ be the concatenation of the randomness which A uses in an execution of the protocol W and his input bit c. We let $P : \{0,1\}^n \to \{0,1\}$ be the predicate which outputs $c = P(x)$.

In order to obtain a contradiction, we fix an adversary B^*_S for the protocol S which violates (6). We would like to apply Theorem 2. For this, we define $C^{(k)}(x_1, b_1, \ldots, x_k, b_k)$ as follows: $C^{(k)}$ first simulates an interaction of B^*_S with A_S, where A_S uses randomness x_i in the ith invocation of the weak protocol W. After this, B^*_S is in some state in which it expects an invocation of the information theoretic protocol. $C^{(k)}$ simulates this information theoretic protocol, but it runs A_I with inputs $b_1, \ldots, b_k$ instead of the actual inputs to the weak protocols. In the end, B^*_S produces a guess for the output bit of A_S, and $C^{(k)}$ outputs 1 if this guess equals the output of $A_I(b_1, \ldots, b_k)$ in the simulation.

In Experiment 1 of Theorem 2, $b_i = P(x_i)$ is used, and so $C^{(k)}$ exactly simulates an execution of the protocol S. Since we assume that B^*_S contradicts (6), we see that the probability that $C^{(k)}$ outputs 1 in Experiment 1 is, for infinitely many n and some constant c at least $\frac{1}{2} + \eta(k) + n^{-c}$.

We now apply Theorem 2 on the circuit $C^{(k)}$ with parameter $n^{-c}/3$. This yields a parameter δ_{T2} (the subscript indicates that it is from Theorem 2). We claim that

$$\delta_{T2} \leq \delta \qquad \text{almost surely.} \tag{7}$$

To see this, we assume otherwise and obtain a contradiction. In Experiment 2, Let Γ_i be the communication produced by the weak protocol W in round i. Assuming all sets S_i^* in the execution are of size at least δ (this happens with probability at least $1 - n^{-c}/3$), the tuples (b_i, Γ_i) are δ-hiding random variables. Consequently, when the circuit $C^{(k)}$ simulates the information theoretic protocol I using bits b_i, it actually simulates it in an instance in which it was designed to be used. Since (5) holds for an arbitrary adversary in this case we get that

$$\Pr[C^{(k)} \text{ outputs 1 in Experiment 2} | \text{No set } S_i^* \text{ was of measure less than } \delta] \\ \leq \frac{1}{2} + \eta(k). \tag{8}$$

Therefore, the probability that $C^{(k)}$ outputs 1 in Experiment 2 is at most $\frac{1}{2} + \eta(k) + \frac{n^{-c}}{3}$, and using "indistinguishability" the probability that $C^{(k)}$ outputs 1 in Experiment 1 is at most $\frac{1}{2} + \eta(k) + \frac{2n^{-c}}{3}$. However, our assumption was that the probability that $C^{(k)}$ outputs 1 is at least $\frac{1}{2} + \eta(k) + n^{-c}$, and so almost surely Gen does not output such a big δ_{T2}, establishing (7).

Theorem 2 also provides us with a non-rewinding circuit Q which treats x obliviously and which satisfies "predictability". We explain how to use Q to break (4), the security property of the weak protocol W.

Since $Q(x)$ is non-rewinding, it uses the input x exclusively in a fixed position i, together with a fixed prefix $(x_1, \ldots, x_{i-1})$, in all calls to $C^{(k)}$. We first extract i and the prefix.

We now explain a crucial point: how to interact with A_W in order to cheat. We simulate the $i-1$ interactions of A_W with B_S^* up to and including round $i-1$ using $(x_1, \ldots, x_{i-1})$ as the input bit and randomness of A. In round i, we continue with the *actual* interaction with A_W. Here, A_W uses randomness x (on which we, however, do not have access).

After this interaction, we need to be able to extract the bit c of A_W. For this, we evaluate $Q(x)$, which we claim is possible. Since Q is oblivious and deterministic, the only difficulty is in evaluating the calls to $C^{(k)}(x_1, b_1, \ldots, x_k, b_k, r)$. All calls use the same values for $x_1, \ldots, x_i$. Recalling how $C^{(k)}$ is defined, we see that we can continue from the state we had after the interaction with A_W in order to evaluate $C^{(k)}$ completely (note that all the b_i are given, so the we can also evaluate the information theoretic protocol I).

We get from Theorem 2 that Q satisfies, almost surely, infinitely often, using (7)

$$\Pr_{x \leftarrow \{0,1\}^n}[Q(x) = P(x)] \geq 1 - \frac{\delta}{2} + \frac{1}{48kn^c} . \tag{9}$$

This therefore gives a contradiction to (4): in order to get rid of the "almost surely", we just consider the algorithm which first runs Gen and then applies the above protocol – this only loses a negligible additive term in the probability.

4 Weakly Verifiable Puzzles

4.1 Interactive Weakly Verifiable Puzzles

Consider a bit commitment protocol, in which a sender commits to a single bit b. In a first phase the sender and the receiver enact an interactive protocol, after which the sender holds some opening information y, and the receiver has some way of checking whether (y, b) is a valid decommitment. If the protocol is secure, then it is a computationally hard problem for the sender to come up with two strings y_0 and y_1 such that both $(y_0, 0)$ and $(y_1, 1)$ are valid decommitments, in addition, he may not even know the function the receiver will use to validate a decommitment pair,[3] and thus in general there is no way for the sender to recognize a valid pair (y_0, y_1). We abstract this situation in the following definition; in it we can say that the solver produces no output because in the security property all efficient algorithms are considered anyhow.

Definition 4. *An* interactive weakly verifiable puzzle *consists of a protocol* (P, S) *and is given by two interactive algorithms* P *and* S, *in which* P *(the problem poser) produces as output a circuit* Γ, *and* S *(the solver) produces no output.*

The success probability *of an interactive algorithm* S^* *in solving a weakly verifiable puzzle* (P, S) *is:*

$$\Pr[y = \langle P, S^* \rangle_{S^*}; \Gamma(y) = 1] \tag{10}$$

The puzzle is non-interactive *if the protocol consists of* P *sending a single message to* S.

Our definition of a non-interactive weakly verifiable puzzle coincides with the usual one [4]. The security property of an interactive weakly verifiable puzzle is that for any algorithm (or circuit) S^* of a restricted class, the success probability of S^* is bounded.

An important property is that S^* does not get access to Γ. Besides bit commitment above, an example of such a puzzle is a CAPTCHA. In both cases it is not obvious whether a given solution is actually a correct solution.

4.2 Strengthening Interactive Weakly Verifiable Puzzles

Suppose that g is a monotone boolean function with k bits of input, and $(P^{(1)}, S^{(1)})$ is a puzzle. We can consider the following new puzzle $(P^{(g)}, S^{(g)})$: the sender and the receiver sequentially create k instances of $(P^{(1)}, S^{(1)})$, which yields circuits

[3] One might want to generalize this by saying that in order to open the commitment, sender and receiver enter yet another interactive protocol. However, our presentation is without loss of generality: the sender can send the randomness he used in the first protocol instead. The receiver then checks, if this randomness together with b indeed produces the communication in the first round, and whether in a simulation of the second protocol he accepts.

$\Gamma^{(1)},\ldots,\Gamma^{(k)}$ for P. Then $P^{(g)}$ outputs the circuit $\Gamma^{(g)}$ which computes $\Gamma^{(g)}(y_1,\ldots,y_k)=g(\Gamma^{(1)}(y_1),\ldots,\Gamma^{(k)}(y_k))$.

Intuitively, if no algorithm solves a single puzzle $(P^{(1)},S^{(1)})$ with higher probability than δ, the probability that an algorithm solves $(P^{(g)},S^{(g)})$ should not be more than approximately $\Pr_{u\leftarrow\mu_\delta^k}[g(u)=1]$. (Recall that μ_δ^k is the distribution on k-bits, where each bit is independent and 1 with probability δ.) The following theorem states exactly this.

Theorem 4. *There exists an algorithm* $\text{Gen}(C,g,\epsilon,\delta,n)$ *which takes as input a circuit* C*, a monotone function* g*, and parameters* ϵ,δ,n*, and produces a circuit* D *such that the following holds. If* C *is such that*

$$\Pr[\Gamma^{(g)}(\langle P^{(g)},C\rangle_C)=1]\geq \Pr_{u\leftarrow\mu_\delta^k}[g(u)=1]+\epsilon, \tag{11}$$

then, D *satisfies almost surely,*

$$\Pr[\Gamma^{(1)}(\langle P^{(1)},D\rangle_D)=1]\geq \delta+\frac{\epsilon}{6k}. \tag{12}$$

Additionally, Gen *and* D *only require oracle access to both* g *and* C*, and* D *is non-rewinding.*

Furthermore, $\text{size}(D)\leq\text{size}(C)\cdot\frac{6k}{\epsilon}\log(\frac{6k}{\epsilon})$ *and* Gen *runs in time* $\text{poly}(k,\frac{1}{\epsilon},n)$ *with oracle calls to* C*.*

The monotone restriction on g in the previous theorem is necessary. For example, consider $g(b)=1-b$. It is possible to satisfy g with probability 1 by producing an incorrect answer, but $\Pr_{u\leftarrow\mu_\delta}[g(u)=1]=1-\delta$.

4.3 Proof of Theorem 4

Algorithm Description. If $k=1$, Gen creates the circuit D which runs C and outputs its answer. Then either g is the identity or a constant function. If g is the identity, the statement is trivial. If g is a constant function, the statement is vacuously true. D is non-rewinding.

In the general case, we need some notation. For $b\in\{0,1\}$, let $\mathcal{G}_b$ denote the set of inputs $\mathcal{G}_b:=\{b_1,\ldots,b_k|g(b,b_2,\ldots,b_k)=1\}$ (i.e., the first input bit is disregarded and replaced by b). We remark that $\mathcal{G}_0\subseteq\mathcal{G}_1$ due to monotonicity of g. We will commonly denote by $u=u_1u_2\cdots u_k\in\{0,1\}^k$ an element drawn from μ_δ^k. After a given interaction of C with $P^{(g)}$, let $c=c_1c_2\cdots c_k\in\{0,1\}^k$ denote the string where c_i is the output of $\Gamma^{(i)}$ on input y_i, which is the ith output of C. We denote the randomness used by $P^{(g)}$ in execution i by π_i.

For $\pi^*,b\in\{0,1\}^n\times\{0,1\}$ we now define the surplus $S_{\pi^*,b}$. It denotes how much better C performs than "it should", in the case where the randomness of $P^{(g)}$ in the first instance is fixed to π^*, and the output of $\Gamma^{(1)}(y_1)$ is ignored (i.e., we don't care whether C solves the first puzzle right), and b is used instead:

$$S_{\pi^*,b}:=\Pr_{\pi^{(k)}}[c\in\mathcal{G}_b|\pi_1=\pi^*]-\Pr_{u\leftarrow\mu_\delta^k}[u\in\mathcal{G}_b], \tag{13}$$

where the first probability is also over the interaction between $P^{(g)}$ and C as well as randomness C uses (if any).

The algorithm then works as follows: first pick $\frac{6k}{\epsilon}\log(n)$ candidates π^* for the randomness of $P^{(g)}$ in the first position. For each of those, simulate the interaction $(P^{(g)}, C)$ and then get estimates $\widetilde{S}_{\pi^*,0}$ and $\widetilde{S}_{\pi^*,1}$ of $S_{\pi^*,0}$ and $S_{\pi^*,1}$ such that $|\widetilde{S}_{\pi^*,b} - S_{\pi^*,b}| \leq \frac{\epsilon}{4k}$ almost surely.

We consider two cases:

- One of the estimates satisfies $\widetilde{S}_{\pi^*,b} \geq (1 - \frac{3}{4k})\epsilon$.
  In this case, we fix $\pi_1 := \pi^*$ and $c_1 := b$, and invoke $\mathrm{Gen}(C', g', (1-\frac{1}{k})\epsilon, \delta, n)$, using the function $g'(b_2, \ldots, b_k) = g(c_1, b_2, \ldots, b_k)$ and circuit C' which is defined as follows: C' first (internally) simulates an interaction of $P^{(1)}$ with C, then follows up with an interaction with $P^{(g')}$.
- For all estimates $\widetilde{S}_{x^*,b} < (1 - \frac{3}{4k})\epsilon$.
  In this case, we output the following circuit D^C: in a first phase, use C to interact with $P^{(1)}$. In the second phase, simulate $k-1$ interactions with $P^{(1)}$ and obtain $(y_1, \ldots, y_k) = C(x, x_2, \ldots, x_k)$. For $i = 2, \ldots, k$ set $c_i = \Gamma_i(y_i)$. If $c = (0, c_2, \ldots, c_k) \in \mathcal{G}_1 \setminus \mathcal{G}_0$, return y_1, otherwise repeat the second phase $\frac{6k}{\epsilon}\log(\frac{6k}{\epsilon})$ times. If all attempts fail, return the special value $\perp$ (or an arbitrary answer).

Due to space constraints, the proof of correctness of the above algorithm is omitted, but can be found in the full version [22].

5 Example: Bit Commitment

Theorems 3 and 4 can be used to show how to strengthen bit commitment protocols, which was the main open problem in [16]. We explain this as an example here. Assume we have given a weak bit protocol, where a cheating receiver can guess a bit after the commitment phase with probability $1 - \frac{\beta}{2}$, and a cheating sender can change the bit he committed to with probability α. We show that such a protocol can be strengthened if $\alpha < \beta - 1/\mathrm{poly}(n)$.

We should point out that a different way to prove a similar theorem exists: one can first show that such a weak bit-commitment protocol implies one-way functions (using the techniques of [27]). The long sequence of works [17,32,36,33,14] imply that one-way functions are sufficient to build bit commitment protocols (the first two papers will yield statistically binding protocols, the last three statistically hiding protocols). However, this will be less efficient and also seems less natural than the method we use here.

In the example, we first define weak bit commitment protocols. We then recall a theorem by Valiant [37], and then show how to use it to strengthen bit commitment. However, due to space constraints, the example only appears in the full version [22].

Acknowledgments

We would like to thank the anonymous referees for useful comments.

References

1. Barak, B., Hardt, M., Kale, S.: The uniform hardcore lemma via approximate bregman projections. In: SODA, pp. 1193–1200 (2009)
2. Barak, B., Shaltiel, R., Wigderson, A.: Computational analogues of entropy. In: Arora, S., Jansen, K., Rolim, J.D.P., Sahai, A. (eds.) RANDOM 2003 and APPROX 2003. LNCS, vol. 2764, pp. 200–215. Springer, Heidelberg (2003)
3. Bellare, M., Impagliazzo, R., Naor, M.: Does parallel repetition lower the error in computationally sound protocols? In: FOCS 1997, pp. 374–383 (1997)
4. Canetti, R., Halevi, S., Steiner, M.: Hardness Amplification of Weakly Verifiable Puzzles. In: Kilian, J. (ed.) TCC 2005. LNCS, vol. 3378, pp. 17–33. Springer, Heidelberg (2005)
5. Chung, K.-M., Liu, F.-H.: Parallel Repetition Theorems for Interactive Arguments. In: Micciancio, D. (ed.) TCC 2010. LNCS, vol. 5978, pp. 19–36. Springer, Heidelberg (2010)
6. Chung, K.-M., Liu, F.-H., Lu, C.-J., Yang, B.-Y.: Efficient string-commitment from weak bit-commitment and full-spectrum theorem for puzzles (2009) (manuscript)
7. Damgård, I., Fehr, S., Morozov, K., Salvail, L.: Unfair Noisy Channels and Oblivious Transfer. In: Naor, M. (ed.) TCC 2004. LNCS, vol. 2951, pp. 355–373. Springer, Heidelberg (2004)
8. Damgård, I., Kilian, J., Salvail, L.: On the (Im)possibility of Basing Oblivious Transfer and Bit Commitment on Weakened Security Assumptions. In: Stern, J. (ed.) EUROCRYPT 1999. LNCS, vol. 1592, pp. 56–73. Springer, Heidelberg (1999)
9. Dodis, Y., Impagliazzo, R., Jaiswal, R., Kabanets, V.: Security Amplification for *Interactive* Cryptographic Primitives. In: Reingold, O. (ed.) TCC 2009. LNCS, vol. 5444, pp. 128–145. Springer, Heidelberg (2009)
10. Dwork, C., Naor, M., Reingold, O.: Immunizing Encryption Schemes from Decryption Errors. In: Cachin, C., Camenisch, J. (eds.) EUROCRYPT 2004. LNCS, vol. 3027, pp. 342–360. Springer, Heidelberg (2004)
11. Goldreich, O., Nisan, N., Wigderson, A.: On Yao's XOR-lemma. Technical Report TR95-050, Electronic Colloquium on Computational Complexity (ECCC) (1995)
12. Haitner, I.: A parallel repetition theorem for any interactive argument. In: FOCS 2009, pp. 241–250 (2009)
13. Haitner, I., Harnik, D., Reingold, O.: On the Power of the Randomized Iterate. In: Dwork, C. (ed.) CRYPTO 2006. LNCS, vol. 4117, pp. 22–40. Springer, Heidelberg (2006)
14. Haitner, I., Reingold, O.: Statistically-hiding commitment from any one-way function. In: Proceedings of the Thirty-Ninth Annual ACM Symposium on Theory of Computing, pp. 1–10 (2007)
15. Haitner, I., Reingold, O., Vadhan, S.: Efficiency improvements in constructing pseudorandom generators from one-way functions. In: Proceedings of the Forty-Second Annual ACM Symposium on Theory of Computing (2010)
16. Halevi, S., Rabin, T.: Degradation and Amplification of Computational Hardness. In: Canetti, R. (ed.) TCC 2008. LNCS, vol. 4948, pp. 626–643. Springer, Heidelberg (2008)
17. Håstad, J., Impagliazzo, R., Levin, L.A., Luby, M.: A pseudorandom generator from any one-way function. SIAM Journal on Computing 28(4), 1364–1396 (1999)

18. Håstad, J., Pass, R., Wikström, D., Pietrzak, K.: An Efficient Parallel Repetition Theorem. In: Micciancio, D. (ed.) TCC 2010. LNCS, vol. 5978, pp. 1–18. Springer, Heidelberg (2010)
19. Holenstein, T.: Key agreement from weak bit agreement. In: Proceedings of the Thirty-Seventh Annual ACM Symposium on Theory of Computing, pp. 664–673 (2005)
20. Holenstein, T.: Strengthening Key Agreement using Hard-Core Sets. PhD thesis, ETH Zürich (2006)
21. Holenstein, T., Renner, R.: One-Way Secret-Key Agreement and Applications to Circuit Polarization and Immunization of Public-Key Encryption. In: Shoup, V. (ed.) CRYPTO 2005. LNCS, vol. 3621, pp. 478–493. Springer, Heidelberg (2005)
22. Holenstein, T., Schoenebeck, G.: General hardness amplification of predicates and puzzles. CoRR, abs/1002.3534 (2010)
23. Impagliazzo, R.: Hard-core distributions for somewhat hard problems. In: The 36th Annual Symposium on Foundations of Computer Science, pp. 538–545 (1995)
24. Impagliazzo, R., Jaiswal, R., Kabanets, V.: Approximately list-decoding direct product codes and uniform hardness amplification. In: The 47th Annual Symposium on Foundations of Computer Science, pp. 187–196 (2006)
25. Impagliazzo, R., Jaiswal, R., Kabanets, V.: Chernoff-type direct product theorems. Journal of Cryptology (2009)
26. Impagliazzo, R., Jaiswal, R., Kabanets, V., Wigderson, A.: Uniform direct product theorems: simplified, optimized, and derandomized. In: Proceedings of the Fortieth Annual ACM Symposium on Theory of Computing, pp. 579–588 (2008)
27. Impagliazzo, R., Luby, M.: One-way functions are essential for complexity based cryptography. In: The 30th Annual Symposium on Foundations of Computer Science, pp. 230–235 (1989)
28. Impagliazzo, R., Wigderson, A.: P = BPP if E requires exponential circuits: Derandomizing the XOR lemma. In: Proceedings of the Twenty-Ninth Annual ACM Symposium on Theory of Computing, pp. 220–229 (1997)
29. Jutla, C.S.: Almost Optimal Bounds for Direct Product Threshold Theorem. In: Micciancio, D. (ed.) TCC 2010. LNCS, vol. 5978, pp. 37–51. Springer, Heidelberg (2010)
30. Klivans, A.R., Servedio, R.A.: Boosting and hard-core sets. In: The 40th Annual Symposium on Foundations of Computer Science, pp. 624–633 (1999)
31. Levin, L.A.: One-way functions and pseudorandom generators. Combinatorica 7(4), 357–363 (1987)
32. Naor, M.: Bit commitment using pseudorandomness. Journal of Cryptology 4(2), 151–158 (1991)
33. Nguyen, M.-H., Ong, S.J., Vadhan, S.P.: Statistical zero-knowledge arguments for NP from any one-way function. In: The 47th Annual Symposium on Foundations of Computer Science, pp. 3–14 (2006)
34. Pass, R., Venkitasubramaniam, M.: An efficient parallel repetition theorem for arthur-merlin games. In: STOC 2007, pp. 420–429 (2007)
35. Pietrzak, K., Wikström, D.: Parallel Repetition of Computationally Sound Protocols Revisited. In: Vadhan, S.P. (ed.) TCC 2007. LNCS, vol. 4392, pp. 86–102. Springer, Heidelberg (2007)
36. Rompel, J.: One-way functions are necessary and sufficient for secure signatures. In: Proceedings of the Twenty-Second Annual ACM Symposium on Theory of Computing, pp. 387–394 (1990)

37. Valiant, L.G.: Short monotone formulae for the majority function. Journal of Algorithms 5, 363–366 (1984)
38. Wullschleger, J.: Oblivious-Transfer Amplification. In: Naor, M. (ed.) EUROCRYPT 2007. LNCS, vol. 4515, pp. 555–572. Springer, Heidelberg (2007)
39. Yao, A.C.: Theory and applications of trapdoor functions (extended abstract). In: The 23rd Annual Symposium on Foundations of Computer Science, pp. 80–91 (1982)

Security Amplification for the Cascade of Arbitrarily Weak PRPs: Tight Bounds via the Interactive Hardcore Lemma

Stefano Tessaro

Department of Computer Science and Engineering
University of California, San Diego
9500 Gilman Drive
La Jolla, CA 92093-0404, USA
stessaro@cs.ucsd.edu

Abstract. We consider the task of amplifying the security of a weak pseudorandom permutation (PRP), called an ε-PRP, for which the computational distinguishing advantage is only guaranteed to be bounded by some (possibly non-negligible) quantity $\varepsilon < 1$. We prove that the cascade (i.e., sequential composition) of m ε-PRPs (with independent keys) is an $((m - (m-1)\varepsilon)\varepsilon^m + \nu)$-PRP, where ν is a negligible function. In the asymptotic setting, this implies security amplification for all $\varepsilon < 1 - \frac{1}{\mathsf{poly}}$, and the result extends to two-sided PRPs, where the inverse of the given permutation is also queried. Furthermore, we show that this result is essentially tight. This settles a long-standing open problem due to Luby and Rackoff (STOC '86).

Our approach relies on the first hardcore lemma for computational indistinguishability of *interactive* systems: Given two systems whose states do not depend on the interaction, and which no efficient adversary can distinguish with advantage better than ε, we show that there exist events on the choices of the respective states, occurring each with probability at least $1 - \varepsilon$, such that the two systems are computationally indistinguishable conditioned on these events.

1 Introduction

1.1 Motivation: Weak PRPs

The security of several cryptographic schemes relies on the assumption that an underlying block cipher is a *pseudorandom permutation* (PRP), a keyed family of permutations $E = \{E_k\}_{k \in \mathcal{K}}$ with the following property: any computationally bounded distinguisher can only decide with negligible advantage over random guessing whether it is given access to E_K (under a random secret key K) or to a uniformly chosen permutation with the same domain.

However, pseudorandomness is a very strong requirement, and continuous progress in cryptanalysis raises some doubts as to whether block-cipher designs such as the *Advanced Encryption Standard* (AES) are indeed secure PRPs. It

Y. Ishai (Ed.): TCC 2011, LNCS 6597, pp. 37–54, 2011.

is therefore a prudent approach, as well as a central question in theoretical cryptography, to investigate *weaker* assumptions on a block cipher which are sufficient to efficiently solve a certain cryptographic task at hand.

A natural weakening of a PRP, considered in this paper, is to only require that the best advantage of a computationally restricted distinguisher is bounded by some given quantity $\varepsilon < 1$; we refer to such a primitive as an ε-PRP. In particular, in the asymptotic setting, ε is not required to be a negligible function. Instead, it may be a constant or even moderately converge to one as a function of the security parameter. For instance, common sense dictates that AES is much more likely to be a 0.99-PRP, rather than a fully secure PRP.

1.2 Our Result: Security Amplification of PRPs by Cascading

We investigate the natural and central problem of finding an efficient construction of a fully secure PRP (i.e., a δ-PRP for a negligible δ) from any ε-PRP $E = \{E_k\}_{k\in\mathcal{K}}$. Such constructions should work for arbitrary $\varepsilon < 1$ and call E as few times as possible (ideally, $\log(1/\delta) \cdot (\log(1/\varepsilon))^{-1}$ times to implement a δ-PRP). This is in the same spirit of the long line of research devoted to security amplification initiated by Yao [16] in the context of one-way functions.

The most natural approach is the *m-fold cascade*, a construction which outputs the value

$$(E_{k_1} \circ \cdots \circ E_{k_m})(x)$$

on input x and keys $k_1, \ldots, k_m$ (which are chosen independently). Here, $\circ$ denotes (sequential) composition of permutations.

Despite its simplicity, proving security amplification for the cascade has been a long-standing open problem. On the one hand, Luby and Rackoff [6] and Myers [12] showed that the c-fold cascade is a $((2-\varepsilon)^{c-1}\varepsilon^c + \nu)$-PRP for any *constant* c, where ν is a negligible additive term, but their results fall short of implying that a sufficiently long cascade yields a fully secure PRP for a non-negligible ε. On the other hand, Maurer and Tessaro [10] showed that the cascade of *arbitrary* (polynomial) length m is a $(2^{m-1}\varepsilon^m + \nu)$-PRP, but their bound, which only implies security amplification when $\varepsilon < \frac{1}{2}$, is clearly not tight in view of the superior result for the constant-length case [6,12].

Our Result on Cascades. This paper closes this gap by providing an exact characterization of the security amplification properties of the cascade: We prove that the cascade of m ε-PRPs (with domain $\mathcal{X}$) is a $((m-(m-1)\varepsilon)\varepsilon^m + \nu)$-PRP, i.e., it is security amplifying for essentially any $\varepsilon < 1 - \frac{1}{|\mathcal{X}|}$.[1] The result extends to *two-sided* ε-PRPs, where the inverse can also be queried, and is shown to be nearly tight. Also, this result arises from the application of new *generic* techniques of independent interest, illustrated in the next section.

[1] This restriction is necessary, as an ε-PRP with a fixed point (independent of the key value) can satisfy $\varepsilon = 1 - \frac{1}{|\mathcal{X}|}$, and the cascade obviously preserves such a fixed point.

FURTHER RELATED WORK. Less efficient constructions of fully secure PRPs from ε-PRPs exist: Maurer and Tessaro [11] showed that XORing two independent keys at both ends of the cascade yields an $(\varepsilon^m + \nu)$-PRP. Alternatively, techniques [13,2,10] for strengthening the security of pseudorandom *functions* (PRF) can be used in conjunction with known PRF-to-PRP conversion techniques such as [7]. However, this paper improves on these works in at least two respects: First, we show that similar amplification is achieved with better efficiency by the most natural construction. Second, our approach entails a new set of generic techniques which promise to be applicable in a wider context.

Additionally, let us point out that cascades have been studied in other contexts and models; a comprehensive discussion is deferred to the full version due to space constraints.

1.3 Our General Paradigm: The Interactive Hardcore Lemma and High-Entropy Permutations

The following technique is well known in the study of random processes: One can always define events $\mathcal{A}$ and $\mathcal{B}$ on any two finite random variables X and Y, by means of conditional probability distributions $\mathsf{P}_{\mathcal{A}|X}$ and $\mathsf{P}_{\mathcal{B}|Y}$, such that:

(i) X and Y are equally distributed conditioned on the respective events, i.e., $\mathsf{P}_{X|\mathcal{A}} = \mathsf{P}_{Y|\mathcal{B}}$,
(ii) $\mathsf{P}[\mathcal{A}] = \mathsf{P}[\mathcal{B}] = 1 - d(X, Y)$, where $d(X, Y)$ is the so called *statistical distance*, which equals the best advantage of a computationally *unbounded* distinguisher in distinguishing X and Y.

A computational version of this statement is due to Maurer and Tessaro [11], and was used to prove security amplification results for PRGs. In this paper, we take this approach one step further by presenting a computational version of the above statement for discrete *interactive* systems.

CC-STATELESS SYSTEMS. We consider the general class of *convex-combination stateless* (or simply *cc-stateless*) interactive systems [10]. Like most cryptographic systems of interest, these systems have the property that the answer of each query *can be seen* as depending solely on the query input and on an *initial state*, but does not depend on previous queries and their answers. A simple example is the cc-stateless system implementing a permutation E_K for a keyed family of permutations $\{E_k\}_{k \in \mathcal{K}}$ and a uniform random key $K \in \mathcal{K}$. A further example is a *uniform random permutation* (URP) $\mathbf{P}$ on a set $\mathcal{X}$, a system choosing a permutation $P : \mathcal{X} \to \mathcal{X}$ uniformly at random, and answering each query x as $P(x)$. Moreover, a randomized encryption scheme where each encryption depends on the random key and some fresh randomness is also cc-stateless.

We stress that being cc-stateless is a property of the input-output behavior of a system, rather than of its actual implementation: Indeed, any implementation using such an initial state may be inefficient (e.g., due to its large size), but at the same time an efficient implementation of a cc-stateless system may be fully stateful. For example, an efficient implementation of a URP keeps an interaction-dependent state (in form of a table of all input-output pairs associated with all previous queries) and employs lazy sampling, returning for each new query a uniformly distributed value among those not returned yet.

THE HARDCORE LEMMA. Our main technical tool is the *Hardcore Lemma (HCL) for computational indistinguishability* (Theorem 2): Informally, it states that if all computationally bounded distinguishers only achieve advantage at most ε in distinguishing two cc-stateless systems $\mathbf{S}$ and $\mathbf{T}$, then there exist events $\mathcal{A}$ and $\mathcal{B}$, defined on the respective initial states of (the cc-stateless representations of) $\mathbf{S}$ and $\mathbf{T}$, such that the following holds:

(i) The (cc-stateless representations of the) systems $\mathbf{S}$ and $\mathbf{T}$ are computationally indistinguishable conditioned on the respective events $\mathcal{A}$ and $\mathcal{B}$.
(ii) Both events occur with probability at least $1 - \varepsilon$.

In addition, some applications of the HCL require the ability to efficiently simulate $\mathbf{S}$ and $\mathbf{T}$ under the assumption that the associated events occur (or do not occur), possibly with the help of some short (but not necessarily efficiently-samplable[2]) advice. In general, it is unclear whether this is possible given any two events satisfying (i) and (ii), even if both systems are efficiently implementable.

As an illustrative example, let $\mathbf{S} = E_K$ and $\mathbf{T} = \mathbf{P}$, where $K \in \mathcal{K}$ is uniformly distributed, $E = \{E_k\}_{k \in \mathcal{K}}$ is an efficiently computable family of permutations, and $\mathbf{P}$ is a URP (all permutations are on the n-bit strings). If E is an ε-PRP, the HCL yields an event $\mathcal{A}$ defined on K and an event $\mathcal{B}$ defined on a uniformly chosen permutation table P, both occurring with probability at least $1 - \varepsilon$, such that $E_{K'}$ (for K' sampled from $\mathsf{P}_{K|\mathcal{A}}$) and a system $\mathbf{P}'$ (implementing a permutation table P' sampled from $\mathsf{P}_{P|\mathcal{B}}$) are computationally indistinguishable. While $E_{K'}$ is efficiently implementable given K', a representation of P' requires $2^{\Theta(n)}$ bits, and it is unclear how to define a short advice (i.e., with length $\mathsf{poly}(n)$) that can be used to efficiently simulate $\mathbf{P}'$. However, quite surprisingly, we will show that one can always find events with short advice as long as $\mathbf{S}$ and $\mathbf{T}$ are efficiently implementable. This will be the major challenge in proving the HCL for the interactive setting.

The core of our proof is a tight generalization (Theorem 1) of Impagliazzo's HCL [5] to the setting of guessing a random bit given access to some interactive system whose behavior is correlated with the bit value.

CASCADE OF PERMUTATIONS WITH HIGH MIN-ENTROPY. We briefly illustrate how the HCL is used to prove our bounds for the cascade of ε-PRPs. The main observation is that P' as above has *min-entropy* at least

$$\begin{aligned} \mathsf{H}_\infty(P') &= \log\left(\min_\pi \frac{1}{\mathsf{P}[P' = \pi]}\right) \\ &= \log\left(\min_\pi \frac{\mathsf{P}[\mathcal{B}]}{\mathsf{P}[P = \pi] \cdot \mathsf{P}\left[\mathcal{B} \,\middle|\, P = \pi\right]}\right) \geq \log(2^n!) - \log\left(\frac{1}{1-\varepsilon}\right), \end{aligned}$$

[2] For now, we only consider the non-uniform setting, thus efficient samplability is not a requirement.

i.e., at most $\log\left((1-\varepsilon)^{-1}\right)$ away from the maximum achievable min-entropy. This gap potentially makes $\mathbf{P}'$ easily distinguishable from a URP. However, we prove (Theorem 3) that the cascade of (at least) two such permutations is indistinguishable from a URP for computationally unbounded distinguishers making at most an exponential number of queries and even when allowing queries to the inverse. (The proof uses techniques from the random systems framework [8], and is of independent interest.)

The main security amplification result (Theorem 4) follows from the observation that by the above at least two (independent) permutations E_{K_i} and E_{K_j} (for $i \neq j$) in the cascade $E_{K_1} \circ \cdots \circ E_{K_m}$ (for independent keys $K_1, \ldots, K_m$) are computationally indistinguishable from $\mathbf{P}'$, except with probability $\varepsilon^m + m(1-\varepsilon)\varepsilon^{m-1}$, and in this case the cascade is computationally indistinguishable from a URP by Theorem 3. The final bound follows from a more fine-grained analysis.

UNIFORM VS. NON-UNIFORM PROOFS. The results of this paper are formulated in a concrete, non-uniform, computational model. This simplifies the presentation considerably and helps conveying the main ideas. In the full version, we highlight the changes required in order to obtain uniform statements and proofs.

2 Preliminaries

Calligraphic letters $\mathcal{X}, \mathcal{Y}, \ldots$ denote sets and events, upper-case letters $X, Y, \ldots$ random variables (with expected values $\mathsf{E}[X], \mathsf{E}[Y], \ldots$), and lower-case letters $x, y, \ldots$ the values they take. Moreover, $\mathsf{P}[\mathcal{A}]$ is the probability of an event $\mathcal{A}$ (we denote as $\overline{\mathcal{A}}$ its complement) and we use the shorthands $\mathsf{P}_X(x) := \mathsf{P}[X = x]$, $\mathsf{P}_{X|Y}(x, y) := \mathsf{P}[X = x|Y = y]$, $\mathsf{P}_{X\mathcal{A}|Y\mathcal{B}}(x, y) := \mathsf{P}[\mathcal{A} \wedge X = x|\mathcal{B} \wedge Y = y]$, etc. Also, P_X, $\mathsf{P}_{X|Y}$, $\mathsf{P}_{\mathcal{A}X|\mathcal{B}Y}$ denote the corresponding (conditional) probability distributions,[3] and $x \stackrel{\$}{\leftarrow} \mathsf{P}_X$ is the action of sampling a value x with distribution P_X. (We use $x \stackrel{\$}{\leftarrow} \mathcal{S}$ to denote the special case where x is drawn uniformly from a finite set $\mathcal{S}$.) The *statistical distance* $d(X, Y)$ (or $d(\mathsf{P}_X, \mathsf{P}_Y)$) of X and Y (both with range $\mathcal{S}$) is defined as $d(X, Y) := \frac{1}{2}\sum_{x\in\mathcal{S}} |\mathsf{P}_X(x) - \mathsf{P}_Y(x)| = \sum_{x:\mathsf{P}_X(x)\geq\mathsf{P}_Y(x)} (\mathsf{P}_X(x) - \mathsf{P}_Y(x))$. Also, recall that a function is *negligible* if it vanishes faster than the inverse of any polynomial.

COMPUTATIONAL MODEL. We consider *interactive* randomized stateful algorithms in some a-priori fixed RAM model of computation. Such an algorithm keeps a state (consisting, say, of the contents of the memory space it employs), and answers each query depending on the input of this query, some coin flips, the current state (which may be updated), and (possibly) one or more queries to an underlying system. It is also convenient to denote by $A[\sigma]$ the algorithm obtained by *setting* the state of A to σ (provided σ is a compatible state), and then behaving according to A's description. We say that A has *time complexity* t_A (where t_A is a function $\mathbb{N} \times \mathbb{N} \to \mathbb{N}$) if the sum of the length of the description of A, of s, and the total number of steps of A is at most $t_A(q, s)$ for all sequences of

[3] In particular, $\mathsf{P}_{X|Y}$ and $\mathsf{P}_{\mathcal{A}X|\mathcal{B}Y}$ take *two* arguments corresponding to all possible values of X and Y, respectively.

q queries, all compatible initial states with size s, and all compatible interactions with an underlying system. We use the shorthand $t_A(q) := t_A(q, 0)$. Furthermore, $s_A : \mathbb{N} \to \mathbb{N}$ is the *space complexity* of A, where $s_A(q)$ is the worst-case amount of memory used by A when answering any q queries.

Systems and Distinguishers. This paper considers abstract discrete interactive systems [8], denoted by bold-face letters $\mathbf{S}, \mathbf{T}, \ldots$, taking as inputs queries $X_1, X_2, \ldots$ and returning outputs $Y_1, Y_2, \ldots$. Such systems may be implemented by an interactive algorithm A (in which case we sometimes write A as a placeholder for the system it implements to explicit this fact), but may also arise from an arbitrary random process. The *input-output behavior* of the system $\mathbf{S}$ is fully described by the (infinite) family of conditional probability distributions $\mathsf{p}^{\mathbf{S}}_{Y_i|X^iY^{i-1}}$ (for $i \geq 1$) of the i-th output Y_i given the first i queries $X^i = [X_1, \ldots, X_i]$, and the first $i-1$ outputs $Y^{i-1} = [Y_1, \ldots, Y_{i-1}]$. In general, every statement that involves a system $\mathbf{S}$ holds for any realization of the system $\mathbf{S}$, i.e., it only depends on its input-output behavior. In particular, we say that two systems $\mathbf{S}$ and $\mathbf{T}$ are *equivalent*, denoted $\mathbf{S} \equiv \mathbf{T}$, if they have the same input-output behavior, i.e., $\mathsf{p}^{\mathbf{S}}_{Y_i|X^iY^{i-1}} = \mathsf{p}^{\mathbf{T}}_{Y_i|X^iY^{i-1}}$ for all $i \geq 1$. Moreover, we say that an algorithm A *implements* the system $\mathbf{S}$ if $A \equiv \mathbf{S}$.

A *distinguisher* $\mathbf{D}$ is a special type of system which interacts with another system $\mathbf{S}$ by means of q queries and outputs a decision bit $\mathbf{D}(\mathbf{S})$ depending on their outputs: Its *advantage* in distinguishing systems $\mathbf{S}$ and $\mathbf{T}$ is

$$\Delta^{\mathbf{D}}(\mathbf{S}, \mathbf{T}) := \left|\mathsf{P}\left[\mathbf{D}(\mathbf{S}) = 1\right] - \mathsf{P}\left[\mathbf{D}(\mathbf{T}) = 1\right]\right|.$$

Moreover, $\Delta_q(\mathbf{S}, \mathbf{T})$ is the best distinguishing advantage $\Delta^{\mathbf{D}}(\mathbf{S}, \mathbf{T})$ over *all* q-query $\mathbf{D}$, whereas $\Delta_{t,q}(\mathbf{S}, \mathbf{T})$ is used when the maximization is restricted to distinguishers implemented by an algorithm with time complexity t.

Stateless Systems. A system $\mathbf{S}$ is called *stateless* if the i-th answer Y_i only depends on the i-th query X_i, that is, there exists a conditional distribution $\mathsf{p}^{\mathbf{S}}_{Y|X}$ such that $\mathsf{p}^{\mathbf{S}}_{Y_i|X^iY^{i-1}}(y_i, x^i, y^{i-1}) = \mathsf{p}^{\mathbf{S}}_{Y|X}(y_i, x_i)$ for all $i \geq 1$, $x^i = [x_1, \ldots, x_i]$, and $y^i = [y_1, \ldots, y_i]$. Furthermore, $\mathbf{S}$ is *convex-combination-stateless* (or simply *cc-stateless*) [10] if there exists a system $\mathbf{T}(\cdot)$ accessing a random variable S (called the *initial state*) such that $\mathbf{S} \equiv \mathbf{T}(S)$ and $\mathbf{T}(s)$ is stateless for all values s taken by S. To save on notation, we usually write $\mathbf{S}(\cdot)$ instead of $\mathbf{T}(\cdot)$, but we stress that $\mathbf{S}(\cdot)$ and $\mathbf{S}$ are different objects, despite their notational similarity. We refer to $\mathbf{S}(S)$ as the *cc-stateless representation* of $\mathbf{S}$.

It is crucial to remark that being cc-stateless is a property of the *input-output behavior* of a system: Its (efficient) implementation may well be stateful, and its cc-stateless representation may be completely inefficient (e.g., because the description of the initial state is even too large to be processed by an efficient algorithm).

Random Functions and Permutations. A system $\mathbf{F}$ taking inputs from a set $\mathcal{X}$ and returning outputs in $\mathcal{Y}$ is a *random function* $\mathcal{X} \to \mathcal{Y}$ if for any two equal queries $X_i = X_j$ we have $Y_i = Y_j$ for the respective answers. Furthermore, if $\mathcal{X} = \mathcal{Y}$, it is called a *random permutation* if $X_i \neq X_j$ also implies $Y_i \neq Y_j$. Typical (cc-stateless) examples are *uniform random function* (URF) $\mathbf{R} : \mathcal{X} \to \mathcal{Y}$, which answers according to a uniformly chosen function $\mathcal{X} \to \mathcal{Y}$, a *uniform random permutation* (URP) $\mathbf{P} : \mathcal{X} \to \mathcal{X}$, implementing a uniformly chosen permutation $\mathcal{X} \to \mathcal{X}$, or E_K for a permutation family $\{E_k\}_{k \in \mathcal{K}}$ and a random $K \in \mathcal{K}$.

The initial state of a cc-stateless random function $\mathbf{F}$ can always be seen without loss of generality as a (randomly chosen) *function table* F according to which $\mathbf{F}$ answers its queries, and usually write $\mathbf{F}(x)$ instead of $F(x)$. In particular, the *inverse* $\mathbf{Q}^{-1}$ of a cc-stateless permutation $\mathbf{Q}$ is well-defined, and $\langle \mathbf{Q} \rangle$ is the *two-sided* random permutation which allows both forward queries $(x, +)$ returning $\mathbf{Q}(x)$ as well as *backward queries* $(y, -)$ returning $\mathbf{Q}^{-1}(y)$. The *cascade* $\mathbf{Q}' \rhd \mathbf{Q}''$ of two random permutations is the system which on input x returns $\mathbf{Q}''(\mathbf{Q}'(x))$, i.e., it implements the composition of the associated permutation tables. (This extends naturally to longer cascades.) Note in particular that for any cascade we have $\mathbf{Q}_1 \rhd \cdots \rhd \mathbf{Q}_m \equiv \mathbf{P}$ whenever there exists i such that $\mathbf{Q}_i \equiv \mathbf{P}$ for a URP $\mathbf{P}$. Moreover, we let $\langle \mathbf{Q}' \rangle \rhd \langle \mathbf{Q}'' \rangle := \langle \mathbf{Q}' \rhd \mathbf{Q}'' \rangle$.

An efficiently implementable family of permutations $E = \{E_k\}_{k \in \mathcal{K}}$ with domain $\mathcal{X}$ and indexed by keys $k \in \mathcal{K}$ is an *ε-pseudorandom permutation* (ε-PRP) if $\Delta_{t,q}(E_K, \mathbf{P}) \leq \varepsilon$ for all polynomially bounded t and q, a uniform $K \in \mathcal{K}$, and a URP $\mathbf{P}$. It is a *two-sided* ε-PRP if $\langle E_K \rangle$ is efficiently implementable and $\Delta_{t,q}(\langle E_K \rangle, \langle \mathbf{P} \rangle) \leq \varepsilon$ for all polynomially bounded t and q.

3 Hardcore Lemmas for Interactive Systems

3.1 System-Bit Pairs, Measures, and State Samplers

We consider the general setting of *system-bit pairs* [10] $(\mathbf{S}, B)$ consisting of a bit B (with an associated probability distribution P_B), and a system $\mathbf{S} = \mathbf{S}(B)$ whose behavior depends on the outcome of the bit B. A system-bit pair $(\mathbf{S}, B)$ is to be interpreted as a system which parallely composes $\mathbf{S}$ and a correlated bit B (which is *initially* chosen, before any interaction with $\mathbf{S}$ has taken place). The notion of a cc-stateless system-bit pair $(\mathbf{S}(S), B(S))$ is obtained naturally. Also, an implementation $A_{(\mathbf{S},B)}$ of a system-bit pair $(\mathbf{S}, B)$ is without loss of generality an algorithm which outputs the bit B and then simulates the system $\mathbf{S}(B)$.

We associate with every system-bit pair $(\mathbf{S}, B)$ a game where an *adversary* $\mathbf{A}$ interacts with $\mathbf{S}(B)$ and outputs a binary guess $\mathbf{A}(\mathbf{S}(B)) \in \{0, 1\}$ for B: Its *guessing advantage* is defined as the quantity

$$\mathrm{Guess}^{\mathbf{A}}(B \,|\, \mathbf{S}) := 2 \cdot \mathsf{P}[\mathbf{A}(\mathbf{S}(B)) = B] - 1 \in [-1, 1].$$

If $\mathrm{Guess}^{\mathbf{A}}(B \,|\, \mathbf{S}) = 1$, then $\mathbf{A}$ always guesses B correctly, whereas $\mathrm{Guess}^{\mathbf{A}}(B \,|\, \mathbf{S}) = -1$ means that $\mathbf{A}$ is always wrong (though flipping $\mathbf{A}$'s output bit yields an adversary which is always correct.) The shorthand $\mathrm{Guess}_{t,q}(B \,|\, \mathbf{S})$ denotes the best guessing advantage taken over all adversaries with time complexity t and issuing at most q queries to $\mathbf{S}$.

Example 1. An example is the (cc-stateless) system-bit pair $(\mathbf{R}, B)$ for a URF $\mathbf{R} : \mathcal{X} \to \{0,1\}$ and $B := \oplus_{x \in \mathcal{X}} \mathbf{R}(x)$ is the parity of its function table. It is easy to see that $\mathrm{Guess}_q(B \,|\, \mathbf{R}) = 0$ for all $q < |\mathcal{X}|$.

Example 2. If $(\mathbf{F}, B)$ is such that B is uniform, and $\mathbf{F}$ behaves as a system $\mathbf{S}$ if $B = 0$, and as another system $\mathbf{T}$ if $B = 1$, then $\mathrm{Guess}^{\mathbf{D}}(B \,|\, \mathbf{F}) = \Delta^{\mathbf{D}}(\mathbf{S}, \mathbf{T})$ for all $\mathbf{D}$ by a standard argument. Note that if both $\mathbf{S}$ and $\mathbf{T}$ are cc-stateless, then $(\mathbf{F}, B)$ is also cc-stateless.

Measures. A *measure* $\mathcal{M}$ for a cc-stateless system $\mathbf{S} \equiv \mathbf{S}(S)$, where $S \in \mathcal{S}$ is the initial state, is a mapping $\mathcal{M} : \mathcal{S} \to [0,1]$. Its *density* is defined as $\mu(\mathcal{M}) := \mathsf{E}[\mathcal{M}(S)] = \sum_{s \in \mathcal{S}} \mathsf{P}_S(s) \cdot \mathcal{M}(s)$. The measure $\mathcal{M}$ is naturally associated with a probability distribution $\mathsf{P}_{\mathcal{M}}$ on $\mathcal{S}$ such that $\mathsf{P}_{\mathcal{M}}(s) := \mathsf{P}_S(s) \cdot \mathcal{M}(s) \cdot \mu(\mathcal{M})^{-1}$ for all $s \in \mathcal{S}$. Also, we define the complement of a measure $\mathcal{M}$ as the measure $\overline{\mathcal{M}}$ such that $\overline{\mathcal{M}}(s) := 1 - \mathcal{M}(s)$ for all $s \in \mathcal{S}$. We repeatedly abuse notation writing $S \xleftarrow{\$} \mathcal{M}$ instead of $S \xleftarrow{\$} \mathsf{P}_{\mathcal{M}}$.

Traditionally, measures are seen as "fuzzy" subsets of $\mathcal{S}$. Alternatively, it is convenient to think of $\mathcal{M}$ in terms of a conditional probability distribution $\mathsf{P}_{\mathcal{A}|S}$ with $\mathsf{P}_{\mathcal{A}|S}(s) := \mathcal{M}(s)$ which adjoins the event $\mathcal{A}$ on the choice of S: In particular, $\mu(\mathcal{M}) = \mathsf{P}[\mathcal{A}]$, $\mathsf{P}_{\mathcal{M}} = \mathsf{P}_{S|\mathcal{A}}$, and $\mathsf{P}_{\overline{\mathcal{M}}} = \mathsf{P}_{S|\overline{\mathcal{A}}}$. In the following, we stick to measures for stating and proving our hardcore lemmas, while an event-based view will be useful when exercising these results.

State Samplers. Ideally, the hardcore lemma for a cc-stateless system-bit pair $(\mathbf{S}, B) \equiv (\mathbf{S}(S), B(S))$ (for initial state $S \in \mathcal{S}$) states that if $\mathrm{Guess}_{t,q}(B \,|\, \mathbf{S}) \leq \varepsilon$, then there exists a measure $\mathcal{M}$ on $\mathcal{S}$ such that (i) $\mu(\mathcal{M}) \geq 1 - \varepsilon$ and (ii) $\mathrm{Guess}_{t',q'}(B(S') \,|\, \mathbf{S}(S')) \approx 0$ for $S' \xleftarrow{\$} \mathcal{M}$ and t', q' as close as possible to t, q. Whenever $\mathbf{S}(S)$ is a random variable, this is equivalent to (a tight) version of Impagliazzo's Hardcore Lemma [4]. However, applications of the hardcore lemma (as the one we give later in this paper) require the ability, possibly given some short advice, to efficiently simulate $(\mathbf{S}(S'), B(S'))$ for $S' \xleftarrow{\$} \mathcal{M}$ or $(\mathbf{S}(S''), B(S''))$ for $S'' \xleftarrow{\$} \overline{\mathcal{M}}$.[4] While in the context of random variables the advice is generally a sample of S' itself, this approach fails in the setting of interactive systems: Recall that the representation $(\mathbf{S}(S), B(S))$ is possibly only a thought experiment, and the description of S' may be of exponential size, or no efficient algorithm implementing $(\mathbf{S}, B)$ from S' exists, even if the system-bit pair itself is efficiently implementable.

To formalize the concept of an advice distribution, we introduce the notion of a state sampler for a cc-stateless system (such as e.g. a system-bit pair).

Definition 1 (State Samplers). *Let* $\mathbf{S} \equiv \mathbf{S}(S)$ *be a cc-stateless system with implementation* $A_{\mathbf{S}}$ *and* $S \in \mathcal{S}$, *let* $\zeta_1, \zeta_2 \in [0,1]$, *and let* $\mathcal{M} : \mathcal{S} \to [0,1]$ *be a*

[4] Formally, one actually needs to prove that $\mathrm{Guess}_{t',q'}(B(S') \,|\, \mathbf{S}(S')) \approx 0$ holds even given access to the advice: While this is implicit in the non-uniform setting (every adversary with advice can be turned in an equally good one without advice), the proof is more challenging in the uniform setting, cf. the full version.

measure for $\mathbf{S}$. *A* (ζ_1, ζ_2)-(state) sampler O *for* $\mathcal{M}$ *and* $A_{\mathbf{S}}$ *with length* ℓ *is a random process* O *such that:*

(i) O *always returns a pair* (σ, z) *with* σ *being a valid state for* $A_{\mathbf{S}}$ *with* $|\sigma| \leq \ell$ *and* $z \in [0,1]$*;*
(ii) For $(\Sigma, Z) \xleftarrow{\$} \mathsf{O}$*, we have*[5]

$$(A_{\mathbf{S}}[\Sigma], Z) \equiv (\mathbf{S}(S), Z'(S)),$$

where $Z'(S) \in [0,1]$ *is a random variable (which only depends on* S*) that differs from* $\mathcal{M}(S)$ *by at most* ζ_1*, except with probability* ζ_2*, for any value taken by* S*.*

Example 3. For all implementations $A_{\mathbf{S}}$ of $\mathbf{S}$, the all-one measure (i.e., $\mathsf{P}_{\mathcal{M}} = \mathsf{P}_S$) admits an error-less sampler O which returns the initial (void) state for $A_{\mathbf{S}}$ and $z = 1$.

Note that O is *not* required to be efficiently implementable, but black-box access to state samplers allow for efficient simulation of $\mathbf{S}(S')$ for $S' \xleftarrow{\$} \mathcal{M}$ using reject-sampling (provided $\mathbf{S}$ admits an efficient implementation): Given the output (Σ, Z) sampled from a (ζ_1, ζ_2)-sampler O, we flip a coin B with $\mathsf{P}_B(1) = Z$: Consider the distribution $\mathsf{P}_{\Sigma|B=1}$ of Σ conditioned on the outcome $B = 1$. If $\zeta_1 = \zeta_2 = 0$, it is not hard to verify that $A_{\mathbf{S}}[\Sigma'] \equiv \mathbf{S}(S')$ for $\Sigma' \xleftarrow{\$} \mathsf{P}_{\Sigma|B=1}$. This is because, by definition, we have $(A_{\mathbf{S}}[\Sigma], Z, B) \equiv (\mathbf{S}(S), \mathcal{M}(S), B')$, where B' is a bit which is 1 with probability $\mathcal{M}(S)$, and thus in particular $A_{\mathbf{S}}[\Sigma'] \equiv \mathbf{S}(S')$ where $S' \xleftarrow{\$} \mathsf{P}_{S|B'=1}$. In addition, since $\mathsf{P}_{B'|S}(1, s) := \mathcal{M}(s)$ and $\mathsf{P}_{B'}(1) := \sum_{s \in \mathcal{S}} \mathsf{P}_S(s) \cdot \mathcal{M}(s) = \mu(\mathcal{M})$,

$$\mathsf{P}_{S|B'}(s, 1) = \mathcal{M}(s) \cdot \mathsf{P}_S(s) \cdot \mu(\mathcal{M})^{-1} = \mathsf{P}_{\mathcal{M}}(s).$$

Of course, one can similarly simulate $\mathbf{S}(S'')$ for $S'' \xleftarrow{\$} \mathsf{P}_{\overline{\mathcal{M}}}$, as we obtain a corresponding sampler by just replacing z by $1 - z$ in the output (σ, z). This approach can be extended to non-zero errors ζ_1 and ζ_2 with some care.

3.2 The Hardcore Lemma for System-Bit Pairs

In the following, for understood parameters γ, ε, ζ_1, and ζ_2, we define

$$\varphi_{\mathsf{hc}} := \frac{6400}{\gamma^2(1-\varepsilon)^4} \cdot \ln\left(\frac{160}{\gamma(1-\varepsilon)^3}\right) \text{ and } \psi_{\mathsf{hc}} := \frac{200}{\gamma^2(1-\varepsilon)^4\zeta_1^2} \cdot \ln\left(\frac{2}{\zeta_2}\right).$$

We now state the HCL for cc-stateless system-bit pairs. Even though we apply the result only in a more restricted setting, we prove a more general statement for arbitrary cc-stateless system-bit pairs.

[5] That is, we consider the parallel composition of a system (either $A_{\mathbf{S}}[\Sigma]$ or $\mathbf{S}(S)$) and a correlated $[0,1]$-valued random variable.

Theorem 1 (HCL for System-Bit Pairs). *Let $(\mathbf{S}, B) \equiv (\mathbf{S}(S), B(S))$ be a cc-stateless system-bit pair admitting an implementation $A_{(\mathbf{S},B)}$ with space complexity $s_{A_{(\mathbf{S},B)}}$. Furthermore, for some integers $t, q > 0$ and some $\varepsilon \in [0, 1)$,*

$$\text{Guess}_{t,q}(B \mid \mathbf{S}) \leq \varepsilon.$$

Then, for all $0 < \zeta_1, \zeta_2 < 1$ and all $0 < \gamma \leq \frac{1}{2}$, there exists a measure $\mathcal{M}$ for $(\mathbf{S}, B)$ with $\mu(\mathcal{M}) \geq 1 - \varepsilon$ such that the following two properties are satisfied:

(i) For $S' \xleftarrow{\$} \mathcal{M}$, $t' := t/\varphi_{hc}$, and $q' := q/\varphi_{hc}$,

$$\text{Guess}_{t',q'}(B(S') \mid \mathbf{S}(S')) \leq \gamma.$$

(ii) There exists a (ζ_1, ζ_2)-sampler for $\mathcal{M}$ and $A_{(\mathbf{S},B)}$ with length $s_{A_{(\mathbf{S},B)}}(\psi_{hc} \cdot q')$. Moreover, if $(\mathbf{S}(s), B(s))$ is deterministic for all s, then there also exists a $(0, 0)$-sampler for $\mathcal{M}$ and $A_{(\mathbf{S},B)}$ with length $s_{A_{(\mathbf{S},B)}}((7 \cdot \gamma^{-2} \cdot (1-\varepsilon)^{-3} + 1) \cdot q')$.

In the remainder of this section, we outline the main ideas behind the proof. The complete proof is found in the full version.

PROOF OUTLINE. The proof is by contradiction: We assume that for all measures $\mathcal{M}$ with $\mu(\mathcal{M}) \geq 1 - \varepsilon$ admitting a (ζ_1, ζ_2)-sampler as in (ii), there exists an adversary $\mathbf{A}$ with time complexity t' and query complexity q' such that $\text{Guess}^{\mathbf{A}}(B(S') \mid \mathbf{S}(S')) > \gamma$ for $S' \xleftarrow{\$} \mathcal{M}$. The core of the proof consists of proving that, under this assumption, there exists a sufficiently small family of adversaries $\mathcal{A}$ (more specifically, $|\mathcal{A}| = 7 \cdot \gamma^{-2} \cdot (1-\varepsilon)^{-3} + 1$) such that either (A) $\alpha(S) > \gamma$ holds with probability higher than $1 - \frac{1-\varepsilon}{4}$ over the choice of S, where $\alpha(s) := \mathsf{E}\left[\text{Guess}^{\mathbf{A}'}(B(s) \mid \mathbf{S}(s))\right]$ for all s, where $\mathbf{A}' \xleftarrow{\$} \mathcal{A}$, or (B) $\mathsf{E}[\alpha(S')] > \Theta\left((1-\varepsilon)^2\gamma\right)$ *for all* measures $\mathcal{M}$ with density $1 - \varepsilon$ and $S' \xleftarrow{\$} \mathcal{M}$.

In Case (A), a simple majority-voting based strategy yields a good adversary breaking the assumed hardness of $(\mathbf{S}, B)$, whereas in Case (B) such an adversary can be built from $\mathcal{A}$ using techniques similar to the case of random variables [5,3]. Both adversaries heavily rely on the cc-stateless property of $(\mathbf{S}, B)$.

To show the existence of an appropriate family, we associate with each family $\mathcal{A}$ and $\tau \in \mathbb{N}$ a measure $\mathcal{M}_{\mathcal{A},\tau}$ such that elements for which $\mathcal{A}$ is worse, i.e., $|\mathcal{A}| \cdot \alpha(s) \leq \tau$, are given high weight (i.e. $\mathcal{M}_{\mathcal{A},\tau}(s) = 1$), whereas elements for which $\mathcal{A}$ performs well, i.e., $|\mathcal{A}| \cdot \alpha(s) \geq \tau + \frac{1}{\gamma(1-\varepsilon)}$, are not chosen ($\mathcal{M}_{\mathcal{A},\tau}(s) = 0$). An intermediate measure value is assigned to states not falling into one of these two categories. In particular, $\mathcal{M}_{\emptyset,0}$ is the all-one measure (i.e., $\mathsf{P}_{\mathcal{M}}$ equals the state distribution P_S), which has density $1 \geq 1 - \varepsilon$. A crucial property is that $\mathcal{M}_{\mathcal{A},\tau}$ admits an (ζ_1, ζ_2)-state sampler for all $\mathcal{A}$ and τ, which is shown by using the observation that $\mathcal{M}_{\mathcal{A},\tau}(S)$ can always be estimated given *black-box* access to $(\mathbf{S}(S), B(S))$. We then consider the following iterative process: It starts with $\mathcal{A} := \emptyset$ and then, at each round, it possibly increases τ to ensure that $\mu(\mathcal{M}_{\mathcal{A},\tau}) \geq 1 - \varepsilon$ and then uses the assumption of the HCL being wrong to find an adversary achieving advantage larger than γ for $\mathcal{M}_{\mathcal{A},\tau}$, and adds it to $\mathcal{A}$. We prove that within $7 \cdot \gamma^{-2} \cdot (1-\varepsilon)^{-3} + 1$ iterations, $\mathcal{A}$ satisfies (A) or (B).

Remark 1. A natural question is whether the HCL can be extended to arbitrary system-bit pairs, where the measure is defined on the randomness of the implementation of the system-bit pair, regardless of the system having a cc-stateless representation. Yet, techniques similar to the ones used in counter-examples to soundness amplification for interactive arguments via parallel repetition [1,14] yield (non cc-stateless) efficiently implementable system-bit pairs for which, given multiple independent instances of the system-bit pair, the probability of guessing all of the bits given access to all of the associated systems in parallel does not decrease with the number of instances. If such a a general HCL were true, then it is not hard to prove that the guessing probability *would* decrease exponentially in the number of instances.

3.3 The Hardcore Lemma for Computational Indistinguishability

This section presents the hardcore lemma for computational indistinguishability of *interactive* systems, which generalizes the statement for random variables previously shown in [11].

Theorem 2 (HCL for Computational Indistinguishability). *Let* $\mathbf{S} \equiv \mathbf{S}(S)$ *and* $\mathbf{T} \equiv \mathbf{T}(T)$ *be cc-stateless systems, with respective implementations* $A_{\mathbf{S}}$ *(with space complexity* $s_{A_{\mathbf{S}}}$*) and* $A_{\mathbf{T}}$ *(with space complexity* $s_{A_{\mathbf{T}}}$*). Furthermore, for some integers* $t, q > 0$ *and some* $\varepsilon \in [0, 1)$,

$$\Delta_{t,q}(\mathbf{S}, \mathbf{T}) \leq \varepsilon.$$

Then, for all $0 < \zeta_1, \zeta_2 < 1$ *and all* $0 < \gamma \leq \frac{1}{2}$*, there exist measures* $\mathcal{M}_{\mathbf{S}}$ *and* $\mathcal{M}_{\mathbf{T}}$ *such that* $\mu(\mathcal{M}_{\mathbf{S}}) \geq 1 - \varepsilon$ *and* $\mu(\mathcal{M}_{\mathbf{T}}) \geq 1 - \varepsilon$ *and the following properties hold:*

(i) For $S' \stackrel{\$}{\leftarrow} \mathcal{M}_{\mathbf{S}}$, $T' \stackrel{\$}{\leftarrow} \mathcal{M}_{\mathbf{T}}$, $t' := t/\varphi_{hc}$*, and* $q' := q/\varphi_{hc}$*, we have*

$$\Delta_{t',q'}(\mathbf{S}(S'), \mathbf{T}(T')) \leq 2\gamma;$$

(ii) There exist a (ζ_1, ζ_2)*-sampler* $\mathsf{O}_{\mathbf{S}}$ *for* $\mathcal{M}_{\mathbf{S}}$ *and* $A_{\mathbf{S}}$ *with length* $s_{A_{\mathbf{S}}}(\psi_{hc} \cdot q')$ *and a* (ζ_1, ζ_2)*-sampler* $\mathsf{O}_{\mathbf{T}}$ *for* $\mathcal{M}_{\mathbf{T}}$ *and* $A_{\mathbf{T}}$ *with length* $s_{A_{\mathbf{T}}}(\psi_{hc} \cdot q')$*. Furthermore, if both* $\mathbf{S}$ *and* $\mathbf{T}$ *are random functions, then both samplers can be made error-less with lengths* $s_{A_{\mathbf{S}}}(\psi \cdot q')$ *and* $s_{A_{\mathbf{T}}}(\psi \cdot q')$*, where* $\psi := 7 \cdot \gamma^{-2} \cdot (1 - \varepsilon)^{-3} + 1$.

We postpone the proof to the full version, which relies on Theorem 1, and only present the main ideas in the following.

Proof Sketch. We define $(\mathbf{F}, B) \equiv (\mathbf{F}(X, B), B)$ to be the cc-stateless system-bit pair with a uniform random bit B and where $\mathbf{F}$ behaves as $\mathbf{S}$ if $B = 0$ and as $\mathbf{T}$ if $B = 1$. In particular, the initial state (X, B) of $(\mathbf{F}, B)$ is sampled by first letting $B \stackrel{\$}{\leftarrow} \{0, 1\}$, and then choosing $X \stackrel{\$}{\leftarrow} \mathsf{P}_S$ if $B = 0$ and $X \stackrel{\$}{\leftarrow} \mathsf{P}_T$ otherwise, and

$$(\mathbf{F}(x, b), B(x, b)) = \begin{cases} (\mathbf{S}(x), 0) & \text{if } b = 0, \\ (\mathbf{T}(x), 1) & \text{if } b = 1. \end{cases}$$

By a standard argument $\Delta_{t,q}(\mathbf{S}, \mathbf{T}) = \text{Guess}_{t,q}(B \,|\, \mathbf{F}) \leq \varepsilon$ holds (also cf. Example 2), and Theorem 1 thus implies that there exists a measure $\mathcal{M}$ for $(\mathbf{F}, B)$ such that $\mu(\mathcal{M}) \geq 1 - \varepsilon$, and $\text{Guess}_{t',q'}(B' \,|\, \mathbf{F}(X')) \leq \gamma$, where $(X', B') \stackrel{\$}{\leftarrow} \mathcal{M}$, $t' = t/\varphi_{\mathsf{hc}}$, and $q' = q/\varphi_{\mathsf{hc}}$. Define $\mathcal{M}_{\mathbf{S}}(s) := \mathcal{M}(s, 0)$ and $\mathcal{M}_{\mathbf{T}}(t) := \mathcal{M}(t, 1)$, and note that

$$\begin{aligned} \mathsf{P}_{X'B'}(s,0) &= \frac{1}{2\mu(\mathcal{M})} \cdot \mathsf{P}_S(s) \cdot \mathcal{M}_{\mathbf{S}}(s), \\ \mathsf{P}_{X'B'}(t,1) &= \frac{1}{2\mu(\mathcal{M})} \cdot \mathsf{P}_T(t) \cdot \mathcal{M}_{\mathbf{T}}(t). \end{aligned} \tag{1}$$

If B' were uniformly distributed (i.e., $\sum_s \mathsf{P}_{X'B'}(s,0) = \sum_t \mathsf{P}_{X'B'}(t,1) = \frac{1}{2}$), we then would have $\mu(\mathcal{M}_{\mathbf{S}}) = \mu(\mathcal{M}_{\mathbf{T}}) = \mu(\mathcal{M}) \geq 1 - \varepsilon$ by (1), and (X', B') could be sampled by choosing B' uniformly, and letting $X' = S' \stackrel{\$}{\leftarrow} \mathcal{M}_{\mathbf{S}}$ if $B' = 0$, and $X' = T' \stackrel{\$}{\leftarrow} \mathcal{M}_{\mathbf{T}}$ if $B' = 1$. This would also yield

$$\Delta_{t',q'}(\mathbf{S}(S'), \mathbf{T}(T')) = \text{Guess}_{t',q'}(B' \,|\, \mathbf{F}(X')) \leq \gamma,$$

concluding the proof. The main challenge in the full proof is dealing with the fact that B' is generally only $\Theta(\gamma)$-close to uniform.

Remark 2. Theorem 2 can be seen as a computational analogue of Lemma 5 in [9], which shows a similar property for information-theoretic indistinguishability (i.e., with respect to computationally unbounded distinguishers). Theorem 2 can of course also be used in the IT setting, and it is somewhat stronger in that it yields events defined on the initial state of the system, instead of interaction-dependent sequences of events as in [9]. However, Lemma 5 in [9] holds for *arbitrary* systems and presents a tight reduction with $q' = q$ and no additive term γ, which we do not know how to achieve in the computational setting.

Connection to Computational Entropy. Let $\mathbf{Q}$ be a cc-stateless random permutation on $\mathcal{X}$ (with $N := |\mathcal{X}|$) with function table Q and such that $\Delta_{t,q}(\mathbf{Q}, \mathbf{P}) \leq \varepsilon$ for a URP $\mathbf{P}$. Theorem 2 yields events $\mathcal{A}$ on Q and $\mathcal{B}$ on a uniform permutation table P such that $\mathsf{P}[\mathcal{A}] \geq 1 - \varepsilon$, $\mathsf{P}[\mathcal{B}] \geq 1 - \varepsilon$, and $\Delta_{t',q'}(\mathbf{Q}', \mathbf{P}') \leq \gamma$, where $\mathbf{Q}'$ and $\mathbf{P}'$ are cc-stateless random functions with function tables $Q' \stackrel{\$}{\leftarrow} \mathsf{P}_{Q|\mathcal{A}}$ and $P' \stackrel{\$}{\leftarrow} \mathsf{P}_{P|\mathcal{B}}$, respectively. In particular, $\mathsf{P}_{P'}(\pi) = \frac{\mathsf{P}_P(\pi) \cdot \mathsf{P}_{\mathcal{B}|P(\pi)}}{\mathsf{P}[\mathcal{B}]} \leq \frac{1}{(1-\varepsilon)\cdot(N!)}$ for all permutations π, and the *min-entropy* $\mathsf{H}_\infty(P') := -\log \max_\pi \mathsf{P}_{P'}(\pi)$ is at least $\log(N!) - \log\left((1-\varepsilon)^{-1}\right)$. Informally, this can be interpreted as Q having *"computational"* min-entropy at most $\log\left((1-\varepsilon)^{-1}\right)$ away from the maximum achievable entropy $\log(N!)$ with probability $1 - \varepsilon$.[6] Clearly, the statement also extends to the two-sided case as well as to other types of systems.

[6] We stress, however, that the distribution P' depends on t, q, as well as on γ.

Remark 3. Another useful fact is that P' has statistical distance ε from P. This follows from the observation that the distribution of P' is a convex combination of flat distributions over subsets of size at least $(1-\varepsilon)\cdot(N!)$: As each such distribution is ε-away from uniform, the bound follows from the convexity of the statistical distance. Therefore, $\Delta_{t,q}(\mathbf{P}', \mathbf{P}) \leq \Delta_{t,q}(\langle \mathbf{P}' \rangle, \langle \mathbf{P} \rangle) \leq d(P', P) \leq \varepsilon$ for all t, q.

4 Cascade of Weak Permutations

4.1 Cascade of Permutations with Large Entropy

Let $\mathbf{Q}_1$ and $\mathbf{Q}_2$ be two independent cc-stateless random permutations on the set $\mathcal{X}$ (with $N := |\mathcal{X}|$) with the property that the min-entropies of their respective function tables Q_1 and Q_2 satisfy $\mathsf{H}_\infty(Q_1) \geq \log(N!) - \log\left((1-\varepsilon)^{-1}\right)$ and $\mathsf{H}_\infty(Q_2) \geq \log(N!) - \log\left((1-\varepsilon)^{-1}\right)$ for some $\varepsilon \in \left[0, 1 - \frac{1}{N}\right)$. We prove that the cascade $\mathbf{Q}_1 \rhd \mathbf{Q}_2$ is indistinguishable from a URP $\mathbf{P}$ for computationally *unbounded* distinguishers, both in the one- and in the two-sided cases.

Theorem 3 (Cascade of Large-Entropy Permutations). *For all* $q, \Lambda \geq 1$,

$$\Delta_q(\langle \mathbf{Q}_1 \rhd \mathbf{Q}_2 \rangle, \langle \mathbf{P} \rangle) \leq \frac{4q\Lambda}{N} + \frac{2\Lambda(q+\Lambda)}{(1-\varepsilon)N} + 2\left(\frac{q\log\left((1-\varepsilon)^{-1}\right)}{\Lambda}\right)^{\frac{1}{2}}.$$

The same bound applies to any cascade $\mathbf{Q}'_1 \rhd \cdots \rhd \mathbf{Q}'_m$ of m independent cc-stateless random permutations such that $\mathbf{Q}'_i \equiv \mathbf{Q}_1$ and $\mathbf{Q}'_j \equiv \mathbf{Q}_2$ for some $i < j$, as such a cascade can be seen as the cascade of two permutations $\overline{\mathbf{Q}}_1 := \mathbf{Q}'_1 \rhd \cdots \rhd \mathbf{Q}'_i$ and $\overline{\mathbf{Q}}_2 := \mathbf{Q}'_{i+1} \rhd \cdots \rhd \mathbf{Q}'_m$ with the same min-entropy guarantees on their function tables. The theorem allows free choice of Λ: For our purposes, it suffices to set $\Lambda := (\log N)^\zeta$ (for a slowly growing $\zeta = \omega(1)$ in the security parameter $\log N$) to achieve indistinguishability for $q = \mathsf{poly}(\log N)$ queries and any $\varepsilon \leq 1 - \frac{(\log N)^{3\zeta}}{N}$.

The core of the proof, which is omitted for lack of space, is a lemma stating that $\langle \mathbf{Q}_i \rangle$ (for $i = 1, 2$) is indistinguishable from a random permutation $\langle \mathbf{Q}_i \rangle_{\overline{\mathbf{D}}_i}$ which is initialized by letting a carefully chosen distinguisher $\overline{\mathbf{D}}_i$ (making Λ queries) interact with $\langle \mathbf{Q}_i \rangle$, and then answering queries according to a randomly chosen permutation consistent with $\overline{\mathbf{D}}_i$'s interaction. (This extends a previous result by Unruh [15] to random permutations.) We employ tools from the random systems framework [8] (including a new lemma) to prove that the cascade of two independent such permutations is indistinguishable from a URP.

4.2 Security Amplification of Weak PRPs

Let $\mathbf{Q}$ be a cc-stateless random permutation with domain $\mathcal{X}$ (for $N := |\mathcal{X}| = 2^n$, where n is the security parameter) such that $\langle \mathbf{Q} \rangle$ is implemented by the algorithm $A_{\langle \mathbf{Q} \rangle}$ with time complexity $t_{A_{\langle \mathbf{Q} \rangle}}$ and space complexity $s_{A_{\langle \mathbf{Q} \rangle}}$. We also consider the canonical (efficient) implementation of a two-sided URP $\langle \mathbf{P} \rangle$ that maintains a table consisting of all input-output pairs (x_i, y_i) of previous queries as its state, and, upon a new query $(x, +)$ or $(y, -)$, it chooses uniformly at random a y' (or x') not appearing as the second (first) element in a previous input-output pair, and adds (x, y') (or (x', y)) to the table. (If a corresponding pair is in the table, it answers accordingly.) Thus each query is answered in time $\mathcal{O}(\log(s))$, where s is the size of the table, and $s = \mathcal{O}(q \cdot n)$ after q queries.

The following is the main security amplification result of this paper.

Theorem 4. *Let $\mathbf{Q}_1, \ldots, \mathbf{Q}_m$ be independent instances of $\mathbf{Q}$ and let $\mathbf{P}$ be a URP, and assume that for some t, q we have $\Delta_{t,q}(\langle \mathbf{Q} \rangle, \langle \mathbf{P} \rangle) \leq \varepsilon$. For all $\gamma > 1$ and $\Lambda > 0$,*

$$\Delta_{t'',q''}(\langle \mathbf{Q}_1 \rhd \cdots \rhd \mathbf{Q}_m \rangle, \langle \mathbf{P} \rangle) \leq (m - (m-1)\varepsilon) \cdot \varepsilon^m + \tfrac{4q''\Lambda}{N} + \tfrac{2\Lambda(q''+\Lambda)}{(1-\varepsilon)N} + 2\left(\frac{q'' \log\left((1-\varepsilon)^{-1}\right)}{\Lambda}\right)^{\frac{1}{2}} + (2m+2)\gamma,$$

where $t'' := t/\varphi_{\mathsf{hc}} - (m-1) \max\left\{t_{A\langle\mathbf{Q}\rangle}(q'', s_{A\langle\mathbf{Q}\rangle}(q'' \cdot \psi)), \mathcal{O}\left(q'' \log(q'' \cdot (\psi + 1)n)\right)\right\}$ and $q'' := q/\varphi_{\mathsf{hc}}$, for $\psi := 7 \cdot \gamma^{-2} \cdot (1-\varepsilon)^{-3} + 1$ and φ_{hc} as in Theorem 2.

Essentially the same result can be proven for the single-sided case. The proof of Theorem 4 follows from the observation that, with very high probability, at least two permutations in the cascade are computational indistinguishable from random permutations with large entropy, allowing application of Theorem 3. Extra work is required to prove a non-trivial bound for the case where *at most* one permutation is guaranteed to have high-entropy. The tightness of these bounds is discussed in Section 4.3.

Proof. Theorem 2 implies that we can define (two-sided) random permutations $\langle \mathbf{Q}' \rangle, \langle \mathbf{Q}'' \rangle$, and $\langle \mathbf{P}' \rangle$ such that the following three properties hold for some $p \leq \varepsilon$: (i) The function table of $\langle \mathbf{P}' \rangle$ has min-entropy at least $\log(N!) - \log\left((1-\varepsilon)^{-1}\right)$, (ii) $\langle \mathbf{Q} \rangle$ behaves as $\langle \mathbf{Q}' \rangle$ with probability $1-p$ and as $\langle \mathbf{Q}'' \rangle$ with probability p, and (iii) $\Delta_{t',q''}(\langle \mathbf{Q}' \rangle, \langle \mathbf{P}' \rangle) \leq 2\gamma$ for $t' := t/\varphi_{\mathsf{hc}}$. Furthermore, $\langle \mathbf{Q}' \rangle$ and $\langle \mathbf{Q}'' \rangle$ can both be perfectly implemented using $A_{\langle \mathbf{Q} \rangle}$ initialized with some appropriately distributed state of length at most $s_{A\langle\mathbf{Q}\rangle}(q'' \cdot \psi)$ given as advice. Similarly, $\langle \mathbf{P}' \rangle$ can be simulated by running the above canonical algorithm initialized with an appropriate state of length $\mathcal{O}(q'' \cdot \psi \cdot n)$. (See the discussion in Section 3.1.)

Additionally, for $\mathcal{I} \subseteq \{1, \ldots, m\}$, let $\mathcal{A}_I$ be the event that $\langle \mathbf{Q}_i \rangle$ behaves as $\langle \mathbf{Q}' \rangle$ for all $i \in \mathcal{I}$ whereas $\langle \mathbf{Q}_i \rangle$ behaves as $\langle \mathbf{Q}'' \rangle$ for all $i \notin \mathcal{I}$. Likewise, for independent instances $\langle \mathbf{Q}'_i \rangle$ and $\langle \mathbf{Q}''_i \rangle$ (for $i = 1, \ldots, m$) of $\langle \mathbf{Q}' \rangle$ and $\langle \mathbf{Q}'' \rangle$, respectively, let $\mathbf{Q}_{\mathcal{I}} := \mathbf{S}_1 \rhd \cdots \rhd \mathbf{S}_m$, where $\mathbf{S}_i := \mathbf{Q}'_i$ for all $i \in \mathcal{I}$ and $\mathbf{S}_i := \mathbf{Q}''_i$ for all $i \notin \mathcal{I}$.

We now fix some distinguisher $\mathbf{D}$ with time complexity t'' and making q'' queries, and we first observe that

$$\delta^{\mathbf{D}}(\langle \mathbf{Q}_1 \rhd \cdots \rhd \mathbf{Q}_m \rangle, \langle \mathbf{P} \rangle) = \sum_{\mathcal{I} \subseteq \{1,\ldots,m\}} q_{\mathcal{I}} \cdot \delta^{\mathbf{D}}(\langle \mathbf{Q}_{\mathcal{I}} \rangle, \langle \mathbf{P} \rangle), \tag{2}$$

where $\delta^{\mathbf{D}}(\mathbf{F},\mathbf{G}) := \mathsf{P}[\mathbf{D}(\mathbf{F})=1] - \mathsf{P}[\mathbf{D}(\mathbf{G})=1]$ and $q_{\mathcal{I}} := \mathsf{P}[\mathcal{A}_{\mathcal{I}}] = (1-p)^{|\mathcal{I}|} \cdot p^{m-|\mathcal{I}|}$. Note that the maximum of $\delta^{\mathbf{D}}(\mathbf{F},\mathbf{G})$ over all distinguishers $\mathbf{D}$ with time complexity t'' and space complexity q'' is $\Delta_{t'',q''}(\mathbf{F},\mathbf{G})$.

We first upper bound the summands corresponding to sets $\mathcal{I}$ with at most one element. To this end, for all $i = 1,\ldots,m$, we define the distinguisher $\mathbf{D}_i$ which, given access to a two-sided random permutation $\langle \mathbf{S} \rangle$, outputs

$$\mathbf{D}(\langle \mathbf{Q}''_1 \rhd \cdots \rhd \mathbf{Q}''_{i-1} \rhd \mathbf{S} \rhd \mathbf{Q}''_{i+1} \rhd \cdots \rhd \mathbf{Q}''_m \rangle),$$

and is implemented with time complexity $t'' + (m-1)t_{A\langle \mathbf{Q} \rangle}(q', s_{A\langle \mathbf{Q} \rangle}(\psi \cdot q')) \leq t'$ given the appropriate advice.

We have $\delta'_i := \delta^{\mathbf{D}_i}(\langle \mathbf{Q}' \rangle, \langle \mathbf{P} \rangle) = \delta^{\mathbf{D}_i}(\langle \mathbf{Q}' \rangle, \langle \mathbf{P}' \rangle) + \delta^{\mathbf{D}_i}(\langle \mathbf{P}' \rangle, \langle \mathbf{P} \rangle) \leq 2\gamma + \varepsilon$, where the bound on the first term follows from the hardcore lemma (for every fixed value of the advice), whereas the bound on the second term follows from Remark 3. Additionally, $\delta^{\mathbf{D}_i}(\langle \mathbf{Q} \rangle, \langle \mathbf{P} \rangle) = (1-p)\cdot\delta'_i + p\cdot\delta''_i \leq \varepsilon$ with $\delta''_i := \delta^{\mathbf{D}_i}(\langle \mathbf{Q}'' \rangle, \langle \mathbf{P} \rangle)$ by the indistinguishability assumption on $\langle \mathbf{Q} \rangle$ and the fact that $t' < t$. Since

$$\langle \mathbf{Q}''_1 \rhd \cdots \rhd \mathbf{Q}''_{i-1} \rhd \mathbf{P} \rhd \mathbf{Q}''_{i+1} \rhd \cdots \rhd \mathbf{Q}''_m \rangle \equiv \langle \mathbf{P} \rangle,$$

we obtain $\delta^{\mathbf{D}}(\langle \mathbf{Q}_{\emptyset} \rangle, \langle \mathbf{P} \rangle) = \delta''_i$ and $\delta^{\mathbf{D}}(\langle \mathbf{Q}_{\{i\}} \rangle, \langle \mathbf{P} \rangle) = \delta'_i$ for all $i \in \{1,\ldots,m\}$, and thus

$$\begin{aligned}\sum_{|\mathcal{I}|\leq 1} q_{\mathcal{I}} \cdot \delta^{\mathbf{D}}(\langle \mathbf{Q}_{\mathcal{I}} \rangle), \langle \mathbf{P} \rangle) &= \sum_{i=1}^{m} \frac{1}{m} \cdot p^m \cdot \delta''_i + p^{m-1}(1-p)\cdot\delta'_i \\ &\leq \max_{i\in\{1,\ldots,m\}} \left\{ p^m \cdot \delta''_i + m \cdot p^{m-1} \cdot (1-p) \cdot \delta'_i \right\}.\end{aligned}$$

However, for all $i \in \{1,\ldots,m\}$, we combine all of the above observations to obtain

$$\begin{aligned}p^m\delta''_i + mp^{m-1}(1-p)\delta'_i &= p^{m-1}(p\delta''_i + (1-p)\delta'_i) + (m-1)p^{m-1}(1-p)\delta'_i \\ &\leq p^{m-1}\varepsilon + (m-1)p^{m-1}(1-p)\varepsilon + 2\gamma \\ &\leq \varepsilon^m + (m-1)\varepsilon^m(1-\varepsilon) + 2\gamma \\ &= \varepsilon^m(m-(m-1)\varepsilon) + 2\gamma,\end{aligned}$$

where we also have used $p \leq \varepsilon$ and the fact that $p^m + (m-1)p^{m-1}(1-p)$ grows monotonically for $p \in [0,1]$.

To bound the remaining summands of Equation (2) with $|\mathcal{I}| \geq 2$, we use a standard hybrid argument and Theorem 3 to obtain

$$\delta^{\mathbf{D}}(\langle \mathbf{Q}_{\mathcal{I}} \rangle, \langle \mathbf{P} \rangle) \leq m\cdot\gamma + \frac{4q''\Lambda}{N} + \frac{2\Lambda(q''+\Lambda)}{(1-\varepsilon)N}.$$

This concludes the proof. □

The following corollary follows by applying the theorem to all $\gamma = 1/p$ (for some polynomial p in n) and to all polynomially bounded t, q, and by choosing an appropriate $\Lambda := n^{\omega(1)}$:

Corollary 1. *Let $E = \{E_k\}_{k\in\mathcal{K}}$ be a (two-sided) ε-PRP for $\varepsilon \le 1 - \frac{1}{\mathsf{poly}(n)}$, where n is the security parameter. Then, for any $m = \mathsf{poly}(n)$, the cascade $\{E_{k_1} \circ \cdots \circ E_{k_m}\}_{k_1,\ldots,k_m\in\mathcal{K}}$ is a (two-sided) $(\varepsilon^m(m-(m-1)\varepsilon)+\nu)$-PRP for some negligible function ν, where $\circ$ denotes permutation composition.*

4.3 Tightness

Let $\varepsilon < 1 - 2^{-n}$ be such that $\log\left((1-\varepsilon)^{-1}\right) \in \{1,\ldots,n\}$. Let $\mathbf{Q} : \{0,1\}^n \to \{0,1\}^n$ be the cc-stateless random permutation which initially chooses $B \in \{0,1\}$ with $\mathsf{P}_B(0) = \varepsilon$. If $B = 0$, then $\mathbf{Q}$ behaves as the identity permutation id, whereas if $B = 1$ it behaves as a uniformly chosen permutation Q' with the constraint that the first $\log\left((1-\varepsilon)^{-1}\right)$ bits of $Q'(0^n)$ are all equal to 0. Clearly, it is possible to give an efficient *stateful* algorithm implementing $\mathbf{Q}$ (or $\langle\mathbf{Q}\rangle$) by using lazy sampling.[7] Also, let $\mathbf{Q}_1,\ldots,\mathbf{Q}_m$ be independent instances of $\mathbf{Q}$. We prove the following two statements:

(i) For all distinguishers $\mathbf{D}$ and an n-bit URP $\mathbf{P}$, we have $\Delta^{\mathbf{D}}(\langle\mathbf{Q}\rangle, \langle\mathbf{P}\rangle) \le \varepsilon$, regardless of their computing power.
(ii) There exists a constant-time distinguisher $\mathbf{D}^*$ making *one* single (forward) query such that

$$\Delta^{\mathbf{D}^*}(\mathbf{Q}_1 \rhd \cdots \rhd \mathbf{Q}_m, \mathbf{P}) \ge (m-(m-1)\varepsilon)\varepsilon^m - \frac{1}{2^n}.$$

Hence, the bound of Theorem 4 cannot be substantially improved, even if allowing a huge security loss (i.e., $t'' << t$ and $q'' << q$). This extends to arbitrary m a previous tightness result given by Myers [12] for the special case $m = 2$.

$\mathbf{Q}$ IS A TWO-SIDED ε-PRP. In the following, let Q and P be random variables representing the distributions of the permutation tables of $\mathbf{Q}$ and $\mathbf{P}$, respectively. There are $(1-\varepsilon)(2^n!)$ permutations π for which the last $\log\left((1-\varepsilon)^{-1}\right)$ bits of $\pi(0^n)$ all equal to 0, and the identity id is one such permutation. Hence,

$$\mathsf{P}_Q(\mathrm{id}) = \varepsilon + (1-\varepsilon)\cdot\frac{1}{(1-\varepsilon)(2^n!)} = \varepsilon + \frac{1}{2^n!} \ge \frac{1}{2^n!} = \mathsf{P}_P(\mathrm{id}).$$

For all $\pi \ne \mathrm{id}$, we have $\mathsf{P}_Q(\pi) \le (1-\varepsilon)\cdot\frac{1}{(1-\varepsilon)(2^n!)} = \frac{1}{2^n!} = \mathsf{P}_P(\pi)$. This yields $\Delta^{\mathbf{D}}(\langle\mathbf{P}\rangle, \langle\mathbf{Q}\rangle) \le d(P,Q) = \mathsf{P}_Q(\mathrm{id}) - \mathsf{P}_P(\mathrm{id}) = \varepsilon$ for *all* distinguishers $\mathbf{D}$.

[7] Also, from any PRP $E = \{E_k\}_{k\in\{0,1\}^n}$ with n-bit string domain, we can define a permutation family $E' = \{E'_{k'}\}_{k'\in\{0,1\}^{\log(1/\varepsilon)+n}}$ which is computationally indistinguishable from $\mathbf{Q}$ under a uniform $(\log(1/\varepsilon)+n)$-bit random key: For all $k' \in \{0,1\}^{\log(1/\varepsilon)}$ and $k \in \{0,1\}^n$, let $E'_{k'\|k}(x) := x$ if $k' = 0^{\log(1/\varepsilon)}$, and $E'_{k'\|k}(x) := E_k(x) \oplus E_k(0^n)|_{\log((1-\varepsilon)^{-1})}$ otherwise, where $z|_r$ sets the last $n-r$ bits of $z \in \{0,1\}^n$ to be 0 (and leaves the first r unchanged) and $\|$ denotes string concatenation.

Lower Bound for Distinguishing the Cascade. We define $\mathbf{D}^*$ as the distinguisher querying 0^n and outputting 1 if and only if the first $\log\left((1-\varepsilon)^{-1}\right)$ bits of the resulting output are all 0, and outputting 0 otherwise. In particular, $\mathsf{P}[\mathbf{D}^*(\mathbf{P}) = 1] = 2^{-\log\left((1-\varepsilon)^{-1}\right)} = 1 - \varepsilon$, as the output of $\mathbf{P}$ on input 0^n is uniform.

Denote as B_i the bit B associated with the i-th instance $\mathbf{Q}_i$, and let $\mathcal{A}_{\mathcal{I}}$ for $\mathcal{I} \subseteq \{1, \ldots, m\}$ be the event that $B_i = 1$ for all $i \in \mathcal{I}$ and $B_i = 0$ for all $i \notin \mathcal{I}$. Furthermore, let $\mathcal{E}$ be the event that $\mathcal{A}_{\mathcal{I}}$ occurs for some $\mathcal{I}$ with $|\mathcal{I}| \leq 1$. Clearly, $\mathsf{P}[\mathcal{E}] = \varepsilon^m + m(1-\varepsilon)\varepsilon^{m-1}$ and $\mathsf{P}\left[\mathbf{D}^*(\mathbf{Q}_1 \rhd \cdots \rhd \mathbf{Q}_m) = 1 \,\middle|\, \mathcal{E}\right] = 1$, since $\mathbf{Q}_1 \rhd \cdots \rhd \mathbf{Q}_m$ under $\mathcal{E}$ behaves either as the identity or as Q', and in both cases the first $\log\left((1-\varepsilon)^{-1}\right)$ output bits are all 0.

Let us fix $\mathcal{I}$ with $k := |\mathcal{I}| \geq 2$, and let $\mathbf{Q}'_1, \ldots, \mathbf{Q}'_k$ be independent random permutations answering according to Q'. Then,

$$\mathsf{P}\left[\mathbf{D}^*(\mathbf{Q}_1 \rhd \cdots \rhd \mathbf{Q}_m) = 1 \,\middle|\, \mathcal{A}_{\mathcal{I}}\right] = \mathsf{P}\left[\mathbf{D}^*(\mathbf{Q}'_1 \rhd \cdots \rhd \mathbf{Q}'_k) = 1\right].$$

For any input $x \neq 0^n$ the probability that the first $\log\left((1-\varepsilon)^{-1}\right)$ output bits of $\mathbf{Q}'_k(x)$ are all 0 is exactly $1-\varepsilon$, whereas the probability that $\mathbf{Q}'_k$ is invoked on 0^n is at most $\frac{1}{(1-\varepsilon)2^n}$ (as regardless of the input, the output $\mathbf{Q}'_{k-1}$ is uniformly distributed on a set of at least size $(1-\varepsilon)2^n$), and therefore

$$\mathsf{P}\left[\mathbf{D}^*(\mathbf{Q}'_1 \rhd \cdots \rhd \mathbf{Q}'_k) = 1\right] \geq \left(1 - \frac{1}{(1-\varepsilon)2^n}\right) \cdot (1-\varepsilon) = 1 - \varepsilon - \frac{1}{2^n},$$

which in turn implies $\mathsf{P}\left[\mathbf{D}^*(\mathbf{Q}_1 \rhd \cdots \rhd \mathbf{Q}_m) = 1 \,\middle|\, \overline{\mathcal{E}}\right] \geq 1 - \varepsilon - \frac{1}{2^n}$. From this, we conclude $\Delta^{\mathbf{D}^*}(\mathbf{Q}_1 \rhd \cdots \rhd \mathbf{Q}_m, \mathbf{P}) \geq (m - (m-1)\varepsilon)\varepsilon^m - \frac{1}{2^n}$.

5 Conclusions and Open Problems

This paper has presented the first tight analysis of the security amplification properties of the cascade of weak PRPs, both in the one- and two-sided cases. Our main tool is a hardcore lemma (Theorem 2) for computational indistinguishability of discrete interactive cc-stateless systems. It is our belief that the generality of this result makes it suitable to the solution of a number of other problems. For instance, an interesting problem is whether *parallel* and *deterministic* security-amplifying constructions for *arbitrarily* weak pseudorandom *functions* exist. To date, the best known constructions are either randomized [13,10], or only work for moderately weak PRFs [2,10]. Also, quantitative improvements of our results should also be of interest. One may try to minimize the length of the state output by the state sampler or to improve the bound of Theorem 3.

Acknowledgments. The author is grateful to Peter Gaži, Thomas Holenstein, Russell Impagliazzo, Ueli Maurer, and Salil Vadhan for helpful discussions and insightful feedback. This work was done while the author was a graduate student at ETH Zurich, supported in part by the Swiss National Science Foundation (SNF), project no. 200020-113700/1. He is currently at UCSD, partially supported by NSF grant CNS-0716790.

References

1. Bellare, M., Impagliazzo, R., Naor, M.: Does parallel repetition lower the error in computationally sound protocols? In: FOCS 1997: Proceedings of the 38th IEEE Annual Symposium on Foundations of Computer Science, pp. 374–383 (1997)
2. Dodis, Y., Impagliazzo, R., Jaiswal, R., Kabanets, V.: Security amplification for *interactive* cryptographic primitives. In: Reingold, O. (ed.) TCC 2009. LNCS, vol. 5444, pp. 128–145. Springer, Heidelberg (2009)
3. Holenstein, T.: Key agreement from weak bit agreement. In: STOC 2005: Proceedings of the 37th Annual ACM Symposium on Theory of Computing, pp. 664–673 (2005)
4. Holenstein, T.: Pseudorandom generators from one-way functions: A simple construction for any hardness. In: Halevi, S., Rabin, T. (eds.) TCC 2006. LNCS, vol. 3876, pp. 443–461. Springer, Heidelberg (2006)
5. Impagliazzo, R.: Hard-core distributions for somewhat hard problems. In: FOCS 1995: Proceedings of the 36th IEEE Annual Symposium on Foundations of Computer Science, pp. 538–545 (1995)
6. Luby, M., Rackoff, C.: Pseudo-random permutation generators and cryptographic composition. In: STOC 1986: Proceedings of the 18th Annual ACM Symposium on Theory of Computing, pp. 356–363 (1986)
7. Luby, M., Rackoff, C.: How to construct pseudorandom permutations from pseudorandom functions. SIAM Journal on Computing 17(2), 373–386 (1988)
8. Maurer, U.: Indistinguishability of random systems. In: Knudsen, L.R. (ed.) EUROCRYPT 2002. LNCS, vol. 2332, pp. 110–132. Springer, Heidelberg (2002)
9. Maurer, U., Pietrzak, K., Renner, R.: Indistinguishability amplification. In: Menezes, A. (ed.) CRYPTO 2007. LNCS, vol. 4622, pp. 130–149. Springer, Heidelberg (2007)
10. Maurer, U., Tessaro, S.: Computational indistinguishability amplification: Tight product theorems for system composition. In: Halevi, S. (ed.) CRYPTO 2009. LNCS, vol. 5677, pp. 355–373. Springer, Heidelberg (2009)
11. Maurer, U., Tessaro, S.: A hardcore lemma for computational indistinguishability: Security amplification for arbitrarily weak prgs with optimal stretch. In: Micciancio, D. (ed.) TCC 2010. LNCS, vol. 5978, pp. 237–254. Springer, Heidelberg (2010)
12. Myers, S.: On the development of block-ciphers and pseudo-random function generators using the composition and XOR operators. Master's thesis, University of Toronto (1999)
13. Myers, S.: Efficient amplification of the security of weak pseudo-random function generators. Journal of Cryptology 16, 1–24 (2003)
14. Pietrzak, K., Wikström, D.: Parallel repetition of computationally sound protocols revisited. In: Vadhan, S.P. (ed.) TCC 2007. LNCS, vol. 4392, pp. 86–102. Springer, Heidelberg (2007)
15. Unruh, D.: Random oracles and auxiliary input. In: Menezes, A. (ed.) CRYPTO 2007. LNCS, vol. 4622, pp. 205–223. Springer, Heidelberg (2007)
16. Yao, A.C.: Theory and applications of trapdoor functions. In: FOCS 1982: Proceedings of the 23rd IEEE Annual Symposium on Foundations of Computer Science, pp. 80–91 (1982)

Dense Model Theorems and Their Applications

Luca Trevisan*

Department of Computer Science, Stanford University,
Stanford CA 94305

In 2004, Ben Green and Terry Tao [6] proved that, for every k, there are infinitely many length-k arithmetic progressions made entirely of prime numbers. This settled a very long-standing open question in number theory that had been open even for the $k = 4$ case.

A key innovation in their proof is a "transference" or "dense model" theorem that, described in a "computer science terminology," states that if R is a "pseudorandom" set of integers, and $P \subseteq R$ is a subset of R that has positive density within R, then there exists a set of integers M that has positive density within all of $\mathbb{N}$ and that is "indistinguishable" from P in the sense that if a statistical property of a certain type is true for M then it must also be true for P.

Green and Tao apply their theorem to the case in which: (i) R are the almost-primes, integers with few, large, prime factors, which were known to have strong pseudorandomness properties; and (ii) P are the primes, which are known to have positive density in R, when the proper quantitative definition of R is given. Because of Szemerédi's Theorem [11, 12], M contains infinitely many arithmetic progressions of any length and, in fact, the probability that a random length-k progression is entirely contained in M is within a constant factor of the probability that are random k-tuple of elements is entirely contained in M. The notion of "indistinguishability" between M and P is such that this statistical property must be true for P too, and so not only it follows that P contains infinitely many arithmetic progressions of any length, but even the stronger statement that there are $\Omega_k(n^2/(\log n)^k)$ length-k progressions entirely composed of primes less than n.

A later paper of Tao and Ziegler [13] presents a more abstract version of the Green-Tao dense model theorem that, *qualitatively*, can be given the following translation in computer science terminology: if R is a pseudorandom distribution over a universe X, and D is a distribution that is c-dominated by R, in the sense that $D(x) \leq c \cdot R(x)$ for a domination parameter c, then there is a model distribution M such that D and M are indistinguishable, and M is $2c$-dominated by the uniform distribution over X. In particular, M has very high entropy $\log_2 |X| - \log_2 c + 1$, and so D has very high *pseudoentropy*. Unfortunately, the proof in [6,13], which is based on the proof of the Szemerédi Regularity Lemma, does not give the right quantitative result, because the complexity of the "security reduction" is exponential in the accuracy of the indistinguishability that one wants to achieve.

* This material is based upon work supported by the National Science Foundation under Grant No. CCF-1017403.

Y. Ishai (Ed.): TCC 2011, LNCS 6597, pp. 55–57, 2011.

Reingold, Trevisan, Tulsiani and Vadhan [10] provide an alternative proof based on duality of linear programming, inspired by Nisan's proof of the Impagliazzo hard-core lemma [7], which has a security reduction of polynomial complexity and hence works with "cryptographically strong" notions of pseudorandomness and indistinguishability. This gave a new characterization of the notion of pseudoentropy, which adds to the study of Barak, Shaltiel and Wigderson [1], who had proved equivalences between various computational analogs of entropy. Gowers [3] independently discovered the same argument based on duality of linear programming, which he has applied to other problems in additive combinatorics in joint work with Wolfe [4, 5].

Dziembowski and Pietrzak [2] formulated the computational dense model theorem in independent work (not inspired by the number-theoretic analog), and proved it based on a result that is attributed to [1], although it is actually first proved in [10]. Dziembowski and Pietrzak apply the dense model theorem to the task of designing cryptosystems that are resilient to key leakage, introducing an approach that has been tremendously influential.

Mironov, Pandey, Reingold and Vadhan [9] give an application of the dense model theorem to the study of *computational differential privacy*.

Impagliazzo [8] showed that the dense model theorem can be proved from a weaker assumption, giving yet another new characterization of pseudoentropy; he also showed that one can derive the dense model theorem in a black box way from any proof of a sufficiently strong version of his hard-core set lemma. This implied that a computational dense model theorem can be derived from an "iterative" approach, which is different from both the original argument of Green and Tao and from the argument based on duality of linear programming. The work of Trevisan, Tulsiani and Vadhan [14] gives a different way of "tying together" the Szemerédi regularity lemma, the dense model theorem, the Impagliazzo hard-core set lemma, and the use of iterative or linear-programming based techniques.

In this talk we will tell the number-theoretic side of this story, and give a quick overview of the several different, but equivalent, ways in which this family of results can be thought about.

References

1. Barak, B., Shaltiel, R., Wigderson, A.: Computational analogues of entropy. In: Arora, S., Jansen, K., Rolim, J.D.P., Sahai, A. (eds.) RANDOM 2003 and APPROX 2003. LNCS, vol. 2764, pp. 200–215. Springer, Heidelberg (2003)
2. Dziembowski, S., Pietrzak, K.: Leakage-resilient cryptography. In: Proceedings of the 49th IEEE Symposium on Foundations of Computer Science, pp. 293–302 (2008)
3. Gowers, T.: Decompositions, approximate structure, transference, and the Hahn-Banach theorem, arXiv:0811.3103 (2008)
4. Gowers, T., Wolf, J.: Linear forms and higher-degree uniformity for functions on $\mathbb{F}_p^n$, arXiv:1002.2208 (2010)
5. Gowers, T., Wolf, J.: Linear forms and quadratic uniformity for functions on $\mathbb{F}_p^n$, arXiv:1002.2209 (2010)

6. Green, B., Tao, T.: The primes contain arbitrarily long arithmetic progressions. Annals of Mathematics 167, 481–547 (2008)
7. Impagliazzo, R.: Hard-core distributions for somewhat hard problems. In: Proceedings of the 36th IEEE Symposium on Foundations of Computer Science, pp. 538–545 (1995)
8. Impagliazzo, R.: Personal Communication (2008)
9. Mironov, I., Pandey, O., Reingold, O., Vadhan, S.P.: Computational differential privacy. In: Halevi, S. (ed.) CRYPTO 2009. LNCS, vol. 5677, pp. 126–142. Springer, Heidelberg (2009)
10. Reingold, O., Trevisan, L., Tulsiani, M., Vadhan, S.: Dense subsets of pseudorandom sets. In: Proceedings of the 49th IEEE Symposium on Foundations of Computer Science, pp. 76–85 (2008)
11. Szemerédi, E.: On sets of integers containing no four elements in arithmetic progression. Acta Math. Acad. Sci. Hung. 20, 89–104 (1969)
12. Szemerédi, E.: On sets of integers containing no k elements in arithmetic progression. Acta Arithmetica 27, 199–245 (1975)
13. Tao, T., Ziegler, T.: The primes contain arbitrarily long polynomial progressions. Acta Mathematica 201, 213–305 (2008)
14. Trevisan, L., Tulsiani, M., Vadhan, S.: Regularity, boosting, and efficiently simulating every high-entropy distribution. In: Proceedings of the 24th IEEE Conference on Computational Complexity (2009)

Parallel Repetition for Leakage Resilience Amplification Revisited

Abhishek Jain[1] and Krzysztof Pietrzak[2]

[1] UCLA
abhishek@cs.ucla.edu
[2] CWI, Amsterdam
pietrzak@cwi.nl

Abstract. If a cryptographic primitive remains secure even if ℓ bits about the secret key are leaked to the adversary, one would expect that at least one of n independent instantiations of the scheme remains secure given $n \cdot \ell$ bits of leakage. This intuition has been proven true for schemes satisfying some special information-theoretic properties by Alwen et al. [Eurocrypt'10]. On the negative side, Lewko and Waters [FOCS'10] construct a CPA secure public-key encryption scheme for which this intuition fails.

The counterexample of Lewko and Waters leaves open the interesting possibility that for any scheme there exists a constant $c > 0$, such that n fold repetition remains secure against $c \cdot n \cdot \ell$ bits of leakage. Furthermore, their counterexample requires the n copies of the encryption scheme to share a common reference parameter, leaving open the possibility that the intuition is true for all schemes without common setup.

In this work we give a stronger counterexample ruling out these possibilities. We construct a signature scheme such that:

1. a single instantiation remains secure given $\ell = \log(k)$ bits of leakage where k is a security parameter.
2. any polynomial number of independent instantiations can be broken (in the strongest sense of key-recovery) given $\ell' = \mathrm{poly}(k)$ bits of leakage. Note that ℓ' does not depend on the number of instances.

The computational assumption underlying our counterexample is that non-interactive computationally sound proofs exist. Moreover, under a stronger (non-standard) assumption about such proofs, our counterexample does not require a common reference parameter.

The underlying idea of our counterexample is rather generic and can be applied to other primitives like encryption schemes.

1 Introduction

In a cryptographic security definition one must precisely specify in which way an anticipated adversary can access the scheme. Classical security definitions usually give the adversary only black-box access, that is she can observe the input/output behavior of the scheme, but nothing else is leaked.

Y. Ishai (Ed.): TCC 2011, LNCS 6597, pp. 58–69, 2011.

Unfortunately, in the last two decades, it has become evident that such notions are often insufficient to guarantee security in the real world where the physical implementation (e.g. on a smart-card) of a scheme is under attack. Two important types on attacks which are not captured by classical security notions are side-channel attacks and malware.

Side-Channel Attacks. Cryptanalytic attacks where the adversary exploits information leakage from the physical implementation of a scheme are called side-channel attacks. Important examples of side-channels that have been exploited include measuring the running-time [24,29], electromagnetic radiation [32,17] or power consumption [25] of a cryptodevice. Cold-boot attacks [19] exploit the fact that memory retains its content for several seconds or even minutes even after being removed from a laptop, allowing the adversary to learn (a noisy version of) the content of the memory. In a probing attack [4], one measures the contents carried by some wires of the circuit.

Malware. Today most computers (and even simpler devices) are connected to the Internet, where they are constantly exposed to attacks by malicious softwares like viruses and Trojans. Even with anti-virus protection in place, it is quite inevitable that a computer is from time to time infected by malware. Such attacks are particularly devastating if cryptographic keys are stored on the computer, as the malware can send them out to the bad guys.

Notions for Key-Leakage. Traditional security notions only consider adversaries having black-box access to the primitive at hand, and thus do not capture side-channel or malware attacks at all. The common approach to protect against side-channel attacks (similarly for malware) is ad-hoc, in the sense that one has to devise a countermeasure against all the known attacks. A more recent approach is to model security against "general" side-channel (or malware) attacks at the definitional level.

Memory-Attacks. A basic such notion is called "security against memory-attacks". A cryptographic scheme is secure against memory attacks, if it remains secure even if a bounded amount of information about the secret key is given to the adversary. In this model, [1,28,7] construct public-key encryption schemes and [22,3] construct signature schemes, identification schemes and key exchange protocols.

Formally, memory attacks are modeled by giving the adversary – on top of the normal black-box access – the power to choose any efficient leakage function f with bounded range ℓ bits; she then gets $f(sk)$ where sk denotes the secret key of the scheme at hand. Of course ℓ must be significantly smaller than $|sk|$ since otherwise the adversary can just leak the entire key.

Although in a memory attack the adversary can learn arbitrary leakage, she is limited to learn a total of ℓ bits. This is not sufficient to protect against most side-channel attacks or malware, where we need stronger models. One way is to consider "continuous" leakage, which requires frequent key-updates, or making the key huge while preserving efficiency of the scheme.

Continuous Leakage Models. Typical side-channel attacks leak small amounts of information per invocation, but since a scheme can typically be invoked many times, no upper bound on the total leakage should be assumed. The notion of "leakage-resilient cryptography" [15,30,16,33,10,18,21] and the recently introduced model of "continuous memory attacks" [12,8] capture such attacks, allowing the adversary to learn a bounded amount of leakage (computed by adaptively chosen leakage functions) *with every invocation*. In the model of leakage-resilience one needs the additional assumption that during each invocation, only the part of the memory that is actually accessed leaks. Continuous memory attacks require (almost) leakage free update phases.

The Bounded Retrieval Model. The bounded-retrieval model (BRM) [13,11] propose a solution to the leakage of cryptographic keys from devices which can get infected by malware as explained above. The idea is to design schemes where the key can be made huge (2GB, say), and the scheme remains secure even if a large amount (1GB, say) of arbitrary information of the key is leaked. The key is protected, even if the computer is infected by malware which can perform arbitrary computation on the infected computer, but can only send out at most 1GB worth of data, either due to low bandwidth or because larger leakage can be detected. This notion is strictly stronger than security against memory attacks, as here one additionally requires that the efficiency of the scheme is basically independent of the size of the secret key. In the BRM model, symmetric authentication schemes [13,11,9], password authentication [11] and secret-sharing [14] were constructed. Recently the first *public-key* primitives in the BRM model were constructed by Alwen at al.[3,2].[1]

Security Amplification by Repetition. Given a scheme that withstands memory attacks with, say $\ell = |sk|/2$ leakage, we can construct a scheme that withstands 1GB of leakage by using a huge key $|sk| =$2GB. This would *not* be considered a solution in the BRM model, as this model requires the efficiency of the scheme to depend only on some security parameter k, but not (or only logarithmically) on the key length (In particular, schemes in the BRM model cannot even read the entire key on a single invocation.)

The approach taken by Alwen et al. [3,2] is to use parallel repetition. They start with a scheme which can tolerate ℓ bits of leakage, and run this scheme n times in parallel. For a signature scheme this means to sample n secret keys $sk_1, \ldots, sk_n$, a signature for the parallel scheme would then consist of n signatures, using the respective keys.[2]

---

[1] The BRM predates the notion of memory-attacks, but public-key cryptography in the BRM came only after it was constructed in the model of memory attacks.

[2] This is still not really a scheme in the BRM model as efficiency of the scheme is linear in n. To actually get signatures, [3] only use a cleverly chosen subset of the keys for signing, and also must "compress" the n public keys in order to make their size independent of n.

The hope is that if the signature scheme is secure against ℓ bits of leakage, the parallel scheme will remain secure given $n \cdot \ell$ bits of leakage. Alwen et al. prove that for their particular scheme this is indeed the case, but their proof exploits information theoretic properties of the scheme, and cannot be generalized to work for any scheme.

The Counterexample of Lewko and Waters. Recently, Lewko and Waters [26] showed that in fact, n-fold parallel repetition does not, in general, amplify leakage resilience from ℓ to $n \cdot \ell$ bits. They construct a scheme (their main example is encryption, but it is outlined how to adapt the counterexample to signatures.) where a single instance is secure given ℓ bits of leakage, but can be broken given $c \cdot n \cdot \ell$ bits of leakage for some constant $c < 1$. Their attack also requires a common setup, in that all the instances of the basic scheme in the parallel system must be over the same group, and the group order must be secret.[3]

Our Results. The counterexample of Lewko and Waters leaves open the possibility that for every scheme, there exists a constant $c > 0$ such that n-fold parallel repetition (even with common setup) amplifies leakage-resilience from ℓ to $c \cdot n \cdot \ell$ bits.

Moreover the common setup seems crucial for their counterexample, and it leaves open the question whether n-fold parallel repetition without common setup amplifies leakage-resilience from ℓ to $n \cdot \ell$ bits for all schemes.

We give a new counterexample that answers both questions in the negative (albeit the 2nd only under a non-standard assumption, details are given below.) More concretely, from any secure signature scheme, we construct a new signature scheme such that:

1. a single instantiation of the scheme remains secure given $\ell = \log(k)$ bits of leakage where k is a security parameter.
2. n *independent* instances (where n can be any polynomial in k) of the scheme can be broken (in the strongest sense of key-recovery) given $\ell' = \text{poly}(k)$ bits of leakage.

Note that ℓ' – the leakage needed to break n instances – does not even depend on n.

Our techniques are quite general and we anticipate that they can be extended to construct counter-examples for other primitives as well. Besides the counterexample for signature schemes just mentioned (which also works for one-time signatures), we provide a similar counterexample for CCA-secure encryption schemes.

We note here that the results of [26] are applicable to even 1-bit (CPA-secure) encryption schemes and signature schemes with "random message unforgeability under no-message attack", while our techniques do not seem to lend themselves to such settings.

[3] The constant c depends on the underlying group, and can be made arbitrary small, but once the group is fixed, so is c.

The main assumption underlying our results is the existence of non-interactive CS proofs (see below). We note that in contrast, the counterexample of [26] is based on a specific-number theoretic, but falsifiable assumption.

Our Techniques. Our negative results make crucial use of computationally sound (CS) proofs as constructed by Micali [27] (using techniques from [23]). Specifically, our counterexample relies on the existence of *non-interactive* CS proofs for NP languages. The usefulness of non-interactive CS proofs lies in the fact that they are extremely short; in particular, their length is only polylogarithmic in the size of the statement and its witness.

Proof Idea. We construct our counterexample by starting with any signature scheme, and extending it as follows. To every secret key we add some random string w, and to the public-key we add the value $G(w)$ where $G(.)$ is a (sufficiently expanding) pseudorandom generator.

The signing algorithm is basically unchanged, except that it additionally checks if the message to be signed contains a CS proof for the fact that $G(w)$ is indeed the the range of $G(.)$. If this is the case, it does something stupid, namely outputting its entire secret key as part of the signature.

The security of the underlying signature scheme and the CS proof system implies that this altered scheme is secure against $\ell = \log(k)$ bit of leakage. On the other hand, a leakage function which has access to n secret keys, can output a short (i.e. of length which is independent of n) CS proof showing that the string $G(w_1), \ldots, G(w_n)$ is indeed the concatenation of n strings in the range of $G(.)$. This proof can then be used (making signing queries) to extract the entire secret key of all the n instances of the scheme.

About the common reference parameter. Non-interactive CS proofs have been shown to exist in the random oracle model (which in practice must be instantiated with an efficient hash-function.) Since a random oracle implies a common reference string, it seems that we haven't improved upon the counterexample of Lewko and Waters [26] as far as the usage of common setup is concerned. Recall that the common setup in [26] contains a number N whose factorisation must remain secret since otherwise the scheme can be trivially broken. In our case, the common setup contains a hash function $h(.)$ (replacing the random oracle). Although for every $h(.)$, there exist CS proofs for invalid statements, this may not necessarily be a problem for us, as all we need is that it is hard to come up with a proof for a *random* word not in the language (the language and exact distribution depends on the PRG we use in our construction.) It is conceivable that for some concrete choices of a PRG and hash function repalcing the random oracle (SHA256, say), it is hard to come up with such invalid proofs even given a polynomial amount of auxiliary information (containing e.g. many invalid proofs.) If we make such a (non-standard) assumption, our counterexample does not need any common setup.

Related Work. We are aware of at least two works where proof systems with short proofs are used to construct counterexamples. Closest to ours is the work of "seed-incompressible functions" of Halevi, Myers and Rackoff [20], who use CS proofs to show that no pseudorandom function exists which remains secure after one leaks a "compressed" key. Another example is the work on parallel repetition of computationally sound proofs [31] (based on [6]), which uses a different kind of proof systems, universal arguments [5], to show that parallel repetition of computationally sound protocols with eight (or more) rounds does in general not reduce the soundness error of the protocol.

2 Preliminaries

2.1 Leakage-Resilient Signatures

Our definition of leakage resilient signatures is essentially the standard notion of existentially unforgeability under adaptive chosen message attacks, except that we allow the adversary to specify an arbitrary function $f(\cdot)$ (whose output length is bounded by the leakage parameter), and obtain the value of f applied to the signing key. Let k be the security parameter, and (KeyGen, Sign, Verify) denote a signature scheme. Let ℓ denote the leakage parameter. In order to define leakage resilient signatures, we consider the following experiment.

1. Compute $(pk, sk) \leftarrow \mathsf{KeyGen}(1^k, \ell)$ and give pk to the adversary.
2. Run the adversary $\mathcal{A}(1^k, pk, \ell)$. The adversary may make adaptive queries to the signing oracle $\textsc{Sign}_{sk}(\cdot)$ and the leakage oracle $\textsc{Leak}_{sk}(\cdot)$, defined as follows:
   - On receiving the ith query m_i, $\textsc{Sign}_{sk}(m_i)$ computes $\sigma_i \leftarrow \mathsf{Sign}(sk, m_i)$ and outputs σ_i.
   - On receiving an input f (where f is a polynomial-time computable function, described as a circuit), $\textsc{Leak}_{sk}(f)$ outputs $f(sk)$ to $\mathcal{A}$. The adversary is allowed only a single query to the leakage oracle. It can choose any function $f(\cdot)$ of the form $f : \{0,1\}^* \to \{0,1\}^\ell$.
3. At some point, $\mathcal{A}$ stops and outputs (m, σ).

We say that $\mathcal{A}$ *succeeds* if (a) $\mathsf{Verify}(pk, \sigma) = 1$, and (b) m was never queried to $\textsc{Sign}_{sk}(\cdot)$.

Definition 1. *A signature scheme* (KeyGen, Sign, Verify) *is ℓ-leakage resilient if all polynomial-time adversaries $\mathcal{A}$ can succeed with only negligible probability in the above experiment.*

2.2 CS Proofs

Our negative result makes crucial use of CS proofs as constructed by Micali [27] (using techniques from [23]). Below, we recall the definition of non-interactive CS proofs. The definition below is taken almost verbatim from [20].

Definition 2 (Non-interactive CS proofs). *A non-interactive CS-proof system for a language $L \in \mathcal{NP}$ (with relation R_L), consists of two deterministic polynomial-time machines, a prover P and a verifier V, operating as follows:*

- *On input* $(1^k, x, w)$ *such that* $(x, w) \in R_L$, *the prover computes a proof* $\pi = P(x, w)$ *such that* $|\pi| \leq poly(k, \log(|x| + |w|))$.
- *On input* $(1^k, x, \pi)$, *the verifier decides whether to accept or reject the proof* π *(i.e.,* $V(x, \pi) \in \{\texttt{accept}, \texttt{reject}\}$*).*

The proof system satisfies the following conditions, where the probabilities are taken over the random coins of the prover and the verifier:

Perfect completeness: *For any* $(x, w) \in R_L$ *and for any* k,

$$\Pr[\pi \leftarrow P(x, w), V(x, \pi) = \texttt{accept}] = 1$$

Computational soundness: *For any polynomial time machine* $\tilde{P}$ *and any input* $x \notin L$, *it holds that*

$$\Pr[\pi \leftarrow \tilde{P}(x), V(x, \pi) = \texttt{accept}] \leq \text{negl}(k)$$

3 A Leakage Resilient Signature Scheme

Let k be the security parameter. Let (KeyGen, Sign, Verify) be any λ-leakage resilient signature scheme where λ is at least logarithmic in the security parameter. Let $G : \{0,1\}^k \rightarrow \{0,1\}^{2k}$ be a pseudo-random generator (PRG). Let L be an $\mathcal{NP}$ language such that a string $y \in L$ iff $y = y_1, y_2, \ldots$ (where $|y_i| = 2k$) and $\forall i$, $\exists w_i$ such that $G(w_i) = y_i$. Let $\langle P, V \rangle$ be a non-interactive CS-proof system for the language L, where both prover and verifier are PPT machines. We now describe a new ℓ-leakage resilient signature scheme (KeyGen, Sign, Verify), where $\ell = \log(k)$.

KeyGen$(1^k, \ell)$: Compute $(pk, sk) \leftarrow \mathsf{KeyGen}(1^k, \lambda)$. Choose random $w \in \{0,1\}^k$ and compute $y = G(w)$. The public key is $PK = (pk, y)$ and the secret key is $SK = (sk, w)$.

Sign(SK, m): To sign message $m = m_1, m_2$ using secret key $SK = (sk, w)$, first parse m_1 as $m_1^1, m_1^2, \ldots$ such that $|m_1^i| = 2k$. If $\exists i$ such that $m_1^i = y$ $(= G(w))$, then run $V(m_1, m_2)$. If it returns accept, then output SK. Otherwise, output $\sigma \leftarrow \mathsf{Sign}(sk, m)$ as the signature.

Verify(PK, m, σ): Given a signature σ on message m with respect to the public key $PK = (pk, y)$, output 1 iff $\mathsf{Verify}(pk, m, \sigma) = 1$.

This completes the description of our signature scheme. We now state our main results.

Theorem 1. *The proposed signature scheme (*KeyGen, Sign, Verify*) is* ℓ*-leakage resilient, where* $\ell = \log(k)$.

Theorem 2. *There exists a fixed polynomial* $q(\cdot)$ *such that any* n*-fold repetition (where* n *can be any polynomial in* k*) of the signature scheme (*KeyGen, Sign, Verify*) can be broken by a key-recovery attack with* $\ell' = q(k)$ *bits of leakage.*

Remark. Note that in Theorem 2, $\ell' = q(k)$ only depends on the security parameter, but does not depend on the number of repetitions n.

We prove Theorem 1 in the next subsection. Theorem 2 is proven in Section 4.

3.1 Leakage Resilience of our Signature Scheme

We will prove theorem 1 by contradiction. Specifically, we will show that given an adversary $\mathcal{A}$ that forges signatures for (KEYGEN, SIGN, VERIFY) with non-negligible probability δ, we can construct an adversary $\mathcal{B}$ that forges signatures for (KeyGen, Sign, Verify) with probability $\delta' = \frac{\delta - \text{negl}(k)}{k}$.

Description of $\mathcal{B}$. Let C denote the challenger for the signature scheme (KeyGen, Sign, Verify). At a high level, $\mathcal{B}$ works by internally running the adversary $\mathcal{A}$; $\mathcal{B}$ answers $\mathcal{A}$'s queries by using the responses (to its own queries) from C, and then outputs the signature forgery created by $\mathcal{A}$. We now give more details.

On receiving a public key pk from C, $\mathcal{B}$ chooses a random $w \in \{0,1\}^k$ and computes $y = G(w)$. It then sends $PK = (pk, y)$ as the public key to $\mathcal{A}$. Now, when $\mathcal{A}$ makes a signature query $m = m_1, m_2$, $\mathcal{B}$ first parses m_1 as $m_1^1, m_1^2, \ldots$ such that $|m_1^i| = 2k$. If $\exists i$ such that $m_1^i = y$, then $\mathcal{B}$ runs $V(m_1, m_2)$. If it returns `accept`, then $\mathcal{B}$ outputs the abort symbol $\perp$ and stops. Otherwise, it obtains a signature σ on message m from C and sends σ to $\mathcal{A}$. Further, when $\mathcal{A}$ makes a leakage query f, $\mathcal{B}$ simply guesses $y = f(sk, w)$ (where sk is the secret key corresponding to pk) and sends y to $\mathcal{A}$. Finally, when $\mathcal{A}$ creates a forgery (m^*, σ^*), $\mathcal{B}$ outputs (m^*, σ^*) and stops. This completes the description of $\mathcal{B}$.

Let us now analyze the interaction between $\mathcal{B}$ and $\mathcal{A}$. First note that since the leakage query function f has a bounded output length $\ell = \log(k)$, $\mathcal{B}$'s response to $\mathcal{A}$'s leakage query is correct with probability $\epsilon = \frac{1}{k}$. Now assuming that $\mathcal{B}$ outputs the abort symbol during its interaction with $\mathcal{A}$ with only negligible probability, we can establish that $\mathcal{B}$ outputs a valid forgery with probability $\delta' = \frac{\delta - \text{negl}(k)}{k}$ which proves our hypothesis. Therefore, in order to complete the proof, we only need to argue that $\mathcal{B}$ outputs the abort symbol with negligible probability.

Lemma 1. *$\mathcal{B}$ outputs the abort symbol $\perp$ with probability* negl(k).

We prove lemma 1 by a simple hybrid experiment. Consider two hybrids $\mathcal{H}_0$ and $\mathcal{H}_1$, described as follows.

$\mathcal{H}_0$: This is the real experiment between $\mathcal{B}$ and $\mathcal{A}$. Here $\mathcal{B}$ uses a pseudo-random generator G to compute y as part of the public key PK.

$\mathcal{H}_1$: Same as $\mathcal{H}_0$, except that $\mathcal{B}$ chooses $y \in \{0,1\}^{2q}$ uniformly at random.

Recall that $\mathcal{B}$ outputs the abort symbol only when $\mathcal{A}$ sends a signature query $m = m_1, m_2$ where $m_1 = m_1^1, m_1^2, \ldots$ ($|m_1^i| = 2q$) and $\exists i$ such that $m_1^i = y$ and $V(m_1, m_2)$ outputs `accept` (i.e., if m contains a non-interactive CS proof that "explains" y). We will denote such a query as a `bad` query. Let p_0 (resp., p_1) be the probability that $\mathcal{A}$ makes a bad query in hybrid $\mathcal{H}_0$ (resp., $\mathcal{H}_1$). Then, from the pseudo-randomness property of G, it follows that:

$$p_0 - p_1 \leq \text{negl}(k) \tag{1}$$

Now, note that since y is chosen uniformly at random in hybrid $\mathcal{H}_1$, $\mathcal{A}$ can make a bad query in $\mathcal{H}_1$ only if it can create a false non-interactive CS proof. Then, from the soundness of the CS proof system $\langle P, V \rangle$, it follows that the probability that $\mathcal{A}$ makes a bad query in $\mathcal{H}_1$ is negligible, i.e., $p_1 = \text{negl}(k)$. Combining this with equation 1, we establish that $p_0 = \text{negl}(k)$. This completes the proof of lemma 1.

4 Attack on Parallel System

Recall from Definition 2 that the length of a non-interactive CS proof is $q'(k, \log(|x|+|w|))$ for some polynomial $q'(.,.)$, where x is the instance and w is the witness for x. Now, note that for any such pair (x, w), if $|x|$ and $|w|$ are polynomial in k, then for sufficiently large k, we have that $|x|+|w| \leq k^{\log(k)}$. Therefore, there exists a fixed polynomial $q(\cdot) \stackrel{def}{=} q'(., \log(k^{\log k}))$ such that for any non-interactive CS proof $\pi = P(x, w)$, we have that $|\pi| \leq q(k)$ for sufficiently large k.

Now consider a parallel system ($\overline{\text{KeyGen}}$, $\overline{\text{Sign}}$, $\overline{\text{Verify}}$) defined as an n-fold repetition of (KeyGen, Sign, Verify), where $n = p(k)$ for any polynomial $p(\cdot)$. $\overline{\text{KeyGen}}$ runs KeyGen n times to generate n key pairs $(PK_1, SK_1), \ldots, (PK_n, SK_n)$. To sign a message m, $\overline{\text{Sign}}$ computes $\sigma_i \leftarrow \text{Sign}(SK_i, m)$ for each $i \in [n]$ and outputs $\sigma = (\sigma_1, \ldots, \sigma_n)$ as the signature. Finally, on input a signature $\sigma = (\sigma_1, \ldots, \sigma_n)$, $\overline{\text{Verify}}$ outputs 1 iff $\forall i$, $\text{Verify}(PK_i, m, \sigma_i) = 1$.

We now describe an adversary $\mathcal{A}$ that can mount a key-recovery attack on ($\overline{\text{KeyGen}}$, $\overline{\text{Sign}}$, $\overline{\text{Verify}}$) given $q(k)$ bits of leakage. The adversary $\mathcal{A}$ receives $(PK_1, \ldots, PK_n)$ from the challenger of the signature scheme. Recall that $\forall i$, key pairs (PK_i, SK_i) are of the form: $PK_i = (pk_i, y_i)$, $SK_i = (sk_i, w_i)$. Let $y = y_1, \ldots, y_n$ and π be a non-interactive CS-proof to prove the membership of y in L. Then, $\mathcal{A}$ makes a leakage query with function f such that $f(SK_1, \ldots, SK_n) = \pi$. As discussed above, it holds that $|\pi| \leq q(k)$. $\mathcal{A}$ now queries the challenger for a signature on the message $m = m_1, m_2$ where $m_1 = y_1, \ldots, y_n$ and $m_2 = \pi$. At this point, $\mathcal{A}$ must receive $SK_1, \ldots, SK_n$ in response since π is a valid proof for the statement $y \in L$.

This completes the description of the attack on the parallel system.

5 Extending Our Techniques to Other Primitives

The techniques used in section 3 are rather general and not specific to signature schemes. In particular, they can be used to construct other leakage resilient crypto primitives that are insecure under parallel repetition. As an example, in this section, we briefly explain how to construct a leakage resilient CCA-secure public-key encryption scheme that is insecure under parallel repetition.

Let (KeyGen, Encrypt, Decrypt) be a CCA-secure public-key encryption scheme that can tolerate at least logarithmic amount of leakage (here the adversary has to choose the leakage function before getting the challange ciphertext). Let G be a length doubling PRG and $\langle P, V \rangle$ be a non-interactive CS-proof system for the language L, as described in section 3. We now describe a new ℓ-leakage resilient CCA-secure public-key encryption scheme (KeyGen, Encrypt, Decrypt), where ℓ is logarithmic in the security parameter.

KeyGen$(1^k, \ell)$: Compute $(pk, sk) \leftarrow \mathsf{KeyGen}(1^k)$. Choose random $w \in \{0,1\}^k$ and compute $y = G(w)$. The public key is $PK = (pk, y)$ and the secret key is $SK = (sk, w)$.

Encrypt(PK, m): To encrypt a message m using public key $PK = (pk, y)$, simply compute $c \leftarrow \mathsf{Encrypt}(pk, m)$.

Decrypt(SK, c): Given a ciphertext $c = c_1, c_2$ and secret key $SK = (sk, w)$, first parse c_1 as $c_1^1, c_1^2, \ldots$ such that $|c_1^i| = 2k$. If $\exists i$ such that $c_1^i = G(w)$, then run $V(c_1, c_2)$. If it returns `accept`, then output SK. Otherwise, output $m \leftarrow \mathsf{Decrypt}(sk, c)$ as the decrypted message.

It is not difficult to see that the new scheme is ℓ-leakage resilient, where $\ell = \log(k)$. Essentially, we can leverage the soundness of CS proof system and the fact that G is a PRG (in the same manner as in the security proof of the signature scheme described in section 3) to reduce the security of (KeyGen, Encrypt, Decrypt) to that of the underlying scheme (KeyGen, Encrypt, Decrypt).

Now, consider an n-fold repetition of the above scheme. Let $PK_i = (pk_i, y_i)$ denote the public key of the ith instance of the above scheme. A natural way to encrypt a message m in the parallel system is to split m into n shares and encrypt each share m_i with PK_i. Now, (as in the proof of Theorem 2) an adversary $\mathcal{A}$ can make a leakage query to obtain a short CS proof π that explains each y_i. Then, since $\mathcal{A}$ has access to a decryption oracle, it can now send a decryption query $c = c_1, c_2$ where $c_1 = y_1, \ldots, y_n$ and $c_2 = \pi$. $\mathcal{A}$ can therefore successfully recover the secret key of the parallel system.

References

1. Akavia, A., Goldwasser, S., Vaikuntanathan, V.: Simultaneous hardcore bits and cryptography against memory attacks. In: Reingold, O. (ed.) TCC 2009. LNCS, vol. 5444, pp. 474–495. Springer, Heidelberg (2009)
2. Alwen, J., Dodis, Y., Naor, M., Segev, G., Walfish, S., Wichs, D.: Public-key encryption in the bounded-retrieval model. In: Gilbert, H. (ed.) EUROCRYPT 2010. LNCS, vol. 6110, pp. 113–134. Springer, Heidelberg (2010)
3. Alwen, J., Dodis, Y., Wichs, D.: Leakage-resilient public-key cryptography in the bounded-retrieval model. In: Halevi, S. (ed.) CRYPTO 2009. LNCS, vol. 5677, pp. 36–54. Springer, Heidelberg (2009)

4. Anderson, R., Kuhn, M.: Tamper resistance: a cautionary note. In: WOEC 1996: Proceedings of the 2nd Conference on Proceedings of the Second USENIX Workshop on Electronic Commerce, pp. 1–11. USENIX Association, Berkeley (1996)
5. Barak, B., Goldreich, O.: Universal arguments and their applications. SIAM J. Comput. 38(5), 1661–1694 (2008)
6. Bellare, M., Impagliazzo, R., Naor, M.: Does parallel repetition lower the error in computationally sound protocols? In: 38th Annual Symposium on Foundations of Computer Science, pp. 374–383. IEEE Computer Society Press, Los Alamitos (October 1997)
7. Brakerski, Z., Goldwasser, S.: Circular and leakage resilient public-key encryption under subgroup indistinguishability (or: Quadratic residuosity strikes back). In: Rabin, T. (ed.) CRYPTO 2010. LNCS, vol. 6223, pp. 1–20. Springer, Heidelberg (2010)
8. Brakerski, Z., Kalai, Y.T., Katz, J., Vaikuntanathan, V.: Overcoming the hole in the bucket: Public-key cryptography resilient to continual memory leakage. In: FOCS (2010)
9. Cash, D., Ding, Y.Z., Dodis, Y., Lee, W., Lipton, R.J., Walfish, S.: Intrusion-resilient key exchange in the bounded retrieval model. In: Vadhan, S.P. (ed.) TCC 2007. LNCS, vol. 4392, pp. 479–498. Springer, Heidelberg (2007)
10. Chow, S.S., Dodis, Y., Rouselakis, Y., Waters, B.: Practical leakage-resilient identity-based encryption from simple assumptions. In: ACM CCS 2010: 17th Conference on Computer and Communications Security. ACM Press, New York (2010)
11. Di Crescenzo, G., Lipton, R.J., Walfish, S.: Perfectly secure password protocols in the bounded retrieval model. In: Halevi, S., Rabin, T. (eds.) TCC 2006. LNCS, vol. 3876, pp. 225–244. Springer, Heidelberg (2006)
12. Dodis, Y., Haralambiev, K., Lopez-Alt, A., Wichs, D.: Cryptography against continuous memory attacks. In: FOCS (2010)
13. Dziembowski, S.: Intrusion-resilience via the bounded-storage model. In: Halevi, S., Rabin, T. (eds.) TCC 2006. LNCS, vol. 3876, pp. 207–224. Springer, Heidelberg (2006)
14. Dziembowski, S., Pietrzak, K.: Intrusion-resilient secret sharing. In: 48th Annual Symposium on Foundations of Computer Science, pp. 227–237. IEEE Computer Society Press, Los Alamitos (October 2007)
15. Dziembowski, S., Pietrzak, K.: Leakage-resilient cryptography. In: 49th Annual Symposium on Foundations of Computer Science, pp. 293–302. IEEE Computer Society Press, Los Alamitos (October 2008)
16. Faust, S., Kiltz, E., Pietrzak, K., Rothblum, G.N.: Leakage-resilient signatures. In: Micciancio, D. (ed.) TCC 2010. LNCS, vol. 5978, pp. 343–360. Springer, Heidelberg (2010)
17. Gandolfi, K., Mourtel, C., Olivier, F.: Electromagnetic analysis: Concrete results. In: Koç, Ç.K., Naccache, D., Paar, C. (eds.) CHES 2001. LNCS, vol. 2162, pp. 251–261. Springer, Heidelberg (2001)
18. Goldwasser, S., Rothblum, G.N.: Securing computation against continuous leakage. In: Rabin, T. (ed.) CRYPTO 2010. LNCS, vol. 6223, pp. 59–79. Springer, Heidelberg (2010)
19. Halderman, J.A., Schoen, S.D., Heninger, N., Clarkson, W., Paul, W., Calandrino, J.A., Feldman, A.J., Appelbaum, J., Felten, E.W.: Lest we remember: Cold boot attacks on encryption keys. In: USENIX Security Symposium, pp. 45–60 (2008)
20. Halevi, S., Myers, S., Rackoff, C.: On seed-incompressible functions. In: Canetti, R. (ed.) TCC 2008. LNCS, vol. 4948, pp. 19–36. Springer, Heidelberg (2008)

21. Juma, A., Vahlis, Y.: Protecting cryptographic keys against continual leakage. In: Rabin, T. (ed.) CRYPTO 2010. LNCS, vol. 6223, pp. 41–58. Springer, Heidelberg (2010)
22. Katz, J., Vaikuntanathan, V.: Signature schemes with bounded leakage resilience. In: Matsui, M. (ed.) ASIACRYPT 2009. LNCS, vol. 5912, pp. 703–720. Springer, Heidelberg (2009)
23. Kilian, J.: A note on efficient zero-knowledge proofs and arguments (extended abstract). In: Proc. 24th STOC, pp. 723–732 (1992)
24. Kocher, P.C.: Timing attacks on implementations of diffie-hellman, RSA, DSS, and other systems. In: Koblitz, N. (ed.) CRYPTO 1996. LNCS, vol. 1109, pp. 104–113. Springer, Heidelberg (1996)
25. Kocher, P.C., Jaffe, J., Jun, B.: Differential power analysis. In: Wiener, M.J. (ed.) CRYPTO 1999. LNCS, vol. 1666, pp. 388–397. Springer, Heidelberg (1999)
26. Lewko, A., Waters, B.: On the insecurity of parallel repetition for leakage resilience. In: FOCS (2010)
27. Micali, S.: A secure and efficient digital signature algorithm. Technical Memo MIT/LCS/TM-501b, Massachusetts Institute of Technology, Laboratory for Computer Science (April 1994)
28. Naor, M., Segev, G.: Public-key cryptosystems resilient to key leakage. In: Halevi, S. (ed.) CRYPTO 2009. LNCS, vol. 5677, pp. 18–35. Springer, Heidelberg (2009)
29. Osvik, D.A., Shamir, A., Tromer, E.: Cache attacks and countermeasures: The case of AES. In: Pointcheval, D. (ed.) CT-RSA 2006. LNCS, vol. 3860, pp. 1–20. Springer, Heidelberg (2006)
30. Pietrzak, K.: A leakage-resilient mode of operation. In: Joux, A. (ed.) EUROCRYPT 2009. LNCS, vol. 5479, pp. 462–482. Springer, Heidelberg (2009)
31. Pietrzak, K., Wikström, D.: Parallel repetition of computationally sound protocols revisited. In: Vadhan, S.P. (ed.) TCC 2007. LNCS, vol. 4392, pp. 86–102. Springer, Heidelberg (2007)
32. Quisquater, J.J., Samyde, D.: ElectroMagnetic analysis (EMA): Measures and counter-measures for smart cards. In: Attali, S., Jensen, T. (eds.) E-smart 2001. LNCS, vol. 2140, pp. 200–210. Springer, Heidelberg (2001)
33. Yu, Y., Standaert, F.X., Pereira, O., Yung, M.: Practical leakage-resilient pseudorandom generators. In: ACM CCS 2010: 17th Conference on Computer and Communications Security. ACM Press, New York (2010)

Achieving Leakage Resilience through Dual System Encryption

Allison Lewko*, Yannis Rouselakis, and Brent Waters**

The University of Texas at Austin
{alewko,jrous,bwaters}@cs.utexas.edu

Abstract. In this work, we show that strong leakage resilience for cryptosystems with advanced functionalities can be obtained quite naturally within the methodology of dual system encryption, recently introduced by Waters. We demonstrate this concretely by providing fully secure IBE, HIBE, and ABE systems which are resilient to bounded leakage from each of many secret keys per user, as well as many master keys. This can be realized as resilience against continual leakage if we assume keys are periodically updated and no (or logarithmic) leakage is allowed during the update process. Our systems are obtained by applying a simple modification to previous dual system encryption constructions: essentially this provides a generic tool for making dual system encryption schemes leakage-resilient.

1 Introduction

Defining and achieving the right security models is crucial to the value of provably secure cryptography. When security definitions fail to encompass *all* of the power of potential attackers, systems which are proven "secure" may actually be vulnerable in practice. It is often not realistic or desirable to address such problems solely at the implementation level. Instead, the ultimate goal of cryptography should be to provide efficient systems which are proven secure against the largest possible class of potential attackers. Additionally, these systems should provide the most advanced functionalities available.

Recently, much progress has been made in obtaining increasingly complex systems with stronger security guarantees. The emergence of *leakage-resilient cryptography* has led to constructions of many cryptographic primitives which can be proven secure even against adversaries who can obtain limited additional information about secret keys and other internal state. This line of research is motivated by a variety of side-channel attacks [46, 13, 7, 12, 53, 8, 47, 58, 33, 41],

* Supported by a National Defense Science and Engineering Graduate Fellowship.
** Supported by NSF CNS-0915361, and CNS-0952692, Air Force Office of Scientific Research (AFO SR) under the MURI award for "Collaborative policies and assured information sharing" (Project PRESIDIO), Department of Homeland Security Grant 2006-CS-001-000001-02 (subaward 641), a Google Faculty Research award, and the Alfred P. Sloan Foundation.

Y. Ishai (Ed.): TCC 2011, LNCS 6597, pp. 70–88, 2011.

which allow attackers to learn partial information about secrets by observing physical properties of a cryptographic execution such as timing, power usage, etc. The cold-boot attack [41] allows an attacker to learn information about memory contents of a machine even after the machine is powered down.

Leakage-resilient cryptography models a large class of side-channel attacks by allowing the attacker to specify an efficiently computable leakage function f and learn the output of f applied to the secret key and possibly other internal state at specified moments in the security game. Clearly, limits must be placed on f to prevent the attacker from obtaining the entire secret key and hence easily winning the game. One approach is to bound the total number of bits leaked over the lifetime of the system to be significantly less than the bit-length of the secret key. Another approach is to continually refresh the secret key and bound the leakage between each update (this is called "continual leakage"). Both of these approaches have been employed successfully in a variety of settings, yielding constructions of stream ciphers, signatures, symmetric key encryption, public key encryption, and identity-based encryption (IBE) which are leakage-resilient under various models of leakage [52, 45, 31, 57, 26, 1, 2, 32, 29, 24, 19, 30, 3, 15, 21, 16, 25].

Concurrently, the methodology of dual system encryption has emerged as a useful tool for improving the security guarantees for efficient cryptosystems with advanced functionalities like identity-based encryption (IBE), hierarchical identity-based encryption (HIBE), attribute-based encryption (ABE) [63,50,48]. These works provide efficient systems with short parameters which are proven fully secure in the standard model under static assumptions. Previous constructions of IBE and HIBE either used random oracles, had large parameters, were only proven selectively secure (a weaker model of security where the attacker must declare its target immediately instead of choosing it adaptively in the course of the security game), or relied on "q-based" assumptions (where the size of the assumption depends on the number of the attacker's queries) [14, 22, 37, 18, 9, 10, 61, 11, 34, 36, 35]. All previous constructions of ABE were only proven selectively secure [59,40,20,6,55,39,62]. Like leakage resilience, moving from selectively secure systems to fully secure systems is important because it results in security against a more powerful class of attackers.

Our Contribution. In this work, we show that the techniques of dual system encryption naturally lead to leakage resilience. We demonstrate this by providing leakage-resilient constructions of IBE, HIBE, and ABE systems which retain all of the desirable features of dual system constructions, like full security from static assumptions and close resemblance to previous selectively secure schemes. We present our combination of dual system encryption and leakage resilience as a convenient abstraction and reduce proving security to the establishment of three properties.

Our approach not only combines the benefits of dual system encryption and leakage resilience, but also qualitatively improves upon the leakage tolerance of previous leakage-resilient IBE schemes [16, 2, 21]. In particular, our IBE system can tolerate leakage on the master key, as well as leakage on several keys for each identity (this can be viewed as continual leakage, where secret keys are periodically updated and leakage is allowed only *between* updates, and not *during* updates).[1] The IBE schemes of [2, 21] only allow bounded leakage on one secret key per identity, and allow no leakage on the master key. The IBE scheme of [16] allows bounded leakage on each of many keys per identity, but allows no leakage on the master key.

We develop a simple and versatile methodology for modifying a dual system encryption construction and proof to incorporate strong leakage resilience guarantees. The change to the constructions is minimal, and can be viewed as the adjoining of a separate piece which does not interfere with the intuitive and efficient structure of the original system. Essentially, we show that dual system encryption and leakage resilience are highly compatible, and their combination results in the strongest security guarantees available for cryptosystems with advanced functionalities, with no sacrifice of efficiency.

Our Techniques. In a dual system encryption scheme, keys and ciphertexts can each take on two forms: normal and semi-functional. Normal keys can decrypt both forms of ciphertexts, while semi-functional keys can only decrypt normal ciphertexts. In the real security game, the ciphertext and all keys are normal. Security is proven by a hybrid argument, where first the ciphertext is changed to semi-functional, and then the keys are changed to semi-functional one by one. We must prove that the attacker cannot detect these changes. Finally, we arrive at a game where the simulator need only produce semi-functional objects, which cannot correctly decrypt. This greatly reduces the burden on the simulator and allows us to now prove security directly.

There is an important challenge inherent in this technique: when we argue the indistinguishability of games where a certain key is changing from normal to semi-functional, it is crucial that the simulator cannot determine the nature of this key for itself by test decrypting a semi-functional ciphertext. However, the simulator should also be prepared to make a semi-functional ciphertext for *any* identity and to use *any* identity for this particular key. This challenge is overcome by allowing the simulator to make *nominal* semi-functional keys: these are keys that are distributed like ordinary semi-functional keys in the attacker's view, but in the simulator's view they are correlated with the challenge ciphertext, so that if the simulator tries to decrypt a semi-functional ciphertext, decryption will always succeed, and hence will not reveal whether the key is normal or nominally semi-functional.

To keep nominal semi-functionality hidden from the attacker's view, previous dual system encryption constructions relied crucially on the fact that the attacker cannot ask for a key capable of decrypting the challenge ciphertext. When we add leakage to this framework, the attacker is now able to ask for leakage on keys which are capable of decrypting the challenge ciphertext: hence we need a

[1] For simplicity, we present our system as allowing no leakage during key updates, but our system can tolerate leakage which is logarithmic in terms of the security parameter using the same methods employed in [16].

new mechanism to hide nominal semi-functionality from attackers who can leak on these keys.

We accomplish this by expanding the semi-functional space to form $n+2$ dimensional vectors, where $n \geq 3$ is a parameter determining the leakage tolerance. Nominality now corresponds to the vector in the semi-functional space of the key being orthogonal to the vector in the semi-functional space of the ciphertext. Because the leakage function on the key must be determined *before* the challenge ciphertext is revealed, an attacker whose leakage is suitably bounded cannot distinguish orthogonal vectors from uniformly random vectors in this context (this is a corollary of the result from [16], which shows that "random subspaces are leakage-resilient"). Hence, the attacker cannot distinguish leakage on a nominally semi-functional key from leakage on an ordinary semi-functional key. This allows us to obtain leakage resilience within the dual system encryption framework.

Comparison to Previous Techniques. One of the leakage-resilient IBE constructions of [21] also applied the dual system encryption methodology, but ultimately relied on the technique of hash proof systems [23, 52, 2] to obtain leakage resilience, instead of deriving leakage resilience from the dual system encryption methodology itself, as we do in this work. More precisely, they used the dual system encryption framework to allow the simulator to produce keys incapable of decrypting the challenge ciphertext, but did not apply dual system encryption to handle leakage on keys which are capable of decrypting the challenge ciphertext. Instead, they relied on a hash proof mechanism for this part of the proof. This leads them to impose the restriction that the attacker can only leak from one key for the challenge identity, and no leakage on the master key is allowed. Essentially, their application of dual system encryption is "orthogonal" to their techniques for achieving leakage resilience. In contrast, our techniques allow us to handle all key generation and leakage queries *within the dual system encryption framework*, eliminating the need for a separate technique to achieve leakage resilience. This enables us to allow leakage from multiple keys which can decrypt the challenge ciphertext, as well as leakage from the master key.

The leakage-resilient IBE construction of [16] in the continual leakage model relies on selective security to allow the simulator to produce the keys incapable of decrypting challenge ciphertext. This is accomplished with a partitioning technique. Their technique for handling leakage on secret keys for the challenge identity is more similar to ours: they produce these keys and ciphertext in such a way that each is independently well-distributed, but the keys for the challenge identity exhibit degenerate behavior relative to the challenge ciphertext. This correlation, however, is information-theoretically hidden from the adversary because the leakage per key is suitably bounded. We employ a similar information-theoretic argument to hide nominal semi-functionality of leaked keys from the attacker's view. However, their technique does not quite fit our dual system encryption framework, and only achieves selective security in their implementation, with no leakage allowed from the master key.

1.1 Related Work

Leakage resilience has been studied in many previous works, under a variety of leakage models [60, 56, 45, 3, 19, 24, 30, 28, 44, 29, 52, 1, 2, 17, 26, 31, 42, 51, 57, 32, 15, 27, 16, 25]. Exposure-resilient cryptography [17,28,44] addressed adversaries who could learn a subset of the bits representing the secret key or internal state. Subsequent works have considered more general leakage functions. Micali and Reyzin [51] introduced the assumption that "only computation leaks information." In other words, one assumes that leakage occurs every time the cryptographic device performs a computation, but that any parts of the memory not involved in the computation do not leak. Under this assumption, leakage-resilient stream ciphers and signatures have been constructed [31, 57, 32]. Additionally, [43,38] have shown how to transform any cryptographic protocol into one that is secure with continual leakage, assuming that only computation leaks information and also relying on a simple, completely non-leaking hardware device.

Since attacks like the cold-boot attack [41] can reveal information about memory contents in the absence of computation, it is desirable to have leakage-resilient constructions that do not rely upon this assumption. Several works have accomplished this by bounding the total amount of leakage over the lifetime of the system, an approach introduced by [1]. This has resulted in constructions of pseudorandom functions, signature schemes, public key encryption, and identity-based encryption [26, 52, 3, 45, 2, 21] which are secure in the presence of suitably bounded leakage. For IBE schemes in particular, this means that an attacker can leak a bounded amount of information from only *one secret key per user*. This does not allow a user to update/re-randomize his secret key during the lifetime of the system.

Recently, two works have achieved continual leakage resilience *without* assuming that only computation leaks information [16, 25]. Dodis, Haralambiev, Lopez-Alt, and Wichs [25] construct one-way relations, signatures, identification schemes, and authenticated key agreement protocols which are secure against attackers who can obtain leakage *between* updates of the secret key. It is assumed the leakage between consecutive updates is bounded in terms of a fraction of the secret key size, and also that there is no leakage during the update process. Brakerski, Kalai, Katz, and Vaikuntanathan [16] construct signatures, public key encryption schemes, and (selectively secure) identity-based encryption schemes which are secure against attackers who can obtain leakage between updates of the secret key, and also a very limited amount of leakage during updates and during the initial setup phase. The leakage between updates is bounded in terms of a fraction of the secret key size, while the leakage during updates and setup is logarithmically small as a function of the security parameter.

The dual system encryption methodology was introduced by Waters in [63]. It has been leveraged to obtain constructions of fully secure IBE and HIBE from simple assumptions [63], fully secure HIBE with short ciphertexts [50], fully secure ABE and Inner Product Encryption (IPE) [48], and fully secure functional encryption combining ABE and IPE [54].

Independently, Alwen and Ibraimi [4] have proposed a leakage resilient system for a special case of Attribute-Based Encryption, where the ciphertext policy is expressed as a DNF. Their work pursues a different technical direction to ours, and provides an interesting application of hash proof systems to the ABE setting. Security is proven from a "q-type" assumption.

2 Preliminaries

Notation. We denote by $s \xleftarrow{\$} S$ the fact that s is picked uniformly at random from a finite set S and by $x, y, z \xleftarrow{\$} S$ that all x, y, z are picked independently and uniformly at random from S. We say that a function is *of constant output size* if the number of bits output by it is independent of the input. By $|x|$, we denote the size/number of bits of term x. Also, the special symbol $\perp$ is meant to serve as a unique dummy value in all our systems. Finally, by PPT we denote a probabilistic polynomial-time algorithm.

Complexity Assumptions. To prove the security of our system, we will use three assumptions in composite order groups, also used in [50, 48]. These are static assumptions, which hold in the generic group model if finding a nontrivial factor of the group order is hard. The proof of this can be found in [50]. The first two of our assumptions belong to the class of General Subgroup Decision Assumptions described in [5]. The specific statement of the assumptions can be found in the full version [49].

2.1 Security Definition

In this section we assume familiarity with the main functionalities of the algorithms of an IBE system. Due to lack of space in this version we included the detailed definition only in the full version [49].

The security of our system is based on a game, called MasterLeak. It is a modified version of the usual IbeCpa security game. In that game, the attacker can make a polynomial number of **Keygen** queries for identities other than the challenge identity. Each of these queries returns a secret key of the requested identity. The main idea of our security game is to allow these queries and in addition allow leakage on the master key and secret keys of the challenge identity. The only restriction we impose is that it can not get leakage of more than ℓ_{MK} bits per master key (remember we can have many master keys) and ℓ_{SK} bits per secret key, where $\ell_{\mathrm{MK}}, \ell_{\mathrm{SK}}$ are parameters of the game.

The game starts with a setup phase, where the challenger runs the setup algorithm and gives the attacker the public parameters. It also gives the attacker a handle (i.e. reference) to the master key. We now allow the attacker to make three kinds of queries, called **Create**, **Leak**, and **Reveal**. With a **Create** query, the attacker asks the challenger to create a key and store it. The attacker supplies a handle that refers to a master key to be used in the key generation algorithm. Each such query returns a unique handle-reference to the generated key, so that the attacker can refer to it later and either apply a leakage function to it and/or ask for the entire key. The original master key (the one created in the **Setup** algorithm) gets a handle of 0.

Using a handle, the attacker can make a leakage query **Leak** on any key of its choice. Since all queries are adaptive (the attacker has the ability to leak from each key a few bits at the time, instead of requiring the leakage to occur all at once) and the total amount of leakage allowed is bounded, the challenger has to keep track of all keys leaked via these queries and the number of leaked bits from each key so far. Thus, it creates a set $\mathcal{T}$ that holds tuples of handles, identities, keys, and the number of leaked bits. Each **Create** query adds a tuple to this set and each **Leak** query updates the number of bits leaked.

The **Reveal** queries allow the attacker to get access to an entire secret key. They get as input a handle to a key and the challenger returns this secret key to the attacker. The obvious restriction is that the attacker cannot get a master key, since it would trivially break the system. For the same reason, no key for the challenge identity should be revealed and thus the challenger has to have another set to keep track of the revealed identities. We will denote this set by $\mathcal{R}$. We also note that the **Reveal** queries model the attacker's ability to "change its mind" in the middle of the game on the challenge identity. Maybe the attacker, after getting leakage from a secret key, decides that it is better to get the entire key via a **Reveal** query. Thus we achieve the maximum level of adaptiveness.

We now define our game formally. The security game is parameterized by a security parameter λ and two leakage bounds $\ell_{\text{MK}} = \ell_{\text{MK}}(\lambda)$, $\ell_{\text{SK}} = \ell_{\text{SK}}(\lambda)$. The master keys', secret keys' and identities' spaces are denoted by $\mathcal{MK}$, $\mathcal{SK}$, and $\mathcal{I}$, respectively. We assume that the handles' space is $\mathcal{H} = \mathbb{N}$. The game MasterLeak consists of the following phases:

Setup: The challenger makes a call to **Setup**(1^λ) and gets a master key MK and the public parameters PP. It gives PP to the attacker. Also, it sets $\mathcal{R} = \emptyset$ and $\mathcal{T} = \{(0, \epsilon, \text{MK}, 0)\}$. Remember that $\mathcal{R} \subseteq \mathcal{I}$ and $\mathcal{T} \subseteq \mathcal{H} \times \mathcal{I} \times (\mathcal{MK} \cup \mathcal{SK}) \times \mathbb{N}$ (handles - identities - keys - leaked bits). Thus initially the set $\mathcal{T}$ holds a record of the original master key (no identity for it and no leakage so far). Also a handle counter H is set to 0.

Phase 1: In this phase, the adversary can make the following queries to the challenger. All of them can be interleaved in any possible way and the input of a query can depend on the outputs of all previous queries (adaptive security).

- **Create**(h, X): h is a handle to a tuple of $\mathcal{T}$ that must refer to a master key and X can be either an identity I or the empty string ϵ.

  The challenger initially scans $\mathcal{T}$ to find the tuple with handle h. If the identity part of the tuple is not ϵ, which means that the tuple holds a secret key of some identity, or if the handle does not exist, it responds with $\perp$.

  Otherwise, the tuple is of the form $(h, \epsilon, \text{MK}', L)$. Then the challenger makes a call to **Keygen**$(\text{MK}', X) \to K$ and adds the tuple $(H+1, X, K, 0)$ to the set $\mathcal{T}$. K is either a secret key for identity I or another master key depending on X. If X is an identity it returns a secret key and if X is the

empty string ϵ it returns another master key. See the full version [49] for a detailed definition. After that, it updates the handle counter to $H \leftarrow H+1$.
- **Leak**(h, f): In this query, the adversary requests leakage from a key that has handle $h \in \mathbb{N}$ with a polynomial-time computable function f of constant output size[2] acting on the set of keys.
  The challenger scans $\mathcal{T}$ to find the tuple with the specified handle. It is either of the form (h, I, SK, L) or $\big(h, \epsilon, \mathrm{MK}', L\big)$[3].
  In the first case, it checks if $L + |f(\mathrm{SK})| \leq \ell_{\mathrm{SK}}$. If this is true, it responds with $f(\mathrm{SK})$ and updates the L in the tuple with $L + |f(\mathrm{SK})|$. If the check fails, it returns $\perp$ to the adversary.
  If the tuple holds a master key MK', it checks if $L + \big|f(\mathrm{MK}')\big| \leq \ell_{\mathrm{MK}}$. If this is true, it responds with $f(\mathrm{MK}')$ and updates the L with $L + \big|f(\mathrm{MK}')\big|$. If the check fails, it returns $\perp$ to the adversary.
- **Reveal**(h): Now the adversary requests the entire key with handle h. The challenger scans $\mathcal{T}$ to find the requested entry. If the handle refers to a master key tuple, then the challenger returns $\perp$. Otherwise, we denote the tuple by (h, I, SK, L). The challenger responds with SK and adds the identity I to the set $\mathcal{R}$.

Challenge: The adversary submits a challenge identity $I^* \notin \mathcal{R}$ and two messages M_0, M_1 of equal size. The challenger flips a uniform coin $c \xleftarrow{\$} \{0,1\}$ and encrypts M_c under I^* with a call to **Encrypt**(M_c, I^*). It sends the resulting ciphertext CT^* to the adversary.

Phase 2: This is the same as **Phase 1** with the restriction that the only queries allowed are **Create** and **Reveal** queries that involve a (non-master) secret key with identity different than I^*. The reason for forbidding **Leak** queries on a master key and on I^* is that the adversary can encode the entire decryption algorithm of CT^* as a function on a secret key, and thus win the game trivially if we allow these queries. For the same reason, the challenger can not give an entire secret key of I^* to the adversary and hence no **Reveal** queries involving I^* are allowed too. **Leak** queries on keys of identities other than I^* are useless, since the adversary can get the entire secret keys.

Guess: The adversary outputs a bit $c' \in \{0,1\}$. We say it succeeds if $c' = c$.

The security definition we will use is the following:

Definition 1. *An IBE encryption system Π is $(\ell_{\mathrm{MK}}, \ell_{\mathrm{SK}})$-master-leakage secure if for all PPT adversaries $\mathcal{A}$ it is true that*

$$\mathsf{Adv}_{\mathcal{A},\Pi}^{\mathsf{MasterLeak}}(\lambda, \ell_{\mathrm{MK}}, \ell_{\mathrm{SK}}) \leq \mathsf{negl}(\lambda)$$

[2] We apply this restriction so that the adversary does not get any "extra" information about the input; only the output bits of the function. This restriction is also present in other works (e.g. in [16] they use circuits as leakage functions).

[3] It can be the case that MK' is the original master key.

where $\mathsf{Adv}_{\mathcal{A},\Pi}^{\mathsf{MasterLeak}}(\lambda, \ell_{\mathrm{MK}}, \ell_{\mathrm{SK}})$ *is the advantage of* $\mathcal{A}$ *in game* MasterLeak *with security parameter* λ *and leakage parameters* $\ell_{\mathrm{MK}} = \ell_{\mathrm{MK}}(\lambda), \ell_{\mathrm{SK}} = \ell_{\mathrm{SK}}(\lambda)$ *and is formally defined as*

$$\mathsf{Adv}_{\mathcal{A},\Pi}^{\mathsf{MasterLeak}}(\lambda, \ell_{\mathrm{MK}}, \ell_{\mathrm{SK}}) = \left|\Pr[\mathcal{A} \text{ succeeds}] - \frac{1}{2}\right|,$$

where the probability is over all random bits used by the challenger and the attacker.

3 Dual System IBE

We now define dual system IBE schemes as an abstraction and define three security properties which will ensure leakage resilience. We show that these properties imply that a dual system IBE scheme is $(\ell_{\mathrm{MK}}, \ell_{\mathrm{SK}})$-master-leakage secure[4].

3.1 Definition

A dual system IBE scheme Π_D has the following algorithms:

Setup$(1^\lambda) \rightarrow (\mathrm{PP}, \mathrm{MK})$**.** The setup algorithm takes in the security parameter, λ, and outputs the public parameters, PP, and a normal master key, MK.

Keygen$(\mathrm{MK}', X) \rightarrow K$**.** The key generation algorithm takes in a normal master key, MK', and either an identity, I, or the empty string ϵ. In the first case, it outputs a normal secret key, SK, for the identity I, and in the second case, it outputs another normal master key, MK''.

Encrypt$(\mathrm{PP}, M, I) \rightarrow \mathrm{CT}$**.** The encryption algorithm takes in the public parameters PP, a message M, and an identity I, and outputs a normal ciphertext, CT.

Decrypt$(\mathrm{CT}, \mathrm{SK}) \rightarrow M$**.** The decryption algorithm takes in a ciphertext CT encrypted to identity I, and a secret key SK for identity I. It outputs the message M, unless both the key and the ciphertext are semi-functional.

KeygenSF$(\mathrm{MK}', X) \rightarrow \widetilde{K}$**.** The semi-functional key generation algorithm works in a similar way to **Keygen** but outputs semi-functional keys. If $X = I$, an identity, it outputs a semi-functional secret key, $\widetilde{SK}$ for identity I. If $X = \epsilon$, the empty string, it outputs a semi-functional master key, $\widetilde{MK}$.

Notice that this algorithm takes in a *normal* master key; not a semi-functional one. Also, this algorithm *need not be polynomial time computable*, in contrast to **Setup**, **Keygen**, **Encrypt**, and **Decrypt**.

EncryptSF$(\mathrm{PP}, M, I) \rightarrow \widetilde{\mathrm{CT}}$**.** The semi-functional encryption algorithm takes in the public parameters PP, a message M, and an identity I, and outputs a semi-functional ciphertext, CT. This algorithm *need not be polynomial time computable.*

[4] We choose not to include nominal semi-functionality as part of our abstraction, since one can use dual system encryption without employing this concept. For example, nominal semi-functionality was not used in [63].

3.2 Security Properties for Leakage Resilience

We now define three security properties for a dual system IBE scheme. For this, we define two additional games which are modifications of the MasterLeak game.

The first game, called MasterLeakC, is exactly the same as the MasterLeak game except that in the **Challenge** phase, the challenger uses **EncryptSF** instead of **Encrypt** to create a semi-functional ciphertext, and returns this to the adversary.

In the second new game, called MasterLeakCK, the challenger again uses **EncryptSF** for the challenge phase. However, the set of tuples $\mathcal{T}$ has a different structure. Each tuple holds for each key (master or secret) a normal and a semi-functional version of it. In this game, all keys leaked or given to the attacker are semi-functional. As we have noted above, the semi-functional key generation algorithm takes as input a normal master key. Thus the challenger stores the normal versions, as well the semi-functional ones so that it can use the normal versions of master keys as input to **Keygen** calls.[5] More precisely, the challenger additionally stores a semi-functional master key in tuple 0 by calling **KeygenSF**(MK, ϵ) after calling **Setup**. Thereafter, for all **Create**(h, X) queries, the challenger makes an additional call to **KeygenSF**(MK', X), where MK' is the *normal* version of the master key stored in tuple h. **Leak** and **Reveal** queries act always on the semi-functional versions of each key.

Finally, notice that the same attackers that play game MasterLeak can play games MasterLeakC and MasterLeakCK without any change in their algorithms - queries etc. The simulator answers them in a different way.

Semi-functional Ciphertext Invariance: We say that a dual system IBE scheme Π_D has *$(\ell_{\mathrm{MK}}, \ell_{\mathrm{SK}})$- semi-functional ciphertext invariance* if for any probabilistic polynomial time algorithm $\mathcal{A}$, the advantage of $\mathcal{A}$ in the MasterLeak game is negligibly close to the advantage of $\mathcal{A}$ in the MasterLeakC game.

Semi-functional Key Invariance: We say that a dual system IBE scheme Π_D has *$(\ell_{\mathrm{MK}}, \ell_{\mathrm{SK}})$-semi-functional key invariance* if for any probabilistic polynomial time algorithm $\mathcal{A}$, the advantage of $\mathcal{A}$ in the MasterLeakC game is negligibly close to the advantage of $\mathcal{A}$ in the MasterLeakCK game.

Semi-functional Security: We say that a dual system IBE scheme Π_D has *$(\ell_{\mathrm{MK}}, \ell_{\mathrm{SK}})$-semi-functional security* if for any probabilistic polynomial time algorithm $\mathcal{A}$, the advantage of $\mathcal{A}$ in the MasterLeakCK game is negligible.

The proof of the following theorem is straightforward, and can be found in the full version [49].

Theorem 1. *If a dual system IBE scheme Π_D =(**Setup**, **Keygen**, **Encrypt**, **Decrypt**, **KeygenSF**, **EncryptSF**) has $(\ell_{\mathrm{MK}}, \ell_{\mathrm{SK}})$-semi-functional ciphertext invariance, $(\ell_{\mathrm{MK}}, \ell_{\mathrm{SK}})$-semi-functional key invariance, and $(\ell_{\mathrm{MK}},$*

---

[5] As one should notice, we will never use the normal versions of non-master keys in this game. However, we have them here because we will need them in the game of the next section and when we move to the HIBE setting.

*ℓ_{SK})-semi-functional security, then Π =(**Setup**, **Keygen**, **Encrypt**, **Decrypt**) is a ($\ell_{\mathrm{MK}}, \ell_{\mathrm{SK}}$)-master-leakage secure IBE scheme.*

3.3 An Alternate Security Property

We additionally define a property called *One Semi-functional Key Invariance.* In the full version we will show that this implies semi-functional key invariance, and so can be substituted for semi-functional key invariance in proving that a system is ($\ell_{\mathrm{MK}}, \ell_{\mathrm{SK}}$)-master-leakage secure. The motivation for this is that proving semi-functional key invariance directly will often involve a hybrid argument, and defining one semi-functional key invariance allows us to include this hybrid as part of our abstraction and hence avoid repeating it for each system.

To define this property, we first define one more variation of our security game, called $\mathsf{MasterLeak}_b$. This is similar to the $\mathsf{MasterLeakCK}$ game, with the main difference being that the attacker can choose on which version of each key to leak or reveal. In other words, on the first leakage or reveal query on a key of the augmented set $\mathcal{T}$, the attacker tells the challenger whether it wants the normal or the semi-functional version of the key. In order for the challenger to keep track of the attacker's choice on each key, we further augment each tuple of $\mathcal{T}$ with a lock-value denoted by $V \in \mathbb{Z}$ that can take one of the three values $\{-1, 0, 1\}$:

- If $V = -1$ the attacker has not made a choice on this key yet and the key is "unlocked". This is the value the tuple gets, in a **Create** query.
- If $V = 0$ the attacker chose to use the normal version of the key on the first leakage or reveal query on it. All subsequent **Leak** and **Reveal** queries act on the normal version.
- If $V = 1$ the attacker chose the semi-functional version and the challenger works as above with the semi-functional version.

To summarize, each tuple is of the form $(h, X, K, \widetilde{K}, L, V)$ i.e. handle - identity or empty string - normal key - semi-functional key - leakage - lock. For example, the original master key is stored at the beginning of the game in the tuple $(0, \epsilon, \mathrm{MK}, \mathbf{KeygenSF}(\mathrm{MK}, \epsilon), 0, -1)$.

At some point, the attacker must decide on a *challenge key* which is "unlocked", $V = -1$, and tell this to the challenger. The challenger samples a uniformly random bit $b \xleftarrow{\$} \{0, 1\}$ and sets $V = b$. Therefore, the attacker has access to either the normal (if $b = 0$) or the semi-functional (if $b = 1$) version of this key via **Leak** and **Reveal** queries. We note that if the attacker did not make a choice for the original master key in tuple 0, it can choose this master key as the challenge key.

The attacker is then allowed to resume queries addressed to either normal or semi-functional keys, with the usual restrictions (i.e. no leakage or reveal queries on keys capable of decrypting the challenge ciphertext after the attacker has seen the challenge ciphertext).

One Semi-functional Key Invariance: We say that a dual system IBE scheme Π_D has $(\ell_{\mathrm{MK}}, \ell_{\mathrm{SK}})$-*one semi-functional key invariance* if, for any probabilistic polynomial time algorithm $\mathcal{A}$, the advantage of $\mathcal{A}$ in the $\mathsf{MasterLeak}_b$ game with $b = 0$ is negligibly close to the advantage of $\mathcal{A}$ in the $\mathsf{MasterLeak}_b$ game with $b = 1$.

The proof of the following theorem can be found in the full version [49].

Theorem 2. *If a dual system IBE scheme Π_D has $(\ell_{\mathrm{MK}}, \ell_{\mathrm{SK}})$-one semi-functional key invariance, then it also has $(\ell_{\mathrm{MK}}, \ell_{\mathrm{SK}})$-semi-functional key invariance.*

4 Master-Leakage Secure IBE Scheme

Our IBE scheme is an augmented version of the Lewko-Waters IBE [50], designed to sustain master and secret key leakage from an arbitrary number of keys. To hide nominal semi-functionality in the attacker's view, we add vectors of dimension n to the front of the ciphertexts and secret keys of the LW system. Notice in the construction below that the last two elements of our secret keys and ciphertexts are very similar to the elements in the LW system (which is essentially a composite order version of the selectively-secure Boneh-Boyen IBE system [9]). Nominal semi-functionality now corresponds to the vector of exponents of the semi-functional components of the key being orthogonal to the vector of exponents of the semi-functional components of the ciphertext. We can use the algebraic lemma of [16] to assert that this orthogonality is hidden from attackers with suitably bounded leakage. Finally, to allow leakage on the master key, we designed the master key to be similar in form to regular secret keys.

Like the original LW scheme, our system uses a bilinear group whose order is the product of three distinct primes (additional background on composite order bilinear groups can be found in the full version [49]). The role of the first prime order subgroup is to "carry" the necessary information of the plaintext message and the secret information of each user or the master authority. The second subgroup is used only in the proof to provide semi-functionality. The third subgroup is used to additionally randomize secret keys. Each of these components is orthogonal to the other two under the pairing operation.

In the construction below, we will use angle brackets to denote vectors and parentheses to denote collections of elements of different types. The dot product of vectors is denoted by $\cdot$ and component-wise multiplication is denoted by $*$. For a group element $u \in \mathbb{G}$ and a vector $\vec{a} \in \mathbb{Z}_N^n$, we define $u^{\vec{a}}$ to be $\langle u^{a_1}, u^{a_2}, \ldots, u^{a_n} \rangle$. We also define a pairing operation of vectors in $\mathbb{G}^n$: For $\vec{v}_1 = \langle v_{11}, v_{12}, \ldots, v_{1n} \rangle \in \mathbb{G}^n$ and $\vec{v}_2 = \langle v_{21}, v_{22}, \ldots, v_{2n} \rangle \in \mathbb{G}^n$, their pairing is $e_n(\vec{v}_1, \vec{v}_2) := \prod_{i=1}^{n} e(v_{1i}, v_{2i}) \in \mathbb{G}_T$, where $e : \mathbb{G} \times \mathbb{G} \to \mathbb{G}_T$ is the bilinear mapping of $\mathbb{G}$ and the product is the group operation of $\mathbb{G}_T$.

4.1 Construction

Our dual system IBE scheme consists of the following algorithms:

Setup(1^λ). The setup algorithm generates a bilinear group $\mathbb{G}$ of composite order $N = p_1 p_2 p_3$, where p_1, p_2, p_3 are three different λ_1,λ_2,λ_3-bit prime numbers respectively[6]. The subgroup of order p_i in $\mathbb{G}$ is denoted by $\mathbb{G}_i$. We assume that the identities of users in our system are elements of $\mathbb{Z}_N$.

We let n be a positive integer greater than or equal to 2. The value of n can be varied - higher values of n will lead to a better fraction of leakage being tolerated (see Section 5), while lower values of n will yield a system with fewer group elements in the keys and ciphertexts.

The algorithm picks 3 random elements $\langle g_1, u_1, h_1 \rangle \in \mathbb{G}_1 \times \mathbb{G}_1 \times \mathbb{G}_1$ and one random element $g_3 \in \mathbb{G}_3$. It also picks $n+1$ random exponents $\langle \alpha, x_1, x_2, \ldots, x_n \rangle \xleftarrow{\$} \mathbb{Z}_N^{n+1}$. It picks $\langle r, y_1, y_2, \ldots, y_n \rangle \xleftarrow{\$} \mathbb{Z}_N^{n+1}$, a random vector $\vec{\rho} = \langle \rho_1, \ldots, \rho_{n+2} \rangle \xleftarrow{\$} \mathbb{Z}_N^{n+2}$, and a random element $\rho_{n+3} \xleftarrow{\$} \mathbb{Z}_N$. It outputs the following public parameters and master key:

$$\mathrm{PP} = \left(N, g_1, g_3, u_1, h_1, e(g_1,g_1)^\alpha, g_1^{x_1}, g_1^{x_2}, \ldots, g_1^{x_n}\right)$$

$$\mathrm{MK} = \left(\vec{K^*}, K^*\right) = \left(\left\langle g_1^{y_1}, \ldots, g_1^{y_n}, g_1^\alpha h_1^{-r} \prod_{i=1}^n g_1^{-x_i y_i}, g_1^r \right\rangle * g_3^{\vec{\rho}}, u_1^r g_3^{\rho_{n+3}}\right)$$

Keygen(MK, PP, X). We first consider when $X = \epsilon$, the empty string. Then this algorithm re-randomizes the master key by picking another $\langle r', y_1', y_2', \ldots, y_n' \rangle \xleftarrow{\$} \mathbb{Z}_N^{n+1}$, a random vector $\vec{\rho'} = \langle \rho_1', \ldots, \rho_{n+2}' \rangle \xleftarrow{\$} \mathbb{Z}_N^{n+2}$, and a random element $\rho_{n+3}' \xleftarrow{\$} \mathbb{Z}_N$. If $\mathrm{MK} = \left(\vec{K^*}, K^*\right)$, it outputs the new (same-sized) master key:

$$\mathrm{MK}' = \left(\vec{K'}, K'\right) = \left(\vec{K^*} * \left\langle g_1^{y_1'}, \ldots, g_1^{y_n'}, h_1^{-r'} \prod_{i=1}^n g_1^{-x_i y_i'}, g_1^{r'} \right\rangle * g_3^{\vec{\rho'}}, K^* u_1^{r'} g_3^{\rho_{n+3}'}\right)$$

If $X = I \in \mathbb{Z}_N$, an identity, the algorithm picks $n+1$ random exponents $\langle r', z_1, z_2, \ldots, z_n \rangle \xleftarrow{\$} \mathbb{Z}_N^{n+1}$. Also it picks $\vec{\rho'} \xleftarrow{\$} \mathbb{Z}_N^{n+2}$ and outputs the secret key:

$$\mathrm{SK} = \vec{K_1} = \vec{K^*} * \left\langle g_1^{z_1}, g_1^{z_2}, \ldots, g_1^{z_n}, (K^*)^{-I} (u_1^I h_1)^{-r'} \prod_{i=1}^n g_1^{-x_i z_i}, g_1^{r'} \right\rangle * g_3^{\vec{\rho'}}$$

The terms $g_1^{-x_i y_i'}$ and $g_1^{-x_i z_i}$ above are calculated by using the g^{x_i} terms of PP.

It is very important to notice that with knowledge of α alone, one can create properly distributed secret keys, because the random terms $r, y_1, \ldots, y_n, \rho_{n+3}, \vec{\rho}$ of the master key are all masked by the random terms $r', z_1, \ldots, z_n, \vec{\rho'}$ generated

[6] The three λ's depend on the security parameter and are chosen appropriately to get a better leakage fraction (see Section 5 for details).

by the algorithm. However, instead of storing α, the master authority now stores $n+3$ elements of $\mathbb{G}$.

Encrypt(M, I)**.** The encryption algorithm picks $s \xleftarrow{\$} \mathbb{Z}_N$ and outputs the ciphertext:

$$\mathrm{CT} = \left(C_0, \vec{C_1}\right) =$$
$$= \left(M \cdot (e(g_1, g_1)^{\alpha})^s, \left\langle (g_1^{x_1})^s, \ldots, (g_1^{x_n})^s, g_1^s, (u_1^I h_1)^s \right\rangle\right) \in \mathbb{G}_T \times \mathbb{G}^{n+2}$$

Decrypt$(\mathrm{CT}, \mathrm{SK})$**.** To calculate the blinding factor $e(g_1, g_1)^{\alpha s}$, one computes $e_{n+2}(\vec{K_1}, \vec{C_1})$ and the message is computed as $M = \frac{C_0}{e_{n+2}(\vec{K_1}, \vec{C_1})}$.

4.2 Semi-functionality

All the ciphertexts, master keys, and secret keys generated by the above algorithms are *normal*, where by normal we mean that they have no $\mathbb{G}_2$ parts. On the other hand, a *semi-functional* key or ciphertext has $\mathbb{G}_2$ parts. We let g_2 denote a generator of $\mathbb{G}_2$. The remaining algorithms of our dual system IBE are the following:

KeygenSF$(\mathrm{MK}, X) \rightarrow \widetilde{K}$**.** This algorithm calls first the normal key generation algorithm **Keygen**(MK, X) to get a normal key $\mathrm{MK} = \left(\vec{K^*}, K^*\right)$ or $\mathrm{SK} = \vec{K_1}$, depending on X.

In the former case, it picks $\vec{\theta} \xleftarrow{\$} \mathbb{Z}_N^{n+2}$ and $\theta \xleftarrow{\$} \mathbb{Z}_N$ and outputs

$$\widetilde{\mathrm{MK}} = \left(\vec{K^*} * g_2^{\vec{\theta}}, K^* g_2^{\theta}\right).$$

In the latter case, it picks $\vec{\gamma} \xleftarrow{\$} \mathbb{Z}_N^{n+2}$ and outputs

$$\widetilde{\mathrm{SK}} = \vec{K_1} * g_2^{\vec{\gamma}}.$$

EncryptSF$(M, I) \rightarrow \widetilde{\mathrm{CT}}$**.** This algorithm calls first the normal encryption algorithm **Encrypt**(M, I) to get the ciphertext $\mathrm{CT} = \left(C_0, \vec{C_1}\right)$. Then it picks $\vec{\delta} \xleftarrow{\$} \mathbb{Z}_N^{n+2}$ and outputs

$$\widetilde{\mathrm{CT}} = \left(C_0, \vec{C_1} * g_2^{\vec{\delta}}\right).$$

We call the three terms $\left(\vec{\theta}, \theta\right), \vec{\gamma}, \vec{\delta}$ the *semi-functional parameters* of the master key, secret key, and ciphertext, respectively. The semi-functional keys are partitioned in *nominal* semi-functional keys and in *truly* semi-functional keys, with respect to a specific semi-functional ciphertext. In short, a nominal secret key can correctly decrypt the ciphertext (by using **Decrypt**), while a nominal master key can generate a semi-functional secret key that correctly decrypts the ciphertext.

As a result, a semi-functional secret key of identity I_k with parameters $\vec{\gamma}$ is nominal with respect to a ciphertext for identity I_c with parameters $\vec{\delta}$ if and only if $\vec{\gamma} \cdot \vec{\delta} = 0 \bmod p_2$ and $I_k = I_c$.

It is easy to see that only then the decryption is correct, because we get an extra term $e(g_2, g_2)^{\vec{\gamma} \cdot \vec{\delta}}$ by the pairing. A semi-functional master key with parameters $\vec{\theta}, \theta$ is nominal with respect to a ciphertext for identity I with parameters $\vec{\delta}$ if and only if $\vec{\delta} \cdot \left(\vec{\theta} + \langle 0, \ldots, 0, -I\theta, 0\rangle\right) = 0 \bmod p_2$.

The proof of the following theorem can be found in the full version [49].

Theorem 3. *Under our complexity assumptions and for* ($\ell_{\text{MK}} = (n - 1 - 2c)\log(p_2), \ell_{\text{SK}} = (n - 1 - 2c)\log(p_2)$)*, where $c > 0$ is a fixed positive constant, our dual system IBE scheme is* ($\ell_{\text{MK}}, \ell_{\text{SK}}$)*-master-leakage secure.*

Our HIBE and ABE constructions as well as their security proofs can also be found in the full version [49].

5 Our Leakage Bound

Our systems allow the same absolute amount of leakage for both the master and the secret keys. That is, $\ell_{\text{MK}} = \ell_{\text{SK}} = (n - 1 - 2c)\log p_2$ bits, where n is an arbitrary integer greater than or equal to 2 and c is a fixed positive constant. Notice that the leakage depends only on the size of the $\mathbb{G}_2$ subgroup, and not on the size of p_1 or p_3. Thus by varying the relative sizes of the $\mathbb{G}_1$, $\mathbb{G}_2$, and $\mathbb{G}_3$ subgroups, we can achieve variable key sizes and allow different fractions of the key size to be leaked. We use the term "leakage fraction" to mean the number of bits allowed to be leaked from a key divided by the number of bits required to represent that key.

Recall that p_1, p_2, p_3 are primes of $\lambda_1, \lambda_2, \lambda_3$ bits, respectively, and $N = p_1p_2p_3$ is the order of our group $\mathbb{G}$. We assume that each group element is represented by approximately $\lambda_1 + \lambda_2 + \lambda_3 = \Theta(\log N)$ bits. Then, by fixing $\lambda_1 = c_1\lambda$, $\lambda_2 = \lambda$, and $\lambda_3 = c_3\lambda$, where λ is the security parameter and c_1, c_3 are arbitrary positive constants, we get that the leakage fractions of our systems are the values presented in the table above.

One notable property of our HIBE scheme is that the higher our keys are in the hierarchy, the less leakage is allowed from them. The master key which is at the top allows for a leakage fraction of $(n-1-2c)/((n+2+D)(1+c_1+c_3))$. This is because the base system we adapted, a HIBE system with short ciphertexts, has keys which contain more group elements for users which are higher in the hierarchy. This feature could be avoided by choosing a different base system.

The leakage fraction can be made arbitrarily close to 1 by modifying n, c_1 and c_3 (if we assume a fixed maximum depth for HIBE and a fixed universe size for ABE). Higher values of n give a better leakage fraction, but larger public parameters, keys, and ciphertexts. Smaller values of c_1, c_3 give a better leakage fraction, but also give fewer bits of security in the $\mathbb{G}_1$ and $\mathbb{G}_3$ subspaces as a function of λ. We must choose λ so that $c_1\lambda$ and $c_3\lambda$ are sufficiently large.

Table 1. c, c_1, c_3 are arbitrary positive constants and n is an integer greater than 2. For the HIBE scheme, D is the maximum depth of the hierarchy and i is the depth of the key in question. The master key is considered to be the root of the hierarchy tree and had depth 0. For the ABE scheme, $|U|$ is the total number of attributes in the system, and $|S|$ is the number of attributes of the key in question. Notice that in the ABE scheme we ignored the size of the representations of U and S. They are included in the keys, but they are considered public; thus not included in the leakage fraction.

Scheme	Master Key	Secret Key
IBE	$\frac{n-1-2c}{n+3} \cdot \frac{1}{1+c_1+c_3}$	$\frac{n-1-2c}{n+2} \cdot \frac{1}{1+c_1+c_3}$
HIBE	$\frac{n-1-2c}{n+2+D-i} \cdot \frac{1}{1+c_1+c_3}$ for key of depth i in the hierarchy	
ABE	$\frac{n-1-2c}{n+2+\|U\|} \cdot \frac{1}{1+c_1+c_3}$	$\frac{n-1-2c}{n+2+\|S\|} \cdot \frac{1}{1+c_1+c_3}$

Acknowledgments

We are thankful to Joël Alwen and Yevgeniy Vahlis for useful observations.

References

1. Akavia, A., Goldwasser, S., Vaikuntanathan, V.: Simultaneous hardcore bits and cryptography against memory attacks. In: Reingold, O. (ed.) TCC 2009. LNCS, vol. 5444, pp. 474–495. Springer, Heidelberg (2009)
2. Alwen, J., Dodis, Y., Naor, M., Segev, G., Walfish, S., Wichs, D.: Public-key encryption in the bounded-retrieval model. In: Gilbert, H. (ed.) EUROCRYPT 2010. LNCS, vol. 6110, pp. 113–134. Springer, Heidelberg (2010)
3. Alwen, J., Dodis, Y., Wichs, D.: Leakage-resilient public-key cryptography in the bounded-retrieval model. In: Halevi, S. (ed.) CRYPTO 2009. LNCS, vol. 5677, pp. 36–54. Springer, Heidelberg (2009)
4. Alwen, J., Ibraimi, L.: Leakage resilient ciphertext-policy attribute-based encryption (2010) (manuscript)
5. Bellare, M., Waters, B., Yilek, S.: Identity-based encryption secure under selective opening attack. In: Ishai, Y. (ed.) TCC 2011. LNCS, vol. 6597. Springer, Heidelberg (2011)
6. Bethencourt, J., Sahai, A., Waters, B.: Ciphertext-policy attribute-based encryption. In: IEEE Symposium on Security and Privacy, pp. 321–334 (2007)
7. Biham, E., Carmeli, Y., Shamir, A.: Bug attacks. In: Wagner, D. (ed.) CRYPTO 2008. LNCS, vol. 5157, pp. 221–240. Springer, Heidelberg (2008)
8. Biham, E., Shamir, A.: Differential fault analysis of secret key cryptosystems. In: Kaliski Jr., B.S. (ed.) CRYPTO 1997. LNCS, vol. 1294, pp. 513–525. Springer, Heidelberg (1997)
9. Boneh, D., Boyen, X.: Efficient selective-ID secure identity-based encryption without random oracles. In: Cachin, C., Camenisch, J.L. (eds.) EUROCRYPT 2004. LNCS, vol. 3027, pp. 223–238. Springer, Heidelberg (2004)

10. Boneh, D., Boyen, X.: Secure identity based encryption without random oracles. In: Franklin, M. (ed.) CRYPTO 2004. LNCS, vol. 3152, pp. 443–459. Springer, Heidelberg (2004)
11. Boneh, D., Boyen, X., Goh, E.: Hierarchical identity based encryption with constant size ciphertext. In: Cramer, R. (ed.) EUROCRYPT 2005. LNCS, vol. 3494, pp. 440–456. Springer, Heidelberg (2005)
12. Boneh, D., Brumley, D.: Remote timing attacks are practical. Computer Networks 48(5), 701–716 (2005)
13. Boneh, D., DeMillo, R.A., Lipton, R.J.: On the importance of checking cryptographic protocols for faults. In: Fumy, W. (ed.) EUROCRYPT 1997. LNCS, vol. 1233, pp. 37–51. Springer, Heidelberg (1997)
14. Boneh, D., Franklin, M.: Identity-based encryption from the weil pairing. In: Kilian, J. (ed.) CRYPTO 2001. LNCS, vol. 2139, pp. 213–229. Springer, Heidelberg (2001)
15. Brakerski, Z., Goldwasser, S.: Circular and leakage resilient public-key encryption under subgroup indistinguishability (or: Quadratic residuosity strikes back). In: Rabin, T. (ed.) CRYPTO 2010. LNCS, vol. 6223, pp. 1–20. Springer, Heidelberg (2010)
16. Brakerski, Z., Kalai, Y.T., Katz, J., Vaikuntanathan, V.: Overcoming the hole in the bucket: Public-key cryptography resilient to continual memory leakage. In: FOCS, pp. 501–510 (2010)
17. Canetti, R., Dodis, Y., Halevi, S., Kushilevitz, E., Sahai, A.: Exposure-resilient functions and all-or-nothing transforms. In: Preneel, B. (ed.) EUROCRYPT 2000. LNCS, vol. 1807, pp. 453–469. Springer, Heidelberg (2000)
18. Canetti, R., Halevi, S., Katz, J.: A forward-secure public-key encryption scheme. In: Biham, E. (ed.) EUROCRYPT 2003. LNCS, vol. 2656, pp. 255–271. Springer, Heidelberg (2003)
19. Cash, D., Ding, Y.Z., Dodis, Y., Lee, W., Lipton, R.J., Walfish, S.: Intrusion-resilient key exchange in the bounded retrieval model. In: Vadhan, S.P. (ed.) TCC 2007. LNCS, vol. 4392, pp. 479–498. Springer, Heidelberg (2007)
20. Cheung, L., Newport, C.: Provably secure ciphertext policy abe. In: ACM Conference on Computer and Communications Security, pp. 456–465 (2007)
21. Chow, S., Dodis, Y., Rouselakis, Y., Waters, B.: Practical leakage-resilient identity-based encryption from simple assumptions. In: ACM Conference on Computer and Communications Security, pp. 152–161 (2010)
22. Cocks, C.: An identity based encryption scheme based on quadratic residues. In: IMA Int. Conf., pp. 360–363 (2001)
23. Cramer, R., Shoup, V.: Universal hash proofs and a paradigm for adaptive chosen ciphertext secure public-key encryption. In: Knudsen, L.R. (ed.) EUROCRYPT 2002. LNCS, vol. 2332, pp. 45–64. Springer, Heidelberg (2002)
24. Di Crescenzo, G., Lipton, R.J., Walfish, S.: Perfectly secure password protocols in the bounded retrieval model. In: Halevi, S., Rabin, T. (eds.) TCC 2006. LNCS, vol. 3876, pp. 225–244. Springer, Heidelberg (2006)
25. Dodis, Y., Haralambiev, K., Lopez-Alt, A., Wichs, D.: Cryptography against continuous memory attacks. In: FOCS, pp. 511–520 (2010)
26. Dodis, Y., Kalai, Y., Lovett, S.: On cryptography with auxiliary input. In: STOC, pp. 621–630 (2009)
27. Dodis, Y., Pietrzak, K.: Leakage-resilient pseudorandom functions and side-channel attacks on feistel networks. In: Rabin, T. (ed.) CRYPTO 2010. LNCS, vol. 6223, pp. 21–40. Springer, Heidelberg (2010)

28. Dodis, Y., Sahai, A., Smith, A.: On perfect and adaptive security in exposure-resilient cryptography. In: Pfitzmann, B. (ed.) EUROCRYPT 2001. LNCS, vol. 2045, pp. 301–324. Springer, Heidelberg (2001)
29. Dziembowski, S.: Intrusion-resilience via the bounded-storage model. In: Halevi, S., Rabin, T. (eds.) TCC 2006. LNCS, vol. 3876, pp. 207–224. Springer, Heidelberg (2006)
30. Dziembowski, S., Pietrzak, K.: Intrusion-resilient secret sharing. In: FOCS, pp. 227–237 (2007)
31. Dziembowski, S., Pietrzak, K.: Leakage-resilient cryptography. In: FOCS, pp. 293–302 (2008)
32. Faust, S., Kiltz, E., Pietrzak, K., Rothblum, G.N.: Leakage-resilient signatures. In: Micciancio, D. (ed.) TCC 2010. LNCS, vol. 5978, pp. 343–360. Springer, Heidelberg (2010)
33. Gandolfi, K., Mourtel, C., Olivier, F.: Electromagnetic analysis: Concrete results. In: Koç, Ç.K., Naccache, D., Paar, C. (eds.) CHES 2001. LNCS, vol. 2162, pp. 251–261. Springer, Heidelberg (2001)
34. Gentry, C.: Practical identity-based encryption without random oracles. In: Vaudenay, S. (ed.) EUROCRYPT 2006. LNCS, vol. 4004, pp. 445–464. Springer, Heidelberg (2006)
35. Gentry, C., Halevi, S.: Hierarchical identity based encryption with polynomially many levels. In: Reingold, O. (ed.) TCC 2009. LNCS, vol. 5444, pp. 437–456. Springer, Heidelberg (2009)
36. Gentry, C., Peikert, C., Vaikuntanathan, V.: Trapdoors for hard lattices and new cryptographic constructions. In: Proceedings of the 40th Annual ACM Symposium on Theory of Computing, pp. 197–206. ACM, New York (2008)
37. Gentry, C., Silverberg, A.: Hierarchical ID-based cryptography. In: Zheng, Y. (ed.) ASIACRYPT 2002. LNCS, vol. 2501, pp. 548–566. Springer, Heidelberg (2002)
38. Goldwasser, S., Rothblum, G.: Securing computation against continuous leakage. In: Rabin, T. (ed.) CRYPTO 2010. LNCS, vol. 6223, pp. 59–79. Springer, Heidelberg (2010)
39. Goyal, V., Jain, A., Pandey, O., Sahai, A.: Bounded ciphertext policy attribute based encryption. In: Aceto, L., Damgård, I., Goldberg, L.A., Halldórsson, M.M., Ingólfsdóttir, A., Walukiewicz, I. (eds.) ICALP 2008, Part II. LNCS, vol. 5126, pp. 579–591. Springer, Heidelberg (2008)
40. Goyal, V., Pandey, O., Sahai, A., Waters, B.: Attribute-based encryption for fine-grained access control of encrypted data. In: ACM Conference on Computer and Communications Security, pp. 89–98 (2006)
41. Halderman, A., Schoen, S., Heninger, N., Clarkson, W., Paul, W., Calandrino, J., Feldman, A., Applebaum, J., Felten, E.: Lest we remember: Cold boot attacks on encryption keys. In: USENIX Security Symposium, pp. 45–60 (2008)
42. Ishai, Y., Sahai, A., Wagner, D.: Private circuits: Securing hardware against probing attacks. In: Boneh, D. (ed.) CRYPTO 2003. LNCS, vol. 2729, pp. 463–481. Springer, Heidelberg (2003)
43. Juma, A., Vahlis, Y.: On protecting cryptographic keys against side-channel attacks. In: Rabin, T. (ed.) CRYPTO 2010. LNCS, vol. 6223, pp. 41–58. Springer, Heidelberg (2010)
44. Kamp, J., Zuckerman, D.: Deterministic extractors for bit-fixing sources and exposure-resilient cryptography. In: FOCS, pp. 92–101 (2003)
45. Katz, J., Vaikuntanathan, V.: Signature schemes with bounded leakage resilience. In: Matsui, M. (ed.) ASIACRYPT 2009. LNCS, vol. 5912, pp. 703–720. Springer, Heidelberg (2009)

46. Kocher, P.C.: Timing attacks on implementations of diffie-hellman, RSA, DSS, and other systems. In: Koblitz, N. (ed.) CRYPTO 1996. LNCS, vol. 1109, pp. 104–113. Springer, Heidelberg (1996)
47. Kocher, P.C., Jaffe, J., Jun, B.: Differential power analysis. In: Wiener, M. (ed.) CRYPTO 1999. LNCS, vol. 1666, pp. 388–397. Springer, Heidelberg (1999)
48. Lewko, A., Okamoto, T., Sahai, A., Takashima, K., Waters, B.: Fully secure functional encryption: Attribute-based encryption and (Hierarchical) inner product encryption. In: Gilbert, H. (ed.) EUROCRYPT 2010. LNCS, vol. 6110, pp. 62–91. Springer, Heidelberg (2010)
49. Lewko, A., Rouselakis, Y., Waters, B.: Achieving leakage resilience through dual system encryption. Cryptology ePrint Archive, Report 2010/438 (2010)
50. Lewko, A., Waters, B.: New techniques for dual system encryption and fully secure HIBE with short ciphertexts. In: Micciancio, D. (ed.) TCC 2010. LNCS, vol. 5978, pp. 455–479. Springer, Heidelberg (2010)
51. Micali, S., Reyzin, L.: Physically observable cryptography. In: Naor, M. (ed.) TCC 2004. LNCS, vol. 2951, pp. 278–296. Springer, Heidelberg (2004)
52. Naor, M., Segev, G.: Public-key cryptosystems resilient to key leakage. In: Halevi, S. (ed.) CRYPTO 2009. LNCS, vol. 5677, pp. 18–35. Springer, Heidelberg (2009)
53. Nguyen, P.Q., Shparlinski, I.: The insecurity of the digital signature algorithm with partially known nonces. J. Cryptology 15(3), 151–176 (2002)
54. Okamoto, T., Takashima, K.: Fully secure functional encryption with general relations from the decisional linear assumption. In: Rabin, T. (ed.) CRYPTO 2010. LNCS, vol. 6223, pp. 191–208. Springer, Heidelberg (2010)
55. Ostrovksy, R., Sahai, A., Waters, B.: Attribute based encryption with non-monotonic access structures. In: ACM Conference on Computer and Communications Security, pp. 195–203 (2007)
56. Petit, C., Standaert, F.X., Pereira, O., Malkin, T., Yung, M.: A block cipher based pseudo random number generator secure against side-channel key recovery. In: ASIACCS, pp. 56–65 (2008)
57. Pietrzak, K.: A leakage-resilient mode of operation. In: Joux, A. (ed.) EUROCRYPT 2009. LNCS, vol. 5479, pp. 462–482. Springer, Heidelberg (2009)
58. Quisquater, J.J., Samyde, D.: ElectroMagnetic analysis (EMA): Measures and counter-measures for smart cards. In: Attali, S., Jensen, T. (eds.) E-smart 2001. LNCS, vol. 2140, pp. 200–210. Springer, Heidelberg (2001)
59. Sahai, A., Waters, B.: Fuzzy identity-based encryption. In: Cramer, R. (ed.) EUROCRYPT 2005. LNCS, vol. 3494, pp. 457–473. Springer, Heidelberg (2005)
60. Standaert, F.X., Malkin, T., Yung, M.: A unified framework for the analysis of side-channel key recovery attacks. In: Joux, A. (ed.) EUROCRYPT 2009. LNCS, vol. 5479, pp. 443–461. Springer, Heidelberg (2009)
61. Waters, B.: Efficient identity-based encryption without random oracles. In: Cramer, R. (ed.) EUROCRYPT 2005. LNCS, vol. 3494, pp. 114–127. Springer, Heidelberg (2005)
62. Waters, B.: Ciphertext-policy attribute-based encryption: An expressive, efficient, and provably secure realization. Cryptology ePrint Archive, Report 2008/290 (2008)
63. Waters, B.: Dual system encryption: Realizing fully secure IBE and HIBE under simple assumptions. In: Halevi, S. (ed.) CRYPTO 2009. LNCS, vol. 5677, pp. 619–636. Springer, Heidelberg (2009)

Signatures Resilient to Continual Leakage on Memory and Computation

Tal Malkin[1], Isamu Teranishi[1,2], Yevgeniy Vahlis[1], and Moti Yung[1,3]

[1] Columbia University
{tal,evahlis}@cs.columbia.edu
[2] NEC Japan
teranisi@ah.jp.nec.com
[3] Google Inc.
moti@cs.columbia.edu

Abstract. Recent breakthrough results by Brakerski *et al* and Dodis *et al* have shown that signature schemes can be made secure even if the adversary *continually* obtains information leakage from the secret key of the scheme. However, the schemes currently do not allow leakage on the secret key and randomness *during signing*, except in the random oracle model. Further, the random oracle based schemes require updates to the secret key in order to maintain security, even when no leakage during computation is present.

We present the first signature scheme that is resilient to full continual leakage: memory leakage as well as leakage from processing during signing (both from the secret key and the randomness), in key generation, and in update. Our scheme can tolerate leakage of a $1 - o(1)$ fraction of the secret key between updates, and is proven secure in the standard model based on the symmetric external DDH (SXDH) assumption in bilinear groups. The time periods between updates are a function of the amount of leakage in the period (and nothing more).

As an additional technical contribution, we introduce a new tool: independent pre-image resistant hash functions, which may be of independent interest.

1 Introduction

Cryptographic schemes have traditionally been modeled under the assumption that the secret key is hidden completely within a device and attacks are of "black box" nature. However, cryptographic engineering has taught us that devices are not perfect and can leak information about the key, primarily via what is known as "side channels." These channels give the adversary some partial information by observing physical leakage correlated with the secret key, such as timing or radiation or power consumption associated with computations (either directly by probing the circuit or via antennas probing unshielded devices), as well as memory leakage that reveals information about the secret key (cf., [BB05, QS01, KJJ99, HSH+08]).

Y. Ishai (Ed.): TCC 2011, LNCS 6597, pp. 89–106, 2011.

The threat of partial leakage of keys has not escaped cryptographers; perhaps the earliest works on the subject are Shamir's secret sharing [S79], and later Rivest's all or nothing transform [R97]. In the last few years a large body of work on leakage-resilient cryptography has emerged (cf., [ISW03,MR04,SMY09,AGV09, DKL09, P09,NS09,ADW09,KV09,FKPR10,DGK+10,FRR+10,BG10,DP10,JV10, GR10, DHLW10,BKKV10]), which provide different leakage models requiring that cryptographic schemes be secure even if partial information about the secret key is leaked to an adversary. In order to have an abstract model that is not directly dependent upon the hardware architecture itself, but rather can produce some general guidelines to systems and hardware designers regarding how much leakage can be tolerated overall (within a given setting), leakage was formally modeled as functions associated with algorithmic steps and states. These leakage functions $f_1(sk), \ldots, f_q(sk)$ of the secret key sk, are from some given family of functions, where the specific f_i is selected by the adversary arbitrarily. The leakage (i.e., the function result which contains only partial information about the key) is applied to the relevant state or computation and is given to the adversary (in addition to the black-box access it gets).

Wiki-Leakage: The variety of Leakage Models. There are many types of leakage sub-models based on various parameters of leakage, which we briefly review here. Adaptive leakage, where the function is chosen dynamically by the adversary, obviously gives it more power than non-adaptive. Such adversaries that rather than observing the device once execute side channel attacks repetitively, are stronger and require the schemes to be *continuous leakage resilient.* Repeating (continual) adaptive leakage requires the scheme to have an update procedure for the secret key. Otherwise, the adversary can eventually leak enough information about the key to break the scheme, even if the amount of leakage per time period is small. However, the public key should *not* be changed by the secret key update. For example, a signature verification key should remain valid throughout the lifetime of the scheme.

Next we note that leakage can be *processing leakage* (i.e., only computation leaks) so that physical computations on the device are the only source of leakage. Procedures such as key generation, key updates, and other kinds of processing (i.e., signing, decryption, etc.) which involve random values may be allowed different amounts of leakage, but are all related to some physics of computations (e.g., timing of operations). The influential works of Ishai, Sahai, and Wagner [ISW03] and Micali and Reyzin [MR04] enable us to construct schemes under the "any computation, and only computation, leak information," model, which has led to many recent achievements. In contrast, *memory leakage* [AGV09] (which, in some sense, can be traced to the original works of Shamir and Rivest mentioned above) are produced as a function of the memory state itself. This type of leakage is orthogonal to computational leakage: an adversary can get memory leakage by probing memories even if the memories are not currently used in any computation (e.g., the cold-boot attacks [HSH+08]). For example, the scheme of [DHLW10] is secure against memory attacks (even continual), but assumes that the signing process leaks no information. The most general model allows *full leakage* which includes leakage both from processing and memory.

The most demanding case for designing digital signature schemes seems to be the case of adaptive and continual full leakage that is available to the adversary from both computational and memory sources (without protection of sub-steps of computations). However, to date, there are no known schemes which achieve a digital signature scheme in this adversarial setting in the standard model. All known schemes with full (memory and processing) leakage either do not have a key update algorithm and thus are not continual (cf., [KV09]), have a key update algorithm but require some restrictions (e.g., [ADW09] which requires an additional leakage-free master key), or are based on the random oracle model (with a relaxation of the definition of a "time period") [BKKV10].

1.1 Our Contributions

We propose the first signature scheme satisfying all of the above requirements, whose security can be proven in the standard model and without further relaxation. Specifically, our scheme protects against (1) continual memory leakages combined with (2) all types of continual processing (computational) leakages, namely leakages from key generation, key updates, and signing.

Moreover, the amount of information that our scheme allows to leak in each time period is optimal, in the sense that our scheme remains secure even if $1 - o(1)$ fraction of the secret key of each time period is leaked. Here "time period" is the period between two consecutive updates of the secret key (the time period is a function of the accumulated leakage itself and not a relaxed notion which depends on other parameters). We stress that our scheme remains secure even when the leakage during signing is a function $f(sk, r)$ of both the secret key and the randomness. Further, the function f can be adaptively chosen by the adversary. Using standard techniques, our scheme also allows $O(\log \kappa)$ leakage in the key generation and in each of the key updates, where κ is a security parameter. (The secret key has $O(\kappa)$ bit length. The fraction of leakage is therefore $O(\log \kappa/\kappa)$).

Comparison with Recent Schemes. Let us compare our scheme with the recent breakthrough results of Dodis, Haralambiev, Lopez-Alt, and Wichs [DHLW10] and Brakerski, Kalai, Katz, and Vaikuntanathan [BKKV10]. The signature scheme of [DHLW10], as noted above, has to assume that there is no leakage from the signing process. The work in [BKKV10] proposed two signature schemes. Their first scheme is secure on if there is no leakage from the signing process. The second scheme relies on random oracle to protect against leakage during signing, and further requires signatures to be treated as leakage. That is, even if there is no actual side-channel leakage during a certain time period, the signing key must be refreshed to preserve security. In contrast, our signature scheme is proved secure in the standard model and the key needs to be refreshed only if leakage occurs (i.e. signatures do not constitute leakage).

Concurrent work. In a concurrent work, Boyle, Segev, and Wichs [BSW10] construct a fully leakage resilient signature scheme using different techniques. [BSW10] take a more generic approach than we do. On the other hand, our scheme is somewhat more efficient.

Other related work. In the recent years a very rich body of research on leakage resilient cryptography has been developed. Dziembowski and Pietrzak [DP08], and Pietrzak [P09] describe the first stream ciphers resilient to continual leakage in the only computation leaks model. Faust *et al* [FKPR10] construct signature schemes resilient to continual leakage in the only computation leaks model. Faust *et al* [FRR+10] give a general compiler using secure hardware that protects an arbitrary circuit against continual leakage that can be modeled as a shallow (AC^0) boolean circuit. Juma and Vahlis [JV10], and separately Goldwasser and Rothblum [GR10], give compilers that protect any algorithm against continual leakage (without complexity restrictions), using secure hardware. Recently, Dodis and Pietrzak [DP10] show how to build continual leakage resilient pseudorandom functions that are secure against non-adaptive leakage.

Discussion on processing leakage. Continual memory attacks are a very powerful leakage model that allows the adversary to continuously obtain leakage from the entire secret key. A natural question to ask is whether adding processing leakage into the mix adds anything to adversarial power. Indeed, the only additional information available to the adversary during processing leakage is ephemeral randomness that is used in the computation. In many cases, such as in the case of public key encryption or decryption (using the corresponding private key), leakage on ephemeral randomness does not provide any useful information to the adversary about the secret key. In fact, in public key encryption and decryption, the adversary can simulate the leakage from the randomness of these algorithms on her own. However, this is not the case for signature schemes.

In a signature scheme, a signature is computed using the secret key, and made public. Consequently, signatures can viewed as a very restricted type of leakage on the secret key. A signature scheme is considered secure if such "signature leakage" is useless to any efficient adversary. When the adversary is given leakage from the randomness of the signing process, she may be able obtain information that will allow her to extract useful information from the accompanying signature. For example, each signature may contain an encryption of the secret key under a random key that is generated during signing, and then forgotten. If the adversary is able to leak this random key, she trivially breaks the security of the scheme.

1.2 The Idea behind Our Construction

We are motivated in our construction by the (non-continual) memory-attack resilient signature schemes of Alwen, Dodis, and Wichs [ADW09], and of Katz and Vaikuntanathan [KV09]. Both [ADW09] and [KV09] use the following high level approach, which is based on the Fiat-Shamir heuristic [FS86]: a signature of a message M relative to a public key pk is an extractable proof of knowledge (in the random oracle model) of a value sk for which $H(sk) = pk$. Here is H is a hash function.

The security proof of these signature schemes relies on the *second preimage resistance* of the hash function H, and the witness extractability of the proofs that are used. That is, they use the property that it is infeasible to find $sk^* \neq sk$ satisfying $H(sk^*) = H(sk)$.

To prove security, they construct a simulator that generates a secret key sk randomly and computes $pk = H(sk)$. The simulator then can answer leakage queries and signing queries using the secret key that it itself has generated. If an adversary can forge a message/signature pair, the simulator extracts the witness sk'. Now, if the fraction of leakage of sk is less than $1 - o(1)$, the exact key sk that is used by the simulator is information theoretically undetermined in the view of the adversary (specifically, there are at least two possible keys, given the leakage). Therefore, with probability at least $1/2$, the witness sk' is different from sk, which breaks the second pre-image resistance of H.

We start with this basic approach, and enhance it along three dimensions. Specifically, we:

1. Remove the requirement for a random oracle, and get a scheme secure in the standard model.
2. Add a key update procedure that refreshes the secret key, while keeping the public key fixed. This yields a signature scheme resilient against *continual* memory attacks [BKKV10].
3. Develop a proof method that allows leakage of randomness used in signing within a period (allowing optimal leakage).

Removing the Random Oracle. The simplest idea to remove the random oracle from the scheme of [ADW09,KV09] is to prove the knowledge of the secret key not by using Fiat-Shamir heuristic, but by using other known non-interactive proof systems in the standard model.

This initial attempt fails for the following reason: the argument of [ADW09, KV09] showing $sk^* \neq sk$ is purely information theoretic. Hence, if we want to use the argument of [ADW09, KV09], the proof systems should hide sk not in a computational sense, but in an information theoretic sense. But if the proof system hides sk information theoretically, the secret key sk^* used by an adversary is also hidden information theoretically (since no random oracle is available). Hence, it is impossible for the simulator to get the second pre-image sk^* from the adversary's forgery, and so we cannot rely on second pre-image resistance.

To overcome the above problem, we use the Waters' function

$$h(\mathcal{H}, M) = H_0 + \sum_k M_k H_k,$$

where M_k is the k-th bit of a message M and $\mathcal{H} = (H_0, \ldots, H_m)$ is a tuple of group elements[1]. The function is introduced in [W05] in order to construct

[1] Here we use *additive notation* to describe the function.

an adaptively secure ID-based encryption scheme. The Waters' function has the property that a simulator can choose the parameters $H_0, \ldots, H_m$ in a special way that defines a subset $\mathcal{M}$ of the range of h. It can then be shown that with non-negligible probability, all signing queries M of the adversary satisfy $h(\mathcal{H}, M) \in \mathcal{M}$, and the forgery M^* satisfies $h(\mathcal{H}, M^*) \notin \mathcal{M}$.

We construct a Waters-like function h' such that $\mathcal{M}$ is a set of all non-DDH tuples in the range of h'. Consequently, we get that with non-negligible probability all signing queries of the adversary map to non-DDH tuples, and the forgery maps to a DDH tuple.

We then combine the above hash function h' with the Groth-Sahai [GS08] proof system. Groth-Sahai is a proof system which uses a common reference string (CRS). The proof system hides the witness information theoretically if the CRS is a non-DDH tuple and hides it only computationally, and the witness therefore is extractable, if the CRS is a DDH tuple.

Hence, by using Groth-Sahai proofs as signatures, and $h'(M)$ as the CRS of the Groth-Sahai proof, we get a proof system that hides the witness sk information theoretically when the simulator generates proofs as answers to signing queries, and allows the witness sk^* to be extracted from the proof generated by an adversary as a forged signature.

The scheme before adding key update. Our scheme therefore has the following basic structure. The private key consists of a vector of group elements $\boldsymbol{W}$, and the public key consists of group elements $\boldsymbol{A}$ and T such that $e(\boldsymbol{A}, \boldsymbol{W}) = T$. The public key also contains the description of a Waters-like hash function h'. To sign a message M, we first use h' to compute a CRS $h'(\mathcal{H}, M)$ for the Groth-Sahai proof system. Then, the signature is a proof, under the CRS $h'(\mathcal{H}, M)$, that $e(\boldsymbol{A}, \boldsymbol{W}) = T$.

Before proceeding to describe our key update procedure, we believe it is instructive to see why the above scheme, without update, is secure. Intuitively, this follows from the second pre-image resistance of the hash function $H_{\boldsymbol{A}}(\boldsymbol{X}) := e(\boldsymbol{A}, \boldsymbol{X})$. From the perfect witness indistinguishability of Groth-Sahai proofs we know that the adversary learns about the specific witness $\boldsymbol{W}$ only from leakage. However, since the amount of leakage is bounded, the actual witness $\boldsymbol{W}$ in the private key remains information theoretically undetermined. Finally, when the adversary submits a successful forgery, we use the indistinguishability of common reference strings technique discussed above to show that, with non-negligible probability, a witness $\boldsymbol{W}'$ can be extracted from the forgery. This witness is likely to be different from $\boldsymbol{W}$, which would give the simulator two inputs $\boldsymbol{W}$ and $\boldsymbol{W}'$ such that $H_{\boldsymbol{A}}(\boldsymbol{W}) = H_{\boldsymbol{A}}(\boldsymbol{W}')$. This in turn violates the second pre-image resistance of $H_{\boldsymbol{A}}$.

We remark that the above technique is somewhat reminiscent of the Feige-Lapidot-Shamir [FLS90] method for using witness indistinguishability to achieve witness hiding.

Adding Key Updates. The approach of Section 1.2 allows us to move from the random oracle to the standard model. However, the above scheme can still tolerate only a bounded amount of leakage. We now describe a method for choosing the private keys of the scheme that allows us to randomize the key without having to issue a new public key.

We start by observing that if our scheme relies for security on the second pre-image resistance of the hash function H, then no key update algorithm can exist. This is because otherwise we could use the key update algorithm itself to break second pre-image resistance as follows:

> If we can get new key $sk^{[i+1]}$ efficiently by updating $sk^{[i]}$, this means that one can get two keys $sk^{[i]}$ and $sk^{[i+1]}$ satisfying of $T = H(sk^{[i]})$ and $T = H(sk^{[i+1]})$, where T is the public key. The function H therefore cannot be second pre-image resistant.

Note that our model does not allow to update the public key for the convenience of verifiers. The above collision of the function H is therefore unavoidable.

(n,k)-independent pre-image resistant (IPIR) hash functions. We overcome the above problem by introducing the following new notion: (n,k)-*independent pre-image resistant hash function* H, where n is an integer and $k \le n-2$. This is a linear function H from an n-dimensional vector space $\mathbb{H}^n$ to a 1-dimensional vector space $\mathbb{T}$, over $\mathbb{Z}_p$. We require that, given certain trapdoor information about H, one can find a tuple $(\boldsymbol{Y}_1, \ldots, \boldsymbol{Y}_{k+1}, T)$ satisfying $T = H(Y_j)$ for all $j \in [k+1]$. However, it must be infeasible to find $\boldsymbol{Y}_* \in \mathbb{H}^n$ satisfying $T = H(\boldsymbol{Y}_*)$ and $\boldsymbol{Y}_* \notin \mathsf{Aff}(\boldsymbol{Y}_1, \ldots, \boldsymbol{Y}_{k+1})$, where $\mathsf{Aff}(\boldsymbol{Y}_1, \ldots, \boldsymbol{Y}_{k+1})$ is the smallest affine space spanned by $\boldsymbol{Y}_1, \ldots, \boldsymbol{Y}_{k+1}$. We call the hardness property (n,k)-*independent preimage resistance.*

We note that although in this paper we give a construction of an (n,k)-IPIR hash function, we do not use it as a black box in our signature scheme. Indeed, the Groth-Sahai proofs that are generated use the parameters of the hash function, which are of a very specific form. However, the notion of IPIR seems useful in itself, and we hope that other applications will be found for it. Furthermore, abstracting the properties that we need from H allows us to present a modular and structured proof of security for our signature scheme.

Generating and updating keys. Using the linearity of H, we can generate any linear sum of $\boldsymbol{Y}_1, \ldots, \boldsymbol{Y}_{k+1}$ in polynomial time. We use this property to perform key update. And we use the IPIR (instead of the second pre-image resistance) when we show that no adversary can forge a signature.

In slightly greater detail, we construct the key generation algorithm and the update algorithm of our scheme as follows. In the key generation, by using the trapdoor of the hash function H, we generate $(\boldsymbol{Y}_1, \boldsymbol{Y}_2, T)$ satisfying $T = H(\boldsymbol{Y}_1) = H(\boldsymbol{Y}_2)$. We then compute $\boldsymbol{Q} \leftarrow \boldsymbol{Y}_1 - \boldsymbol{Y}_2$ and publish $\boldsymbol{Q}$ as a part of the public key. The secret key is $\boldsymbol{W}^{[0]} \leftarrow \boldsymbol{Y}_2$. Note that $H(\boldsymbol{Q}) = H(\boldsymbol{Y}_1) - H(\boldsymbol{Y}_2) = 0$ holds.

Key update then works as follows: in the $(i+1)$-st round, we select $s^{[i]} \xleftarrow{\$} \mathbb{Z}_p$ randomly and compute $\boldsymbol{W}^{[i+1]} \leftarrow \boldsymbol{W}^{[i]} + s^{[i]}\boldsymbol{Q}$. Based on the linearity of H, and the equality $H(\boldsymbol{Q}) = 0$, one can easily show that $H(\boldsymbol{W}^{[i+1]}) = H(\boldsymbol{W}^{[i]}) = \cdots H(\boldsymbol{W}^{[0]}) = H(\boldsymbol{Y}_2) = T$ holds, and that $\boldsymbol{W}^{[i]}$ is an element of $\mathsf{Aff}(\boldsymbol{Y}_1, \boldsymbol{Y}_2)$. The latter holds because $\boldsymbol{W}^{[i]}$ are linear sums of $\boldsymbol{W}^{[0]} = \boldsymbol{Y}_2$ and $\boldsymbol{Q} = \boldsymbol{Y}_1 - \boldsymbol{Y}_2$. We then use an adaptation of the "leakage resilient subspace" technique from [BKKV10] to show that the affine subspace $\mathsf{Aff}(\boldsymbol{Y}_1, \boldsymbol{Y}_2)$ (or even $\mathsf{Aff}(\boldsymbol{Y}_1, \ldots, \boldsymbol{Y}_{k+1})$) is hidden from the adversary, even given continual leakage (assuming the refresh procedure above is periodically performed). Given the hiding property of the affine space, it follows that if the adversary forges a signature $\sigma^* = \mathsf{Prf}_{M^*}(\boldsymbol{W}^*)$ for some message M^*, she likely uses a witness $\boldsymbol{W}^* \notin \mathsf{Aff}(\boldsymbol{Y}_1, \boldsymbol{Y}_2)$. However, this violates the IPIR of H. The security of the scheme follows.

Security Against Leakage in Signing. The main challenge in achieving security under leakage from the signing process is that the signature and the secret key are correlated through the randomness that was used to produce the signature. When the adversary obtains leakage on the randomness, the signature may become much more valuable, and potentially allow the adversary to break the scheme (as we discussed in the introduction).

In the proof based signature schemes we described above, there is no guarantee that leakage on the randomness of the prover does not break the zero-knowledge or witness indistinguishability property of the proof system. We solve this problem through a combination of several tools: first, as described above, we rely on Groth-Sahai proofs, which have a dual mode – witness hiding and witness binding. When the proof is (perfectly) hiding, it is information theoretically independent from the witness, which is the secret key, if there is no leakage on the randomness of the prover.

We use the above fact to "invert" the order in which components of the signature are generated: first the GS proof σ in the signature is generated using some globally fixed witness $\mathcal{Y}$ (note that this is only done by the simulator in the analysis, and so there is no leakage on the witness $\mathcal{Y}$). Then, given an actual witness $\boldsymbol{W}$ for the proof, we "reverse engineer" the randomness R that would yield the same proof σ, and compute leakage on $(\boldsymbol{W}, R)$. We use an additional property of Groth-Sahai that for every pair of proof and witness $(\sigma, \boldsymbol{W})$ there exists a unique randomness $R_{\sigma,\boldsymbol{W}}$ that causes the prover to output σ given witness $\boldsymbol{W}$. Moreover, since the proof is perfectly witness hiding, for all witnesses $\boldsymbol{W}$, the distribution on the tuple $(\sigma, R_{\sigma,\boldsymbol{W}})$ are identical whether we first generate the proof using witness $\boldsymbol{V}$ and then determine the randomness, or choose the randomness uniformly, and compute the proof directly.

The above approach, however, does not work as is, since the process of finding R may not be efficiently computable! We therefore rely on an information theoretic leakage resilience property of random subspaces that was shown in [BKKV10] (in fact, we prove a slightly stronger version that suits our construction). We combine both techniques together, simultaneously using the randomness reverse engineering technique described above to handle leakage from signing, and information theoretic indistinguishability of random subspaces under leakage. Using these techniques together, we show that the adversary is unable to gain information about the subspace from which we generate private keys during update, even if leakage on the signing process is available. We then use independent pre-image resistance to conclude that our scheme is secure.

2 Preliminaries

Notations. Let $[n]$ denote $\{1,\ldots,n\}$ and $[k..n]$ denote $\{k,\ldots,n\}$. For two random variables X and Y, $\mathsf{dist}(X,Y)$ denote the statistical distance between X and Y.

Linear Algebra. Unless otherwise stated, vectors in this paper are column vectors. We represent a row vector as a transpose of a column vector in this paper. For natural numbers n and m, let $\mathbb{Z}_p^{n\times m}$ denote the set of $n\times m$ matrices over $\mathbb{Z}_p$. Let A^T denote the transposed of a matrix $A=(a_{i,j})_{i,j}$, that is, $A^T=(a_{j,i})_{i,j}$. For two vectors $\boldsymbol{u}=(u_1,\ldots,u_n)$, $\boldsymbol{v}=(v_1,\ldots,v_n)\in\mathbb{Z}_p^n$, we let $\langle\boldsymbol{u},\boldsymbol{v}\rangle$ denote the inner product of them in $\mathbb{Z}_p$, that is, $\langle\boldsymbol{u},\boldsymbol{v}\rangle=\sum_i u_iv_i \mod p$.

For a vector $\boldsymbol{v}=(v_1,\ldots,v_n)\in\mathbb{Z}_p^n$ and an element W of the group $\mathbb{G}$ or $\mathbb{H}$, let $\boldsymbol{v}W$ denote the vector $(v_1W,\ldots,v_nW)$.

For a vector space $\mathcal{V}$ and vectors $\boldsymbol{Y}_1,\ldots,\boldsymbol{Y}_{k+1}\in\mathcal{V}$, let $\mathsf{Span}(\boldsymbol{Y}_1,\ldots,\boldsymbol{Y}_k)$ denote the smallest vector subspace of $\mathcal{V}^n$ which contains all of $(\boldsymbol{Y}_j)_{j\in[k]}$, that is,

$$\mathsf{Span}(\boldsymbol{Y}_1,\ldots,\boldsymbol{Y}_k)=\{\boldsymbol{Y}\in\mathcal{V}^n \mid \exists s_1,\ldots,s_k\in\mathbb{Z}_p:\boldsymbol{Y}=\sum_{j\in[k]}s_j\boldsymbol{Y}_j\}.$$

Similarly, let $\mathsf{Aff}(\boldsymbol{Y}_1,\ldots,\boldsymbol{Y}_{k+1})$ is the smallest affine subspace of $\mathcal{V}^n$ which contains all of $(\boldsymbol{Y}_j)_{j\in[k+1]}$, that is,

$$\mathsf{Aff}(\boldsymbol{Y}_1,\ldots,\boldsymbol{Y}_{k+1})=\{\boldsymbol{Y}\in\mathcal{V}^n \mid \exists s_1,\ldots,s_{k+1}\in\mathbb{Z}_p:\boldsymbol{Y}=\sum_{j\in[k+1]}s_j\boldsymbol{Y}_j,\ \sum_{j\in[k+1]}s_j=1\}.$$

Note that the space $\mathsf{Aff}(\boldsymbol{Y}_1,\ldots,\boldsymbol{Y}_{k+1})$ becomes k dimensional, when $k+1$ vectors $\boldsymbol{Y}_1,\ldots,\boldsymbol{Y}_{k+1}$ are linear independent.

2.1 Signature Schemes Resilient against Continual Leakage

A *signature scheme with key update* $\mathcal{SGN}$ consists of four algorithms Kg, Sig, Ver, and Update. The inputs and outputs of Kg, Sig, and Ver are the same as in standard signature schemes. Update takes as input a secret key and a public key and outputs a new element of the secret key space. $\mathcal{SGN}=(\mathsf{Kg},\mathsf{Sig},\mathsf{Ver},\mathsf{Update})$ has to satisfy the following property:

- **(Correctness).** For any integer $n\geq 0$ and any message M, if we compute $(pk,sk_0)\leftarrow\mathsf{Gen}(1^\kappa)$, $sk_1\leftarrow\mathsf{Update}_{pk}(sk_0)$, $\ldots$, $sk_n\leftarrow\mathsf{Update}_{pk}(sk_{n-1})$, and $\sigma\leftarrow\mathsf{Sig}(sk_n,M)$, $\mathsf{Ver}(pk,M,\sigma)=1$ always holds.

We follow the definition of [BKKV10] of leakage resilient signatures.

Setup. $\mathcal{A}(1^\kappa)$ sends to the challenger a function f satisfying $|f(R)| \le \rho_G|R|$ for all R. The challenger then selects $R \xleftarrow{\$} \mathsf{Rnd}[\mathsf{Gen}]$ computes $(pk, sk_0) \leftarrow \mathsf{Gen}(1^\kappa; R)$ sends $(pk, f(R))$ to $\mathcal{A}$, and initializes $i \leftarrow 0$ and $L_i \leftarrow |f(R)|$. Here i represents the number of updates and L_i denote the bit length of all leakages about the i-th secret key.

Queries. $\mathcal{A}$ makes queries of the following three types polynomial number of times:

- Update queries (update, f) where f is a circuit satisfying $|f(sk, R)| \le \rho_U(|sk|+|R|)$ for any (sk, R). If $L_i + |f(sk_i, R)| \le \rho_M|sk_i|$ holds, the challenger chooses $R \xleftarrow{\$} \mathsf{Rnd}[\mathsf{Update}]$ randomly, computes $sk_{i+1} \leftarrow \mathsf{Update}_{pk}(sk_i)$ and sends $f(sk_i, R)$ back to $\mathcal{A}$ and resets $i \leftarrow i+1$ and $L_i \leftarrow |f(sk_i, R)|$. Otherwise, the challenger aborts.
- (Memory) leak queries (leak, f), where f is a circuit. If $L_i + |f(sk_i)| \le \rho_M|sk_i|$ holds, the challenger sends $f(sk_i)$ to adversary and resets $L_i \leftarrow L_i + |f(sk_i)|$. Otherwise, the challenger aborts.
- Signing queries (sig, M, f) where f is a circuit with $|f(sk, R)| \le \rho_S(|sk|+|R|)$ for any (sk, R). The challenger chooses $R \leftarrow \mathsf{Rnd}[\mathsf{Sig}]$ randomly, computes $\sigma \leftarrow \mathsf{Sig}(sk_i, M; R)$ and sends $(\sigma, f(sk_i, R))$ back to $\mathcal{A}$.

Challenge. Assuming the challenger did not aborts, $\mathcal{A}$ outputs (M_*, σ_*). It succeeds if $\mathsf{Ver}(pk, M_*, \sigma_*) = 1$ holds and $\mathcal{A}$ never made query (sig, M_*).

Fig. 1. Game of $(\rho_G, \rho_U, \rho_M, \rho_S)$-EU-CMA-CML secure

Definition 1 ([BKKV10]). Let ρ_G, ρ_U, ρ_M, and ρ_S be elements of the real range $[0, 1]$. We say that $\mathcal{SGN} = (\mathsf{Gen}, \mathsf{Sig}, \mathsf{Ver}, \mathsf{Update})$ is $(\rho_G, \rho_U, \rho_M, \rho_S)$- *EU-CMA-CML secure* (stand for existentially unforgeable under chosen message attack in the CML model) if no PPT adversary $\mathcal{A}$ succeeds in the game of Fig.1 with non-negligible probability. Here $\mathsf{Rnd}[\mathsf{Algo}]$ denote the set of randomnesses for algorithm Algo.

2.2 Bilinear Pairings

In our paper, we are working on a bilinear pairing, $\mathsf{e} : \mathbb{G} \times \mathbb{H} \to \mathbb{T}$ with prime order p, where $\mathbb{G} \ne \mathbb{H}$ holds and there is no efficiently computable homomorphism between two groups $\mathbb{G}$ and $\mathbb{H}$ (Such a pairing is called Type III [GPS08]). We denote by gk bilinear map parameters of the form $(p, \mathbb{G}, \mathbb{H}, \mathbb{T}, \mathsf{e})$. We will occasionally refer to gk as *group description.*

Our proofs rely on basic properties of linear algebra. We therefore find it convenient to use *additive notation* for pairings. For example, we write $\mathsf{e}((a+b)A, W) = a \cdot \mathsf{e}(A, W) + b \cdot \mathsf{e}(A, W)$. For two (column) vectors $\boldsymbol{A} = (A_1, \ldots, A_n)^\top \in \mathbb{G}$ and $\boldsymbol{W} = (W_1, \ldots, W_n)^\top \in \mathbb{H}$, we denote

$$\mathsf{e}(\boldsymbol{A}^\top, \boldsymbol{W}) = \sum_{i \in [n]} \mathsf{e}(A_i, W_i)$$

Assumption 2 (SXDH assumption [GS08]). *We say that* $gk = (p, \mathbb{G}, \mathbb{H}, \mathbb{T}, \mathsf{e})$ *satisfies the Symmetric external Diffie-Hellman (SXDH) assumption, if the*

DDH assumption holds both over $\mathbb{G}$ and $\mathbb{H}$ (note that this is possible in type III pairings).

2.3 Groth-Sahai Proofs

Groth and Sahai [GS08] proposed efficient non-interactive witness indistinguishable proof systems for settings where a bilinear pairing is available. Their system allows efficient proofs of various statements about the groups of the pairing, and the pairing relation.

In this work we prove statements of the form $\mathsf{e}(\boldsymbol{A}^\top, \boldsymbol{W}) = T$, where an instance $(\boldsymbol{A}, T) \in \mathbb{G}^n \times \mathbb{T}$ is the input to the verifier, and $\boldsymbol{W}$ is the witness used by the prover.

Let $gk = (p, \mathbb{G}, \mathbb{H}, \mathbb{T}, \mathsf{e})$ be a group description and $crs = (\boldsymbol{G}, \boldsymbol{H}) \in \mathbb{H}^2$. The Groth-Sahai proof using crs as CRS (Common Reference String) is as follows.

- $\mathsf{Prf}(gk, crs, (T, \boldsymbol{A}), \boldsymbol{W})$: Parse gk and crs as $(p, \mathbb{G}, \mathbb{H}, \mathbb{T}, \mathsf{e})$ and $(\boldsymbol{G}, \boldsymbol{H})$ respectively. Select $R \stackrel{\$}{\leftarrow} \mathbb{Z}_p^{n \times 2}$ randomly, and compute
$$(\boldsymbol{C}, \boldsymbol{D}) \leftarrow (R \cdot \boldsymbol{G}, \boldsymbol{W} + R \cdot \boldsymbol{H})$$
$$\boldsymbol{\Pi} \leftarrow R^\top \boldsymbol{A}$$
and output $\sigma = (\boldsymbol{C}, \boldsymbol{D}, \boldsymbol{\Pi})$.
- $\mathsf{Vrf}(gk, crs, (\boldsymbol{A}, T), \sigma)$: Parse gk, crs, and σ as $(p, \mathbb{G}, \mathbb{H}, \mathbb{T}, \mathsf{e})$, $(\boldsymbol{G}, \boldsymbol{H})$ and $(\boldsymbol{C}, \boldsymbol{D}, \boldsymbol{\Pi})$ respectively.
Output 1 iff the following equality holds.
$$(\mathsf{e}(\boldsymbol{A}^\top, \boldsymbol{C}), \mathsf{e}(\boldsymbol{A}^\top, \boldsymbol{D})) \stackrel{?}{=} (\mathsf{e}(\boldsymbol{\Pi}^\top, \boldsymbol{G}), T + \mathsf{e}(\boldsymbol{\Pi}^\top, \boldsymbol{H}))$$

One can easily show that a correctly generated proof is always accepted by Vrf.

Groth and Sahai [GS08] gave the following two algorithms $\mathsf{HideCRS}$ and $\mathsf{BindCRS}$ to generate two different types of CRS: hiding and binding. When a hiding CRS is used for the proof, the witness is perfectly (information theoretically) hidden. When a binding CRS is used, $\mathsf{BindCRS}$ provides a trapdoor along with the CRS, and the trapdoor can be used to extract the witness from any proof. Finally, it is required that the two types of CRS are computationally indistinguishable.

For the above proof system, let $gk = (p, \mathbb{G}, \mathbb{H}, \mathbb{T}, \mathsf{e})$. The algorithms $\mathsf{HideCRS}$ and $\mathsf{BindCRS}$ are defined as follows:

- $\mathsf{HideCRS}(gk)$: Select $\boldsymbol{G}, \boldsymbol{H} \stackrel{\$}{\leftarrow} \mathbb{H}^2$ randomly and output $crs \leftarrow (\boldsymbol{G}, \boldsymbol{H})$.
- $\mathsf{BindCRS}(gk)$: Select $\boldsymbol{G} \stackrel{\$}{\leftarrow} \mathbb{H}^2$ and $\alpha \stackrel{\$}{\leftarrow} \mathbb{Z}_p$ randomly, compute $\boldsymbol{H} \leftarrow \alpha \boldsymbol{G}$ and output $crs \leftarrow (\boldsymbol{G}, \boldsymbol{H})$ and the trapdoor α.

Groth-Sahai show that in the above proof system, a hiding CRS and a binding CRS are indistinguishable under the SXDH assumption. Formally, the perfect witness indistinguishability, and the witness extractability properties are defined as follows [GS08]. Below, $\mathsf{Setup}(1^\kappa)$ be an algorithm which generates a group description $gk = (p, \mathbb{G}, \mathbb{H}, \mathbb{T}, \mathsf{e})$.

– **(Composable Perfect Witness Indistinguishability).** For all (possibly unbounded) adversaries $\mathcal{A}$

$$\Pr\left[\begin{array}{l} gk \leftarrow \mathsf{Setup}(1^\kappa),\ crs \leftarrow \mathsf{HideCRS}(gk), \\ (\boldsymbol{A}, T, \boldsymbol{W}_0, \boldsymbol{W}_1, st) \leftarrow \mathcal{A}(gk, crs), \sigma \leftarrow \mathsf{Prf}(gk, crs, (\boldsymbol{A}, T), \boldsymbol{W}_0), b \leftarrow \mathcal{A}(\sigma, st) \end{array} : b = 1\right]$$

$$= \Pr\left[\begin{array}{l} gk \leftarrow \mathsf{Setup}(1^\kappa),\ crs \leftarrow \mathsf{HideCRS}(gk), \\ (\boldsymbol{A}, T, \boldsymbol{W}_0, \boldsymbol{W}_1, st) \leftarrow \mathcal{A}(gk, crs), \sigma \leftarrow \mathsf{Prf}(gk, crs, (\boldsymbol{A}, T), \boldsymbol{W}_1), b \leftarrow \mathcal{A}(\sigma, st) \end{array} : b = 1\right]$$

where we require $\mathsf{e}(\boldsymbol{A}, \boldsymbol{W}^{[0]}) = \mathsf{e}(\boldsymbol{A}, \boldsymbol{W}^{[1]}) = T$.

– **(Perfect Extractability).** For all possible output $gk = (p, \mathbb{G}, \mathbb{H}, \mathbb{T}, \mathsf{e})$ of $\mathsf{Setup}(1^\kappa)$, all possible output (crs, α) of $\mathsf{BindCRS}(gk)$, all $(\boldsymbol{A}, T) \in \mathbb{G}^n \times \mathbb{T}$ and all $\sigma = (\boldsymbol{C}, \boldsymbol{D}, \boldsymbol{\Pi}) \in \mathbb{H}^2 \times \mathbb{H}^2 \times \mathbb{G}^2$ satisfying $\mathsf{Vrf}(gk, crs, (\boldsymbol{A}, T), \sigma) = 1$, if we set

$$\boldsymbol{W}^* \leftarrow \boldsymbol{D} - \alpha \boldsymbol{C},$$

the equality $\mathsf{e}(\boldsymbol{A}, \boldsymbol{W}^*) = T$ always holds.

3 Independent Preimage Resistant (IPIR) Hash Functions

We introduce a new notion: independent pre-image resistance (IPIR), that we use in the construction and analysis of our scheme. As we have already mentioned in the introduction, our construction does not use the IPIR hash function described below in a black box way. Nevertheless, we believe it to be instructive to define this notion separately, both to conceptually isolate the properties of the hash function that we use, and for potential future use.

Definition 3 (Independent Preimage Resistant Hash Function). Let n be a positive number, and let $\mathbb{H}$ and $\mathbb{T}$ be cyclic groups of order p. Let $\mathsf{Gen}, H, \mathsf{GenSim}, \mathsf{Check}$ be polynomial time algorithms whose inputs and outputs are as follows. Below, k is the parameter representing the dimension. Note that a k-dimensional *affine* (not linear) subspace contains $k+1$ vectors and GenSim of the below therefore outputs $k+1$ vectors $\mathcal{Y}_j$.

– $\mathsf{Gen}(1^\kappa)$ outputs some public information P.
– For any possible output P of Gen, H_P is a deterministic linear function from $\mathbb{H}^n$ to $\mathbb{T}$.
– GenSim takes an integer $k \in [n-2]$ as an input and outputs a public information P, a trapdoor td, $(T, \boldsymbol{Y}_1, \ldots, \boldsymbol{Y}_{k+1}) \in \mathbb{T} \times (\mathbb{H}^n)^{k+1}$ satisfying $T = H_P(\boldsymbol{Y}_i)$ for all $i \in [k+1]$.
– Check takes P, $(T, \boldsymbol{Y}_1, \ldots, \boldsymbol{Y}_{k+1})$, an element $\boldsymbol{Y}'$ of $\mathbb{H}^n$, and a trapdoor td as inputs and outputs 1 or 0.

For integers n and $k \in [n-2]$, we say that H is (n,k)-*independent pre-image resistant* with respect to $(\mathsf{Gen}, \mathsf{GenSim}, \mathsf{Check})$ if it satisfies the following properties.

- **(Correctness).** For all outputs $(P, td, (T, \boldsymbol{Y}_1, \ldots, \boldsymbol{Y}_{k+1}))$ of GenSim and all $\boldsymbol{Y}' \in \mathbb{H}^n$,
  $\mathsf{Check}(P, td, (T, \boldsymbol{Y}_1, \ldots, \boldsymbol{Y}_{k+1}), \boldsymbol{Y}') = 1$ holds iff $T = H_P(\boldsymbol{Y}')$ and $\boldsymbol{Y}' \in \mathsf{Aff}(\boldsymbol{Y}_1, \ldots, \boldsymbol{Y}_{k+1})$ holds.
  Moreover, for an output $(P, td, (T, \boldsymbol{Y}_1, \ldots, \boldsymbol{Y}_{k+1}))$ of GenSim, P have the same distribution as an output of $\mathsf{Gen}(1^\kappa)$ and $(\boldsymbol{Y}_1, \ldots, \boldsymbol{Y}_{k+1})$ uniformly distributed on $\{\boldsymbol{Y} \in \mathbb{H}^n \mid H_P(\boldsymbol{Y}) = T\}^{k+1}$.
- **((n,k)-Independent Preimage Resistance).** For any polytime adversary $\mathcal{A}$,

$$\Pr\left[\begin{array}{l}(P, td, (T, \boldsymbol{Y}_1, \ldots, \boldsymbol{Y}_{k+1})) \leftarrow \mathsf{GenSim}(1^\kappa), \\ \boldsymbol{Y}_* \leftarrow \mathcal{A}(P, T, \boldsymbol{Y}_1, \ldots, \boldsymbol{Y}_{k+1})\end{array} : \begin{array}{l} T = H_P(\boldsymbol{Y}_*) \\ \boldsymbol{Y}_* \notin \mathsf{Aff}(\boldsymbol{Y}_1, \ldots, \boldsymbol{Y}_{k+1})\end{array}\right]$$

  is negligible.
  Note that one can check whether $\boldsymbol{Y}_* \notin \mathsf{Aff}(\boldsymbol{Y}_1, \ldots, \boldsymbol{Y}_{k+1})$ holds or not in polytime by using Check and the a trapdoor td.

Construction of an independent pre-image resistant function. In our signature scheme we use the function $H_{\boldsymbol{A}}(\boldsymbol{Y}) = \mathsf{e}(\boldsymbol{A}, \boldsymbol{Y})$. The algorithms Gen, GenSim, and Check for the function H are as follows. Below, $\mathsf{Setup}(1^\kappa)$ is an algorithm which generates a group description $gk = (p, \mathbb{G}, \mathbb{H}, \mathbb{T}, \mathsf{e})$.

- $\mathsf{Gen}(1^\kappa)$: Compute $gk \leftarrow \mathsf{Setup}(1^\kappa)$ and generates $\boldsymbol{A} \overset{\$}{\leftarrow} \mathbb{G}^n$ randomly, and output $P \leftarrow (gk, \boldsymbol{A})$.
- $\mathsf{GenSim}(1^\kappa)$: Compute $gk \leftarrow \mathsf{Setup}(1^\kappa)$, and choose randomly $A \overset{\$}{\leftarrow} \mathbb{G}$, and $\boldsymbol{a} \overset{\$}{\leftarrow} \mathbb{Z}_p^n$. Compute $\boldsymbol{A} \leftarrow \boldsymbol{a}A$. Choose randomly $\boldsymbol{Y} \overset{\$}{\leftarrow} \mathbb{H}$, $t \overset{\$}{\leftarrow} \mathbb{Z}_p$, and $\boldsymbol{y}_j \in \mathbb{Z}_p^n$ satisfying $\langle \boldsymbol{a}, \boldsymbol{y}_j \rangle = t$ for $j \in [k+1]$. Choose $n-k$ linearly independent vectors $\boldsymbol{e}_1, \ldots, \boldsymbol{e}_{n-k}$ satisfying $\langle \boldsymbol{e}_i, \boldsymbol{y}_j \rangle = 0$ for all $i \in [n-k]$, $j \in [k]$. Output $P = (gk, \boldsymbol{A})$, $td \leftarrow (\boldsymbol{e}_i)_{i \in [n-k]}$, and $\boldsymbol{Y}_j \leftarrow \boldsymbol{y}_j \boldsymbol{Y}$ for $j \in [k]$, $T \leftarrow t\boldsymbol{Y}$.
- $\mathsf{Check}(P, td, (T, \boldsymbol{Y}_1, \ldots, \boldsymbol{Y}_{k+1}), \boldsymbol{Y}')$: Parse P and td as $(gk, \boldsymbol{A})$ and $(\boldsymbol{e}_i)_{i \in [n-k]}$ respectively. Output 1 iff $\mathsf{e}(\boldsymbol{A}, \boldsymbol{Y}') = T$ and $\langle \boldsymbol{e}_i, \boldsymbol{Y}' \rangle = 0$ holds for all $i \in [n-k]$.

Proposition 4. *For any n and $k \leq n-2$, the scheme* $(\mathsf{Setup}, H, \mathsf{GenSim}, \mathsf{Check})$ *is (n,k)-independent pre-image resistant under the SXDH assumption.*

Correctness is straightforward. We give the proof of independent pre-image resistance in the full paper.

4 Proposed Scheme

Let $n \geq 3$ and m be integers. Let Setup be a polytime algorithm that generates a group description $gk = (p, \mathbb{G}, \mathbb{H}, \mathbb{T}, \mathsf{e})$, as discussed above, where $\mathsf{e} : \mathbb{G} \times \mathbb{H} \to \mathbb{T}$. For $\mathcal{H} = (\boldsymbol{H}_0, \boldsymbol{H}_1, \ldots, \boldsymbol{H}_m) \in (\mathbb{H}^2)^{m+1}$ and $M \in \{0,1\}^m$, we define a Water's hash function h as

$$h_{gk}(\mathcal{H}, M) = \boldsymbol{H}_0 + \sum_{k \in [m]} M_k \boldsymbol{H}_k,$$

where M_k is the k-th bit of M. Let Prf and Vrf be the proof algorithm and the verification algorithm of the Groth-Sahai proof system reviewed in Section 2.3. Our signature scheme $\mathcal{SGN} = (\mathsf{Kg}, \mathsf{Update}, \mathsf{Sig}, \mathsf{Ver})$ works as follows.

Key Generation $\mathsf{Kg}(1^\kappa)$**:** $gk \leftarrow (p, \mathbb{G}, \mathbb{H}, \mathbb{T}, \mathsf{e}) \leftarrow \mathsf{Setup}(1^\kappa)$, $\boldsymbol{G} \leftarrow \mathbb{H}^2, \mathcal{H} \leftarrow (\boldsymbol{H}_0, \boldsymbol{H}_1, \ldots, \boldsymbol{H}_m) \leftarrow (\mathbb{H}^2)^{m+1}$.

Randomly select $A \stackrel{\$}{\leftarrow} \mathbb{G}$, $Q \stackrel{\$}{\leftarrow} \mathbb{H}$, and $\boldsymbol{a}, \boldsymbol{q} \stackrel{\$}{\leftarrow} \mathbb{Z}_p^n$ satisfying $\langle \boldsymbol{a}, \boldsymbol{q} \rangle = 0$ and compute $\boldsymbol{A} \leftarrow \boldsymbol{a}A$, $\boldsymbol{Q} \leftarrow \boldsymbol{q}Q$. Select $\boldsymbol{W}^{[0]} \stackrel{\$}{\leftarrow} \mathbb{H}^n$ randomly, compute $T \leftarrow \mathsf{e}(\boldsymbol{A}, \boldsymbol{W}^{[0]})$, and outputs $pk \leftarrow (gk, \boldsymbol{G}, \mathcal{H}, \boldsymbol{A}, T, \boldsymbol{Q})$ and $sk^{[0]} \leftarrow \boldsymbol{W}^{[0]}$.

Key Update $\mathsf{Update}_{pk}(sk^{[i]})$**:** Parse pk and $sk^{[i]}$ as $(gk, \boldsymbol{G}, \mathcal{H}, \boldsymbol{A}, T, \boldsymbol{Q})$ and $\boldsymbol{W}^{[i]}$ respectively, select $s \stackrel{\$}{\leftarrow} \mathbb{Z}_p$ randomly, and output $sk^{[i+1]} \leftarrow \boldsymbol{W}^{[i+1]} \leftarrow \boldsymbol{W}^{[i]} + s\boldsymbol{Q}$.

Signing $\mathsf{Sig}(sk^{[i]}, M)$ **for** $M \in \{0,1\}^m$**:** Parse pk and $sk^{[i]}$ as $(gk, \boldsymbol{G}, \mathcal{H}, \boldsymbol{A}, T, \boldsymbol{Q})$ and $\boldsymbol{W}^{[i]}$. Compute $\boldsymbol{H}_M \leftarrow h_{gk}(\mathcal{H}, M)$, set $crs_M \leftarrow (\boldsymbol{G}, \boldsymbol{H}_M)$, and $\sigma \leftarrow \mathsf{Prf}(gk, crs_M, (\boldsymbol{A}, T), \boldsymbol{W}^{[i]})$ and output σ.

Verification $\mathsf{Ver}(pk, M, \sigma)$**:** Parse pk as $(gk, \boldsymbol{G}, \mathcal{H}, \boldsymbol{A}, T, \boldsymbol{Q})$, compute $\boldsymbol{H}_M \leftarrow h_{gk}(\mathcal{H}, M)$, and set $crs_M \leftarrow (\boldsymbol{G}, \boldsymbol{H}_M)$. If $\mathsf{Ver}(gk, crs_M, (\boldsymbol{A}, T), \sigma) = 1$, output 1. Otherwise, output 0.

Theorem 5. *For any constants $c > 0$ and any $\gamma = \Theta(1/\sqrt{\kappa})$, the proposed scheme $\mathcal{SIG}$ is $(\rho_G, \rho_U, \rho_M, \rho_S)$-EU-CMA-CML secure under the SXDH assumption. Here*

$$(\rho_G, \rho_U, \rho_M, \rho_S) = \left(\frac{c \cdot \log k}{n \log p}, \frac{c \cdot \log k}{n \log p}, 1 - \frac{2+\gamma}{n}, 1 - \frac{2+\gamma}{n} \right).$$

We can achieve the fraction $1 - o(1)$ of leakage in signing and in memory by setting $n = \kappa$.

4.1 Overview of Security Analysis

Our proof starts with a reduction that shows how to convert any adversary that obtains leakage on key generation and updates, to an adversary that does not require such leakage. This follows from the lemma of Brakerski *et al* [BKKV10]. (See our full paper for the proof.)

Lemma 1 (Informally given in [BKKV10]). *Let $\mathcal{SGN} = (\mathsf{Kg}, \mathsf{Sig}, \mathsf{Ver}, \mathsf{Update})$ be a $(0, 0, \rho_M, \rho_S)$-EU-CMA-CML secure signature scheme. Then if $\rho_M = \omega(\log n)$, it is also $(c \log \kappa/m, c \log \kappa/m, \rho_M, \rho_S)$-EU-CMA-CML secure for any c. Here m is the maximum of the length of secret key.*

The proof of Theorem 5 then proceeds by a sequence of games (depicted in Fig.2):

1. In games 1-3 we follow a standard argument about the properties of Waters' hash. Specifically, we show that with non-negligible probability the common reference strings determined by the messages that the adversary submits in

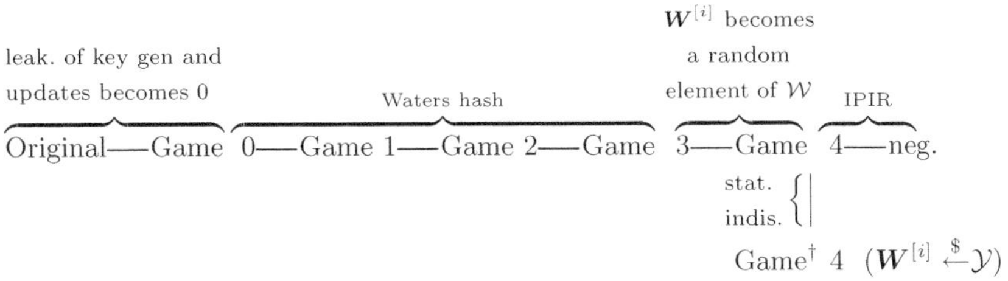

Fig. 2. Games in the reduction

signing queries are hiding CRS (and therefore hide the witness perfectly), and the CRS of the forgery is binding (and therefore the witness can be extracted).

This part of discussion of the our proof is essentially the same as that of [W05] (simplified by [BR09]).

2. In game 4, we change the way that secret keys are generated. Instead of being generated by the update algorithm, the secret key is now randomly chosen from an n-3 dimensional subspace $\mathcal{W}$ of the set $\mathcal{Y} = \{W | \mathsf{e}(A, W) = T\}$ of valid secret keys (which is of dimension $n - 1$).

   Indistinguishability with game 3 follows from the DDH assumption on the group $\mathbb{H}$.
3. Game† 4 is described only to prove a specific property of Game 4 that we need. In Game† 4 the keys are chosen randomly from the space $\mathcal{Y}$ of all valid secret keys.

   We rely on the perfect witness hiding of Groth-Sahai proof and a lemma from [BKKV10] to show that game 4 and Game† 4 are *statistically* indistinguishable.

   We then obtain the property of Game† 4 that the subspace $\mathcal{W}$ is information theoretically hidden, and this property transfers to Game 4 due to the indistinguishability of the two games.
4. Finally, in Game 4 we use the fact that $\mathcal{W}$ is information theoretically hidden from the adversary to argue that the witness W^* extracted from the forgery the an adversary will almost certainly be an element of $\mathcal{Y} \setminus \mathcal{W}$.

   This allows us to violate the independent pre-image resistance of the hash function $H_A(Y) = \mathsf{e}(A, Y)$, because we can find a pre-image W^* of T under H_A, and that pre-image is independent from the set of known vectors $\mathcal{W}$.

5 Conclusion

In this work, we propose a signature scheme that protects against (1) continual memory leakage combined with (2) all types of continual processing (computational) leakage, namely leakage from key generation, key updates, and signing. Our scheme remains secure even when the leakage during signing is a function $f(sk, r)$ of both the secret key and the randomness.

The security of our scheme is proven in the standard model. Moreover, the amount of information that our scheme is allowed to leak during each time period is optimal, in the sense that our scheme remains secure even if $1 - o(1)$ fraction of the secret key of each time period is leaked.

References

[AFGKHO10] Abe, M., Fuchsbauer, G., Groth, J., Haralambiev, K., Ohkubo, M.: Structure-Preserving Signatures and Commitments to Group Elements. In: Rabin, T. (ed.) CRYPTO 2010. LNCS, vol. 6223, pp. 209–236. Springer, Heidelberg (2010)

[AGV09] Akavia, A., Goldwasser, S., Vaikuntanathan, V.: Simultaneous Hardcore Bits and Cryptography against Memory Attacks. In: Reingold, O. (ed.) TCC 2009. LNCS, vol. 5444, pp. 474–495. Springer, Heidelberg (2009)

[ADW09] Alwen, J., Dodis, Y., Wichs, D.: Leakage-Resilient Public-Key Cryptography in the Bounded-Retrieval Model. In: Halevi, S. (ed.) CRYPTO 2009. LNCS, vol. 5677, pp. 36–54. Springer, Heidelberg (2009)

[BSW10] Boyle, E., Segev, G., Wichs, D.: Fully Leakage-Resilient Signatures. eprint archive (2010/488) (2010)

[BB05] Brumley, D., Boneh, D.: Remote timing attacks are practical. Computer Networks 48(5), 701–716 (2005)

[BR09] Bellare, M., Ristenpart, T.: Simulation without the Artificial Abort: Simplified Proof and Improved Concrete Security for Waters' IBE Scheme. In: Joux, A. (ed.) EUROCRYPT 2009. LNCS, vol. 5479, pp. 407–424. Springer, Heidelberg (2009)

[BFO08] Boldyreva, A., Fehr, S., O'Neill, A.: On Notions of Security for Deterministic Encryption, and Efficient Constructions without Random Oracles. In: Wagner, D. (ed.) CRYPTO 2008. LNCS, vol. 5157, pp. 335–359. Springer, Heidelberg (2008)

[BSS99] Blake, I.F., Seroussi, G., Smart, N.P.: Elliptic Curves in Cryptography. London Mathematical Society, vol. 265. Cambridge University Press, Cambridge (1999)

[BB04] Boneh, D., Boyen, X.: Secure Identity Based Encryption Without Random Oracles. In: Franklin, M. (ed.) CRYPTO 2004. LNCS, vol. 3152, pp. 443–459. Springer, Heidelberg (2004)

[BG10] Brakerski, Z., Goldwasser, S.: Circular and Leakage Resilient Public-Key Encryption under Subgroup Indistinguishability. In: Rabin, T. (ed.) CRYPTO 2010. LNCS, vol. 6223, pp. 1–20. Springer, Heidelberg (2010)

[BKKV10] Brakerski, Z., Kalai, Y.T., Katz, J., Vaikuntanathan, V.: Overcoming the Hole in the Bucket: Public-Key Cryptography Resilient to Continual Memory Leakage. In: FOCS 2010 (2010)

[DKL09] Dodis, Y., Kalai, Y.T., Lovett, S.: On cryptography with auxiliary input. In: STOC 2009, pp. 621–630 (2009)

[DHLW10] Dodis, Y., Haralambiev, K., Lopez-Alt, A., Wichs, D.: Cryptography Against Continuous Memory Attacks. In: FOCS 2010 (2010)

[DHLW10b] Dodis, Y., Haralambiev, K., Lopez-Alt, A., Wichs, D.: Efficient Public-Key Cryptography in the Presence of Key Leakage. Cryptology ePrint Archive, Report 2010/154

[DGK+10] Dodis, Y., Goldwasser, S., Kalai, Y.T., Peikert, C., Vaikuntanathan, V.: Public-Key Encryption Schemes with Auxiliary Inputs. In: Micciancio, D. (ed.) TCC 2010. LNCS, vol. 5978, pp. 361–381. Springer, Heidelberg (2010)

[DP08] Dziembowski, S., Pietrzak, K.: Leakage-Resilient Cryptography. In: FOCS 2008, pp. 293–302 (2008)

[DS05] Dodis, Y., Smith, A.: Correcting errors without leaking partial information. In: STOC 2005, pp.654–663 (2005)

[DP10] Dodis, Y., Pietrzak, K.: Leakage-Resilient Pseudorandom Functions and Side-Channel Attacks on Feistel Networks. In: Rabin, T. (ed.) CRYPTO 2010. LNCS, vol. 6223, pp. 21–40. Springer, Heidelberg (2010)

[FKPR10] Faust, S., Kiltz, E., Pietrzak, K., Rothblum, G.N.: Leakage-Resilient Signatures. In: Micciancio, D. (ed.) TCC 2010. LNCS, vol. 5978, pp. 343–360. Springer, Heidelberg (2010)

[FRR+10] Faust, S., Rabin, T., Reyzin, L., Tromer, E., Vaikuntanathan, V.: Protecting Circuits from Leakage: the Computationally-Bounded and Noisy Cases. In: Gilbert, H. (ed.) EUROCRYPT 2010. LNCS, vol. 6110, pp. 135–156. Springer, Heidelberg (2010)

[FS86] Fiat, A., Shamir, A.: How to prove yourself: Practical solutions to identification and signature problems. In: Odlyzko, A.M. (ed.) CRYPTO 1986. LNCS, vol. 263, pp. 186–194. Springer, Heidelberg (1987)

[FLS90] Feige, U., Lapidot, D., Shamir, A.: Multiple Non-Interactive Zero Knowledge Proofs Based on a Single Random String (Extended Abstract). In: FOCS 1990, pp. 308–317 (1990)

[GPS08] Galbraith, S.D., Paterson, K.G., Smart, N.P.: Smart: Pairings for cryptographers. Discrete Applied Mathematics 156(16), 3113–3121 (2008)

[GR10] Goldwasser, S., Rothblum, G.N.: Securing Computation against Continuous Leakage. In: Rabin, T. (ed.) CRYPTO 2010. LNCS, vol. 6223, pp. 59–79. Springer, Heidelberg (2010)

[GSW10] Ghadafi, E., Smart, N.P., Warinschi, B.: Groth–sahai proofs revisited. In: Nguyen, P.Q., Pointcheval, D. (eds.) PKC 2010. LNCS, vol. 6056, pp. 177–192. Springer, Heidelberg (2010)

[GS08] Groth, J., Sahai, A.: Efficient Non-interactive Proof Systems for Bilinear Groups. In: Smart, N.P. (ed.) EUROCRYPT 2008. LNCS, vol. 4965, pp. 415–432. Springer, Heidelberg (2008)

[HSH+08] Halderman, J.A., Schoen, S.D., Heninger, N., Clarkson, W., Paul, W., Calandrino, J.A., Feldman, A.J., Appelbaum, J., Felten, E.W.: Lest We Remember: Cold Boot Attacks on Encryption Keys. In: USENIX Security Symposium 2008, pp.45–60 (2008)

[JV10] Juma, A., Vahlis, Y.: Protecting Cryptographic Keys against Continual Leakage. In: Rabin, T. (ed.) CRYPTO 2010. LNCS, vol. 6223, pp. 41–58. Springer, Heidelberg (2010)

[ISW03] Ishai, Y., Sahai, A., Wagner, D.: Private Circuits: Securing Hardware against Probing Attacks. In: Boneh, D. (ed.) CRYPTO 2003. LNCS, vol. 2729, pp. 463–481. Springer, Heidelberg (2003)

[KJJ98] Kocher, P., Jaffe, J., Jun, B.: Introduction to Differential Power Analysis and Related Attacks (1998), `http://www.cryptography.com/dpa/technical/`

[KJJ99] Kocher, P.C., Jaffe, J., Jun, B.: Differential Power Analysis. In: Wiener, M. (ed.) CRYPTO 1999. LNCS, vol. 1666, pp. 388–397. Springer, Heidelberg (1999)

[KV09] Katz, J., Vaikuntanathan, V.: Signature Schemes with Bounded Leakage Resilience. In: Matsui, M. (ed.) ASIACRYPT 2009. LNCS, vol. 5912, pp. 703–720. Springer, Heidelberg (2009)

[MR04] Micali, S., Reyzin, L.: Physically Observable Cryptography (Extended Abstract). In: Naor, M. (ed.) TCC 2004. LNCS, vol. 2951, pp. 278–296. Springer, Heidelberg (2004)

[NS09] Naor, M., Segev, G.: Public-Key Cryptosystems Resilient to Key Leakage. In: Halevi, S. (ed.) CRYPTO 2009. LNCS, vol. 5677, pp. 18–35. Springer, Heidelberg (2009)

[P09] Pietrzak, K.: A Leakage-Resilient Mode of Operation. In: Joux, A. (ed.) EUROCRYPT 2009. LNCS, vol. 5479, pp. 462–482. Springer, Heidelberg (2009)

[QS01] Quisquater, J.-J., Samyde, D.: ElectroMagnetic Analysis (EMA): Measures and Counter-Measures for Smart Cards. In: Attali, S., Jensen, T. (eds.) E-smart 2001. LNCS, vol. 2140, pp. 200–210. Springer, Heidelberg (2001)

[R97] Rivest, R.L.: All-or-Nothing Encryption and the Package Transform. In: Biham, E. (ed.) FSE 1997. LNCS, vol. 1267, pp. 210–218. Springer, Heidelberg (1997)

[S79] Shamir, A.: How to Share a Secret. Commun. ACM 22(11), 612–613 (1979)

[SMY09] Standaert, F.-X., Malkin, T., Yung, M.: A Unified Framework for the Analysis of Side-Channel Key Recovery Attacks. In: Joux, A. (ed.) EUROCRYPT 2009. LNCS, vol. 5479, pp. 443–461. Springer, Heidelberg (2009)

[W05] Waters, B.: Efficient Identity-Based Encryption Without Random Oracles. In: Cramer, R. (ed.) EUROCRYPT 2005. LNCS, vol. 3494, pp. 114–127. Springer, Heidelberg (2005)

After-the-Fact Leakage in Public-Key Encryption

Shai Halevi[1] and Huijia Lin[2,⋆]

[1] IBM Research
shaih@alum.mit.edu
[2] Cornell University
huijia@cs.cornell.edu

Abstract. What does it mean for an encryption scheme to be leakage-resilient? Prior formulations require that the scheme remains semantically secure even in the presence of leakage, but only considered leakage that occurs *before the challenge ciphertext is generated*. Although seemingly necessary, this restriction severely limits the usefulness of the resulting notion.

In this work we study after-the-fact leakage, namely leakage that the adversary obtains after seeing the challenge ciphertext. We seek a "natural" and realizable notion of security, which is usable in higher-level protocols and applications. To this end, we formulate *entropic leakage-resilient PKE*. This notion captures the intuition that as long as the entropy of the encrypted message is higher than the amount of leakage, the message still has some (pseudo) entropy left. We show that this notion is realized by the Naor-Segev constructions (using hash proof systems).

We demonstrate that entropic leakage-resilience is useful by showing a simple construction that uses it to get semantic security in the presence of after-the-fact leakage, in a model of bounded memory leakage from a split state.

1 Introduction

In the traditional view of cryptography, some parts of the system are designated as secret, and these parts are kept beyond reach for the attackers and only interact with the non-secret parts via well defined interfaces under the control of the designers. In contrast, in reality many times attackers can "design their own interfaces" for accessing the secret state. For example, they may get parts of the secret state via a myriad of side-channels (e.g., timing, radiation, etc.), read it off some backup tape or physical memory, or maybe bribe people who have access to parts of the secret state (or install a virus on their machines). Recent years saw many advances in our ability to reason formally about such unintended leakage and to construct schemes that resist broad class of leakage attacks (e.g., [13,15,9,1,16,3,14,10] and others). This line of work is typically referred to as *leakage-resilient cryptography*.

⋆ Lin is supported by a Microsoft Research Fellowship.

Y. Ishai (Ed.): TCC 2011, LNCS 6597, pp. 107–124, 2011.

The general theme in formulating leakage resilience of some primitive is that in addition to the usual interfaces that are available by design, the adversary can also choose arbitrary leakage functions (from some broad class) and get the result of applying these functions to the secret state of the scheme. We then require that the scheme still meets the original notion of security, even in the presence of this more powerful adversary. This approach was successfully applied to model leakage resilience of many cryptographic schemes, such as pseudorandom generators, signature schemes, etc.

The same approach was also applied to leakage-resilience of encryption schemes, e.g., in [1,16,3], but here there seems to be a problem: Basic notions of security for encryption schemes require that they hide the content of the plaintext even from an adversary that knows many things about that plaintext. In particular, semantic security of encryption requires that an adversary that knows two messages m_0, m_1 and sees a ciphertext c encrypting one of them, will not be able to tell which of the two messages is encrypted in c. But if we let the adversary learn arbitrary functions of the secret key, then it could ask for a function that decrypts the "challenge ciphertext" c and outputs 0 if it is decrypted to m_0 and 1 otherwise. In other words, *given the challenge ciphertext the adversary can design a leakage function that will leak to it exactly the one bit that we try to hide using encryption.*

Prior work on leakage-resilient PKE [1,16,3] bypassed this definitional difficulty by only considering before-the-fact leakage. Namely, the adversary could only ask for leakage on the secret key before it sees the challenge ciphertext, and the scheme was deemed leakage resilient if it remained semantically secure in face of such leakage. This approach indeed bypasses the technical problem, but pays dearly in terms of the meaning and applicability of the resulting notions. Indeed this solution means that as soon as even one bit of the secret key is leaked, we cannot say anything about the secrecy of any message that was encrypted before that bit was leaked.

Consider for example trying to address memory leakage (such as the "cold boot attacks" [11]) by using leakage-resilient encryption. In a memory-leakage attack, an attacker may get a laptop where the disk is encrypted. Lacking the password to access the decryption functionality, the attacker may try to read the decryption key directly off the physical memory. In this setting, the adversary could first see the encrypted disk (hence getting access to the ciphertext), and then try to design a method of measuring the memory specifically for the purpose of decrypting this ciphertext. Existing notions of before-the-fact leakage-resilient encryption say nothing about the secrecy of the plaintext under this attack. This definitional problem was acknowledged in prior work, but no solutions were offered. For example, Naor and Segev wrote in [16] that "It will be very interesting to find an appropriate framework that allows a certain form of challenge-dependent leakage."

1.1 Our Contributions

In this work we study after-the-fact leakage, where the adversary obtains leakage information after seeing the challenge ciphertext. Our main contribution is formulating the notion of *entropic leakage-resilient PKE* and showing how to meet it. Intuitively, this notion says that even if the adversary designs its leakage function according to the challenge ciphertext to leak the things it wants to know, if it only leaks k bits then it cannot "amplify" them to learn more than k bits about the plaintext. Technically, our notion can be viewed as an extension of HILL entropy [12] to the interactive setting. Namely, our notion would say that the message still looks like it has some min-entropy, even to the interactive adversary that participated in the game of semantic-security with leakage.

We remark that this notion is not trivial: Indeed it is not hard to construct "contrived" encryption schemes that are semantically secure (even with respect to before-the-fact leakage), but such that leaking (say) $\sqrt{n}$ bits after the fact lets the adversary recover n bits of plaintext. On the other hand, we show that the same construction that Naor and Segev used in [16] for obtaining leakage-resilient encryption from hash proof systems, in fact realizes also the stronger notion of entropic leakage-resilience relative to after-the-fact leakage.

To demonstrate the usefulness of entropic leakage-resilience we show that in some cases it can be used to get full semantic security, even in the presence of after-the-fact leakage. For this, we of course have to limit the type of leakage functions that the adversary has access to. Specifically, we consider a model where the key is broken into several parts, and the adversary can get access to leakage from every part separately, but not to a global leakage from the entire secret state. (This model is often used in conjunction with the only-computation-leaks axiom of Micali and Reyzin [15].)

To get semantic security in that model, we use two instances of an entropic leakage resilient encryption scheme. To encrypt a message m, we choose two random long strings x_1, x_2, encrypt each x_i under a different copy of the entropic scheme, and use a two-source extractor to compute $Ext2(x_1, x_2) \oplus m$. To decrypt we recover the two strings x_1, x_2 (which we can do by working separately with the two secret keys) and then recover m. The intuition is that as long as the adversary can only get leakage functions from the two keys separately, the entropic leakage resilience of the underlying scheme implies that x_1, x_2 still have a lot of entropy, and hence $Ext2(x_1, x_2)$ still hides the message. (We remark that we view this construction more as an example to the usefulness of our new notion than as a stand-alone application.)

Discussion: on defining useful leakage primitives. On some level, our notion of entropic leakage-resilience departs from the usual theme described above for defining leakage-resilience primitives. Namely, we no longer insist that the scheme retains its original security properties even in the face of leakage. In the face of the impossibility of achieving the strong notion of semantic security, we are willing to settle on a weaker achievable notion so long as it is useful in higher-level applications. It is interesting to formulate such useful weakened notions also for other primitives, such as commitment, key-agreement, etc.

In this context we note that when thinking about encryption as part of a communication system, our notion only captures leakage at the receiver side (i.e., from the secret key) and not at the sender side (i.e., from the encryption randomness). It is interesting to find ways of simultaneously addressing leakage at both ends.

1.2 Recent Related Work

We mention that two recent works by Goldwasser and Rothblum [10], and Juma and Vahlis [14] implicitly also considered after-the-fact leakage for encryption schemes. They presented general methods for compiling any circuit with secret components into one that resists continuous leakage (in the only-computation-leaks model), using leakage-free hardware. Their transformations use as a technical tool encryptions schemes that remain semantically secure even at the presence of after-the-fact leakage of the secret key, provided that the adversary sees *only part of the challenge ciphertext.* (Such notion of semantic-security with respect to adversaries that cannot see the entire challenge ciphertext, if defined as a stand-alone primitive, could be another example of a useful weaker leakage primitive.)

2 Preliminaries

We denote random variables by uppercase English letters. For a random variable A we (slightly) abuse notation and denote by A also the probability distribution on the support of this variable. We write $A \in D$ to denote that A is drawn from domain D. We use U_t to denote the uniform distribution on t-bit binary strings. We write $x \leftarrow A$ to denote the random variable A assuming the value x. We will rely on the following fact about independent random variables, which is proved in the full version of [8].

Lemma 1. *Let A, B be independent random variables and consider a sequence $V_1, \ldots, V_m$ of random variables, where for some function ϕ, $V_i = \phi(V_1, \ldots, V_{i-1}, C_i)$, with each C_i being either A or B. Then A and B are independent conditioned on $V_1, \ldots, V_m$.*

2.1 Min-entropy and Average Min-entropy

The statistical distance between two random variables A and B over a finite domain Ω is $\mathsf{SD}(A,\ B) = \frac{1}{2}\sum_{\omega \in \Omega} |\Pr[A = \omega] - \Pr[B = \omega]|$. Two variables are ε-close if their statistical distance is at most ε. The min-entropy of a random variable A is $H_\infty(A) = -\log(\max_x \Pr[A = x])$. The notion of average min-entropy, formalized in [6], captures the remaining unpredictability of the random variable A conditioned on the value of another random variable B. Formally,

$$\begin{aligned} \tilde{H}_\infty(A|B) &= -\log\left(\mathbb{E}_{y \leftarrow B}\left[\max_x \Pr\left[A = x \mid B = y\right]\right]\right) \\ &= -\log\left(\mathbb{E}_{y \leftarrow B}\left[2^{-H_\infty(A|B=y)}\right]\right) \end{aligned}$$

We will rely on the following useful properties of average min-entropy.

Lemma 2 ([6]). *Let A, B, C be random variables. If B has at most 2^λ possible values, then $\tilde{H}_\infty(A|(B,C)) \geq \tilde{H}_\infty(A|C) - \lambda$.*

Lemma 3. *Let A be a random variable with domain Ω, and U the random variable describing a uniformly sampled element from Ω; and let B a random variable. For any $\varepsilon \in [0,1]$, if $\mathsf{SD}\big((A,B),\ (U,B)\big) \leq \varepsilon$, then $\tilde{H}_\infty(A|B) \geq -\log\left(\frac{1}{|\Omega|} + \varepsilon\right)$.*

This lemma follows directly from the definition of average min-entropy; we omit the proof here.

2.2 Seeded Extractors

Definition 1 ([18]). *A function $Ext : \{0,1\}^n \times \{0,1\}^r \to \{0,1\}^m$ is a (worst-case) (k,ε)-strong extractor if for every random variable $A \in \{0,1\}^n$, such that, $H_\infty(A) \geq k$, it holds that, $\mathsf{SD}\big((Ext(A,S),S),\ (U_m,S)\big) \leq \varepsilon$, where S is uniform on $\{0,1\}^r$.*

Dodis et al. [6] generalized the definition above to the setting of average min-entropy, and showed the following generalized variant of the leftover hash lemma, stating that any family of pairwise independent hash functions is an average-case strong extractor.

Definition 2. *A function $Ext : \{0,1\}^n \times \{0,1\}^r \to \{0,1\}^m$ is an average-case (k,ε)-strong extractor if for all pairs of random variables A and B, such that, $A \in \{0,1\}^n$ and $\tilde{H}_\infty(A|B) \geq k$, it holds that, $\mathsf{SD}\big((Ext(A,S),S,B),\ (U_m,S,B)\big) \leq \varepsilon$, where S is uniform on $\{0,1\}^r$.*

Lemma 4. *Assume $\big\{H_x : \{0,1\}^n \to \{0,1\}^l\big\}$ is a family of universal hash functions. Then for any random variables A and B, such that $A \in \{0,1\}^n$ and $\tilde{H}_\infty(A|B) \geq m$, $\mathsf{SD}\big((H_x(A),X,B),\ (U_l,X,B)\big) \leq \varepsilon$ whenever $l \leq m - 2\log\frac{1}{\varepsilon}$.*

2.3 Two-Source Extractors

The extractors defined in the last section require the use of a short but *truly random* seed, which sometimes might be hard to obtain. The notion of two-source extractor [19,20,4] eliminates the use of a truly random seed, and instead extracts random bits from two *independent* sources of randomness.

Definition 3. *A function $Ext2 : (\{0,1\}^t)^2 \to \{0,1\}^m$ is a (worst-case) (s,ε) two-source extractor, if for all independent random variables $A, B \in \{0,1\}^t$ with min-entropy s, it holds that $\mathsf{SD}\big(Ext2(A,B),\ U_m\big) \leq \varepsilon$.*

Definition 4. *A function $Ext2 : (\{0,1\}^t)^2 \to \{0,1\}^m$ is an average-case (s,ε)-two source extractor, if for all random variables $A, B \in \{0,1\}^t$ and C, such that, conditioned on C, A and B are independent and have average min-entropy s, it holds that $\mathsf{SD}\big((Ext2(A,B),C),\ (U_m,C)\big) \leq \varepsilon$.*

It follows from the same proof of Lemma 2.3 in [6] that any worst-case two-source extractor is also an average-case two-source extractor.

Lemma 5. *For any $\delta > 0$, if $Ext2 : (\{0,1\}^t)^2 \to \{0,1\}^m$ is a (worst-case) $(s - \log\frac{1}{\delta}, \varepsilon)$-two-source extractor, then $Ext2$ is also an average-case $(s, \varepsilon + 2\delta)$-two-source extractor.*

2.4 Hash Proof Systems

Hash proof systems were introduced by Cramer and Shoup [5]. We briefly recall the presentation in [16], which views the hash proof systems as key encapsulation mechanisms.

Smooth Projective Hashing. All the notations below should be thought of as relying on an implicit security parameter (and maybe some other system parameters, such as the underlying algebraic groups). Let $\mathcal{SK}, \mathcal{PK}$ be the domains of secret and public keys, let $\mathcal{K}$ be the space of encapsulated symmetric keys, $\mathcal{C}$ be the space of ciphertexts and $\mathcal{V} \subset \mathcal{C}$ be the space of valid ciphertexts.

Let $F = \{F_{sk} : \mathcal{C} \to \mathcal{K}\}_{sk \in \mathcal{SK}}$ be a collection of hash functions with domain $\mathcal{C}$ and range $\mathcal{K}$, and let $\mu : \mathcal{SK} \to \mathcal{PK}$ be a projection function. Let $\mathcal{F}$ be the construction which is described by all these sets, $\mathcal{F} = (\mathcal{SK}, \mathcal{PK}, \mathcal{C}, \mathcal{V}, \mathcal{K}, F, \mu)$.

Definition 5. *The construction $\mathcal{F}$ is a projective hash family if for all $v \in \mathcal{V}$, and for all $sk_1, sk_2 \in \mathcal{SK}$ such that $\mu(sk_1) = \mu(sk_2)$, it holds that $F_{sk_1}(v) = F_{sk_2}(v)$.*

In other words, for all indexes $sk \in \mathcal{SK}$, the actions of F_{sk} on elements in $\mathcal{V}$ are uniquely determined by $\mu(sk)$. On the other hand, for elements not in $\mathcal{V}$, we require the hash function F_{sk} to behave almost "randomly". Formally,

Definition 6. *The construction $\mathcal{F}$ is ε-smooth, if it holds that*

$$\mathsf{SD}\big((pk, c, F_{sk}(c)),\ (pk, c, k)\big) \leq \varepsilon,$$

where $sk \in \mathcal{SK}$, $c \in \mathcal{C}/\mathcal{V}$, and $k \in \mathcal{K}$ are sampled uniformly at random and $pk = \mu(sk)$.

Hash Proof System. A hash proof system is roughly a construction of smooth-projective hash functions with several efficient associated algorithms. Specifically, we assume an efficient parameter-generating algorithm *Param* that given the security parameter outputs the description of $\mathcal{F} = (\mathcal{SK}, \mathcal{PK}, \mathcal{C}, \mathcal{V}, \mathcal{K}, F, \mu)$, such that $\mathcal{V}$ is an NP language and there is an efficient algorithm for sampling $c \leftarrow \mathcal{V}$ together with a witness w. We also assume that there are efficient algorithms for sampling $sk \leftarrow \mathcal{SK}$ and $c \leftarrow \mathcal{C} \setminus \mathcal{V}$.

We also have two algorithms for computing the hash function F_{sk}. One is a *private evaluation algorithm* $Priv(sk, c)$ that on input a secret key $sk \in \mathcal{SK}$ and a ciphertext $c \in \mathcal{C}$, outputs the encapsulated key $k = F_{sk}(c)$. The other is a *public evaluation algorithm* Pub that computes the same given the public key, but only on valid ciphertexts and only when it is also given a witness of validity. Namely, for every $sk \in SK$ and $pk = \mu(sk)$ and for every $c \in \mathcal{V}$ with witness w, it holds that $Pub(pk, c, w) = F_{sk}(c)$.

Cramer and Shoup noted that a hash proof system immediately implies a KEM mechanism, where key-generation consists of running the parameter generating routine, then choosing a random secret key $sk \leftarrow \mathcal{SK}$ and computing the corresponding public key $pk = \mu(sk)$. Encapsulating a key is done by choosing at random $c \leftarrow \mathcal{V}$ together with a witness w. Then the ciphertext is c and the corresponding encapsulated key is computed by the sender using the public evaluation algorithm, setting $k = Pub(pk, c, w)$. On the receiving side, the same key is recovered using the private evaluation algorithm, setting $k = Priv(sk, c)$. Security of this scheme follows from the smoothness of the construction, in conjunction with the hardness subset membership problem, as defined below.

Subset Membership Problem. A hash proof system as above is said to have a hard subset membership problem if a randomly generated valid ciphertext is computationally indistinguishable from a randomly generated invalid ciphertext. Formally, the following two ensembles are indistinguishable

$$\mathsf{VALID} = \{\mathcal{F} = (\mathcal{SK}, \mathcal{PK}, \mathcal{C}, \mathcal{V}, \mathcal{K}, F, \mu) \leftarrow Param(1^n),\ c \leftarrow \mathcal{V}\ :\ (\mathcal{F}, c)\}_{n \in N}$$
$$\mathsf{INVALID} = \{\mathcal{F} = (\mathcal{SK}, \mathcal{PK}, \mathcal{C}, \mathcal{V}, \mathcal{K}, F, \mu) \leftarrow Param(1^n),\ c \leftarrow \mathcal{C} \backslash \mathcal{V}\ :\ (\mathcal{F}, c)\}_{n \in N}$$

3 Entropic Security against After-the-Fact Leakage

Roughly speaking, we say that an encryption scheme is entropic leakage resilient if a message M with high min-entropy still "looks random" to the adversary even after it sees an encryption of it and some leakage information (even if this leakage was obtained after seeing the ciphertext). This is formulated by postulating the existence of a simulator that generates a view, which on one hand is indistinguishable from the real adversary view and on the other hand still leaves M with high min-entropy.

More formally, we define two games, one "real" and the other "simulated". Both games depend on several parameters: k is the a-priory min-entropy of the message, and $\ell_{\text{pre}}, \ell_{\text{post}}$ control the amount of leakage in various parts of the games (namely the pre- and post- challenge-ciphertext leakage). All of these parameters are of course functions of the security parameter n. (For simplicity, in the definition below we assume that the message M is a uniform random k-bit string. This is all we need for our application in Section 4 to the only-computation-leak model, and extending the definition to arbitrary high-min-entropy distributions is easy.)

The "real" game. Given the parameters $(k, \ell_{\text{pre}}, \ell_{\text{post}})$ and the encryption scheme $\Psi = (Gen, Enc, Dec)$, the real game is defined as follows:

Key Generation: The challenger chooses at random a message $m \leftarrow U_k$. The challenger also generates $(sk, pk) \leftarrow Gen(1^n)$, and sends pk to the adversary.

Pre-Challenge Leakage: The adversary makes a pre-challenge leakage query, specifying a function $f^{\mathrm{pre}}(\cdot)$. If the output length of f^{pre} is at most ℓ_{pre} then the challenger replies with $f^{\mathrm{pre}}(sk)$. (Else the challenger ignores the query.)

Challenge: Upon a challenge query, the challenger encrypts the message m and sends the ciphertext $c = Enc(pk, m)$ to the adversary.

Post-Challenge Leakage: The adversary makes a post-challenge leakage query, specifying another function $f^{\mathrm{post}}(\cdot)$. If the output length of f^{post} is at most ℓ_{post} then the challenger replies with $f^{\mathrm{post}}(sk)$. (Else the challenger ignores the query.)

We let $\mathsf{View}_A^{\mathrm{rl}}(\Psi) = (\text{randomness}, pk, f^{\mathrm{pre}}(sk), c, f^{\mathrm{post}}(sk))$ be the random variable describing the view of the adversary A in the game above, and by M^{rl} we denote the message that was chosen at the onset of this game. (We view them as correlated random variables, namely when we write $(M^{\mathrm{rl}}, \mathsf{View}_A^{\mathrm{rl}}(\Psi))$ we mean the joint distribution of the message M^{rl} and A's view in a real game with M^{rl}).

The "simulated" game. In the simulated game we replace the challenger from above by a simulator Simu that interacts with A in any way that it sees fit. Simu gets a uniformly chosen message M^{sm} as input, and it needs to simulate the interaction conditioned on this M^{sm}. The view of A when interacting with S is denoted $\mathsf{View}_A^{\mathrm{sm}}(\mathsf{Simu})$.

Below we say that Ψ is entropic leakage-resilient (with respect to all the parameters) if on one hand the distributions $\mathsf{View}_A^{\mathrm{rl}}(\Psi)$, $\mathsf{View}_A^{\mathrm{sm}}(\mathsf{Simu})$ are indistinguishable even given the message M, and on the other hand M^{sm} has high min-entropy given $\mathsf{View}_A^{\mathrm{sm}}(\mathsf{Simu})$.

Definition 7. *Let $k, \ell_{\mathrm{pre}}, \ell_{\mathrm{post}}$ be parameters as above, and let δ be another "slackness parameter." A public-key encryption scheme $\Psi = (Gen, Enc, Dec)$ is entropic leakage resilient with respect to these parameters if there exists a simulator* Simu, *such that, for every $\mathcal{PPT}$ adversary A the following two conditions hold:*

- *The two ensembles $\left(M^{\mathrm{rl}}, \mathsf{View}_A^{\mathrm{rl}}(\Psi)\right)$, $\left(M^{\mathrm{sm}}, \mathsf{View}_A^{\mathrm{sm}}(\mathsf{Simu})\right)$ (indexed by the security parameter) are computationally indistinguishable.*
- *The average min-entropy of M^{sm} given $\mathsf{View}_A^{\mathrm{sm}}(\mathsf{Simu})$ is*

$$\tilde{H}_\infty(M^{\mathrm{sm}} \mid \mathsf{View}_A^{\mathrm{sm}}(\mathsf{Simu})) \geq k - \ell_{\mathrm{post}} - \delta.$$

Intuitively, in the simulated game the message M^{sm} retains its initial entropy, except for the ℓ_{post} post-challenge leakage bits and possibly also some "overhead" of δ bits. And since the simulated game cannot be distinguished from the real game, then M^{rl} has the same number of pseudo-entropy bits also in the real game.

What happened to ℓ_{pre}? Note that the min-entropy of M^{sm} is only reduced by the amount of the post-challenge leakage (and "overhead") irrespective of the pre-challenge leakage. This is reminiscent of the prior results that can tolerate pre-challenge leakage while maintaining semantic security (hence "not losing any entropy" of the message). Indeed, the security notions from [1,16] can be obtained as a special case of our definition with $\ell_{\mathrm{post}} = \delta = 0$.

Adaptiveness. It was pointed out in [1] that the pre-challenge leakage can be made adaptive without effecting the definition. The same holds also for the post-challenge leakage.

3.1 Constructing Entropic Leakage-Resilient Scheme

We show that the generic construction of Naor and Segev for pre-challenge leakage resilient encryption from hash proof systems [16], is actually entropic secure against bounded after-the-fact leakage. The encryption algorithm (overly simplified) samples a valid ciphertext c of the hash proofs system, and uses the key encapsulated in c to hide the message. To show entropic security, the entropic simulator proceed the same as the encryption algorithm except that it uses invalid ciphertexts. It follows from the indistinguishability of the valid and invalid ciphertexts that the real and the simulated games are indistinguishable. Furthermore, due to smoothness the key encapsulated in an invalid ciphertext has high min-entropy, and hence the message is well "hidden", and has high average min-entropy.

In more details, we need an ε-smooth hash proof system $\mathcal{F} = (\mathcal{SK}, \mathcal{PK}, \mathcal{C}, \mathcal{V}, \mathcal{K}, F, \mu)$, where the symmetric encapsulated keys are assumed (w.l.o.g.) to be just t_1-bit strings, $\mathcal{K} = \{0,1\}^{t_1}$. We also need a function $Ext : \{0,1\}^{t_1} \times \{0,1\}^{t_2} \to \{0,1\}^{t_3}$ which is an average-case (t_4, δ) strong extractor. Namely, it has t_1-bit inputs, t_2-bit seeds and t_3-bit outputs, and for a random seed and input with t_4 bits of min entropy, the output is δ-away from a uniform t_3-bit string. Then, the encryption scheme $\Psi = (Gen, Enc, Dec)$ proceeds as follows:

Key Generation: The key generation algorithm, on input a security parameter 1^n, generates an instance of a projective hash family $\mathcal{F} = (\mathcal{SK}, \mathcal{PK}, \mathcal{C}, \mathcal{V}, \mathcal{K}, F, \mu) \leftarrow Param(1^n)$, samples a secret key $sk \leftarrow \mathcal{SK}$, and computes the corresponding public key $pk = \mu(sk)$.

Encryption: The encryption algorithm, on input a message $m \in \{0,1\}^{t_3}$, samples a valid ciphertext together with a corresponding witness $(c, w) \leftarrow \mathcal{V}$, and computes the encapsulated key k using the public evaluation algorithm, i.e., $k = Pub(pk, c, w)$. It then samples a random seed $s \in \{0,1\}^{t_2}$, and computes $\psi = Ext(k, s) \oplus m$. Finally, it outputs the ciphertext $\hat{c} = (c, s, \psi)$.

Decryption: The decryption algorithm on input a ciphertext (c, s, ψ), computes the encapsulated key k using the private evaluation algorithm, i.e., $k = Priv(sk, c)$, and outputs the message $m = Ext(k, s) \oplus \psi$.

It follows using the same proof as in [16] that the encryption scheme Ψ is a correct public-key encryption scheme. Namely the decryption algorithm always recovers the original message m correctly. Next, we proceed to prove the entropic leakage resilience of Ψ against after-the-fact leakage.

Lemma 6. *The public-key encryption scheme Ψ from above is entropic leakage-resilient with respect to leakage $\ell_{\text{pre}}, \ell_{\text{post}}$ and "overhead" δ', as long as these parameters satisfy the following constraints:*

$$\ell_{\text{pre}} \leq \log\left(\frac{1}{\frac{1}{|\mathcal{K}|} + \varepsilon}\right) - t_4 \text{ and } \delta' \leq t_3 - \log\frac{1}{2^{-t_3} + \delta}$$

To interpret these parameters, it is useful to think of a "very smooth" hash proof system ($\varepsilon \ll 1/|\mathcal{K}| = 2^{-t_1}$), and a very good extractor that can work with inputs that have min-entropy $t_4 \ll \log |\mathcal{K}| = t_1$ and produces outputs whose distance from uniform t_3-bit strings is $\delta < 2^{-t_3}$. For such building blocks we can tolerate pre-challenge leakage of $\ell_{\text{pre}} \approx t_1 - t_4 = t_1(1 - o(1))$, and our overhead is $\delta' < 1$ bits.

Proof. To prove Lemma 6 we need to describe a simulator, whose answers to the adversary are indistinguishable from the real game but at the same time leave many bits of min-entropy in the message m.

In our case, the simulator S proceeds almost identically to the challenger in the real game, except that to generate the ciphertext $\hat{c} = (c, s, \psi)$ it samples an *invalid ciphertext* for the hash-prof system, $c \leftarrow \mathcal{C}/\mathcal{V}$, then it computes $k = Priv(sk, c)$ using the secret key sk that it knows, and outputs the ciphertext $\hat{c} = (c, s, \psi)$, where $\psi = Ext(k, s) \oplus m$.

It follows directly from the indistinguishability of the valid and invalid ciphertexts of the hash proof system that the simulated view is indistinguishable from the real one even given the message m. It only remains to show the min-entropy condition.

On a high-level, the proof consists of two steps. The first step shows that conditioned on all the information that the adversary receives till the end of the Challenge Phase, the message m still has high average min-entropy, namely at least $t_3 - \delta'$.

To see this, note that by ε-smoothness the encapsulated key k has almost t_1 bits of min-entropy even given pk and c, and therefore almost $t_1 - \ell_{\text{pre}}$ bits of min-entropy even given pk, c and the pre-challenge leakage. Specifically k has at least $\log\left(\frac{1}{\frac{1}{|\mathcal{K}|}+\varepsilon}\right) - \ell_{\text{pre}} \geq t_4$ bits of min-entropy, and therefore the bits extracted from k using the extractor Ext are statistically close to random (even given pk, c the pre-challenge leakage and the seed s). Thus the message m is δ-close to a uniform t_3-bit string, even given pk, c, the pre-challenge leakage, the seed s, and the value ψ. (So far this is exactly the same argument as in the proof of the Naor-Segev construction.) Hence upto this phase, the message m has at least $t_3 - \delta'$ bits of min entropy.

Next, by further relying on the fact that the post-challenge leakage is bounded by ℓ_{post} bits, the min-entropy of m is reduced by at most this much, so it retains at least $t_3 - \ell_{\text{post}} - \delta'$ bits of average min-entropy.

4 Semantic Security in a Split-State Model

We next demonstrate how Definition 7 can be used in a "higher level protocol". Specifically, we consider a split-state model, where the secret state of the cryptosystem at hand is partitioned to a few parts and the adversary can obtain leakage of its choice on every part separately but not a global leakage function from the entire secret state.

This model is often used in conjunction with the only-computation-leaks axiom (OCL) of Micali and Reyzin [15]. In our case we talk only about CPA security and there is no decryption oracle, hence after-the-fact there isn't any computation to leak from and talking about only-computation-leaks does not make sense. (An extension of this construction to get CCA-security may be applicable to the OCL model, but this is beyond the scope of the current work.)

Definition 8. *A 2-split-state encryption scheme is a public-key encryption scheme* $\Pi = (Gen, Enc, Dec)$ *that has the following structure:*

- *The secret key consists of a pair of strings* $S = (S_1, S_2)$*, and similarly the public key consists of a pair* $P = (P_1, P_2)$*.*
- *The key generation algorithm Gen consists of two subroutines* Gen_1 *and* Gen_2*, where* Gen_i *for* $i \in \{1,2\}$ *outputs* (P_i, S_i)*.*
- *The decryption algorithm Dec also consists of two partial decryption subroutines* Dec_1 *and* Dec_2 *and a combining subroutine Comb. Each* Dec_i *takes as input the ciphertext and* S_i *and outputs partial decryption* t_i*, and the combining subroutines Comb takes the ciphertext and the pair* (t_1, t_2) *and recovers the plaintext.*

In the split-state model, we assume that information is leaked independently from the two parts. Semantic security for such a 2-split-state scheme in the presence of After-the-Fact Leakage in this model is defined below. Let $\ell_{\text{pre}}, \ell_{\text{post}}$ be parameters as before, and we consider the following game:

Key Generation: The challenger chooses $r_1, r_2 \in \{0,1\}^*$ at random, generates $(sk_b, pk_b) \leftarrow Gen(1^n, r_b)$ for $b = 1, 2$, and sends (pk_1, pk_2) to the adversary.
Pre-Challenge Leakage: The adversary makes an arbitrary number of leakage queries $(f_{1,i}^{\text{pre}}, f_{2,i}^{\text{pre}})$ adaptively. Upon receiving the ith leakage query the challenger sends back $(f_{1,i}^{\text{pre}}(sk_1), f_{2,i}^{\text{pre}}(sk_2))$, provided that the total output length of all the pre-challenge queries so far does not exceed ℓ_{pre} in each coordinate. (Otherwise the challenger ignores the query.)
Challenge: The adversary sends two messages $m_0, m_1 \in \{0,1\}^n$. The challenger chooses a random bit σ, encrypts the message m_σ, and returns the ciphertext $c = Enc(pk, m_\sigma)$.
Post-Challenge Leakage: The adversary can submit an arbitrary number of leakage queries $(f_{1,i}^{\text{post}}, f_{2,i}^{\text{post}})$ adaptively. Upon receiving the ith leakage query the challenger sends back $(f_{1,i}^{\text{post}}(sk_1), f_{2,i}^{\text{post}}(sk_2))$, provided that the total output length of all the post-challenge queries so far does not exceed ℓ_{post} in each coordinate. (Otherwise the challenger ignores the query.)
Output: The adversary outputs a bit σ'.

Definition 9. *A 2-split-state encryption scheme* $\Psi = (Gen, Enc, Dec)$ *is* resilient to $(\ell_{\text{pre}}, \ell_{\text{post}})$ leakage in the split-state model, *if for every* $\mathcal{PPT}$ *adversary A that participates in an experiment as above, there is a negligible function* negl *such that* $\Pr[\sigma' = \sigma] < 1/2 + \mathsf{negl}(n)$.

4.1 Our Construction

As defined above, our 2-split-state scheme maintains a split secret key (S_1, S_2) (and a corresponding split public key (P_1, P_2)), where S_i is generated by Gen_i

and used by Dec_i. Due to the the restriction on the leakage in the split-state model, the adversary can never obtain leakage on S_1 and S_2 jointly, so even after leakage we can hope that each of the two parts still has sufficient entropy *and moreover they are independent.* Hence we can use two-source extractors to get a close-to-uniform string from these two parts, and use it to mask the message.

In the scheme below, we do not try to extract from the secret keys themselves, but rather in each encryption we encrypt two random strings, one under each of the keys, and extract randomness from these two ephemeral random strings. We argue that if the two copies are implemented using an entropic leakage-resilient scheme, then we get semantic security in the split-state model. Intuitively, the reason is that the entropic security ensures that the two ephemeral strings still have high (pseudo)entropy even given the leakage, and the split-state model ensures that they are independent, so the two-source extraction should give us what we want.

The formal proof roughly follows this intuitive reasoning, with just one additional complication, related to adaptivity: In the post-challenge leakage, the adversary can choose the leakage functions from the two key parts after it already saw the value that was extracted from the two random strings, causing a circularity in the argument. We solve this issue essentially by "brute force": We argue below that if the extracted value has only u bits, then the adaptivity issue can increase the advantage of the adversary by at most a 2^u factor, and set our parameters to get around this factor[1].

The construction. Let n be the security parameter, and let u be the bit-length of the messages that we want to encrypt. Also let $t, v, \ell_{\text{pre}}, \ell_{\text{post}}$ be some other parameters (to be defined later). Let $\Psi = (Gen^{\mathsf{Ent}}, Enc^{\mathsf{Ent}}, Dec^{\mathsf{Ent}})$ be a entropic secure encryption for t-bit messages, resilient to leakage $(\ell_{\text{pre}}, \ell_{\text{post}})$, with overhead of one bit.

Also, let $Ext2 : \{0,1\}^t \times \{0,1\}^t \to \{0,1\}^u$ be an average-case (v, ε)-two-source extractor, with $\varepsilon = 2^{-u-\omega(\log n)}$. Namely both inputs to $Ext2$ are of length t, and as long as they are independent and both have more than v min entropy, the output of $Ext2$ is at most ε away from a uniform u-bit string[2]. Given these ingredients, our 2-split-state encryption scheme $\Pi = (Gen, Enc, Dec)$ proceeds as follows:

Key Generation: The key generation algorithm runs two subroutines Gen_1 and Gen_2, where Gen_i for $i \in \{1,2\}$ on input 1^n generates a public and secret key pair $(S_i, P_i) \leftarrow Gen^{\mathsf{Ent}}(1^n)$ of the entropic encryption scheme Ψ. The public key is $P = (P_1, P_2)$ and the secret key is $S = (S_1, S_2)$.

Encryption: The encryption algorithm, on input a message $m \in \{0,1\}^u$, chooses two random strings $x_1, x_2 \in \{0,1\}^t$ and encrypts the two strings using the two public keys P_1 and P_2 respectively; set $c_i = Enc^{\mathsf{Ent}}(P_i, x_i)$. Then, it computes $\psi = Ext2(x_1, x_2) \oplus m$, and outputs the ciphertext $\hat{c} = (c_1, c_2, \psi)$.

[1] We remark that we *do not* make exponential hardness assumptions to achieve this. See proof of Claim 8 for more details.

[2] Note that we set ε so that it remain negligible even if we multiply it by 2^u, this is needed for the adaptivity issue in the proof.

Decryption: The decryption algorithm, on input a ciphertext (c_1, c_2, ψ), executes the following three subroutines sequentially.

- The subroutine Dec_1 decrypts c_1 using S_1 and outputs the plaintext $x_1 = Dec^{\mathsf{Ent}}(S_1, c_1)$.
- The subroutine Dec_2 decrypts c_2 using S_2 and outputs the plaintext $x_2 = Dec^{\mathsf{Ent}}(S_2, c_2)$.
- The subroutine $Comb$ on input x_1, x_2 and ψ, outputs the message $M = Ext2(x_1, x_2) \oplus \psi$.

Lemma 7. *The 2-split-state scheme Π from above is semantically secure with respect to leakage $(\ell'_{\text{pre}}, \ell'_{\text{post}})$ in the split-state model, as long as the parameters satisfy the following constraints:*

$$\ell'_{\text{pre}} \leq \ell_{\text{pre}} \text{ and } \ell'_{\text{post}} \leq \min(\ell_{\text{post}} - u,\ t - v - 1).$$

Proof. We need to show that $\mathcal{PPT}$ adversaries only have negligible advantage in guessing the choice bit σ in the semantic security game. To that end, fix a semantic-security adversary $\mathcal{A}^{\mathsf{ss}}$, and let Simu be the entropic simulator that exists for the underlying entropic scheme Ψ[3]. We now consider the hybrid experiments hyb_1 and hyb_2, which are defined as follows:

Hybrid hyb_1: In this hybrid, the challenger generates the ciphertext c_2 and answers leakage queries against S_2 just as in the real game. However, it uses the entropic simulator Simu to generate the ciphertext c_1 and to answer leakage queries against S_1.

In more details, the challenger chooses x_1, x_2 at random, then generates (S_2, P_2) using the key-generation of Ψ, but it gets P_1 by running $\mathsf{Simu}(x_1)$. (Recall that the entropic simulator expects a random plaintext string in its input.) Then to answer a pre-challenge query $(f_{1,i}^{\text{pre}}, f_{2,i}^{\text{pre}})$, the challenger forward $f_{1,i}^{\text{pre}}$ to Simu and gets the answer from it, computes the answer $f_{2,i}^{\text{pre}}(S_2)$ by itself, and send both answer to $\mathcal{A}^{\mathsf{ss}}$.

When $\mathcal{A}^{\mathsf{ss}}$ makes a challenge query (m_0, m_1), the challenger asks Simu for the first ciphertext c_1, and computes c_2 by itself $c_2 = Enc(P_2, x_2)$. (Recall again that Simu was run with input x_1, so the ciphertext that it returns is supposed to simulate an encryption of x_1.)

Next, **the challenger makes a direct post-challenge leakage query to** Simu, querying with the function $h_1(S_1) = Ext2(Dec(S_1, c_1), x_2)$ (that has u bits of output). Getting some answer r', the challenger just discards that answer, instead computing $r = Ext2(x_1, x_2)$, choosing a random bit σ, setting $\psi = r \oplus m_\sigma$ and sending (c_1, c_2, ψ) to $\mathcal{A}^{\mathsf{ss}}$.

After that, post-challenge queries of $\mathcal{A}^{\mathsf{ss}}$ are handled just like the pre-challenge queries, with the challenger asking Simu for the first part of the answer (for the query against S_1) and computing the answer to the query against S_2 by itself.

---

[3] Our Definition 7 has only one pre- and one post-challenge query. Below we assume for convenience that the entropic-security adversary can make adaptive queries, it was noted in [1] that these definition are equivalent.

Hybrid hyb_2**:** In this hybrid the challenger still chooses x_1, x_2 at random, but now both parts of the game are handled by the simulator, running as $\mathsf{Simu}(x_1)$ to answer the first part of all the queries (and to get c_1) and as $\mathsf{Simu}(x_2)$ to answer the second part of all the queries (and to get c_2).

The challenger makes direct post-challenge queries to both copies of the simulator, asking the first for $r' = h_1(S_1) = Ext2(Dec(S_1, c_1), x_2)$ and the second for $r'' = h_2(S_2) = Ext2(x_1, Dec(S_2, c_2))$. The challenger still ignores both answers, computing instead $r = Ext2(x_1, x_2)$ and setting the ψ component of the Π-ciphertext as $r \oplus m_\sigma$.

Before proceeding with the proof, we point out that the direct post-challenge leakage queries that the challenger makes are expected to return the same value that the challenger computes itself, $r' = r'' = r$. (Indeed we prove below that they almost always do). The reason that the challenger still makes them is to ensure that the entropic simulators see the same queries in these hybrids as in the reductions that we use below. One consequence of these direct queries is that the entropic simulators need to answer more post-challenge queries than what the semantic-security adversary asks. Specifically, it needs to answer u more bits, hence the constraint $\ell'_{\text{post}} \le \ell_{\text{post}} - u$.

We now prove that the event $\sigma' = \sigma$ holds in the hybrids with essentially the same probability as in the real game, by reducing to the indistinguishability property of the entropic simulator.

The hybrid hyb_1. Assume toward contradiction that the event $\sigma' = \sigma$ happens in the real game with probability which is larger than in the first hybrid hyb_1 by a noticeable amount ρ. We describe an entropic adversary $\mathcal{A}^{\mathsf{ent}}$ and distinguisher $\mathcal{D}^{\mathsf{ent}}$ that break this indistinguishability property. (In fact, for the same entropic adversary $\mathcal{A}^{\mathsf{ent}}$ we describe two distinguishes $\mathcal{D}_1^{\mathsf{ent}}$, $\mathcal{D}_2^{\mathsf{ent}}$, and prove that at least one of them has advantage $\rho/2$ or more.)

- The entropic adversary $\mathcal{A}^{\mathsf{ent}}$, on input public key P_1, chooses (P_2, S_2) and x_2 in the same way as the hyb_1 challenger, and sends (P_1, P_2) to the semantic-security adversary $\mathcal{A}^{\mathsf{ss}}$. It then proceeds similarly to the hyb_1 challenger, answering the first part of every query using its oracle and computing the answer to the second part by itself.

  The only difference between $\mathcal{A}^{\mathsf{ent}}$ and the hyb_1 challenger is in the way that the ψ component of the ciphertext is computed. Once $\mathcal{A}^{\mathsf{ent}}$ gets c_1 from its oracle and computes $c_2 = Enc(P_2, x_2)$, it makes a post-challenge leakage query to its oracle asking for $r' = h_1(S_1) = Ext2(Dec(S_1, c_1), x_2)$. Since $\mathcal{A}^{\mathsf{ent}}$ does not have x_1, it does not discard the answer but rather uses it for setting $\psi = r' \oplus m_\sigma$.
- The first distinguisher $\mathcal{D}_1^{\mathsf{ent}}$ gets the view of $\mathcal{A}^{\mathsf{ent}}$, which includes x_2 and r', and also the string x_1 (which was supposed to be encrypted in c_1). $\mathcal{D}_1^{\mathsf{ent}}$ simply verifies that $r' = Ext2(x_1, x_2)$, outputting 1 if they are equal and 0 otherwise.
- The second distinguisher $\mathcal{D}_2^{\mathsf{ent}}$ gets the view of $\mathcal{A}^{\mathsf{ent}}$, which includes σ and σ', and outputs 1 if they are equal and 0 otherwise.

Clearly, if the oracle of $\mathcal{A}^{\mathsf{ent}}$ is the real encryption scheme Ψ then the transcript that $\mathcal{A}^{\mathsf{ss}}$ sees is identical to the real semantic-security game. In particular, the ciphertext c_1 is indeed an encryption of x_1, and therefore we have $r' = Ext2(x_1, x_2)$ with probability 1.

If the oracle of $\mathcal{A}^{\mathsf{ent}}$ is the simulator $\mathsf{Simu}(x_1)$, then we have two possible cases: either the event $r' \neq Ext2(x_1, x_2)$ happens with probability at least $\rho/2$, or it happens with smaller probability. In the first case, the distinguisher $\mathcal{D}_1^{\mathsf{ent}}$ clearly has an advantage at least $\rho/2$ in distinguishing between the real scheme Ψ and the simulator Simu.

In the second case, the transcript that $\mathcal{A}^{\mathsf{ss}}$ sees is the same as in the hybrid hyb_1, except for an event of probability less than $\rho/2$. Since the probability of $\sigma = \sigma'$ in the real game is larger by ρ than this probability in hyb_1, then it is larger by more than $\rho/2$ than this probability in the interaction with $\mathcal{A}^{\mathsf{ent}}$. Hence the distinguisher $\mathcal{D}_2^{\mathsf{ent}}$ has advantage more than $\rho/2$. ■

The hybrid hyb_2. The proof of indistinguishability between hyb_1 and hyb_2 is essentially the same as the proof of indistinguishability between the real game and hyb_1, and is omitted here. ■

The advantage in hyb_2. Having shown that the probability of $\sigma = \sigma'$ in the second hybrid hyb_2 is negligibly close to the probability in the real game, we now proceed to bound it. For that purpose, we consider another mental experiment $\widetilde{\mathsf{hyb}}$ as follows:

Hybrid $\widetilde{\mathsf{hyb}}$**:** hybrid $\widetilde{\mathsf{hyb}}$ proceeds the same as hyb_2, except that, in the Challenge Phase, instead of sending the adversary $\mathcal{A}^{\mathsf{ss}}$ the complete ciphertext $\hat{c} = (c_1, c_2, \psi)$, the challenger sends only c_1 and c_2, and defers sending ψ until after the Post-Challenge Leakage Phase.

Of course, the mental experiment $\widetilde{\mathsf{hyb}}$ is very much distinguishable from hyb_2. Moreover, compared with the adversary in $\widetilde{\mathsf{hyb}}$, the adversary in hyb_2 has the advantage of choosing the leakage functions in the Post-Challenge Leakage Phase based on ψ. Still, we argue that this advantage is limited, up to an exponential factor in u. Namely, we show in Claim 8 that if $\mathcal{A}^{\mathsf{ss}}$ has advantage α in guessing the bit σ in hyb^2, then there is another adversary $\tilde{A}$ that has advantage at least $\alpha/2^u$ in guessing the bit σ in the mental experiment $\widetilde{\mathsf{hyb}}$.

Claim 8. *If for some* $\alpha > 0$ *there exists an adversary* $\mathcal{A}^{\mathsf{ss}}$ *for which* $\Pr_{\mathsf{hyb}_2}[\sigma = \sigma'] \geq \frac{1}{2} + \alpha$*, then there exists another adversary* $\tilde{A}$ *for which* $\Pr_{\widetilde{\mathsf{hyb}}}[\sigma = \sigma'] \geq \frac{1}{2} + \frac{\alpha}{2^u}$.

Proof. We present a generic construction of $\tilde{A}$ given $\mathcal{A}^{\mathsf{ss}}$. The adversary $\tilde{A}$ in $\widetilde{\mathsf{hyb}}$ runs $\mathcal{A}^{\mathsf{ss}}$ internally, and forward messages externally to the Challenger in $\widetilde{\mathsf{hyb}}$. Except that in the Challenge Phase, $\tilde{A}$ randomly chooses some string $\psi' \in \{0,1\}^u$ and sends it to $\mathcal{A}^{\mathsf{ss}}$ in lieu of ψ. Later, when $\tilde{A}$ gets the "real ψ" from the challenger, it aborts if it guessed wrong, $\psi' \neq \psi$, and proceeds just like $\mathcal{A}^{\mathsf{ss}}$ if the guess was correct. Since the guess is correct with probability 2^{-u}, it follows that the advantage of $\tilde{A}$ is exactly $\alpha/2^u$.

The advantage in $\widetilde{\mathsf{hyb}}$. We are now ready to use the min-entropy property of the simulator S to prove that the advantage pf $\tilde{A}$ in $\widetilde{\mathsf{hyb}}$ is at most 2ε. Since we set $\varepsilon = 2^{-u-\omega(\log n)}$, then by Claim 8 it follows that the advantage of $\mathcal{A}^{\mathsf{ss}}$ in hyb_2 is at most $2\varepsilon \cdot 2^u = 2^{1-\omega(\log n)} = \mathsf{negl}(n)$, as needed.

In the mental experiment $\widetilde{\mathsf{hyb}}$, let Γ be the (partial) transcript of messages that $\tilde{A}$ receives till the end of the Post-Challenge Leakage Phase (i.e., before it gets ψ). We show that the average min-entropy of each of the two seeds x_1, x_2, conditioned on Γ is at least v. Let $\Gamma = (\Gamma_1, \Gamma_2)$, where Γ_1 denote the partial transcript including the public key P_1, the simulated encryption c_1 of x_1, and all the leakage on S_1, and Γ_2 the partial transcript including P_2, c_2 and all the leakage on S_2. By the entropic security of Ψ in the simulated game, and the fact that $\ell'_{\text{post}} \leq t - v - 1$, we have that $\tilde{H}_\infty(x_1|\Gamma_1) \geq t - \ell'_{\text{post}} - 1 \geq v$. Furthermore, since conditioned on Γ_1, x_1 and Γ_2 are independent, we get $\tilde{H}_\infty(x_1|\Gamma) \geq v$. Similarly, it also holds that $\tilde{H}_\infty(x_2|\Gamma) \geq v$.

Since both x_1, x_2 have min-entropy more than v, and furthermore, by Lemma 1, are independent conditioned on Γ (as in Γ no function computes on both x_1 and x_2), the output of the average-case (v, ε) two-source extractor $Ext2(x_1, x_2)$ is at most ε away from uniform. Therefore the two distribution $Ext2(x_1, x_2) \oplus m_0$ and $Ext2(x_1, x_2) \oplus m_1$ are at most 2ε apart (since each is at most ε away from uniform). Therefore the advantage of $\tilde{A}$ is at most 2ε.

4.2 Instantiations and Parameters

Naor and Segev presented some instances of their construction [16] based on the DDH assumption (or DDH and the d-linear assumption), and the same constructions work for our case too. This gives entropic leakage resilient scheme Ψ with respect to any leakage $\ell_{\text{pre}}, \ell_{\text{post}}$ and overhead 1, as long as ℓ_{pre} is bounded by $(1 - o(1))L' - 3t$, where L' and t are respectively the lengths of the secret key and plaintext of the scheme Ψ. Therefore, we only need to focus on instantiating the two-source extractor $Ext2$ with exponentially small error $\varepsilon = 2^{-u-\omega(\log n)}$ in the length u of the output. In the work of Bouragin [2] it was shown how to extract randomness from two independent sources with min-entropy rate slightly less than half.

Theorem 9 ([2]). *There exists a universal constant* $\gamma < 1/2$ *and a polynomial time computable function* $\mathsf{Bou} : \{0,1\}^t \times \{0,1\}^t \to \{0,1\}^{u'}$ *that is a* (v, ε)*-two-source extractor, with* $v = \gamma t$, $\varepsilon = 2^{-\Omega(u')}$, *and* $u' = \Omega(t)$.

It follow from Lemma 5 that Bou is also an average-case extractor as needed. Furthermore, this construction lets us get two-source extractors with statistical distance as small as we want. Namely, to get $\varepsilon = 2^{-u-\omega(\log n)}$ we simply use it with u' sufficiently larger than u. Then we can truncate the output to length u without increasing the statistical distance, thus getting the parameters that we need.

Remark 1. The scheme Π uses a two-source extractor. We show that the construction can be easily modified to use a c-source extractor, for any $c > 2$:

instead of having two secret keys, maintain c secret keys $S_1, \ldots, S_c$; each key S_i is used to encrypt and decrypt a random seed X_i sampled independently, and the message is hidden using the random bits extracted from X_i's—we call it a c-split-state encryption scheme. It follows from the same proof as above that, this scheme is secure in the split-state model. This can be used to improve the parameters of our construction.

5 Conclusion and Future Work

In this paper, we study after-the-fact leakage for public-key encryption schemes. We show that a meaningful notion of security, namely, entropic security, can be achieved even at the presence of arbitrary (but bounded) leakage after the ciphertext is generated, and furthermore, the full fledged semantic security can be retained if considering some restricted form of leakage, namely a split-state model.

It is, of course, very interesting to explore other notions of security and other models in the context of after-the-fact leakage. For instance, Naor and Segev [16] showed that PKE that is semantically secure resilient to before-the-fact leakage can be transformed into a scheme that is CCA2-secure resilient to before-the-fact leakage, following the Naor-Yung "double encryption" paradigm [7,17]. It is interesting to see if a similar transformation can be done even with after-the-fact leakage.

Furthermore, recently, there has been some developments in leakage resilient cryptography in the continuous leakage model. One question studied in [14,10] is how to transform any circuit with a secret hard-coded, into another one that hides the secret even at the presence of arbitrary leakage during the computation of the circuit, *in the OCL model*. It would be interesting to investigate if their techniques can be applied to our scheme to make it secure even in the continuous leakage model.

Another interesting question is to handle leakage from the encryption randomness, not just the secret key. Perhaps the dense-model theorem from [9] can be used to prove ressitance at least to logarithmically many leakage bits.

Beyond just encryption, it is interesting to see if there are "natural" and useful relaxations of other primitives that can be achieved in the presence of After-the-Fact Leakage, for example commitment, key-agreement, etc.

Acknowledgement. We would like to thank Daniel Wichs for many delightful discussions and the anonymous referees of TCC 2011 for their helpful comments and suggestions.

References

1. Akavia, A., Goldwasser, S., Vaikuntanathan, V.: Simultaneous hardcore bits and cryptography against memory attacks. In: Reingold, O. (ed.) TCC 2009. LNCS, vol. 5444, pp. 474–495. Springer, Heidelberg (2009)
2. Bourgain, J.: More on the sum-product phenomenon in prime fields and its applications. International Journal of Number Theory 1, 1–32 (2005)

3. Brakerski, Z., Kalai, Y.T., Katz, J., Vaikuntanathan, V.: Overcoming the Hole in the Bucket: Public-Key Cryptography Resilient to Continual Memory Leakage. In: FOCS 2010, pp. 501–510. IEEE Computer Society, Las Vegas (2010)
4. Chor, B., Goldreich, O.: Unbiased bits from sources of weak randomness and probabilistic communication complexity (extended abstract). In: FOCS 1985, pp. 429–442. IEEE, Los Alamitos (1985)
5. Cramer, R., Shoup, V.: Design and analysis of practical public-key encryption schemes secure against adaptive chosen ciphertext attack. SIAM Journal on Computing 33(1), 167–226 (2003); Preliminary version in CRYPTO 1998
6. Dodis, Y., Ostrovsky, R., Reyzin, L., Smith, A.: Fuzzy extractors: How to generate strong keys from biometrics and other noisy data. SIAM J. Comput. 38(1), 97–139 (2008)
7. Dolev, D., Dwork, C., Naor, M.: Non-malleable cryptography (extended abstract). In: Proceedings of the 23rd Annual ACM Symposium on Theory of Computing, New Orleans, Louisiana, pp. 542–552. ACM, New York (May 1991)
8. Dziembowski, S., Pietrzak, K.: Intrusion-resilient secret sharing. In: FOCS, pp. 227–237 (2007)
9. Dziembowski, S., Pietrzak, K.: Leakage-resilient cryptography. In: FOCS 2008, pp. 293–302. IEEE Computer Society, Los Alamitos (2008)
10. Goldwasser, S., Rothblum, G.N.: Securing computation against continuous leakage. In: Rabin, T. (ed.) CRYPTO 2010. LNCS, vol. 6223, pp. 59–79. Springer, Heidelberg (2010)
11. Halderman, J.A., Schoen, S.D., Heninger, N., Clarkson, W., Paul, W., Calandrino, J.A., Feldman, A.J., Appelbaum, J., Felten, E.W.: Lest we remember: cold-boot attacks on encryption keys. Commun. ACM 52(5), 91–98 (2009)
12. Håstad, J., Impagliazzo, R., Levin, L.A., Luby, M.: A pseudorandom generator from any one-way function. SIAM J. Comput. 28(4), 1364–1396 (1999)
13. Ishai, Y., Sahai, A., Wagner, D.: Private circuits: Securing hardware against probing attacks. In: Boneh, D. (ed.) CRYPTO 2003. LNCS, vol. 2729, pp. 463–481. Springer, Heidelberg (2003)
14. Juma, A., Vahlis, Y.: Protecting cryptographic keys against continual leakage. In: Rabin, T. (ed.) CRYPTO 2010. LNCS, vol. 6223, pp. 41–58. Springer, Heidelberg (2010)
15. Micali, S., Reyzin, L.: Physically observable cryptography. In: Naor, M. (ed.) TCC 2004. LNCS, vol. 2951, pp. 278–296. Springer, Heidelberg (2004)
16. Naor, M., Segev, G.: Public-key cryptosystems resilient to key leakage. In: Halevi, S. (ed.) CRYPTO 2009. LNCS, vol. 5677, pp. 18–35. Springer, Heidelberg (2009)
17. Naor, M., Yung, M.: Public-key cryptosystems provably secure against chosen ciphertext attacks. In: Proceedings of the 22nd Annual ACM Symposium on Theory of Computing, pp. 427–437 (1990)
18. Nisan, N., Zuckerman, D.: Randomness is linear in space. J. Comput. Syst. Sci. 52(1), 43–52 (1996)
19. Santha, M., Vazirani, U.V.: Generating quasi-random sequences from semi-random sources. J. Comput. Syst. Sci. 33(1), 75–87 (1986)
20. Vazirani, U.V.: Strong communication complexity or generating quasirandom sequences form two communicating semi-random sources. Combinatorica 7(4), 375–392 (1987)

One-Time Computable Self-erasing Functions*

Stefan Dziembowski**, Tomasz Kazana***, and Daniel Wichs†

Abstract. This paper studies the design of cryptographic schemes that are secure even if implemented on untrusted machines that fall under adversarial control. For example, this includes machines that are infected by a software virus.

We introduce a new cryptographic notion that we call a *one-time computable pseudorandom function (PRF)*, which is a PRF $F_K(\cdot)$ that can be evaluated *on at most one input*, even by an adversary who controls the device storing the key K, as long as: (1) the adversary cannot "leak" the key K out of the device completely (this is similar to the assumptions made in the *Bounded-Retrieval Model*), and (2) the local read/write memory of the machine is restricted, and not too much larger than the size of K. In particular, the *only way* to evaluate $F_K(x)$ on such device, is to overwrite part of the key K during the computation, thus preventing all future evaluations of $F_K(\cdot)$ at any other point $x' \neq x$. We show that this primitive can be used to construct schemes for *password protected storage* that are secure against dictionary attacks, even by a virus that infects the machine. Our constructions rely on the random-oracle model, and lower-bounds for *graphs pebbling* problems.

We show that our techniques can also be used to construct another primitive, called *uncomputable hash functions*, which are hash functions that have a short description but require a large amount of space to compute on any input. We show that this tool can be used to improve the communication complexity of *proofs-of-erasure* schemes, introduced recently by Perito and Tsudik (ESORICS 2010).

1 Introduction

A recent trend in cryptographic research is to construct cryptographic schemes that have some provable security properties, even when they are implemented on devices that are not fully trusted. In general, two types of adversarial models are considered in this area. In the *passive* one, the adversary can get some partial information ("leakage") about the internal data stored on a cryptographic machine $\mathcal{M}$. This line of research, motivated by various *side-channel attacks* [24] was initiated in the seminal papers of Ishai et al. [28] and Micali and Reyzin [32], and followed by many recent works [22,1,35,29,33,12,37,13,14,25,8,7]. Some

* This work was supported by the ERC Starting Grant, agreement CNTM-207908.

** University of Warsaw and Sapienza University of Rome.

*** BioInfoBank Research Institute and University of Warsaw.

† New York University, part of this work was done while this author was visiting Sapienza University of Rome.

Y. Ishai (Ed.): TCC 2011, LNCS 6597, pp. 125–143, 2011.

papers have also been motivated by the attacks of the malicious software (like viruses) against computers [11,18,17,10,21,3,2]. What all these works have in common is that they provide a formal model for reasoning about the adversary that can obtain some information about the cryptographic secrets stored on $\mathcal{M}$. It is easy to see that some restrictions on the information that the adversary can learn is necessary, as the adversary that has an unlimited access to the internal data of $\mathcal{M}$ can simply "leak" the internals in their entirety, which is usually enough to completely break any type of security. A common assumption in this area is the *bounded-retrievability property*, which states that the adversary can retrieve at most some *input-shrinking* function f of the secret K stored on the device, i.e. he can learn a value $f(K)$ such that $|f(K)| \ll |K|$. The second class of models considered in the literature [27,26,23,30] are those where the adversary is *active*, which corresponds to the so-called *tampering attacks*. In these models the adversary is allowed to maliciously modify the internals of the device. For example, in the model of [27] the adversary that can tamper a restricted number of wires of the circuit that performs the computation (in a given time-frame), and in [26] it is assumed that a device is equipped with a small tamper-free component.

The above discussion motivates the following question:

> Can we achieve security against an adversary that has *complete active/passive control* over a device $\mathcal{M}$ storing cryptographic secrets, by only relying on simple physical characteristics of the device? For which cryptographic primitives can this be achieved and what characteristics are needed?

In this work, we focus on answering the above question for new primitive called a *one-time computable pseudorandom function*. That is, we consider a device that stores a key K for a function $F_K(\cdot)$ and allows the user to evaluate the function at a single arbitrary input x. Moreover, even if an adversary gains complete control of the device, he should be unable to learn any information, beyond a single value $F_K(x)$ at a single point x. We rely on the following two physical characteristics of the device $\mathcal{M}$ on which the secret key K is stored: (1) $\mathcal{M}$ satisfies the bounded-retrievability property, and (2) the read/write memory of $\mathcal{M}$ is restricted in size and not much larger than the size of the key K. Intuitively, the first property ensures that the attacker cannot leak the key K out of the device, while the second property will prevent the attacker from evaluating $F_K(\cdot)$ at multiple inputs using the resources of the device itself. The main application of this primitive is a scheme for password-protected storage. We also construct another, related primitive, that we call *uncomputable hash functions*, and use it construct an improved protocol for *proofs-of-erasure*, a concept recently introduced in [34].

1.1 One-Time Computable Functions

In this section we informally define the concept of a one-time computable PRF $F_K(\cdot)$ implemented on a resource-constrained device $\mathcal{M}$. Firstly, for correctness/functinality, we need to ensure that the key K of the PRF can be stored on the device $\mathcal{M}$ and that there is a method for honestly evaluating $F_K(\cdot)$ on a single arbitrary input x, using the resources of the device $\mathcal{M}$. Secondly, for security, we consider an attacker that gains control of the device $\mathcal{M}$. Such an adversary may learn the value $F_K(x)$ for some arbitrary point x, but should not have any other information about $F_K(x')$ for any other point $x' \neq x$.

So far we have not been very specific about the type of constraints placed on the resources of the device $\mathcal{M}$, and the type of control that the adversary gets over the device. One could imagine settings in which the above would be easy to implement. For example, if the adversary only gets black-box access to $\mathcal{M}$ then we can use an arbitrary PRF to achieve the above goal; simply have the device only perform only one evaluation of the PRF and then set a flag to stop responding to all future inputs. However, if the adversary can perform even relatively simple tampering attacks, it may be possible for it to "reset" the flag on the device and thus break security of the above solution.

In this work, we consider an adversary that has *complete control* over the device $\mathcal{M}$. That is, for the purpose of security, we can consider the device $\mathcal{M}$ itself to be a resource-constrained adversarial entity that gets the key K, and can communicate with an external unconstrained adversary $\mathcal{A}$. In this case, we must place some constraints on the resources of $\mathcal{M}$. Firstly, we must bound the amount of outgoing communication available to the device $\mathcal{M}$, as otherwise the $\mathcal{M}$ can just "leak" the entire key K to the external adversary $\mathcal{A}$, who can then evaluate $F_K(\cdot)$ at arbitrarily many points. Secondly, we must also place some limits on the computational resources available to $\mathcal{M}$, to prevent it from e.g. evaluating $F_K(x_0), F_K(x_1)$ at two points $x_0 \neq x_1$ and leaking the first bit of each output to the external adversary $\mathcal{A}$. In this work, we will assume that the amount of read/write memory available to the device $\mathcal{M}$ is bounded, and not much larger than the size of the key K. (The device can have arbitrary additional read-only or write-only memory).

Putting this together, our goal is to design a PRF $F_K(\cdot)$ which can be efficiently evaluated at any single point x on a memory-constrained device, but cannot be evaluated at any additional point $x' \neq x$ afterward. Roughly, our main idea is to construct F_K in such a way that any computation of $F_K(x)$ has to (at least partially) destroy K, by overwriting it, and thus prevents future computations of the function. That is, we assume that the key K itself is stored on the read/write memory of the device and takes up $m = |K|$ bits of space, which is a large fraction of the total. We design the PRF in such a way that there is an honest computation of $F_K(x)$ that uses (almost) no extra space beyond that on which K was originally stored, but overwrites K with various intermediate values during the computation. On the other hand, assuming the total memory on the device is $s < 2m$, we show that there is *no* method for computing of $F_K(x)$ at any single point x, without erasing part of the key K and preventing evaluation at any other input. Note that it is necessary for us to require that the key takes up more than half of the available read/write memory of the device, as otherwise it is possible to make a "copy" of the key that does not get damaged during the computation $F_K(x)$. In fact, we show a stronger result along these lines, where we also allow the adversarial memory-constrained device $\mathcal{M}$ to communicate up to $c < m$ bits to an external unconstrained adversary $\mathcal{A}$ (and we allow unbounded communication from $\mathcal{A}$ to the device).

One-time computable functions – a generalization. We also construct a generalization of the concept described above, where a single key K defines T different pseudorandom functions: $(F_{1,K}, \ldots . F_{T,K})$. Using the resources of the device, the honest user can evaluate each of the function $F_{i,K}$ at a single arbitrary point (i.e. the user first chooses an arbitrary x_1 and evaluates $F_{1,K}(x_1)$, then adaptively chooses x_2 and evaluates $F_{2,K}(x_2) \ldots$). However, even if the device is under full adversarial control, the attacker cannot get information about any of the T functions at more than one point – i.e. the attacker cannot get information about $F_{i,K}(x), F_{i,K}(x')$ for any two distinct points $x \neq x'$ and the same index i. The construction is given in Section 5. The maximal T that we can have is approximately equal to $\frac{c+s}{2(c+s-m)}$ (cf. (3)).

Application: Password-protected storage. Let us now describe an application of the primitives described above. Our application is related to *password-based cryptography*, which is an area that deals with the protocols where the secrets used by the parties are human-memorizable passwords. The crucial difference between a password and a cryptographic key is that the latter is usually assumed to be chosen uniformly at random from a large domain, while the former may come from some relatively small (polynomial sized) *dictionary set* $\mathcal{D}$. One of the main problems in constructing the password-based protocols is that one needs to consider the so-called *offline dictionary attacks*, where the adversary simply tries to break the scheme by analyzing all of the passwords from $\mathcal{D}$ one-by-one.

In this paper we are particularly interested in designing schemes for *password-protected storage*, which are schemes for secure encryption of data using passwords. A typical scheme of this type works as follows: let $\pi \in \mathcal{D}$ be a password. To encrypt a message X we apply a *key-derivation function* H to π and then encrypt X with $H(\pi)$ using some standard symmetric encryption scheme (Enc, Dec). To decrypt a ciphertext $C = Enc(H(\pi), X)$ one simply calculates $Dec(H(\pi), C)$.

A typical choice for H is a hash function. This solution is vulnerable to a following offline dictionary attack. An attacker simply tries, for every $\pi' \in \mathcal{D}$ to decrypt C until he finds π' such that $Dec(H(\pi'), C)$ "makes sense". Most likely there will be only one such π', and hence, with a good probability, this will be the correct π that the user has chosen to compute C.

A common way to make this attack harder is to design H in such a way that it is moderately expensive to compute it. The time needed to compute H should be acceptable for a legitimate user, and to high for the adversary if he has to do it for all passwords in $\mathcal{D}$. A drawback of this solution is that it depends on the amount of computing power available to the adversary. Moreover, the algorithm of the adversary can be easily parallelized.

An interesting solution to this problem was proposed in [9]. Here, a computation of H requires the user to solve the CAPTCHA puzzles [38], which are small puzzles that are easy to solve by a human, and hard to solve be a machine. A disadvantage of this solution is that it imposes additional burden on the user (he needs to solve the CAPTCHAs when he wants to access his data). Moreover, experience shows that designing secure CAPTCHAs gets increasingly difficult.

In this paper we show an alternative solution to this problem. Our solution works in a model where the data is stored on some machine that can be infected by a virus. In this model, the virus can get a total control over the machine, but he can retrieve only c bits from it. The main idea is that we will use a one-time computable function F (secure against an adversary with c-bounded communication and s-bounded storage) as the key-derivation function. To encrypt a message X with a password π we first choose randomly a key R for a one-time computable PRF. We then calculate $K = F_R(\pi)$. The ciphertext stored on the machine is $Enc(K, X)$. It is now clear that the honest user can easily compute K in space bounded by $c - \delta$. On the other hand, the adversary can compute K only once, even if he has space c. Of course, the adversary could use a part of the ciphertext $Enc(K, X)$ as his additional storage. This is not a problem if X is short (shorter than δ). If X is long, we can solve this problem by assuming that $Enc(K, X)$ is stored on a read-only memory.

A problem with this solution is that if an honest user makes an error and types in a wrong password then he does not have a chance to try another password. This can be solved by using the generalized version of the one-time computable functions. The scheme works as follows. First, we choose a key K for symmetric encryption. Then, we choose randomly R and for each $i = 1, \ldots, T$ we calculate $K_i = F_{R_i}(\pi) \oplus K$ (where the keys R_i are derived from R). The values that are stored on the machine are $(R, (K_1, \ldots, K_T), Enc(K, M))$. Now, to decrypt the message, the user first calculates $K = F_{R_1}(\pi) \oplus K_1$, and then decrypts $Enc(K, M)$ using K. If a user makes an error and calculates K_1 using a wrong π he still has a chance to calculate K_2, and so on.

1.2 Uncomputable Functions

We also introduce a notion of *uncomputable* hash functions, which we explain here informally. A hash function H is (s, ϵ)*-uncomputable*, if any machine that uses space s and takes a random input $x \in \{0,1\}^*$ outputs $H(x)$ with probability at most ϵ. We say that H is s'*-computable* if it *can* be computed in space s'. Note that in this case we assume that the adversary cannot use any external help to compute H (using the terminology from the previous sections: his communication is 0-bounded). Informally, we are interested in constructing (s, ϵ)-uncomputable, s'-computable functions for a small ϵ and s' being only slightly larger than s.

This notion can be used to construct an improved scheme for the *proof of erasure*, a concept recently introduced in [34]. Essentially the proof of erasure is a protocol between two parties: a powerful verifier $\mathcal{V}$ and a weak prover $\mathcal{P}$ (that can be, e.g., an embedded device). The goal of the verifier is to ensure that the prover has erased all the data that he stores in his RAM (we assume that $\mathcal{P}$ can also have a small ROM). This is done by forcing $\mathcal{P}$ to overwrite his RAM. Let

m be the size of RAM. Then, a simple proof of erasure consists of $\mathcal{V}$ sending to $\mathcal{P}$ a random string R such that $|R| = m$, and then $\mathcal{V}$ replying with R. In [34] the authors observe that the communication from $\mathcal{P}$ to $\mathcal{V}$ can be reduced in the following way: instead of asking $\mathcal{P}$ to send the entire R, we can just verify his knowledge of R using a protocol for the "proof of data possession" (see, e.g., [4]). Such a protocol still requires the verifier to send a large string R to the prover, hence the communication from the verifier to the prover is m. Using our uncomputable functions we can reduce this communication significantly.

Our idea as follows. Suppose we have a function H that is m-computable and $(m - \delta, \epsilon)$-uncomputable (for some small $\delta \in \mathcal{N}$ and a negligible $\epsilon \in [0, 1]$). Moreover, assume that H has a short domain and co-domain, say: $H : \{0,1\}^w \rightarrow \{0,1\}^w$ for some $w \ll m$. We can now design the following protocol:

1. $\mathcal{V}$ selects $X \leftarrow \{0,1\}^w$ at random and sends it to $\mathcal{P}$,
2. $\mathcal{P}$ calculates $Y = H(X)$ and sends it back to $\mathcal{V}$,
3. $\mathcal{V}$ accepts if $Y = H(X)$.

Clearly, an honest prover can calculate Y, since he has enough memory for this. On the other hand, from the $(m - \delta, \epsilon)$-uncomputability of H we get that a cheating prover cannot calculate Y with probability greater than ϵ without overwriting $m - \delta$ bits. The total communication between $\mathcal{P}$ and $\mathcal{V}$ has length $2w$. Note, that we need to assume that an adversary that controls the prover cannot communicate any data outside of the machine (therefore we are interested only in protocols with 0-bounded communication). This is because otherwise he could simply forward X to some external party that has more memory. The same assumption needs to be made in the protocols of [34]. What remains is to show a construction of such an H. We do it in Section 6.

1.3 Related Work

Most of the related work was already described in the previous sections. In our paper we will use a technique called *graph pebbling* (see e.g. [36]). This technique has already been used in cryptography in an important work of [16], some of our methods were inspired by this paper. The assumption that the adversary is memory-bounded has been used in the so-called *bounded-storage model* [31,5,20]. As similar assumption was also used in [15]. The proof of erasures can be viewed as a special case of the *remote attestation protocols* (see [34] for a list of relevant references).

1.4 Notation

For a sequence $R = (R_1, \ldots, R_N)$ and for indices i, j such that $1 \leq i \leq j \leq N$, we define $R[i, \ldots, j] = (R_i, \ldots, R_j)$.

2 Model of Computation

To make our statements precise, we must fix a model of computation. We will usually consider an adversary that consists of two parts: a "space-bounded" component $\mathcal{A}_{small}$ which gets access to the internals of an attacked device and has "bounded communication" to an external, and otherwise unrestricted, adversary $\mathcal{A}_{big}$.

Since the lower bounds on the computational complexity of functions are usually hard to prove, it seems difficult to show any meaningful statements in this model using purely complexity-theoretic settings. We will therefore use the *random-oracle model* [6]. Recall, that in this case a hash function is modeled as an external oracle containing a random function, and the oracle can be queried by all the parties in the protocol (including the adversary).

Using the random-oracle model in our case is a little bit tricky. To illustrate the problem consider a following protocol for the proof of erasure: (1) $\mathcal{V}$ sends to $\mathcal{P}$ a long random string R, (2) $\mathcal{P}$ replies with $H(R)$, where H is a hash function. Now, this protocol is obviously not secure for most of the real-life hash functions. For example, if H is designed using the Merkle-Damgård paradigm, then it can be computed "on fly", and hence there is no need to store the whole R before starting the computation of H.

On the other hand, if we model H as a random oracle, then the protocol described above can be proven secure, as the adversary has to wait until he gets the complete R before sending it to the oracle. We solve this problem in the following way: we will require that the only way in which the hash function is used is that it is applied to small inputs, i.e. if w is the length of the output of a hash function (e.g.: $w = 128$) then the hash function will have a type $H : \{0,1\}^{\xi w} \to \{0,1\}^w$, for some small ξ. Observe that if $\xi = 2$ then the function H can simply be a *compression function* used to construct the hash function).

We model our adversary $\mathcal{A} = (\mathcal{A}_{big}, \mathcal{A}_{small})$ as a pair of interactive algorithms[1] with oracle-access to a random-oracle $H(\cdot)$. The algorithm $\mathcal{A}_{big}$ will only be restricted in the number of oracle calls made. On the other hand, we impose the following additional restrictions on $\mathcal{A}_{small}$:

- s-bounded space: The total amount of space used by $\mathcal{A}_{small}$ is bounded by s. That is, we can accurately describe the entire configuration of $\mathcal{A}_{small}$ at any point in time using s bits[2].
- c-bounded communication: The total number of outgoing bits communicated by $\mathcal{A}_{small}$ is bounded by c[3].

We use the notation $\mathcal{A}^{H(\cdot)}(R) = \left(\mathcal{A}_{big}^{H(\cdot)}() \leftrightarrows \mathcal{A}_{small}^{H(\cdot)}(R)\right)$ to denote the interactive execution of $\mathcal{A}_{big}$ and $\mathcal{A}_{small}$, where $\mathcal{A}_{small}$ gets input R and both machines have access to the oracle $H(\cdot)$.

---

[1] Say ITMs, interactive RAMs, ... The exact model will not matter.

[2] This is somewhat different than standard space-complexity considered in complexity theory, even when we restrict the discussion to ITMs. Firstly, the configuration of $\mathcal{A}_{small}$ includes the value of *all* tapes, including the input tape. Secondly, it includes the current state that the machine is in and the position of all the tape heads.

[3] To be precise, we assume that we can completely describe the patters of outgoing communication of $\mathcal{A}_{small}$ using c bits. That is, $\mathcal{A}_{small}$ cannot convey additional information in when it sends these bits, how many bits are sent at a given time and so on...

3 Definitions

Let $W^{H(\cdot)}$ be an algorithm that takes as input $R \in \{0,1\}^m$ and has access to the oracle H. Let $(F_{1,R}^H, \ldots .F_{T,R}^H)$ be sequence of functions that depend on H and R. Assume that $W^{H(\cdot)}$ is interactive, i.e. it may receive queries from the outside. Let $x_1, \ldots, x_T$ be the sequence of queries that $W^{H(\cdot)}$ received. The algorithm $W^{H(\cdot)}$ replies to such a query by issuing a special *output query* to the oracle H. We assume that after receiving each $x_i \in \{0,1\}^*$ the algorithm $W^{H(\cdot)}$ always issues an output query to H of a form $((F_{i,R}^H(x_i), (i, x_i)), \texttt{out})$. We say that $W^{H(\cdot)}$ is a $(c, s, m, q, \epsilon, T)$-*onetime computable PRF* if:

- $W^{H(\cdot)}$ has m-bounded storage, and 0-bounded communication.
- for any $\mathcal{A}^{H(\cdot)}(R)$ that makes at most q queries to H and has s-bounded storage and c-bounded communication, the probability that $\mathcal{A}^{H(\cdot)}(R)$ (for a randomly chosen $R \leftarrow \{0,1\}^m$) issues two queries $((F_{i,R}^H(x), (i, x)), \texttt{out})$ and $((F_{i,R}^H(x'), (i, x')), \texttt{out})$, for $x \neq x'$, is at most ϵ.

Basically, what this definition states is that no adversary with s-bounded storage and c-bounded communication can compute the value of any $F_{i,R}$ on two different inputs. It may look suspicious that we defined the secrecy of a value in terms of the hardness of guessing it, instead of using the indistinguishability paradigm. We now argue why our approach is ok. There are two reasons for this. The first one is that in the schemes that we construct that output of each F^H is always equal to some output of H (i.e. the algorithm F simply outputs on the the responses he got from H). Hence $\mathcal{A}$ cannot have a "partial knowledge" of the output (either he was lucky and he queried H on the "right" inputs, or not – in the latter case the output is indistinguishable from random, from his point of view).

The second reason is that, even if it was not the case — i.e. even if F^H outputted some value y that is a more complicated function of the responses he got from H — we could modify F^H by hashing y with H (and hence if y is "hard to guess" then $H(y)$ would be completely random, with a high probability).

Now, suppose that $V^{H(\cdot)}$ is defined identically to $W^{H(\cdot)}$ with the only difference that it receives just one query $x \in \{0,1\}^*$, and afterwards it issues one output query $((F^H(x), x), \texttt{out})$ (for some function F that depends on H). We say that $V^{H(\cdot)}$ is an $(s, w, q, \delta, \epsilon)$-*uncomputable hash function* if:

- $V^{H(\cdot)}$ has s-bounded storage, and 0-bounded communication.
- for any $\mathcal{A}^{H(\cdot)}(R)$ that makes at most q queries to H and has $(s-\delta)$-bounded storage and c-bounded communication, the probability that $\mathcal{A}^{H(\cdot)}(R)$ (for a randomly chosen $R \leftarrow \{0,1\}^w$) issues a query $((F^H(x), x_i), \texttt{out})$ is at most ϵ.

4 Random Oracle Graphs and the Pebbling Game

We show a connection between an adversary computing a "random oracle graph" and a pebbling strategy for the corresponding graph. A similar connection appears in [16].

4.1 Random-Oracle Labeling of a Graph

Let $G = (V, E)$ be a DAG with $|V| = N$ vertices. Without loss of generality, we will just assume that $V = \{1, \ldots, N\}$ (we will also consider infinite graphs, in which case we will have $N = \infty$). We call vertices with no incoming edges *input vertices*, and will assume there are $M \le N$ of them. A *labeling* of G is a function $\texttt{label}(\cdot)$, which assigns values $\texttt{label}(v) \in \{0,1\}^w$ to vertices $v \in V$. We call w the *label-length*. For any function $H : \{0,1\}^* \to \{0,1\}^w$ and input-labels $R = (R_1, \ldots, R_M)$ with $R_i \in \{0,1\}^w$, we define the (H, R)-labeling of G as follows:

- The labels of the M distinct input vertices $v_1 < v_2 < \ldots < v_M$ are given by $\texttt{label}(v_i) \stackrel{\text{def}}{=} R_i$.
- The label of every other vertex v is defined recursively by

  $$\texttt{label}(v) \stackrel{\text{def}}{=} H(\texttt{label}(v_1), \ldots, \texttt{label}(v_d), v)$$

  where $v_1 < \ldots < v_d$ are the d parents of v.

A *random oracle labeling* of G is an (H, R)-labeling of G where H is a random-function and R is chosen uniformly at random.

For convenience, we also define $\texttt{preLabel}(v) \stackrel{\text{def}}{=} (\texttt{label}(v_1), \ldots, \texttt{label}(v_d), v)$, where $v_1 < \ldots < v_d$ are the parents of v, so that $\texttt{label}(v) = H(\texttt{preLabel}(v))$.

The *output vertices* of G are the vertices that have no children. Let $v_1, \ldots, v_K$ be the output vertices of G. Let $Eval(G, H, (R_1, \ldots, R_M))$ denote the sequence of labels $(\texttt{label}(v_1), \ldots, \texttt{label}(v_K))$ of the output vertices calculated with the procedure described above (with $R_1, \ldots, R_M$ being the labels of the input vertices $v_1, \ldots, v_M$ and H being the hash function).

Our main goal is to show that computing the labeling of a graph G requires a large amount of resources in the random-oracle model, and is therefore difficult. We will usually (only) care about the list of random-oracle calls made by $\mathcal{A}_{big}$ and $\mathcal{A}_{small}$ during such an execution. We say that an execution $\mathcal{A}^{H(\cdot)}(R)$ *labels* a vertex v, if a random-oracle call to $\texttt{preLabel}(v)$, is made by either $\mathcal{A}_{big}$ or $\mathcal{A}_{small}$.

4.2 Pebbling Game

We will consider a new variant of the pebble game that we call the "red-black" pebble game over a graph G. Each vertex of the graph G can either be empty, contain a red pebble, contain a black pebble, or contain both types of pebbles. An initial configuration consists of (only) a black pebble placed on each input vertex of G. The game proceeds in steps where, in each step, one of the following four actions is taken:

1. A red pebble can be placed on any vertex already containing a black pebble.
2. If all the parents of a vertex v have a red pebble on them, a red pebble can be placed on v.

3. If all the parents of v have *some* pebble on them (red or black), a black pebble can be placed on v.
4. A black pebble can be removed from any vertex.

We define the *black-pebble complexity* of a pebbling strategy to be the maximum number of *black pebbles* in use at any given time. We define the *red-pebble complexity* of a pebbling strategy to be the total number of steps in which action 1 is taken. We also define the *all-pebble complexity* of a pebbling strategy to be the sum of its black- and red-pebble complexities. By *heavy-pebbles* we will mean the black pebbles, or the red-pebbles that appeared on the graph because of action 1. Note, that these are exactly the pebbles that count when we calculate the all-pebble complexity of a strategy.

Remark 1. Let G be a graph with N vertices and M input vertices. Let v be an output vertex of G and let $v_{i_1}, \ldots, v_{i_d}$ be a subset of the set of input vertices. Suppose there exists a pebbling strategy that (1) pebbles v while keeping the pebbles on the vertices $v_{i_1}, \ldots, v_{i_d}$, and (2) has black-pebble complexity b and it does not use the red pebbles, i.e. its red-pebble complexity 0. Then the value of $Eval(G, H, (R_1, \ldots, R_M))$ can be computed by a machine with bw-bounded storage and an access to a random oracle that computes H. This is because the only thing that the machine needs to remember are the labels of at most b vertices (each of those labels has length at most w). The computation may overwrite some part of the input $(R_1, \ldots, R_M)$, however, it does not overwrite the input corresponding to the vertices $v_{i_1}, \ldots, v_{i_d}$, i.e.: $(R_{v_1}, \ldots, R_{v_d})$.

It is more complicated to show a connection in the opposite direction, namely to prove that if a graph cannot be pebbled with a strategy with low black- and red-complexities, then it cannot be computed by a machine with a restricted storage and communication. The following lemma establishes such a connection (its proof, that appears in [19], is an extension of the proof of Lemma 1 in [16].)

Lemma 1. *Let G be a DAG with M input vertices and K output vertices. Let r and b be arbitrary parameters. Suppose that G is such that there does not exist a pebbling strategy such that (1) its all-pebble complexity is at most a, and that (2) pebbles at least α output vertices (for some $\alpha \in \{1, \ldots, K\}$).*

Then, for any c, s, w such that $\frac{c+s+w}{w-\log(q)} < a$, and for any $\mathcal{A} = (\mathcal{A}_{big}, \mathcal{A}_{small})$ that makes at most q oracle calls and has s-bounded space and c-bounded communication the probability that $\mathcal{A}$ labels more than $\alpha - 1$ output vertices is at most $(q+1) \cdot 2^{-w}$ (where the probability is taken over the randomness of $\mathcal{A}$ and the random choice of H and R).

5 One-Time Computable Functions

In this section we show specific examples of graphs that are hard to pebble in limited space and bounded communication. Let M, M' be a parameters such that $M' < M$. The (M, M')-*lambda graph* (denoted $Lam^{M}_{M'}$) is defined in the

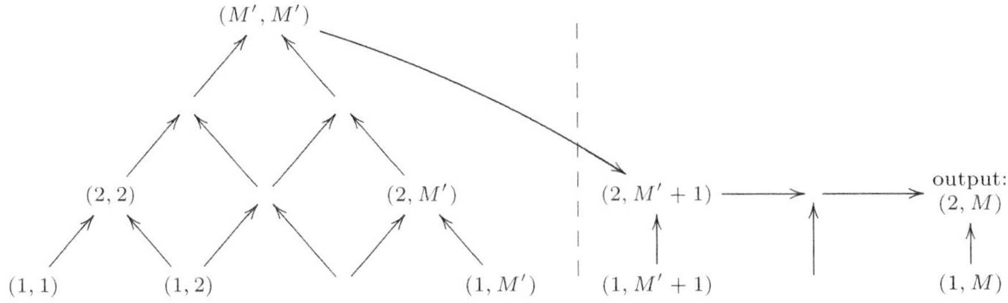

Fig. 1. An (M, M')-lambda graph for $M' = 4$ and $M = 7$. The sub-graph on the left-hand side of the dashed line is an M'-pyramid.

following way (cf. Fig. 1). Its set of vertices is equal to $V_0 \cup V_1$, where $V_0 = \{(i,j) : 1 \leq i \leq j \leq M'\}$ and $V_1 = \{1,2\} \times \{M'+1, \ldots, M\}$. The set of input vertices is equal to $\{1\} \times \{1, \ldots, M\}$. The output vertex is $(2, M)$. The set of edges is equal to the following sum: $\{((i-1, j-1), (i,j)) : (i-1, j-1), (i,j) \in V_0\} \cup \{((i-1, j), (i,j)) : (i-1, j), (i,j) \in V_0\} \cup \{((M', M'), (2, M'+1))\} \cup \{((1, j-1), (1, j)) : (1, j-1), (1, j) \in V_1\} \cup \{((1, j), (2, j)) : (1, j), (2, j) \in V_1\}$. If $M' = M$ then a (M, M)-lambda graph is defined as above, with $V_1 = \emptyset$ and with the set of edges consisting only of the first two summands of the sum above. Its output vertex is (M, M). Such a graph is also called an *M-pyramid graph.*

Lemma 2. *For any $X < M' - 1$ there exists a strategy that pebbles the output vertex of $\mathrm{Lam}_{M'}^{M}$ that satisfies the following:*

- *it uses $M + M' - 1 - X$ black pebbles (remember that all the M input vertices are initially pebbled with a black pebble, and therefore using $M + M' - 1 - X$ means having $M' - 1 - X$* extra *pebbles),*
- *it uses no red pebbles, and*
- *at the moment when the output vertex is pebbled there are still pebbles on the last $M - X$ input vertices, i.e.: vertices from the set $\{1\} \times \{X+1, \ldots, M\}$.*

Proof. The pebbling strategy consists of the following steps:

pebble the second row of the pyramid. In this step we pebble the second row of the pyramid, i.e. the vertices from the set $\{2\} \times \{2, \ldots, M'\}$. We do it by removing X pebbles from the input of the pyramid, and by using the $M' - 1 - X$ extra pebbles that we have. The procedure is as follows:

1. First, we put pebbles on the vertices from the set $\{2\} \times \{2, \ldots, X+1\}$. We do it in the following way: for $j = 2, \ldots, X'+1$ we put a pebble on $(2, j)$ and remove it from $(1, j-1)$.
2. We then put pebbles on the vertices from the set $\{2\} \times \{X+2, M'\}$. We do it just using the extra pebbles, without removing any pebble from the input. Clearly we have enough extra pebbles, since $|\{2\} \times \{X+2, M'\}| = M' - 1 - X$.

pebble the rest of the pyramid. In this step we pebble the pyramid row-by-row, starting from the third row, and ending with the top of the pyramid (M', M'). We do it in the by executing the following procedure for $i = 3, \ldots, M'$:
  - for $j = i, \ldots, M'$ do the following: put a pebble on (i, j) and remove it from $(i-1, j-1)$.

pebble the rest of the graph. We now pebble the rest of the graph in the following way. First, we put a pebble on $(2, M'+1)$ and remove it from (M', M'). Then, for $j = M'+2, \ldots, M$ we put a pebble on $(2, j)$ and remove it from $(2, j-1)$. At the end of this loop there output vertex is pebbled.

It is easy to see that the above procedure results in a correct pebbling strategy. Moreover, it uses only $M' - 1 - X$ extra pebbles, and it removes the pebbles only from the first X vertices of the input. □

5.1 Hardness of Pebbling

Consider a configuration of the red and black pebbles on some DAG G. Let v be a vertex of G. We say that v *is input-dependent in this configuration* if, after removing all the pebbles from the input it is impossible to pebble the vertex v. If v is not input-dependent then we say that it is *input-independent*.

Lemma 3. *For $M \geq 2$ consider an M-pyramid graph Lam_M^M and some configuration of pebbles on it. If the output vertex (M, M) is input-dependent then the number of heavy pebbles is at least M.*

Proof. We prove it by induction on $M = 2, 3, \ldots$. To root the induction we first consider the case when $M = 2$. In this case the graph consists of 3 vertices only: 2 input vertices, and 1 output vertex. If it is input-dependent then the output vertex is not pebbled. Hence both input vertices need to have a pebble.

Now, let us assume the hypothesis for $M-1$ and consider $G_M = Lam_M^M$. Take some configuration γ of pebbles. Denote the set of heavy pebbles in γ by $\mathcal{X}$. Let G_{M-1} be a subgraph of G_M induced by all the vertices of G_M except of the input row (in other words: G_{M-1} is equal to G_M with the bottom row "cut"). Of course G_{M-1} is Lam_{M-1}^{M-1}.

Now, put black pebbles on the vertices of the input row of G_{M-1} in the following way: put a pebble on a vertex v whenever v has both parents in $\mathcal{X}$ (and keep the old pebbles from the configuration γ). It is easy to see that the number of black pebbles in this new configuration is at most $|\mathcal{X}| - 1$.

Clearly the resulting configuration of pebbles on G_{M-1} satisfies the following: (1) the output vertex can be pebbled from this configuration, and (2) the output vertex on G_{M-1} is input-dependent (if it was not input-dependent then also the configuration γ would not be input-dependent). Hence, by the induction hypothesis $|\mathcal{X}| - 1 \geq M - 1$, which implies that $|\mathcal{X}| \geq M$. □

Lemma 4. *Suppose $M > 2$. Consider a pebbling strategy for Lam_M^M that pebbles the vertex (M, M). In the first configuration in which (M, M) is input-independent we have that: (1) the total number of the heavy pebbles that are not on the input row is at least $M - 1$, and (2) there is no pebble on (M, M).*

Proof. Let $G_M = Lam_M^M$, and let G_{M-1} be defined as in the proof of Lemma 3. Let γ be the first configuration in which (M, M) is input-independent, and let γ' be the configuration that directly precedes γ, i.e. the last configuration that is input-dependent. Keep on the vertices of G_{M-1} all the pebbles from the configuration γ. We now show that in such a configuration of the pebbles on G_{M-1} the output of G_{M-1} is input-dependent. After showing it we will be done: part (1) will follow directly from Lemma 3 (applied to G^{M-1}), and part (2) will follow from the fact that (for $M - 1 > 1$) if the output vertex is input-dependent then it cannot be pebbled.

To finish the proof assume that the output of G_{M-1} is input-independent. We obtain contradiction by showing that in this case also G_M needs to be input-independent. Clearly the only way in which γ' was transformed into γ was that a pebble was added on the input row of G_{M-1}. However, by our assumption the output of G_{M-1} (and hence also of G_M) does not depend on this row. Therefore also in the configuration γ' the output cannot depend on the two bottom rows of G_M. This gives us a contradiction. □

Lemma 5. *Consider a pebbling strategy that pebbles the output of $Lam_{M'}^M$. As long as the vertex (M', M') has not been pebbled, there has to be a heavy pebble on every input vertex on the second part of $Lam_{M'}^M$ (i.e. the vertices $(1, j)$ such that $j \in \{M' + 1, \ldots, M\}$).*

Proof. This follows easily from the construction of the $Lam_{M'}^M$ graph: if one removes a pebble from any vertex $(1, j)$ such that $j \in \{M' + 1, \ldots, M\}$ then one cannot put a pebble on it in the future. Therefore it will never be possible to pebble $(2, j)$, and hence also $(2, M)$. □

For $\ell \in \mathcal{N} \cup \{\infty\}$ consider a family of ℓ DAGs $\{G_k = (V_k, E_k)\}_{k=1}^{\ell}$ such that every DAG in this family has the same set of input V_I of input vertices. Define $V_k' = V_k \setminus V_I$. The graph $G = (V, E)$ is a *sum of* $\{G_k = (V_k, E_k)\}_{k=1}^{\ell}$ if is defined as follows: the set of vertices V is equal to V_I plus the disjoint sum of the sets V_k'. More precisely: $V := V_I \cup \bigcup_{k=1}^{\ell} \{k\} \times V_k$. The set E of edges is defined as: $E := \{((k, v), (k, v')) : v, v' \in V_k' \text{ and } (v, v') \in E_k\} \cup \{(v, (k, v')) : v \in V_I \text{ and } v' \in V_k' \text{ and } (v, v') \in E_k\}$. The set of input vertices of G is equal to V_I, and the set of the output vertices is equal to $V_{O,L} \cup V_{O,R}$, where $V_{O,L}$ and $V_{O,R}$ are the sets of the output verices of G_L and G_R, respectively.

Lemma 6. *Consider a family $\{G_k\}_{k=1}^{\ell}$ of (M, M')-lambda graphs. Let G be a sum of the graphs in this family. Then there does not exist a pebbling strategy with all-pebble complexity bounded by $M + M' - 2$ that pebbles more than one output of G.*

Proof. For the sake of contradiction suppose that such a strategy exists. Pebbling the output of $Lam^{M}_{M'}$ requires first pebbling the top of the pyramid graph that is a part of $Lam^{M}_{M'}$. Therefore there has to exist a pebbling strategy with all-pebble complexity bounded by $M + M' - 2$ that pebbles two different vertices that are the tops of the pyramids in some G_k and G_h (i.e. vertices $(k, (M', M'))$ and $(h, (M', M'))$).

Clearly, at the beginning of any pebbling strategy the top of each pyramid is dependent on the input of this pyramid. Consider the first configuration where the top of one of the pyramids, the one belonging to G_k, say, gets independent from the input of this pyramid. In this moment, by Lemma 4 the total number of the heavy pebbles that are not on the input row of G_k is at least $M' - 1$. Since in this moment the top vertex of G_h is still dependent on the input, hence, by Lemma 3 the total number of the heavy primary red pebbles and the black pebbles on G_k is at least M'. Therefore the number of the heavy pebbles on the two pyramids is at least $2M' - 1$.

On the other hand, by the second part of Lemma 4 the vertex (M', M') is not yet pebbled in this configuration. Hence, by Lemma 5, there needs to be a heavy pebble on every vertex from the second part of the input of G_h and G_k, i.e. on the vertices $(1, j))$ such that $j \in \{M' + 1, \ldots, M\}$. Therefore altogether we have $2M' - 1 + (M - M') = M + M' - 1$ pebbles on the sum of G_h and G_k. This yields a contradiction with the assumption that the all-pebble complexity of the strategy is bounded by $M + M' - 2$. □

Combining Lemma 1 with Lemma 6 we get the following.

Corollary 1. *Consider a family $\{G_k\}_{k=1}^{\ell}$ of (M, M')-lambda graphs. Let G be a sum of the graphs in this family. Then, for any s, c, w and q, such that $\frac{c+s+w}{w-\log(q)} < M + M' - 2$, and any adversary $\mathcal{A}$ that has s-bounded storage and c-bounded communication, and makes at most q queries to the oracle, the probability that $\mathcal{A}$ labels more than one output of G is at most $(q+1) \cdot 2^{-w}$.*

5.2 The Construction

In our construction the hash function will depend on an additional parameter a. Formally, let $H : \{0,1\}^* \times \{0,1\}^* \to \{0,1\}^w$ be a function that is modeled as a random oracle. For any $a \in \{0,1\}^*$ let H^a denote a function defined as $H^a(z) = H(a, z)$. Let M, U and T be some positive integer parameters such that

$$T < \frac{U - 1}{2\Delta} \tag{1}$$

where $\Delta := U - M$. We now construct an interactive algorithm $COMP^{H}_{U,M,T,w}$ that has access to a random oracle H and stores a key consisting of M blocks (of length w). That is, the input to the algorithm is $R = (R_1, \ldots, R_M)$, and it behaves as follows. Suppose it is queried on some inputs $x_1, \ldots, x_T$. Then, after receiving each x_i it computes the value of

$$Eval(Lam^{M-(i-1)\Delta}_{i\Delta+2}, H^{(i,x_i)}, (R[1 + (i-1)\Delta, \ldots, M]). \tag{2}$$

The algorithm $COMP^H_{U,M,T,w}(R)$ simply computes each (2) one-by-one for $i = 1, \ldots, T$. Each of these steps destroys Δ values R_j from the input. Thus, before the i-th step we keep in the memory only $R[1 + (i-1)\Delta, \ldots, M]$. This means that the space used by the remaining part of the input ($R[1, \ldots, (i-1)\Delta]$) is now free and it can be used as additional storage for computation. So, just before the beginning of the i-th step the free storage (i.e. storage not including kept fragment of the input) is bounded by $e_i = E_i \cdot w$, where $E_i := 1 + (i-1)\Delta$. The algorithm in the i-th step is just a simple application of Remark 1 from Section 4.2. Observe that from Lemma 2 we get a pebbling strategy that pebbles output vertex of $Lam^{M-(i-1)\Delta}_{i\Delta+2}$ using E_i extra pebbles and removes the first $(i\Delta+2)-1-E_i$ input pebbles. From the definition of E_i we have $(i\Delta+2)-1-E_i = \Delta$. So, from Remark 1 we get that there is an algorithm that computes (2) overwriting $\Delta \cdot w$ first bits of remaining input. So, after this step the algorithm can keep $R[1 + i\Delta, \ldots, M]$ to be used in the next steps.

Theorem 1. *For any integers c, s, m, w, let $U \stackrel{def}{=} \lfloor \frac{c+s+w}{w-\log(q)} \rfloor$ and $M \stackrel{def}{=} \lfloor \frac{m}{w} \rfloor - 1$. Then, for any integer $T < \frac{U-1}{2(U-M)}$ the algorithm $COMP^H_{U,M,T,w}$ is a $(c, s, m, q, (q+1) \cdot 2^{-w}, T)$-one-time computable PRF.*

Asymptotically, as $c, s, m \gg w \gg \log(q)$, the maximal T becomes

$$T \approx \frac{c+s}{2(c+s-m)}. \tag{3}$$

Proof (of Theorem 1). Suppose $(R_1, \ldots, R_M)$ is chosen uniformly at random. Let $\mathcal{A} = (\mathcal{A}_{big}, \mathcal{A}_{small})$ be an arbitrary adversary with oracle access to H that has s-bounded space and c-bounded communication and makes at most q oracle calls. Consider an execution $\mathcal{A}^{H(\cdot)}(R)$. Let $\mathcal{E}$ be an event that for some i and for two different x and x' the adversary labeled the output vertex of $Lam^{M-(i-1)\Delta}_{i\Delta+2}$ in the $(H^{(i,x)}, R)$-labeling and $(H^{(i,x')}, R)$-labeling. To prove the theorem we need to show the following.

$$P(\mathcal{E}) \leq T \cdot (q+1) \cdot 2^{-w}. \tag{4}$$

Fix some $\tilde{i}$, and let $\mathcal{E}_{\tilde{i}}$ denote the event that $\mathcal{E}$ happened for $i = \tilde{i}$. Let G be equal to the sum of following infinite sequence of graphs $\left\{ Lam^{M-(\tilde{i}-1)\Delta}_{\tilde{i}\Delta+2} \right\}_{x \in \{0,1\}^*}$.
We now show an adversary $\tilde{\mathcal{A}}$ with an s-bounded space and c-bounded communication that has access to an oracle $\tilde{H}$ and makes at most q queries to it, and satisfies the following: for a randomly-chosen $\tilde{R} = (R_{1+(\tilde{i}-1)\Delta}, \ldots, R_M) \in (\{0,1\}^w)^{M-(\tilde{i}-1)\Delta}$ in the execution $\tilde{\mathcal{A}}^{\tilde{H}(\cdot)}(\tilde{R})$ the probability that the adversary labels at least two different output vertices of G is equal to $P(\mathcal{E}_{\tilde{i}})$.

The adversary $\tilde{\mathcal{A}}$ simulates $\mathcal{A}$ in the following way. First, since $\mathcal{A}$ "expects" the input to have length M, it fills-in the "missing" elements of $\tilde{R}$, i.e. he selects randomly $(R_1, \ldots, R_{(\tilde{i}-1)\Delta})$ and sets $R = (R_1, \ldots, R_{(\tilde{i}-1)\Delta})||\tilde{R}$. Next, it simply runs $\mathcal{A}$. The only thing that we need to take care of is to "translate" the oracle queries issued by $\mathcal{A}$ to H into oracle queries issued by $\tilde{\mathcal{A}}$ to $\tilde{H}$. Let Q be a query issued by $\mathcal{A}$. Consider the following cases:

- Q has a form $((\tilde{i}, x), (\texttt{label}_1, \ldots, \texttt{label}_d, v))$ (for some $x, \texttt{label}_1, \ldots, \texttt{label}_d$) — in this case we translate it into a a query $(\texttt{label}_1, \ldots, \texttt{label}_d, ((i, x), v)$,
- Q has a form or $((\tilde{i}, x), (\texttt{label}, v, \texttt{out}))$ (for some x and $\texttt{label}$) — in this case we translate it into a a query $(\texttt{label}, ((\tilde{i}, x), v), \texttt{out})$.
- if Q does not have any of the forms above — we translate it in some arbitrary (deterministic and injective) way.

It is easy to see that $\tilde{\mathcal{A}}$ labels an output vertex $((\tilde{i}, x), v)$ of G if and only if his simulated copy of $\mathcal{A}$ labeled v in the graph $Lam_{\tilde{i}\Delta+2}^{M-(\tilde{i}-1)\Delta}$. Therefore the probability that $\tilde{\mathcal{A}}$ labeled two output vertices of G is equal to $P(E_{\tilde{i}})$. Now, by Corollary 1 we get that this probability is at most $(q+1) \cdot 2^{-w}$ as long as $\frac{s+c+w}{w-\log(q)} < M - (\tilde{i}-1)\Delta + 1 + \tilde{i}\Delta + 1 - 2 = M + \Delta = U$, which is exactly the assumption that we made in the statement of the lemma. Since $\mathcal{E} = \cup_{i=1}^{T} \mathcal{E}_i$, therefore, by the union-bound we get that $P(\mathcal{E}) \leq T \cdot ((q+1) \cdot 2^{-w})$. Therefore (4) is proven.

6 Arrowhead Functions

In this section we define a class of DAGs that we call the *arrowhead graphs*. For every $M \in \mathcal{N}$ let Arr_M be a graph consisting of defined previously M-pyramid with one additional vertex $(0,0)$ and additional edge from $(0,0)$ to $(1,x)$ for $x \in 1, \ldots M$. More precisely, $Arr_M = (V_M, E_M)$, where $V = \{(0,0)\} \cup \{(i,j) : 1 \leq i \leq j \leq M$. A graph Arr_M consists of one input vertex $(0,0)$ and one output vertex (M,M). The follwoing figure shows an example of an M-arrowhead graph for $M = 4$. The subgraph on the upper side of the dashed line is an M-pyramid.

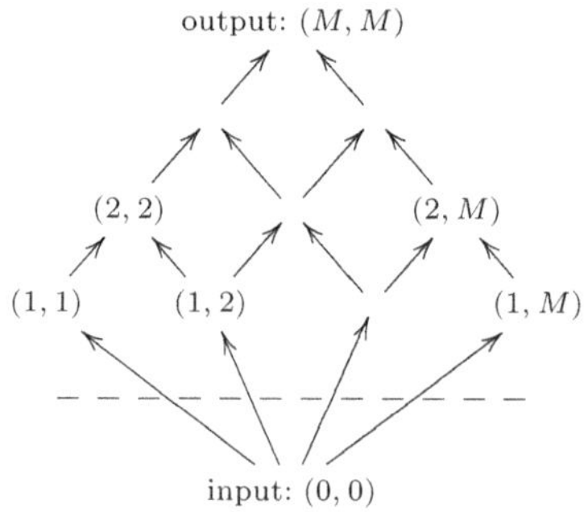

Lemma 7. *For any a and $R = (R_1, \ldots, R_M)$ the value of $Eval(Arr_M, H, R)$ can be computed by an algorithm that has access to a random oracle H and has $(M+1) \cdot w$-bounded storage.*

Proof. Using Remark 1 it suffices to show a strategy that pebbles Arr_M using $M+1$ pebbles. Such a strategy is straighforward and appears in the full version of this paper [19].

The following lemma shows the optimality of the algorithm given in Lemma 7.

Lemma 8. *For any s, λ and q, such that $\frac{s+\lambda}{w-\log(q)} < M+1$, and any adversary $\mathcal{A}$ that has s-bounded storage and 0-bounded communication, and makes at most q queries to the oracle, the probability that $\mathcal{A}$ labels the output of Arr_M is at most $q \cdot 2^{-w} + 2^{-\lambda}$.*

This lemma follows from the fact that every strategy that pebbles the output of Arr_M, and does not use the red pebbles, must use at least $M-1$ black pebbles. The proof of this fact is very similar to the proof of Lemma 10.2.1 in the book of John Savage ([36]), and it appears in the full version of this paper [19]. Lemma 7 and 8 imply the following.

Theorem 2. *The hash function that takes as input R and outputs $Eval(Arr_M, H, R)$ is $((M+1) \cdot w, w, q, \log(q)(M+1) + \lambda, q \cdot 2^{-w} + 2^{-\lambda})$-uncomputable.*

Acknowledgments

The authors wish to thank Yevgeniy Dodis for helpful discussions.

References

1. Akavia, A., Goldwasser, S., Vaikuntanathan, V.: Simultaneous hardcore bits and cryptography against memory attacks. In: Reingold, O. (ed.) TCC 2009. LNCS, vol. 5444, pp. 474–495. Springer, Heidelberg (2009)
2. Alwen, J., Dodis, Y., Naor, M., Segev, G., Walfish, S., Wichs, D.: Public-key encryption in the bounded-retrieval model. In: Gilbert, H. (ed.) EUROCRYPT 2010. LNCS, vol. 6110, pp. 113–134. Springer, Heidelberg (2010)
3. Alwen, J., Dodis, Y., Wichs, D.: Leakage-resilient public-key cryptography in the bounded-retrieval model. In: Halevi, S. (ed.) CRYPTO 2009. LNCS, vol. 5677, pp. 36–54. Springer, Heidelberg (2009)
4. Ateniese, G., Di Pietro, R., Mancini, L., Tsudik, G.: Scalable and efficient provable data possession. In: SecureComm (2008)
5. Aumann, Y., Ding, Y.Z., Rabin, M.O.: Everlasting security in the bounded storage model. IEEE Transactions on Information Theory 48(6) (2002)
6. Bellare, M., Rogaway, P.: Random oracles are practical: A paradigm for designing efficient protocols. In: ACM Conference on Computer and Communications Security (1993)
7. Brakerski, Z., Goldwasser, S.: Circular and leakage resilient public-key encryption under subgroup indistinguishability. In: Rabin, T. (ed.) CRYPTO 2010. LNCS, vol. 6223, pp. 1–20. Springer, Heidelberg (2010)
8. Brakerski, Z., Tauman Kalai, Y., Katz, J., Vaikuntanathan, V.: Cryptography resilient to continual memory leakage. In: FOCS (2010)
9. Canetti, R., Halevi, S., Steiner, M.: Mitigating dictionary attacks on password-protected local storage. In: Dwork, C. (ed.) CRYPTO 2006. LNCS, vol. 4117, pp. 160–179. Springer, Heidelberg (2006)
10. Cash, D.M., Ding, Y.Z., Dodis, Y., Lee, W., Lipton, R.J., Walfish, S.: Intrusion-Resilient Key Exchange in the Bounded Retrieval Model. In: Vadhan, S.P. (ed.) TCC 2007. LNCS, vol. 4392, pp. 479–498. Springer, Heidelberg (2007)

11. Di Crescenzo, G., Lipton, R.J., Walfish, S.: Perfectly secure password protocols in the bounded retrieval model. In: Halevi, S., Rabin, T. (eds.) TCC 2006. LNCS, vol. 3876, pp. 225–244. Springer, Heidelberg (2006)
12. Davì, F., Dziembowski, S., Venturi, D.: Leakage-resilient storage. In: Garay, J.A., De Prisco, R. (eds.) SCN 2010. LNCS, vol. 6280, pp. 121–137. Springer, Heidelberg (2010)
13. Dodis, Y., Goldwasser, S., Tauman Kalai, Y., Peikert, C., Vaikuntanathan, V.: Public-key encryption schemes with auxiliary inputs. In: Micciancio, D. (ed.) TCC 2010. LNCS, vol. 5978, pp. 361–381. Springer, Heidelberg (2010)
14. Dodis, Y., Haralambiev, K., Lopez-Alt, A., Wichs, D.: Cryptography against continuous memory attacks. In: FOCS (2010)
15. Dwork, C., Goldberg, A., Naor, M.: On memory-bound functions for fighting spam. In: Boneh, D. (ed.) CRYPTO 2003. LNCS, vol. 2729, pp. 426–444. Springer, Heidelberg (2003)
16. Dwork, C., Naor, M., Wee, H.: Pebbling and proofs of work. In: Shoup, V. (ed.) CRYPTO 2005. LNCS, vol. 3621, pp. 37–54. Springer, Heidelberg (2005)
17. Dziembowski, S.: Intrusion-resilience via the bounded-storage model. In: Halevi, S., Rabin, T. (eds.) TCC 2006. LNCS, vol. 3876, pp. 207–224. Springer, Heidelberg (2006)
18. Dziembowski, S.: On forward-secure storage. In: Dwork, C. (ed.) CRYPTO 2006. LNCS, vol. 4117, pp. 251–270. Springer, Heidelberg (2006)
19. Dziembowski, S., Kazana, T., Wichs, D.: One-Time Computable Self-Erasing Functions (2010), Cryptology ePrint Archive
20. Dziembowski, S., Maurer, U.M.: Optimal randomizer efficiency in the bounded-storage model. J. Cryptology 17(1) (2004)
21. Dziembowski, S., Pietrzak, K.: Intrusion-resilient secret sharing. In: FOCS (2007)
22. Dziembowski, S., Pietrzak, K.: Leakage-resilient cryptography. In: FOCS (2008)
23. Dziembowski, S., Pietrzak, K., Wichs, D.: Non-malleable codes. In: ICS (2010)
24. ECRYPT. The Side Channel Cryptanalysis Lounge, `http://www.crypto.rub.de/en_sclounge.html`
25. Faust, S., Kiltz, E., Pietrzak, K., Rothblum, G.N.: Leakage-resilient signatures. In: Micciancio, D. (ed.) TCC 2010. LNCS, vol. 5978, pp. 343–360. Springer, Heidelberg (2010)
26. Gennaro, R., Lysyanskaya, A., Malkin, T., Micali, S., Rabin, T.: Algorithmic tamper-proof (ATP) security: Theoretical foundations for security against hardware tampering. In: Naor, M. (ed.) TCC 2004. LNCS, vol. 2951, pp. 258–277. Springer, Heidelberg (2004)
27. Ishai, Y., Prabhakaran, M., Sahai, A., Wagner, D.: Private Circuits II: Keeping Secrets in Tamperable Circuits. In: Vaudenay, S. (ed.) EUROCRYPT 2006. LNCS, vol. 4004, pp. 308–327. Springer, Heidelberg (2006)
28. Ishai, Y., Sahai, A., Wagner, D.: Private Circuits: Securing Hardware against Probing Attacks. In: Boneh, D. (ed.) CRYPTO 2003. LNCS, vol. 2729, pp. 463–481. Springer, Heidelberg (2003)
29. Katz, J., Vaikuntanathan, V.: Signature schemes with bounded leakage resilience. In: Matsui, M. (ed.) ASIACRYPT 2009. LNCS, vol. 5912, pp. 703–720. Springer, Heidelberg (2009)
30. Liu, F.-H., Lysyanskaya, A.: Algorithmic tamper-proof security under probing attacks. In: Garay, J.A., De Prisco, R. (eds.) SCN 2010. LNCS, vol. 6280, pp. 106–120. Springer, Heidelberg (2010)
31. Maurer, U.M.: Conditionally-perfect secrecy and a provably-secure randomized cipher. J. Cryptology 5(1) (1992)

32. Micali, S., Reyzin, L.: Physically observable cryptography. In: Naor, M. (ed.) TCC 2004. LNCS, vol. 2951, pp. 278–296. Springer, Heidelberg (2004)
33. Naor, M., Segev, G.: Public-key cryptosystems resilient to key leakage. In: Halevi, S. (ed.) CRYPTO 2009. LNCS, vol. 5677, pp. 18–35. Springer, Heidelberg (2009)
34. Perito, D., Tsudik, G.: Secure code update for embedded devices via proofs of secure erasure. In: Gritzalis, D., Preneel, B., Theoharidou, M. (eds.) ESORICS 2010. LNCS, vol. 6345, pp. 643–662. Springer, Heidelberg (2010)
35. Pietrzak, K.: A leakage-resilient mode of operation. In: Joux, A. (ed.) EUROCRYPT 2009. LNCS, vol. 5479, pp. 462–482. Springer, Heidelberg (2009)
36. Savage, J.E.: Models of Computation: Exploring the Power of Computing (1997)
37. Standaert, F.-X., Malkin, T., Yung, M.: A unified framework for the analysis of side-channel key recovery attacks. In: Joux, A. (ed.) EUROCRYPT 2009. LNCS, vol. 5479, pp. 443–461. Springer, Heidelberg (2009)
38. von Ahn, L., Blum, M., Hopper, N.J., Langford, J.: Captcha: Using hard ai problems for security. In: Biham, E. (ed.) EUROCRYPT 2003. LNCS, vol. 2656. Springer, Heidelberg (2003)

Perfectly Secure Oblivious RAM without Random Oracles

Ivan Damgård, Sigurd Meldgaard, and Jesper Buus Nielsen

Department of Computer Science, Aarhus University

Abstract. We present an algorithm for implementing a secure oblivious RAM where the access pattern is perfectly hidden in the information theoretic sense, without assuming that the CPU has access to a random oracle. In addition we prove a lower bound on the amount of randomness needed for implementing an information theoretically secure oblivious RAM.

1 Introduction

In many cases it is attractive to store data at an untrusted place, and only retrieve the parts of it you need. Encryption can help to ensure that the party storing the data has no idea of what he is storing, but it may still be possible to get information about the stored data by analyzing the access pattern.

A trivial solution is to access all the data every time one piece of data is needed. However, many algorithms are designed for being efficient in the RAM-model, where access to any word of the memory takes constant time, and so accessing all data for every data access gives an overhead that is linear in the size of the used memory.

This poses the question: is there any way to perfectly hide which data is accessed, while paying a lower overhead cost than for the trivial solution?

Goldreich and Ostrovsky [6] solved this problem in a model with a secure CPU that is equipped with a random oracle and small (constant size) memory. The CPU runs a program while using a (large) RAM that is observed by the adversary. The results from [6] show that any program in the standard RAM model can be transformed using an "'oblivious RAM simulator" into a program for the oblivious RAM model, where the access pattern is information theoretically hidden. The overhead of this transformation is polylogarithmic in the size of the memory.

Whereas it is not reasonable to assume a random oracle in a real implementation, Goldreich and Ostrovski point out that one can replace it by a pseudorandom function (PRF) that only depends on a short key stored by the CPU. This way, one obtains a solution that is only computationally secure. Moreover, in applications related to secure multiparty computation (see below), one would need to securely compute the PRF, which introduces a very significant overhead.

It is a natural question whether one can completely avoid the random oracle/PRF. One obvious approach is to look for a solution that uses a very small

Y. Ishai (Ed.): TCC 2011, LNCS 6597, pp. 144–163, 2011.

number of random bits. But this is not possible: in this paper we show a lower bound on the number of secret random bits that an oblivious RAM simulator must use to hide the access pattern information theoretically: the number of random bits used must grow linearly with the number of memory accesses and logarithmically with the memory size. The natural alternative is to generate the random bits on the fly as you need them, and store those you need to remember in the external RAM. This assumes, of course, that the adversary observes only the access pattern and not the data written to the RAM. However, as we discuss below, there are several natural scenarios where this can be assumed, including applications for secure multiparty computation. The advantage of this approach is that it only assumes a source that delivers random bits on demand, which is clearly a more reasonable assumption than a random oracle and much easier to implement in a secure multiparty computation setting.

Using this approach, we construct an oblivious RAM simulator where we can make an access with an amortized $\log^3(N)$ overhead, where N is the size of the memory provided. The result remains the same, even if the adversary is told which operations the program is executing, and even if the simulator has no internal memory.

In recent concurrent and independent work [2] Ajtai also deals with oblivious RAM and unconditional security. His result solves essentially the same problem as we do, but using a completely different technique that does not lead to an error-free solution. In Ajtai's algorithm, a certain error event must not occur and Ajtai shows that this event happens with probability only $n^{-\log n}$. If the error event happens, then the simulation reveals information it shouldn't, so one cannot get an error-free solution from Ajtai's by, for instance, rebuilding the entire data structure if things go wrong. In comparison, our algorithm fails with probability zero and is simpler to describe and to prove, being a more "direct fit" to the model.

In work following ours [4], Beame and Machmouchi prove that an oblivious RAM simulation always requires a superlogarithmic overhead in both time and space, but in contrast to our work, their result says nothing about the amount of randomness needed.

2 Applications

Software protection: This was the main original application of Goldreich and Ostrovsky. A tamper-proof CPU with an internal secret key and randomness could run an encrypted program stored in an untrusted memory. Now using an oblivious RAM, the observer would only learn the running time and the required memory of the program, and nothing else. Note that, while the adversary would usually be able to see the data written to RAM in such a scenario, this does not have to be the case: if the adversary is doing a side channel attack where he is timing the memory accesses to see if the program hits or misses the cache, he is exactly in a situation where only information on the access pattern leaks, and our solution would give unconditional security.

Secure multiparty computation: If secure multiparty computation is implemented by secret sharing, each player will have a share of the inputs, and computations can be done on the shares. We can use the oblivious RAM model to structure the computation by thinking of the players as jointly implementing the secure CPU, while each cell in the RAM is represented as a secret shared value. This is again a case where an adversary can observe the access pattern (since the protocol must reveal which shared values we access) but not the data[1]. Using an oblivious RAM, we can hide the access pattern and this allows us, for instance, to do array indexing with secret indices much more efficiently than if we had used the standard approach of writing the desired computation as an arithmetic circuit.

Note that players can generate random shared values very efficiently, so that our solution fits this application much better than an approach where a PRF is used and must be securely computed by the players.

Cloud computing: It is becoming more and more common to outsource data storage to untrusted third parties. And even if the user keeps all data encrypted, analysis of the access patterns can still reveal information about the stored data. Oblivious RAM eliminates this problem, leaving the untrusted party only with knowledge about the size of the stored data, and the access frequency.

3 The Model

An oblivious RAM simulator is a functionality that implements the interface of a RAM, using auxiliary access to another RAM (the physical RAM). We say that such a simulation securely implements an oblivious RAM, if for any two access patterns to the simulated RAM, the respective access patterns that the simulation makes to the physical RAM are indistinguishable.

To simplify notation we assume that the interface of a RAM has only one operation, which writes a new value to a memory position and returns the previous value.

We model that the identity of the instructions performed by the simulator leak, but not the operands. We assume that the indices of the memory positions updated by the simulation leak, but not the values being retrieved and stored.

A RAM is modeled as an interactive machine behaving as follows:

1. Set $\mathcal{N}[i] = 0$ for $i \in \mathbb{Z}_N$.
2. On each subsequent input $(\texttt{update}, i, v)$ on the input tape, where $i \in \mathbb{Z}_N$ and $v \in \mathbb{Z}_q$, let $w = \mathcal{N}[i]$, set $\mathcal{N}[i] = v$ and output w on the output tape.

We consider lists of form $U = ((\texttt{update}, i_1, v_1), \ldots, (\texttt{update}, i_\ell, v_\ell))$ with $i_j \in \mathbb{Z}_N$ and $v_j \in \mathbb{Z}_q$. Let $\mathrm{IO}_{\mathrm{RAM}}(U) = ((\texttt{update}, i_1, v_1), w_1, \ldots, (\texttt{update}, i_\ell, v_\ell), w_\ell)$ denote the sequence of inputs and outputs on the input tape and the output tape when the interactive machine is run on the update sequence U, where $|U| = \ell$.

[1] This type of application was also already proposed by Goldreich and Ostrovsky.

Formally, an ORAM simulator is a tuple $\mathcal{S} = (\mathcal{C}, N, M, q)$, where $\mathcal{C} = (\mathcal{C}[0], \ldots, \mathcal{C}[|\mathcal{C}| - 1])$ is the finite program code where each $\mathcal{C}[j]$ is one of $(\texttt{random}, i)$, $(\texttt{const}, i, v)$, $(\texttt{+}, t, l, r)$, $(\texttt{-}, t, l, r)$, $(\texttt{*}, t, l, r)$, $(\texttt{=}, t, l, r)$, $(\texttt{<}, t, l, r)$, $(\texttt{goto}, i)$ with $i, t, l, r \in \mathbb{Z}_M$ and $v \in \mathbb{Z}_q$, and $N \in \mathbb{N}$ is the size of the simulated RAM, $M \in \mathbb{N}$ is the size of the physical RAM, and $q \in \mathbb{N}$ indicates the word size of the RAMs: the simulated and the physical RAM store elements of $\mathbb{Z}_q$. We require that $q > \max(N, M)$ so a word can store a pointer to any address. We denote the simulated memory by $\mathcal{N} \in \mathbb{Z}_q^N$, indexed by $\{0, 1, \ldots, N-1\} \subseteq \mathbb{Z}_q$. We denote the physical memory by $\mathcal{M} \in \mathbb{Z}_q^M$, indexed by $\{0, 1, \ldots, M-1\} \subseteq \mathbb{Z}_q$.

The simulation can be interpreted as an interactive machine. It has a register $\mathbf{c}$ and a special *leakage tape*, which we use for modeling purposes. The machine behaves as follows:

1. Set $\mathcal{M}[i] = 0$ for $i \in \mathbb{Z}_M$.
2. Set $\mathbf{c} = 0$.
3. On each subsequent input $(\texttt{update}, i, v)$ on the input tape, where $i \in \mathbb{Z}_M$ and $v \in \mathbb{Z}_q$, proceed as follows:
   (a) Set $\mathcal{M}[0] = i$ and $\mathcal{M}[1] = v$.
   (b) If $\mathbf{c} > |\mathcal{C}|$, then output $\mathcal{M}[2]$ on the output tape, append $(\texttt{return})$ to the leakage tape and halt. Otherwise, execute the instruction $C = \mathcal{C}[\mathbf{c}]$ as described below, let $\mathbf{c} = \mathbf{c} + 1$ and go to Step 3b. Each instruction C is executed as follows:
      - If $C = (\texttt{random}, i)$, then sample a uniformly random $r \in \mathbb{Z}_q$, set $\mathcal{M}[i] = r$ and append $(\texttt{random}, i)$ to the leakage tape.
      - If $C = (\texttt{const}, i, v)$, set $\mathcal{M}[i] = v$ and append $(\texttt{const}, i, v)$ to the leakage tape.
      - If $C = (\texttt{+}, t, l, r)$, set $\mathcal{M}[t] = \mathcal{M}[l] + \mathcal{M}[r] \bmod q$ and append $(\texttt{+}, t, l, r)$ to the leakage tape. The commands - and * are handled similarly.
      - If $C = (\texttt{=}, t, l, r)$, set $\mathcal{M}[t] = 1$ if $\mathcal{M}[l] = \mathcal{M}[r]$ and set $\mathcal{M}[t] = 0$ otherwise, and append $(\texttt{=}, t, l, r)$ to the leakage tape.
      - If $C = (\texttt{<}, t, l, r)$, set $\mathcal{M}[t] = 1$ if $\mathcal{M}[l] < \mathcal{M}[r]$ as residues in $\{0, \ldots, q-1\}$ and set $\mathcal{M}[t] = 0$ otherwise, and append $(\texttt{<}, t, l, r)$ to the leakage tape.
      - If $C = (\texttt{goto}, i)$, let $\mathbf{c} = \mathcal{M}[i]$ and append $(\texttt{goto}, i, \mathbf{c})$ to the leakage tape.

By $\mathrm{IO}_{\mathcal{S}}(U)$ we denote the random variable describing the sequence of inputs and outputs on the input and output tapes when $\mathcal{S}$ is executed as above on the update sequence U, where the randomness is taken over the values r sampled by the random-commands. By $\mathrm{LEAK}_{\mathcal{S}}(U)$ we denote the random variable describing the outputs on the leakage tape.

Definition 1. *$\mathcal{S}$ is an ϵ-secure ORAM simulator if for all update sequences U, the statistical difference $\Delta(\mathrm{IO}_{\mathcal{S}}(U), \mathrm{IO}_{\mathrm{RAM}}(U)) \leq \epsilon$ and it holds for all update sequences U_0 and U_1 with $|U_0| = |U_1|$ that $\Delta(\mathrm{LEAK}_{\mathcal{S}}(U_0), \mathrm{LEAK}_{\mathcal{S}}(U_1)) \leq \epsilon$. We say that $\mathcal{S}$ is a perfectly secure ORAM simulator if it is a 0-secure ORAM simulator.*

4 Oblivious Sorting and Shuffling

In the following, we will need to shuffle a list of records obliviously. One way to do this, is to assign a random number to each, and sort them according to this number. If the numbers are large enough we choose distinct numbers for each value with very high probability, and then the permutation we obtain is uniformly chosen among all permutations.

This issue is, in fact, the only source of error in our solution.

If we want to make sure we succeed, we can simply run through the records after sorting to see if all random numbers were different. If not, we choose a new random set of numbers and do another sorting. This will succeed in expected $O(1)$ attempts each taking $O(n)$ time, and so in asymptotically the same (expected) time, we can have a perfect solution.

We can sort obliviously by means of a sorting network. This can be done with $O(n \log n)$ compare-and-switch operations, but a very high constant overhead [1], or more practically with a Batcher's network [3] using $O(n \log^2 n)$ switches.

Each switch can be implemented with two reads and two writes to the memory, and a constant number of primitive operations. Sorting in this way is oblivious because the accesses are fixed by the size of the data, and therefore independent of the data stored.

By storing the choice bits of each switch we can arrange according to the inverse permutation by running the elements through the switches of the sorting network in backwards order while switching in the opposite direction from before.

5 A Solution with Polylogarithmic Overhead

Like the original algorithm of Goldreich and Ostrovsky in [6], the algorithm works by randomly permuting the data entries, and then storing the permutation in a dictionary data structure used for lookups. Previously accessed elements are put into a cache to ensure we do not have to look at the same place twice, as that would reveal patterns in the accesses. The dictionary also contains dummy elements we can look at, to hide whether we found an element in the cache or in the dictionary. We amortize the time used for searching the increasingly large cache by making levels of caches of growing sizes that are reshuffled when they have been used enough.

The construction in [6] used a hash-table as the dictionary, and security was proven in the random oracle model. A random oracle allows to remember and randomly access a large number of random bits "for free", but we want to do without this, so instead we save our randomness in physical memory using a binary tree for the dictionary.

We now present a solution with an amortized overhead per access of $\log^3 N$. The solution proceeds somewhat like the protocol of [6] with polylogarithmic overhead.

As in [6] the idea is to have $\log_2(N) + 1$ levels of simulated RAMs each level being a cache for the one below. The cache at level 0 has size 1, and the one at level $\log_2(N)$ size N, and in general the cache at level ℓ stores $O(2^\ell)$ elements.

As in [5] it is also possible to make a conceptually simpler solution with a higher asymptotic overhead, this might be useful for some input sizes, and also may aid in the understanding of the full algorithm, we describe that in section 6.

5.1 Shuffled Trees

A shuffled tree is a complete binary search tree with all nodes (except the root node) permuted randomly in the RAM. So when you follow a child-pointer, it will give you a uniformly random location in the RAM where the child-node is stored. The actual data is stored at the leafs.

Deciding between the left and right child can be done obliviously by looking at both pointers. In this way a full lookup in a shuffled tree is oblivious: an adversary observing the access pattern to the RAM storing the shuffled tree, sees only $\log N$ random accesses to nodes, and thus learns nothing about what element we find.

In the algorithm described below there is a single logical tree. The nodes of this tree are represented by records in physical memory.

We will have several levels of caches, each can store a number of records representing nodes from the tree. Pointers in the tree will be represented as a pair: the number of the level and an offset into the memory block for that level. The division of the physical RAM into levels is public information that does not have to be hidden.

During a lookup we look at some nodes from the logical tree, this results in touching the records representing these nodes. However, we will never again read those records, we now call them *dead* records. Instead new records representing the same nodes of the tree will be created on lower levels. In this way a node of the tree might be represented by several records on different levels, but only one will ever be alive. Specifically every logical node will always be represented by a single live record at some level. In section 5.3 we show how to construct such a tree obliviously.

5.2 Overview of the Construction

The main ideas behind the solution are as follows:

Initial setup. We start with a random shuffled tree at level $\log N$ with all data in this tree and all other levels (the caches) being empty.

Each internal node of the tree stores: a left and a right child-pointer, each storing both the level, and the index of that level where the child can be found, a bound for dispatching during the lookup (this bound can actually be inferred from the path we are following, and is mostly included for easing the presentation), and a tag telling if it is a dummy node or not. The leaf nodes store a data-item and the original index of the item. The smallest kind of node has to be padded to make them have the same size.

Lookup and caching. When a lookup is made, we start at the root node, and follow child-pointers to a leaf, like when looking up in a binary tree. But now a child pointer can point to a node at the same or any higher cache-level.

When a path $(\nabla_1, \ldots, \nabla_m)$ is followed, we first try to put the nodes $\{\nabla_i\}$ in the bottom level, i.e. level 0. If this level is empty, the path is shuffled (such that the nodes ∇_i are stored in random positions at level 0 and such that ∇_i points to the new physical location of ∇_{i+1} at the bottom level) and inserted here. Otherwise it is merged with the tree already on this level, and we try to put the result on the next level and so on until we find an empty level. Below we describe a procedure for merging two trees while shuffling them so all records end up being permuted randomly.

The pointers from ∇_i to the physical addresses of the siblings of the ∇_i which were not on the path $(\nabla_1, \ldots, \nabla_m)$, and hence were not moved, are just copied along with the nodes during the shuffling, so that the nodes ∇_i at the bottom level might now have pointers to untouched nodes in the trees at higher levels.

The root. The only node whose true position is not hidden is the root node. Because it is touched as the first node for every look-up, it will always be at the first position of the youngest tree. We take special precautions to make sure it is not moved during the shuffling procedure. This is still oblivious because the node is touched for every lookup independently of the index we search for.

Dummy nodes. If we were just to start each search at the root node and then following the updated points to physical addresses it is true that we would never touch the same node twice in the same position, as a touched node is moved down and then shuffled up. The pattern of how the path jumps between the levels would however leak information[2]. We therefore make sure to touch each level once every time we follow one pointer by doing *dummy reads* at all other levels. If we are following a pointer to level ℓ we do dummy read at levels $i = \log_2 N, \ldots, \ell + 1$, then we read the node at level ℓ, and then we do dummy reads at levels $i = \ell - 1, \ldots, 1$. Only the real node ∇ read at level ℓ is moved to the bottom level.

This requires the tree at each level to contain as many dummy nodes as it has real nodes. These will be chained by their child-pointers, so we can chose for each level to either look at the real node or the next dummy. The location of the head of the dummy chain for each level is stored publicly, like the root node.

---

[2] If, e.g., we look up the same leaf node twice, the path to the node is moved to the bottom level in the first update, so the second update would only touch nodes at the bottom level. If we look up a distinct leaf in the second update the pointer jumping would at some point take us back to the bottom level. This would allow the adversary to distinguish.

Algorithm 5.1: LOOKUP(*key*)

```
output: Value stored in the ORAM at index key
global: live — The set of levels containing records
(current-level, current-offset) ← (min(live), 0) — Start at root
for i ← 0 to log2(N)
  do
    for each level ∈ live
      do
        if level = current-level
          then
            ∇ ← physical[level, current-offset]
            if key < ∇.bound
              then
                (next-level, next-offset) ← (∇.left.level, ∇.left.offset)
                (∇.left.level, ∇.left.offset) ← (0, i)
                physical[level].dummy ← physical[level].dummy
              else {...same thing for right side...
          else
            Dummy lookup:
            t ← physical[level, physical[level].dummy]
            physical[level].dummy ← t.left.offset
    path ← path · ∇
    (current-level, current-offset) ← (next-level, next-offset)
INSERTPATH(path)
return (∇.data)
```

Algorithm 5.2: INSERTPATH(*path*)

```
global: live — The set of levels containing records
level ← 0
while level ∈ live — Cascade down until we find a not-live level
  do
    path ← MERGELAYERS(path, physical[level])
    live ← live − {level} — Level is no longer live
    level ← level + 1
physical[level] ← path — Insert the merged nodes
live ← live ∪ {level}
```

A sequence of accesses is illustrated in Fig. 1. The procedure `MergeLayers` is not described in pseudocode, but it is explained in detail in the following section.

5.3 Merging and Shuffling of Two Unbalanced Trees

Each time a lookup is made, we build a chain of records representing nodes on a path from the root of the tree to the leaf-node we are looking up by copying the nodes from the used path, and making sure that the child pointer we follow points to the next node.

This chain will go to the bottom level cache. If the bottom level already is filled, it will be merged with the new one path and they will be put one level higher, etc.

The records we merge from two levels represent some subset of the nodes from the logical tree. This subset is always forming a tree but not in general a full tree, but an unbalanced tree.

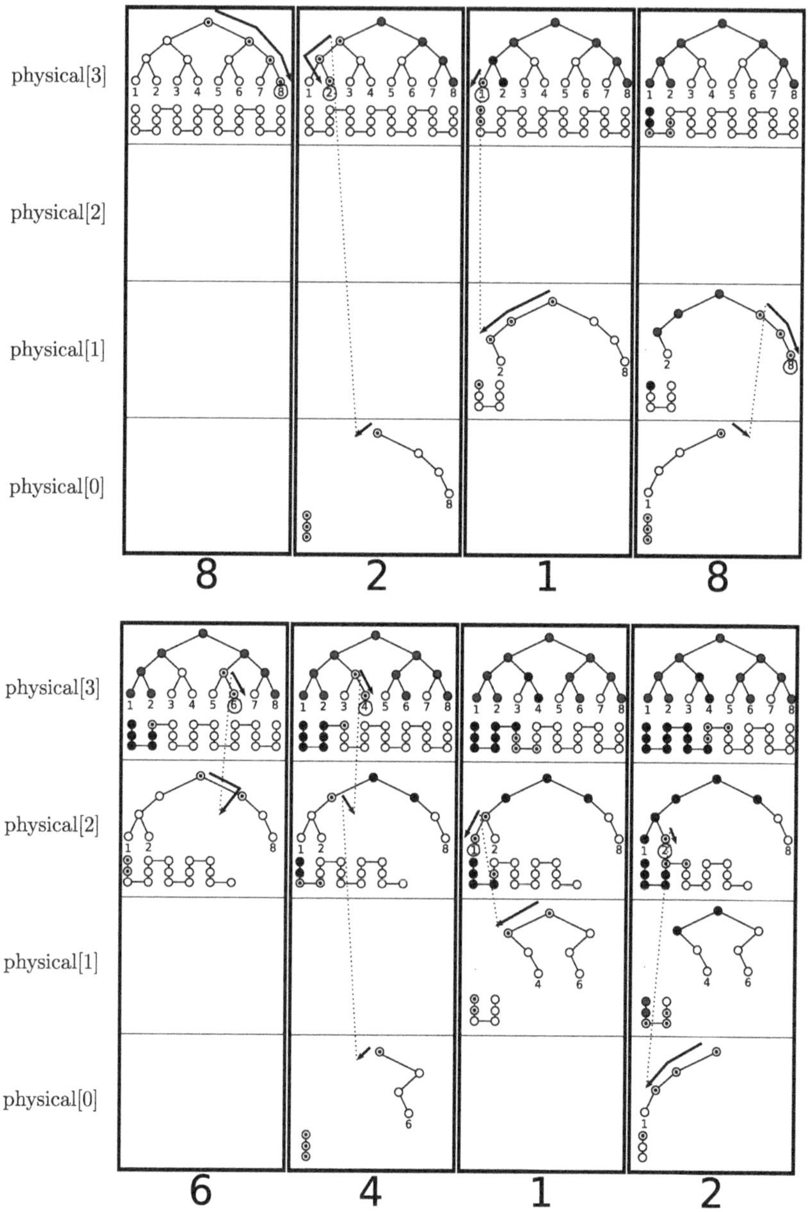

Fig. 1. To be read as a comic strip. Each frame shows the logical (unshuffled) view of the different layers of each step of a sequence of accesses to an oblivious RAM storing 8 elements. Dummy elements at each level are shown at the bottom. The arrow shows how the lookup progresses, and nodes with a dot are those read at this step. Used nodes are shown in gray. Notice how no node is ever touched twice, and how, besides the root node, $\log n$ nodes are read from each live level per lookup.

We now describe how to merge and shuffle two such trees while updating the pointers so they point to the right new place.

We make a crucial observation about the algorithm; that the two merged levels are always two youngest existing levels, this in turn gives us the following:

- There will never be pointers from other levels into the merged records so nothing outside the level has to be updated.
- The child-pointers that are not internal to the two levels will point out of the local level and into an older level, as older levels are not updated, these pointers do not have to be changed.
- The root of the tree will always be in one of the merged levels (the youngest one).

We will copy all the records from both trees that we have not touched yet (the live records) into a new list in the same order as they where before (but with the dead records removed). We run through the second list, and obliviously add to each internal child pointer the number of live records before it in the list, so all internal child-pointers are unique.

Also we connect the two dummy chains, by running through all the records of the first tree, and for each obliviously checking if it is the end of a dummy chain, if so, it is set to point at the child of the head of the dummy chain of the second tree.

In the two lists there will be a number of live dummies already, and they will be reused, but for the resulting list we need more dummies (as many as there are already), so we make a chain of new dummies and make the end of that point to the head of the old dummy chain. These new records are put at the end of the list of nodes, and we shuffle them together with the rest.

We know that every live node is a unique representative of a node in the logical tree. It always (except for the root) has exactly one incoming pointer from the other nodes in the list. And it has one or two children among the copied nodes, the pointers to these needs to be updated, while the (one or zero) child-pointers pointing *out of the tree* to an older tree from where it was copied should stay unchanged.

The root-node and the head of the dummy chain are exceptions to this invariant as they have no incoming pointers, on the other hand their location is not secret, and we take special care of these two nodes.

Note that we cannot allow the dummy nodes to have both child-pointers pointing to the next node because that would give dummy nodes two incoming nodes as well. Instead they must have one forward pointer, an nil-pointer and a tag so we can recognize them as dummy nodes, and when following such one, we always obliviously choose to follow the real pointer.

We now collect this into a subroutine for updating the pointers while we merge and shuffle the two lists with a total of p nodes at the same time. The process is illustrated in Fig. 2

1. For each child pointer in the list in order, we write in a new list a pointer y that, if the pointer is internal, is a copy of the child pointer and ∞ otherwise

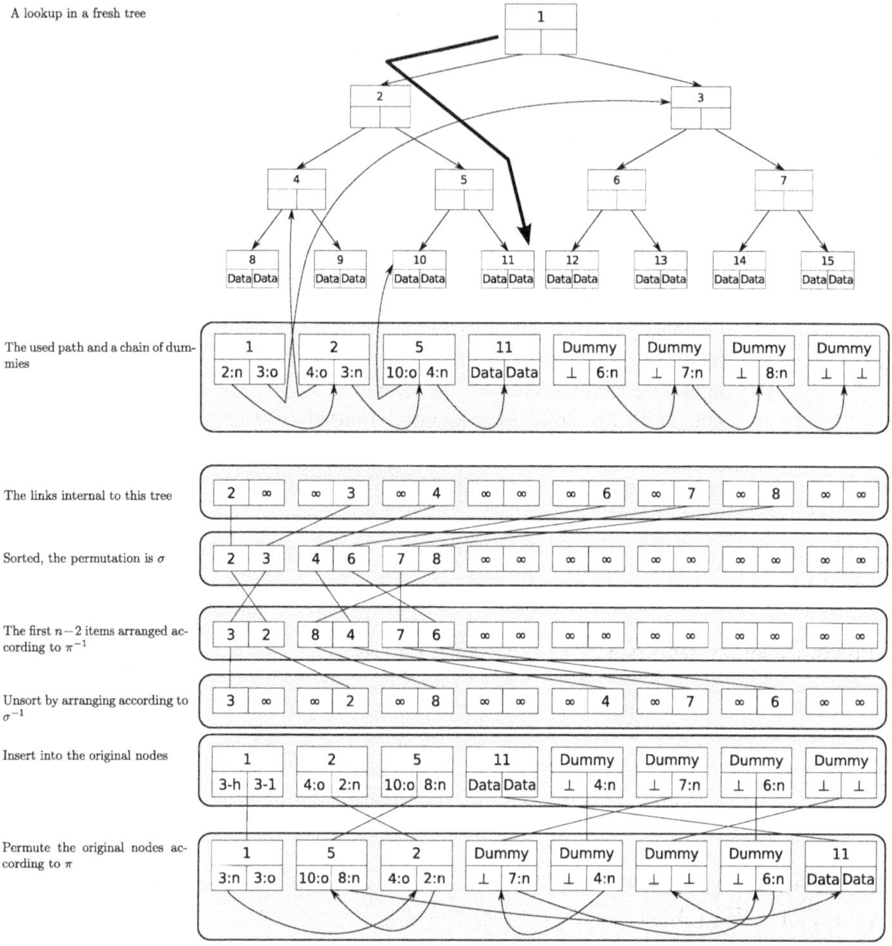

Fig. 2. Illustration of the shuffling process. ":n" here indicates a pointer inside the level that is shuffled, ":o" is a pointer to the old level where the node was copied from.

so it will be sorted last. These represent pointers out of the tree, and should not be updated.

When shuffling a list with p live records, the list of pointers will have $2p$ entries. Approximately half of these will be external pointers, and nil-pointers from dummies. It might be more than half, because the paths represented by the two trees might overlap. This does not affect the shuffling procedure, only the dummy chains will end up longer than what can ever be used before reshuffling, but this knowledge is never revealed to the adversary.

2. Sort the row of y's obliviously, and save the sorting permutation as σ. Because each node (except the two special nodes) has exactly one incoming pointer, the first $p-2$ nodes will be internal, and those are the ones we want to update, so they point to the new locations of those nodes.

3. Now we create the permutation, π, that we want to shuffle the final list in. This permutation should leave location 1 (the root) and the location of the head of the dummy chain the same place, and be uniformly random for all other indices.
4. Now permute a list containing the numbers from 1 to p according to π^{-1}, and remove the first element, and the dummy-head pointer (they stay in place under π). Take this list of $p-2$ elements and concatenate them with $p+2$ ∞'s.
5. Now first undo the sorting by applying σ^{-1} to this list, and use the unsorted list to overwrite the pointers of the original records where they came from (obliviously choosing to change the value only when the new value is $\neq \infty$), now we have updated all the internal pointers, and left the external pointers untouched.
6. Finally shuffle the nodes according to π. We now have a shuffled list with all nodes pointing to the right places.

The main trick behind this procedure is that $\pi(\mathcal{M})[\pi^{-1}(x)] = \mathcal{M}[x]$ (where $\mathcal{M}$ denotes the physical memory, seen as an array). The time the procedure takes is dominated by the time it takes to sort the list of pointers.

5.4 Security

The transcript of a lookup, as seen by the adversary, always consists of the following parts:

- An access to the root node, it will always be the first entry of the youngest tree
- $\log n$ accesses at uniformly random records at each live cache-level of the tree
- A fixed amount of merging and reshuffling, done obliviously

This transcript is independent from the simulated access, and therefore the algorithm is secure.

5.5 Performance

To see how many log factors we use, we look at the two parts of the algorithm (that takes time). That is the operations for doing the *lookup*, and the operations for doing the *reshuffling*.

Following a path of length $\log N$ requires to look up $\log N$ nodes. For each node we have to touch each level for obliviousness, which gives a total of $\log^2 N$ reads.

For reshuffling we saw that the time for shuffling a level is dominated by sorting it. We have $\log N$ cache-levels. Level i is of size $O(2^i \cdot \log N)$, so that takes $O(2^i \cdot \log N \cdot \log(2^i \cdot \log N)) = O(2^i \cdot \log N \cdot i)$ operations to sort (because

i is in $O(\log N)$, so it dominates $\log\log N$). But that level is only shuffled every 2^{i-1}th. lookup. On average we therefore use time for all levels:

$$O\left(\sum_{i=1}^{\log N} \log N \cdot i\right) = O(\log^3 N) .$$

Therefore the total amortized time spent per lookup is in $O(\log^3 N)$.

This gives rise to the following theorem:

Theorem 1. *The described algorithm implements a perfectly secure ORAM simulation in the standard model with a time and memory overhead in* $O(\log^3 N)$.

6 The Square Root Algorithm

In this section we describe a simpler algorithm implementing an oblivious RAM using memory and amortized time in $O(\sqrt{N}\log^2 N)$. The algorithm assumes access to a functionality that shuffles n elements of the physical RAM in time $O(n\log^2 n)$.

In this solution the tree is stored with each level shuffled individually, so each child pointer from a node points to a random location of the next level. Also we make $\sqrt{N}$ dummy chains, that are also shuffled into the tree. Only the root level is (trivially) not shuffled. The shuffling is depicted in Fig. 3.

Making a lookup. A lookup in the simulated RAM is implemented by making a lookup in the binary search tree. In order to touch every node in the tree only once, we do a (possibly dummy) lookup in the physical RAM on each level of the tree, and for each level we also linearly scan through all of the cache to see if we have accessed the same node earlier. If we found the element in the cache, the next access in the tree will still be to a dummy node.

The physical memory is split up into $\log_2(N)$ parts, the i'th part is again split in two; $physical[i]$ storing $2^i + \sqrt{N}$ records (the tree, and the dummy chains), and $cache[i]$ storing $\sqrt{N}$ records (the cache). Each node in the three record stores a .left and .right field for pointing to the next level, and a .bound for directing the search in the tree.

The leaf-records at the lowermost level are different, they contain the .data that are stored, an .index field naming the original index where the data is stored in the simulated RAM, and a .used field that is used for reshuffling the data as described below.

An invariant for the outer loop in the Lookup-algorithm below can be phrased:

1. *next* is the real index we are going to look for at *level*
2. *next_from_tree* is the index of the tree where we will look if we do not find the item in the cache. If this is different from next, it is still pointing at a dummy chain.

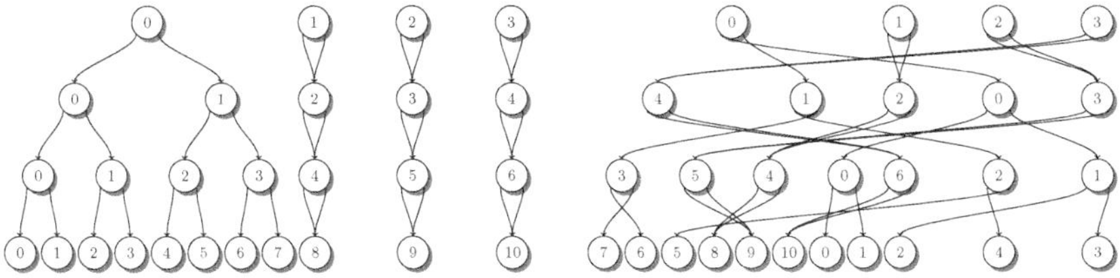

(a) Before (b) After

Fig. 3. Visualization of the memory layout of a tree storing 8 elements, before and after shuffling the tree. The edges indicate child-pointers.

By changing the index of the cached value to ∞ when it is found at 1 we implicitly invalidate it; it will be sorted last and therefore thrown away when we reshuffle. This is only necessary to do for the cache of the last level of the tree.

Obliviously shuffling a tree. This is a somewhat simpler algorithm for shuffling a tree that the one described in the main part of the paper: it works by shuffling one level at a time, starting at the bottom of the tree, and ending with the root-level. Because we always have all nodes present, we can allow ourselves to consider only one level at a time. After shuffling a level L with permutation π (so $L'[i] = L[\pi(i)]$), we apply π^{-1} on a sequence $[1, \ldots, n]$, and copy these numbers into the child-pointer fields of the level above. This gives the right result, because what before pointed at k will now point at $\pi^{-1}(k)$. But looking up in the shuffled layer yields the record at $L'[\pi^{-1}(k)] = L[\pi(\pi^{-1}(k))] = L[k]$. We take special care to also shuffle the dummy chains, and ensuring that their left and right pointers point to the same next node in the chain.

Algorithm 6.1: UNMINGLE( )

$a \leftarrow$ Filter out any physical[$\log_2 N$] record which has `.used`= **true**
$b \leftarrow a$ concatenated with cache[$\log_2 N$].
Obliviously sort b according to the original *index* of the records
Remove the last $\sqrt{N}$ records of b
$physical[log_2 N] \leftarrow b$

Algorithm 6.2: SHUFFLE( )

for $level \leftarrow log_2(n)$ **downto** 1
do
  Choose a permutation π uniformly at random
  Shuffle physical[level] according to π
  $temp = [1, 2, \ldots, 2^{level} + \sqrt{N}]$
  Shuffle temp according to π^{-1}
  for $i \leftarrow 0$ **to** $2^{level-1}$
  do $physical[level-1, i]$.left $\leftarrow temp[2i]$
     $physical[level-1, i]$.right $\leftarrow temp[2i+1]$
  for $i \leftarrow 2^{level-1}$ **to** $2^{level-1} + \sqrt{N}$
  do $physical[level-1, i]$.left $\leftarrow temp[2^{level} + i]$
     $physical[level-1, i]$.right $\leftarrow temp[2^{level} + i]$

Algorithm 6.3: DISPATCH($key, record$)

output: The left or right child of record, depending on key
if $key < record.bound$
  then return ($record$.left)
  else return ($record$.right)

Algorithm 6.4: LOOKUP(key)

input: key
output: Value stored at key
if $count \geq \sqrt{n}$
  then $\begin{cases} \text{UNMINGLE}() \\ \text{SHUFFLE}() \\ count \leftarrow 0 \end{cases}$
  else $count \leftarrow count + 1$
$next \leftarrow count$
$next\_from\_tree \leftarrow count$
for $level \leftarrow 0$ **to** $\log_2(N) - 1$
  do
    $found \leftarrow False$
    for $i \leftarrow 0$ **to** $count - 1$
      do
        $k \leftarrow cache[level, i]$
        if k.index = next
          then $\begin{cases} k\_from\_cache \leftarrow k \\ found \leftarrow True \\ k.\text{index} = \infty \\ cache[level, i] \leftarrow k \end{cases}$ (1)
    if $found$
      then $next \leftarrow next\_from\_tree$
    $k\_from\_tree \leftarrow physical[level, next]$
    $physical[level, next]$.used $= True$
    $next\_from\_tree \leftarrow$ DISPATCH($index, k\_from\_tree$)
    $next \leftarrow$ DISPATCH($index, k\_from\_tree$)
    if $found$
      then $next \leftarrow$ DISPATCH($index, k\_from\_tree$)
    $cache[level, count] \leftarrow (next,$ UPDATE$(k))$

Security. The transcript of a lookup, as seen by the adversary, always consists of the following parts:

- An access at index $count$ of the first level of the tree
- An access at a uniformly random location at each lower level of the tree
- A scan of the full cache of each level
- For each $\sqrt{N}$ accesses, the tree is reshuffled

All these are independent of the access pattern to the original RAM, and thus an eavesdropper will learn nothing whatsoever about the access pattern.

Performance. The time for a single lookup (without the shuffling) is dominated by accessing $\log_2 N$ caches, each of size $O(\sqrt{N})$. For each $\sqrt{N}$ lookups, we perform a shuffle taking time $O(N \log^2 N)$. Giving an amortized running time of $O(\sqrt{N} \log^2 N)$ per lookup.

7 Lower Bounds on Randomness

An equivalent way to state the definition of a secure oblivious RAM simulation is that, for simulation of any program running on an standard RAM, the resulting distribution of accesses to physical memory is the same for every choice of input. In particular, if we choose (part of) the input at random, the distribution of memory accesses must remain the same.

Our goal in this section will be to show a lower bound on the number of random bits that a simulation must use in order to be secure. We will for the moment assume that the simulation is *data-oblivious*, i.e., the simulation treats every word it is asked to read or write as a black-box and does not look at the data it contains. Any such word is called a data-word. All known oblivious RAM simulators are data oblivious. Indeed, letting anything the simulator does depend on the content of a data-word would only seem to introduce extra difficulties, since at the end of the day the access pattern must not depend on this content. Nevertheless, we show in section 8 a weaker lower bound for data non-oblivious simulation.

To show the lower bound, we consider the program that first writes random data to all N locations in RAM. It then executes d read operations from randomly chosen locations. Let U denote this sequence of update instructions.

Let P be the random variable describing the choice of locations to read from. Clearly $H(P) = d \log N$. Let $C = \text{LEAK}_{\mathcal{S}}(U)$ be the random variable describing the history of the simulation as seen by the adversary. Finally, let K be the random variable describing the random choices made by the simulation during the execution of the program, the r-values of the `random`-commands. We will show a bound on $H(K|C)$, that is, a bound on the number of random bits that are unknown to the adversary.

By construction of the random experiment, K is independent of P and, assuming the simulation is perfectly secure, C and P are independent.

From this it follows by elementary information theory that

$$H(K|C) \geq H(P) - H(P|C,K) = d \log N - H(P|C,K) . \tag{1}$$

Now, let us assume that each read operation causes the simulation to access at most n locations in physical RAM. From this we will show an upper bound on $H(P|C,K)$.

Let P_i be the random variable describing the choice of location to read from in the i 'th read. Then we can write $P = (P_d, \ldots, P_1)$, and we have

$$\begin{aligned} H(P|C,K) &= H(P_1|C,K) + H(P_2|P_1,C,K) + \cdots + H(P_d|P_{d-1}..P_1,C,K) \quad (2) \\ &\leq H(P_1|C,K) + H(P_2|C,K) + \cdots + H(P_d|C,K) . \quad (3) \end{aligned}$$

The plan is now to bound $H(P_i|C = c, K = k)$ for each i and arbitrary, fixed values c, k, from this will follow a bound on $H(P|C, K)$. We will write $H(P_i|C = c, K = k) = H(P_i|c, k)$ for short in the following.

Note first that once we fix $K = k, C = c$, in particular the value of c specifies a choice of at most n locations that are accessed during the i'th read operation. This will constrain the distribution of P_i to be only over values that cause these locations to be accessed. Let w be the number of remaining possible choices for P_i. This is a set of addresses that we call the *relevant* addresses. Since there are w relevant addresses, we have $H(P_i|c, k) \leq \log w$.

Let $a = \log_2 q$ be the number of bits in a memory location, and recall that the program initially writes random data to all addresses. Let the random variable $D_{c,k}$ represent the choice of data originally written to the w relevant addresses. Since the simulation is data oblivious, fixing $C = c, K = k$ does not constrain the data stored in the relevant addresses, and hence $D_{c,k}$ is uniform over the 2^{aw} possible values, or equivalently $H(D_{c,k}) = aw$.

Let $R_{c,k}$ represent all the data the simulator accesses while it executes the i'th read operation given the constraints we have defined. Since at most n words from external memory are accessed, we have $H(R_{c,k}) \leq an$.

But on the other hand, $H(R_{c,k})$ must be at least $H(D_{c,k})$, since otherwise the simulator does not have enough information to return a correct result of the read operation. More precisely, since the simulation always returns correct results, we can reconstruct the exact value of $D_{c,k}$ as a deterministic function of $R_{c,k}$ by letting the value of P_i run through all w possibilities and computing in each case what the simulation would return. Since applying a deterministic function can only decrease entropy, it must be that $H(D_{c,k}) \leq H(R_{c,k})$.

We therefore conclude that $aw \leq an$, and from this and $H(P_i|c, k) \leq \log w$ it follows immediately that $H(P_i|c, k) \leq \log n$. By definition of conditional entropy we have $H(P_i|C, K) \leq \log n$ as well, and hence by (3) that $H(P|K, C) \leq d \log n$. Combining this with (1) we see that $H(K|C) \geq d \log(N/n)$, when d read operations are executed. Thus we have:

Theorem 2. *Suppose we are given a perfectly secure oblivious RAM simulation of a memory of size N. Assume it accesses at most n locations in physical RAM per read operation and is is data-oblivious. Then there exist programs such that the simulator must, on average, use at least* $\log(N/n)$ *secret random bits per read operation.*

Extensions. Suppose the simulation is not perfect, but leaks a small amount of information. This means that P and C are not independent, rather the distributions of C caused by different choices of P are statistically close. Therefore the information overlap $I(C; P)$ is negligible as a function of the security parameter. If we drop the assumption that P, C are independent (3) becomes $H(K|C) \geq d \log N - H(P|C, K) - I(P; C)$. The rest of proof does not depend on C, P being independent, so the lower bound changes by only a negligible amount.

The result also holds even if the adversary does not know which instructions are executed by the simulated program, as the proof does not exploit the fact that in our model, he knows this information.

Another remark is that the result assumes that every read operation causes at most n locations to be accessed. This does not, actually, cover the known solutions because they may at times access a very large number of locations, but do so seldom enough to keep the amortized overhead small. So we would like to show that even if we only assume the amortized overhead to be small, a large number of secret random bits is still needed. To this end, consider the same program as before, and suppose we have a simulation that accesses at most n_i bits during the i'th read operation. Let $\bar{n}$ be the average, $\bar{n} = \frac{1}{d}\sum_i n_i$. Then we have

Theorem 3. *Suppose we are given a perfectly secure oblivious RAM simulation of a memory of size N. Assume it accesses at most n_i locations in physical RAM during the i'th read operation and is data-oblivious. Then there exist programs such that the simulator must, on average, use at least* $\log(N/\bar{n})$ *secret random bits per read operation.*

Proof. Observe first that the argument in the proof for Theorem 2 for equations (3) and (1) still holds here, and that furthermore the argument for $H(P_i|C,K) \leq \log n$ does not depend on the bound n being the same for every read operation. Hence, under the assumption given here, we get $H(P_i|C,K) \leq \log n_i$, and hence by (3), (1) that

$$
\begin{aligned}
H(K|C) &\geq d\log N - \sum_{i=1}^{d} \log n_i = d\log N - d\frac{1}{d}\sum_{i=1}^{d}\log n_i \\
&\geq d\log N - d\log(\frac{1}{d}\sum_{i=1}^{d} n_i) = d\log N - d\log\bar{n} = d\log(N/\bar{n}) \ .
\end{aligned}
$$

where the last inequality follows from Jensen's inequality.

8 Data Non-oblivious Simulation

The argument used in section 7 breaks down if the simulation is not data-oblivious. The problem comes from the fact that then, even if security demands that C is independent of the data D that is written, this may no longer be the case if K is given. And then, if we fix C, this may mean that the data written in the w locations in the proof above are no longer uniformly distributed.

We can nevertheless prove a different lower bound. Since we assume that $q \geq N$ we have that the word size $a = \log_2 q$ is $\theta(\log N)$. Our result would get stronger if a was larger. The goal will be to show that when executing $\theta(N)$ read operations, security implies that $H(K|C)$ must be at least $\theta(N)$, and so one must use a constant number of secret random bits per read operation.

We will consider the same program as in the previous subsection, and we will reuse the notation. We will, however, choose d to be $d = \alpha N$ where α is a constant.

Observe first that since both C and K are independent of D, it follows that $I(K; D|C) = I(C; D|K)$. It is also straightforward to see that $H(K|C) \geq I(K; D|C)$. So if $I(C; D|K) \geq d$ we are already done. So, we may assume $I(C; D|K) < d$. We have that $H(D) = Na$, and also it holds that

$$H(D|C, K) \geq H(D) - I(K; D|C) - I(C; D|K) \geq Na - 2d,$$

where the first inequality holds for any 3 random variables D, K, C. Now by our choice of d, it is much smaller than Na, so $H(D|C, K)$ is almost as large as it can be. So since $H(D|K, C)$ is the average of the values $H(D|K = k, C = c)$ we want to claim that $H(D|K = k, C = c)$ is "large most of the time". Concretely, let p be the probability that $H(D|K = k, C = c) \leq 3Na/4$, taken over the choices of K, C. Then it is straightforward to argue that

$$(1 - p) \cdot Na + p \cdot 3Na/4 \geq Na - 2d = Na - 2\alpha N,$$

from which we conclude that $p \leq 8\alpha/a$, and so is $\theta(1/\log N)$.

We will now do exactly the same argument as for the previous theorem, so we will want to bound $H(P_i|c, k) \leq \log w$, by looking at the number w of possible choices for P_i given that $C = c, K = k$.

We will use the fact we established above, that with probability $1 - p$, $H(D|c, k) \geq 3Na/4$. Values c, k for which this holds will be called a *good* choice of c, k.

So we will assume for the moment that we are in the good case. Now, there are two sub-cases: first, if $w \leq N/2$ we already know enough, as we shall see in a moment. In the second sub-case $w > N/2$, and we now use that $H(D|c, k) \geq 3Na/4$: there are at most $N/2$ words that are *not* among those w we consider, so their entropy can be at most $Na/2$. It follows that the joint entropy of the w words we do consider is at least $Na/4 \geq wa/4$.

We now argue as before, that since the simulator only has access to at most na bits in order to answer the read request, it must be that $na \geq wa/4$, or $w \leq 4n$.

Hence, for a good choice of c, k we have

$$H(P_i|c, k) \leq \max(\log(N/2), \log(4n)).$$

We now make the reasonable assumption that $4n \leq N/2$, and note that even for bad values of c, k $H(P_i|c, k) \leq \log N$. From this we conclude that

$$H(P_i|C, K) \leq p \log N + (1 - p) \log(N/2) = \log(N/2) + p \log 2.$$

Plugging this into (1), we finally obtain that

$$H(K|C) \geq d \log N - d \log(N/2) - dp \log 2 = d(1 - p \log 2),$$

which is indeed $\theta(N)$, since p is $\theta(1/\log N)$, and is in fact essentially d for large N. We therefore have the following:

Theorem 4. *Suppose we are given a perfectly secure oblivious RAM simulation of a memory of size N, accessing at most n locations in physical RAM per read operation. Even if the simulation is data non-oblivious, if $n \leq N/8$, such a simulation must use, on average, at least $\theta(1)$ secret random bits per read operation.*

The analysis in fact shows that one must use (essentially) 1 bit per read operation. This bound is almost certainly not tight, but we have not spent much effort in optimizing it, since data oblivious simulation clearly is the way to construct actual solutions.

The result can easily be extended to the case where the simulation does not access the same number of locations in every operation, as long as the bound $N/8$ is always observed.

References

1. Ajtai, M., Komlós, J., Szemerédi, E.: An 0(n log n) sorting network. In: STOC 1983: Proceedings of the Fifteenth Annual ACM Symposium on Theory of Computing, pp. 1–9. ACM, New York (1983)
2. Ajtai, M.: Oblivious rams without cryptographic assumptions. In: STOC 2010: Proceedings of the 42nd Annual ACM Symposium on Theory of Computing (2010) (to be published at STOC)
3. Batcher, K.E.: Sorting networks and their applications. In: AFIPS 1968 (Spring): Proceedings of the Spring Joint Computer Conference, April 30-May 2, pp. 307–314. ACM, New York (1968)
4. Beame, P., Machmouchi, W.: Making RAMs Oblivious Requires Superlogarithmic Overhead. Electronic Colloquium on Computational Complexity (ECCC) 10(104) (2010)
5. Goldreich, O.: Towards a theory of software protection and simulation by oblivious rams. In: STOC 1987: Proceedings of the Nineteenth Annual ACM Symposium on Theory of Computing, pp. 182–194. ACM, New York (1987)
6. Goldreich, O., Ostrovsky, R.: Software protection and simulation on oblivious rams. J. ACM 43(3), 431–473 (1996)

Unconditional and Composable Security Using a Single Stateful Tamper-Proof Hardware Token

Nico Döttling, Daniel Kraschewski, and Jörn Müller-Quade

Institute of Cryptography and Security, Faculty of Informatics,
Karlsruhe Institute of Technology, Germany
doettling@kit.edu, kraschewski@kit.edu, mueller-quade@kit.edu

Abstract. Cryptographic assumptions regarding tamper proof hardware tokens have gained increasing attention. Even if the tamper-proof hardware is issued by one of the parties, and hence not necessarily trusted by the other, many tasks become possible: Tamper proof hardware is sufficient for universally composable protocols, for information-theoretically secure protocols, and even allow to create software which can only be used once (One-Time-Programs). However, all known protocols employing tamper-proof hardware are either indirect, i.e., additional computational assumptions must be used to obtain general two party computations or a large number of devices must be used. In this work we present the first protocol realizing universally composable two-party computations (and even trusted One-Time-Programs) with information-theoretic security using only one single tamper-proof device issued by one of the mutually distrusting parties.

Keywords: Secure Two-Party Computation, Universal Composability, Tamper-Proof Hardware, Information-Theoretical Security.

1 Introduction

Recently, tamper-proof hardware tokens have received increasing attention. Tamper-proof hardware tokens allow information-theoretically secure protocols which are universally composable (UC) [4], they can be employed for protocols in the *globalized UC* framework [5,17], and they even allow for One-Time-Programs, i.e. circuits which can be evaluated only once [13]. However, all known protocols employing tamper-proof hardware are either indirect, i.e., the secure hardware is used to implement commitments or Zero Knowledge proofs and additional computational assumptions must be used to obtain general, composable two party computations [19,22,8,7], or a large number of devices must be used [13,15]. However, issuing multiple independent tamper-proof devices requires much stronger isolation assumptions. Not only the communication between the device and the issuer must be prevented, but also the many devices must be mutually isolated. This is especially difficult as the devices are not necessarily trusted (see [3] for the difficulty of isolating two devices in one location).

Y. Ishai (Ed.): TCC 2011, LNCS 6597, pp. 164–181, 2011.

In this work we present the first protocol realizing universally composable two-party computations (and even trusted One-Time-Programs) with information-theoretic security using only one single (untrusted) tamper-proof device. One of the main challenges is to prevent a corrupted token from encoding previous inputs in subsequent outputs.

2 Related Work

The idea of secure computation based on separation assumptions was introduced by Ben-Or et al. [2] to construct multi-prover interactive proof systems. Ben-Or et al. [2] construct an unconditionally secure protocol for Rabin-OT [23] between two provers and a verifier. Even though this result is not explicitly stated in the context of tamper-proof hardware[1] and is proven secure in a standalone, synchronous model, we suppose that an amplified variant of the protocol of [2] can be proven UC-secure if the sender is allowed to issue two tamper-proof hardware tokens.

The idea of explicitly using tamper-proof hardware for cryptographic purposes was introduced by Goldreich and Ostrovsky [12]. They showed that tamper-proof hardware can be used for the purpose of software-protection.

The interest in secure hardware and separation assumptions was renewed, when it was realized that universally secure multi-party computation can be based on the setup assumption of tamper-proof hardware tokens. The tamper-proof hardware must suffice strong separation conditions, even if a more recent result showed that the assumptions about the physical separation can be relaxed to some extent [8,7].

Generally, the work on secure multi-party computation with tamper-proof hardware assumption can be divided in works dealing with either stateful or stateless hardware-tokens. Katz [19] considers a scenario where all parties can create and issue stateful tamper-proof hardware tokens. Using additional number theoretic assumptions, [19] implements a multiple commitment functionality in this scenario. Subsequently Moran and Segev [22] improved upon Katz result, by constructing commitments in an asymmetric scenario, where only one out of two parties is able to issue stateful tamper-proof hardware tokens. Hofheinz, Mller-Quade and Unruh [17] use (stateless) signature cards, issued by a trusted authority to achieve universal composability with respect to global setup assumptions [5]. Fischlin et al. [10] show how set intersection can be computed securely using a single untrusted tamper-proof hardware token and additional computational assumptions.

Goldwasser et al. [13] show that using a minimalistic stateful tamper-proof hardware assumption called One-Time-Memory, a new cryptographic primitive called One-Time-Program can be implemented. Recently, Kolesnikov [21] implemented string oblivious transfer with stateless tamper-proof hardware tokens. Goyal et al. [15] consider a unified treatment of tamper-proof hardware assumptions. Important in the context of this work, they show that in a mutually

[1] [2] mention that the provers in their protocol might be implemented as bank-cards.

mistrusting setting, trusted One-Time-Programs can be implemented statistically secure from a polynomial number of OTMs. Goyal et al. [14] show that a single stateless tamper-proof hardware token is sufficient to implement statistically secure commitments and statistical zero-knowledge. Furthermore, if stateless tokens can be encapsulated into other stateless tokens, general statistically secure composable multi-party computation is possible in this setting. [14] also show that unconditionally secure Oblivious Transfer cannot be realized from stateless tamper-proof hardware alone.

With the exception of [2], all of the above schemes based on stateful tamper-proof hardware either use additional complexity assumptions to achieve secure two party computations [17,19,22,13,8,7,21] or a large number of hardware tokens must be issued [13,15]. The question if one single tamper-proof device, issued by one of two mistrusting parties, suffices for information-theoretically secure two-party computations remains open in the literature.

3 Our Contribution

In this paper, we show that general, statistically secure, composable two-party computations are possible in a setting where a single untrusted stateful tamper-proof hardware token can be issued by one party. All previous solutions supposed that either the creator of the tamper-proof hardware is honest, that additional complexity assumptions are used, or that a larger number of independent tamper-proof hardware tokens is issued. As a reasonable abstraction for the primitives that can be implemented in our setting, we introduce a new primitive which we call *Sequential-One-Time-Memory*. Sequential-One-Time-Memories provide a large amount of single One-Time-Memories, with the additional guarantee that the memory cells can only be queried sequentially. Just like One-Time-Memories, Sequential-One-Time-Memories have the property of being non-signaling to their issuer once they have been sent. We show that the Oblivious Transfer (OT) functionality can be realized straightforwardly using Sequential-One-Time-Memories, thus our results for statistically secure, composable two party computations follow immediately by [18,20]. Our main contribution is a statistically secure, universally composable protocol that realizes the Sequential-One-Time-Memory functionality with a single, untrusted, tamper-proof hardware token. Our protocol construction is efficient, it realizes m sequential n-bit-string-OTMs while having a communication overhead of $O(n^2m)$ bits in its interactive send phase, where n is the security parameter. This is achieved by using tokens that compute blinded tensor-product functions. The algebraic structure of these functions allows the token-sender to encapsulate two random keys, one of which the token-receiver can learn. In turn, the token-receiver has the promise that the token will not learn his choice-bit from his function input and that the key he computes from the function-output solely depends on his choice-bit. Moreover, these functions can be computed by simple linear algebra operations. The technical challenge in the security-proof, compared to issuing a large number of independent devices, is that an untrusted, adversarial token might deviate arbitrarily from the specification of an honest token. It may thus try to correlate its output or abort-behavior with previous inputs. Our protocol is semi-interactive as it requires an interactive send-phase. This interaction is necessary, as we can further show that any protocol that implements Sequential-OTMs using a single untrusted device without any further interaction is insecure. Furthermore, a simple modification of our protocol yields a completely non-interactive protocol that implements Sequential-One-Time-Memories from two tamper-proof hardware tokens. Finally, we can resolve a question left open by [22] positively. Moran and Segev asked if multiple commitments from the receiver of a token to its issuer can be realized with a single untrusted tamper-proof hardware token. As we can realize Oblivious Transfer, standard techniques (e.g. [9,18]) can be used to implement commitments.

4 Framework

We state and prove our results in the Universal-Composability (UC) framework of [4]. In this framework security is defined by comparison of an *ideal model* and a *real model*. The protocol of interest is running in the latter, where an adversary $\mathcal{A}$ coordinates the behavior of all corrupted parties. In the ideal model, which is secure by definition, a simulator $\mathcal{S}$ tries to mimic the actions of $\mathcal{A}$. An environment $\mathcal{Z}$ is plugged either to the ideal or the real model and has to guess, which model it is actually plugged to.

By the random variable $\mathrm{View}_{\Pi}^{\mathcal{A}}(\mathcal{Z})$ we denote the complete view of the environment $\mathcal{Z}$ when plugged to the real model with protocol Π and adversary $\mathcal{A}$. Analogously, by the random variable $\mathrm{View}_{\mathcal{F}}^{\mathcal{S}}(\mathcal{Z})$ we denote the view of $\mathcal{Z}$ when plugged to the ideal model, where the functionality $\mathcal{F}$ realizes the protocol task and the simulator $\mathcal{S}$ coordinates the behavior of the corrupted parties. Therefore Π is a (statistically) UC-secure implementation of $\mathcal{F}$, if for every adversary $\mathcal{A}$ there exists a simulator $\mathcal{S}$, such that for all environments $\mathcal{Z}$ the random variables $\mathrm{View}_{\Pi}^{\mathcal{A}}(\mathcal{Z})$ and $\mathrm{View}_{\mathcal{F}}^{\mathcal{S}}(\mathcal{Z})$ are (statistically) close.

In our case the adversarial entities $\mathcal{A}, \mathcal{S}$ and the environment $\mathcal{Z}$ are computationally unbounded and a hybrid functionality $\mathcal{F}_{\mathrm{wrap}}^{\mathrm{stateful}}$ models our tamper-proof hardware assumption (cf. Section 6). Moreover, as it was proven sufficient in [4], we always focus on the dummy adversary $\tilde{\mathcal{A}}$, which is completely controlled by the environment $\mathcal{Z}$.

5 Preliminaries

We use the notation $[m] := \{1, \ldots, m\}$. To formulate our protocol, we need some elementary concepts of linear algebra over finite fields. $\mathbb{F}_2$ is the finite field with two elements. We can canonically identify the vector-space $\mathbb{F}_2^n$ with the set $\{0,1\}^n$ of strings of length n. For two column-vectors $a \in \mathbb{F}_2^n$ and $b \in \mathbb{F}_2^k$ we write $ab^T = (a_i b_j)_{ij} \in \mathbb{F}_2^{n \times k}$ for the outer product (or tensor-product) of a and b. Similarly, for $a, b \in \mathbf{F}_2^n$ we denote the inner product by $a^T b = \sum_{i=1}^n a_i b_i \in \mathbb{F}_2$. Let $C \in \mathbb{F}_2^{n \times 2n}$. Then $\dim(\ker(C)) \geq n$. Let $B = \{b_1, \ldots, b_n\} \subseteq \ker(C)$ be a linearly independent set. We can choose a set $B^* = \{b_{n+1}, \ldots, b_{2n}\}$ such that $B \cup B^*$ is a basis of $\mathbb{F}_2^{2n}$. Let $e_i \in \mathbb{F}_2^n$ be the i-th unit-vector. Then there exists a matrix $G \in \mathbb{F}_2^{n \times 2n}$ such that $Gb_i = e_i$ for $i = 1, \ldots, n$ and $Gb_i = 0$ for $i = n+1, \ldots, 2n$. We call such a G a *complementary-matrix* to C. It holds that $\operatorname{rank}(G) = n$ and $B^* \subseteq \ker(G)$. For such C and G, we can always solve linear equation systems $Cx = r$, $Gx = s$ by solving $Cx_{B^*} = r$ and $Gx_B = s$ independently for some $x_B \in \operatorname{span}(B)$ and $x_{B^*} \in \operatorname{span}(B^*)$. We can then set $x = x_B + x_{B^*}$. It holds $Cx = r$ and $Gx = s$, as $x_B \in \ker(C)$ and $x_{B^*} \in \ker(G)$.

6 Modeling Tamper-Proof Hardware

Our formulation of general stateful tamper-proof hardware resembles the meanwhile standard definitions of [19] and [22]. To model tamper-proof hardware, we employ the $\mathcal{F}_{\text{wrap}}^{\text{stateful}}$ wrapper-functionality. A sender-party G (Goliath) provides as input a Turing-machine $\mathcal{M}$ to $\mathcal{F}_{\text{wrap}}^{\text{stateful}}$. The receiver party D (David) can now query $\mathcal{F}_{\text{wrap}}^{\text{stateful}}$ on inputs w, whereupon $\mathcal{F}_{\text{wrap}}^{\text{stateful}}$ runs $\mathcal{M}$ on input w, sends the output y that $\mathcal{M}$ produced to D and stores the new state of $\mathcal{M}$. Every time D sends a new query w' to $\mathcal{F}_{\text{wrap}}^{\text{stateful}}$, it resumes simulating $\mathcal{M}$ with its most recent state, sends the output to D and updates the stored state of $\mathcal{M}$. This captures the following properties one expects from tamper-proof hardware. First, G is unable to revoke $\mathcal{M}$ once it has sent a token to D. Second, D can run $\mathcal{M}$ on inputs of his choice, but the program-code and state of $\mathcal{M}$ are out of reach for D, due to the tokens tamper-proofness. Note that stateful tokens don't need a trusted source of randomness, as $\mathcal{M}$ can be provided with a sufficiently long hard-coded random-tape. Thus we can, w.l.o.g restrict $\mathcal{M}$ to be deterministic. See Figure 1.

A very basic tamper-proof hardware primitive called One-Time-Memory (OTM) (Figure 2), that can be considered minimal in the sense that it only needs to be able to erase its contents after being queried, was defined in [13].

Functionality $\mathcal{F}_{\text{wrap}}^{\text{stateful}}$

Parametrized by an implicit security parameter n and a (polynomial) runtime bound $p(\cdot)$.

Creation. Upon receiving a message (`create`,sid,P_i,P_j,$\mathcal{M}$) from party P_i, where $\mathcal{M}$ is a deterministic Turing-machine, go to state `ready` and store $(\mathcal{M}, \perp)$, where $\perp$ is the initial state of $\mathcal{M}$. Send (`ready`,sid,P_i,P_j) to party P_j.

Execution. Upon receiving a message (`run`,sid,P_i,P_j,in) where in is an input word, first check if current state is `ready`. If so, run $\mathcal{M}(\text{state}, in)$ for at most $p(k)$ steps. Read out from the output tape of $\mathcal{M}$ and save its new state. Send (`out`,sid,P_i,P_j,out) to P_j.

Fig. 1. The wrapper functionality that models general stateful tamper-proof hardware

OTMs resemble the well known Oblivious Transfer (OT) functionality [23,9], with just a slight difference. While OT notifies the sender when the receiver queries the primitive, this is not the case for OTM. Moreover, OTMs have the property of immediate delivery, which means that even a malicious sender cannot revoke or abort the functionality once it has been sent. Together with One-Time-Memories, [13] defined the notion of One-Time-Programs (OTP), and its generalization, k-Time-Programs (k-TP). OTP is a functionality that allows a receiver to evaluate a circuit on exactly one input, then it shuts down. Similarly, a k-TP can be evaluated exactly k-times, then shuts down. While [13] showed that OTPs can be realized with OTM and computational assumptions and assumed that the sender of the OTPs is trusted, [15] provided an unconditionally secure implementation of *trusted* OTPs using only OTMs, where the sender of the OTPs is untrusted.

Functionality $\mathcal{F}^{\mathrm{OTM}}$

Parametrized by a security parameter n.

Creation. Upon receiving a message (`create`,sid,P_i,P_j,(s_0, s_1)) from party P_i, go to state `ready` and store (s_0, s_1). Send (`ready`,sid,P_i,P_j) to party P_j.

Choice. Upon receiving a message (`choice`,sid,P_i,P_j,x) from P_j, check if current state is `ready`. If so, send (`out`,sid,P_i,P_j,s_x) to P_j and go to state `dead`.

Fig. 2. The One-Time-Memory Functionality

We introduce a new variant of the One-Time-Memory functionality that we call Sequential-One-Time-Memory (Seq-OTM) (Figure 3). Seq-OTM constitutes a set of m single OTM functionalities, with the restriction that they can only be queried in a fixed order. We refer to the single OTMs of Seq-OTM as its stages. We impose the sequential ordering to the stages, to model that the untrusted token might stop answering queries after an arbitrary stage. The limitation to sequential access makes applications impossible where random access is essential. All constructions where sequential access is sufficient can be implemented with Seq-OTM as well. As the construction of unconditionally secure trusted OTPs [15] using multiple OTM tokens is quite involved, we cannot present it here. However, we observe that the construction of [15] works using Seq-OTM instead of multiple independent OTM tokens. This holds because in this construction, an honest receiver queries all the OTM primitives he received in a fixed order anyway. Interestingly, the technical challenges [15] deals with in constructing unconditionally secure OTPs arises from the fact that a malicious receiver might query the OTMs *out of order*. Finally, a remark is in place. Even though Seq-OTM can be used to implement several OTPs, the sequential nature of Seq-OTM demands that those OTPs can only be executed in a fixed order. If one wishes to execute several OTPs in random order, multiple Seq-OTMs (and thus multiple hardware tokens) have to be issued.

The restriction to sequential access also bears advantages in certain applications. In [15], a construction was provided that realizes a single OT from polynomially many OTM tokens. The construction is rather expensive regarding the number of OTMs used, as it involves One-Time-Programs to issue a proof that the primitive has been queried. Such a proof is necessary to convince the OT-sender that the receiver has provided his input. Again, what makes this construction technically challenging is that the single OTM tokens can be queried in an arbitrary order.

Functionality $\mathcal{F}^{\text{Seq-OTM}}$

Parametrized by a security parameter n and a parameter $m = \text{poly}(n)$, which is the number of single OTMs that can be stored.

Creation. Upon receiving a message (`create`,sid,P_i,P_j,$((s_{1,0}, s_{1,1}), \ldots, (s_{m,0}, s_{m,1}))$) from party P_i, go to state `ready` and store $((s_{1,0}, s_{1,1}), \ldots, (s_{m,0}, s_{m,1}))$. Set a counter $t = 1$. Send (`ready`,sid,P_i,P_j) to party P_j.

Choice. Upon receiving a message (`choice`,sid,P_i,P_j,x) from P_j, check if current state is `ready` and $t \leq m$. If so, send (`out`,sid,P_i,P_j,$s_{t,x}$) to P_j and increment t by 1. If $t > m$, go to state `dead`.

Fig. 3. The Sequential-One-Time-Memory Functionality

We will now briefly outline how a polynomial number of OTs can be implemented using the Seq-OTM functionality. Given a Seq-OTM primitive with $2m$ single OTMs, $\text{OTM}_1, \ldots, \text{OTM}_{2m}$, the sender programs OTM_{2i-1} with his inputs for OT_i and programs a random-string r_i into OTM_{2i}. In the choice-phase, the receiver queries OTM_{2i-1} with his i-th choice-bit x_i and OTM_{2i} with choice-bit 0. To prove to the sender that OTM_{2i-1} has already been queried, the receiver sends r_i to the sender. As Seq-OTM forces the receiver to open $\text{OTM}_1, \ldots, \text{OTM}_{2m}$ sequentially, he has only negligible chance to guess r_i correctly unless he has opened OTM_{2i-1} already. Thus, this reduction is perfectly secure against the sender of the OT and statistically secure against the receiver of the OT. Noting that OT can be stored and reversed [1,24,25], we conclude that in the Seq-OTM hybrid-model, OT can be implemented in both ways (from the Seq-OTM sender to the Seq-OTM receiver and from the Seq-OTM receiver to the Seq-OTM sender).

7 The Necessity of Interaction

We will now show that any protocol, which implements a Seq-OTM with 2 or more stages from a single hardware token, requires some interaction between sender and receiver. The proof is a variant of the impossibility result of [6], which states that Bit-Commitments cannot be realized in the plain UC-model. We show that any non-interactive protocol, realizing Seq-OTM from a single tamper-proof device, that is secure against a corrupted receiver is necessarily insecure against a corrupted sender. More precisely, once there exists a simulator against a corrupted receiver, a corrupted sender can use this simulator to construct a token that learns the receiver's inputs. Such a token can make the output of the second stage dependent on the input of the first stage. However, this behavior cannot be simulated, because in the ideal experiment, the sender's inputs to $\mathcal{F}^{\text{Seq-OTM}}$ have to be provided before the choice-phase begins, and can thus not depend on any choice-bits.

Remark 1. This argument assumes that both the receiver and $\mathcal{T}$ are equivalent in their computational resources, as it uses the simulator against a corrupted receiver to construct a corrupted token. Thus, it only holds if the receiver is subject to the same computational restrictions as the token. If we allow the receiver (and thus the simulator against a corrupted receiver) to be computationally unbounded but require the token to run in polynomial time, there may exist non-interactive UC-secure protocols for $\mathcal{F}^{\text{Seq-OTM}}$ from $\mathcal{F}^{\text{stateful}}_{\text{wrap}}$ relative to some computational assumption.

Theorem 1. *There is no non-interactive protocol which UC-realizes $\mathcal{F}^{\text{Seq-OTM}}$ with two or more stages from a single tamper-proof hardware instance $\mathcal{F}^{\text{stateful}}_{\text{wrap}}$, given that the receiver D and the token $\mathcal{T}$ ran by $\mathcal{F}^{\text{stateful}}_{\text{wrap}}$ are bounded by the same computational restrictions.*

Proof. Assume there exists a protocol Π that UC-realizes a Seq-OTM with two stages, which requires no further interaction between sender and receiver. For simplicity, we assume that the Seq-OTM stages are bit-OTMs. As Π is UC-secure, there exists a receiver-simulator $\mathcal{S}_D$ which extracts the choice-bits x_1 and x_2 from the dummy-adversary $\tilde{\mathcal{A}}_D$. As Π is non-interactive, $\tilde{\mathcal{A}}_D$ only interacts with a token $\mathcal{T}$. But that means that $\mathcal{S}_D$ must be able to extract $\tilde{\mathcal{A}}_D$'s input from the messages $\tilde{\mathcal{A}}_D$ has provided to $\mathcal{T}$. Consider the case of a corrupted sender. We will provide an environment $\mathcal{Z}'$ that distinguishes real and ideal with probability $\geq \frac{1}{2}$. $\mathcal{Z}'$ constructs a token $\mathcal{T}$, which does the following. The token $\mathcal{T}$ internally simulates the simulator $\mathcal{S}_D$ and some functionality $\mathcal{F}'$. $\mathcal{S}_D$ is wired to $\mathcal{F}'$ instead of $\mathcal{F}^{\text{Seq-OTM}}$. $\mathcal{F}'$ is a "corrupted" version of $\mathcal{F}^{\text{Seq-OTM}}$, which always gives 0 as its first-stage OTM-output and x_1 (its first-stage input) as its second-stage output. $\mathcal{Z}'$ instructs $\tilde{A}_G$ to input $\mathcal{T}$ into $\mathcal{F}^{\text{wrap}}_{\text{stateful}}$. In the choice-phase, $\mathcal{Z}'$ sets the receiver's first choice-bit x_1 to a uniformly random value and the second choice-bit x_2 to 0. This concludes the description of $\mathcal{Z}'$. We claim that for every simulator $\mathcal{S}_G$, the statistical distance between $\text{View}^{\mathcal{A}}_{\Pi}(\mathcal{Z}')$ and $\text{View}^{\mathcal{S}}_{\mathcal{F}}(\mathcal{Z}')$ is at least $\frac{1}{2}$. In the real experiment, the output s_2 that the receiver gets will always be the same as his first input x_1. In the ideal experiment however, the simulator $\mathcal{S}_G$ has to guess the choice-bit x_1 in advance. As x_1 depends solely on an internal coin-toss of $\mathcal{Z}'$, the chance of $\mathcal{S}_G$ guessing x_1 correctly is $\leq \frac{1}{2}$. Thus, the probability that D gives an incorrect output to $\mathcal{Z}'$ is $\geq \frac{1}{2}$.

8 The David-and-Goliath Sequential-OTM Protocol

In this Section, we will describe our protocol $\Pi_{\text{Seq-OTM}}$ for David-and-Goliath Sequential-OTM and give a sketch of the security proof. We will refer to the sender of the Seq-OTM as Goliath and to the receiver as David. Let n be the statistical security parameter. We first explain the token program that will be used in the protocol. The token is created with m affine functions stored in it. It has a counter t, which is initialized with 1. In the execution phase, if the token is queried with an input x, it will evaluate its t-th affine function on x and output the result. After each query the counter t is incremented by 1. When the counter reaches m, the token stops functioning.

Before we describe in more detail how these tokens can be used to implement Sequential-OTM, we will explain more precisely which kind of affine functions are stored on the token. Let $a \in \mathbb{F}_2^{2n}$, $B \in \mathbb{F}_2^{2n \times 2n}$ and $z \in \mathbb{F}_2^{2n}$ be an input vector. The functions have the form $V(z) = az^T + B$. Thus, $V(z)$ is an outer product az^T blinded by a matrix B. See Figure 4.

Program: Seq-OTM Token

Parametrized by a security parameter n.

Hardwired Inputs

- A parameter $m \in \mathbb{N}$, the maximum number of stages.
- $(a_1, B_1), \ldots, (a_m, B_m)$, where $a_i \in \mathbb{F}_2^{2n}$ and $B_i \in \mathbb{F}_2^{2n \times 2n}$ for $i \in [m]$

Stateful Variables

- A counter variable t that is, upon creation, initialized with $t = 1$

Execution Upon receiving an input $z \in \mathbb{F}_2^{2n}$

- Check if $t \leq m$, if not abort.
- Compute $V_t = a_t z^T + B_t$.
- Set $t \leftarrow t + 1$.
- Output V_t and wait for further inputs.

Fig. 4. The program that is run by an honest token for Sequential-One-Time-Memories

We will first give a high-level picture of protocol $\Pi_{\text{Seq-OTM}}$. The protocol realizes a Seq-OTM with $m = \text{poly}(n)$ stages, where each stage is an n-bit-string OTM. It has three phases. In the first phase, Goliath creates a token with random affine functions stored on it, then sends the wrapped token to David. Hereafter Goliath is physically committed to the token. The second phase is the only interactive part of the protocol. It involves the sending of check-values and one-time-pad-encrypted OTM inputs. In the third phase, which is non-interactive,

David queries the token to get his desired one-time-pad keys. David's input to the token is a randomized value, in which his choice-bit is hidden. After receiving output from the token, David checks the output with the check-values provided by Goliath and aborts if the output does not pass the check. If the output passes the check, David computes a key and decrypts his desired OTM output.

We will now describe the protocol in detail. In the creation-phase, Goliath programs the token $\mathcal{T}$ with uniformly chosen $(a_i, B_i) \in \mathbb{F}_2^{2n} \times \mathbb{F}_2^{2n \times 2n}$, for $i = 1, \ldots, m$. Goliath then sends the wrapped token $\mathcal{T}$ to David.

The send-phase proceeds as follows. Let $(s_{i,0}, s_{i,1}) \in \mathbb{F}_2^n \times \mathbb{F}_2^n$ be Goliath's i-th Seq-OTM input. First, David announces a randomly chosen check-matrix $C \in \mathbb{F}_2^{n \times 2n}$ to Goliath. Then for $i = 1, \ldots, m$, Goliath computes $(\tilde{a}_i, \tilde{B}_i) \in \mathbb{F}_2^n \times \mathbb{F}_2^{n \times 2n}$ by $\tilde{a}_i = Ca_i$ and $\tilde{B}_i = CB_i$. Goliath further chooses a matrix $G \in \mathbb{F}_2^{n \times 2n}$ such that G is a *complementary*-matrix of C (cf. Section 5). Goliath now sends the $(\tilde{a}_i, \tilde{B}_i)_{i \in [m]}$ and G to David. The $(\tilde{a}_i, \tilde{B}_i)$ can be seen as commitments that commit the token $\mathcal{T}$ to a unique output behavior, but reveal no meaningful information to David. After David has received $(\tilde{a}_i, \tilde{B}_i)_{i \in [m]}$ and G, he chooses a random $\mathbf{h} = (h_1, \ldots, h_m) \in \left(\mathbb{F}_2^{2n} \backslash \{0\}\right)^m$ and sends $\mathbf{h}$ to Goliath. Goliath computes *ciphertexts* $\tilde{s}_{i,0} = s_{i,0} + GB_i h_i$ and $\tilde{s}_{i,1} = s_{i,1} + GB_i h_i + Ga_i$ for $i = 1, \ldots, m$ and sends $((\tilde{s}_{i,0}, \tilde{s}_{i,1}))_{i \in [m]}$ to David. This concludes the send-phase.

The choice-phase is non-interactive and has m stages. In stage i David queries the i-th OTM of the Seq-OTM functionality being implemented. Let $x_i \in \mathbb{F}_2$ be David's i-th choice-bit. David uniformly samples an element $z_i \in \mathbb{F}_2^{2n}$ such that $z_i^T h_i = x_i$. That is, if $x_i = 0$ David samples z_i from the $2n - 1$ dimensional hyperplane $A_{i,0} = \{z \in \mathbb{F}_2^{2n} : z^T h_i = 0\}$. Likewise, if $x_i = 1$, then z_i is sampled from the hyperplane $A_{i,1} = \{z \in \mathbb{F}_2^{2n} : z^T h_i = 1\}$. Let V_i be the output of the token when input z_i in stage i. If the token was created honestly, it holds that $V_i = a_i z_i^T + B_i$. David checks if $CV_i \stackrel{?}{=} \tilde{a}_i z^T + \tilde{B}_i$. If not, then the token computed a different function than the one Goliath committed to and David aborts the protocol. If the check is passed, David computes $s_{i,x_i} = \tilde{s}_{i,x_i} + GV_i h_i$ and outputs s_{i,x_i}. This concludes the description of the protocol. The full protocol is given in Figure 5.

Discussion. We will shortly sketch the ideas behind this construction. First notice, that from the view of the token, David's inputs look almost uniform. This comes from the fact that the token is oblivious of the randomly chosen vector h_i, that defines $A_{i,0}$ and $A_{i,1}$. In fact, it can be shown, that a complete input history $\mathbf{z} = (z_1, \ldots, z_m) \in \left(\mathbb{F}_2^{2n}\right)^m$, for adversarially chosen choice-bits of David, is statistically close to a uniformly chosen input history $\mathbf{u} = (u_1, \ldots, u_m) \in \left(\mathbb{F}_2^{2n}\right)^m$. This, in turn, means that a dishonest token is oblivious of David's input. The check-operation is performed to make sure that the token responds to queries in an unambiguous way and that it does not encode further information into Davids output. If the token answers dishonestly, David will notice that with overwhelming probability. The token's output V_i needs to be projected with the matrix G, so that David cannot learn anything about Goliath's inputs $s_{i,0}$ and $s_{i,1}$ from $\tilde{a}_i$ and $\tilde{B}_i$. Notice that Goliath commits to the check-values $(\tilde{a}_1, \tilde{B}_1), \ldots, (\tilde{a}_m, \tilde{B}_m)$

Protocol $\Pi_{\text{Seq-OTM}}$: David & Goliath Sequential OTM

Let n be the statistical security parameter and let $m = \text{poly}(n)$ be the number of sequential OTMs. Further let $\mathcal{F}_{\text{wrap}}^{\text{stateful}}$ be the wrapper-functionality that models the tamper-proof hardware.

Creation-Phase: (**only Goliath**)

- For every index $i \in [m]$, choose $a_i \in \mathbb{F}_2^{2n}$ and $B_i \in \mathbb{F}_2^{2n \times 2n}$ uniformly at random.
- Program a Seq-OTM token $\mathcal{T}$ with hardwired inputs m and (a_i, B_i) for $i \in [m]$.
- Send $\mathcal{T}$ to $\mathcal{F}_{\text{wrap}}^{\text{stateful}}$.

Send-Phase: Let $((s_{1,0}, s_{1,1}), \ldots, (s_{m,0}, s_{m,1})) \in (\mathbb{F}_2^n \times \mathbb{F}_2^n)^m$ be Goliath's Seq-OTM input.

1. (**David**) Wait until the `ready` message from $\mathcal{F}_{\text{wrap}}^{\text{stateful}}$ was received. Then, choose a random matrix $C \in \mathbb{F}_2^{n \times 2n}$ and send C to Goliath.
2. (**Goliath**) Compute a matrix $G \in \mathbb{F}_2^{n \times 2n}$, such that G is complementary to C. For all $i \in [m]$, set $\tilde{a}_i = Ca_i$ and $\tilde{B}_i = CB_i$. Send $(G, (\tilde{a}_i, \tilde{B}_i)_{i \in [m]})$ to David.
3. (**David**) Choose $\mathbf{h} = (h_1, \ldots, h_m) \in (\mathbb{F}_2^{2n} \setminus \{0\})^m$ uniformly a random and send $\mathbf{h}$ to Goliath.
4. (**Goliath**) For each $i \in [m]$, set $\tilde{s}_{i,0} = s_{i,0} + GB_i h_i$ and $\tilde{s}_{i,1} = s_{i,1} + GB_i h_i + Ga_i$. Send $(\tilde{s}_{i,0}, \tilde{s}_{i,1})_{i \in [m]}$ to David.

Choice-Phase (Stage $i \in [m]$): (**only David**) Let $x_i \in \mathbb{F}_2$ be David's i-th Seq-OTM input.

- Choose $z_i \in \mathbb{F}_2^{2n}$ uniformly at random such that $z_i^T h_i = x_i$ and input z_i into $\mathcal{F}_{\text{wrap}}^{\text{stateful}}$.
- Let V_i be the output of $\mathcal{F}_{\text{wrap}}^{\text{stateful}}$. Check if $CV_i = \tilde{a}_i z_i^T + \tilde{B}_i$. If not, go to abort state and output $\perp$ for this query and for all further queries.
- If the check is passed, output $s_{i,x_i} = \tilde{s}_{i,x_i} + GV_i h_i$.

Fig. 5. A protocol for sequential OTM from a single stateful token

before he sees the vectors $(h_1, \ldots, h_m)$. Thus, the condition on which David aborts is independent of the choice of $(h_1, \ldots, h_m)$. The use of the outer-product operation for the affine functions stems from a subtle issue. We need to ensure that the environment $\mathcal{Z}$ does not see any of the random coins David used to sample the z_i. If a corrupted Goliath could create a token that encodes several bits of z_i in its output, the environment might be able to tell apart real and ideal model in a later stage. Such a token might, for instance, abort in a later stage if some of the bits of z_i fulfill a certain condition (e.g. have odd parity or some special hash value). Given that the token actually computes the specified function (which is enforced by the check-operation), the outer-product form ensures that there are only two different outputs an honest David might produce in each stage. Particularly, let $V(z) = az^T + B$ be one of the functions computed by the token. Basically, the outer product form allows David to derandomize the token's output. If x is such that $z^T h = x$, then the randomness of z is removed by $V(z)h = (az^T + B)h = az^T h + Bh = ax + Bh$. The term $ax + Bh$ is independent of the random coins used to sample z. Consequently, David's outputs s_{i,x_i} are (with overwhelming probability) independent of the random coins used to sample z.

Correctness of the Protocol. If both parties are honest, the correctness of the protocol can be seen straightforwardly. Let s'_{i,x_i} be one of David's outputs.

$$s'_{i,x_i} = \tilde{s}_{i,x_i} + GV_i h_i = \tilde{s}_{i,x_i} + G(a_i z_i^T + B_i)h_i = \tilde{s}_{i,x_i} + Ga_i z_i^T h_i + GB_i h_i$$
$$= \tilde{s}_{i,x_i} + Ga_i x_i + GB_i h_i = s_{i,x_i}$$

8.1 Security of $\Pi_{\text{Seq-OTM}}$

The security of protocol $\Pi_{\text{Seq-OTM}}$ is summarized in Theorem 2.

Theorem 2. *Protocol $\Pi_{Seq\text{-}OTM}$ UC-realizes the $\mathcal{F}^{\text{Seq}-\text{OTM}}$ functionality, with perfect security against a corrupted receiver and statistical security against a corrupted sender.*

We will provide the security-proof against a corrupted receiver and outline the security-proof against a corrupted sender

Corrupted Receiver. We will first consider the case of a corrupted receiver, as this is the easy case. Let $\tilde{\mathcal{A}}$ be the dummy adversary for David. We will provide a simulator $\mathcal{S}_D$ and show, that for any environment $\mathcal{Z}$ the distributions $\text{View}^{\tilde{\mathcal{A}}}_{\Pi_{Seq-OTM}}(\mathcal{Z})$ and $\text{View}^{\mathcal{S}_D}_{\mathcal{F}_{Seq-OTM}}(\mathcal{Z})$ are identical. Simulator $\mathcal{S}_D$ runs in strict polynomial time. The simulator $\mathcal{S}_D$ is given in Figure 6.

First notice that there is always a solution V_i for $CV_i = \tilde{a}_i z_i^T + \tilde{B}_i$ and $GV_i h_i = \tilde{s}_{i,x_i} + s_{i,x_i}$, as G is a complementary matrix of C and has rank n. Furthermore, from $\tilde{\mathcal{A}}$'s view, each V_i with $CV_i = \tilde{a}_i z_i^T + \tilde{B}_i$ and $GV_i h_i = \tilde{s}_{i,x_i} + s_{i,x_i}$ is equally likely. Thus we obtain perfect indistinguishability for $\text{View}^{\tilde{\mathcal{A}}}_{\Pi_{Seq-OTM}}(\mathcal{Z})$ and $\text{View}^{\mathcal{S}_D}_{\mathcal{F}_{Seq-OTM}}(\mathcal{Z})$.

Corrupted Sender. The case of a corrupted sender is the technically challenging part of this proof. Let $\tilde{\mathcal{A}}$ be the dummy adversary for Goliath. We will provide a simulator $\mathcal{S}_G$ and show, that for any environment $\mathcal{Z}$ the distributions $\text{View}^{\tilde{\mathcal{A}}}_{\Pi_{Seq-OTM}}(\mathcal{Z})$ and $\text{View}^{\mathcal{S}_G}_{\mathcal{F}_{Seq-OTM}}(\mathcal{Z})$ are statistically close. Simulator $\mathcal{S}_G$ runs in expected polynomial time. The simulator $\mathcal{S}_G$ is given in Figure 7.

Remark 2. Let $\mathcal{T}$ be a (possibly malicious) token program. Write $(V_1, \ldots, V_m) = \mathcal{T}(z_1, \ldots, z_m)$ for a token run with inputs $(z_1, \ldots, z_m)$ and outputs $(V_1, \ldots, V_m)$.

The proof proceeds roughly as follows. In a hybrid argument, we show for each stage i, that if David does not abort in stage i, then for both inputs $x_i = 0$ and $x_i = 1$, the simulator can extract outputs $\hat{V}_{i,0}$ and $\hat{V}_{i,1}$ from $\mathcal{T}$ that David both

Simulator $\mathcal{S}_D$

- Setup a Goliath-machine $\mathcal{G}$ and provide $\mathcal{G}$ with a random tape.
- Simulate the creation phase with machine $\mathcal{G}$ and extract a Seq-OTM token program $\mathcal{T}$
- Send the `ready` message to $\tilde{\mathcal{A}}$.
- If $\tilde{\mathcal{A}}$ queries $\mathcal{F}_{\mathrm{wrap}}^{\mathrm{stateful}}$ with a z_i, before h_i has been received, wire the input to $\mathcal{T}$ and forward $\mathcal{T}$'s output V_i to $\tilde{\mathcal{A}}$.
- Simulate the send phase between $\mathcal{G}$ and $\tilde{\mathcal{A}}$. For all indices $i \in [m]$ such that $\tilde{\mathcal{A}}$ has queried $\mathcal{F}_{\mathrm{wrap}}^{\mathrm{stateful}}$ before sending h_i, upon receiving h_i, set $x_i = z_i^T h_i$. Query $\mathcal{F}^{\mathrm{Seq-OTM}}$ with x_i and receive s_{i,x_i}. Set $\tilde{s}_{i,x_i} = s_{i,x_i} + GV_i h_i$. Except for that, continue the interaction between $\mathcal{G}$ and $\tilde{\mathcal{A}}$ normally.
- For every index i such that $\tilde{\mathcal{A}}$ queries $\mathcal{F}_{\mathrm{wrap}}^{\mathrm{stateful}}$ with input z_i after h_i has been received, compute $x_i = z_i^T h_i$. Query $\mathcal{F}^{\mathrm{Seq-OTM}}$ with x_i and receive s_{i,x_i}. Uniformly sample $V_i \in \mathbb{F}_2^{2n \times 2n}$ such that $CV_i = \tilde{a}_i z_i^T + \tilde{B}_i$ and $GV_i h_i = \tilde{s}_{i,x_i} + s_{i,x_i}$. Forward V_i to $\tilde{\mathcal{A}}$.

Fig. 6. The simulator for a corrupted receiver

would accept. The extraction is done using a rewind-to-the-beginning technique. This means, instead of rewinding an existing run to the previous stage, we sample a completely new run. We can show that the event that the simulator fails to extract happens only with negligibly small probability. Next, we show that the token's output in the hybrid real part matches the corresponding extracted output with overwhelming probability (over the simulators coins). We can thus modify the simulator to use the extracted token outputs instead of the token output from the hybrid real part. After this transformation, David's outputs are ideal. Thus, only the stage at which David aborts, supposed that he aborts, depends on the hybrid real part (and thus on input made by $\mathcal{Z}$). We will call the random variable that describes at which stage David aborts the stoptime S. To determine the stage at which to abort in the ideal part, the simulator runs the token with purely random inputs $u_1^*, \ldots, u_m^*$. Let $(V_1^*, \ldots, V_m^*) = \mathcal{T}(u_1^*, \ldots, u_m^*)$. Let i_0 be the first index such that $CV_i \neq \tilde{a}_i u_i^{*T} + \tilde{B}_i$. The simulator sets David's outputs to be $\perp$ for indices $i_0, \ldots, m$. We can show that from $\mathcal{Z}$'s view, the stoptime S of the hybrid real part is statistically close to the stoptime S' of the hybrid ideal part.

Consider the following sequence of games. In game i, the environment $\mathcal{Z}$ interacts with simulator $\mathcal{S}_i$. As the send-phase is identical for the real protocol and the simulation, we only care about the choice-phase. Recall that it has stages $1, \ldots, m$.

Game 0. Simulator $\mathcal{S}_0$ simulates the real protocol $\Pi_{\mathrm{Seq-OTM}}$.

Game i (for $i = 1, \ldots, m$). $\mathcal{S}_i$ does the same as $\mathcal{S}_{i-1}$, except for the following. In stage i, if V_i passes the test $CV_i = \tilde{a}_i z_i^T + \tilde{B}_i$ sample $\hat{V}_{i,0}$ and $\hat{V}_{i,1}$ in the following manner. For both $x = 0$ and $x = 1$ repeat for at most $2^{\frac{n}{4}}$ times:

Simulator $\mathcal{S}_G$

Simulation

- Setup a David-machine $\mathcal{D}$ and provide $\mathcal{D}$ with a random tape.
- Wait until $\tilde{\mathcal{A}}$ sends a token $\mathcal{T}$ to $\mathcal{F}_{\text{wrap}}^{\text{stateful}}$.
- Simulate the send phase between $\tilde{\mathcal{A}}$ and $\mathcal{D}$

Extraction. Once the send-phase is complete, proceed as follows. The variables $C, G, h_i, \tilde{a}_i, \tilde{B}_i, \tilde{s}_{i,0}, \tilde{s}_{i,1}$ are taken from the view of $\mathcal{D}$.

- Choose $\mathbf{u}^* = (u_1^*, \ldots, u_m^*) \in (\mathbb{F}_2^{2n})^m$ uniformly. Compute $(V_1^*, \ldots, V_m^*) = \mathcal{T}(\mathbf{u}^*)$.
- For $i = \{1, \ldots, m\}$
  - If $CV_i^* \neq \tilde{a}_i u_i^{*T} + \tilde{B}_i$, then set $s_{j,0} = s_{j,1} = \perp$ for $j \geq i$ and abort loop.
  - Otherwise, for $x = 0$ and $x = 1$, try for $2^{\frac{n}{4}}$ times
    - Choose $\mathbf{u} \in (\mathbb{F}_2^{2n})^{i-1}$ uniformly at random.
    - Choose $z \in \mathbb{F}_2^{2n}$ uniformly at random such that $z^T h_i = x$.
    - Let $(V_1', \ldots, V_i') = \mathcal{T}(\mathbf{u}, z)$. If for all $j \in [i-1]$ it holds that $CV_j' = \tilde{a}_j u_j^T + \tilde{B}_j$ and $CV_i' = \tilde{a}_i z^T + \tilde{B}_i$, then set $s_{i,x} = \tilde{s}_{i,x} + GV_i' h_i$ and exit loop, otherwise repeat trying.
- Input $((s_{1,0}, s_{1,1}), \ldots, (s_{m,0}, s_{m,1}))$ into $\mathcal{F}^{\text{Seq-OTM}}$.

Fig. 7. The simulator for a corrupted sender

- Uniformly choose input-history $(u_1, \ldots, u_{i-1}) \in (\mathbb{F}_2^{2n})^{i-1}$.
- Uniformly choose $z \in \mathbb{F}_2^{2n}$ such that $z^T h_i = x$.
- Let $(V_1', \ldots, V_i') = \mathcal{T}(u_1, \ldots, u_i, z)$. If for all $j \in [i]$ it holds that $CV_j' = \tilde{a}_j u_j^T + \tilde{B}_j$ and $CV_i' = \tilde{a}_i z^T + \tilde{B}_i$, then set $\hat{V}_{i,x} = V_i'$ and exit loop.

If $\mathcal{S}_i$ fails to sample either $\hat{V}_{i,0}$ or $\hat{V}_{i,1}$ after the $2^{\frac{n}{4}}$ iterations, $\mathcal{S}_i$ halts.

Game $m + i$ (for $i = 1, \ldots, m$). The same as game $m + i - 1$, except that now David's i-th output is $G\hat{V}_{i,x_i} h_i + \tilde{s}_{i,x_i}$ instead of $GV_i h_i + \tilde{s}_{i,x_i}$.

Game $2m + 1$. The same as game $2m$, except that now the stage at which David aborts is determined differently. $\mathcal{S}_{2m+1}$ samples $(u_1^*, \ldots, u_m^*) \in (\mathbb{F}_2^{2n})^m$ uniformly at random. Let $(V_1^*, \ldots, V_m^*) = \mathcal{T}(u_1^*, \ldots, u_m^*)$. For every stage $i \in [m]$, $\mathcal{S}_{2m+1}$ replaces the abort condition $CV_i = \tilde{a}_i z_i^T + \tilde{B}_i$ by $CV_i^* = \tilde{a}_i u_i^{*T} + \tilde{B}_i$. This is the ideal game.

Proof Techniques. We will briefly sketch the proofs that establish indistinguishability between successive games. The indistinguishability of games $i - 1$ and i, for $i = 1, \ldots, m$ can be established as follows. Once a token $\mathcal{T}$ and a check-matrix C (together with $(\tilde{a}_1, \tilde{B}_1), \ldots, (\tilde{a}_m, \tilde{B}_m)$) are fixed, we can define a set $D_{\mathcal{T},C}$ of accepting input-histories. That is, $D_{\mathcal{T},C}$ consists of all input histories $(z_1, \ldots, z_i)$ for which David accepts $\mathcal{T}$'s corresponding outputs. As h_i is chosen independently of $D_{\mathcal{T},C}$, almost every choice of h_i is *good*, in the sense that it partitions $D_{\mathcal{T},C}$ in two sets of almost (up to negligible) the same size, one for which $z_i^T h_i = 0$ and $z_i^T h_i = 1$ for the other. As an analogy, h_i can be thought of as a universal hash function. If we fix h_i to be good in the above-mentioned sense, the chance to find an accepting run that belongs to choice-bit 0 roughly equals the chance to find an accepting run for choice bit 1. It can be shown that the probability of the hybrid real part producing a run that lies in $D_{\mathcal{T},C}$ is statistically close to a uniformly chosen run lying in $D_{\mathcal{T},C}$. Hence, the simulator's probability of extraction-failure is negligible (over its random coins).

The indistinguishability for games $m+i-1$ and $m+i$, for $i = 1,\dots,m$ is proven as follows. We need to show that for a fixed h_i and a fixed choice-bit x_i, Davids output $\tilde{s}_{i,x_i} + GV_i h_i$ in the hybrid real part of game $m+i$ and the extracted output $\tilde{s}_{i,x_i} + G\hat{V}_{i,x_i} h_i$ are identical. Assume there was a token $\mathcal{T}$ so that $\mathcal{T}$ succeeds to output different V_i and $\hat{V}_{i,x_i}$ such that $V_i h_i \neq \hat{V}_{i,x_i} h_i$, that are both accepted by David. Let z be the token-input corresponding to $\hat{V}_{i,x_i}$. If David accepts both runs, it must hold that

$$CV_i = \tilde{a}_i z_i^T + \tilde{B}_i \tag{1}$$

$$C\hat{V}_{i,x_i} = \tilde{a}_i z^T + \tilde{B}_i. \tag{2}$$

This implies that

$$CV_i h_i = \tilde{a}_i z_i^T h_i + \tilde{B}_i h_i = \tilde{a}_i x_i + \tilde{B}_i h_i \tag{3}$$

$$C\hat{V}_{i,x_i} h_i = \tilde{a}_i z^T h_i + \tilde{B}_i h_i = \tilde{a}_i x_i + \tilde{B}_i h_i, \tag{4}$$

and thus $CV_i h_i = C\hat{V}_{i,x_i} h_i$ holds. But this means that $\mathcal{T}$ could as well form hash-collisions for the universal hash-function C, of which it is oblivious. However, this event has only negligible probability.

Finally, in game $2m$, all of David's output values are determined in the hybrid ideal part. From $\mathcal{Z}$'s view, the probability that either of the simulators $\mathcal{S}_{2m}$ or $\mathcal{S}_{2m+1}$ halts due to an extraction error is negligible. Thus we need to show that the stoptimes S and S' for game $2m$ and game $2m+1$ are statistically close from the view of $\mathcal{Z}$. The proof needs to take into account that all the C and h_i are contained in $\mathcal{Z}$'s view. Again, it can be shown that almost all choices for the h_i are *good*, which shows that even for fixed C and h_i the stoptimes S and S' in game $2m$ and S' in game $2m+1$ are statistically close.

8.2 A Non-interactive Protocol Using Two Tokens

The send-phase in protocol $\Pi_{\text{Seq-OTM}}$ (Figure 5) is interactive. We have proven this interaction to be necessary, given that only one token is issued (Theorem 1). However, if we allow the sender Goliath to issue a second stateful token $\mathcal{T}'$ to David, then we obtain a non-interactive protocol $\Pi'_{\text{Seq-OTM}}$ for Seq-OTM.

The second token $\mathcal{T}'$ runs Goliath's program for the send-phase of $\Pi_{\text{Seq-OTM}}$. Instead of running the send-phase interactively with David, Goliath sends $\mathcal{T}'$ to a second instance of $\mathcal{F}_{\text{stateful}}^{\text{wrap}}$ in the creation-phase. David then runs the send-phase with $\mathcal{T}'$. The security-proof for $\Pi_{\text{Seq-OTM}}$ still holds for $\Pi'_{\text{Seq-OTM}}$, due to the following observations. In protocol $\Pi_{\text{Seq-OTM}}$, a corrupted sender Goliath is controlled by the environment $\mathcal{Z}$. In $\Pi'_{\text{Seq-OTM}}$, $\mathcal{Z}$ can only control Goliath during the creation-phase. Afterwards, the tokens $\mathcal{T}$ and $\mathcal{T}'$ cannot communicate with $\mathcal{Z}$ anymore. Thus a corrupted sender in $\Pi'_{\text{Seq-OTM}}$ is strictly less powerful than in $\Pi_{\text{Seq-OTM}}$, and we can conclude that $\Pi'_{\text{Seq-OTM}}$ is UC-secure against a corrupted sender. It is also UC-secure against a corrupted receiver David, as from the view of a corrupted receiver, $\Pi_{\text{Seq-OTM}}$ and $\Pi'_{\text{Seq-OTM}}$ are the same (it makes no difference if David interacts with Goliath or $\mathcal{T}'$).

9 Memory Limited Tokens

The protocol presented in the last Section guarantees perfect security against David. However, to achieve this, the token needs to be able to store $O(n^2m)$ bits of information (for $m = \text{poly}(n)$). For large m, this contradicts the idea of a tamper-proof hardware token being a small and simple device. Moran and Segev [22] noted, that if David is computationally bounded, then the functions stored on the token could be chosen to be pseudorandom [11,16]. The same is true for our construction. It suffices that the token stores a succinct seed of length $O(n)$ for a pseudorandom function F. The token can answer queries z by temporarily computing $(a_t, B_t) = F(t)$ and outputting $V_t = a_t z^T + B_t$.

10 Conclusion

In this paper, we showed that a single (untrusted) tamper-proof hardware token is sufficient for non-interactive, composable computation. We require no additional complexity assumptions. As we only need a single device and no zero-knowledge proofs, our approach is more efficient than previous solutions. Our new primitive, Sequential-One-Time-Memory, is sufficient to realize unconditionally secure One- and k-Time-Programs. However, Sequential-One-Time-Memory is restricted to sequential-access, thus it cannot be used directly to implement several independent One-Time-Programs that can be executed in random order. We consider it an interesting open problem whether it is possible to implement multiple random-access One-Time-Memories with a single untrusted tamper-proof hardware token. However, this seems improbable. Any protocol realizing multiple random-access One-Time-Memories with a single token needs to effectively hide the order in which the One-Time-Memories are queried from the token.

Acknowledgments. We would like to thank the anonymous reviewers of TCC 2011 for their helpful comments.

References

1. Beaver, D.: Correlated pseudorandomness and the complexity of private computations. In: STOC, pp. 479–488 (1996)
2. Ben-Or, M., Goldwasser, S., Kilian, J., Wigderson, A.: Multi-prover interactive proofs: How to remove intractability assumptions. In: STOC, pp. 113–131 (1988)
3. Brouchier, J., Kean, T., Marsh, C., Naccache, D.: Temperature attacks. IEEE Security & Privacy 7(2), 79–82 (2009)
4. Canetti, R.: Universally composable security: A new paradigm for cryptographic protocols. In: FOCS, pp. 136–145 (2001)
5. Canetti, R., Dodis, Y., Pass, R., Walfish, S.: Universally composable security with global setup. In: Vadhan, S.P. (ed.) TCC 2007. LNCS, vol. 4392, pp. 61–85. Springer, Heidelberg (2007)
6. Canetti, R., Fischlin, M.: Universally Composable Commitments. In: Kilian, J. (ed.) CRYPTO 2001. LNCS, vol. 2139, pp. 19–40. Springer, Heidelberg (2001)
7. Damgård, I., Nielsen, J.B., Wichs, D.: Isolated proofs of knowledge and isolated zero knowledge. In: Smart, N.P. (ed.) EUROCRYPT 2008. LNCS, vol. 4965, pp. 509–526. Springer, Heidelberg (2008)
8. Damgård, I., Nielsen, J.B., Wichs, D.: Universally composable multiparty computation with partially isolated parties (2007)
9. Even, S., Goldreich, O., Lempel, A.: A randomized protocol for signing contracts. Commun. ACM 28(6), 637–647 (1985)
10. Fischlin, M., Pinkas, B., Sadeghi, A.-R., Schneider, T., Visconti, I.: Secure set intersection with untrusted hardware tokens. In: Kiayias, A. (ed.) CT-RSA 2011. LNCS, vol. 6558, pp. 1–16. Springer, Heidelberg (2011)
11. Goldreich, O., Goldwasser, S., Micali, S.: How to construct random functions. J. ACM 33(4), 792–807 (1986)
12. Goldreich, O., Ostrovsky, R.: Software protection and simulation on oblivious rams. J. ACM 43(3), 431–473 (1996)
13. Goldwasser, S., Kalai, Y.T., Rothblum, G.N.: One-time programs. In: Wagner, D. (ed.) CRYPTO 2008. LNCS, vol. 5157, pp. 39–56. Springer, Heidelberg (2008)
14. Goyal, V., Ishai, Y., Mahmoody, M., Sahai, A.: Interactive locking, zero-knowledge PCPs, and unconditional cryptography. In: Rabin, T. (ed.) CRYPTO 2010. LNCS, vol. 6223, pp. 173–190. Springer, Heidelberg (2010)
15. Goyal, V., Ishai, Y., Sahai, A., Venkatesan, R., Wadia, A.: Founding cryptography on tamper-proof hardware tokens. In: Micciancio, D. (ed.) TCC 2010. LNCS, vol. 5978, pp. 308–326. Springer, Heidelberg (2010)
16. Håstad, J., Impagliazzo, R., Levin, L.A., Luby, M.: A pseudorandom generator from any one-way function. SIAM J. Comput. 28(4), 1364–1396 (1999)
17. Hofheinz, D., Müller-Quade, J., Unruh, D.: Universally composable zero-knowledge arguments and commitments from signature cards. In: In Proc. of the 5th Central European Conference on Cryptology MoraviaCrypt 2005 (2005)
18. Ishai, Y., Prabhakaran, M., Sahai, A.: Founding cryptography on oblivious transfer – efficiently. In: Wagner, D. (ed.) CRYPTO 2008. LNCS, vol. 5157, pp. 572–591. Springer, Heidelberg (2008)
19. Katz, J.: Universally composable multi-party computation using tamper-proof hardware. In: Naor, M. (ed.) EUROCRYPT 2007. LNCS, vol. 4515, pp. 115–128. Springer, Heidelberg (2007)
20. Kilian, J.: Founding cryptography on oblivious transfer. In: STOC, pp. 20–31 (1988)

21. Kolesnikov, V.: Truly efficient string oblivious transfer using resettable tamper-proof tokens. In: Micciancio, D. (ed.) TCC 2010. LNCS, vol. 5978, pp. 327–342. Springer, Heidelberg (2010)
22. Moran, T., Segev, G.: David and goliath commitments: UC computation for asymmetric parties using tamper-proof hardware. In: Smart, N.P. (ed.) EUROCRYPT 2008. LNCS, vol. 4965, pp. 527–544. Springer, Heidelberg (2008)
23. Rabin, M.O.: How to exchange secrets by oblivious transfer. technical report tr-81. Technical report, Aiken Computation Laboratory, Harvard University (1981)
24. Wolf, S., Wullschleger, J.: Oblivious transfer is symmetric. In: Vaudenay, S. (ed.) EUROCRYPT 2006. LNCS, vol. 4004, pp. 222–232. Springer, Heidelberg (2006)
25. Wullschleger, J.: Oblivious-transfer amplification. CoRR, abs/cs/0608076 (2006)

Correlated-Input Secure Hash Functions

Vipul Goyal[1], Adam O'Neill[2,⋆], and Vanishree Rao[3,⋆]

[1] Microsoft Research, India
vipul@microsoft.com
[2] University of Texas at Austin
adamo@cs.utexas.edu
[3] University of California, Los Angeles
vanishri@cs.ucla.edu

Abstract. We undertake a general study of hash functions secure under *correlated inputs*, meaning that security should be maintained when the adversary sees hash values of many related high-entropy inputs. Such a property is satisfied by a random oracle, and its importance is illustrated by study of the "avalanche effect," a well-known heuristic in cryptographic hash function design. One can interpret "security" in different ways: e.g., asking for one-wayness or that the hash values look uniformly and independently random; the latter case can be seen as a generalization of correlation-robustness introduced by Ishai et al. (CRYPTO 2003). We give specific applications of these notions to password-based login and efficient search on encrypted data. Our main construction achieves them (without random oracles) for inputs related by *polynomials* over the input space (namely $\mathbb{Z}_p$), based on corresponding variants of the q-Diffie Hellman Inversion assumption. Additionally, we show relations between correlated-input secure hash functions and cryptographic primitives secure under related-key attacks. Using our techniques, we are also able to obtain a host of new results for such related-key attack secure cryptographic primitives.

1 Introduction

In practice is often useful to view a cryptographic hash function like a random oracle, as formalized in the random oracle model [6]. However, as random oracles do not exist in reality (and indeed, in general the random oracle model may lead to insecure schemes [13]), an important line of research suggested by [13] seeks to formalize various useful properties satisfied by a random oracle and construct hash functions meeting them under standard assumptions. In this paper, we do so for what we call *correlated-input security*, meaning that (various notions of) security should be maintained when the adversary sees hash values of many related high-entropy inputs.

The importance of correlated-input security in practice is illustrated by the so-called *avalanche effect*, a well-known heuristic in cryptographic hash function

⋆ Work done in part while at Microsoft Research, India.

Y. Ishai (Ed.): TCC 2011, LNCS 6597, pp. 182–200, 2011.

design. (The name "avalanche effect" was coined by Feistel [17], although the idea goes back to Shannon's notion of diffusion [26].) Roughly, the avalanche effect states that making any change to an input should result in a drastically different hash value. Clearly, such a hash function should satisfy a notion of correlated-input security. Our results help to shed light on whether or not this is feasible from a theoretical perspective.

1.1 Notions of Correlated-Input Security

We get different specific notions of correlated-input security depending on how we interpret "security." We first discuss the different interpretations we consider and and how we formalize the resulting notions.

Three Notions. The first and most basic interpretation we consider is "one-wayness." To formalize one-wayness under correlated-inputs, we consider a hash function H and circuits $C_1, \ldots, C_n$, where each C_i takes as input some random coins and outputs a point in the domain of H. The adversary is given hash values $H(x_1), \ldots, H(x_n)$ where each x_i is the output of $C_i(r)$ for random coins r. Note that each C_i is run on the *same* random coins. Therefore the x_i are correlated[1]. The adversary's goal is to output any one of the x_i. Informally, we say that H is one-way under correlated inputs for a class of circuits $\{C\}$ if for any n and any choice of $C_1, \ldots, C_n$ from $\{C\}$, any efficient adversary succeeds with at most negligible probability.

The next interpretation we consider is "unpredictability." To formalize unpredictabililty under correlated-inputs, we consider a hash function H and circuits $C_1, \ldots, C_{n+1}$, where each C_i is as before. Now the adversary is given hash values $H(x_1), \ldots, H(x_n)$ and tries to output $H(x_{n+1})$, where each x_i is the output of $C_i(r)$ as before. The notion is defined for a class of circuits $\{C\}$ analogously to the one-wayness case. It mainly serves as a stepping-stone to our final notion.

Finally, the last interpretation we consider is "pseudorandomness." To formalize pseudorandomness under correlated-inputs, we consider a hash function H and circuits $C_1, \ldots, C_{n+1}$, each C_i is as before. Now the adversary is given hash values $H(x_1), \ldots, H(x_n)$ as well as a "challenge" value that is either $H(x_{n+1})$ or a random string of appropriate length, where each x_i is the output of $C_i(r)$ as before. (This of course requires the circuits to have distinct outputs.) Again, the notion is defined for a class of circuits $\{C\}$ analogously.

Discussion. We make a few observations about these notions. One is that they are only achievable for a class of circuits $\{C\}$ such that $C(r)$ for random r has sufficient min-entropy for any C in the class. In fact, it is not hard to show that a random oracle satisfies our notions for the class of all such circuits. However, in the standard model they are in general only achievable by a *keyed* hash function H. To see this, fix an unkeyed hash function H and consider circuits C_1, C_2 where

[1] For example, the x_i's might agree in most bit positions but vary in the others. It may even be the case that a single input x_i completely determines the rest.

$C_1(r)$ outputs r and $C_1(r)$ outputs $H(r)$. Clearly, no fixed H is even one-way under correlated-inputs for these circuits. By a similar argument, the circuits must not depend on the choice of the hash key. We stress that in our notions the hash function key is *public*. Similar counter-examples show that in general pseudorandom generators and functions do not meet our notions (note that the latter has a *secret* key); we give details in the full version.

Another point is that when considering non-uniform inputs, these notions are non-trivial to achieve even in the case of a single input. However, this can be done; for example [27] considers one-way functions for a non-uniform input and [16] considers pseudorandom generators (that expand the input, which we do not require) for non-uniform seed. Additionally, we note for any *a priori bounded* number of circuits, pseudorandomness under correlated-input can be met even under information-theoretic indistinguishability (where the key-size depends on the bound). This follows from a generalization of the Leftover Hash Lemma [24, Lemma 6.1] (which follows [22, Lemma 3.2]). On the other hand, the focus of our work is on *correlations* among an *unbounded* number of inputs.

1.2 Applications

We now discuss some specific practical applications for our new notions.

Password-based login. An application of our notion of one-wayness under correlated-inputs is password-based login. For example, UNIX maintains a "password" file that, for each user in the system, stores a hash of their password that is compared against the hash of a candidate password supplied at login by someone claiming to be this user. The goal is to prevent an adversary with access to the password file from gaining the ability to impersonate a user. Informally, it is often said that the property of the hash function needed to ensure this is one-wayness. But the standard notion of one-wayness is obviously insufficient here. Passwords, while they should contain entropy, are certainly not uniformly random. Moreover, passwords are typically *correlated*, both across different users, and across the same user on different systems (and the adversary may recover the password file for multiple systems).

This issue seems to be largely ignored in prior work. A paper (which we already mentioned) that considers the relevance of one-way functions for high entropy inputs to this application is [27]; however, they do not consider multiple related inputs and relations among them. Our notion of one-wayness under correlated input seems to be an appropriate security notion for this application[2].

[2] For simplicity, this ignores "salting" the passwords, which can be viewed as considering a randomized hash function. This may make the problem easier for an approach based on the Leftover Hash Lemma (using a similar argument to [24, Lemma 6.1]), but as discussed in [27] such an approach is impractical due to its "entropy loss." For our approach it does not make the problem significantly easier, and anyway it circumvents the core issue that in practice a (deterministic) cryptographic hash function is assumed to satisfy correlated-input security; indeed, salting passwords is only meant to slow down dictionary attacks.

Efficient search on encrypted data. An application of our notion of pseudorandomness under correlated-inputs is efficient search on encrypted data. It is becoming increasingly common for companies to store large amounts of data remotely on servers maintained by an untrusted third party. To provide privacy for the client, the data should be encrypted. However, we still want to allow search on the data without retrieving and decrypting the entire database. Techniques like public-key encryption with keyword search [11] make search possible, but it takes linear time in the database size. On the other hand, practitioners require search time to be comparable to that for unencrypted data.

This problem was first studied from a cryptographic perspective by Bellare et al. [2], who introduced deterministic encryption and the more general concept of efficiently searchable encryption (ESE) as a solution. The basic idea is to attach a hash of each keyword to an encrypted file. Keywords are obviously not uncorrelated, and thus our notions are natural to apply. In fact, if a hash function meets our notion of pseudorandomness under correlated-inputs, then it can be "bootstrapped" to hide all partial information information by encrypting a hash value under the one-way trapdoor permutation based deterministic encryption scheme of [4] (actually, a one-way permutation without a trapdoor suffices here, since we do not need to decrypt).

1.3 Our Construction and Its Security

Next we turn to whether our security definitions can be achieved and under what cryptographic assumptions.

Our Construction. We propose the following construction: Letting G be a group of prime order p, the hash key is a random generator $g \in G$ and random $c \in \mathbb{Z}_p$, and the evaluation of the hash on input $x \in \mathbb{Z}_p$ is $g^{1/(x+c)}$ where $1/(x+c)$ denotes the inverse of $x + c$ modulo p.

Security Analysis. We show that this construction is secure under each of our three notions of security assuming (appropriate variants of) the *q-Diffie-Hellman inversion assumption* (q-DHI). Roughly speaking, this assumption says that given $g^{\alpha}, g^{\alpha^2}, \ldots, g^{\alpha^q}$, it is hard to compute $g^{1/\alpha}$. An assumption of this form was first introduced by Boneh and Boyen [9], who considered it in groups with a *a bilinear map* (pairing) e and asked that it be hard to compute (or distinguish from random) $e(g,g)^{1/x}$ instead of $g^{1/x}$. However, since our hash function is deterministic and thus automatically publically verifiable, we do not need bilinear maps here.

The class of circuits we consider in our proofs are the ones that are (efficiently) representable by a polynomial over $\mathbb{Z}_p$ (the input space of the hash function). In other words, each $x_i = p_i(x, y, z, \ldots)$ where p_i is a polynomial over $\mathbb{Z}_p$ the $x, y, z, \ldots$ are randomly chosen but fixed for all i. In our main theorems, we treat the case of univariate polynomials p_i over $\mathbb{Z}_p$, but it is easy to extend our proofs to multivariate polynomials as well. This is quite a broad class, in particular just considering univariate polynomials and taking p_1 to be the identity polynomial covers the well-known attacks on RSA [15].

First, we show that our inversion hash is one-way under correlated inputs for this class of circuits assuming the q-strong discrete logarithm (q-SDL) assumption (which is weaker than q-DHI in that it need only be hard to compute α itself)[3]. For unpredictability and psuedorandomness, we require an additional assumption that each input to the hash function is individually uniform[4]. However we stress that given other inputs, an input may even be fully computable. For this class of polynomials, we show the inversion hash is unpredictable under correlated inputs assuming q-DHI. Using standard hardcore bit techniques this already gives a construction with small output length achieving pseudorandomness under correlated inputs with the same assumption. However, we directly show pseudorandomness under correlated inputs of our construction for the same class of polynomials assuming the decisional version of q-DHI.

Discussion. One may notice the similarity of our construction to the Boneh-Boyen short signature scheme [10]. Our proof techniques build on those introduced in [10], but note that, as opposed to the latter, we focus on a setting where there is no secret key, and we use the q-DHI assumption instead of the q-SDH assumption of [10]. Indeed, our proofs use some new ideas. In particular, the role of c in our reductions for unpredictability and pseudorandomness is completely different from that of the message in [10].

We also note that our security proofs are under a notion of "selective" security where the circuits that sample the inputs do not depend on the public hash key. As we mentioned, in general this restriction is inherent. However, for restricted classes of circuits (such as arithmetic circuits we consider) which are not able to efficiently compute the hash function in question, it may be possible to achieve an adaptive notion of security even by using an *unkeyed* hash function. We discuss a positive result for this case below.

Finally, we note that our construction as defined is not compressing. However, once we obtain a construction meeting any of our notions, it is easy to obtain one which is also compressing. In the case of one-wayness we can apply a collision-resistant hash to the output, and in the case of pseudorandomness we can truncate the output. It can be shown that the resulting (compressing) hash function retains correlated-input security.

1.4 Relations to Related-Key Attacks

Security under related-key attacks (RKA), first formalized by [5] in the context of pseudorandom functions/permutations, is a well-established notion that, like correlated-input security, asks for security to be maintained under related values of a "secret" input. We explore elations between pseudorandomness under

[3] In fact, for this result the c component of the hash key can be any fixed element in $\mathbb{Z}_p$ (for instance, set $c = 0$).

[4] This translates to the requirement that the polynomials individually have uniform output on uniform input (e.g., this is the case for permutation polynomials). By making non-standard assumptions, it may be possible to drop this restriction. However, considering individually uniform but correlated inputs to the hash function is natural, and we focus on results under standard assumptions.

correlated-inputs secure hash (which we simply call CI-secure hash below) and RKA-security of various cryptographic primitives.

Equivalence to One-Input RKA-Scure wPRF. We first observe that a CI-secure hash is in some sense equivalent to what we call a "one-input RKA-secure weak pseudorandom function (wPRF)." To define such a wPRF F, the adversary is given an input-output pair $x, F(K, x)$ for random x and may query for for other outputs of the form $F(\phi(K), x)$ for relations ϕ of its choosing. (Note that the same x is re-used each time.) Following [20], we note that as compared to a CI-secure hash, the role of the key and the input are simply "switched"[5]. Note that a one-input RKA-secure wPRF is implied by an RKA-secure PRF; in fact, if we start with an RKA-secure PRF (rather than wPRF), the resulting CI-secure hash *does not need a public key.* The latter is significant because [3] gives RKA-secure PRFs, in particular one under the decisional Diffie-Hellman assumption for adaptively chosen, group-induced relations (i.e., multiplication by a constant). We thus obtain an unkeyed adaptively-secure CI-secure hash for the corresponding class of circuits.

A General Transformation for RKA Security. Addtionally, we propose a transformation to "bootstrap" any cryptographic primitive to one that is RKA-secure: simply hash the coins used to generate the secret key for the former. (This can be seen as replacing a RO in this transformation with a CI-secure hash.) Note, however, that in the case the CI-secure hash has a public key, an authentic version of the latter is then needed by any algorithm that uses the secret key (e.g., the signing algorithm for a signature scheme), which may not always be practical. Additionally, while our main construction of CI-secure hash is selectively secure, leading to selective security under RKA (i.e., relations chosen before seeing the public parameters). However, by using our techniques in a non-blackbox way, we can sometimes achieve adaptive security instead and without any public key. In particular, we show how to do this for RKA-secure symmetric encryption, a primitive introduced concurrently to our work in [1];[6]. In this case, we are even able to handle the entire class of (non-zero) polynomial relations.

1.5 Other Related Work

Our work is related to several other lines of research, as we now discuss.

Realizing Random Oracles. As we mentioned, our work can be seen as extending a research agenda proposed by Canetti, Halevi, and Goldriech [13], in which one identifies and realizes useful properties of a random oracle (RO). Indeed, by using the techniques of [2, Theorem 5.1] one can show that a RO meets all the security definitions we consider (which is why we have sought realizations under standard cryptographic assumptions). Other useful properties of a RO that

[5] Note, however, that CI-input security is more general than RKA-security in that it considers inputs sampled from a common "history."

[6] We obtained this result after seeing [1], but the rest of our work was concurrent.

have undergone a similar treatment include perfect one-wayness [12,14] and non-malleability [7] We note that none of these works address security under multiple correlated inputs.

In fact, a significant prior work of Ishai et al. [21, Definition 1] considered a notion of "correlation-robustness" for pseudorandom hash functions, with the motivation of instantiatng the RO in their oblivious transfer protocols. Their notion is more restrictive than our notion of pseudorandomness under correlated-input, as the former is defined only for hash functions that output a single bit and considers inputs obtained by computing the exclusive-or's of a "master" random input s with public random values.

Finally, we note that while realizing the "avalanche effect" satisfied by a RO forms a major motivation for considering correlated-input security, it is not the only way the latter could be formalized. In particular, it talks only about the change in the output behavior relative to any change to an *unknown* input. The notion of "(multiple input) correlation intractability" due to [13] is a possible formalization the effect without this restriction. On the other hand, the latter notion seems harder to work with and more difficult to achieve.

Deterministic Encryption. Our security notions can be viewed as relaxations or variants of the notions of privacy proposed for deterministic encryption (DE) in [2,8,4,24] (that seek to hide partial information) in the case of hash functions rather than encryption schemes. Indeed, the results of of [2,8,4,24] show that in some sense the "hard part" of realizing DE without random oracles is dealing with correlations among the inputs. Our work studies this issue at a more basic level, asking whether it is feasible even without supporting decryption and for weaker security notions like one-wayness.

Related Security Notions. Recently, Rosen and Segev [25] introduced the notion of *correlated-product secure* trapdoor functions (TDFs). Correlated-product security was later considered for hash functions in [20]. Correlated-product security differs from our notions in that the former refers to security when related inputs are evaluated under *independent instances* of the function; in other words, there does not exist a single function which is evaluated on related inputs (as is the case for correlated-input security). Indeed, our techniques are quite different and unrelated to those in [25,20].

The recent work of Goldenberg and Liskov [19] also considers a form of correlated-input security (which they call "related-secret" security) for various primitives. While their work has some similarities to ours (for example, they consider "related-secret" one-way functions), there are some important differences. Namely, they focus on hardcore bits and pseudorandom functions rather than hash functions, and they follow the definitional framework for RKA-security introduced in [5]. As mentioned above, this definitional framework is less general than ours. Additionally, their results are mainly negative.

2 Preliminaries

Notations. Let $x \xleftarrow{\$} X$ denote the operation of selecting a random element x from X. Let by $x \leftarrow y$ denote the assignment of a value y to x. Let $|X|$ denote the size of the set X, and $|x|$ denote the length of the string x. Let λ denote the security parameter. Let $\mathsf{FF}(\mathcal{K}, \mathcal{D}, \mathcal{R})$ be the set of all families of functions $F : \mathcal{K} \times \mathcal{D} \longrightarrow \mathcal{R}$. For sets X, Y, let $\mathsf{Fun}(X, Y)$ be the set of all functions mapping X to Y. For brevity, we say that an algorithm outputs a set/function as a shorthand to mean that it outputs their descriptions.

Complexity Assumptions. We first state our complexity assumptions, namely q-DHI [9,10], which is weaker than q-BDHI [9] as well as q-SDH [10], and we introduce what we call the q-Strong Discrete Logarithm (q-SDL) assumption which is weaker than all of q-DHI, q-BDHI, and q-SDH assumptions.

Let GrpGen be a PPT algorithm that takes as input the security parameter 1^λ and outputs parameters for some cyclic multiplicative group $\mathbb{G}$, including the group order p which is a $poly(\lambda)$-bit integer, a generator g, and an efficient algorithm (e.g., circuit) for multiplication (and thus also exponentiation). We denote it as $(\mathbb{G}, p, g) \leftarrow \mathsf{GrpGen}(1^\lambda)$.

q-Strong Discrete Logarithm (q-SDL) Problem. The q-SDL problem in $\mathbb{G}$ is defined as follows: given a $(q+1)$-tuple $(g, g^x, g^{x^2}, \ldots, g^{x^q}) \in (\mathbb{G}^\star)^{q+1}$ for some unknown $x \in \mathbb{Z}_p^\star$, output x.

An algorithm $\mathcal{A}$ solves the q-SDL problem in the group $\mathbb{G}$ with advantage ϵ if

$$\mathsf{SDL\ Adv}_{\mathcal{A},q} := \Pr[\mathcal{A}(g, g^x, g^{x^2}, \ldots, g^{x^q}) = x] \geq \epsilon$$

where the probability is over the random choice of generator $g \in \mathbb{G}^\star$, the random choice of $x \in \mathbb{Z}_p^\star$, and the random bits consumed by $\mathcal{A}$.

Definition 1. *We say that the (q, t, ϵ)-SDL assumption holds in $\mathbb{G}$ (or GrpGen satisfies the (q, t, ϵ)-SDL assumption) if no probabilistic t-time algorithm has advantage at least ϵ in solving the q-SDL problem in $\mathbb{G}$.*

It is easy to see that the 1-SDL assumption is equivalent to the standard Discrete Logarithm assumption.

q-Diffie-Hellman Inversion (q-DHI) Problem. The q-DHI problem [9,10] in $\mathbb{G}$ is defined as follows: given a $(q+1)$-tuple $(g, g^x, g^{x^2}, \ldots, g^{x^q}) \in (\mathbb{G}^\star)^{q+1}$ for some unknown $x \in \mathbb{Z}_p^\star$, output $g^{\frac{1}{x}} \in \mathbb{G}$.

An algorithm $\mathcal{A}$ solves the q-DHI problem in the group $\mathbb{G}$ with advantage ϵ if

$$\mathsf{DHI\ Adv}_{\mathcal{A},q} := \Pr[\mathcal{A}(g, g^x, g^{x^2}, \ldots, g^{x^q}) = g^{\frac{1}{x}}] \geq \epsilon$$

where the probability is over the random choice of generator $g \in \mathbb{G}^\star$, the random choice of $x \in \mathbb{Z}_p^\star$, and the random bits consumed by $\mathcal{A}$.

Definition 2. *We say that the (q, t, ϵ)-DHI assumption holds in $\mathbb{G}$ (or GrpGen satisfies the (q, t, ϵ)-DHI assumption) if no probabilistic t-time algorithm has advantage at least ϵ in solving the q-DHI problem in $\mathbb{G}$.*

q-DHI Problem can be equivalently stated as follows [10]: given a $(q+2)$-tuple $(g, g^x, g^{x^2}, \ldots, g^{x^q}, c) \in (\mathbb{G}^\star)^{q+1} \times \mathbb{Z}_p \backslash \{x\}$ for some unknown $x \in \mathbb{Z}_p^\star$, output $g^{\frac{1}{x+c}} \in \mathbb{G}$. This definition also clearly points out the distinction between the q-DHI problem and the q-SDH problem: in case of the q-DHI problem, the value of c is prescribed in the problem instance itself, whereas in case of the q-SDH problem, the solver is free to choose $c \in \mathbb{Z}_p$ and output a pair, $(c, g^{\frac{1}{x+c}})$. Obviously, the q-DHI assumption is weaker than the q-SDH assumption.

Decisional q-Diffie-Hellman Inversion (Decisional q-DHI) Problem. The decisional q-DHI problem in $\mathbb{G}$ is defined as follows: given a $(q+1)$-tuple $(g, g^x, g^{x^2}, \ldots, g^{x^q}) \in (\mathbb{G}^\star)^{q+1}$ for some unknown $x \in \mathbb{Z}_p^\star$, distinguish between $g^{\frac{1}{x}}$ and a random element $R \xleftarrow{\$} \mathbb{G}$.

An algorithm $\mathcal{A}$ solves the decisional q-DHI problem in $\mathbb{G}$ with advantage ϵ if

$$\begin{aligned}\mathsf{DDHI\ Adv}_{\mathcal{A},q} := \big|\Pr[\mathcal{A}(g, g^x, g^{x^2}, \ldots, g^{x^q}, g^{\frac{1}{x}}) = 1] \\ - \Pr[\mathcal{A}(g, g^x, g^{x^2}, \ldots, g^{x^q}, R) = 1]\big| \geq \epsilon\end{aligned}$$

where the probability is over the random choice of generator $g \in \mathbb{G}^\star$, the random choice of $x \in \mathbb{Z}_p^\star$, the random choice of $R \in \mathbb{G}^\star$, and the random bits consumed by $\mathcal{A}$. The distribution on the left is referred to as $\mathcal{P_{DDHI}}$ and the distribution on the right as $\mathcal{R_{DDHI}}$.

Definition 3. *We say that the decisional (q, t, ϵ)-DHI assumption holds in $\mathbb{G}$ (or* GrpGen *satisfies the decisional (q, t, ϵ)-DHI assumption) if no probabilistic t-time algorithm has advantage at ϵ in solving the decisional q-DHI problem in $\mathbb{G}$.*

3 Our Model: Correlated Input Security

In this section, we define our new notion of security for cryptographic hash functions. We would be interested in preserving various properties of hash functions (like one-wayness and pseudo-randomness) when the function maybe evaluated on a tuple of inputs which maybe be *correlated* in an arbitrary way. Standard notions of security do not provide any guarantee in such a setting. In the full version, we discuss examples of functions which are secure in the standard sense but may be completely insecure when evaluated on multiple inputs which are correlated.

Before we go further, we first discuss how we represent correlations among a tuple of inputs $(m_1, \ldots, m_n)$. In general, such an input tuple maybe generated by a polynomial-size sampler circuit *Samp*. In other words, *Samp* takes a random tape r (of appropriate length) as input such that $(m_1, \ldots, m_n) \leftarrow Samp(r)$. Note such a sampler circuit can generate the input tuple for any type of polynomial-time computable correlations. Equivalently, one can generate the (correlated) tuple of inputs using a *tuple* of polynomial-size circuits $(C_1, \ldots, C_n)$ when initialized on the same random tape. In other words, fix a random string r and set $m_i \leftarrow C_i(r)$. It is easy to see that both these sampling procedures are equivalent. For the rest of the paper, we shall stick to the latter mode of using a tuple of circuits (for convenience, as will be clear later on).

Also, it will be understood that the range of every circuit considered is a subset of the input-space (or keyspace) in question. We refer to an adversary as $\{C\}$-restricted, if its correlated-input circuits are restricted to the class of circuits $\{C\}$.

We first define the syntax for a general function family (or a hash function family if the functions are compressing). We will then move on to formalize the various security properties such a function family might satisfy.

Definition 4 (Function Family). *A family of deterministic functions $\mathcal{H}$ is specified by a PPT algorithm Gen. The algorithm Gen, given input 1^λ, outputs a parameter set $\mathcal{I}_h$, domain $\mathcal{D}_h$, and range $\mathcal{R}_h$, and outputs $c \in \mathcal{I}_h$ as a description of a function $h_c : \mathcal{D}_h \longrightarrow \mathcal{R}_h$. The sizes of the domain and range sets are each exponential in the security parameter.*

Now, we shall discuss our first notion of security called *correlated-input one-wayness*. Informally, we consider a function $h(\cdot)$ such that given $(h(m_1), \ldots, h(m_n))$, where inputs $(m_1, \ldots, m_n)$ maybe correlated, it is hard for any PPT adversary to output any valid preimage m_i. This can be viewed as a generalization of the standard notion of one-way functions. We allow the adversary to specify the correlations to the challenger by giving a tuple of circuits $(C_1, \ldots, C_n)$, where each circuit is from a class of correlated-input circuits $\{C\}$. Note that, for this definition to be satisfiable, each circuit $C_i, \in \{C\}$ individually should have high min-entropy output distribution for uniform random input distribution[7]. Thus, we quantify only over such circuits in our definition. We discuss it under both selective and adaptive security notions. More details follow.

In the following we shall only formalize the selective notion, while we refer the reader to the full version for definitions of the adaptive notions.

The Selective Correlated-Input Inverting experiment $\mathsf{Exp}^{sCI-inv}_{\mathcal{A},\mathcal{H},\{C\}}$. For a family of deterministic functions $\mathcal{H}$, an adversary $\mathcal{A}$, and a family of efficiently-computable correlated-input circuits $\{C\}$, we define the following game between a challenger and the adversary $\mathcal{A}$.

- **Setup Phase 1.** Challenger runs the Gen algorithm of $\mathcal{H}$ for a security parameter input 1^λ and gets $h_c : \mathcal{D}_h \longrightarrow \mathcal{R}_h$. Challenger gives $\mathcal{D}_h$ to $\mathcal{A}$.
- **Query Phase.** $\mathcal{A}$ chooses a positive integer n $(= poly(\lambda))$, and gives to the challenger n circuits $\{C_i\}_{i\in[n]} \subset \{C\}$.
- **Setup Phase 2.** Challenger gives $h_c(\cdot)$ to $\mathcal{A}$ and chooses r, a uniform random string of appropriate length.
- **Response Phase.** $\forall i \in [n]$, challenger responds via $h_c(C_i(r))$.
- **Invert Phase.** $\mathcal{A}$ outputs $(\hat{k}, \hat{y})$ for $\hat{k} \in [n]$ and $\hat{y} \in \mathcal{D}_h$.

The output of the experiment is defined to be 1 if $h_c(\hat{y}) = h_c(C_{\hat{k}}(r))$ and 0 otherwise.

[7] This requirement is similar to one in the standard notion of one-way functions. If the input does not have sufficient min-entropy, it is easy to see that an adversary can guess a preimage and succeed with noticeable probability.

We define the advantage of an adversary $\mathcal{A}$ in the above game as:

$$\mathsf{Adv}^{sCI-inv}_{\mathcal{A},\mathcal{H},\{C\}}(\lambda) = \Pr[\mathsf{Exp}^{sCI-inv}_{\mathcal{A},\mathcal{H},\{C\}} = 1]$$

The probability is over the random bits used by the challenger and the adversary.

Definition 5. *A family of functions $\mathcal{H}$ is said to be selective correlated-input one-way with respect to a family of correlated-input circuits $\{C\}$, if for all $\mathcal{A} \in$ PPT there exists a negligible function negl, such that:*

$$Adv^{sCI-inv}_{\mathcal{A},\mathcal{H},\{C\}}(\lambda) \leq negl(\lambda)$$

We now consider two more correlated-input security notions where we talk about the unpredictability of the *output* as opposed to that of the input. Informally, we consider a function $h_c : \mathcal{D}_h \longrightarrow \mathcal{R}_h$ with the following properties. Consider a tuple of correlated inputs $(m_1, \ldots, m_{n+1})$. The adversary is given the function outputs $(h_c(m_1), \ldots, h_c(m_n))$ and it tries to compute $h_c(m_{n+1})$. In the first security notion called *correlated-input unpredictability* (CI-unpredictability), we require that it should be hard for the adversary to output $h_c(m_{n+1})$. In the next notion called *correlated-input pseudorandomness* (CI-pseudorandomness), we require that the adversary should not be able to distinguish $h_c(m_{n+1})$ from a random element in $\mathcal{R}_h$, given $(h_c(m_1), \ldots, h_c(m_n))$. It is easy to show that this notion of CI-pseudorandomness is equivalent to a notion where an adversary gets either $(h_c(m_1), \ldots, h_c(m_{n+1}))$ or $n+1$ independent random elements in $\mathcal{R}_h$ and is required to distinguish the two cases. Note that, for any of these notions to be satisfiable, besides the requirement that each circuit $C_i \in \{C\}$ individually should have high min-entropy output distribution for uniform random input distribution, we also require that, for every two distinct circuits C_i and C_j in $\{C\}$, and for a uniform random input r of appropriate length, $C_i(r) = C_j(r)$ happens only with negligible probability over the choice of r.

In trying to give more power to the adversary (thus making our definition stronger), we allow the adversary to specify the correlation by giving a tuple of circuits $(C_1, \ldots, C_{n+1})$, where $C_i \in \{C\}$, to the challenger, as before. In addition, for the selective case, for a tuple of inputs $(m_1, \ldots, m_{n+1})$, we allow the adversary to get outputs on n *adaptively chosen* input indices of its choice before trying to predict the remaining output[8]. More details follow.

The definitions of the selective CI-pseudorandomness and adaptive notions of all the three notions are given in the full version.

The Selective Correlated-Input Predicting *experiment* $\mathsf{Exp}^{sCI-pred}_{\mathcal{A},\mathcal{H},\{C\}}$. For a family of deterministic functions $\mathcal{H}$, an adversary $\mathcal{A}$, and a family of efficiently-computable correlated-input circuits $\{C\}$, we define the following game between a challenger and the adversary $\mathcal{A}$.

[8] By incurring a security loss of a factor of n, this definition can actually be shown to be equivalent to a weaker definition where the adversary is required to predict the output specifically on input m_{n+1} fixed after it presents its queries but before it reveives the responses. However, working directly with this definition might lead to better concrete security guarantees in the real world.

- **Setup Phase 1.** Challenger runs the Gen algorithm of $\mathcal{H}$ for a security parameter input 1^λ and gets $h_c : \mathcal{D}_h \longrightarrow \mathcal{R}_h$. Challenger gives $\mathcal{D}_h$ to $\mathcal{A}$.
- **Query Phase.** $\mathcal{A}$ chooses a positive integer n $(= poly(\lambda))$, and gives to the challenger $n+1$ distinct circuits $\{C_i\}_{i\in[n+1]} \subset \{C\}$.
- **Setup Phase 2.** Challenger gives $h_c(\cdot)$ to $\mathcal{A}$ and chooses r, a uniform random string of appropriate length.
- **Partially Adaptive Query-Response Phase.** $\mathcal{A}$ presents n queries, where an ith query is $k_i \in [n+1]$. Challenger responds to it via $h_c(C_{k_i}(r))$.
- **Predict Phase.** The adversary outputs $\hat{y} \in \mathcal{R}_h$.

Let $k_{n+1} \in [n+1]$ be such that $k_{n+1} \neq k_i \ \forall\, i \in [n]$. The output of the experiment is defined to be 1 if $\hat{y} = h_c(C_{k_{n+1}}(r))$ and 0 otherwise.

We define the advantage of an adversary $\mathcal{A}$ in the above game as:

$$\mathsf{Adv}^{sCI-pred}_{\mathcal{A},\mathcal{H},\{C\}}(\lambda) = \Pr[\mathsf{Exp}^{sCI-pred}_{\mathcal{A},\mathcal{H},\{C\}} = 1]$$

The probability is over the random bits used by the challenger and the adversary.

Definition 6. *A family of functions $\mathcal{H}$ is said to be selective correlated-input unpredictable with respect to a family of correlated-input circuits $\{C\}$, if for all $\mathcal{A} \in \mathsf{PPT}$ there exists a negligible function negl, such that:*

$$\mathit{Adv}^{sCI-pred}_{\mathcal{A},\mathcal{H},\{C\}}(\lambda) \leq negl(\lambda)$$

4 Proposed Construction

In the sequel, we give the construction of our function and prove that it is correlated-input secure for a class of polynomials over $\mathbb{Z}_p$ (where p is a prime number) in the sense of each of the three selective security models defined above.

Our construction is given in Figure 1.

Our proposed function is extremely simple and efficient to compute. The cost of computation is dominated by a single exponentiation operation. The construction can be seen as similar to a short signature scheme by Boneh and Boyen [10]. Our main novelty can be seen in the proofs of security. Indeed, interestingly, our proofs show that the original signature scheme of Boneh and Boyen is secure even if an adversary is allowed to obtain messages signed by various *correlated secret signing keys*, where the correlations are from a set of polynomials over $\mathbb{Z}_p$. We refer the reader to the full version for a detailed description of this implication.

$\mathsf{Gen}(1^\lambda)$. Run GrpGen: $(\mathbb{G}, p, g) \leftarrow \mathsf{GrpGen}(1^\lambda)$, where p is a prime number. Gen uniformly samples a random element c from $\mathbb{Z}_p$ and a random generator g of group $\mathbb{G}$. It outputs g, c and a function $h : \mathbb{Z}_p \longrightarrow \mathbb{G}$ defined by,

$$h(m) := g^{\frac{1}{m+c}}$$

for any $m \in \mathbb{Z}_p$ (where $\frac{1}{m+c}$ is computed mod p).

Fig. 1. Our Construction

4.1 Analysis of the above Construction

We prove that our construction is selectively secure against a class of correlations computable by polynomials over $\mathbb{Z}_p$. In what follows, we introduce some more notations that will be used in the rest of the paper and then discuss the three security games with correlated-input circuits computing polynomials over $\mathbb{Z}_p$.

Notations. Let $deg[f(X)]$ denote the degree of a polynomial $f(X)$ over $\mathbb{Z}_p$. If the output distribution of a polynomial is uniform in $\mathbb{Z}_p$ (or, in other words, if the range of the polynomial is $\mathbb{Z}_p$ itself), then we refer to the polynomial as uniform-output polynomial. On the other hand, if the output distribution of a polynomial has high min-entropy in $\mathbb{Z}_p$, then we refer to the polynomial as high-min-entropy-output polynomial (any non-zero polynomial of degree polynomial in the security parameter has a range of size exponential in the security parameter). We shall only consider polynomials of degree is at least 1 and polynomial in λ. We only state our theorems in the following and give the proofs in the full version.

4.2 Selective Correlated-Input One-Wayness

Theorem 1. *Suppose that (q, t', ϵ)-SDL assumption holds in $\mathbb{G}$. Let $\{C\}$ be a set of non-zero polynomials over $\mathbb{Z}_p$. Then, for $\mathcal{H}$ as in Figure 1, there exists no probabilistic t-time adversary $\mathcal{A}$ for which $\mathsf{Adv}^{sCI-inv}_{\mathcal{A},\mathcal{H},\{C\}}(\lambda)$ is at least ϵ provided that*

$$d \leq q \text{ and } t \leq t' - \Theta(nq\tau)$$

where $d = d(\lambda)$ upper bounds the sum of the degrees of the polynomials that $\mathcal{A}$ queries upon and τ is the maximum time for an exponentiation in $\mathbb{G}$ and $\mathbb{Z}_p$.

4.3 Selective Correlated-Input Unpredictability

Theorem 2. *Suppose that (q, t', ϵ')-DHI assumption holds in $\mathbb{G}$. Let $\{C\}$ be a set of uniform-output polynomials over $\mathbb{Z}_p$. Then, for $\mathcal{H}$ as in Figure 1, there exists no probabilistic t-time adversary $\mathcal{A}$ for which $\mathsf{Adv}^{sCI-pred}_{\mathcal{A},\mathcal{H},\{C\}}$ is at least ϵ provided that*

$$d \leq q + 1,\ \epsilon \geq 2(n+1)\epsilon' \text{ and } t \leq t' - \Theta(nq\tau)$$

where $d = d(\lambda)$ upper bounds the sum of the degrees the polynomials that $\mathcal{A}$ queries upon and τ is the maximum time for an exponentiation in $\mathbb{G}$ and $\mathbb{Z}_p$.

The proof for the above theorem and for CI-pseudorandomness is given in the full version.

5 Relations between CI-Security and RKA-Security

We conclude by examining relations beween correlated-input secure hash functions and security under related-key attacks, whose formal treatement was initiated by [5]. The latter asks that security of a cryptographic primitive (e.g., a pseudorandom function) maintains security when used with *related secret keys*. In this section, we only consider our notion of pseudorandomness under correlated-inputs, so "CI-security" below refers by default to this notion.

5.1 Relations to RKA-Secure Weak PRFs

We start by showing an equivalence between CI-secure hash functions and some form of RKA-security for *weak* PRFs we introduce. The idea, following [20], is to "switch" the input and key for these primitives.

RKA-Secure Weak PRFs. Recall that weak PRFs [23], as opposed to normal ones, handle only random inputs. In defining RKA-security for this primitive, a modeling choice we need to consider is whether, when the adversary queries for a value of the function under a related key, a *new* random input is chosen (or the previous random-input re-used). We give general definitions that capture the possibilities. (However, for our results we only use a notion where the same random input is re-used.) We also consider both "selective" and "adaptive" security; in the former, the adversary chooses the relations applied to the secret key before receiving any responses.

In defining RKA security for various primitives, we use a non-standard formalization based on our framework of circuits, which in this case sample keys. This is for ease of comparison to our notion of CI-security. For example, by $\{C\}$-RKA-PRF we refer to an RKA-PRF where the secret keys are sampled according to circuits in $\{C\}$ (executed on a common random input). Note that, in this framework, RKA-security corresponds to a special case of CI-security where the first circuit samples a random key K and and the remaining circuits operate only on K (and not the coins used to sample it) to produce a related key. (The latter is sufficient in the context of RKA-security.)

We also note that a PRF function family is specified by an efficient probabilistic parameter-generation algorithm Gen_{prf} which takes as input a security parameter 1^λ and outputs the description of a function including a description of its keyspace, domain and range. However, for simplicity of exposition, we only consider a single PRF function in most part of the following discussion as long as there is no ambiguity.

Definition 7 (**$q_{input} - \{C\}$-aRKA-wPRF.**). *Let $F : \mathcal{K} \times \mathcal{D} \longrightarrow \mathcal{R}$ be an efficiently computable function. Let $\{C\} \subseteq \mathsf{Fun}(\mathcal{K}, \mathcal{K})$ be a set of RKD circuits. F is said to be $q_{input} - \{C\}$-aRKA-wPRF, if, $\forall \mathcal{A} \in PPT$:*
$\mathsf{Adv}^{aRKA-wPRF}_{\mathcal{A},F,\{C\},q_{input}}(\lambda) := |\Pr[k \xleftarrow{\$} \mathcal{K} : \mathcal{A}^{\mathcal{O}^{weak}_{F(k,\cdot)}(\cdot,\cdot)} \longrightarrow 1] - \Pr[k \xleftarrow{\$} \mathcal{K}, G \xleftarrow{\$} \mathsf{FF}(\mathcal{K}, \mathcal{D}, \mathcal{R}) : \mathcal{A}^{\mathcal{O}^{weak}_{G(k,\cdot)}(\cdot,\cdot)} \longrightarrow 1]|$ is negligible in λ, where the related-key-wprf oracle $\mathcal{O}^{weak}_{f(k,\cdot)}(\cdot)$ takes as input $(index_i, C_i) \in ([q_{input}], \{C\})$, and outputs $(x_{index_i}, f(C_i(k), x_{index_i}))$, where $x_{index_i} \xleftarrow{\$} \mathcal{D}$.

The definition for selective version appears in the full version.

Let $\mathcal{H}$ be a family of functions specified by Gen. A family of functions $\{F\}$ is defined by the following parameter-generation algorithm:
$\mathsf{Gen}_{prf}(1^\lambda)$. Gen_{prf} runs $\mathsf{Gen}(1^\lambda)$ that outputs the description of a parameter set $\mathcal{D}$, $c \in \mathcal{D}$, and a function $h_c : \mathcal{K} \longrightarrow \mathcal{R}$. Gen_{prf} outputs $\mathcal{K}$, $\mathcal{D}$ and $\mathcal{R}$ for keyspace, domain and range, respectively, of a function F defined by,

$$F(k, x) := h_x(k)$$

for any $k \in \mathcal{K}$ and $x \in \mathcal{D}$.

Fig. 2. Construction of $1 - \{C\}$-(s/a)RKA-wPRF Family from $\{C\}$-(s/a)CI-pseudorandom Function Family

The Equivalence. In the following, for a class of circuits $\{C\}$ we show an equivalence between a one-input RKA-Secure wPRF for $\{C\}$ and a CI-secure hash function for $\{C\}$. (The equivalence holds respectively in the cases of selective and adaptive security notions.) First, we construct a family of functions $\{F\}$ from a family of functions $\mathcal{H}$ and show that if $\mathcal{H}$ is a $\{C\}$-aCI-pseudorandom function family (resp. $\{C\}$-sCI-pseudorandom function family) then $\{F\}$ is a $1 - \{C\}$-aRKA-wPRF (resp. $1 - \{C\}$-sRKA-wPRF).

Theorem 3. *$\{C\}$-aCI-pseudorandom function family (resp. $\{C\}$-sCI-pseudorandom function family) implies $1 - \{C\}$-aRKA-wPRF (resp. $1 - \{C\}$-sRKA-wPRF).*

The proof is given in the full version.

In the full version, we also give a construction of $1 - \{C\}$-aRKA-wPRF (resp. $1 - \{C\}$-sRKA-wPRF) from $\{C\}$-aCI-pseudorandom function family (resp. $\{C\}$-sCI-pseudorandom function family).

New CI-Secure Hash Functions. As an application of the above equivalence, we obtain new CI-secure hash functions. In particular, note that an adaptive one-input RKA-secure wPRF for a class of circuits $\{C\}$ is trivially implied by an RKA-Secure PRF for $\{C\}$. (Here, the latter is defined as expected, namely as in prior work except cast in our framework of circuits; see the full version for more details.) We can therefore use the recent constructions of RKA-Secure PRFs by Bellare and Cash [3] to obtain adaptive CI-secure hash functions. Namely, the latter are secure under the standard DDH assumption for class of circuits computing multiplication by a group element or under exponential-hardness of DDH for addition by a group element.

Though the resulting CI-secure hash functions are secure for much weaker classes of relations as compared to our main construction, they are remarkable in that they are both adaptively secure and do not need a public key. (They do not even need any randomly-generated global parameters, as the constructions of [3] work in a fixed group.) The latter is because in the case that we start with an RKA-secure PRF (rather than wPRF), our construction of CI-secure hash can be modified by applying the PRF to any fixed value in the domain of the latter (still using the input as the key).

5.2 CI-Secure Functions Imply Other RKA-Secure Primitives

In this section, we discuss a more general technique for building RKA-secure cryptographic primitives from a CI-secure hash function. The basic idea is to hash the coins used to generate keys for the the former, using a CI-secure hash function.

Informally, let Ψ be a scheme for a cryptographic primitive. Let KeyGen be a PPT algorithm for Ψ, and let $l(\lambda)$ be the length of the random string used by it. Our transformation uses a $\{C\}$-pseudorandom function family $\mathcal{H}$ specified by Gen, and the transformation involves modifying $\mathsf{KeyGen}(1^\lambda; r)$ where $r \xleftarrow{\$} \{0,1\}^{l(\lambda)}$ to $\mathsf{KeyGen}(1^\lambda; h(r'))$ where $r' \xleftarrow{\$} \{0,1\}^{t(\lambda)}$ and $(h : \{0,1\}^{t(\lambda)} \longrightarrow \{0,1\}^{l(\lambda)}) \leftarrow \mathsf{Gen}(1^\lambda)$. The resulting scheme is then expected to be " $\{C\}$-RKA-secure" besides preserving the security properties of the underlying untransformed scheme.

More concretely, we exemplify the above technique for digital signatures. We give our formalization of RKA-security for signatures in the full version.

In what follows, we show that $\{C\}$-aCI-pseudorandom function family (resp. $\{C\}$-sCI-pseudorandom function family) implies $\{C\}$-aRKA-unforgeable scheme (resp. $\{C\}$-sRKA-unforgeable scheme). The transformation is given in Figure 3.

Theorem 4. *$\{C\}$-aCI-pseudorandom function family (resp. $\{C\}$-sCI-pseudorandom function family) implies $\{C\}$-aRKA-unforgeable scheme (resp. $\{C\}$-sRKA-unforgeable scheme).*

The proof is given in the full version.

Discussion. In the case that the starting CI-secure hash function has a public key, the above transformation results in a cryptographic primitive for which algorithms operating on the secret key also need to access an authentic public key. In some scenarios, e.g. smart cards, this may not always be practical. Moreover,

Let $\mathcal{H}$ be a function family specified by Gen. Let $\Sigma' = (\mathsf{KeyGen}', \mathsf{Sign}', \mathsf{Verify}')$ be a signature scheme and let $l(\lambda)$ be the length of the randomness used in the KeyGen'. The signature scheme $\Sigma = (\mathsf{KeyGen}, \mathsf{Sign}, \mathsf{Verify})$ is defined by:

- $\mathsf{KeyGen}(1^\lambda)$: $(h : \{0,1\}^{t(\lambda)} \longrightarrow \{0,1\}^{l(\lambda)}) \leftarrow \mathsf{Gen}(1^\lambda)$; $sk \xleftarrow{\$} \{0,1\}^{t(\lambda)}$; $(sk', pk') \leftarrow \mathsf{KeyGen}'(1^\lambda; h(sk))$; output sk as the secret key, and $pk := (h, pk')$ as the public key.
- $\mathsf{Sign}(sk, m)$: Run $\mathsf{KeyGen}'(1^\lambda, h(sk))$ to obtain sk'. The signature on message m is set as $\sigma \leftarrow \mathsf{Sign}'(sk', m)$.
- $\mathsf{Verify}(pk, m, \sigma)$: Output valid if $\mathsf{Verify}'(pk', m, \sigma) = \mathsf{valid}$ and output $\mathsf{invalid}$ otherwise.

Fig. 3. Construction of RKA-secure Signature Scheme from CI-secure Pseudorandom Functions our main construction of CI-secure hash is only selectively-secure, resulting in a selectively RKA-secure the cryptographic primitive. On the other hand, it is sometimes possible to use our techniques (in a "non-blackbox" way) to design an RKA-secure scheme without a public key and that is adaptively secure. In particular, we show this for RKA-secure symmetric-key encryption recently introduced in [1] in the full version. We also mention that using the CI-secure hash functions derived from the Bellare-Cash RKA-secure PRFs [3] avoid these issues, but for a much weaker class of relations.

Relation to Tampering Attacks. We also note that RKA-security for a cryptographic primitive can also be used to to protect against tampering attacks [18], where, for instance, the secret key stored by a smart card is tampered with and its behavior is observed while it acts using the tampered secret, with an objective of gaining advantage against the security of the functionality of the smart card when using the original secret. However, as discussed in [1], security against tampering attacks is easier to achieve in general, through some kind of "sanity check" on the secret key (for instance, by including a signature on the secret key as a part of the public key, which is verified by any algorithm using the former); although, as discussed above, this approach may not always be practical. This does not work for RKA-security, since we actually want related secret keys to function like *independently generated* ones.

Acknolwedgements

We thank the anonymous reviewers of TCC 2011 for detailed comments and suggestions. The second author thanks David Cash for discussions about related-key security, and was supported in part by Brent Waters' grants NSF CNS-0915361 and CNS-0952692.

References

1. Applebaum, B., Harnik, D., Ishai, Y.: Semantic security under related-key attacks and applications. Cryptology ePrint Archive, Report 2010/544 (2010), `http://eprint.iacr.org/`
2. Bellare, M., Boldyreva, A., O'Neill, A.: Deterministic and efficiently searchable encryption. In: Menezes, A. (ed.) CRYPTO 2007. LNCS, vol. 4622, pp. 535–552. Springer, Heidelberg (2007)
3. Bellare, M., Cash, D.: Pseudorandom functions and permutations provably secure against related-key attacks. In: Rabin, T. (ed.) CRYPTO 2010. LNCS, vol. 6223, pp. 666–684. Springer, Heidelberg (2010)
4. Bellare, M., Fischlin, M., O'Neill, A., Ristenpart, T.: Deterministic encryption: Definitional equivalences and constructions without random oracles. In: Wagner, D. (ed.) CRYPTO 2008. LNCS, vol. 5157, pp. 360–378. Springer, Heidelberg (2008)
5. Bellare, M., Kohno, T.: A theoretical treatment of related-key attacks: Rka-prps, rka-prfs, and applications. In: Biham, E. (ed.) EUROCRYPT 2003. LNCS, vol. 2656, pp. 491–506. Springer, Heidelberg (2003)

6. Bellare, M., Rogaway, P.: Random oracles are practical: A paradigm for designing efficient protocols. In: ACM Conference on Computer and Communications Security, pp. 62–73 (1993)
7. Boldyreva, A., Cash, D., Fischlin, M., Warinschi, B.: Foundations of non-malleable hash and one-way functions. In: Matsui, M. (ed.) ASIACRYPT 2009. LNCS, vol. 5912, pp. 524–541. Springer, Heidelberg (2009)
8. Boldyreva, A., Fehr, S., O'Neill, A.: On notions of security for deterministic encryption, and efficient constructions without random oracles. In: Wagner, D. (ed.) CRYPTO 2008. LNCS, vol. 5157, pp. 335–359. Springer, Heidelberg (2008)
9. Boneh, D., Boyen, X.: Efficient Selective-ID Secure Identity-Based Encryption Without Random Oracles. In: Cachin, C., Camenisch, J. (eds.) EUROCRYPT 2004. LNCS, vol. 3027, pp. 223–238. Springer, Heidelberg (2004)
10. Boneh, D., Boyen, X.: Short signatures without random oracles. In: Cachin, C., Camenisch, J. (eds.) EUROCRYPT 2004. LNCS, vol. 3027, pp. 56–73. Springer, Heidelberg (2004)
11. Boneh, D., Di Crescenzo, G., Ostrovsky, R., Persiano, G.: Public key encryption with keyword search. In: Cachin, C., Camenisch, J.L. (eds.) EUROCRYPT 2004. LNCS, vol. 3027, pp. 506–522. Springer, Heidelberg (2004)
12. Canetti, R.: Towards realizing random oracles: Hash functions that hide all partial information. In: Kaliski Jr., B.S. (ed.) CRYPTO 1997. LNCS, vol. 1294, pp. 455–469. Springer, Heidelberg (1997)
13. Canetti, R., Goldreich, O., Halevi, S.: The random oracle methodology, revisited. J. ACM 51(4), 557–594 (2004)
14. Canetti, R., Micciancio, D., Reingold, O.: Perfectly one-way probabilistic hash functions (preliminary version). In: STOC, pp. 131–140 (1998)
15. Coppersmith, D., Franklin, M.K., Patarin, J., Reiter, M.K.: Low-exponent RSA with related messages. In: Maurer, U.M. (ed.) EUROCRYPT 1996. LNCS, vol. 1070, pp. 1–9. Springer, Heidelberg (1996)
16. Dziembowski, S., Pietrzak, K.: Leakage-resilient cryptography. In: FOCS, pp. 293–302 (2008)
17. Feistel, H.: Cryptography and computer privacy. Scientific American 228(5) (1973)
18. Gennaro, R., Lysyanskaya, A., Malkin, T., Micali, S., Rabin, T.: Algorithmic tamper-proof (ATP) security: Theoretical foundations for security against hardware tampering. In: Naor, M. (ed.) TCC 2004. LNCS, vol. 2951, pp. 258–277. Springer, Heidelberg (2004)
19. Goldenberg, D., Liskov, M.: On related-secret pseudorandomness. In: Micciancio, D. (ed.) TCC 2010. LNCS, vol. 5978, pp. 255–272. Springer, Heidelberg (2010)
20. Hemenway, B., Lu, S., Ostrovsky, R.: Correlated product security from any one-way function and the new notion of decisional correlated product security. Cryptology ePrint Archive, Report 2010/100 (2010), http://eprint.iacr.org/
21. Ishai, Y., Kilian, J., Nissim, K., Petrank, E.: Extending oblivious transfers efficiently. In: Boneh, D. (ed.) CRYPTO 2003. LNCS, vol. 2729, pp. 145–161. Springer, Heidelberg (2003)
22. Kiltz, E., Pietrzak, K., Stam, M., Yung, M.: A new randomness extraction paradigm for hybrid encryption. In: Joux, A. (ed.) EUROCRYPT 2009. LNCS, vol. 5479, pp. 590–609. Springer, Heidelberg (2009)
23. Naor, M., Reingold, O.: Synthesizers and their application to the parallel construction of pseudo-random functions. J. Comput. Syst. Sci. 58(2), 336–375 (1999)
24. O'Neill, A.: Deterministic public-key encryption revisited. Cryptology ePrint Archive, Report 2010/533 (2010), http://eprint.iacr.org/

25. Rosen, A., Segev, G.: Chosen-ciphertext security via correlated products. In: Reingold, O. (ed.) TCC 2009. LNCS, vol. 5444, pp. 419–436. Springer, Heidelberg (2009)
26. Shannon, C.E.: Communication theory of secrecy systems. Bell System Technical Journal 28(4), 656–715 (1949)
27. Wagner, D., Goldberg, I.: Proofs of security for the unix password hashing algorithm. In: Okamoto, T. (ed.) ASIACRYPT 2000. LNCS, vol. 1976, pp. 560–572. Springer, Heidelberg (2000)

Black-Box Circular-Secure Encryption beyond Affine Functions

Zvika Brakerski[1], Shafi Goldwasser[2], and Yael Tauman Kalai[3]

[1] Weizmann Institute of Science
zvika.brakerski@weizmann.ac.il
[2] Weizmann Institute of Science and MIT
shafi@theory.csail.mit.edu
[3] Microsoft Research
yael@microsoft.com

Abstract. We show how to achieve public-key encryption schemes that can securely encrypt nonlinear functions of their own secret key. Specifically, we show that for any constant $d \in \mathbb{N}$, there exists a public-key encryption scheme that can securely encrypt any function f of its own secret key, assuming f can be expressed as a polynomial of total degree d. Such a scheme is said to be key-dependent message (KDM) secure w.r.t. degree-d polynomials. We also show that for any constants c, e, there exists a public-key encryption scheme that is KDM secure w.r.t. all Turing machines with description size $c \log \lambda$ and running time λ^e, where λ is the security parameter. The security of such public-key schemes can be based either on the standard decision Diffie-Hellman (DDH) assumption or on the learning with errors (LWE) assumption (with certain parameters settings).

In the case of functions that can be expressed as degree-d polynomials, we show that the resulting schemes are also secure with respect to *key cycles* of any length. Specifically, for any polynomial number n of key pairs, our schemes can securely encrypt a degree-d polynomial whose variables are the collection of coordinates of all n secret keys. Prior to this work, it was not known how to achieve this for nonlinear functions.

Our key idea is a general transformation that amplifies KDM security. The transformation takes an encryption scheme that is KDM secure w.r.t. *some* functions even when the secret keys are weak (i.e. chosen from an arbitrary distribution with entropy k), and outputs a scheme that is KDM secure w.r.t. a *richer* class of functions. The resulting scheme may no longer be secure with weak keys. Thus, in some sense, this transformation converts security with weak keys into amplified KDM security.

1 Introduction

Secure encryption is one of the most fundamental tasks in cryptography, and significant work has gone into defining and attaining it. In many classical notions of secure encryption, it is assumed that the plaintext messages to be encrypted are independent of the secret decryption keys. However, over the years, it was observed that in some situations the plaintext messages *do* depend on the secret keys. This more demanding setting, termed *key-dependent message security*

Y. Ishai (Ed.): TCC 2011, LNCS 6597, pp. 201–218, 2011.

(KDM security) by Black, Rogoway and Shrimpton [7], has received much attention in recent years [12, 1, 21, 19, 5, 4, 20, 8, 17, 11, 3, 6, 9].

KDM security w.r.t. a class $\mathcal{F}$ of efficiently computable functions is modeled as follows[1]. An adversary is given public keys $\mathrm{pk}_1, \ldots, \mathrm{pk}_n$ and can access an oracle $\mathcal{O}$ that upon receiving a query (i, f), where f is a function in the class $\mathcal{F}$, and $i \in [n]$ is an index, returns an encryption of $f(\mathrm{sk}_1, \ldots, \mathrm{sk}_n)$ under the public key pk_i. The scheme is $\mathrm{KDM}^{(n)}$ secure w.r.t. $\mathcal{F}$, where n is the number of public keys, if the adversary cannot distinguish between the oracle $\mathcal{O}$ and an oracle that always returns an encryption of (say) the all-zero string. In particular, in $\mathrm{KDM}^{(1)}$ security, the adversary is given a single public key pk and can ask for encryptions (under pk) of functions of the corresponding secret key sk.

Starting with the breakthrough work of Boneh, Halevi, Hamburg and Ostrovsky [8] and continuing with the work of Applebaum, Cash, Peikert and Sahai [3], it is known how to achieve KDM security under a variety of computational assumptions. However, the above works achieve security only w.r.t. *affine* functions of the secret key, leaving unanswered the question of achieving security w.r.t. richer classes of functions.

The motivation to explore beyond affine functions is a straightforward extension of that provided in [8]: Assume a secret key is stored on a hard drive which is being encrypted as a part of a backup process. The encrypted contents thus depend on the secret key in a way that may not necessarily be affine (conditioned on the file type and the file system used).

Heitner and Holenstein [17] gave impossibility results with regards to black-box constructions of $\mathrm{KDM}^{(1)}$-secure encryption (even in the symmetric case). They showed that $\mathrm{KDM}^{(1)}$ security w.r.t. poly-wise independent functions is not black-box reducible to one-way trapdoor permutations, and also that $\mathrm{KDM}^{(1)}$ security w.r.t. *all* functions is not black-box reducible to essentially any cryptographic assumption.

In a work independent and concurrent to ours, Barak, Haitner, Hofheinz and Ishai [6] show how to overcome the latter black-box separation of [17]. They give a very strong positive result, showing that for any polynomial p there exists a $\mathrm{KDM}^{(1)}$-secure schemes w.r.t. all functions computable by circuits of size at most p, based on either the DDH or LWE assumptions. They also achieve $\mathrm{KDM}^{(n)}$ security at the cost of having the ciphertext length depend on the number of users n. Altogether, our work and theirs are complementary and achieve incomparable results. See a detailed comparison at the end of Section 1.1 below.

1.1 Our Results

We provide a general transformation that amplifies KDM security. Throughout this work, we restrict our attention to public-key encryption schemes in which the key-generation algorithm works by first sampling a secret key and then applying some, possibly randomized, function to produce the public key. Many known encryption

[1] We define KDM security in the public-key setting since this is the focus of this work. A similar definition can be provided for the symmetric setting.

schemes have this property, e.g. [26, 14, 24, 8, 3] and others. We say that an encryption scheme is *entropy-k KDM-secure* if it is KDM-secure even when the secret key is sampled from an arbitrary distribution with min-entropy k, and the computation of the public key is performed with perfect randomness[2]. Our transformation starts with an encryption scheme $\mathcal{E} = (G, E, D)$ that is entropy-k KDM$^{(n)}$-secure w.r.t. some class of functions $\mathcal{F}$, and converts it into another scheme $\mathcal{E}^* = (G^*, E^*, D^*)$, which is KDM$^{(n)}$ secure w.r.t. a larger class of functions $\mathcal{F}'$.

Theorem 1.1 (informal). *Let $\mathcal{E} = (G, E, D)$ be a public-key encryption scheme that is entropy-k KDM$^{(n)}$-secure w.r.t. a function class $\mathcal{F}$. Let $\mathcal{S}$ denote the space of the secret keys of $\mathcal{E}$, and let $\mathcal{K}$ be any set of size at least 2^k. Then for every deterministic, efficiently computable and injective mapping $\alpha : \mathcal{K} \to \mathcal{S}$ there exists an encryption scheme $\mathcal{E}^*_\alpha = (G^*, E^*, D^*)$, whose secret key, sk^*, is chosen at random from $\mathcal{K}$, such that $\mathcal{E}^*_\alpha$ is KDM$^{(n)}$ secure w.r.t. the function class $\mathcal{F}' = \mathcal{F} \circ \alpha = \{(f \circ \alpha)(\mathrm{sk}^*_1, \ldots, \mathrm{sk}^*_n) = f(\alpha(\mathrm{sk}^*_1), \ldots, \alpha(\mathrm{sk}^*_n)) : f \in \mathcal{F}\}$.*

For example, one can think of $\alpha(\mathrm{sk})$ as the vector of all monomials of degree d; namely, $\alpha(x_1, \ldots, x_k) = (\prod_{i \in I} x_i)_{|I| \leq d}$, where $\mathrm{sk} = (x_1, \ldots, x_k) \in \{0,1\}^k$. Another example is where $\alpha(\mathrm{sk})$ is the vector of all Turing machines with description length $O(\log k)$ and running time at most t (for some polynomial t), applied to sk. Namely, $\alpha(\mathrm{sk}) = \langle M(\mathrm{sk}) \rangle_M$, where M is a Turing machine with description length $O(\log k)$ that runs for at most t steps on sk.

In the first example, if $\mathcal{F}$ is the class of all linear functions, then $\mathcal{F}' = \mathcal{F} \circ \alpha$ is the class of all degree $\leq d$ polynomials. In the the second example, if $\mathcal{F}$ contains the identity function, then $\mathcal{F}' = \mathcal{F} \circ \alpha$ contains all the Turing machines with description length $O(\log k)$ and running time at most t.

We emphasize that in Theorem 1.1, we start with a scheme $\mathcal{E}$ that is entropy-k KDM$^{(n)}$-secure w.r.t. a function class $\mathcal{F}$, and end up with a scheme $\mathcal{E}^*_\alpha$ that is not necessarily entropy-k secure anymore. However, it is KDM$^{(n)}$-secure w.r.t. a (supposedly richer) function class $\mathcal{F}'$. Therefore this theorem gives a way to convert security with weak keys, into enhanced KDM security, thus showing a formal connection between the two notions[3]. Another connection between these notions in the symmetric encryption case, was shown by Canetti et. al. [13], in the context of obfuscation of multi-bit point functions: Loosely speaking, they show that an encryption scheme that is entropy-k KDM-secure implies a multi-bit point obfuscators, and vice versa. However, showing a direct implication between the notions (or showing that one does not exist) remains an interesting open problem.

We apply Theorem 1.1 to the schemes of [8] and [3] to obtain Theorems 1.2 and 1.3, respectively, presented below. In order to do that, we will argue that these schemes (or rather, a slight modification thereof) are entropy-k KDM$^{(1)}$-secure. In what follows, λ denotes the security parameter.

---

[2] This notion is different from security with key-leakage, where the leakage may depend on the public key.

[3] We stress that our reduction does not yield that leakage resiliency by itself implies KDM security; rather, we show that leakage resiliency on top of KDM security enables amplifying the KDM security property.

Theorem 1.2 (informal). *Assume the* DDH *assumption in a group* $\mathbb{G}$ *of order* q*, and let* g *be any generator of* $\mathbb{G}$*. Then, for any class* $\mathcal{H} = \{h_1, \ldots, h_\ell : h_i \in \{0,1\}^k \to \{0,1\}\}$ *of* $\text{poly}(\lambda)$*-time computable functions, with cardinality* $\ell = \text{poly}(\lambda)$*, there exists a* $\text{KDM}^{(1)}$*-secure encryption scheme w.r.t. the class of functions*

$$\mathcal{F}_{\mathcal{H}} = \left\{ f(g^{\mathbf{x}}) = g^{\sum_{i \in [\ell]} t_i h_i(\mathbf{x}) + w} : \mathbf{x} \in \{0,1\}^k, (\mathbf{t}, w) \in \mathbb{Z}_q^\ell \times \mathbb{Z}_q \right\} .$$

In this scheme, the secret key is a vector in $\mathbb{G}^k$ whose i^{th} coordinate is $g^{x_i} \in \{1, g\}$. Theorem 1.2 is obtained by applying Theorem 1.1 to the public-key encryption scheme of [8], which is KDM secure w.r.t. affine functions in the exponent, using the mapping $\alpha(g^{\mathbf{x}}) = (g^{h_1(\mathbf{x})}, \ldots, g^{h_\ell(\mathbf{x})})$.

In particular, taking $\mathcal{H}$ to be the class of all degree-d monomials, we show that for any constant $d \in \mathbb{N}$, there exists a public-key encryption scheme that is $\text{KDM}^{(1)}$-secure w.r.t. all polynomials of total degree d (in the exponent). This is because degree-d polynomials over k variables can be viewed as affine functions applied to the vector of degree-d monomials. A different selection of $\mathcal{H}$ implies that for any polynomial t, there exists a public-key scheme that is $\text{KDM}^{(1)}$-secure w.r.t. all Turing machines of description length bounded by $\log t$ and running time bounded by t[4].

Theorem 1.3 (informal). *Under the* LWE *assumption with modulus* $q = p^2$*, for a prime* p*, for any class* $\mathcal{H} = \{h_1, \ldots, h_\ell : h_i \in \{0,1\}^k \to \{0,1\}\}$ *of* $\text{poly}(\lambda)$*-time computable functions, with cardinality* $\ell = \text{poly}(\lambda)$*, there exists a* $\text{KDM}^{(1)}$*-secure encryption scheme w.r.t. the class of functions*

$$\mathcal{F}_{\mathcal{H}} = \left\{ f(\mathbf{x}) = \sum_{i \in [\ell]} t_i h_i(\mathbf{x}) + w \pmod{p} : (\mathbf{t}, w) \in \mathbb{Z}_p^\ell \times \mathbb{Z}_p \right\} .$$

The secret key space in this scheme is $\{0,1\}^k$. The result is obtained by applying Theorem 1.1 to (a variant of) the public-key encryption scheme of [3], which is KDM secure w.r.t. affine functions, using the mapping $\alpha(\mathbf{x}) = (h_1(\mathbf{x}), \ldots, h_\ell(\mathbf{x}))$.

In a similar manner to the DDH based result, appropriate selections of $\mathcal{H}$ imply a $\text{KDM}^{(1)}$-secure scheme w.r.t. all polynomials of total degree d and a $\text{KDM}^{(1)}$-secure scheme w.r.t. all Turing machines of description length bounded by $\log t$ and running time bounded by t, for $t = \text{poly}(\lambda)$.

This ability, to tailor an encryption scheme to the required set of functions, can be useful when, as a part of a cryptographic protocol, encryptions of certain functions of the secret key need to be transmitted.

We are able to extend the above results, using additional techniques (Theorem 1.1 will not suffice), and show that for the case of degree-d polynomials, both schemes obtained above are in fact $\text{KDM}^{(n)}$-secure, based on their respective assumptions. These results are stated in the theorems below.

[4] Bear in mind that any *uniform* function can be represented by a Turing machine of *constant* description. This means that for any uniform function f (computable in time t), our scheme becomes secure asymptotically with the security parameter.

Theorem 1.4 (informal). *Under the* DDH *assumption, for any constant* $d \in \mathbb{N}$*, there exists a public-key encryption scheme that is* $\text{KDM}^{(n)}$*-secure w.r.t. degree-*d *polynomials in the exponent, for any* $n = \text{poly}(\lambda)$.

Theorem 1.5 (informal). *Under the* LWE *assumption, for any constant* $d \in \mathbb{N}$*, there exists a public-key encryption scheme that is* $\text{KDM}^{(n)}$*-secure w.r.t. degree-*d *polynomials, for any* $n = \text{poly}(\lambda)$.

Additional Uses for Our Amplification Theorem. While Theorem 1.1 is stated in terms of public-key encryption, it in fact also works, in a straightforward manner, for other primitives such as symmetric encryption or pseudo-random functions (under an appropriate adaptation of the definitions of KDM and entropy-k security). In this paper, though, we focus on public-key encryption.

One could also consider applying the theorem to the OAEP (i.e. random oracle) based scheme of Backes, Dürmuth and Unruh [4]. However, in order to do that, entropy-k secure one-way trapdoor functions are required. Such are currently not known to exist, to the best of our knowledge, and thus we do not elaborate on this scheme.

Comparison With [6]. As mentioned above, a recent independent work of [6] achieves KDM security for a very rich class of functions: the class of functions computable by circuits of polynomial size p (the polynomial p affects the parameters of the scheme as we explain below). Their main technique is a non black-box use of the functions in the class, resulting in the ciphertext's containing a garbled circuit corresponding to a size-p circuit. Our implementation, in contrast, makes black-box use of the functions and does not require garbled circuits. The downside is that the size of the function class has to be limited (as demonstrated by the negative result of [17]). Another difference is that in the $\text{KDM}^{(n)}$ scheme of [6], the ciphertext size depends on n, unlike our schemes.

We also note that while the [6] framework applies only for public-key encryption, ours can be applied to symmetric encryption as well as other primitives.

1.2 Our Techniques

Let us present the intuition behind the KDM amplification theorem (Theorem 1.1). Given an encryption scheme $\mathcal{E}$ that is entropy-k $\text{KDM}^{(n)}$-secure w.r.t. a function class $\mathcal{F}$, we construct the encryption scheme $\mathcal{E}^*$ as follows: The key generation algorithm G^*, rather than choosing the secret key from $\mathcal{S}$, chooses $\text{sk} \xleftarrow{\$} \mathcal{K}$, and sets pk to be the public key corresponding to the secret key $\alpha(\text{sk})$. As an example, one can think of $\mathcal{K} = \{0,1\}^k$, $\mathcal{S} = \{0,1\}^\ell$ where $\ell = \sum_{i=0}^{d} \binom{k}{i}$, and $\alpha(\text{sk})$ is the vector of all monomials of degree d; namely, $\alpha(x_1, \ldots, x_k) = (\prod_{i \in I} x_i)_{|I| \le d}$, where $\text{sk} = (x_1, \ldots, x_k) \in \{0,1\}^k$. Another example is where $\mathcal{K} = \{0,1\}^k$, $\mathcal{S} = \{0,1\}^{\text{poly}(k)}$, and $\alpha(\text{sk})$ as being the vector of all Turing machines with description length $O(\log k)$ and running time at most t (for some polynomial t), applied to sk. Namely, $\alpha(\text{sk}) = \langle M(\text{sk}) \rangle_M$, where M is a Turing machine with description length $O(\log k)$ that runs for at most t steps on sk.

The encryption algorithm E^* is identical to E. The decryption algorithm D^* takes the secret key sk, computes $\alpha(\mathrm{sk})$, and decrypts the ciphertext by applying the decryption algorithm D with the secret key $\alpha(\mathrm{sk})$[5].

We next exemplify why the scheme $\mathcal{E}^*$ has amplified KDM security. Assume, for example, that $\mathcal{E}$ was entropy-k KDM$^{(1)}$ secure w.r.t. all affine functions. Consider, as in the example above, $\alpha(\mathrm{sk})$ that is the vector of all monomials of degree d. Then $\mathcal{E}^*$ is still secure, because it applies the scheme $\mathcal{E}$ with a weak secret key of min-entropy k. Moreover, the fact that $\mathcal{E}$ is entropy-k KDM$^{(1)}$-secure w.r.t. all affine functions, implies that the scheme $\mathcal{E}^*$ is secure w.r.t. all affine functions of $\alpha(\mathrm{sk})$, i.e. all degree d polynomials of sk. Similarly, if $\alpha(\mathrm{sk})$ is the vector of all Turing machines with description length $O(\log k)$ and with running time at most t, applied to sk, then $\mathcal{E}^*$ is KDM$^{(1)}$ secure w.r.t. all functions computed by these Turing machines.

Thus, Theorem 1.1 provides us with a generic tool to amplify KDM security of schemes that are entropy-k KDM-secure to begin with. However, the question that remains is: *Do there exist entropy-k KDM-secure schemes?*

KDM$^{(1)}$ *Security.* [8, 3] presented encryption schemes that are KDM$^{(1)}$-secure w.r.t. some classes of functions. We argue that these schemes are in fact entropy-k KDM$^{(1)}$-secure (for some setting of parameters). This enables us to apply Theorem 1.1 and amplify KDM$^{(1)}$ security "for free". Specifically, this implies KDM$^{(1)}$-secure schemes w.r.t. degree-d polynomials or bounded description and bounded running time Turing machines.

KDM$^{(n)}$ *security.* Two problems arise when trying to utilize Theorem 1.1 to obtain KDM$^{(n)}$ security. First, a direct application of Theorem 1.1 may not produce the strongest result. Consider, for example, the case of bounded degree polynomials. Even if we had a scheme that was entropy-k KDM$^{(n)}$-secure w.r.t. affine functions, Theorem 1.1 would only imply a scheme that is KDM$^{(n)}$-secure w.r.t. bounded-degree polynomials where each monomial only contains variables of the same secret key. Second, we are not able to show entropy-k KDM$^{(n)}$ security for any scheme and therefore cannot satisfy the conditions of the theorem.

To obtain Theorems 1.4 and 1.5, therefore, additional ideas are required. Rather than applying Theorem 1.1 directly for KDM$^{(n)}$, we consider the schemes obtained by Theorems 1.2 and 1.3 for the specific case where $\mathcal{H}$ is the class of all degree-d monomials. We then show that these schemes are not only KDM$^{(1)}$-secure w.r.t. degree-d polynomials, but are also KDM$^{(n)}$-secure w.r.t. the same class. We emphasize that monomials can contain variables from all secret keys in the system. This part contains the bulk of technical difficulty of this work.

While the proof for each scheme requires special treatment, the crux of the idea in both cases is similar. We use the "linear" behavior exhibited by both underlying schemes (in the DDH-based scheme, linearity is in the exponent) which

---

[5] We must require that α is deterministic so that $\alpha(\mathrm{sk})$ evaluates to the same value at each invocation (and thus is consistent with pk). It is interesting to explore whether similar techniques can be used when α is a randomized mapping (and thus can even increase the entropy of $\alpha(\mathrm{sk})$ compared to sk).

enables the following form of homomorphism: starting from a single public key, that corresponds to a secret key sk, it is possible to generate a public key that corresponds to a linearly-related secret key. This is done without knowing the original secret key sk, only the (linear) relation. We need to be careful in utilizing this property: as it turns out (and hinted by the intuition of Theorem 1.1 provided above), we need to apply this homomorphism on secret keys whose coordinates are low-degree monomials. Therefore we cannot use arbitrary linear transformations to "switch" between secret keys. We solve this problem by presenting a class of linear transformations that do preserve the structure of the input secret key.

1.3 Other Related Works and Notions

One can consider an "entropy-k" variant for any security measure for public-key encryption, analogously to our definition of entropy-k KDM security; i.e., requiring that the scheme remains secure, in the relative measure, even when the secret key is sampled from an arbitrary entropy-k distribution. This notion is incomparable to that of key-leakage resilience, defined by Akavia, Goldwasser and Vaikuntanathan [2]. On the one hand, the notion of entropy-k security is weaker since imperfect randomness is only used to generate the secret key, while the computation of the corresponding public key uses perfect randomness. On the other hand, key-leakage resilience is weaker since it requires security to hold, with high probability, over *some* family of distributions, whereas entropy-k security requires security to hold for *all* high min-entropy distributions.

In this work, we restructure the secret key of a public-key encryption scheme in order to achieve additional properties. Previous works also used a key distribution other than the obvious one to obtain stronger results. In the KDM-secure scheme of [8], binary vectors in the exponent of a group generator are used as secret keys, instead of the more natural selection of vectors in $\mathbb{Z}_q$. This is done in order to achieve KDM security w.r.t. the desired function class. In [22], the secret key distribution of the [8] scheme is again modified, this time using vectors of higher dimension than required, thus achieving security against key-leakage. The KDM-secure public-key scheme of [3] is very similar to that of [24], with one of the changes being that the secret key distribution is selected from a narrow Gaussian rather than being uniform. This is done, again, in order for KDM security to apply w.r.t. the desired set of functions.

In a followup work, Brakerski and Goldwasser [9] present a KDM (and memory leakage resilient) secure scheme based on the quadratic residuosity assumption. They then use our techniques to amplify the KDM security of their scheme by showing that it is entropy-k KDM secure.

1.4 Paper Organization

We provide notation and standard definitions in Section 2, new definitions and tools appear in Section 3. The KDM amplification theorem (Theorem 1.1) is formally restated and proven in Section 4, with examples of applying it to specific function classes. Due to space limitations, we omit the discussion of our DDH based solution and refer the reader to the full version [10] for the full details. Our LWE based construction appears in Section 5, where Theorems 1.3 and 1.5 are formally restated. Many proofs are omitted, see full version [10].

2 Notation and Definitions

We denote scalars in plain lowercase ($x \in \{0,1\}$), vectors in bold lowercase ($\mathbf{x} \in \{0,1\}^k$) and matrices in bold uppercase ($\mathbf{X} \in \{0,1\}^{k\times k}$). All vectors are column vectors by default. The ith coordinate of $\mathbf{x}$ is denoted x_i. For a set I, we use $\mathbf{x} = \langle x_i \rangle_{i\in I}$ to denote a vector that is indexed by elements in I.

Vectors in $\{0,1\}^k$ are treated both as elements in $\mathbb{Z}_q^k$ (for an appropriately defined $q \in \mathbb{N}$) and as elements in $\mathbb{Z}_2^k$. We use standard arithmetic notation for arithmetics over $\mathbb{Z}_q^k$ and use $\mathbf{x} \oplus \mathbf{y}$ to denote the addition in $\mathbb{Z}_2^k$ (i.e. bitwise XOR operation).

Let X be a probability distribution. We write $x \stackrel{\$}{\leftarrow} X$ to indicate that x is sampled from X. X^n denotes the n-fold product distribution of X. The uniform distribution over a set S is denoted $U(S)$ but we also denote $x \stackrel{\$}{\leftarrow} S$ to abbreviate $x \stackrel{\$}{\leftarrow} U(S)$. The *min entropy* of a random variable X is denoted $\mathbf{H}_\infty(X)$. For any function f, $f(X)$ denotes the random variable (or corresponding distribution) obtained by sampling $x \stackrel{\$}{\leftarrow} X$ and outputting $f(x)$.

We write $\text{negl}(n)$ to denote an arbitrary *negligible* function, i.e. one that vanishes faster than the inverse of any polynomial.

The *statistical distance* between two distributions X, Y is denoted $\mathsf{SD}(X, Y)$. Two ensembles $\{X_n\}_n$, $\{Y_n\}_n$ are *statistically indistinguishable* if $\mathsf{SD}(X_n, Y_n) = \text{negl}(n)$, and are *computationally indistinguishable* if for every $\text{poly}(n)$-time adversary $\mathcal{A}$ it holds that $|\Pr[\mathcal{A}(X_n) = 1] - \Pr[\mathcal{A}(Y_n) = 1]| = \text{negl}(n)$.

Let M be a deterministic Turing Machine. We use $|M|$ to denote the description length of M and use $\mathsf{execute}(M, 1^t, x)$ to denote the content of M's output tape after running on x for t computation steps. Clearly $\mathsf{execute}(M, 1^t, x)$ is computable in time $\text{poly}(|M|, t)$.

2.1 Learning with Errors (LWE)

We use the *decisional* version of the LWE ([24]) assumption. For any $m, n, q \in \mathbb{N}$ such that $q > 2$, all functions of the security parameter λ, and any probability distribution χ on $\mathbb{Z}_q$, the $\text{LWE}_{q,m,n,\chi}$ assumption is that the distributions $(\mathbf{A}, \mathbf{A}\mathbf{s}+\mathbf{x})$ and $(\mathbf{A}, \mathbf{u})$ are computationally indistinguishable, where $\mathbf{A} \stackrel{\$}{\leftarrow} \mathbb{Z}_q^{m\times n}$, $\mathbf{s} \stackrel{\$}{\leftarrow} \mathbb{Z}_q^n$, $\mathbf{x} \stackrel{\$}{\leftarrow} \chi^m$, $\mathbf{u} \stackrel{\$}{\leftarrow} \mathbb{Z}_q^m$.

We remark that the *search* version of the assumption, where the challenge is to find $\mathbf{s}$, is equivalent to the decisional version, for prime q, under $\text{poly}(q)$-time reductions. It is shown in [3] that this equivalence also holds for $q = p^e$, for integer constant e and prime p, provided that χ is a distribution over $\mathbb{Z}_q$ that produces an element in $\{-\frac{p-1}{2}, \ldots, \frac{p-1}{2}\}$ with all but negligible probability.

Worst-case to average-case reductions ([24, 23]) can be used to obtain a connection between LWE instances and worst case lattice problems, for some (Gaussian like) distribution χ.

Noise Distributions. In our construction, we use distributions that are derived from Gaussians. For any $\sigma > 0$, we denote $D_\sigma(x) = e^{-\pi(x/\sigma)^2}/\sigma$, the (scaled) density function of the one dimensional Gaussian distribution. For any $q \in \mathbb{N}$ and $\sigma > 0$ we define $\bar{\Psi}_\sigma$ to be the distribution over $\mathbb{Z}_q$ obtained by sampling $y \xleftarrow{\$} D_\sigma$ and outputting $\lfloor q \cdot y \rceil \pmod{q}$. We define $D_{\mathbb{Z}^m,\sigma}$ to be the distribution over all $\mathbf{x} \in \mathbb{Z}^m$ such that $\Pr[\mathbf{x}]$ is proportional to $\prod_{i\in[m]} D_\sigma(x_i)$. We note that this distribution is efficiently sampleable for any $\sigma > 0$.

2.2 KDM Security

A public-key encryption scheme $\mathcal{E} = (G, E, D)$ is defined by its key generation, encryption and decryption algorithms. The key generation algorithm G takes as input the unary vector 1^λ, where λ is called the *security parameter* of the scheme. All other parameters of the scheme are parameterized by λ. We let $\mathcal{S} = \{\mathcal{S}_\lambda\}$ denote the space of secret keys and $\mathcal{M} = \{\mathcal{M}_\lambda\}$ denote the message space of the encryption scheme. We refer the reader to [15] for a formal definition of encryption schemes and their security.

In the scenario of key-dependent messages, we wish to model the case where functions of the secret key can be encrypted, and require that the resulting ciphertexts are indistinguishable from encryptions of 0. We want our definition to apply also for the case of "key cycles" where a function of one user's secret key is encrypted by another's public key and vice versa. The most inclusive definition, therefore, is parameterized by the number of users n and allows encrypting a function of the entire vector of n secret keys under any of the corresponding public keys (this is sometimes referred to as "clique security"). An additional parameter to be considered is the set of functions of the secret key that we allow to encrypt. We use the definition presented in [8].

Formally, let $\mathcal{E} = (G, E, D)$ be a public key encryption scheme, $n > 0$ be an integer, $\mathcal{S} = \{\mathcal{S}_\lambda\}$ be the space of secret keys, and let $\mathcal{F} = \{\mathcal{F}_\lambda\}$ be a class of functions such that $\mathcal{F}_\lambda \subseteq \mathcal{S}_\lambda^n \to \mathcal{M}_\lambda$.

We define the KDM$^{(n)}$ game, w.r.t. the function class $\mathcal{F}$, played between a challenger and an adversary $\mathcal{A}$, as follows.

- **Initialize.** The challenger selects $b \xleftarrow{\$} \{0, 1\}$ and generates, for all $i \in [n]$, key pairs $(\mathrm{sk}_i, \mathrm{pk}_i) \xleftarrow{\$} G(1^\lambda)$. The challenger then sends $\{\mathrm{pk}_i\}_{i\in[n]}$ to $\mathcal{A}$.
- **Query.** The adversary makes queries of the form $(i, f) \in [n] \times \mathcal{F}_\lambda$. For each query, the challenger computes $y \leftarrow f(\mathrm{sk}_1, \ldots, \mathrm{sk}_n)$ and sends the following ciphertext to $\mathcal{A}$.
$$c \leftarrow \begin{cases} E_{\mathrm{pk}_i}(y) \text{ if } b = 0 \\ E_{\mathrm{pk}_i}(0) \text{ if } b = 1. \end{cases}$$
- **Finish.** $\mathcal{A}$ outputs a guess $b' \in \{0, 1\}$.

Adversary $\mathcal{A}$ *wins* the game if $b' = b$. The scheme is secure if the *advantage* $\text{KDM}^{(n)}\text{Adv}[\mathcal{A}, \mathcal{E}](\lambda) = |\Pr[b' = b] - 1/2|$ of any polynomial time $\mathcal{A}$ is negligible.

We use $\text{KDM}_{\mathcal{F}}^{(n)}$ to denote KDM security w.r.t. the function class $\mathcal{F}$.

3 New Definitions and Tools

3.1 Projective Encryption Schemes and Weak Keys

Projection. Throughout this paper, we only consider encryption schemes that have a *projection* between the secret and public key. Namely, the key generation can be described as first sampling the secret key from some set and then applying an efficiently computable projection function (which can be randomized) to generate the public key.

Definition 3.1 (projection). *Let $\mathcal{E} = (G, E, D)$ be a public-key encryption scheme. $\mathcal{E}$ is* projective *if $G(1^\lambda) = (\text{sk}, \text{pk} = \mathsf{Proj}(\text{sk}))$ where $\text{sk} \xleftarrow{\$} \mathcal{S}$ and $\mathsf{Proj}(\cdot)$ is an efficiently computable (possibly randomized) function.*

We remark that many known encryption schemes are indeed projective, e.g. [26, 14, 24, 8, 3] and others. We further remark that any secure scheme can be modified to be projective by using the randomness of the key generation as the secret key. However such transformation does not preserve KDM security and thus we will need to require projection explicitly.

Weak Keys and Entropy-k Security. We are also interested in a more specific case where a (projective) scheme remains secure even when the key generation is "improper": the secret key is sampled from an arbitrary distribution on $\mathcal{S}$ that has min-entropy k. The projection is then applied to the sampled value.

We can think of an "entropy-k variant" of any security notion σ, we thus provide a general definition. In this work, however, we instantiate this definition with σ being KDM security.

Definition 3.2 (entropy-k security). *Let $\mathcal{E} = (G, E, D)$ be a projective public-key encryption scheme and let σ be some security notion. Consider a distribution ensemble $\mathcal{D} = \{\mathcal{D}_\lambda\}$ over $\mathcal{S} = \{\mathcal{S}_\lambda\}$. Let $G_{\mathcal{D}}$ denote the following key-generator: $G_{\mathcal{D}}(1^\lambda) = (\text{sk}, \mathsf{Proj}(\text{sk}))$ where $\text{sk} \leftarrow \mathcal{D}_\lambda$.*

Let $k : \mathbb{N} \to \mathbb{R}^+$ be some function. $\mathcal{E}$ is entropy-k σ-secure *if for any ensemble $\mathcal{D}$ with $\mathbf{H}_\infty(\mathcal{D}_\lambda) \geq k(\lambda)$ it holds that $\mathcal{E}_{\mathcal{D}}(G_{\mathcal{D}}, E, D)$ is σ-secure.*

We stress that entropy-k security, as defined above, is a notion incomparable to that of key-leakage resilience (as defined in [2, 22]). On the one hand, the notion of entropy-k security is weaker since imperfect randomness is only used to generate the secret key, while the projection $\mathsf{Proj}(\cdot)$ uses perfect randomness to compute the corresponding public key. On the other hand, key-leakage resilience is weaker since it requires security to hold with high probability over *some* family of distributions, whereas entropy-k security requires security to hold for *all* high min-entropy distributions.

3.2 Transformations on Expanded Secret Keys

Let q be some modulus. The set of *affine functions* modulo q on $\mathbb{Z}_q^k$ is

$$\mathcal{F}_{\text{aff}} = \{f_{\mathbf{t},w}(\mathbf{x}) = \mathbf{t}^T\mathbf{x} + w : (\mathbf{t}, w) \in \mathbb{Z}_q^k \times \mathbb{Z}_q\} .$$

Degree-d polynomials over k variables can be viewed as affine functions applied to the vector of degree-d monomials. While we consider polynomials over $\mathbb{Z}_q$, we only apply them to binary variables, $\mathbf{x} \in \{0,1\}^k$. We define a mapping $\gamma_{k,d}$ that maps $\mathbf{x} \in \{0,1\}^k$ into the vector containing all monomials of degree d of the variables of $\mathbf{x}$.

Definition 3.3 (the vector of monomials $\gamma_{k,d}$). *For all $k, d \in \mathbb{N}$ and $\mathbf{x} \in \{0,1\}^k$, we define the vector of all degree-d monomials in $\mathbf{x}$ by*

$$\gamma_{k,d}(\mathbf{x}) = \left\langle \prod_{j \in J} x_j \right\rangle_{\substack{J \subseteq [k], \\ |J| \le d}} .$$

In other words, letting $\nu_{k,d} = \sum_{j=0}^{d} \binom{k}{j}$ denote the number of such degree-d monomials, $\gamma_{k,d} : \{0,1\}^k \to \{0,1\}^{\nu_{k,d}}$ is a mapping between vectors. We denote its image by $\Gamma_{k,d} = \{\gamma_{k,d}(\mathbf{x}) : \mathbf{x} \in \{0,1\}^k\}$.

It follows immediately from the definition that $\gamma_{k,d}$ is injective, since $(\gamma_{k,d}(\mathbf{x}))_{\{i\}} = x_i$, and thus that $|\Gamma_{k,d}| = 2^k$.

Intuitively, in the context of KDM security amplification, $\mathbf{x}$ is our "real" secret key, whereas $\gamma_{k,d}(\mathbf{x})$, the expanded version of $\mathbf{x}$, is used as a "secret key" for a scheme that is KDM-secure w.r.t. affine functions. This results in a KDM-secure scheme w.r.t. degree-d polynomials.

We denote the set of all *degree-d polynomials* over $\mathbb{Z}_q$ with binary variables $\mathbf{x} \in \{0,1\}^k$ by

$$\mathcal{F}_d = \{f_{\mathbf{t}}(\mathbf{x}) = \mathbf{t}^T \cdot \gamma_{k,d}(\mathbf{x}) : \mathbf{t} \in \mathbb{Z}_q^\ell\} .$$

Note that $\gamma_{k,d}(\mathbf{x})_\emptyset = 1$, i.e. the vector of monomials contains the empty monomial that always evaluates to 1. Therefore there is no need for an additional free term w as in the definition of affine functions[6].

The following lemma states that that given $\mathbf{y} \in \{0,1\}^k$, we can efficiently compute a matrix $\mathbf{T} \in \mathbb{Z}_q^{\ell \times \ell}$ such that for all $\mathbf{x} \in \{0,1\}^k$ it holds that $\mathbf{T} \cdot \gamma_{k,d}(\mathbf{x}) = \gamma_{k,d}(\mathbf{x} \oplus \mathbf{y})$. We think of $\mathbf{y}$ as the known relation between secret keys $\mathbf{x}$ and $\mathbf{x} \oplus \mathbf{y}$. The transformation $\mathbf{T}$ allows us to convert the expanded version of $\mathbf{x}$ to the expanded version of $\mathbf{x} \oplus \mathbf{y}$, i.e. to convert $\gamma_{k,d}(\mathbf{x})$ into $\gamma_{k,d}(\mathbf{x} \oplus \mathbf{y})$. For proof of the lemma, see the full version [10].

[6] We remark, for the interested reader, that for our DDH based scheme, described in the full version [10], we need to define the analogous sets of *affine functions in the exponent* and *degree-d polynomials in exponent* over $\mathbb{G}^k$:

$$\hat{\mathcal{F}}_{\text{aff}} = \{h_{\mathbf{t},w}(g^{\mathbf{x}}) = g^{\mathbf{t}^T\mathbf{x}+w} : (\mathbf{t}, w) \in \mathbb{Z}_q^k \times \mathbb{Z}_q\}$$
$$\hat{\mathcal{F}}_d = \{h_{\mathbf{t}}(g^{\mathbf{x}}) = g^{\mathbf{t}^T \cdot \gamma_{k,d}(\mathbf{x})} : \mathbf{t} \in \mathbb{Z}_q^\ell\} ,$$

where $\mathbb{G}$ is a group of order q and g is a generator of $\mathbb{G}$.

Lemma 3.4. *For all $k, d, q \in \mathbb{N}$ such that $q > 2$, there exists an efficiently computable function $\mathsf{T}_{k,d,q} : \{0,1\}^k \to \mathbb{Z}_q^{\ell \times \ell}$, where $\ell = \nu_{k,d}$, such that setting $\mathbf{T} = \mathsf{T}_{k,d,q}(\mathbf{y})$, for all $\mathbf{x} \in \{0,1\}^k$ it holds that $\mathbf{T} \cdot \gamma_{k,d}(\mathbf{x}) = \gamma_{k,d}(\mathbf{x} \oplus \mathbf{y})$. Moreover $\mathbf{T}$ is an involution, i.e. $\mathbf{T}^2$ is the identity matrix.*

4 Amplification of KDM Security

In this section we give a general result: We show that an entropy-k KDM-secure scheme, w.r.t. a certain class of functions, can be converted into various schemes that are KDM-secure w.r.t. richer classes. We start by stating the general result in Section 4.1 and then, in Section 4.2, we present corollaries for specific classes of functions.

4.1 Main Theorem

Before stating our theorem, let us give some intuition for how KDM security can be amplified for projective entropy-k schemes (as defined in Section 3.1).

Consider, for example, a projective encryption scheme $\mathcal{E}$ that is entropy-k KDM-secure w.r.t. the class of indexing functions $\mathcal{I} = \{h_i(\mathbf{s}) = s_i\}$ or, in other words, a bit by bit encryption of the secret key is secure. Entropy-k security in particular means that we can sample the secret key $\mathrm{sk} = \mathbf{s} \in \{0,1\}^\ell$ as follows: first, sample the first k bits uniformly, call this part $\mathbf{x}$; then, set the remaining bits of $\mathbf{s}$ to $s_i = f_i(\mathbf{x})$, where $\{f_i\}_{i=k+1,\ldots,\ell}$ is an arbitrary class of efficiently computable deterministic functions. The resulting secret key distribution has min-entropy k and thus $\mathcal{E}$ is still KDM-secure w.r.t. $\mathcal{I}$ with the resulting secret key distribution. Namely, $\mathcal{E}$ is secure w.r.t. the functions $h_i(\mathbf{s}) = s_i = f_i(\mathbf{x})$. Therefore, we can convert $\mathcal{E}$ into a scheme $\mathcal{E}^*$ by setting the secret key in $\mathcal{E}^*$ to be $\mathbf{x}$. This $\mathcal{E}^*$ is KDM-secure w.r.t. indexing functions as well as the functions $\{f_i\}_{i=k+1,\ldots,\ell}$.

Theorem 1.1 (restated). *Let $\mathcal{E} = (G, E, D)$ be a projective public-key encryption scheme that is entropy-k KDM$^{(n)}$-secure w.r.t. a function class $\mathcal{F}$. Let $\mathcal{S} = \{\mathcal{S}_\lambda\}$ be the space of secret keys.*

Let $\mathcal{K} = \{\mathcal{K}_\lambda\}$ be a family of sets such that $|\mathcal{K}| \geq 2^k$ and let $\alpha : \mathcal{K} \to \mathcal{S}$ be a deterministic, efficiently computable and injective function. Then there exists a projective encryption scheme $\mathcal{E}^*_\alpha = (G^*, E^*, D^*)$ with secret key space $\mathcal{K}$ that is KDM$^{(n)}$ secure w.r.t. $\mathcal{F} \circ \alpha = \{(f \circ \alpha)(\mathrm{sk}_1, \ldots, \mathrm{sk}_n) = f(\alpha(\mathrm{sk}_1), \ldots, \alpha(\mathrm{sk}_n)) : f \in \mathcal{F}\}$.

Proof. Consider the ensemble $\mathcal{D}$ where $\mathcal{D}_\lambda = \alpha(U(\mathcal{K}_\lambda))$ and consider the scheme $\mathcal{E}_\mathcal{D} = (G_\mathcal{D}, E, D)$ as in Definition 3.2. $\mathcal{E}^*_\alpha$ is similar to $\mathcal{E}_\mathcal{D}$ with the following modifications. $G^*(1^\lambda)$ first samples $\mathrm{sk}^* \xleftarrow{\$} \mathcal{K}$ and then computes $\mathrm{pk} = \mathsf{Proj}^*(\mathrm{sk}^*) = \mathsf{Proj}(\alpha(\mathrm{sk}^*))$. Note that the distribution of the public keys is identical to that of $\mathcal{E}_\mathcal{D}$ while the distributions of secret keys differ. The encryption E^* is performed identically to E. The decryption $D^*_{\mathrm{sk}^*}(c)$ is performed by first computing $\mathrm{sk} = \alpha(\mathrm{sk}^*)$ and then outputting $D_{\mathrm{sk}}(c)$.

Since α is injective, it holds that $\mathbf{H}_\infty(\mathcal{D}_\lambda) \geq k$, and thus by definition, $\mathcal{E}_\mathcal{D}$ is KDM$^{(n)}$-secure w.r.t. $\mathcal{F}$.

We next show that for any adversary $\mathcal{A}^*$ for the KDM$^{(n)}$ game with $\mathcal{E}^*_\alpha$, there exists an adversary $\mathcal{A}$ for the KDM$^{(n)}$ game with $\mathcal{E}_\mathcal{D}$ such that

$$\mathrm{KDM}^{(n)}_{\mathcal{F}}\mathrm{Adv}[\mathcal{A}, \mathcal{E}_\mathcal{D}](\lambda) = \mathrm{KDM}^{(n)}_{\mathcal{F}\circ\alpha}\mathrm{Adv}[\mathcal{A}^*, \mathcal{E}^*_\alpha](\lambda) ,$$

completing the proof of the theorem. Our adversary $\mathcal{A}$ will simulate $\mathcal{A}^*$ as follows.

- **Initialize.** Since the public key distributions of $\mathcal{E}_\mathcal{D}$ and $\mathcal{E}^*_\alpha$ are identical, $\mathcal{A}$ forwards its input $\mathrm{pk}_1, \ldots, \mathrm{pk}_n$ to $\mathcal{A}^*$.
- **Queries.** When $\mathcal{A}^*$ sends the query $(i, f \circ \alpha) \in [n] \times (\mathcal{F} \circ \alpha)$, $\mathcal{A}$ sends the query (i, f)[7]. Let sk^*_i denote the secret key corresponding to pk_i in $\mathcal{E}^*_\alpha$, then by definition $\mathrm{sk}_i = \alpha(\mathrm{sk}^*_i)$ is the secret key corresponding to pk_i in $\mathcal{E}_\mathcal{D}$. Therefore $f(\mathrm{sk}_1, \ldots, \mathrm{sk}_n) = (f \circ \alpha)(\mathrm{sk}^*_1, \ldots, \mathrm{sk}^*_n)$, and $\mathcal{A}$ can forward the answer to $\mathcal{A}^*$. Thus, $\mathcal{A}$ can simulate any query made by $\mathcal{A}^*$ during the game.
- **Finish.** When $\mathcal{A}^*$ returns b', $\mathcal{A}$ also terminates and returns the same b'.

Since $\mathcal{A}$ simulates $\mathcal{A}^*$ exactly, it follows that $\mathcal{A}$ achieves the same advantage in the KDM$^{(n)}$ game with $\mathcal{E}_\mathcal{D}$ as $\mathcal{A}^*$ does with $\mathcal{E}^*_\alpha$. □

4.2 Exemplifying for Specific Function Classes

We demonstrate specific cases where Theorem 1.1 amplifies KDM security. We restrict our attention to KDM$^{(1)}$ security (see discussion below).

- *Bounded description functions.* We first show how to amplify the class of indexing functions $\mathcal{I} = \{h_i(\mathbf{s}) = s_i\}$ into the class of all functions computable by a Turing machine with bounded description length and bounded running time. Let $\mathcal{E}$ be an entropy-k KDM$^{(1)}$-secure encryption scheme w.r.t. the class of indexing functions, with message space $\mathcal{M} = \{0,1\}$ and secret key space $\mathcal{S} = \{0,1\}^\ell$. Let $\mathcal{K} = \{0,1\}^k$ and $\alpha(\mathbf{x}) = \langle \mathsf{execute}(M, 1^{t(\lambda)}, \mathbf{x}) \rangle_{|M| \leq \log \ell}$ where $t(\cdot)$ is some (fixed) polynomial. Then $\mathcal{E}^*_\alpha$, defined in the proof of Theorem 1.1, is KDM$^{(1)}$-secure w.r.t. all functions computable by a Turing machine with description length $\log \ell$ and running time $t(\lambda)$[8].
- *Bounded degree polynomials.* We now show how to amplify the class of affine functions into the class of bounded degree polynomials. Let $\mathcal{E}$ be an entropy-k KDM$^{(1)}$-secure encryption scheme w.r.t. the class of affine functions $\mathbb{F}^\ell \to \mathbb{F}$, with $\mathcal{M} = \mathbb{F}$ and $\mathcal{S} \subseteq \mathbb{F}^\ell$, for a finite field $\mathbb{F}$. Let $\mathcal{K} = \{0,1\}^k \subseteq \mathbb{F}^k$ and let d be such that $\ell = \nu_{k,d}$ (see Definition 3.3), this implies that d is at least $\frac{\log \ell}{\log(k+1)}$. Consider $\alpha(\mathbf{x}) = \gamma_{k,d}(\mathbf{x})$, i.e. α contains all degree d monomials. Then $\mathcal{E}^*_\alpha$, defined in the proof of Theorem 1.1, is KDM$^{(1)}$-secure w.r.t. all degree-d polynomials $\mathbb{F}^k \to \mathbb{F}$.

[7] We represent $f \circ \alpha$ in such a way that enables to derive f.

[8] One has to be careful when showing that α is injective. We can either assume that the first k coordinates of the output contain the input, or, if ℓ is sufficiently larger than k, we can rely on the short description and running time of the indexing functions.

We provided examples only for the case of $\text{KDM}^{(1)}$ security for two reasons. First, while in Section 5.2 we present entropy-k $\text{KDM}^{(1)}$-secure schemes based on the LWE assumption[9], we are unable to obtain entropy-k $\text{KDM}^{(n)}$-secure schemes for $n > 1$. Second, even if such exist, the result of applying Theorem 1.1 for the classes above would be weaker than expected. This is because while the functions in the class $\mathcal{F}$ are applied to the vector of n secret keys, the mapping α is only applied to one secret key at a time. Therefore, the first example above would imply $\text{KDM}^{(n)}$ security w.r.t. Turing machines that only take one of the secret keys as input; the second would imply $\text{KDM}^{(n)}$ security w.r.t. degree-d polynomials where each monomial only contains variables from one secret key.

5 LWE Based KDM Security

For any constant d, we present a scheme that is $\text{KDM}^{(n)}$ secure w.r.t. all degree-d polynomials, $\mathcal{F}_d$. We also present a scheme that is $\text{KDM}^{(1)}$-secure w.r.t. the class of all functions computed by Turing machines with description length at most $\log t$ and running time t, for some polynomial t (more generally, w.r.t. any class of efficiently computable functions of polynomial cardinality). Our starting point is the LWE based scheme presented in [3], which we denote $\mathcal{E}_{\text{ACPS}}$, which is extended using ideas from Section 4.

First, in Section 5.1, we present the relevant previous work, in this case - the scheme of [3], denoted $\mathcal{E}_{\text{ACPS}}$. Then, in Section 5.2, we prove the entropy-k $\text{KDM}^{(1)}$ security of $\mathcal{E}_{\text{ACPS}}$ w.r.t. affine functions $\mathcal{F}_{\text{aff}}$, and present the consequences of applying Theorem 1.1 to $\mathcal{E}_{\text{ACPS}}$. Finally, in Section 5.3, we show that in the special case of degree-d polynomials, we can in fact prove $\text{KDM}^{(n)}$ security of the scheme obtained from Theorem 1.1.

5.1 Scheme $\mathcal{E}_{\text{ACPS}}$

We present the $\mathcal{E}_{\text{ACPS}}[\mathcal{S}]$ scheme which is similar to the scheme presented in [3]. The only difference is that we take the distribution of secret keys as a parameter. We also use slightly different notation for consistency with the rest of this paper.

- **Parameters.** Let p be a prime and $q = p^2$. We set $\ell, m \in \mathbb{N}$ to be polynomial functions of λ such that $m \geq 2(\ell+1)\log q$. Let $\chi = \bar{\Psi}_\sigma$ for $\sigma = \sigma(\lambda) \in (0,1)$ such that $\sigma \leq \frac{1}{p\cdot\sqrt{m}\cdot\omega(\log\lambda)}$. We also fix some $\tau = \omega(\sqrt{\log\lambda})$. Finally, let $\mathcal{S} \subseteq \mathbb{Z}_p^\ell$. The secret key space is $\mathcal{S}$ and the message space is $\mathbb{Z}_p$.
- **Key Generation.** On input 1^λ, sample $\mathbf{s} \xleftarrow{\$} \mathcal{S}$ and set $\text{sk} = \mathbf{s}$[10]. Then, sample $\mathbf{A} \xleftarrow{\$} \mathbb{Z}_q^{m\times\ell}$ and $\boldsymbol{\eta} \xleftarrow{\$} \chi^m$ and set $\text{pk} = (\mathbf{A}, \mathbf{A}\cdot\mathbf{s}+\boldsymbol{\eta}) \in \mathbb{Z}_q^{m\times\ell} \times \mathbb{Z}_q^m$.
- **Encryption.** Define the distribution $E_{\mathbf{A},\mathbf{b}}$ in $\mathbb{Z}_q^\ell \times \mathbb{Z}_q$ as follows. $E_{\mathbf{A},\mathbf{b}}$ samples $\mathbf{r} \xleftarrow{\$} D_{\mathbb{Z}^m,\tau}$, $e \xleftarrow{\$} \bar{\Psi}_{\tau'}$ where $\tau' = \tau\sqrt{m}(\sigma + \frac{1}{2q})$ and outputs $(\mathbf{A}^T\cdot\mathbf{r}, \mathbf{b}^T\cdot\mathbf{r}+e) \in \mathbb{Z}_q^\ell \times \mathbb{Z}_q$.

---

9 An additional DDH based example is provided in the full version [10].

10 In [3], $\mathbf{s}$ is sampled from the distribution χ^ℓ.

On input a public key $\mathrm{pk} = (\mathbf{A}, \mathbf{b})$ and a message $w \in \mathbb{Z}_p$, the encryption algorithm samples $(\mathbf{u}, v) \xleftarrow{\$} E_{\mathbf{A},\mathbf{b}}$ and outputs

$$(\mathbf{u}, v + w \cdot p) \ .$$

– **Decryption.** On input a secret key $\mathbf{s}$ and a ciphertext $(\mathbf{u}, c)$, the decryption algorithm outputs

$$\left\lfloor \left(c - \mathbf{u}^T \cdot \mathbf{s} \pmod q\right) / p \right\rceil \pmod p \ .$$

When $\sigma \le \frac{1}{p \cdot \sqrt{m} \cdot \omega(\log \lambda)}$, correctness (for any $\mathbf{s} \in \mathbb{Z}_p^\ell$) follows directly from [3].

5.2 Amplification of KDM$^{(1)}$ Security

We use Theorem 1.1 to amplify the KDM$^{(1)}$ security of $\mathcal{E}_{\text{ACPS}}$. We say that a finite set of functions, $\mathcal{H} = \{h_1, \ldots, h_\ell\}$, with a common domain, is *entropy preserving* if $\alpha_{\mathcal{H}}(x) = (h_1(x), \cdots h_\ell(x))$ is an injective function.

Theorem 1.3 (restated). *Let p be a prime number that is super-polynomial in λ and denote $q = p^2$. Let m, ℓ, σ, χ be as in the parameters of $\mathcal{E}_{\text{ACPS}}$. Let $k \le \ell$ and set $k' = \frac{k - \omega(\log \lambda)}{\log q}$. Let $\beta = \beta(\lambda) \in (0, 1)$ be such that $\frac{\beta}{\sigma} = \mathrm{negl}(\lambda)$ and denote $\chi' = \bar{\Psi}_\beta$. Let $\mathcal{H} = \{h_1, \ldots, h_\ell : h_i \in \{0,1\}^k \to \{0,1\}\}$ be an entropy preserving class of efficiently computable functions with cardinality $\ell = \mathrm{poly}(\lambda)$. Then under the $\mathrm{LWE}_{q,m,k',\chi'}$ assumption, there exists a public-key encryption scheme that is KDM$^{(1)}$ secure w.r.t. function class*

$$\mathcal{F}_{\mathcal{H}} = \left\{ f(\mathbf{x}) = \sum_{i \in [\ell]} t_i h_i(\mathbf{x}) + w \pmod p : (\mathbf{t}, w) \in \mathbb{Z}_p^\ell \times \mathbb{Z}_p \right\} .$$

Before we outline the proof, let us discuss the parameters of our hardness assumption. The decisional $\mathrm{LWE}_{q,m,k',\chi'}$ assumption (see Section 2.1) is equivalent to the search version under a $\mathrm{poly}(q)$-time reduction. The search version, in turn, is shown in [24] to correspond to worst-case lattice problems, under quantum reductions. In [23], a classical reduction from other worst-case lattice problems to search LWE is shown. Thus, we can set p and q to be quasi-polynomial in λ, set $\beta \ge n/q$ and set $\frac{\sigma}{\beta}$ to be quasi-polynomial in λ as well (recall that for correctness we must take $\sigma \le \frac{1}{p \cdot \sqrt{m} \cdot \omega(\log \lambda)}$, so we cannot set σ to be too large, but one can verify that a proper selection of parameters exists). Using such parameters we can relate the security of our scheme to either the worst case hardness of obtaining a quasi-polynomial approximation factor for a lattice problem such as GapSVP, using quasi-polynomial time quantum algorithms, or to the worst case hardness of obtaining a classical quasi-polynomial time algorithm for a lattice problem such as $\mathrm{GapSVP}_{\zeta,\gamma}$ with quasi-polynomial ζ.

To prove Theorem 1.3, we employ Theorem 1.1. As a precondition, we will need to establish entropy-k KDM$^{(1)}$ security for $\mathcal{E}_{\text{ACPS}}$, which is not straightforward. We do this in two steps. First, we prove KDM$^{(1)}$ security based on a nonstandard assumption. Then, we use a result of Goldwasser, Kalai, Peikert and Vaikuntanathan [16] that implies that for the parameters of Theorem 1.3, LWE reduces to our new assumption, thus ultimately basing our scheme on standard decisional LWE. We remark that it may be possible to achieve better parameters than stated in Theorem 1.3 using a more efficient reduction, if such exists. See full version [10] for details and proof.

In the specific case of using the set of all degree-d monomials as the function class $\mathcal{H}$, we obtain a KDM$^{(1)}$-secure scheme w.r.t. $\mathcal{F}_d$, all degree-d polynomials modulo p. We describe this scheme, $\mathcal{E}_2$[11], explicitly. In Section 5.3 we show that $\mathcal{E}_2$ is KDM$^{(n)}$-secure w.r.t. $\mathcal{F}_d$. ($\gamma_{k,d}$, $\nu_{k,d}$, $\Gamma_{k,d}$ were defined in Definition 3.3.)

Encryption Scheme $\mathcal{E}_2$. Let $k, d \in \mathbb{N}$ and consider $p, q, m, \sigma, \chi, \tau$, as in the definition of $\mathcal{E}_{\text{ACPS}}[\Gamma_{k,d}]$, specifically let $\ell = \nu_{k,d}$. The secret key space of $\mathcal{E}_2$ is $\{0,1\}^k$ and the message space is $\mathbb{Z}_p$.

– **Key Generation.** On input 1^λ, select $\mathbf{x} \xleftarrow{\$} \{0,1\}^k$ and set $\text{sk} = \mathbf{x}$. We denote $\mathbf{s} = \gamma_{k,d}(\mathbf{x})$ and note that $\mathbf{s}$ is uniform in $\Gamma_{k,d}$. The public key pk is generated as in $\mathcal{E}_{\text{ACPS}}[\Gamma_{k,d}]$. Namely, $\text{pk} = (\mathbf{A}, \mathbf{A} \cdot \mathbf{s} + \boldsymbol{\eta}) \in \mathbb{Z}_q^{m \times \ell} \times \mathbb{Z}_q^m$. Note that the distributions of the public keys in $\mathcal{E}_2$ and $\mathcal{E}_{\text{ACPS}}[\Gamma_{k,d}]$ are identical.
– **Encryption.** On inputs a public key pk and message w, the encryption algorithm runs the encryption algorithm of $\mathcal{E}_{\text{ACPS}}[\Gamma_{k,d}]$ with the same inputs.
– **Decryption.** On inputs a secret key $\text{sk} = \mathbf{x} \in \{0,1\}^k$ and a ciphertext $(\mathbf{u}, c)$, the decryption algorithm uses $\mathbf{x}$ to obtain $\mathbf{s} = \gamma_{k,d}(\mathbf{x})$. Decryption proceeds as in $\mathcal{E}_{\text{ACPS}}[\Gamma_{k,d}]$, with inputs a secret key $\mathbf{s}$ and a ciphertext $(\mathbf{u}, c)$.

5.3 KDM$^{(n)}$ Security w.r.t. Degree-d Polynomials

We show that $\mathcal{E}_2$ is KDM$^{(n)}$-secure w.r.t. $\mathcal{F}_d$. For proof, see full version [10].

Theorem 1.5 (restated). *Consider the scheme $\mathcal{E}_2$ with p being super-polynomial in λ. Let $k' = \frac{k-\omega(\log \lambda)}{\log q}$ and let $\beta = \beta(\lambda) \in (0,1)$ be such that $\frac{\beta}{\sigma} = \text{negl}(\lambda)$. Define $\chi' = \bar{\Psi}_\beta$. Under the $\text{LWE}_{q,m\cdot n,k',\chi'}$ assumption, $\mathcal{E}_2$ is KDM$^{(n)}$-secure w.r.t. the class of degree-d polynomials modulo p.*

Note that if $\text{LWE}_{q,m\cdot n,k',\chi'}$ is hard for all $n = \text{poly}(\lambda)$, then $\mathcal{E}_2$ is KDM$^{(n)}$-secure for any polynomial number of "users". We also note that as in Theorem 1.3, the LWE assumption we rely on is related to worst-case lattice problems. See discussion in Section 5.2 for more details.

References

[1] Adão, P., Bana, G., Herzog, J., Scedrov, A.: Soundness of formal encryption in the presence of key-cycles. In: di Vimercati, S.D.C., Syverson, P.F., Gollmann, D. (eds.) ESORICS 2005. LNCS, vol. 3679, pp. 374–396. Springer, Heidelberg (2005)

[11] This scheme is denoted $\mathcal{E}_2$ for consistency with the full version [10] where $\mathcal{E}_1$ denotes our DDH based scheme.

[2] Akavia, A., Goldwasser, S., Vaikuntanathan, V.: Simultaneous hardcore bits and cryptography against memory attacks. In: Reingold [25], pp. 474–495
[3] Applebaum, B., Cash, D., Peikert, C., Sahai, A.: Fast cryptographic primitives and circular-secure encryption based on hard learning problems. In: Halevi [18], pp. 595–618
[4] Backes, M., Dürmuth, M., Unruh, D.: OAEP is secure under key-dependent messages. In: Pieprzyk, J. (ed.) ASIACRYPT 2008. LNCS, vol. 5350, pp. 506–523. Springer, Heidelberg (2008)
[5] Backes, M., Pfitzmann, B., Scedrov, A.: Key-dependent message security under active attacks - brsim/uc-soundness of symbolic encryption with key cycles. In: CSF, pp. 112–124. IEEE Computer Society, Los Alamitos (2007)
[6] Barak, B., Haitner, I., Hofheinz, D., Ishai, Y.: Bounded key-dependent message security. In: Gilbert, H. (ed.) EUROCRYPT 2010. LNCS, vol. 6110, pp. 423–444. Springer, Heidelberg (2010)
[7] Black, J., Rogaway, P., Shrimpton, T.: Encryption-scheme security in the presence of key-dependent messages. In: Nyberg, K., Heys, H.M. (eds.) SAC 2002. LNCS, vol. 2595, pp. 62–75. Springer, Heidelberg (2003)
[8] Boneh, D., Halevi, S., Hamburg, M., Ostrovsky, R.: Circular-secure encryption from decision diffie-hellman. In: Wagner, D. (ed.) CRYPTO 2008. LNCS, vol. 5157, pp. 108–125. Springer, Heidelberg (2008)
[9] Brakerski, Z., Goldwasser, S.: Circular and leakage resilient public-key encryption under subgroup indistinguishability - (or: Quadratic residuosity strikes back). In: Rabin, T. (ed.) CRYPTO 2010. LNCS, vol. 6223, pp. 1–20. Springer, Heidelberg (2010)
[10] Brakerski, Z., Goldwasser, S., Kalai, Y.: Black-box circular-secure encryption beyond affine functions (full version of this paper). Cryptology ePrint Archive, Report 2009/485 (2009), `http://eprint.iacr.org/`
[11] Camenisch, J., Chandran, N., Shoup, V.: A public key encryption scheme secure against key dependent chosen plaintext and adaptive chosen ciphertext attacks. In: Joux, A. (ed.) EUROCRYPT 2009. LNCS, vol. 5479, pp. 351–368. Springer, Heidelberg (2009)
[12] Camenisch, J.L., Lysyanskaya, A.: An efficient system for non-transferable anonymous credentials with optional anonymity revocation. In: Pfitzmann, B. (ed.) EUROCRYPT 2001. LNCS, vol. 2045, pp. 93–118. Springer, Heidelberg (2001)
[13] Canetti, R., Tauman Kalai, Y., Varia, M., Wichs, D.: On symmetric encryption and point obfuscation. In: Micciancio, D. (ed.) TCC 2010. LNCS, vol. 5978, pp. 52–71. Springer, Heidelberg (2010)
[14] El Gamal, T.: A public key cryptosystem and a signature scheme based on discrete logarithms. In: Blakely, G.R., Chaum, D. (eds.) CRYPTO 1984. LNCS, vol. 196, pp. 10–18. Springer, Heidelberg (1985)
[15] Goldreich, O.: Foundations of Cryptography - Basic Applications. Cambridge University Press, Cambridge (2004)
[16] Goldwasser, S., Kalai, Y., Peikert, C., Vaikuntanathan, V.: Robustness of the learning with errors assumption (2009) (manuscript)
[17] Haitner, I., Holenstein, T.: On the (im)possibility of key dependent encryption. In: Reingold [25], pp. 202–219
[18] Halevi, S. (ed.): CRYPTO 2009. LNCS, vol. 5677. Springer, Heidelberg (2009)
[19] Halevi, S., Krawczyk, H.: Security under key-dependent inputs. In: Ning, P., di Vimercati, S.D.C., Syverson, P.F. (eds.) ACM Conference on Computer and Communications Security, pp. 466–475. ACM, New York (2007)

[20] Hofheinz, D., Unruh, D.: Towards key-dependent message security in the standard model. In: Smart, N.P. (ed.) EUROCRYPT 2008. LNCS, vol. 4965, pp. 108–126. Springer, Heidelberg (2008)
[21] Laud, P., Corin, R.: Sound computational interpretation of formal encryption with composed keys. In: Lim, J.-I., Lee, D.-H. (eds.) ICISC 2003. LNCS, vol. 2971, pp. 55–66. Springer, Heidelberg (2004)
[22] Naor, M., Segev, G.: Public-key cryptosystems resilient to key leakage. In: Halevi [18], pp. 18–35
[23] Peikert, C.: Public-key cryptosystems from the worst-case shortest vector problem: extended abstract. In: Mitzenmacher, M. (ed.) STOC, pp. 333–342. ACM, New York (2009)
[24] Regev, O.: On lattices, learning with errors, random linear codes, and cryptography. In: Gabow, H.N., Fagin, R. (eds.) STOC, pp. 84–93. ACM, New York (2005)
[25] Reingold, O. (ed.): TCC 2009. LNCS, vol. 5444. Springer, Heidelberg (2009)
[26] Rivest, R.L., Shamir, A., Adleman, L.M.: A method for obtaining digital signatures and public-key cryptosystems. Commun. ACM 21(2), 120–126 (1978)

Homomorphic Encryption: From Private-Key to Public-Key

Ron Rothblum*

Weizmann Institute of Science, Rehovot, Israel
ron.rothblum@weizmann.ac.il

Abstract. We show how to transform any additively homomorphic private-key encryption scheme that is compact, into a public-key encryption scheme. By compact we mean that the length of a homomorphically generated encryption is independent of the number of ciphertexts from which it was created. We do not require anything else on the distribution of homomorphically generated encryptions (in particular, we do not require them to be distributed like real ciphertexts).

Our resulting public-key scheme is homomorphic in the following sense. If the private-key scheme is $i+1$-hop homomorphic with respect to some set of operations then the public-key scheme we construct is i-hop homomorphic with respect to the same set of operations.

1 Introduction

Homomorphic encryption is a paradigm that refers to the ability, given encryptions of some messages, to generate an encryption of a value that is related to the original messages. Specifically, this ability means that from encryptions of k messages $m_1, \ldots, m_k$ it is possible to generate an encryption of $m^* = f(m_1, \ldots, m_k)$ for some (efficiently computable) function f. Ideally, one may want the homomorphically generated encryption of m^* to be distributed identically (or statistically close) to a standard encryption of m^* (even given the original ciphertexts). We call schemes that have this property distribution-preserving homomorphic encryption schemes. Indeed, some proposed homomorphic encryption schemes are distribution-preserving w.r.t some algebraic operations such as addition or multiplication (e.g. Goldwasser-Micali [10], El-Gamal [5]).

For some applications, it seems as though distribution-preserving homomorphic encryption is an overkill. There are weaker notions of homomorphic encryption that might be easier to construct and still suffice for these applications. The very minimal requirement is that a homomorphically generated encryption decrypts correctly to the corresponding message. Alas, this minimalistic requirement does not seem to be useful as is, because it captures schemes that we do

* This research was partially supported by the Israel Science Foundation (grant No. 1041/08).

Y. Ishai (Ed.): TCC 2011, LNCS 6597, pp. 219–234, 2011.

not really consider to be homomorphic: Actually, *any* encryption scheme can be slightly modified to satisfy this requirement w.r.t *any* efficient operation[1]. A more meaningful notion is obtained by restricting the length of the homomorphically generated encryption. Specifically, we call a homomorphic encryption scheme compact if homomorphically generated encryptions properly decrypt to the correct message *and* their lengths depend *only* on the security parameter and the message length (and not on the number of input ciphertexts). Note that every distribution preserving homomorphic scheme is compact whereas the converse does not hold. Thus, the compactness property is strictly weaker than being distribution-preserving.

1.1 Private-Key vs. Public-Key

When discussing homomorphic encryption, we did not specify whether we consider private-key or public-key encryption schemes. Indeed, one can define homomorphic encryption in both settings (with only minor differences). The focus of this paper is showing the connection between public-key and private-key homomorphic encryption.

The easy direction is showing that a public-key homomorphic encryption scheme can be transformed into a private-key homomorphic scheme. This transformation is quite simple and involves only a minor issue. Intuitively, it seems as though any public-key homomorphic scheme *is* a private-key homomorphic scheme. The only problem is that in the public-key setting (in contrast to the private-key one), the homomorphic evaluation algorithm is also given the encryption-key. A simple transformation that addresses this issue is to append the encryption-key to each ciphertext. The resulting private-key scheme clearly retains the homomorphic properties of the public-key scheme (this holds for both distribution-preserving or merely compact homomorphic schemes).

The harder direction is showing that a private-key homomorphic encryption scheme implies a public-key one. This direction will be addressed by our main result, Theorem 2, which basically states that any *compact* additively homomorphic *private-key* encryption scheme can be transformed into a public-key encryption scheme. We present two such constructions both of which partially retain the homomorphic properties of the underlying private-key scheme (see Section 1.2).

We note that it is quite easy to transform a *distribution preserving* homomorphic private-key scheme into a distribution preserving homomorphic public-key one. In fact, this transformation was used by Barak [1] in his exposition of the work of van Dijk et al. [3]. For further discussion, see Section 1.4.

---

[1] Consider implementing the homomorphic evaluation algorithm as the identity function. That is, given ciphertexts and a description of an operation, just output both. Then, modify the decryption algorithm to first decrypt all the ciphertexts and then apply the operation to the decrypted messages. Thus, homomorphic evaluation is delegated to the decryption algorithm that, using the decryption key, can trivially evaluate the required operation.

1.2 Homomorphic Properties of the Public-Key Scheme

So far we have described homomorphic evaluation as a one-shot process, however one can consider repeated applications of the homomorphic evaluation algorithm. For *distribution-preserving* homomorphic encryption it is possible to do this because homomorphically generated values are identical (or statistically close) to real ciphertexts. For *compact* homomorphic encryption, the homomorphically generated encryptions can completely differ from actual ciphertexts, hence it is unclear that it is possible to keep computing on such homomorphically generated data. Gentry et al. [7] called a scheme that supports i such repeated applications an i-hop homomorphic encryption scheme.

The public-key schemes that we construct are homomorphic in the following sense. If the original private-key scheme is $(i+1)$-hop homomorphic w.r.t some set of operations (which must include addition modulo 2), then the public-key schemes are i-hop homomorphic w.r.t the same set of operations. That is, we lose one application of the homomorphic operation in the construction.

1.3 Connection to Prior Work

It is possible to combine previous results to construct a public-key encryption scheme from a compact additively homomorphic private-key scheme. However, the resulting public-key scheme does not (necessarily) retain the homomorphic properties of the private-key scheme. The indirect construction works as follows.

Kushilevitz and Ostrovsky [12] show that a compact additively homomorphic *public-key* scheme can be used to construct a two-message private information retrieval (PIR) protocol but their construction also works when using a private-key scheme (that is compact additively homomorphic). Di Crescenzo et al. [2] show that such a PIR protocol implies a two-message oblivious transfer (OT) protocol which in turn easily implies a public-key encryption scheme.

The public-key scheme constructed by combining these results is not necessarily homomorphic. Our simpler direct constructions retain the homomorphic properties of the private-key scheme (in the sense outlined in Section 1.2). Additionally, our schemes are fairly efficient and do require the amplification step used in [2] (to construct OT from PIR).

1.4 Technique

The intuition for how to move from private to public-key can be seen in a more straightforward manner in the case of *distribution preserving* homomorphic encryption. The following construction was suggested implicitly in [1].

Let E and D be the respective encryption and decryption algorithm of a private-key encryption scheme. Suppose that the scheme is distribution-preserving homomorphic w.r.t the identity function. That is, it is possible to "re-randomize" ciphertexts[2]. Such a scheme can be used to construct a public-key bit-encryption

[2] This means that there exists an algorithm RR such that for *any* encryption c of a bit b, the output of $RR(c)$ is distributed identically to $E_e(b)$.

scheme[3] as follows. The (private) decryption-key is a key k of the private-key scheme and the (public) encryption-key consists of an encryption of 0 and an encryption of 1 (i.e. $E_k(0)$ and $E_k(1)$). To encrypt a bit σ just re-randomize the ciphertext corresponding to σ. To decrypt, apply the private-key decryption algorithm using k (i.e. D_k).

The security of this construction follows from the fact that after re-randomization, all information on the original ciphertext, which was re-randomized, is completely lost. However, if the private-key scheme is only *compactly* homomorphic then we do not have a guarantee on the distribution of the homomorphically generated ciphertext and the above transformation fails. Hence, we use more complicated constructions, outlined next.

We present two constructions of public-key bit-encryption schemes based on any private-key scheme that is compactly homomorphic w.r.t addition modulo 2. The first construction was suggested to us by Yuval Ishai after we discovered the second construction. Both constructions are fairly straightforward but the first construction has a very simple proof. The second construction has a more complex proof based on an information-theoretic theorem, which may be of independent interest.

For both constructions the basic idea is to run the homomorphic algorithm, which outputs at most m bits, on more than m ciphertexts, and so forcing the algorithm to somehow actually compress the input ciphertexts. In the first construction, the decryption-key is once again a key k of the private-key scheme. The encryption-key consists of a random ℓ bit string r, where $\ell \gg m$, together with a sequence of encryptions of the bits of r using the key k. To encrypt a bit σ, a random vector $s \in \{0,1\}^\ell$ is selected such that the inner product of r and s equals σ. The homomorphic operation is then applied to the subset of the ℓ ciphertexts in the encryption-key, that correspond to coordinates in which s equals 1. Since the ciphertexts in the public-key are encryptions of the bits of r, the encryption process produces a homomorphically generated encryption of the inner product of r and s, which equals σ.

To show that this construction is semantically secure we consider, as a mental experiment, changing the key generation algorithm to encrypt zeros instead of the bits of r in the public encryption-key. We then consider an adversary that is given an encryption of a random bit (under this new scheme) and is asked to guess the bit. Observe that the encryption process now depends solely on s and does not contain any information on r. Thus, in essence, the adversary is given the ℓ bit string r and m bits of information on s, where $m \ll \ell$ and is asked to find $\langle r, s\rangle$. Using the Leftover Hash Lemma, we show that it is impossible to predict $\langle r, s\rangle$ with probability that is noticeably greater than $\frac{1}{2}$. Thus, an adversary for the proposed public-key scheme would imply a distinguisher for the underlying private-key scheme that distinguishes between ℓ encryptions of random bits and ℓ encryptions of 0.

---

[3] A bit-encryption scheme is a public-key encryption scheme that only handles single-bit messages. Such schemes suffice to construct full-fledged public-key encryption schemes (see [8]).

The second construction is somewhat similar, however its proof of security is more complex and is based on an information-theoretic theorem, which may be of independent interest. Again, the decryption-key is a key k of the private-key scheme but the public-key consists of two lists of ciphertexts; the first is a list of ℓ encryptions of 0 and the second is a list of ℓ encryptions of 1. To encrypt a bit σ we choose a *random subset* $S \subseteq [\ell]$ that has parity σ (i.e. $|S| \equiv \sigma \bmod 2$). We use S to select ℓ ciphertexts from the public-key by selecting the i-th ciphertext from the first list if $i \notin S$ (and from the second if $i \in S$). By homomorphically adding the selected ciphertexts modulo 2, we obtain a ciphertext that correctly decrypts to σ.

To prove security, once again we consider a mental experiment in which both lists in the public-key are encryptions of 0. Because the mental experiment is computationally indistinguishable from the actual scheme, proving that the original scheme is secure reduces to showing that when *both* lists consist of encryptions of 0, it is essentially impossible to find the parity of the random subset used in the homomorphic encryption process.

We prove the latter via the following information-theoretic theorem: Let $X_1, \ldots, X_\ell$ and $Y_1, \ldots, Y_\ell$ be independent and identically distributed over a finite set Ω and let S be a random subset of $[\ell]$. We consider the list Z, defined as $Z_i = X_i$ for $i \notin S$ and $Z_i = Y_i$ for $i \in S$. The theorem states that it is essentially impossible to guess the parity of S based on X, Y and m bits of information on Z. That is, any such guess will be correct with probability that is bounded by (roughly) $\frac{1}{2} + 2^{-\Omega(\ell-m)}$. The proof of the information-theoretic theorem makes use of the Efron-Stein decomposition [4], an extension of Fourier analysis for product distributions.

Remarks. First we mention that both of our constructions are secure even if we use a weaker definition of *compact* homomorphic encryption. Specifically, when homomorphically adding ciphertexts, the output can be of length that is a sub-linear function of the number of input ciphertexts (rather than being independent of it).

We also mention that while our definition of *compact* homomorphic encryption considers homomorphic operations over *arbitrarily* many ciphertexts, one might instead consider compact homomorphic encryption over just two ciphertexts. However, this definition can be implemented trivially[4] and a more meaningful notion is obtained by requiring that the homomorphic operation support multiple hops. Our constructions can be implemented by a scheme that satisfies the new definition by implementing the homomorphic addition of ℓ ciphertexts using a logarithmic number of hops.

1.5 Application of Our Construction to Fully-Homomorphic Encryption

Our generic transformation from private-key to public-key encryption can be used as a general methodology for constructing (compact) homomorphic public-key

[4] As in Footnote 1 while keeping the ciphertexts sufficiently short.

encryption. One application of this methodology, which actually motivated this work, is to simplify the presentation of the DGHV fully-homomorphic encryption scheme [3].

A fully-homomorphic encryption scheme is an encryption scheme that is homomorphic w.r.t any (efficiently computable) function. The concept of fully-homomorphic encryption was first proposed by Rivest et al. [13] in the 70's, but the first concrete proposal was only made recently in the breakthrough work of Gentry [6].

Building on the work of Gentry [6], van Dijk et al. [3], proposed a simpler fully-homomorphic public-key scheme. From a high-level view, the DGHV fully homomorphic scheme is constructed by first proposing a simple *private-key* homomorphic scheme that is only "somewhat" homomorphic (that is, homomorphic w.r.t some restricted functions), and then showing how to modify this scheme into a somewhat homomorphic *public-key* scheme. Finally, using the bootstrapping technique of [6] the somewhat homomorphic public-key scheme is transformed into a fully-homomorphic public-key scheme.

One way in which our transformation can be used is to replace the aforementioned modification (i.e. from private-key to public-key) that uses specific properties of the DGHV scheme. The advantage is that our transformation is generic. Although the somewhat homomorphic public-key scheme constructed by our transformation is slightly different from the one of [3], the final steps of bootstrapping (see [6]) and reducing the (multiplicative) depth of the decryption circuit can still be applied to both of our constructions.

An alternate way to use our transformation is to first construct a compact fully-homomorphic *private-key* scheme and then, using our generic transformation, to obtain a compact fully-homomorphic *public-key* scheme. We believe that this approach simplifies the presentation of the scheme because the bootstrapping step is done on the simpler private-key scheme. This approach was suggested by Barak [1] for one of the variants of DGHV that is actually *distribution-preserving*. However, using our transformation, the approach can be extended to the compact variants as well.

The two approaches to construct fully homomorphic encryption based on the DGHV scheme are depicted in Figure 1.

2 Preliminaries

For a set S, we denote by $x \in_R S$ a random element uniformly distributed in S. Similarly, we denote by $X \subseteq_R S$ a uniformly distributed random subset of S. For a vector $X = X_1, \ldots, X_\ell$ and a set $I \subseteq [\ell]$, we denote by X_I the projection of X to coordinates in I; i.e. if $I = \{i_1, \ldots, i_m\}$, where $i_1 < \cdots < i_m$, then $X_I = X_{i_1}, \ldots, X_{i_m}$.

Non-Standard Notation. For every $\ell \in \mathbb{N}$, random variables $X = X_1, \ldots, X_\ell$ and $Y = Y_1, \ldots, Y_\ell$ and set $S \subseteq [\ell]$, we denote by $X_{\overline{S}}Y_S$, the random variable $Z = Z_1, \ldots, Z_\ell$ where $Z_i = X_i$ for $i \notin S$ and $Z_i = Y_i$ for $i \in S$.

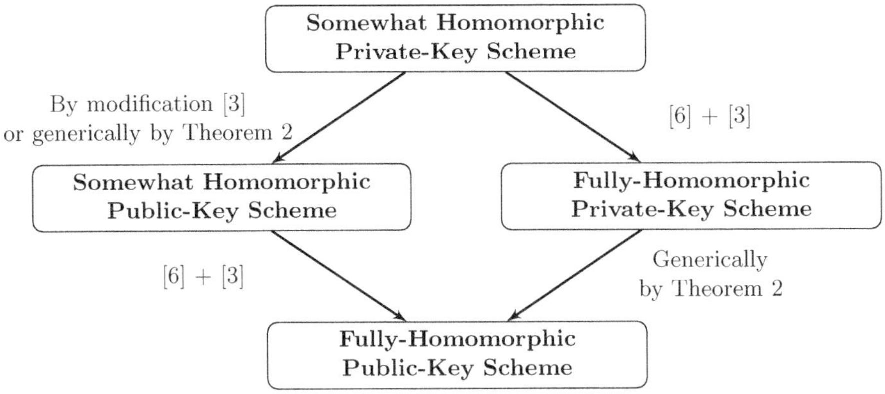

Fig. 1. Constructing the compact homomorphic variant of the DGHV fully-homomorphic public-key scheme

2.1 Encryption Schemes

We follow notations and definitions of [8]. In particular we use their definition of semantically secure encryption schemes, both in the private-key and public-key settings. Throughout this paper we restrict our attention to bit-encryption schemes, i.e., schemes that encrypt a single bit. For simplicity, we say public-key (resp. private-key) encryption when we actually mean public-key (resp. private-key) bit-encryption.

When discussing private-key schemes, we consider schemes with *multiple-message* security, i.e., semantic security w.r.t to an adversary that gets encryptions of polynomially many messages. Recall that in the private-key setting (in contrast to public-key one), multiple-message security does not follow from single-message security (see [8, Chapter 5]).

2.2 Homomorphic Encryption

Since we only consider *compact* homomorphic encryption, from here on, when we say homomorphic we always mean in the *compact* sense as defined next.

Definition 1. (G, E, D, H) *is a* homomorphic public-key encryption scheme *with respect to a set of families of polynomial-sized circuits* $\mathcal{C}$ *if* (G, E, D) *are a public-key encryption scheme,* H *is a probabilistic polynomial-time algorithm and there exists a polynomial*[5] $m(\cdot)$ *such that for every circuit family* $\{C_k\}_{k\in\mathbb{N}} \in \mathcal{C}$*, polynomial* $\ell(\cdot)$*, for every* $n \in \mathbb{N}$*, keys* $(e, d) \leftarrow G(1^n)$*, and* $\ell = \ell(n)$ *single bit messages* $b_1, \ldots, b_\ell \in \{0, 1\}$ *the following holds:*

- *Correct decryption of homomorphically generated encryptions:*

$$D_d\left(H\left(e, C_\ell, E_e(b_1), \ldots, E_e(b_\ell)\right)\right) = C_\ell\left(b_1, \ldots, b_\ell\right). \tag{1}$$

[5] For convenience we assume that m is at least linear.

– *The length of homomorphically generated encryptions is independent of ℓ:*

$$|H(e, C_\ell, E_e(b_1), \ldots, E_e(b_\ell))| \leq m(n). \tag{2}$$

Homomorphic *private-key* encryption is defined analogously (with the modification that H does not get the encryption key as part of its input).

2.3 i-Hop Homomorphic Encryption

The homomorphic evaluation algorithm in Definition 1 is only required to operate on ciphertexts that were output by the encryption algorithm. The definition does not specify what happens if the homomorphic evaluation algorithm is applied to its own output. Gentry et al. [7] defined an i-hop homomorphic encryption scheme as a scheme for which it is possible to apply the homomorphic evaluation algorithm consecutively i times.

Let G, E, D, H be a homomorphic encryption scheme w.r.t to a set of circuit families $\mathcal{C}$. For a given encryption key e, we denote by $W_0(e)$ the set of all valid ciphertexts of the encryption scheme, i.e., all possible outputs of the encryption algorithm E_e applied to a single bit message. For $j \geq 1$, we define $W_j(e)$ to be the set of all possible outputs of the homomorphic evaluation algorithm H when applied to a sequence of ciphertexts in $W_{j-1}(e)$ and any circuit $C \in \mathcal{C}$. We say that elements in $W_j(e)$ are j-th level ciphertexts and define i-hop homomorphic encryption (in both the public and private-key settings) by requiring that Equations (1) and (2) of Definition 1 hold not only for standard ciphertexts (i.e., in $W_0(e)$) but also for j-th level ciphertexts for $j \leq i$ (i.e., in $W_j(e)$).

3 Constructing a Public-Key Scheme from a Homomorphic Private-Key Scheme

In this section we prove our main theorem:

Theorem 2. *Any multiple-message semantically secure private-key encryption scheme that is compactly homomorphic with respect to addition modulo 2 can be transformed into a semantically secure public-key encryption scheme. Furthermore, if the private-key scheme is $(i+1)$-hop homomorphic w.r.t to a set of circuit families, then the constructed public-key scheme is i-hop homomorphic w.r.t to the same set.*

We present two alternate constructions. Both constructions are fairly straightforward but the proof of the first construction (Construction 3) is more simple. However, the tools used in the proof of the second construction (Construction 7) may be of independent interest.

To prove Theorem 2, we assume the existence of a private-key scheme (G, E, D, H) that is compactly homomorphic with respect to addition modulo 2 and the polynomial $m(\cdot)$ as in Definition 1. We denote by $H_\oplus$ the algorithm H when applied to the circuit family that computes addition modulo 2. The discussion on the homomorphic properties of the schemes (i.e. the furthermore part of Theorem 2) is presented in Section 4.

3.1 First Construction

Construction 3. *The public-key encryption scheme (G', E', D', H') is defined as follows:*

Key Generation - $G'(1^n)$:
Select $k \leftarrow G(1^n)$ *and* $r \in_R \{0,1\}^\ell$ *where* $\ell = 4m(n)$.
Set $X = (X_1, \ldots, X_\ell)$ *where* $X_i \leftarrow E_k(r_i)$.
Output (X, r) *as the public-key and* k *as the private-key.*

Encryption - $E'_{X,r}(\sigma)$:
Select at random a vector $s \in \{0,1\}^\ell$ *such that* $\langle s, r\rangle = \sigma$.[6] *At convenience, we identity the set* S *with the natural representation of the vector* s *as a set, i.e.,* $S = \{i \colon s_i = 1\}$.
Output $H_\oplus(X_S)$ *where* X_S *is the projection of* X *to coordinates in* S.

Decryption - $D'_k(c)$:
Output $D_k(c)$.

Homomorphic Evaluation - $H'(C, (X, r), c_1, \ldots, c_\ell)$:
Output $H(C, c_1, \ldots, c_\ell)$.

We start by showing that the decryption algorithm correctly decrypts proper ciphertexts. We then proceed to the main part of the proof, showing that Construction 3 is indeed semantically secure. In Section 4 we discuss the homomorphic properties of the scheme.

Proposition 4. *For every* $n \in \mathbb{N}$, $\sigma \in \{0,1\}$ *and* $((X, r), k) \leftarrow G'(1^n)$ *it holds that*

$$D'_k\left(E'_{X,r}(\sigma)\right) = \sigma.$$

Proof. Based on the first property of homomorphic encryption (Definition 1),

$$D'_k\left(E'_{X,r}(\sigma)\right) = D_k\left(H_\oplus\left(X_S\right)\right) = \bigoplus_{i \in S} D_k(X_i) = \bigoplus_{i \in S} D_k(E_k(r_i))$$

where S is the random subset selected in the encryption algorithm E' and $\bigoplus$ denotes addition modulo 2. Since D decrypts correctly, $D_k(E_k(r_i) = r_i$. Therefore, $D'_k\left(E'_{X,r}(\sigma)\right) = \bigoplus_{i \in S} r_i = \langle s, r\rangle = \sigma$. □

We proceed to the main part of the proof, showing that Construction 3 is semantically secure.

Proposition 5. *If* (G, E, D) *is a multiple-message semantically secure private-key scheme, then* (G', E', D') *is a semantically secure public-key scheme.*

[6] If $r = 0^\ell$ then such a vector s does not necessarily exist. However, this case only happens with exponentially vanishing probability and can be handled by choosing $r \neq 0^\ell$ in the key-generation process.

Proof. Suppose toward a contradiction that there exists a probabilistic polynomial-time adversary A that can predict the value of a random bit σ based on an encryption of σ. That is, there exists a polynomial $p(\cdot)$ and infinitely many $n \in \mathbb{N}$ for which:

$$\Pr_{\substack{(X,r),k \leftarrow G'(1^n) \\ \sigma \in_R \{0,1\}}} \left[A\left((X,r), E'_{X,r}(\sigma)\right) = \sigma\right] > \frac{1}{2} + \frac{1}{p(n)} \tag{3}$$

where the probability is also over the coin tosses of A and E'. Based on the definitions of G' and E' this implies that:

$$\Pr_{\substack{s,r \in_R \{0,1\}^\ell,\ k \leftarrow G(1^n) \\ X_i \leftarrow E_k(r_i)}} \left[A\left(X, r, H_\oplus\left(X_S\right)\right) = \langle s, r\rangle\right] > \frac{1}{2} + \frac{1}{p(n)} \tag{4}$$

where $S = \{i \colon s_i = 1\}$.

Consider X' which is distributed as ℓ encryptions of 0 under k (in contrast to X which is distributed as an encryption of the bits of r). We claim that for every (computationally unbounded) algorithm A and for every $n \in \mathbb{N}$,

$$\Pr_{\substack{s,r \in_R \{0,1\}^\ell,\ k \leftarrow G(1^n) \\ X'_i \leftarrow E_k(0)}} \left[A\left(X', r, H_\oplus\left(X'_S\right)\right) = \langle s, r\rangle\right] \le \frac{1}{2} + 3 \cdot 2^{-\frac{\ell}{2}+m(n)} . \tag{5}$$

Equation (5) implies a simple distinguisher for the private-key scheme. Specifically, we refer to a distinguisher that gets $r \in_R \{0,1\}^\ell$ and Y which is either an encryption of the bits of r or ℓ encryptions of zero. The distinguisher chooses a random vector $s \in_R \{0,1\}^\ell$, computes $A(Y, r, H_\oplus(Y_S))$ and outputs 1 if it equals $\langle s, r\rangle$ and 0 otherwise. Based on Equations (4) and (5), this distinguisher distinguishes between the two cases with a noticeable gap. Thus, we only need to show that Eq. (5) holds.

To prove that Eq. (5) holds, we first view it as follows:

$$\Pr_{\substack{s,r \in_R \{0,1\}^\ell,\ k \leftarrow G(1^n) \\ X'_i \leftarrow E_k(0)}} \left[A_{X'}\left(r, H_{X'}(s)\right) = \langle s, r\rangle\right] \le \frac{1}{2} + 3 \cdot 2^{-\frac{\ell}{2}+m(n)} . \tag{6}$$

Observe that since $H_{X'}$ does not depend on r, the adversary $A_{X'}$ needs to predict $\langle s, r\rangle$ based on r and m bits of information on s. The following proposition shows that $A_{X'}$ can only succeed with a negligible advantage, for *every* k and X' (hence, also for k and X' distributed as above):

Proposition 6. *For every function $f : \{0,1\}^\ell \to \{0,1\}^m$ and every (computationally unbounded) algorithm A,*

$$\Pr_{r,s \in \{0,1\}^\ell} \left[A(r, f(s)) = \langle s, r\rangle\right] \le \frac{1}{2} + 3 \cdot 2^{-\frac{\ell}{2}+m} . \tag{7}$$

Proof. By the Leftover Hash Lemma[7] [11], because $h_r(s) = \langle s, r\rangle$ is a universal hash function family, for every $\alpha \in \{0,1\}^m$ the distribution $(r, \langle r, s\rangle)$ conditioned on $f(s) = \alpha$ and the distribution (r, σ) for $\sigma \in_R \{0,1\}$ are $2\sqrt{\frac{2}{|f^{-1}(\alpha)|}}$-close in statistical distance. Thus:

$$\begin{aligned}
&\Pr_{r,s\in_R\{0,1\}^\ell}[A(r, f(s)) = \langle r, s\rangle] \\
&= \sum_{\alpha\in\{0,1\}^m} \Pr_{s\in_R\{0,1\}^\ell}[f(s) = \alpha] \cdot \Pr_{r,s\in_R\{0,1\}^\ell}[A(r,\alpha) = \langle r, s\rangle | f(s) = \alpha] \\
&\le \sum_{\alpha\in\{0,1\}^m} \frac{|f^{-1}(\alpha)|}{2^\ell} \cdot \left(\Pr_{\substack{r\in_R\{0,1\}^\ell \\ \sigma\in_R\{0,1\}}} [A(r,\alpha) = \sigma] + 2 \cdot \sqrt{\frac{2}{|f^{-1}(\alpha)|}} \right) \\
&\le \frac{1}{2} \cdot \sum_{\alpha\in\{0,1\}^m} \frac{|f^{-1}(\alpha)|}{2^\ell} + \sum_{\alpha\in\{0,1\}^m} \frac{2\sqrt{2|f^{-1}(\alpha)|}}{2^\ell} \\
&\le \frac{1}{2} + 3 \cdot 2^{-\frac{\ell}{2}+m} .
\end{aligned}$$

□

3.2 Second Construction

We proceed to present the second construction. Recall that (G, E, D, H) is a homomorphic private-key scheme with respect to addition modulo 2 and the polynomial $m(\cdot)$ as in Definition 1.

Construction 7. *The public-key encryption scheme (G'', E'', D'', H'') is defined as follows:*

Key Generation - $G''(1^n)$:
Select $k \leftarrow G(1^n)$, $X = (X_1, \ldots, X_\ell)$ and $Y = (Y_1, \ldots, Y_\ell)$ where $\ell = 10m(n)$, such that $X_i \leftarrow E_k(0)$ and $Y_i \leftarrow E_k(1)$ (with fresh random coins for each i).
Output (X, Y) as the public-key and k as the private-key.

Encryption - $E''_{X,Y}(\sigma)$:
Select at random a subset $S \subseteq [\ell]$ that has size of parity σ (i.e. $|S| \equiv \sigma \bmod 2$).
Output $H_\oplus(X_{\overline{S}}Y_S)$ (recall that $X_{\overline{S}}Y_S$ is a list of ℓ ciphertexts that are encryptions of 1 for coordinates in S and encryptions of 0 elsewhere).

Decryption - $D''_k(c)$:
Output $D_k(c)$.

---

[7] The Leftover Hash Lemma states that if h is selected at random from a universal hash function family from $\{0,1\}^\ell$ to $\{0,1\}^k$ and t is selected uniformly from a set $T \subseteq \{0,1\}^\ell$ then the distribution $h, h(t)$ and the distribution h, u, where u is distributed uniformly in $\{0,1\}^k$ are $2\sqrt{\frac{2^k}{|S|}}$-close (see, e.g. [9, Appendix D]).

Homomorphic Evaluation - $H''(C, (X,Y), c_1, \ldots, c_\ell)$**:**
Output $H(C, c_1, \ldots, c_\ell)$.

As in Construction 3, we first show that the decryption algorithm works and then move on to the main part, showing that the construction is indeed semantically secure.

Proposition 8. *For every* $n \in \mathbb{N}$, $\sigma \in \{0,1\}$ *and* $((X,Y), k) \leftarrow G''(1^n)$*:*

$$D''_k\left(E''_{X,Y}(\sigma)\right) = \sigma.$$

Proof. Based on the first property of homomorphic encryption (Definition 1),

$$D''_k\left(E''_{X,Y}(\sigma)\right) = D_k\left(H_\oplus\left(X_{\overline{S}} Y_S\right)\right) = \bigoplus_{i=1}^{\ell} D_k(Z_i)$$

where S is the random subset selected in the encryption process, $Z_i = Y_i$ for $i \in S$ and $Z_i = X_i$ otherwise. Since D decrypts correctly, $D_k(X_i) = 0$ and $D_k(Y_i) = 1$. Therefore, $D''_k\left(E''_{X,Y}(\sigma)\right) = \bigoplus_{i \in S} 1 = |S| \bmod 2 = \sigma$. □

We proceed to the main part of the proof, showing that Construction 7 is semantically secure.

Proposition 9. *If* (G, E, D) *is a multiple-message semantically secure private-key scheme then* (G'', E'', D'') *is a semantically secure public-key scheme.*

Proof. Assume toward a contradiction that (G'', E'', D'') is not semantically secure. This means that there exists a probabilistic polynomial-time adversary A'' and a polynomial $p(\cdot)$ such that for infinitely many $n \in \mathbb{N}$:

$$\Pr_{\substack{(X,Y),k \leftarrow G''(1^n) \\ \sigma \in_R \{0,1\}}}\left[A''\left(X, Y, E''_{X,Y}(\sigma)\right) = \sigma\right] > \frac{1}{2} + \frac{1}{p(n)} . \tag{8}$$

To derive a contradiction, we consider n from this infinite set and construct a probabilistic polynomial-time adversary A for the underlying private-key scheme. The adversary A receives 2ℓ ciphertexts $(\alpha_1, \ldots, \alpha_\ell, \beta_1, \ldots, \beta_\ell)$ and will be shown to distinguish between the following two cases:

- $\alpha_1, \ldots, \alpha_\ell$ are encryptions of 0 and $\beta_1, \ldots, \beta_\ell$ are encryptions of 1.
- $\alpha_1, \ldots, \alpha_\ell, \beta_1, \ldots, \beta_\ell$ are encryptions of 0.

Algorithm A operates as follows:

1. Set $X = (\alpha_1, \ldots, \alpha_\ell)$ and $Y = (\beta_1, \ldots, \beta_\ell)$.
2. Select $S \subseteq_R [\ell]$.
3. Output 1 if $A''(X, Y, H_\oplus(X_{\overline{S}} Y_S)) = |S| \bmod 2$ and 0 otherwise.

Accordingly,

$$\Pr_{\substack{k\leftarrow G(1^n)\\ \alpha_j,\beta_j}}\left[A\left(\alpha_1,\ldots,\alpha_\ell,\beta_1,\ldots,\beta_\ell\right)=1\right]=\Pr_{\substack{k\leftarrow G(1^n)\\ X,Y,S}}\left[A''\left(X,Y,H_\oplus\left(X_{\overline{S}}Y_S\right)\right)=|S| \bmod 2\right].$$

We proceed by analyzing A's behavior in the two different cases. In the first case, $\alpha_i = E_k(0)$ and $\beta_i = E_k(1)$. Consequently, $H_\oplus(X_{\overline{S}}Y_S)$ is distributed identically to an encryption of a random bit under E'' and so, by Eq. (8), it holds that

$$\Pr_{\substack{k\leftarrow G(1^n)\\ X,Y,S}}\left[A''\left(X,Y,H_\oplus\left(X_{\overline{S}}Y_S\right)\right)=|S| \bmod 2\right]=\Pr_{\substack{(X,Y),k\leftarrow G''(1^n)\\ \sigma\in_R\{0,1\}}}\left[A''\left(X,Y,E''_{X,Y}\left(\sigma\right)\right)=\sigma\right]$$
$$> \frac{1}{2}+\frac{1}{p(n)}\ .$$

In the second case, $\alpha_i = \beta_i = E_k(0)$. We argue that in this case for every $n \in \mathbb{N}$ and even for an unbounded adversary A',

$$\Pr_{\substack{k\leftarrow G(1^n)\\ X,Y,S}}\left[A''\left(X,Y,H_\oplus\left(X_{\overline{S}},Y_S\right)\right) = |S| \bmod 2\right] < \frac{1}{2} + 2^{-0.2\ell+m(n)+1}. \qquad (9)$$

Equation (9) follows from an information-theoretic theorem (Theorem 10) that will be stated next. See the full version of this paper [14] for the proof of Theorem 10.

Using Theorem 10, we conclude that A distinguishes between the two cases with non-negligible probability, in contradiction to the multiple-message security of (G, E, D), □

Information-Theoretic Theorem. Let Ω be a finite non-empty set and $\ell \in \mathbb{N}$. Let $\mu_1, \ldots, \mu_\ell$ be distributions over Ω and $\mu = \mu_1 \times \cdots \times \mu_\ell$ be a product distribution over Ω^ℓ. Let X and Y be independent random variables identically distributed according to μ over Ω^ℓ.

Theorem 10. *For any $\ell, m \in \mathbb{N}$ and any functions $h\colon \Omega^\ell \to \{0,1\}^m$ and $g\colon \Omega^\ell \times \Omega^\ell \times \{0,1\}^m \to \{0,1\}$, it holds that*

$$\Pr_{X,Y,S\subseteq_R[\ell]}\left[g\left(X,Y,h(X_{\overline{S}}Y_S)\right) = |S| \bmod 2\right] < \frac{1}{2} + 2^{-0.2\ell+m+1}.$$

Equation (9) seems to follow immediately from Theorem 10 by setting A'' as g, $H_\oplus$ as h and having X and Y distributed as ℓ independent encryptions of 0 each. However, there is a small subtlety - Theorem 10 addresses g and h that are *deterministic* functions, in contrast to A'' and H that are *probabilistic* algorithms. Additionally, since X and Y are distributed w.r.t to the same randomly chosen key, they are not product distributions as required by Theorem 10.

Both issues are resolved by an averaging argument. If Eq. (9) does not hold for some $n \in \mathbb{N}$, then there exist random coins for A'', H and a fixed private-key k for which it does not hold. Once we fix these coins, A'' and H become deterministic functions. Additionally, we set X and Y to each be distributed as ℓ encryptions of 0 under the fixed key k, which is in particular a product distribution. Thus, the hypothesis that Eq. (9) does not hold contradicts Theorem 10. The proof of Theorem 10 uses the Efron-Stein decomposition [4], an extension of Fourier analysis for general product distributions and appears in the full version of this paper [14] (see an outline next).

Outline of the Proof of Theorem 10. Theorem 10 considers a game in which a computationally unbounded adversary sees X, Y and m bits of information on $X_{\overline{S}}Y_S$ and needs to decide whether S is of even or odd cardinality. That is, the adversary specifies a function $h : \Omega^\ell \to \{0,1\}^m$ and based on $X, Y, h(X_{\overline{S}}Y_S)$ needs to find $|S| \bmod 2$. Theorem 10 states that winning this game with probability noticeably better than $\frac{1}{2}$ is impossible as long as m is sufficiently smaller than ℓ. Note that winning the game becomes easy if m is sufficiently larger[8] than ℓ (as long as the probability of a collision in each coordinate, i.e. $\Pr[X_i = Y_i]$, is sufficiently small).

To prove Theorem 10, we would like to that show that for a *typical* $\gamma \in \{0,1\}^m$, the number of odd S that map to γ (i.e., $h(X_{\overline{S}}Y_S) = \gamma$) and the number of even S are roughly the same. This would imply that any adversary, which sees only X, Y and γ, cannot guess whether γ was produced from an odd or even S, which is exactly what we are looking to prove.

The proof is composed of two lemmas. The main lemma states that for *every* $\gamma \in \{0,1\}^m$, w.h.p, the number of odd S that map to γ is fairly close to the number of even S (in absolute terms). We prove this lemma by showing that the probability of a collision of two random sets S and T of the same parity (i.e. the probability that $h(X_{\overline{S}}Y_S) = h(X_{\overline{T}}Y_T)$ where $|T| = |S|$) is roughly the same as the collision probability of two sets of different parity. We use the Efron-Stein decomposition to express the collision probability and bound it. The second lemma is more straightforward and states that for a *typical* γ the total number of S that map to it is very large. Combining these two lemmas we prove Theorem 10. See the full version of the paper for details.

4 Homomorphic Properties of the Public-Key Scheme

In this section, we discuss the homomorphic properties of the public-key schemes presented in Section 3. We show that if the underlying private-key scheme supports $i+1$ repeated homomorphic operations then both Construction 3 and Construction 7 support i such operations. Intuitively, this follows by the fact that in both constructions, the encryption algorithm applies a single homomorphic operation (see Fact 12), thus exactly one hop is lost.

Proposition 11. *Suppose G, E, D, H are an $(i+1)$-hop homomorphic private-key scheme w.r.t to a set of circuit families $\mathcal{C}$ that includes addition modulo 2. Then Constructions 3 and Construction 7 are i-hop homomorphic w.r.t the set $\mathcal{C}$.*

[8] If $m \geq \ell \log(|\Omega|)$ just take h to be the identity function.

We prove that Proposition 11 holds for Construction 3 while noting that the proof for Construction 7 is completely analogous. Thus, we refer to G', E', D', H' as in Construction 3.

Let $(X, r), k$ be a pair of encryption/decryption keys for Construction 3 (w.r.t to the security parameter n). We denote the j-th level ciphertexts of the private-key scheme by $W_j(k)$ and the j-th level ciphertexts of the public-key scheme by $W'_j(X, r)$.

Fact 12. *For every $j \in \mathbb{N}$, $W'_j(X, r) \subseteq W_{j+1}(k)$.*

Proof. By induction on j.

Let $\{C_k\}_k \in \mathcal{C}$, $0 \leq j \leq i$, $\ell = \ell(n)$ and $w_1, \ldots, w_\ell$ be j-th level ciphertexts of the public-key scheme (i.e., in $W'_j(X, Y)$). We proceed by showing that it j-th level ciphertexts decrypt properly. By Fact 12, it holds that $w_1, \ldots, w_\ell \in W_{j+1}(k)$ and thus,

$$\begin{aligned} H'(C_\ell, (X, r), w_1, \ldots, w_\ell) &= H(C_\ell, w_1, \ldots, w_\ell) \\ &= C_\ell(D_k(w_1), \ldots, D_k(w_\ell)) \\ &= C_\ell(D'_k(w_1), \ldots, D'_k(w_\ell)). \end{aligned}$$

where the first and third equalities follow from the definition of H' and D' respectively and the second equality follows from the fact that (G, E, D, H) are $i+1$-hop homomorphic and that $w_1, \ldots, w_\ell$ are ciphertexts of level $j+1 \leq i+1$ of the private-key scheme.

A similar argument shows that the scheme is also compact. Indeed, since $w_1, \ldots, w_\ell \in W'_j(X, Y) \subseteq W_{j+1}(k)$ it holds that,

$$|H'(C_\ell, (X, r), w_1, \ldots, w_\ell)| = |H(C_\ell, w_1, \ldots, w_\ell)| \leq m(n)$$

for every $0 \leq j \leq i$.

Acknowledgments

I would like to express my thanks and appreciation to my M.Sc. advisor, Oded Goldreich, for his encouragement and guidance in completing this work. In particular, I would like to thank him for many helpful discussions and constructive comments that helped present this work in a more coherent way.

I would like to thank Benny Applebaum for useful conversations. I am also extremely grateful to Yuval Ishai for suggesting Construction 3 and for allowing me to include it in this paper.

References

1. Barak, B.: Cryptography course - Lecture notes, COS 433. Princeton University, Computer Science Department (Spring (2010), `http://www.cs.princeton.edu/courses/archive/spring10/cos433`

2. Di Crescenzo, G., Malkin, T., Ostrovsky, R.: Single database private information retrieval implies oblivious transfer. In: Preneel, B. (ed.) EUROCRYPT 2000. LNCS, vol. 1807, p. 122. Springer, Heidelberg (2000)
3. van Dijk, M., Gentry, C., Halevi, S., Vaikuntanathan, V.: Fully homomorphic encryption over the integers. In: Gilbert, H. (ed.) EUROCRYPT 2010. LNCS, vol. 6110, pp. 24–43. Springer, Heidelberg (2010), http://dx.doi.org/10.1007/978-3-642-13190-5
4. Efron, B., Stein, C.: The jackknife estimate of variance. The Annals of Statistics 9(3), 586–596 (1981), http://www.jstor.org/stable/2240822
5. El Gamal, T.: A public key cryptosystem and a signature scheme based on discrete logarithms. In: Blakely, G.R., Chaum, D. (eds.) CRYPTO 1984. LNCS, vol. 196, pp. 10–18. Springer, Heidelberg (1985), http://dx.doi.org/10.1007/3-540-39568-7_2
6. Gentry, C.: Fully homomorphic encryption using ideal lattices. In: Proceedings of the 41st Annual ACM Symposium on Theory of Computing, STOC 2009, pp. 169–178. ACM, New York (2009), http://doi.acm.org/10.1145/1536414.1536440
7. Gentry, C., Halevi, S., Vaikuntanathan, V.: i-hop homomorphic encryption and rerandomizable yao circuits. In: Rabin, T. (ed.) CRYPTO 2010. LNCS, vol. 6223, pp. 155–172. Springer, Heidelberg (2010), http://dx.doi.org/10.1007/978-3-642-14623-7
8. Goldreich, O.: Foundations of Cryptography. Basic Applications, vol. 2. Cambridge University Press, Cambridge (2004)
9. Goldreich, O.: Computational complexity: a conceptual perspective. SIGACT News 39(3), 35–39 (2008)
10. Goldwasser, S., Micali, S.: Probabilistic encryption. Journal of Computer and System Sciences 28(2), 270–299 (1984)
11. Håstad, J., Impagliazzo, R., Levin, L.A., Luby, M.: A pseudorandom generator from any one-way function. SIAM Journal on Computing 28, 12–24 (1999)
12. Kushilevitz, E., Ostrovsky, R.: Replication is not needed: single database, computationally-private information retrieval. In: Proceedings of the 38th Annual Symposium on Foundations of Computer Science, FOCS 1997, p. 364. IEEE Computer Society, Los Alamitos (1997)
13. Rivest, R.L., Adleman, L., Dertouzos, M.L.: On data banks and privacy homomorphisms. In: Foundations of Secure Computation, pp. 169–180. Academic Press, London (1978)
14. Rothblum, R.: Homomorphic encryption: from private-key to public-key. Electronic Colloquium on Computational Complexity (ECCC) 17, 146 (2010), http://eccc.hpi-web.de/report/2010/146

Identity-Based Encryption Secure against Selective Opening Attack

Mihir Bellare[1], Brent Waters[2], and Scott Yilek[3]

[1] Department of Computer Science and Engineering,
University of California San Diego, La Jolla CA, USA
http://cseweb.ucsd.edu/~mihir/

[2] Department of Computer Science,
University of Texas at Austin, Austin TX, USA
http://www.cs.utexas.edu/~bwaters/

[3] Department of Computer and Information Sciences,
University of St. Thomas, St. Paul, MN, USA
http://personal.stthomas.edu/yile5901/

Abstract. We present the first IBE schemes that are proven secure against selective opening attack (SOA). This means that if an adversary, given a vector of ciphertexts, adaptively corrupts some fraction of the senders, exposing not only their messages *but also their coins*, the privacy of the unopened messages is guaranteed. Achieving security against such attacks is well-known to be challenging and was only recently done in the PKE case. We show that IBE schemes having a property we call 1-sided public openability (1SPO) yield SOA secure IBE schemes and then provide two 1SPO IBE schemes, the first based on the Boyen-Waters anonymous IBE and the second on Waters' dual-system approach.

1 Introduction

Security against selective-opening attack (SOA) is arguably the most paradoxical and vexing open question in the theory of encryption. Recently (and 10 years after the problem was identified), we have seen solutions [2]. These and followups [24,22], however, have been for the case of Public-Key Encryption (PKE). Another domain where the problem arises, and is important for applications, is Identity-Based Encryption (IBE). The techniques used for PKE do not yield solutions here.This paper initiates a treatment of IBE secure under SOA, providing definitions of security and the first schemes achieving them. Our schemes do not use random oracles.

Background. A selective-opening attack on a PKE scheme imagines n senders and receivers. Sender i encrypts a message $\mathbf{m}[i]$ under fresh, random coins $\mathbf{r}[i]$ and the public key $\mathbf{pk}[i]$ of the i-th receiver to get a ciphertext $\mathbf{c}[i]$. An adversary given the vector $\mathbf{c}$ corrupts some subset of the senders and learns not only their messages but also their coins. SOA-security requires that the remaining, unopened messages retain their privacy. SOA-security is required when implementing the

Y. Ishai (Ed.): TCC 2011, LNCS 6597, pp. 235–252, 2011.

assumed secure channels in an adaptively-secure multi-party computation protocol. More pragmatically, it would be required to distribute shares in a distributed file-system that is using secret-sharing for privacy.

IND-CPA and IND-CCA, widely-accepted as the "right" notions of encryption privacy, are not known to imply security under SOA. The difficulty of establishing SOA-security stems from the fact that the adversary gets the coins and also that the messages $\mathbf{m}[1], \ldots, \mathbf{m}[n]$ may be related. Constructions of SOA secure schemes also remained elusive, the area colored by negative results for commitment schemes [21,2,28]. Finally, Bellare, Hofheinz, and Yilek (BHY) [2] showed a large class of encryption schemes, which they call *lossy* [2,25,30], are SOA secure. Schemes they show to be lossy include variants of El Gamal [27], the IND-CPA scheme built from lossy trapdoor functions by Peikert and Waters [31], and even the original Goldwasser-Micali encryption scheme [23]. Hemenway, Libert, Ostrovsky and Vergnaud [24] showed that re-randomizable encryption and statistically hiding, two-round oblivious transfer imply lossy encryption, yielding still more examples of SOA secure PKE schemes via the lossy-implies-SOA-secure connection of BHY. Fehr, Hofheinz, Kiltz, and Wee (FHKW) [22] use a deniable encryption [13] approach to achieve CC-SOA (Chosen-Ciphertext SOA) secure PKE.

SOA FOR IBE. We can adapt the SOA framework to IBE in a natural way. A vector $\mathbf{id}$ of adversarially-chosen target receiver identities replaces the vector $\mathbf{pk}$ of public receiver keys. Sender i encrypts message $\mathbf{m}[i]$ under coins $\mathbf{r}[i]$ for identity $\mathbf{id}[i]$ to get a ciphertext $\mathbf{c}[i]$. As before the adversary, given $\mathbf{c}$, corrupts a subset of the senders and learns their messages and coins, and SOA-security requires that the unopened messages are secure. At any time, the adversary can query **Extract** with any identity not in the vector $\mathbf{id}$ and obtain its decryption key.

There are two elements here, new compared to PKE, that will be central to the technical challenges in achieving the goal. The first is the **Extract** oracle, a feature of IBE security formalizations since the pioneering work of Boneh and Franklin [9], that allows the adversary to obtain the decryption key of any (non-target) receiver of its choice. The second is that the target identities are chosen by the adversary. (We will achieve full, rather than selective-id security [15].)

IBE can conveniently replace PKE in applications such as those mentioned above, making its SOA-security important. Beyond this, we feel that determining whether SOA-secure IBE is possible is a question of both foundational and technical interest.

CONTRIBUTIONS IN BRIEF. We provide a simulation-based, semantic security formalization of SOA-secure IBE. (This means our results do not need to assume conditional re-samplability of message spaces, in contrast to some of the results of [2] for IND-style notions.) We provide a general paradigm to achieve SOA-secure IBE based on IBE schemes that are IND-CPA and have a property we call 1-Sided Public Openability (1SPO). We discuss why obtaining 1SPO IND-CPA IBE schemes without random oracles is not immediate and then illustrate two ways to do it.

Scheme	Pars	Ctxt	Keys	Enc	Dec	F/S	Assumption
LoR	$n+6$	5	5	5 exp	5 pr	F	DLIN
BBoR	4	2	2	2 exp	2 pr	F	GSD

Fig. 1. Our 1SPO IND-CPA IBE schemes. These encrypt 1-bit messages. Bit-by-bit encryption yields SOA-secure IBE schemes encrypting full messages. "Pars" is the size of the public parameters, "Ctxt" of the ciphertext and "Keys" of the decryption keys, all in group elements, with n the length of identities. (In practice $n = 160$ by hashing identities.) "Enc" and "Dec" are the encryption and decryption costs with "exp" standing for an exponentiation or multi-exponentiation and "pr" for a pairing. "F/S" indicates whether we get Full or Selective-id security. "GSD" stands for the General Subgroup Decision assumption.

The first, adapting the anonymous IBE scheme of Boyen and Waters [12], yields a SOA-secure IBE scheme based on the DLIN (Decision Linear) assumption of [7]. The second, using the dual-system approach of [32], yields a SOA-secure IBE scheme in the Boneh-Boyen style [6] based on a subgroup decision assumption in composite order groups. Attributes of the schemes are summarized in Figure 1. We now expand on these contributions.

1SPO IBE IMPLIES SOA-SECURE IBE. There are fundamental obstacles to extending BHY's lossy-implies-SOA-secure approach, that worked for SOA-secure PKE, to the IBE setting. (Briefly, we cannot make the encryption undetectably lossy on all challenge identities because the adversary has an **Extract** oracle and we wish to achieve full, not selective-id [15] security.) Instead we return to ideas from non-committing [14] and deniable [13] encryption. We define IBE schemes that have a property we call *one-sided public openability* (1SPO) and is an IBE-analogue of a weak form of deniable PKE [13]. In short, an IBE scheme for 1-bit messages is 1SPO if it is possible, given the public parameters par, an identity id, and the encryption c of message 1 under par and id, to efficiently open the encryption, meaning find correctly-distributed randomness r such that encrypting a 1 using $\mathsf{par}, \mathsf{id}, r$ results in the ciphertext c. We emphasize that this opening must be done without the aid of any secret information. Bit-by-bit encryption then results in a scheme that can encrypt long messages. We show in Theorem 1 that if the starting 1-bit 1SPO scheme is also IND-CPA secure then the constructed IBE scheme is SOA-secure. This reduces the task of obtaining SOA-secure IBE schemes to obtaining IND-CPA secure 1SPO schemes.

FHKY [22] develop a similar approach in the PKE setting. Their work and ours are concurrent and independent. (Both were submitted to Eurocrypt 2010 but only theirs was accepted.)

FINDING 1SPO IBE SCHEMES. Known Random Oracle (RO) model IBE schemes [9,19] can be adapted to be 1SPO secure, yielding SOA-secure IBE in the RO model. Achieving it without ROs, however, turns out not to be straightforward. The natural approach, extending that used for PKE [13,22], is to build IBE schemes that are what we call 1SIS (1-sided invertibly samplable). Here, encryptions of 0 to a certain identity would have a certain structure. This structure should be detectable with the secret key associated to the identity, but *not* without it, and thus not by an attacker. On the other hand, encryptions of 1 would be random, but in a special way, namely there is a public procedure that given an encryption c of a 1 can compute randomness (coins) under which the encryption algorithm applied to 1 would produce c. Any such scheme is 1SPO. The challenge that emerges is to find 1SIS IND-CPA IBE schemes. Existing IBE schemes do not have the property, and nor do direct adaptations work. The Boneh-Boyen approach [6] is probably the most widely used in IBE design. (Waters' IBE scheme [33] is one instance.) However, ciphertexts in BB-schemes contain group elements that obey relations an attacker can test and thus cannot be undetectably replaced with random group elements. We will obtain our first solution by a different approach. Then, however, we will go back to show how the dual-system approach can be used to make a BB-style scheme work if we use composite order groups.

The linear scheme. In our "Linear or Random" (LoR) scheme, an encryption of 0 to a given identity, id, is done using (a modification of) the Boyen-Waters (BW) encryption algorithm [12]. This output of the encryption will be five group elements that share a certain structure that is only detectable to a user with the private key for id. To encrypt a 1 we simply choose five random group elements. This, however, must be done using what we call a publicly invertible process (see below). The main feature of this encryption scheme is that an encryption of 0 can always be claimed as just five random group elements, and thus as an encryption of 1. This reveals why we choose to build of the BW anonymous IBE scheme as opposed to other simpler IBE systems without random oracles. The main feature of the BW ciphertexts is that they have no detectable structure from an attacker that does not have a private key for id. In contrast, in BB-style IBE systems [6,33] the attacker can test for structure between two group elements in well formed ciphertexts. Therefore we cannot create a secure encryption system simply by replacing these with random group elements.

We prove LoR is 1SPO directly. We must also, however, prove it is IND-CPA. We adapt techniques from [12,3] to do this under the DLIN assumption. The proof technique of [3] allows us to avoid Waters' artificial abort step [33] thereby resulting in a more efficient reduction.

The dual-system scheme. Waters introduced a new approach to IBE called the dual system approach in which both the challenge ciphertext and keys are replaced in the proof by "semi-functional" versions [32]. We adapt this approach to get a 1SIS (and thus 1SPO) IND-CPA IBE scheme and thus a SOA-secure IBE scheme. An interesting feature of the scheme is that ciphertexts have a BB-form, showing that the dual-system approach can surmount the above-mentioned difficulties in making BB-style systems 1SIS. We accordingly call the scheme BBoR (BB or Random). In addition this is interesting because it illustrates a quite different technique and yields a scheme based on a different assumption (subgroup decision in a composite group, not known to imply or be implied by DLIN in a prime-order group). As Figure 1 shows, the main pragmatic difference compared to LoR is short public parameters. (Those of LoR are long due to the Waters' hash function [33] which is required to get full security.) Others costs have dropped as well (from 5 to 2) but the group is larger so a closer analysis would be needed to determine whether this translates to actual efficiency gains.

Our starting points are the dual-system based Lewko-Waters (LW) IBE scheme [26] and its anonymous extension by De Caro, Iovino and Persiano (DIP) [17]. We modify these to get a 1SIS scheme where an encryption of a 0 is BB-ciphertext but in a subgroup while an encryption of a 1 is a pair of random points in the full group. While extending these schemes we manage simultaneously to make the assumptions simpler, more natural and fewer. Specifically, all these schemes rely on a pairing $\mathrm{e}\colon \mathbb{G} \times \mathbb{G} \to \mathbb{G}_T$ where $\mathbb{G}, \mathbb{G}_T$ have composite order N. LW make three different assumptions (numbered 1,2,3), the first two being about subgroup decision in $\mathbb{G}$ and the third in $\mathbb{G}_T$. DIP also make three assumptions, with the third being quite ad hoc and tailored to the scheme. We eliminate the third assumption in both cases and unify the rest, formulating what we call the general subgroup decision assumption, which is only in $\mathbb{G}$, and basing the proof solely on this single assumption.

PUBLICLY INVERTIBLE SAMPLING. We have said that encryptions of a 1 in our 1SIS schemes are random group elements. This, however, is not enough. They have to be invertibly sampled. As we explained above, this means there is a public procedure that given an encryption c of a 1 can compute randomness (coins) under which the encryption algorithm applied to 1 would produce c. To illustrate the subtleties of the notion, consider a scheme in which the encryption c of a 1 is computed by picking an exponent x at random and returning g^x where g is a generator of a group $\mathbb{G}$. Although the ciphertext is random, this is not invertibly samplable since we cannot recover x from c. Instead, a ciphertext must be sampled "directly" as $c \leftarrow_{\$} \mathbb{G}$. The difficulty is that whether or not this is possible depends on the group. In the PKE case, it is possible to stay within simple groups such as $\mathbb{Z}_p^*$ for prime p, where such sampling is easy. (Pick a a random integer in the range $1, \ldots, p-1$.) In our case, however, $\mathbb{G}$ is a complex group, namely a subgroup of the points on an elliptic curve. We show how to sample invertibly nonetheless, relying on the structure of the elliptic curve groups in question. Specifically, we modify some methods used to implement the hash function of the BLS signature scheme [11].

EXTENSIONS AND OPEN PROBLEMS. After seeing a preliminary version of our work, Peikert [29] has said that the lattice-based IBE schemes of [18,1] adapt to yield 1SPO schemes, whence, by our results, SOA-secure IBE. One interesting direction for further work is to define SOA-secure HIBE and then extend our schemes (the second in particular) as well as the lattice ones to achieve it. Another direction is to define deniable IBE and then fuse our approaches with those of [13] to achieve it. We remark that even given SOA-secure HIBE, we do not directly achieve CC-SOA (Chosen-ciphertext SOA) secure IBE because the BCHK transform [8] does not work in the SOA setting. (The problem is opening the randomness used in creating the one-time-signature key.) Achieving CC-SOA secure IBE is another interesting open question.

RELATED WORK. Canetti, Feige, Goldreich and Naor [14] introduced non committing encryption (NCE) to achieve adaptively secure multi-party computation in the computational (as opposed to secure channels) setting without erasures. In their treatment, NCE is an interactive protocol, and their definition of security is in the MPC framework. The model allows corruption of both senders and receivers. They show how to achieve NCE but, viewed as a public-key system, they would have keys larger than the total number of message bits that may be securely encrypted. Damgård and Nielsen [20] introduced more efficient schemes but this restriction remained, and Nielsen [28] showed it was necessary. With partial erasures, more efficient solutions were provided by Canetti, Halevi and Katz [16].

Dwork, Naor, Reingold and Stockmeyer [21] extracted out a stand-alone notion of commitment secure against selective opening defined directly by a game rather than via the MPC framework. Corruptions allow the adversary to obtain the committer's coins along with its message. This was adapted to public-key encryption in [2], who focused on sender (as opposed to receiver) corruptions and were then able to obtain solutions based on lossy encryption.

Canetti, Dwork, Naor and Ostrovsky [13] introduced deniable encryption, where a sender may open a ciphertext to an arbitrary message by providing coins produced by a faking algorithm. The authors explain that this is stronger than NCE because in the latter only a simulator can open in this way. A weak form of their requirement is that encryptions of 1 can be opened as encryptions of 0 even if not vice versa. 1SPO IBE is an IBE analogue of this notion.

SENDER VERSUS RECEIVER CORRUPTIONS. We clarify that our model and results are for adaptive sender corruptions, not adaptive receiver corruptions. (The latter would correspond to being allowed to query to **Extract** identities in the challenge vector **id**.) Security against adaptive receiver corruptions seems out of reach of current techniques for PKE let alone for IBE. (Without either erasures or keys as long as the total number of messages bits ever encrypted.) We do allow receiver corruptions via the **Extract** oracle but these are non-adaptive. We view this as retaining (meaning neither weakening nor strengthening) the guarantees against receiver corruption already provided by the basic definition of IND-CPA-secure IBE [9]. Our notion, security against adaptive sender and non-adaptive receiver corruptions, is still very strong.

2 Preliminaries

NOTATION. We use boldface to denote vectors, i.e., $\mathbf{m}$. For vector $\mathbf{m}$, we let $|\mathbf{m}|$ denote the number of components in the vector. When $\mathbf{m}[i] \in \{0,1\}^*$, we denote by $\mathbf{m}[i][j]$ the jth bit of the ith component of $\mathbf{m}$, i.e., the jth bit of $\mathbf{m}[i]$. On the other hand, when $\mathbf{c}[i]$ is a sequence, we let $\mathbf{c}[i][j]$ denote the jth value in the sequence $\mathbf{c}[i]$. We sometimes abuse notation and treat vectors as sets. Specifically, if S is a set we may write $S \cup \mathbf{m}$ to denote $S \cup \{\mathbf{m}[1]\} \cup \{\mathbf{m}[2]\} \ldots$. If two adversaries $\mathcal{A}$ and $\mathcal{B}$ have access to different oracles with the same name (e.g., **NewMesg**) we sometimes write $\mathbf{NewMesg}_{\mathcal{B}}$ to mean $\mathcal{B}$'s version of the oracle. For $n \in \mathbb{N}$ we denote by $[n]$ the set $\{1, \ldots, n\}$.

We fix pairing parameters $\mathsf{GP} = (\mathbb{G}, \mathbb{G}_T, p, \mathrm{e})$ where $\mathbb{G}, \mathbb{G}_T$ are groups of order prime p and the map $\mathrm{e} : \mathbb{G} \times \mathbb{G} \to \mathbb{G}_T$ is an efficiently computable non-degenerate bilinear map. We let $\mathsf{T}_{\mathrm{exp}}(\mathbb{G})$ be the time to compute an exponentiation in the group $\mathbb{G}$. We let $\mathsf{T}_{\mathrm{op}}(\mathbb{G})$ be the time to compute a group operation in $\mathbb{G}$. For any group $\mathbb{G}$, let $\mathbb{G}^*$ denote the generators of $\mathbb{G}$.

CODE-BASED GAMES. We use code based games [4] for our security definitions. A game consists of numerous procedures including an **Initialize** procedure and a **Finalize** procedure. When an adversary $\mathcal{A}$ executes with the game, the **Initialize** procedure is executed first and its outputs are the initial inputs to adversary $\mathcal{A}$. Then $\mathcal{A}$ executes and its oracle queries are answered by the corresponding procedures of the game. When the adversary halts with some final output, this output is given as input to the **Finalize** procedure. The output of the **Finalize** procedure is then considered the output of the game. We let $G^{\mathcal{A}} \Rightarrow y$ be the event that game G, when executed with adversary $\mathcal{A}$, has output y. We abbreviate "$G^{\mathcal{A}} \Rightarrow \mathsf{true}$" by "$G^{\mathcal{A}}$". The running time of the adversary while playing the game is considered to be the running time of the adversary while playing the game plus the time to execute all of the game procedures during the execution.

RANDOMIZED ALGORITHMS AND SAMPLING FROM GROUPS. We have to model randomized algorithms carefully and in a particular way to define invertible sampling. We assume that all algorithms have access to a RNG Rand that is the only source of randomness in the system. On input a positive integer n, function Rand returns a value uniformly distributed in $\mathbb{Z}_n$. We stress that Rand is not viewed as having an underlying source of coins in the form of bits as in complexity-theoretic/Turing machine models. Rather, its operation is atomic and its outputs *are* the coins.

When we write $a \leftarrow_\$ \mathbb{G}$ we mean that we run $i \leftarrow_\$ \mathsf{Rand}(p)$, where $p = |\mathbb{G}|$, and let $a = g^i$ where g is a generator of $\mathbb{G}$. However, we also want to use publicly reversible sampling. A publicly reversible (PR) sampler Samp takes no input and, via access to Rand, outputs a point in $\mathbb{G}$ or the failure symbol $\perp$. It has sampling failure probability ζ if the probability that it outputs $\perp$ is at most ζ. We require that $\Pr[a' = a \mid a' \neq \perp] = 1/|\mathbb{G}|$ for all $a \in \mathbb{G}$, where the probability is over $a' \leftarrow_\$ \mathsf{Samp}$.

If $(r_1, \ldots, r_s)$ is a sequence of non-negative integers, we let $\mathsf{Samp}[r_1, \ldots, r_s]$ be the result of running Samp with Rand replaced by the subroutine that returns r_i in response to the i-th query made to it, for $1 \le i \le s$. We require that there is an algorithm Samp^{-1} which on input $a \in \mathbb{G}$ outputs a sequence $(r_1, \ldots, r_s)$ such that $\mathsf{Samp}[r_1, \ldots, r_s] = a$. ($\mathsf{Samp}^{-1}$, as with any other algorithm, has access to

Rand.) Samp^{-1} also might fail (and output $\perp$). We call this the reverse sampling failure probability and denote it with θ.

IDENTITY-BASED ENCRYPTION. An Identity-based encryption scheme (IBE) is a tuple of algorithms $\Pi = (\mathsf{Pg}, \mathsf{Kg}, \mathsf{Enc}, \mathsf{Dec})$ with identity space IdSp, message space MsgSp, and the following properties. The parameter generation algorithm Pg takes no input and outputs a public parameter string par and a master secret key msk. The identity key generation algorithm Kg takes as input the public parameter string par, the master secret key msk, and an identity id, and outputs a secret key sk for identity id. The encryption algorithm Enc takes as input the public parameters par, an identity id, and a message M, and outputs a ciphertext C. Lastly, the decryption algorithm Dec takes as input the public parameters par, an identity secret key sk, and a ciphertext C, and outputs either a message M or a failure symbol $\perp$. We say that an IBE scheme has completeness error ϵ if the probability that $\mathsf{Dec}(\mathsf{par}, sk, \mathsf{id}, \mathsf{Enc}(\mathsf{par}, \mathsf{id}, M)) = M$ is $\geq 1 - \epsilon$ for all $\mathsf{id} \in \mathsf{IdSp}$, all $M \in \mathsf{MsgSp}$, all $(\mathsf{par}, \mathsf{msk}) \in [\mathsf{Pg}]$, and all $sk \in [\mathsf{Kg}(\mathsf{par}, \mathsf{msk}, \mathsf{id})]$, where the probability is taken over the coins used in encryption.

A one-bit IBE scheme $\Pi = (\mathsf{Pg}, \mathsf{Kg}, \mathsf{Enc}, \mathsf{Dec})$ is one with $\mathsf{MsgSp} = \{0, 1\}$, while an ℓ-bit IBE scheme has $\mathsf{MsgSp} = \{0, 1\}^\ell$. We will build ℓ-bit IBE schemes from one-bit IBE schemes as follows. Given one-bit IBE scheme Π as above, let $\Pi^\ell = (\mathsf{Pg}^\ell, \mathsf{Kg}^\ell, \mathsf{Enc}^\ell, \mathsf{Dec}^\ell)$ be an ℓ-bit IBE scheme defined as follows: parameter and key generation are unchanged, i.e., $\mathsf{Pg}^\ell = \mathsf{Pg}$ and $\mathsf{Kg}^\ell = \mathsf{Kg}$. The encryption algorithm Enc^ℓ, on input par, id, $M \in \{0, 1\}^\ell$, outputs $\mathsf{Enc}(\mathsf{par}, \mathsf{id}, M[1]) \parallel \ldots \parallel \mathsf{Enc}(\mathsf{par}, \mathsf{id}, M[\ell])$, where $M[i]$ is the ith bit of M. In other words, encryption encrypts each bit separately and concatenates the resulting ciphertexts. Decryption works in the obvious way: decrypt each ciphertext component separately to learn individual bits. It is easy to see that if Π has ϵ completeness error, then the resulting ℓ-bit scheme has completeness error at most $\ell \cdot \epsilon$.

The standard notion of security for IBE schemes is indistinguishability under chosen plaintext attack (IND-CPA) [9]. We define the IND-CPA advantage of an IND-CPA adversary A against IBE scheme Π to be $\mathbf{Adv}_{\Pi}^{\text{ind-cpa}}(A) = 2 \cdot \Pr\left[\, \text{INDCPA}_{\Pi}^{A} \Rightarrow \mathsf{true} \,\right] - 1$, where game INDCPA can be found in Figure 2. An IND-CPA adversary interacts with game INDCPA, querying **LR** only once and on an identity $\mathsf{id}^* \in \mathsf{IdSp}$ that is never queried to **Extract** and on equal length messages $M_0, M_1 \in \mathsf{MsgSp}$. We note that adversaries may query the same identity id to **Extract** multiple times, since key generation is randomized.

We associate to encryption algorithm Enc the set $\mathsf{Coins}(\mathsf{par}, m)$. This is the set from which Enc draws its coins when encrypting message m using parameters par. Similarly, we let $\mathsf{Coins}(\mathsf{par}, \mathsf{id}, c, 1)$ be the set of coins $\{r \mid c = \mathsf{Enc}(\mathsf{par}, \mathsf{id}, 1; r)\}$.

3 Security against Selective Opening Attacks

In this section we formalize SOA security for IBE, closely following the formalizations from [2]. Before proceeding, we need two definitions. A (k, ℓ)-message sampler is a randomized algorithm $\mathcal{M}$ that on input string $\alpha \in \{0, 1\}^*$ outputs

proc. Initialize:	**proc. LR**(id, M_0, M_1): IND-CPA$_\Pi$
(par, msk) $\leftarrow_\$$ Pg ; $b \leftarrow_\$ \{0,1\}$	Return Enc(par, id, M_b)
Return par	**proc. Finalize**(b'):
proc. Extract(id):	Return ($b = b'$)
Return Kg(par, msk, id)	

Fig. 2. The IBE IND-CPA Game

a vector of messages $\mathbf{m}$ such that $|\mathbf{m}| = k$ and each $\mathbf{m}[i] \in \{0,1\}^\ell$. A relation $\mathcal{R}$ is any randomized algorithm that outputs a single bit.

An soa-adversary is one that runs with game REAL making one query to **NewMesg** before making one query to **Corrupt**; it may make one or more queries to **Extract** at any time during the game. An soa-simulator is an adversary that runs with game SIM, makes one query to **NewMesg** and later makes one query to **Corrupt**. It makes no **Extract** queries. We define the soa-advantage of soa-adversary $\mathcal{A}$ against an IBE scheme Π with respect to a (k,ℓ)-message sampler $\mathcal{M}$, relation $\mathcal{R}$, and soa-simulator $\mathcal{S}$ as

$$\mathbf{Adv}^{\mathrm{soa}}_{\Pi,k,\mathcal{S},\mathcal{M},\mathcal{R}}(\mathcal{A}) = \Pr\left[\,\mathrm{REAL}^{\mathcal{A}}_{\Pi,k,\mathcal{M},\mathcal{R}} \Rightarrow 1\,\right] - \Pr\left[\,\mathrm{SIM}^{\mathcal{S}}_{\Pi,k,\mathcal{M},\mathcal{R}} \Rightarrow 1\,\right] .$$

Discussion. In game REAL (shown in Figure 3), the **Initialize** procedure runs the parameter generation algorithm and returns the scheme parameters to the adversary. The adversary then runs with oracles **NewMesg**, **Corrupt**, and **Extract**. The adversary may never query an identity to **Extract** that appears in a query to **NewMesg**.

The adversary may query the **NewMesg** oracle once with a vector of identities **id** and a string α that is meant to capture state to pass on to the message sampler. Procedure **NewMesg**, on input **id** and α, samples a vector of messages from the message sampling algorithm $\mathcal{M}$ and encrypts the entire vector using independent coins to the identities specified in **id**. This means that the ith component of the resulting ciphertext vector $\mathbf{c}$ is $\mathsf{Enc}(\mathsf{par}, \mathbf{id}[i], \mathbf{m}[i]; \mathbf{r}[i])$, the encryption of the ith message to the ith identity with the ith coins.

After querying the **NewMesg** oracle, the adversary may make one query to **Corrupt** with a set of indices $I \subseteq [k]$. These indices specify which ciphertexts from the vector $\mathbf{c}$ returned by **NewMesg** the adversary would like opened. The **Corrupt** procedure returns the messages and randomness used in **NewMesg** corresponding to indices in I. Additionally, at any time the adversary may query the **Extract** oracle on an identity of its choice and learn a secret key for that identity. We do not allow the adversary to query **Extract** on any identity appearing in the vector id queried to **NewMesg**.

Finally, the adversary halts with output *out* and the output of the game is the relation $\mathcal{R}$ applied to the message vector $\mathbf{m}$, the set of challenge IDs ChID, the corrupt set I, and the output *out*.

In game SIM (shown in Figure 4), the **Initialize** procedure does nothing and returns $\perp$ to the simulator. The simulator then runs with two oracles, **NewMesg**

proc. Initialize:
$(\mathsf{par}, \mathsf{msk}) \leftarrow_{\$} \mathsf{Pg}$
Return par

proc. NewMesg$(\mathbf{id}, \alpha)$:
If $\mathbf{id} \cap \mathrm{ExID} \neq \emptyset$ then return $\bot$
$\mathrm{ChID} \leftarrow \mathrm{ChID} \cup \mathbf{id}$; $\mathbf{m} \leftarrow_{\$} \mathcal{M}(\alpha)$
For i in 1 to k
  $\mathbf{r}[i] \leftarrow_{\$} \mathsf{Coins}(\mathsf{par}, \mathbf{m}[i])$
  $\mathbf{c}[i] \leftarrow \mathsf{Enc}(\mathsf{par}, \mathbf{id}[i], \mathbf{m}[i]; \mathbf{r}[i])$
Return $\mathbf{c}$

proc. Extract(id):
If $\mathsf{id} \in \mathrm{ChID}$ then return $\bot$
$\mathrm{ExID} \leftarrow \mathrm{ExID} \cup \{\mathsf{id}\}$
$sk \leftarrow_{\$} \mathsf{Kg}(\mathsf{par}, \mathsf{msk}, \mathsf{id})$
Return sk

proc. Corrupt(I):
Return $\mathbf{r}[I]$, $\mathbf{m}[I]$

proc. Finalize(out):
Return $\mathcal{R}(\mathbf{m}, \mathrm{ChID}, I, out)$

Fig. 3. Game $\mathrm{REAL}_{\Pi,k,\mathcal{M},\mathcal{R}}$

proc. Initialize:
Return $\bot$

proc. NewMesg$(\mathbf{id}, \alpha)$:
$\mathrm{ChID} \leftarrow \mathrm{ChID} \cup \mathbf{id}$; $\mathbf{m} \leftarrow_{\$} \mathcal{M}(\alpha)$
Return $\bot$

proc. Corrupt(I):
Return $\mathbf{m}[I]$

proc. Finalize(out):
Return $\mathcal{R}(\mathbf{m}, \mathrm{ChID}, I, out)$

Fig. 4. Game $\mathrm{SIM}_{\Pi,k,\mathcal{M},\mathcal{R}}$

and **Corrupt**. On input an identity vector **id** and a string α, oracle **NewMesg** samples a vector **m** of messages using the message sampling algorithm $\mathcal{M}$ applied to the state string α. Nothing is returned to the simulator. The simulator is only allowed one **NewMesg** query. At a later time, the simulator may then make a single query to oracle **Corrupt** with a set of indices I and as a result will learn the messages in **m** corresponding to I. Finally, the simulator halts with output out and the output of the game is the relation $\mathcal{R}$ applied to the message vector **m**, the set of challenge IDs ChID, the corrupt set I, and the output out.

As noted at the end of Section 1, we model adaptive sender corruptions while retaining standard IBE security against non-adaptive receiver corruptions. (Adaptive receiver corruptions would correspond to removing the restriction that **Extract** return $\bot$ when queried on a challenge identity.) Security against adaptive receiver corruptions seems out of reach of current techniques for PKE let alone IBE.

4 SOA Secure IBE from 1SPO IBE

A perfect one-sided public (1SP) opener for one-bit IBE scheme $\Pi = (\mathsf{Pg}, \mathsf{Kg}, \mathsf{Enc}, \mathsf{Dec})$ is an algorithm $\mathsf{OpToOne}$ that takes input parameters par, identity id, and ciphertext c, and has the following property: for all $\mathsf{par} \in [\mathsf{Pg}]$, all $\mathsf{id} \in \mathsf{IdSp}$, every $c \in [\mathsf{Enc}(\mathsf{par}, \mathsf{id}, 1)]$, and every $\bar{r} \in \mathsf{Coins}(\mathsf{par}, \mathsf{id}, c, 1)$,

$$\Pr\left[\, r \leftarrow_{\$} \mathsf{OpToOne}(\mathsf{par}, \mathsf{id}, c) \;:\; r = \bar{r} \,\right] = \frac{1}{|\mathsf{Coins}(\mathsf{par}, \mathsf{id}, c, 1)|} \,.$$

We can weaken this definition slightly by considering opening algorithms that can fail with some probability δ, but in the case of success their output distribution is identical to the actual coin distribution. This is reflected as for all $\mathsf{par} \in [\mathsf{Pg}]$, all $\mathsf{id} \in \mathsf{IdSp}$, every $c \in [\mathsf{Enc}(\mathsf{par}, \mathsf{id}, 1)]$, and every $\bar{r} \in \mathsf{Coins}(\mathsf{par}, \mathsf{id}, c, 1)$,

$$\Pr\left[\, r \leftarrow_{\$} \mathsf{OpToOne}(\mathsf{par}, \mathsf{id}, c) \;:\; r = \bar{r} \mid r \neq \bot \,\right] = \frac{1}{|\mathsf{Coins}(\mathsf{par}, \mathsf{id}, c, 1)|} \,.$$

Notice that the probability is only over the coins used by OpToOne. We call such an OpToOne algorithm a δ-1SP opener and we also call an IBE scheme with a δ-1SP opener δ-one-sided publicly openable (δ-1SPO).

The idea of constructing encryption schemes with such one-sided opening originates with Canetti, Dwork, Naor, and Ostrovsky [13], who used PKE schemes with this property to build deniable PKE schemes. From translucent sets they get PKE schemes where an encryption of a 1 is pseudorandom while an encryption of a 0 is random. It is then possible to claim the encryption of a 1 was random and thus open it to a 0. More recently, in independent work, Fehr, Hofheinz, Kiltz, and Wee [22] used PKE schemes with a 1SPO property as a building block to achieve CC-SOA public-key encryption security. Of course, both of these works focus on PKE while we focus on IBE.

From 1SPO to SOA. We now state our main result: IND-CPA 1SPO one-bit IBE schemes lead to many-bit SOA-secure IBE schemes. Let $\Pi = (\mathsf{Pg}, \mathsf{Kg}, \mathsf{Enc}, \mathsf{Dec})$ denote a one-bit IBE scheme that is δ-one sided openable and let $\Pi^\ell = (\mathsf{Pg}^\ell, \mathsf{Kg}^\ell, \mathsf{Enc}^\ell, \mathsf{Dec}^\ell)$ the ℓ-bit IBE scheme built from Π as described in Section 2.

Theorem 1. *Let Π be a one-bit IBE scheme with a δ one-sided opener* OpToOne, *and let Π^ℓ be the ℓ-bit scheme built from it. Let k be an integer, $\mathcal{A}$ an soa-adversary making at most q queries to **Extract**, $\mathcal{R}$ a relation, and $\mathcal{M}$ a (k, ℓ)-message sampler. Then there exists an soa-simulator $\mathcal{S}$ and an IND-CPA adversary $\mathcal{B}$ such that*

$$\mathbf{Adv}^{\mathrm{soa}}_{\Pi^\ell, k, \mathcal{M}, \mathcal{R}, \mathcal{S}}(\mathcal{A}) \leq k\ell \cdot \mathbf{Adv}^{\mathrm{ind\text{-}cpa}}_{\Pi}(\mathcal{B}) + k\ell \cdot \delta \,,$$

where $\mathsf{T}(\mathcal{S}) = \mathcal{O}(\mathsf{T}(\mathcal{A}) + k\ell \cdot \mathsf{T}(\mathsf{OpToOne}) + q \cdot \mathsf{T}(\mathsf{Kg}^\ell) + k \cdot \mathsf{T}(\mathsf{Enc}^\ell) + \mathsf{T}(\mathsf{Pg}^\ell))$ *and* $\mathsf{T}(\mathcal{B}) = \mathcal{O}(\mathsf{T}(\mathcal{A}) + \mathsf{T}(\mathcal{M}) + k\ell \cdot \mathsf{T}(\mathsf{Enc}) + k\ell \cdot \mathsf{T}(\mathsf{OpToOne}) + \mathsf{T}(\mathcal{R}))$. □

The full proof is in [5]. We briefly sketch the ideas here. Simulator $\mathcal{S}$ runs $\mathcal{A}$ and gives it encryptions of all 0s. When $\mathcal{A}$ asks for some of the ciphertexts to be opened, $\mathcal{S}$ queries its own **Corrupt** oracle, learns the messages it needs to open the ciphertexts to, and then opens bit-by-bit. If it needs to open a ciphertext component to a 0, it simply gives $\mathcal{A}$ the coins it used when originally creating the ciphertext. If it needs to open a ciphertext to a 1, it uses the scheme's OpToOne algorithm to find the coins. The simulator then outputs the same output as $\mathcal{A}$. The IND-CPA security of the scheme will allow us to argue that the simulator is successful. We will do a hybrid over the ℓ individual components of the k messages sampled from $\mathcal{M}$, where in the ith hybrid game the first i bits sampled from $\mathcal{M}$ are ignored and 0s are encrypted in their place. Thus, in the first hybrid game all bits sampled from $\mathcal{M}$ are accurately encrypted, while in the last hybrid game only 0s are encrypted. This hybrid causes the loss of a factor $k \cdot \ell$ in the theorem.

A natural question is why not prove Theorem 1 for $\ell = 1$ (meaning, show that any 1-bit 1SPO IBE scheme is SOA-secure) and then prove the general result that if 1-bit Π is SOA-secure then so is Π^{ℓ} for any ℓ? The answer is that we do not know how to prove this general result.

5 A First Attempt

As a first attempt at constructing a 1SPO IBE scheme, we try to adapt techniques from deniable PKE [13], in particular the idea that an encryption of a 0 should be "pseudorandom" while the encryption of a 1 is "random". A typical IBE scheme has ciphertexts consisting of a tuple of group elements with some structure. To make such a scheme 1SPO, a natural idea is to make the honestly-generated ciphertext tuple the encryption of a 0, and a tuple of random group elements an encryption of a 1. The secret key for an identity then contains information which helps test for this structure. Let us see what happens if we apply this idea to the IBE scheme of Boneh and Boyen (BB) [6], which has been the basis for many other IBE schemes including the Waters (W) IBE [33].

Recall in the BB scheme (and its variants) a ciphertext has the form (C_1, $C_2, C_3) = (\mathrm{e}(g_1, g_2)^s \cdot M, g^s, H(\mathbf{u}, \mathsf{id})^s)$, where different variants define the hash function H differently, $g, g_1, g_2, \mathbf{u}$ are part of the parameters, and s is chosen randomly by the encryptor. Now, if we follow the ideas described above for making the scheme 1SPO, encryptions of 0 would be $(C, C') = (g^s, H(\mathbf{u}, \mathsf{id})^s)$, while encryptions of 1 would be a pair of group elements chosen uniformly at random from $\mathbb{G} \times \mathbb{G}$.

Syntactically the scheme works, but it is unfortunately not IND-CPA secure. The reason is that distinguishing an encryption of a 0 from an encryption of a 1 is exactly the DDH problem in $\mathbb{G}$ and hence easy given the pairing. Given a ciphertext (C, C'), we can output 0 if $\mathrm{e}(g, C') = \mathrm{e}(C, H(\mathbf{u}, \mathsf{id}))$ and 1 otherwise. This is (with high probability) a correct decryption.

The fundamental issue is that in the BB scheme, the structure of the ciphertexts can be detected given only public parameters. The key idea in our two schemes is to destroy any structure that is publicly detectable.

6 A 1SPO IBE Scheme Based on the DLIN Assumption

The security of our first scheme will rely on the Decisional Linear Assumption [7]. The decisional linear game DLIN is found in Figure 5. We say the DLIN-advantage of an adversary A against GP is

$$\mathbf{Adv}^{\mathrm{dlin}}_{\mathsf{GP}}(A) = 2 \cdot \Pr\left[\, \mathrm{DLIN}^{A}_{\mathsf{GP}} \Rightarrow \mathsf{true} \,\right] - 1 \, .$$

proc. Initialize:	$\mathrm{DLIN}_{\mathbb{G}}$
$g, g_1, g_2 \leftarrow_{\$} \mathbb{G}^*$; $a_1, a_2 \leftarrow_{\$} \mathbb{Z}_p$; $d \leftarrow_{\$} \{0,1\}$ $h_1 \leftarrow g_1^{a_1}$; $h_2 \leftarrow g_2^{a_2}$ If $d = 1$ then $W = g^{a_1+a_2}$ Else $W \leftarrow_{\$} \mathbb{G}$ Return $(g, g_1, g_2, h_1, h_2, W)$	**proc. Finalize**(d'): Return $(d = d')$

Fig. 5. The DLIN game for the decisional linear assumption

SCHEME DESCRIPTION. Our IBE scheme $\mathsf{LoR} = (\mathsf{Pg}, \mathsf{Kg}, \mathsf{Enc}, \mathsf{Dec})$ is a one-bit version of the anonymous IBE scheme from Boyen and Waters [12] but using the Waters' hash function [33] for adaptive security. The name is short for "Linear or Random" to represent the fact that the encryption of a 0 consists of five group elements whose relationship is similar to that of the group elements in the decisional linear assumption, while an encryption of a 1 consists of five random group elements. The scheme will use a cyclic group $\mathbb{G}$ of prime order p with an efficiently computable pairing $\mathrm{e} : \mathbb{G} \times \mathbb{G} \to \mathbb{G}_T$. We also require the group has a PR sampler Samp with failure probability ζ and corresponding inverse sampler Samp^{-1} with reverse failure probability θ. (The full version [5] gives details on how to instantiate such groups.) Let $\mathbb{G}^*$ denote the generators of $\mathbb{G}$. Let $1_{\mathbb{G}_T}$ be the identity element of the target group $\mathbb{G}_T$. Define hash function $H : \mathbb{G}^{n+1} \times \{0,1\}^n \to \mathbb{G}$ as $H(\mathbf{u}, \mathsf{id}) = \mathbf{u}[0] \prod_{i=1}^{n} \mathbf{u}[i]^{\mathsf{id}[i]}$, where $\mathsf{id}[i]$ is the ith bit of string id. This is the Waters' hash function [33]. The scheme LoR, shown in Figure 6, has message space $\{0,1\}$ and identity space $\{0,1\}^n$. The scheme has completeness error $1/p^2$.

We claim the scheme is δ-1SPO where $\delta \leq 5\theta$. The algorithm $\mathsf{OpToOne}$ simply runs Samp^{-1} on each of the five ciphertext components with independent failure probabilities θ.

The following says LoR is IND-CPA-secure based on DLIN. The proof combines techniques from [12,33,3] and can be found in the full version [5].

Theorem 2. Fix pairing parameters $\mathsf{GP} = (\mathbb{G}, \mathbb{G}_T, p, \mathrm{e})$ and an integer $n \geq 1$, and let $\mathsf{LoR} = (\mathsf{Pg}, \mathsf{Kg}, \mathsf{Enc}, \mathsf{Dec})$ be the one-bit IBE scheme associated to GP and $\mathsf{IdSp} = \{0,1\}^n$. Assume $\mathbb{G}$ is PR-samplable with sampling failure probability ζ. Let $\mathcal{A}$ be an IND-CPA adversary against LoR which has advantage $\epsilon = \mathbf{Adv}^{\text{ind-cpa}}_{\mathsf{LoR}}(\mathcal{A}) > 2^{n+1}/p + 5\zeta$ and makes at most $q \in [1 \,..\, p\epsilon/9n]$ queries to its **Extract** oracle. Let

$$\delta = \frac{1}{2}\left(\frac{\epsilon}{2} - \frac{2^n}{p} - 5\zeta\right) .$$

Then there is a DLIN-adversary $\mathcal{B}$ such that

$$\mathbf{Adv}^{\mathrm{dlin}}_{\mathsf{GP}}(\mathcal{B}) \geq \frac{\delta^2}{9qn + 3\delta} \quad \text{and} \quad \mathsf{T}(\mathcal{B}) = \mathsf{T}(\mathcal{A}) + \mathsf{T}_{\mathrm{sim}}(n, q) \tag{1}$$

where $\mathsf{T}_{\mathrm{sim}}(n,q) = \mathcal{O}(qn + (n+q)\mathsf{T}_{\exp}(\mathbb{G}))$. □

Alg. Pg:

$g \leftarrow_\$ \mathbb{G}^*$; $\mathbf{u} \leftarrow_\$ \mathbb{G}^{n+1}$
$t_1, t_2, t_3, t_4 \leftarrow_\$ \mathbb{Z}_p^*$
$v_1 \leftarrow g^{t_1}$; $v_2 \leftarrow g^{t_2}$
$v_3 \leftarrow g^{t_3}$; $v_4 \leftarrow g^{t_4}$
$\mathsf{par} \leftarrow (g, \mathbf{u}, v_1, v_2, v_3, v_4)$
$\mathsf{msk} \leftarrow (t_1, t_2, t_3, t_4)$
Return $(\mathsf{par}, \mathsf{msk})$

Alg. Kg(par, msk, id):

$(g, \mathbf{u}, v_1, v_2, v_3, v_4) \leftarrow \mathsf{par}$
$(t_1, t_2, t_3, t_4) \leftarrow \mathsf{msk}$
$r_1, r_2 \leftarrow_\$ \mathbb{Z}_p$; $d_0 \leftarrow g^{r_1 t_1 t_2 + r_2 t_3 t_4}$
$d_1 \leftarrow H(\mathbf{u}, \mathsf{id})^{-r_1 t_2}$
$d_2 \leftarrow H(\mathbf{u}, \mathsf{id})^{-r_1 t_1}$
$d_3 \leftarrow H(\mathbf{u}, \mathsf{id})^{-r_2 t_4}$
$d_4 \leftarrow H(\mathbf{u}, \mathsf{id})^{-r_2 t_3}$
Return $(d_0, d_1, d_2, d_3, d_4)$

Alg. Enc(par, id, M):

$(g, \mathbf{u}, v_1, v_2, v_3, v_4) \leftarrow \mathsf{par}$
If $M = 0$ then
  $s, s_1, s_2 \leftarrow_\$ \mathbb{Z}_p$; $C_0 \leftarrow H(\mathbf{u}, \mathsf{id})^s$
  $C_1 \leftarrow v_1^{s - s_1}$; $C_2 \leftarrow v_2^{s_1}$
  $C_3 \leftarrow v_3^{s - s_2}$; $C_4 \leftarrow v_4^{s_2}$
Else
  For $i = 0$ to 4 do $C_i \leftarrow_\$ \mathsf{Samp}_{\mathbb{G}}()$
Return $(C_0, C_1, C_2, C_3, C_4)$

Alg. Dec(par, sk, C):

$(g, \mathbf{u}, v_1, v_2, v_3, v_4) \leftarrow \mathsf{par}$
$(d_0, d_1, d_2, d_3, d_4) \leftarrow sk$
$(C_0, C_1, C_2, C_3, C_4) \leftarrow C$
If $\prod_{i=0}^4 \mathrm{e}(C_i, d_i) = 1_{\mathbb{G}_T}$ then
  Return 0
Else return 1

Fig. 6. Scheme LoR based on Boyen-Waters IBE

7 A Scheme Based on Dual System IBE

General Subgroup Decision. We introduce the general subgroup decision problem and assumption as a generalization of several assumptions in the literature. An order-n group generator with security parameter k is an algorithm Gen that returns a pair $(\pi, \bar{\pi})$, where $\pi = (\langle \mathbb{G} \rangle, \langle \mathbb{G}_T \rangle, \langle \mathrm{e} \rangle, N)$ and $\bar{\pi} = (p_1, \ldots, p_n)$ with $p_1 < \ldots < p_n$ primes; $\mathbb{G}, \mathbb{G}_T$ groups and $\mathrm{e} : \mathbb{G} \times \mathbb{G} \to \mathbb{G}_T$ a non-degenerate bilinear map; $p_i \in \{2^{k-1}, \ldots, 2^k - 1\}$ for $1 \leq i \leq n$; $N = p_1 \cdots p_n = |\mathbb{G}| = |\mathbb{G}_T|$. For $S \subseteq [n]$ we let $\mathbb{G}(S)$ denote the unique subgroup of $\mathbb{G}$ of order $\prod_{i \in S} p_i$. By $\mathbb{H}^*$ we denote the set of generators of a cyclic group $\mathbb{H}$.

The orthogonality property is that if $S_1, S_2 \subseteq [n]$ are disjoint and $g_i \in \mathbb{G}(S_i)$ $(i = 1, 2)$, then $\mathrm{e}(g_1, g_2) = 1_{\mathbb{G}_T}$. Now suppose $S_0, S_1 \subseteq [n]$ and given $T \in \mathbb{G}(S_b)$ we wish to determine $b \in \{0, 1\}$. Orthogonality makes this easy if we possess $g \in \mathbb{G}(S)$ where one of $S \cap S_0, S \cap S_1$ is empty and the other is not. The general subgroup decision assumption is that it is hard without such a g, even if we possess elements of any $G(S)$ for which $S \cap S_0, S \cap S_1$ are both empty or both not empty. Our formalization uses game $\mathrm{GSD}_{\mathsf{Gen}}$ of Figure 7. Adversary $\mathcal{A}$ must make exactly one **Ch** query, consisting of a pair $S_0, S_1 \subseteq [n]$, and this must be its first oracle query. Subsequently it can query **Gen**(S) on any $S \subseteq [n]$ and is allowed multiple queries to this oracle. It terminates by outputting a bit b' and its advantage is

$$\mathbf{Adv}^{\mathrm{gsd}}_{\mathsf{Gen}}(\mathcal{A}) = 2 \cdot \Pr\left[\, \mathrm{GSD}^{\mathcal{A}}_{\mathsf{Gen}} \Rightarrow \mathsf{true} \,\right] - 1 \,.$$

proc. Initialize:
$(\pi, \bar{\pi}) \leftarrow_{\$} \mathsf{Gen}$; $b \leftarrow_{\$} \{0,1\}$
$(\langle \mathbb{G} \rangle, \langle \mathbb{G}_T \rangle, \langle \mathrm{e} \rangle, N) \leftarrow \pi$
$(p_1, \ldots, p_n) \leftarrow \bar{\pi}$
Return π

proc. Ch(S_0, S_1):
If $(S_0 = \emptyset$ or $S_1 = \emptyset)$ then return $\perp$
$T \leftarrow_{\$} \mathbb{G}(S_b)$
Return T

proc. Gen(S):
If $(S \cap S_0 = \emptyset) \wedge (S \cap S_1 \neq \emptyset)$
  then return $\perp$
If $(S \cap S_0 \neq \emptyset) \wedge (S \cap S_1 = \emptyset)$
  then return $\perp$
$g \leftarrow_{\$} \mathbb{G}(S)$
Return g

proc. Finalize(b'):
Return $(b = b')$

Fig. 7. Game $\mathrm{GSD}_{\mathsf{Gen}}$

DISCUSSION. Lewko and Waters [26] make several different subgroup decision assumptions about order-n group generators with $n = 3$, and [17] do the same with $n = 4$. Each of these corresponds to a particular choice of S_0, S_1 queried to **Ch**, and particular queries to **Gen**, in our game. (And hence can be formulated without these oracles. We note these papers also make other assumptions, some pertaining to $\mathbb{G}_T$, that we will not need or consider.) Although the authors make only a few specific assumptions, it is apparent that they would be willing to make any "allowed" one in the family, where "allowed" means that the adversary can get elements of $\mathbb{G}(S)$ only as long as $S \cap S_0, S \cap S_1$ are both empty or both not empty. Our aim in formulating GSD has been to make this more transparent, namely, to make the full family of potential choices explicit, thereby generalizing, unifying and explaining subgroup decisions assumptions from [10,26,17].

GSD may at first glance look like an "interactive" assumption. It isn't. The value n will be a fixed constant, eg. $n = 3$ for [26] and $n = 4$ for us. The GSD assumption is then just a compact way of stating a constant number —one for each subset $\{S_0, S_1\}$ of $2^{[n]}$ with $S_0, S_1 \neq \emptyset$— of non-interactive assumptions. (By non-interactive we mean the game has only **Initialize** and **Finalize** procedures, no oracles.)

We don't really need the full strength of GSD. As in previous works, we only need a few special cases, namely a few particular choices of queries S_0, S_1 to **Ch** and queries S to **Gen**. But we feel that stating GSD better elucidates the source of the assumptions, and it will allow more compact assumption and theorem statements.

SCHEME DESCRIPTION. For our scheme we require a 4 group generator Gen with the property that the group $\mathbb{G}$ described by the first output of Gen has a PR sampler Samp with failure probability ζ and corresponding inverse sampler Samp^{-1} with reverse failure probability θ. (The full version [5] describes how we can instantiate such groups.) The scheme $\mathsf{BBoR} = (\mathsf{Pg}, \mathsf{Kg}, \mathsf{Enc}, \mathsf{Dec})$ associated to a order 4 group generator Gen is shown in Figure 8, where $\mathsf{IdSp} = \mathbb{Z}_{2^{4k-4}}$ and $\mathsf{MsgSp} = \{0,1\}$. (We use identity space $\mathbb{Z}_{2^{4k-4}}$ since N will vary but will always be at least 2^{4k-4}.) If $(C, C') \in [\mathsf{Enc}(\mathsf{par}, \mathsf{id}, 0)]$ and $(K, K') \in [\mathsf{Kg}(\mathsf{par}, \mathsf{msk}, \mathsf{id})]$,

Alg. Pg:

$(\pi, \bar{\pi}) \leftarrow_{\$} \mathsf{Gen}$
$(\langle\mathbb{G}\rangle, \langle\mathbb{G}_T\rangle, \langle \mathrm{e} \rangle, N) \leftarrow \pi$
$(p_1, p_2, p_3, p_4) \leftarrow \bar{\pi}$
$g_1 \leftarrow_{\$} \mathbb{G}(\{1\})^*$
$g_3 \leftarrow_{\$} \mathbb{G}(\{3\})^*$; $g_4 \leftarrow_{\$} \mathbb{G}(\{4\})^*$
$u_1 \leftarrow_{\$} \mathbb{Z}_N$; $U_1 \leftarrow g_1^{u_1}$
$u_4 \leftarrow_{\$} \mathbb{Z}_N$; $U_4 \leftarrow g_4^{u_4}$
$x_1 \leftarrow_{\$} \mathbb{Z}_N$; $X_1 \leftarrow g_1^{x_1}$
$x_4 \leftarrow_{\$} \mathbb{Z}_N$; $X_4 \leftarrow g_4^{x_4}$
$w_4 \leftarrow_{\$} \mathbb{Z}_N$; $W_4 \leftarrow g_4^{w_4}$; $U_{14} \leftarrow U_1 U_4$
$W_{14} \leftarrow g_1 W_4$; $X_{14} \leftarrow X_1 X_4$
$\mathsf{par} \leftarrow (\pi, U_{14}, X_{14}, W_{14}, g_4)$
$\mathsf{msk} \leftarrow (g_1, U_1, X_1, g_3)$

Alg. $\mathsf{Dec}(\mathsf{par}, (K, K'), (C, C'))$:

$(\pi, U_{14}, X_{14}, W_{14}, g_4) \leftarrow \mathsf{par}$
If $\mathrm{e}(C, K) = \mathrm{e}(C', K')$ then return 0
Else return 1

Alg. $\mathsf{Kg}(\mathsf{par}, \mathsf{msk}, \mathsf{id})$:

// $\mathsf{id} \in \mathbb{Z}_{2^{4k-4}} \subseteq Z_N$
$(\pi, U_{14}, X_{14}, W_{14}, g_4) \leftarrow \mathsf{par}$
$(g_1, U_1, X_1, g_3) \leftarrow \mathsf{msk}$
$r, r_3, r_3' \leftarrow_{\$} \mathbb{Z}_N$
$K \leftarrow g_1^r g_3^{r_3}$; $K' \leftarrow (U_1^{\mathsf{id}} X_1)^r g_3^{r_3'}$
Return (K, K')

Alg. $\mathsf{Enc}(\mathsf{par}, \mathsf{id}, M)$:

$(\pi, U_{14}, X_{14}, W_{14}, g_4) \leftarrow \mathsf{par}$
If $M = 0$ then
  $s \leftarrow_{\$} \mathbb{Z}_N$; $t_4, t_4' \leftarrow_{\$} \mathbb{Z}_N$
  $C \leftarrow (U_{14}^{\mathsf{id}} X_{14})^s g_4^{t_4}$
  $C' \leftarrow W_{14}^s g_4^{t_4'}$
Else $C, C' \leftarrow_{\$} \mathsf{Samp}_{\mathbb{G}}()$
Return (C, C')

Fig. 8. Scheme BBoR based on composite order pairing groups

then decryption always succeeds. On the other hand, if $(C, C') \leftarrow_{\$} \mathsf{Enc}(\mathsf{par}, \mathsf{id}, 1)$ and $(K, K') \leftarrow_{\$} \mathsf{Kg}(\mathsf{par}, \mathsf{msk}, \mathsf{id})$ then $\Pr[\mathrm{e}(C, K) = \mathrm{e}(C', K')] \le 8 \cdot 2^{-2k}$ where k is the security parameter associated to Gen.

We claim the scheme is δ-1SPO with $\delta \le 2\theta$. The algorithm OpToOne runs Samp^{-1} on each of the two ciphertext components. Each component will give independent reverse sample failure probability of θ. The IND-CPA security of the scheme is captured by the following theorem, proven in our full version [5].

Theorem 3. *Let* Gen *be an order 4 group generator and let the resulting group* $\mathbb{G}$ *be PR-samplable with sampling failure probability* ζ. *Let* BBoR = (Pg, Kg, Enc, Dec) *the associated IBE scheme defined above. For all adversaries* $\mathcal{A}'$ *making* q ***Extract*** *queries there exists an adversary* $\mathcal{B}$ *such that*

$$\mathbf{Adv}^{\text{ind-cpa}}_{\mathsf{BBoR}}(\mathcal{A}') \le (9 + 2q) \cdot \mathbf{Adv}^{\text{gsd}}_{\mathsf{Gen}}(\mathcal{B}) + 4 \cdot \zeta\,.$$

Adversary $\mathcal{B}$ *makes at most 5 queries to* ***Gen*** *and runs in time at most* $\mathsf{T}(\mathcal{B}) = \mathsf{T}(\mathcal{A}') + \mathcal{O}(q \cdot \mathsf{T}_{\exp}(\mathbb{G}) + q \cdot \mathsf{T}(\gcd))$. □

Acknowledgements

Bellare is supported in part by NSF grants CNS-0627779, CCF-0915675 and CCF-0904380. Waters is supported in part by NSF grants CNS-0915361 and CNS-0952692, an AFOSR MURI award for "Collaborative policies and assured information sharing" (Project PRESIDIO), Department of Homeland Security

Grant 2006-CS-001-000001-02 (subaward 641), a Google Faculty Research award, and the Alfred P. Sloan Foundation. Part of Yilek's work was done at UCSD, supported in part by NSF grants CNS-0831536 and CNS-0627779.

References

1. Agrawal, S., Boneh, D., Boyen, X.: Efficient lattice (H)IBE in the standard model. In: Gilbert, H. (ed.) EUROCRYPT 2010. LNCS, vol. 6110, pp. 553–572. Springer, Heidelberg (2010)
2. Bellare, M., Hofheinz, D., Yilek, S.: Possibility and impossibility results for encryption and commitment secure under selective opening. In: Joux, A. (ed.) EUROCRYPT 2009. LNCS, vol. 5479, pp. 1–35. Springer, Heidelberg (2009)
3. Bellare, M., Ristenpart, T.: Simulation without the artificial abort: Simplified proof and improved concrete security for waters' IBE scheme. In: Joux, A. (ed.) EUROCRYPT 2009. LNCS, vol. 5479, pp. 407–424. Springer, Heidelberg (2009)
4. Bellare, M., Rogaway, P.: The security of triple encryption and a framework for code-based game-playing proofs. In: Vaudenay, S. (ed.) EUROCRYPT 2006. LNCS, vol. 4004, pp. 409–426. Springer, Heidelberg (2006)
5. Bellare, M., Waters, B., Yilek, S.: Identity-based encryption secure against selective opening attack. IACR ePrint Archive Report 2010/159
6. Boneh, D., Boyen, X.: Efficient selective-ID secure identity-based encryption without random oracles. In: Cachin, C., Camenisch, J. (eds.) EUROCRYPT 2004. LNCS, vol. 3027, pp. 223–238. Springer, Heidelberg (2004)
7. Boneh, D., Boyen, X., Shacham, H.: Short group signatures. In: Franklin, M. (ed.) CRYPTO 2004. LNCS, vol. 3152, pp. 41–55. Springer, Heidelberg (2004)
8. Boneh, D., Canetti, R., Halevi, S., Katz, J.: Chosen-ciphertext security from identity-based encryption. SIAM Journal on Computing 36(5), 915–942 (2006)
9. Boneh, D., Franklin, M.K.: Identity based encryption from the Weil pairing. SIAM Journal on Computing 32(3), 586–615 (2003)
10. Boneh, D., Goh, E.-J., Nissim, K.: Evaluating 2-DNF formulas on ciphertexts. In: Kilian, J. (ed.) TCC 2005. LNCS, vol. 3378, pp. 325–341. Springer, Heidelberg (2005)
11. Boneh, D., Lynn, B., Shacham, H.: Short signatures from the Weil pairing. Journal of Cryptology 17(4), 297–319 (2004)
12. Boyen, X., Waters, B.: Anonymous hierarchical identity-based encryption (Without random oracles). In: Dwork, C. (ed.) CRYPTO 2006. LNCS, vol. 4117, pp. 290–307. Springer, Heidelberg (2006)
13. Canetti, R., Dwork, C., Naor, M., Ostrovsky, R.: Deniable encryption. In: Kaliski Jr., B.S. (ed.) CRYPTO 1997. LNCS, vol. 1294, pp. 90–104. Springer, Heidelberg (1997)
14. Canetti, R., Feige, U., Goldreich, O., Naor, M.: Adaptively secure multi-party computation. In: 28th ACM STOC, pp. 639–648. ACM Press, New York (May 1996)
15. Canetti, R., Halevi, S., Katz, J.: A forward-secure public-key encryption scheme. In: Biham, E. (ed.) EUROCRYPT 2003. LNCS, vol. 2656, pp. 255–271. Springer, Heidelberg (2003)
16. Canetti, R., Halevi, S., Katz, J.: Adaptively-secure, non-interactive public-key encryption. In: Kilian, J. (ed.) TCC 2005. LNCS, vol. 3378, pp. 150–168. Springer, Heidelberg (2005)

17. Caro, A.D., Iovino, V., Persiano, G.: Fully secure anonymous HIBE with short ciphertexts. IACR ePrint Archive Report 2010/197
18. Cash, D., Hofheinz, D., Kiltz, E., Peikert, C.: Bonsai trees, or how to delegate a lattice basis. In: Gilbert, H. (ed.) EUROCRYPT 2010. LNCS, vol. 6110, pp. 523–552. Springer, Heidelberg (2010)
19. Cocks, C.: An identity based encryption scheme based on quadratic residues. In: Honary, B. (ed.) Cryptography and Coding 2001. LNCS, vol. 2260, pp. 360–363. Springer, Heidelberg (2001)
20. Damgård, I., Nielsen, J.B.: Improved non-committing encryption schemes based on a general complexity assumption. In: Bellare, M. (ed.) CRYPTO 2000. LNCS, vol. 1880, pp. 432–450. Springer, Heidelberg (2000)
21. Dwork, C., Naor, M., Reingold, O., Stockmeyer, L.: Magic functions. Journal of the ACM 50(6), 852–921 (2003)
22. Fehr, S., Hofheinz, D., Kiltz, E., Wee, H.: Encryption schemes secure against chosen-ciphertext selective opening attacks. In: Gilbert, H. (ed.) EUROCRYPT 2010. LNCS, vol. 6110, pp. 381–402. Springer, Heidelberg (2010)
23. Goldwasser, S., Micali, S.: Probabilistic encryption. Journal of Computer and System Sciences 28(2), 270–299 (1984)
24. Hemenway, B., Libert, B., Ostrovsky, R., Vergnaud, D.: Lossy encryption: Constructions from general assumptions and efficient selective opening chosen ciphertext security. IACR ePrint Archive Report 2009/088
25. Kol, G., Naor, M.: Cryptography and game theory: Designing protocols for exchanging information. In: Canetti, R. (ed.) TCC 2008. LNCS, vol. 4948, pp. 320–339. Springer, Heidelberg (2008)
26. Lewko, A.B., Waters, B.: New techniques for dual system encryption and fully secure HIBE with short ciphertexts. In: Micciancio, D. (ed.) TCC 2010. LNCS, vol. 5978, pp. 455–479. Springer, Heidelberg (2010)
27. Naor, M., Pinkas, B.: Efficient oblivious transfer protocols. In: 12th SODA, pp. 448–457. ACM-SIAM (January 2001)
28. Nielsen, J.B.: Separating random oracle proofs from complexity theoretic proofs: The non-committing encryption case. In: Yung, M. (ed.) CRYPTO 2002. LNCS, vol. 2442, pp. 111–126. Springer, Heidelberg (2002)
29. Peikert, C.: Private Communication (May 2010)
30. Peikert, C., Vaikuntanathan, V., Waters, B.: A framework for efficient and composable oblivious transfer. In: Wagner, D. (ed.) CRYPTO 2008. LNCS, vol. 5157, pp. 554–571. Springer, Heidelberg (2008)
31. Peikert, C., Waters, B.: Lossy trapdoor functions and their applications. In: Ladner, R.E., Dwork, C. (eds.) 40th ACM STOC, pp. 187–196. ACM Press, New York (May 2008)
32. Waters, B.: Dual system encryption: Realizing fully secure IBE and HIBE under simple assumptions. In: Halevi, S. (ed.) CRYPTO 2009. LNCS, vol. 5677, pp. 619–636. Springer, Heidelberg (2009)
33. Waters, B.: Efficient identity-based encryption without random oracles. In: Cramer, R. (ed.) EUROCRYPT 2005. LNCS, vol. 3494, pp. 114–127. Springer, Heidelberg (2005)

Functional Encryption: Definitions and Challenges

Dan Boneh[1,⋆], Amit Sahai[2], and Brent Waters[3,⋆⋆]

[1] Stanford University
[2] University of California at Los Angeles
[3] University of Texas at Austin

Abstract. We initiate the formal study of functional encryption by giving precise definitions of the concept and its security. Roughly speaking, functional encryption supports restricted secret keys that enable a key holder to learn a specific function of encrypted data, but learn nothing else about the data. For example, given an encrypted program the secret key may enable the key holder to learn the output of the program on a specific input without learning anything else about the program.

We show that defining security for functional encryption is non-trivial. First, we show that a natural game-based definition is inadequate for some functionalities. We then present a natural simulation-based definition and show that it (provably) cannot be satisfied in the standard model, but can be satisfied in the random oracle model. We show how to map many existing concepts to our formalization of functional encryption and conclude with several interesting open problems in this young area.

1 Introduction

Encryption is a method for a user to securely share data over an insecure network or storage site. Before the advent of public key cryptography, a widely held view was that for two users to communicate data confidentially they would need to a priori establish a mutually held secret key k. While this might be acceptable for some small or tightly knit organizations, such a solution was clearly infeasible for larger networks such as today's Internet consisting of billions of users. Over thirty years ago, Diffie and Hellman [DH76a, DH76b] put forth a radically new idea in the concept of public key cryptography, where two parties can securely communicate with each other *without* having an a prior mutual secret — radically challenging the conventional wisdom of the time.

Today public key encryption is an invaluable tool and its use is ubiquitous in building tools from secure web communication (e.g., SSH, SSL), to disk encryption, and secure software patch distribution. However, there is an ingrained view that: (1) *Encryption is a method to send a message or data to a single entity holding a secret key*, and (2) *Access to the encrypted data is all or nothing – one can either decrypt and read the entire plaintext or one learns nothing at all about the plaintext other than its length.*

⋆ Supported by NSF, MURI, and the Packard foundation.

⋆⋆ Supported by NSF CNS-0716199, CNS-0915361, and CNS-0952692, Air Force Office of Scientific Research (AFO SR) under the MURI award for "Collaborative policies and assured information sharing" (Project PRESIDIO), Department of Homeland Security Grant 2006-CS-001-000001-02 (subaward 641), and the Alfred P. Sloan Foundation.

Y. Ishai (Ed.): TCC 2011, LNCS 6597, pp. 253–273, 2011.

For many emerging applications such as "cloud" services this notion of public-key encryption is insufficient. For example, there is often a need to specify a decryption policy in the ciphertext and only individuals who satisfy the policy can decrypt. More generally, we may want to only give access to a function of the plaintext, depending on the decryptor's authorization. As a concrete example, consider a cloud service storing encrypted images. Law enforcement may require the cloud to search for images containing a particular face. Thus, the cloud needs a restricted secret key that decrypts images that contain the target face, but reveals nothing about other images. More generally, the secret key may only reveal a function of the plaintext image, for example an image that is blurred everywhere except for the target face. Traditional public-key cryptography cannot help with such tasks.

We believe that it is time to adopt a new broad vision of encryption systems. To this end, we explore the concept of *functional encryption*. In a functional encryption system, a decryption key allows a user to learn a *function* of the encrypted data. Briefly, in a functional encryption system for functionality $F(\cdot,\cdot)$ (modeled as a Turing Machine) an authority holding a master secret key can generate a key sk_k that enables the computation of the function $F(k,\cdot)$ on encrypted data. More precisely, using sk_k the decryptor can compute $F(k,x)$ from an encryption of x. Intuitively, the security of the system guarantees that one cannot learn anything more about x, but as we shall see, capturing this rigorously is quite challenging.

We can now see the power of functional encryption. Let us consider what can be achieved if we could realize functional encryption for any polynomial-time Turing Machine $F(\cdot,\cdot)$. In applications of access control, one could let $x = (\mathsf{ind}, m)$ encode a message m as well as an arbitrarily complex access control program ind that will act over the description of a user's credentials. The functionality F would interpret the program ind over k and output the message m if and only if ind accepts on input k. Moreover, the program ind would be hidden and thus one would not necessarily know why decryption was successful or what other keys would satisfy ind. We give many more examples in Section 3.

Our Contributions. Recently, there have been multiple systems that suggest moving beyond the traditional boundaries of encryption. Some examples include Identity-Based Encryption [Sha84, BF03, Coc01], searchable encryption [BCOP04] and Attribute-Based Encryption [SW05]. These and other related works such as [BW07, KSW08] propose specific new systems for problems ranging from expressive access control to searching on encrypted data. In the last few years, the term "functional encryption[1]" was adopted to describe this new area [LOS $^+$10, OT10, AL10].

While these results contain special cases of functional encryption, the general concept has never been formally defined or studied. In this paper we put forth a formal treatment of the subject and discuss many of the remaining challenges. We begin with a general framework and syntax for functional encryption and show how existing

[1] We note that both the term "functional encryption" and its underlying concept were introduced by the authors of this paper. This term was first publicly used to describe the line of work starting with [SW05] in a talk "Functional Encryption: Beyond Public Key Cryptography" [Wat08] in 2008, given by one of the authors of this paper.

encryption concepts, such as attribute based encryption and many others, can be elegantly expressed as particular functionalities of functional encryption.

Defining security of abstract functional encryption turns out to be highly non-trivial. We begin with a natural indistinguishability game-based definition (based on a definition of secure predicate encryption from [BW07, KSW08]). Unfortunately, we show that this simple definition is inadequate for certain functionalities since trivially insecure constructions may satisfy it.

Given the inadequacy of game-based definitions we move to simulation-based definitions in the spirit of the original notion of semantic security of Goldwasser and Micali [GM84]. The goal is to capture the notion that the adversary learns nothing about the plaintext other than functions $F(k,\cdot)$ of the plaintext for which he has a secret key. Somewhat surprisingly, we show a connection to non-committing encryption [CFGN96, Nie02] which proves that our definition cannot be satisfied for the same reason that non-interactive non-committing encryption is impossible. However, we show that our definition can be satisfied in the random oracle model, and we exhibit constructions for interesting functionalities that can be shown to be secure. (Independently, O'Neill [O'N10] also observed a gap between simulation and game-based definitions and a connection to non-committing encryption.)

Functional encryption is still in its infancy and many fascinating open problems remain. We conclude with several directions for future work. The key challenge is the construction of functional encryption for more general functionalities. Another important question is understanding the relative power of functionalities: when does one functionality imply another and when can functionalities be black-box separated?

2 Functional Encryption Syntax

We begin by describing the syntactic definition of functional encryption (FE) for a functionality F. The functionality F describes the functions of a plaintext that can be learned from the ciphertext. More precisely, a functionality is defined as follows.

Definition 1. *A functionality F defined over (K,X) is a function $F: K \times X \to \{0,1\}^*$ described as a (deterministic) Turing Machine. The set K is called the* key space *and the set X is called the* plaintext space. *We require that the key space K contain a special key called the* empty key *denoted ϵ.*

A functional encryption scheme for the functionality F enables one to evaluate $F(k,x)$ given the encryption of x and a secret key sk_k for k. The algorithm for evaluation $F(k,x)$ using sk_k is called *decrypt*. More precisely, a functional encryption scheme is defined as follows.

Definition 2. *A functional encryption scheme (FE) for a functionality F defined over (K,X) is a tuple of four PPT algorithms* (setup, keygen, enc, dec) *satisfying the following correctness condition for all $k \in K$ and $x \in X$:*

$(pp, mk) \leftarrow \text{setup}(1^\lambda)$ *(generate a public and master secret key pair)*

$sk \leftarrow \text{keygen}(mk, k)$ *(generate secret key for k)*

$c \leftarrow \text{enc}(pp, x)$ *(encrypt message x)*

$y \leftarrow \text{dec}(sk, c)$ *(use sk to compute $F(k, x)$ from c)*

then we require that $y = F(k, x)$ *with probability* 1.

We define security of a functional encryption scheme in Section 4. For now, we briefly show that standard public-key encryption is a simple example of functional encryption. Let $K := \{1, \epsilon\}$ and consider the following functionality F defined over (K, X) for some plaintext space X:

$$F(k, x) := \begin{cases} x & \text{if } k = 1 \\ \text{len(x)} & \text{if } k = \epsilon \end{cases}$$

A secret key for $k = 1$ fully decrypts valid ciphertexts, while the empty key $k = \epsilon$ simply returns the bit length of the plaintext. Hence, this functionality syntactically defines standard public-key encryption.

The empty key ϵ: The special key ϵ in K captures all the information about the plaintext that intentionally leaks from the ciphertext, such as the length of the encrypted plaintext. The secret key for ϵ is empty and also denoted by ϵ. Thus, anyone can run $\text{dec}(\epsilon, c)$ on a ciphertext $c \stackrel{R}{\leftarrow} \text{enc}(\text{pp}, x)$ and obtain all the information about x that intentionally leaks from c.

Further parametrization. In some cases the key space K and plaintext space X are further parametrized by quantities generated by the setup algorithm. For example, setup may output an RSA modulus N in which case the sets K and X and the functionality F are defined as tuples over $\mathbb{Z}_N$. More generally, we allow setup to output a third parameter π and we denote the key and plaintext space by K_π and X_π. The functionality F is the defined as

$$F_\pi : K_\pi \times X_\pi \to \{0, 1\}^* .$$

When π is clear from context, we avoid writing it as an explicit subscript.

2.1 Sub-classes of Functional Encryption

So far we defined the most general syntax for a functional encryption scheme. For the applications we have in mind it is convenient to define two sub-classes of functional encryption where the plaintext space X has additional structure.

Predicate encryption [BW07, KSW08]. In many applications a plaintext $x \in X$ is itself a pair $(\mathsf{ind}, m) \in I \times M$ where ind is called an index and m is called the payload message. For example, in an email system the index might be the sender's name while the payload is the email contents.

In this context, an FE scheme is defined in terms of a polynomial-time predicate $P : K \times I \to \{0, 1\}$ where K is the key space. More precisely, the FE functionality over $(\, K \cup \{\epsilon\},\ (I \times M)\,)$ is defined as

$$F\big(k \in K,\ (\mathsf{ind}, m) \in X\,\big) := \begin{cases} m & \text{if } P(k, \mathsf{ind}) = 1, \text{ and} \\ \perp & \text{if } P(k, \mathsf{ind}) = 0 \end{cases}$$

Consequently, let c be an encryption of (ind, m) and let sk_k be a secret key for $k \in K$. Then $\mathsf{dec}(\mathsf{sk}_k, c)$ reveals the payload in c when $P(k, \mathsf{ind}) = 1$ and reveals nothing new about m otherwise.

Predicate encryption with public index. A sub-class of predicate encryption makes the plaintext index easily readable from the ciphertext. In particular, in this type of FE the empty key ϵ explicitly reveals the index ind, namely

$$F\big(\epsilon,\ (\mathsf{ind}, m)\ \big) = (\mathsf{ind},\ \mathsf{len}(m)\ )$$

Hence, $\mathsf{dec}(\epsilon, c)$ gives anyone the index component of the plaintext as well as the bit length of m.

3 Capturing Cryptosystems in the Context of Functional Encryption

Many recent encryption concepts and constructions can be viewed as special cases of Functional Encryption. In this section we give a few examples to show how functional encryption captures these encryption concepts. Security of these schemes is captured by the general security definitions of functional encryption in the next section.

3.1 Predicate Encryption Systems with Public Index

The first class of systems that we consider are Predicate encryption schemes with public index[2]. We begin our study with the simplest interesting case of Identity-Based Encryption and then advance to more expressive methods of access formulas. We will describe these systems using the notation for predicate encryption defined in Subsection 2.1[3].

Identity-Based Encryption. In Identity-Based Encryption (IBE) [Sha84] ciphertexts and private keys are associated with strings (a.k.a identities) and a key can decrypt a ciphertext if the two strings are equal. IBE represents the first functionality that is not directly realizable from public key encryption [BPR$^+$08]. IBE is formally described as a Predicate Encryption scheme where:

1. The key space K is $K := \{0,1\}^* \cup \{\epsilon\}$.
2. The plaintext is a pair (ind, m) where the index space $I := \{0,1\}^*$.
3. The predicate P on $K \times I$ is defined as

$$P\big(k \in K \setminus \{\epsilon\},\ \mathsf{ind} \in I\ \big) := \begin{cases} 1 & \text{if } k = \mathsf{ind}, \text{ and} \\ 0 & \text{otherwise} \end{cases}$$

[2] This class has also been informally referred to as "payload hiding" [BW07, KSW08] in the literature.

[3] Recall that for all predicate encryption schemes with public index we have that $F\big(\epsilon,\ (\mathsf{ind}, m)\ \big) = (\mathsf{ind},\ \mathsf{len}(m)\ )$.

Boneh and Franklin [BF03] and Cocks [Coc01] construct the first practical IBE systems, which were proven secure according to an indistinguishability definition that is a special case of our definition of functional encryption security (Definition 3 in Section 4). These first schemes were proven secure in the random oracle model. Subsequent schemes were proven secure in the standard model, but under a weaker notion known as selective security [CHK03, BB04a], and further subsequent systems were proven adaptively secure [BB04b, Wat05, Gen06]. Recently, there have been multiple lattice-based constructions of IBE systems [GPV08, CHKP10, ABB10].

For these systems to properly support the empty key ϵ function, the ciphertext must explicitly include ind and the length of the message in the clear.

Attribute-Based Encryption. Sahai and Waters [SW05] proposed a notion of encryption, called Attribute-Based Encryption (ABE), where one could express complex access policies. Subsequently, Goyal, Pandey, Sahai and Waters [GPSW06] refined this concept into two different formulations of ABE: Key Policy ABE and Ciphertext-Policy ABE.

We first describe Key-Policy ABE for boolean formulas, as was realized by Goyal et. al. [GPSW06][4]. A Key-Policy ABE system over n variables can be described as a predicate encryption scheme (with public index) for the predicate $P_n : K \times I \to \{0,1\}$ where:

1. The key space K is the set of all poly-sized boolean formulas ϕ in n variables $\boldsymbol{z} = (z_1, \dots, z_n) \in \{0,1\}^n$. We let $\phi(\boldsymbol{z})$ denote the value of the formula ϕ at $\boldsymbol{z}$.
2. The plaintext is a pair $(\mathsf{ind} = \boldsymbol{z},\ m)$ where the index space is $I := \{0,1\}^n$, and where we interpret $\boldsymbol{z}$ as a bit vector representing the boolean values $z_1, \dots z_n$.
3. The predicate P_n on $K \times I$ is defined as

$$P_n\big(\phi \in K \setminus \{\epsilon\},\ \mathsf{ind} = \boldsymbol{z} \in I\big) := \begin{cases} 1 & \text{if } \phi(\boldsymbol{z}) = 1, \text{and} \\ 0 & \text{otherwise} \end{cases}$$

In these systems the key provides an access formula that operates over a set of n attributes that must evaluate to true for decryption to yield the message m. Goyal et al. also describe how to construct a "Large Universe" construction where ind can be viewed as a set of strings. Then K consists of all monotone boolean formulas over strings. To evaluate $\phi(\mathsf{ind})$ we evaluate a leaf labeled with string x in ϕ as '0' if $x \notin \mathsf{ind}$.

Ciphertext-Policy ABE. A dual concept of Attribute-Based Encryption is Ciphertext-Policy Attribute-Based Encryption (CP-ABE), where the roles of the ciphertext and key are essentially reversed. A Ciphertext-Policy ABE system over n variables can be described as predicate encryption scheme (with public index) for the predicate $P_n : K \times I \to \{0,1\}$ where:

1. The key space $K := \{0,1\}^n$ is the set of all n bit strings representing n boolean variables $\boldsymbol{z} = (z_1, \dots, z_n) \in \{0,1\}^n$.

[4] The ABE solutions of Goyal et. al. and others [BSW07, OSW07, GJPS08, Wat11] actually extend to formulas over threshold gates and to Monotone Span Programs; however, we restrict our description to Boolean formulas for simplicity.

2. The plaintext is a pair (ind $= \phi,\ m$) where the index space I is the set of all poly-sized boolean formulas ϕ over n variables.
3. The predicate P_n on $K \times I$ is defined as

$$P_n\big(\boldsymbol{z} \in K \setminus \{\epsilon\},\ \mathsf{ind} = \phi \in I\ \big) := \begin{cases} 1 & \text{if } \phi(\boldsymbol{z}) = 1, \text{ and} \\ 0 & \text{otherwise} \end{cases}$$

CP-ABE systems are constructed in [BSW07, GJPS08, Wat11]. Most constructions of ABE (both Ciphertext-Policy and Key-Policy) were proven secure in a weaker selective model of security. Recently Lewko et. al. [LOS$^+$10] showed how to give fully secure realizations meeting our security definition.

3.2 Predicate Encryption Systems

While the previous systems described allow for expressive forms of access control, they are limited in two ways. First, the policy ind is given in the clear as part of the empty functionality — often this in itself can be considered sensitive. Second, it does not allow for computation on the encrypted data, which might include such applications as search. Here we describe current Predicate Encryption systems that do not leak the index ind.

Anonymous Identity-Based Encryption. The problem of Anonymous Identity-Based Encryption was first proposed by Boneh et. al. [BCOP04] and later formalized by Abdalla et. al. [ABC$^+$08]. Other constructions include [BW06, Gen06, CHKP10, ABB10]. The functionality of Anonymous IBE is similar to IBE except that the string representing the ciphertext identity is hidden and one can only determine it if they have the corresponding private key. Therefore, we can describe Anonymous IBE in the exact same manner as above, except we have that $F\big(\epsilon,\ (\mathsf{ind}, m)\ \big) = \mathsf{len}(m)$. The empty functionality only gives the message length, but ind stays hidden.

Hidden Vector Encryption. Boneh and Waters [BW07] proposed what they called a hidden vector encryption system. In such a system a ciphertext contains a vector of n elements in $\{0,1\}^*$ and a private key contains of a vector of n elements in $\{*\} \cup \{0,1\}^*$ where we refer to $*$ as a wildcard character. More precisely,

1. The key space K is all $(v_1, \ldots v_n)$ where each $v_i \in \{*\} \cup \{0,1\}^*$.
2. The plaintext is a pair (ind $= (w_1, \ldots, w_n),\ m$) where each $w_i \in \{0,1\}^*$. The index space in $I := (\{0,1\}^*)^n$.
3. The predicate P_n on $K \times I$ is defined as

$$P_n\big(\ (v_1, \ldots, v_n) \in K \setminus \{\epsilon\},\ \mathsf{ind} = (w_1, \ldots, w_n)\ \big) := \begin{cases} 1 & \text{if } v_i = w_i \text{ whenever } v_i \neq *, \text{ and} \\ 0 & \text{otherwise} \end{cases}$$

Applications of the predicate include conjunctive and range searches. Independently, Shi et. al. [SBC$^+$07] proposed a related system in a weaker security model. Again we note that $F\big(\epsilon,\ (\mathsf{ind}, m)\ \big) = \mathsf{len}(m)$ so that ciphertexts do not reveal ind.

Inner Product Predicate. The previous system was limited to conjunctive searches. Katz, Sahai and Waters [KSW08] proposed a system for testing if a dot product operation over the ring Z_N is equal to 0, where N is the product of three random primes chosen by the setup algorithm. This enabled more complex evaluations on disjunctions, polynomials, and CNF/DNF formulae. Subsequently, Okamoto and Takashima [OT09] and Lewko et. al. [LOS$^+$10] gave constructions over the field F_p. We describe this predicate for vectors of length n.

1. The setup algorithm defines a randomly chosen prime p of length κ, where κ is the security parameter.
2. The key space K is all $\boldsymbol{v} = (v_1, \ldots v_n)$ where each $v_i \in F_p$.
3. The plaintext is a pair $(\mathsf{ind} = (w_1, \ldots, w_n),\ m)$ where each $w_i \in F_p$. The index space is $I := (F_p)^n$.
4. The predicate $P_{n,p}$ on $K \times I$ is defined as

$$P_{n,p}\big(\,(v_1, \ldots, v_n) \in K \setminus \{\epsilon\},\ \mathsf{ind} = (w_1, \ldots, w_n)\,\big) := \begin{cases} 1 & \text{if } \sum_{i=1,\ldots,n} v_i \cdot w_i = 0 \\ 0 & \text{otherwise} \end{cases}$$

3.3 Other Systems and Combinations

Different researchers have realized different combinations of the above core systems. Examples of these include combinations of: Attribute-Based Encryption and Broadcast Encryption [AI09], Identity-Based Broadcast Encryption [Del07, DPP07, SF07, GW09], broadcast HIBE [BH08], and Inner-Product Encryption and ABE [OT10]. All are captured as special cases of functional encryption.

4 Security Definitions

Given the syntactic definitions of Functional Encryption (FE) from Section 2 we now turn to defining security of an FE scheme. In this section we give game based definitions. In Section 5 we discuss simulation-based definitions.

Let $\mathcal{E}$ be an FE scheme for functionality F defined over (K, X). Our goal is to define security against an adaptive adversary that repeatedly asks for secret keys sk_k for keys $k \in K$ of the attacker's choice. As we shall see, defining security against such attackers is more delicate than one might first expect. The problem is how to define the challenge ciphertext in a semantic security game. As usual, once the attacker obtains all the secret keys he desires, he outputs two challenge messages $m_0, m_1 \in X$ and expects to get back an encryption c of m_0 or m_1 chosen at random by the challenger. Clearly, if the attacker has a secret key sk_k for some $k \in K$ for which $F(k, m_0) \neq F(k, m_1)$ then he can easily answer the challenge c by outputting

$$\begin{cases} 0 & \text{if } \mathrm{dec}(\mathsf{sk}_k, c) = F(k, m_0), \text{ and} \\ 1 & \text{otherwise} \end{cases}$$

Hence, for the definition to be satisfiable we must severely restrict the attacker's choice of m_0, m_1 and require that they satisfy

$$F(k, m_0) = F(k, m_1) \quad \text{for all } k \text{ for which the attacker has } \mathrm{sk}_k. \tag{1}$$

Since the empty key ϵ reveals the plaintext length, condition (1) ensures that $|m_0| = |m_1|$, as in the standard PKE definition of semantic security.

Security definition. With requirement (1) in place we obtain a natural game for defining security of an FE scheme $\mathcal{E}$. For $b = 0, 1$ define experiment b for an adversary $\mathcal{A}$ as follows:

- Setup: run $(\mathrm{pp}, \mathrm{mk}) \leftarrow \mathrm{setup}(1^\lambda)$ and give pp to $\mathcal{A}$.
- Query: $\mathcal{A}$ adaptively submits queries k_i in K for $i = 1, 2, \ldots$ and is given $\mathrm{sk}_i \leftarrow \mathrm{keygen}(\mathrm{mk}, k_i)$.
- Challenge: $\mathcal{A}$ submits two messages $m_0, m_1 \in X$ satisfying (1) and is given $\mathrm{enc}(\mathrm{pp}, m_b)$.
- $\mathcal{A}$ continues to issue key queries as before subject to (1) and eventually outputs a bit in $\{0, 1\}$.

For $b = 0, 1$ let W_b be the event that the adversary outputs 1 in Experiment b and define

$$\mathrm{FE}_{\mathrm{adv}}[\mathcal{E}, \mathcal{A}](\lambda) := \big| \Pr[W_0] - \Pr[W_1] \big|$$

Definition 3. *An FE scheme $\mathcal{E}$ is secure if for all PPT $\mathcal{A}$ the function $FE_{adv}[\mathcal{E}, \mathcal{A}](\lambda)$ is negligible.*

Definition 3 is a generalization of related definitions from [BW07, KSW08].

4.1 A "Brute Force" Construction

We briefly show that any functionality F where the key space K has polynomial size can be easily realized. Write $s = |K| - 1$ and $K = \{\epsilon, k_1, \ldots, k_s\}$. In this brute force construction, the size of public parameters, secret key, and ciphertext are all proportional to s. A closely related construction is given in [BW07].

The brute force FE scheme realizing F uses a semantically secure public-key encryption scheme (G, E, D) and works as follows:

- $\mathrm{setup}(1^\lambda)$: for $i = 1, \ldots, s$ run $(\mathrm{pp}_i, \mathrm{mk}_i) \leftarrow G(1^\lambda)$.

$$\text{output:} \qquad \mathrm{pp} := (\mathrm{pp}_1, \ldots, \mathrm{pp}_s) \qquad \text{and} \qquad \mathrm{mk} := (\mathrm{mk}_1, \ldots, \mathrm{mk}_s)$$

- $\mathrm{keygen}(\mathrm{mk}, k_i)$: output $\mathrm{sk}_i := \mathrm{mk}_i$.
- $\mathrm{enc}(\mathrm{pp}, x)$: output $\mathbf{c} := \big(F(\epsilon, x),\ E(\mathrm{pp}_1, F(k_i, x)),\ \ldots,\ E(\mathrm{pp}_s, F(k_s, x)) \big)$.
- $\mathrm{dec}(\mathrm{sk}_i, \mathbf{c})$: output c_0 if $\mathrm{sk}_i = \epsilon$, and output $D(\mathrm{sk}_i, c_i)$ otherwise.

Clearly, a ciphertext $\mathbf{c}$ leaks the bit lengths of $F(k_i, x)$ for $i = 1, \ldots, s$. Therefore, for this construction to be secure we must assume that this information is already leaked by the empty functionality $F(\epsilon, \cdot)$, namely that $|F(k_i, x)|$ for $i = 1, \ldots, s$ is contained in $F(\epsilon, x)$. If so then we say that F reveals functional bit lengths.

Theorem 1. *Let F be a functionality that reveals functional bit lengths. If (G, E, D) is a semantically secure public-key encryption scheme then the brute force FE system implementing F is secure.*

Proof (Proof Sketch). The proof is by a standard hybrid argument across the s components of the challenge ciphertext.

4.2 Insufficiency of the Game-Based Security Definition

We will now show that for certain complex functionalities Definition 3 is too weak. For these functionalities we construct systems that are secure under Definition 3, but should not be considered secure. Nevertheless, for functionalities such as predicate encryption *with public index* we show in Section 5 that Definition 3 is adequate.

We give a simple example of a functionality for which the game-based Definition 3 is insufficient. Let π be a one-way permutation and consider the functionality F that only admits the trivial key ϵ, defined as follows:

$$F(\epsilon, x) = \pi(x)$$

It is clear that the "right" way to achieve functional encryption for this very simple functionality is to have the functional encryption algorithm itself simply output $\pi(x)$ on input x, namely $\text{enc}(\text{pp}, x) = \pi(x)$. This scheme would also clearly achieve the simulation-based definition of security presented in Section 5.

However, consider an "incorrect" realization of this functionality where the functional encryption algorithm outputs x on input x, namely $\text{enc}(\text{pp}, x) = x$. Clearly this system leaks more information about the plaintext than needed. Nevertheless, it is easy to verify that this construction satisfies the game-based definition from Section 4. This is because for any two values x and y, it is the case that $F(\epsilon, x) = F(\epsilon, y)$ if and only if $x = y$ and therefore the attacker can only issue challenge messages m_0, m_1 where $m_0 = m_1$.

This problematic system, however, would clearly not achieve the simulation-based definition of security presented in Section 5, since if x is chosen at random, the real-life adversary would be able to recover x always, while the simulator would not be able to recover x without breaking the one-wayness of the permutation π.

While the simple example above may seem to be "abusing" the role of the trivial key ϵ, it is easy to modify the functionality example F above so that there is exactly one non-trivial key $k \in K$ that outputs $\pi(x)$. The only difference to the construction above would be that the functional encryption algorithm would output a public-key encryption[5] of either $\pi(x)$ (in the "correct" implementation) or x (in the "incorrect" implementation), and the secret key for key k would be the secret key of the public-key encryption scheme. Again, it is easy to verify that the incorrect implementation satisfies the game-based definition.

Discussion. What does this separation show? While this is a subjective question, our view is that it shows that if the output of the functionality is supposed to have some

[5] The public-key encryption would need to be non-committing to achieve the simulation-based definition of security for the good case.

computational hiding properties – that is, security of your application is not only based on the information-theoretic properties of the function, but also on the computational properties of the function – then there is a real problem with the game-based formulation of security. The game-based formulation essentially ignores any computational hiding properties of the function, and therefore offers no security guarantees that could be meaningfully combined with such computational considerations.

5 Simulation Based Definitions

In this section, we explore security definitions for functional encryption that arise from the simulation paradigm [GM84, GMR85, GMW86] that has served us so well, especially in the closely related context of secure computation protocols.

We begin by considering a simulation-based definition[6] of security for functional encryption that captures the most basic intuition we have: That getting the secret key sk_k corresponding to the key $k \in K$ should only reveal $F(k, x)$ when given an encryption of x.

It turns out that we can achieve this simulation-based definition for natural functionalities in the *random oracle model*, where in the ideal model the random oracle would also be simulated. We argue that in fact this (very strong) random oracle model seems necessary for a meaningful simulation-based definition of security for functional encryption: we show that even in the *non-programmable* random oracle model (where the simulator, too, only has oracle access to the same random oracle that is provided to the distinguisher), simulation-secure functional encryption (for a seemingly "minimal" formulation of simulation-security) even just for the IBE functionality is impossible to achieve. At a high level, this is because any simulation-based definition that allows the adversary to query for secret keys after seeing the challenge ciphertext must achieve something very similar in spirit to *non-interactive non-committing encryption*, where exactly these kinds of impossibility and possibility results are known [Nie02].

In our main definition below (that we will achieve in our positive results), we will use some non-standard syntax for representing a *stateful oracle*[7]. When we write $A^{B(\cdot)[[x]]}$, we mean that the algorithm A can issue a query q to its oracle, at which point $B(q, x)$ will be executed and output a pair (y, x'). The value y is then communicated to A as the response to its query, and the variable x is set to x', and this updated value is fed to the algorithm B the next time it is queried as an oracle, and fed to any algorithms executed later in an experiment that want x as an input. Also, if we write $A^{B^{\circ}(\cdot)}$, we mean that A can send a query q to its oracle, at which point $B^{\circ}(q)$ is executed, and any oracle queries that B makes are answered by A.

Definition 4. *An FE scheme $\mathcal{E}$ is simulation-secure if there exists an (oracle) PPT algorithm $Sim = (Sim_1, Sim_O, Sim_2)$ such that for any (oracle) PPT algorithms*

[6] We note that there are several natural variants possible for such a definition. We have chosen a definition that is strong in the sense that it requires a universal black-box simulator. We will later discuss some weaker formulations.

[7] The more standard way to formalize this communication structure would be through interactive Turing Machines, but we find this notation to be simpler to parse.

$Message$ and Adv, we have that the following two distribution ensembles (over the security parameter λ) are computationally indistinguishable:

Real Distribution:

1. $(pp, mk) \leftarrow \mathrm{setup}(1^\lambda)$
2. $(\boldsymbol{x}, \tau) \leftarrow Message^{\mathrm{keygen}(mk,\cdot)}(pp)$
3. $\mathbf{c} \leftarrow \mathrm{enc}(pp, \boldsymbol{x})$
4. $\alpha \leftarrow Adv^{\mathrm{keygen}(mk,\cdot)}(pp, \mathbf{c}, \tau)$
5. *Let $y_1, \ldots, y_\ell$ be the queries to* keygen *made by $Message$ and Adv in the previous steps.*
6. *Output $(pp, \boldsymbol{x}, \tau, \alpha, y_1, \ldots, y_\ell)$*

Ideal Distribution:

1. $(pp, \sigma) \leftarrow Sim_1(1^\lambda)$
2. $(\boldsymbol{x}, \tau) \leftarrow Message^{SimO(\cdot)[[\sigma]]}(pp)$
3. $\alpha \leftarrow Sim_2^{F(\cdot,\boldsymbol{x}),\ Adv^{\circ}(pp,\cdot,\tau)}(\sigma, F(\epsilon, \boldsymbol{x}))$
4. *Let $y_1, \ldots, y_\ell$ be the queries to F made by Sim in the previous steps*[8].
5. *Output $(pp, \boldsymbol{x}, \tau, \alpha, y_1, \ldots, y_\ell)$*

We note that this definition can be extended further to allow the adversary to receive challenge ciphertexts adaptively (instead of as a single vector), and all our positive results below would extend to this setting. We omit this generalization due to the notational complexity that would be required to formulate such a definition.

5.1 Impossibility of Simulation-Secure Functional Encryption

In this section, we briefly sketch the impossibility result for simulation-secure functional encryption, even for a quite simple functionality (the functionality corresponding to IBE), in the non-programmable random oracle model. As the proof closely mirrors the argument of Nielsen [Nie02] for non-interactive non-committing encryption, we give only a high-level overview of the proof.

We note that our impossibility result in fact holds for much less stringent formulations of simulation security for functional encryption. In particular, we consider the following weaker version of our main definition:

Definition 5. *An FE scheme $\mathcal{E}$ is weakly simulation-secure if for any (oracle) PPT algorithms $Message$ and Adv, there exists an (oracle) PPT algorithm Sim such that we have that the following two distribution ensembles (over the security parameter λ) are computationally indistinguishable:*

Real Distribution:

1. $(pp, mk) \leftarrow \mathrm{setup}(1^\lambda)$
2. $(\boldsymbol{x}, \tau) \leftarrow Message(1^\lambda)$

[8] Note that Sim does not need to query the oracle for $F(\epsilon, \boldsymbol{x})$, as this is provided as an explicit input to Sim_2. We choose this formulation since in the real distribution, the Adversary does not explicitly need to ask keygen for the key corresponding to ϵ in order to gain knowledge about $F(\epsilon, \boldsymbol{x})$.

3. $\mathbf{c} \leftarrow \text{enc}(pp, \boldsymbol{x})$
4. $\alpha \leftarrow Adv^{\text{keygen}(mk,\cdot)}(pp, \mathbf{c}, \tau)$
5. *Let* $y_1, \ldots, y_\ell$ *be the queries to* keygen *made by* Adv *in the previous steps.*
6. *Output* $(\boldsymbol{x}, \tau, \alpha, y_1, \ldots, y_\ell)$

Ideal Distribution:

1. $(\boldsymbol{x}, \tau) \leftarrow Message(1^\lambda)$
2. $\alpha \leftarrow Sim^{F(\cdot,\boldsymbol{x})}(1^\lambda, \tau, F(\epsilon, \boldsymbol{x}))$
3. *Let* $y_1, \ldots, y_\ell$ *be the queries to* F *made by* Sim *in the previous step.*
4. *Output* $(\boldsymbol{x}, \tau, \alpha, y_1, \ldots, y_\ell)$

We note another weakening of the definition above would be to have the distributions output the queries $y_1, \ldots, y_\ell$ as an unordered set, instead of an ordered tuple. Our impossibility proof can be extended to this weakening as well. We now sketch the proof of the following theorem.

Theorem 2. *Let* F *be the functionality for IBE. There does not exist any weakly simulation-secure FE scheme for* F *in the non-programmable random oracle model.*

Proof (Brief Proof Sketch). The overall idea of this proof is almost identical to the impossibility proof of Nielsen [Nie02] for non-interactive non-committing encryption. Let H represent the random oracle. Consider the following concrete adversary algorithms:

$Message(1^\lambda)$ works as follows: Let len_{sk} be the maximum bit length produced by the keygen algorithm for the key 0 for security parameter λ. Then the vector $\boldsymbol{x}$ consists of the following elements: for $i = 1, \ldots, len_{sk} + \lambda$, the element $(r_i, 0)$ where r_i is a randomly and independently chosen bit for each i. The value τ is empty.

$Adv^{\text{keygen}(\text{mk},\cdot)}(\text{pp}, \mathbf{c}, \tau)$ works as follows: Call the random oracle H on the input $(\text{pp}, \mathbf{c})$ to obtain a string w of length λ. Now request the secret key for the identity (w) first, and then for the identity 0. Use the key for identity 0 to decrypt the entire ciphertext. Output a full transcript of the entire computation done by Adv, including all calls to the random oracle and the interaction with the keygen oracle.

Now consider what Sim must do in order to output a distribution indistinguishable from the real interaction. Because Adv only makes a single key query of the form (w), it is the case that Sim must make exactly one query – its first query – to F of this form as well. Furthermore, the distinguisher can check if this w is the output of H applied to some string of the form $(\text{pp}, \mathbf{c})$. Thus, the simulator must perform this query to H before making any queries to F. The simulator at this point has no information whatsoever about the plaintexts r_i (which is only revealed when the simulator queries F for identity 0 afterwards). Thus, this fixed string $z = (\text{pp}, \mathbf{c})$ has the (impossible) property that after receiving only len_{sk} bits of information, it can deterministically "decode" z to be a an arbitrary string of length $len_{sk} + \lambda$.

We remark that the proof above made use of the fact that the simulator's queries to F are recorded *in order*. However, we note that the same impossibility result would hold even if the security definition only recorded the *unordered set* of queries to F, but using a slightly more involved adversary and message distribution. Roughly speaking, the only identities in the system would be of the form $(i, 0)$ and $(i, 1)$ for $i = 1, \ldots, \lambda$, and the messages to be encrypted would be random long messages for each identity. The adversary would apply the random oracle to $(\mathrm{pp}, \mathbf{c})$ to obtain a single string w of length λ exactly as above, but it would now use this string to obtain keys (i, w_i) for $i = 1, \ldots, \lambda$. The argument would now proceed by looking at the point when the simulator has made at least $\lambda/2$ queries to F. By now with overwhelming probability, a single query $(\mathrm{pp}, \mathbf{c})$ to H would be compatible with these queries, and that could be used to define the "impossible string" needed above.

5.2 A Simulation-Based Brute Force Scheme

We now consider FE schemes that are simulation-secure in the random oracle model (where the scheme algorithms and the $Message$ and Adv algorithms all have oracle access to a random oracle, but the simulator algorithms can emulate the random oracle itself). We note that this is the standard formulation of the random oracle model, more recently called the "full" or "programmable" random oracle model.

The modified "brute-force" construction. We first consider the following slight modification of the brute-force construction given earlier. The modification just uses the random oracle to randomly mask the output values of the function.

We will make use of a random oracle $H : \{0,1\}^* \to \{0,1\}$. Note that we will abuse notation and also write $H(x)$ to produce strings of arbitrary length (which will be clear from context). This can be accomplished by interpreting $H(x)$ to mean the concatenation of $H((x,1)), \ldots, H((x,\ell))$ to produce strings of length ℓ.

Recall that we write $s = |K|-1$ and $K = \{\epsilon, k_1, \ldots, k_s\}$. The brute force FE scheme realizing F uses a semantically secure public-key encryption scheme (G, E, D), and works as follows:

- setup(1^λ): for $i = 1, \ldots, s$ run $(\mathrm{pp}_i, \mathrm{mk}_i) \leftarrow G(1^\lambda)$.

$$\text{output:} \qquad \mathrm{pp} := (\mathrm{pp}_1, \ldots, \mathrm{pp}_s) \qquad \text{and} \qquad \mathrm{mk} := (\mathrm{mk}_1, \ldots, \mathrm{mk}_s)$$

- keygen(mk, k_i): output $\mathrm{sk}_i := \mathrm{mk}_i$.
- enc(pp, x): choose random values $r_1, \ldots, r_s \in_R \{0,1\}^\lambda$. output

$$\mathbf{c} := \Big(F(\epsilon, x),\ E(\mathrm{pp}_1, r_1),\ H(r_1) \oplus F(k_1, x),\ \ldots,\ E(\mathrm{pp}_s, r_s),\ H(r_s) \oplus F(k_s, x) \Big).$$

- dec$(\mathrm{sk}_i, \mathbf{c})$: output c_0 if $\mathrm{sk}_i = \epsilon$, and output $H\big(D(\mathrm{sk}_i, c_{2i-1})\big) \oplus c_{2i}$ otherwise.

A proof sketch of the following theorem is given in the full version of the paper [BSW10].

Theorem 3. *Let F be a functionality that reveals functional bit lengths. If (G, E, D) is a semantically secure public-key encryption scheme then the modified brute force FE system above implementing F is simulation-secure in the random oracle model.*

5.3 An Equivalence for Public Index Schemes

We show that any predicate encryption system with public index that is secure under the game-based Definition 3 also satisfies the simulation based Definition 4 in the random oracle model. This result shows equivalence (in the random oracle model) for the large class of public index schemes including various forms of Attribute-Based encryption.

Let $\mathcal{E} :=$ (setup, keygen, enc, dec) be an FE predicate encryption system with public index for predicate $P : K \times I \to \{0,1\}$. We convert the system into a scheme $\mathcal{E}_H :=$ (setup, keygen, enc_H, dec_H) where encryption is done using a random oracle H:

- $\text{enc}_H(\text{pp}, (\text{ind}, m))$: choose a random value $r \in \{0,1\}^\lambda$ and output

$$c := \big(\, \text{enc}(\text{pp}, (\text{ind}, r)\,),\;\; H(r) \oplus m \,\big)\,.$$

- $\text{dec}_H(\text{sk}, (c_1, c_2))$: if $\text{dec}(\text{sk}, c_1) = \bot$ output $\bot$, otherwise output $\text{dec}(\text{sk}, c_1) \oplus c_2$.

The following theorem shows that this construction is simulation secure.

Theorem 4. *If the system $\mathcal{E}$ is game-secure (Definition 3) then $\mathcal{E}_H$ is simulation secure (Definition 4) in the random oracle model.*

Proof (sketch). We construct the universal simulators Sim_1, $SimO$, and Sim_2 needed for simulation security. These algorithms work as follows:

- $Sim_1(1^\lambda)$ simply executes $\text{setup}(1^\lambda)$ to obtain pp and mk. It outputs pp unmodified, and outputs $\sigma = (\text{mk}, O, \kappa)$, where O and κ are empty lists. This list O will keep track of the simulated random oracle, and κ will keep track of key queries.
- $SimO(\cdot)[[\sigma]]$ works as follows: It responds to random oracle queries and keygen queries the adversary $Message$ makes as follows:
    - **Random Oracle Queries:** On query q, the simulator first checks to see if a pair (q, y) already exists in the list O. If so, it provides y as the response to the adversary's query. If not, the simulator chooses a fresh random string y, adds the pair (q, y) to the list O, and provides y as the response to the adversary's query. This list O is updated in the state variable σ.
    - **Key Queries:** When the adversary asks for the key k, the simulator sends the secret key $\text{sk} \leftarrow \text{keygen}(\text{mk}, k)$ to the adversary. The simulator adds k to the list κ. This list κ is updated in the state variable σ.
- $Sim_2^{F(\boldsymbol{x},\cdot),\ Adv^\circ(\text{pp},\cdot,\tau)}(\sigma, F(\boldsymbol{x}, \epsilon))$ works as follows:
    1. The algorithm begins by preparing a "fake" vector of ciphertexts as follows: Let n be the number of elements in $\boldsymbol{x}$ and let $\text{ind}_1, \ldots, \text{ind}_n$ be the indices in $\boldsymbol{x}$. Sim_2 obtains n and these indices by querying its F oracle at $F(\epsilon, \boldsymbol{x})$.

       Now, for $i = 1 \ldots n$, it chooses random strings $r_1, \ldots, r_n$ and $R_1, \ldots R_n$, and creates the ciphertext components

$$c_{i,1} := \text{enc}\big(\text{pp}, (\text{ind}_i, r_i)\,\big) \quad \text{and} \quad c_{i,2} = R_i \qquad \text{for } i = 1, \ldots, n.$$

       Let $\mathbf{c}$ be the vector of n ciphertexts $\mathbf{c} := (c_{i,1}, c_{i,2})_{i=1,\ldots,n}$.

2. For each key k in the list κ of keys already queried, the simulator does the following:
(1) it invokes the F oracle and obtains $F(k, \boldsymbol{x}) = (z_1, \ldots, z_n)$,
(2) for $i = 1, \ldots, n$ if $z_i \neq \perp$ it adds the pair $(r_i, R_i \oplus z_i)$ to the list O. If any of these r_i values were already in the list O, the simulation aborts.
3. Then it invokes $Adv(\text{pp}, \mathbf{c}, \tau)$ using this "fake" ciphertext vector $\mathbf{c}$ created above.
4. It now monitors which random oracle queries and keygen queries the adversary Adv makes. It responds to these queries as follows:

   - **Random Oracle Queries:** On query q, the simulator first checks to see if a pair (q, y) already exists in the list O. If so, it provides y as the response to the adversary's query. If not, the simulator chooses a fresh random string y, adds the pair (q, y) to the list O, and provides y as the response to the adversary's query.
   - **Key Queries:** If the adversary asks for the key k, then the simulator invokes the F oracle and obtains $F(k, \boldsymbol{x}) = (z_1, \ldots, z_n)$. For $i = 1, \ldots, n$ if $z_i \neq \perp$ it adds the pair $(r_i, \ R_i \oplus z_i)$ to the list O. If for any i there is already a pair (r_i, R) in the list O with $R \neq R_i \oplus z_i$ then the simulation aborts. Finally, it sends the secret key sk $\leftarrow$ keygen(mk, k) to the adversary. It is easy to confirm that the decryption procedure will work as it should after we have modified the random oracle as detailed above.

5. When the adversary terminates and outputs α, then the simulator outputs this α as well, finishing the simulation.

The same argument as in the proof of Theorem 3 shows that the simulator aborts with negligible probability and that the distribution generated by these simulators is statistically close to the real distribution. In particular, the negligible probability of abort follows from the game-based security of $\mathcal{E}$, since game-based security implies one-way security for encrypting random values, which implies that the adversary is extremely unlikely to query the random oracle on the r_i values prior to obtaining a secret key that can open the i'th ciphertext.

Other Simulation-Secure Functional Encryption Schemes. Since the above equivalence only applies to public index schemes, an interesting question is whether we can achieve simulation security for more general systems. Intuitively this is more challenging, since it goes beyond just hiding a payload, to "hiding a computation" and is arguably closer to our counter example of Section 4.2.

In the full version of the paper [BSW10] we prove the simulation security of the Boneh-Franklin construction for the anonymous IBE functionality. An interesting direction is to prove simulation security for systems with more functionality. One challenge is that it is not completely clear how to apply the random oracle heuristic to these systems, as the correctness of such schemes typically relies on structure that a hash function might break.

6 Extending Functional Encryption

In this work, we focus on the "core" case of functional encryption. However, there are multiple ways to extend the concept. We briefly outline these here. We hope future work will develop these extensions and give precise definitions of security and constructions.

Delegation. Delegation is the ability of an algorithm to transform a key k in a functional encryption system to another key k'. For example, one might want to share the ability to decrypt all messages of a certain subject to another user. Typically, we think of the resulting key k' as being more restricted than the source key k. We observe that the set of allowed delegations must respect the security definition of the system. Delegation in functional encryption systems is typically associated with Hierarchical Identity-Based Encryption [HL02, GS02], but was also considered in Attribute-Based Encryption [GPSW06] and other predicate encryption systems [SW08].

Encryption over Multiple Parameters and Multiple Systems. Our functional encryption systems allow for functionalities $F : K \times X \to \{0,1\}^*$ that take in a single key and plaintext as inputs. However, we could extend our system to allows for functionalities that take in multiple keys $F : (K_1, \dots, K_n) \times X \to \{0,1\}^*$. This can be useful in applications where we want users to combine their capabilities in a specified manner or when one of the keys can represent an event such as a certain time period arriving, or publication of a revocation list [BGK08].Another interesting direction is to allow for a functional encryption system that operates over multiple ciphertexts.

Taking things further we could consider an encryption system where encryption takes in multiple public parameters each from *different* authorities and where the functionality is evaluated over private keys generated by different master secret keys. One notable application of this is Attribute-Based Encryption where multiple authorities are used [Cha07, CC09].

Hiding Information about capabilities of the key. One consistent feature of all systems is that there do not exist any security notions about the attacker's inability to distinguish what type of key k he is given a secret key for. A natural reason for this is that in a public key system, he can distinguish whether he has the capability for k_0, k_1 by simply encrypting an $x \in X$ such that $F(k_0, x) \neq F(k_1, x)$. However, one might try to consider such a problem when encryption is not public key [SSW09, BIP10].

7 Future Directions in Functional Encryption

The results to date scratch the surface of functional encryption and only implement relatively simple functionalities. Here we list a few fascinating open problems that remain.

- The grand challenge is to construct a secure functional encryption scheme for all polynomial-time functionalities. A more modest goal is to do the same for predicate encryption for all polynomial-time predicates. Currently, the best we can do is predicates defined by inner products [KSW08]. The inner product construction uses bilinear maps and our inability to move beyond inner products is due to the "bi" in bilinear maps. Other tools, perhaps borrowing from fully homomorphic encryption [Gen09], may lead to a more general class of predicates.

- If not all polynomial time functionalities, can we realize complex interesting ones such as data-mining functionalities? That is, can we build a secret key that given an encrypted data set will produce a cleartext model (e.g. a decision tree) for the data set, but reveal nothing else about the data? Nothing in this vain is currently known.
- Are there black box separations between different functionalities? Currently, the only result in this direction separates IBE from public-key encryption [BPR+08]. Is there a generic separation result that separates any two functionalities that are not trivially implied one by the other?

References

[ABB10] Agrawal, S., Boneh, D., Boyen, X.: Efficient lattice (H)IBE in the standard model. In: Gilbert, H. (ed.) EUROCRYPT 2010. LNCS, vol. 6110, pp. 553–572. Springer, Heidelberg (2010)

[ABC+08] Abdalla, M., Bellare, M., Catalano, D., Kiltz, E., Kohno, T., Lange, T., Malone-Lee, J., Neven, G., Paillier, P., Shi, H.: Searchable encryption revisited: Consistency properties, relation to anonymous ibe, and extensions. J. Cryptology 21(3), 350–391 (2008)

[AI09] Attrapadung, N., Imai, H.: Conjunctive broadcast and attribute-based encryption. In: Shacham, H., Waters, B. (eds.) Pairing 2009. LNCS, vol. 5671, pp. 248–265. Springer, Heidelberg (2009)

[AL10] Attrapadung, N., Libert, B.: Functional encryption for inner product: Achieving constant-size ciphertexts with adaptive security or support for negation. In: Nguyen, P.Q., Pointcheval, D. (eds.) PKC 2010. LNCS, vol. 6056, pp. 384–402. Springer, Heidelberg (2010)

[BB04a] Boneh, D., Boyen, X.: Efficient selective-ID secure identity-based encryption without random oracles. In: Cachin, C., Camenisch, J.L. (eds.) EUROCRYPT 2004. LNCS, vol. 3027, pp. 223–238. Springer, Heidelberg (2004)

[BB04b] Boneh, D., Boyen, X.: Secure identity based encryption without random oracles. In: Franklin, M. (ed.) CRYPTO 2004. LNCS, vol. 3152, pp. 443–459. Springer, Heidelberg (2004)

[BCOP04] Boneh, D., Di Crescenzo, G., Ostrovsky, R., Persiano, G.: Public key encryption with keyword search. In: Cachin, C., Camenisch, J.L. (eds.) EUROCRYPT 2004. LNCS, vol. 3027, pp. 506–522. Springer, Heidelberg (2004)

[BF03] Boneh, D., Franklin, M.K.: Identity-based encryption from the weil pairing. SIAM J. Comput. 32(3), 586–615 (2003); Extended abstract In: Kilian, J. (ed.) CRYPTO 2001. LNCS, vol. 2139, p. 213. Springer, Heidelberg (2001)

[BGK08] Boldyreva, A., Goyal, V., Kumar, V.: Identity-based encryption with efficient revocation. In: ACM Conference on Computer and Communications Security, pp. 417–426 (2008)

[BH08] Boneh, D., Hamburg, M.: Generalized identity based and broadcast encryption schemes. In: Pieprzyk, J. (ed.) ASIACRYPT 2008. LNCS, vol. 5350, pp. 455–470. Springer, Heidelberg (2008)

[BIP10] Blundo, C., Iovino, V., Persiano, G.: Predicate encryption with partial public keys. Cryptology ePrint Archive, Report 2010/476 (2010), `http://eprint.iacr.org/`

[BPR+08] Boneh, D., Papakonstantinou, P.A., Rackoff, C., Vahlis, Y., Waters, B.: On the impossibility of basing identity based encryption on trapdoor permutations. In: FOCS, pp. 283–292 (2008)

[BSW07] Bethencourt, J., Sahai, A., Waters, B.: Ciphertext-policy attribute-based encryption. In: IEEE Symposium on Security and Privacy, pp. 321–334 (2007)

[BSW10] Boneh, D., Sahai, A., Waters, B.: Functional encryption: Definitions and challenges. Cryptology ePrint Archive, Report 2010/543 (2010), `http://eprint.iacr.org/2010/543`

[BW06] Boyen, X., Waters, B.: Anonymous hierarchical identity-based encryption (Without random oracles). In: Dwork, C. (ed.) CRYPTO 2006. LNCS, vol. 4117, pp. 290–307. Springer, Heidelberg (2006)

[BW07] Boneh, D., Waters, B.: Conjunctive, subset, and range queries on encrypted data. In: Vadhan, S.P. (ed.) TCC 2007. LNCS, vol. 4392, pp. 535–554. Springer, Heidelberg (2007)

[CC09] Chase, M., Chow, S.S.M.: Improving privacy and security in multi-authority attribute-based encryption. In: ACM Conference on Computer and Communications Security, pp. 121–130 (2009)

[CFGN96] Canetti, R., Feige, U., Goldreich, O., Naor, M.: Adaptively secure multi-party computation. In: STOC 1996, pp. 639–648 (1996)

[Cha07] Chase, M.: Multi-authority attribute based encryption. In: Vadhan, S.P. (ed.) TCC 2007. LNCS, vol. 4392, pp. 515–534. Springer, Heidelberg (2007)

[CHK03] Canetti, R., Halevi, S., Katz, J.: A forward-secure public-key encryption scheme. In: Biham, E. (ed.) EUROCRYPT 2003. LNCS, vol. 2656, pp. 255–271. Springer, Heidelberg (2003)

[CHKP10] Cash, D., Hofheinz, D., Kiltz, E., Peikert, C.: Bonsai trees, or how to delegate a lattice basis. In: Gilbert, H. (ed.) EUROCRYPT 2010. LNCS, vol. 6110, pp. 523–552. Springer, Heidelberg (2010)

[Coc01] Cocks, C.: An identity based encryption scheme based on quadratic residues. In: IMA Int. Conf., pp. 360–363 (2001)

[Del07] Delerablée, C.: Identity-based broadcast encryption with constant size ciphertexts and private keys. In: Kurosawa, K. (ed.) ASIACRYPT 2007. LNCS, vol. 4833, pp. 200–215. Springer, Heidelberg (2007)

[DH76a] Diffie, W., Hellman, M.E.: Multiuser cryptographic techniques. In: AFIPS National Computer Conference, pp. 109–112 (1976)

[DH76b] Diffie, W., Hellman, M.E.: New directions in cryptography. IEEE Transactions on Information Theory IT-22(6), 644–654 (1976)

[DPP07] Delerablée, C., Paillier, P., Pointcheval, D.: Fully collusion secure dynamic broadcast encryption with constant-size ciphertexts or decryption keys. In: Takagi, T., Okamoto, T., Okamoto, E., Okamoto, T. (eds.) Pairing 2007. LNCS, vol. 4575, pp. 39–59. Springer, Heidelberg (2007)

[Gen06] Gentry, C.: Practical identity-based encryption without random oracles. In: Vaudenay, S. (ed.) EUROCRYPT 2006. LNCS, vol. 4004, pp. 445–464. Springer, Heidelberg (2006)

[Gen09] Gentry, C.: A fully homomorphic encryption scheme. PhD thesis, Stanford University (2009), `http://crypto.stanford.edu/craig`

[GJPS08] Goyal, V., Jain, A., Pandey, O., Sahai, A.: Bounded ciphertext policy attribute based encryption. In: Aceto, L., Damgård, I., Goldberg, L.A., Halldórsson, M.M., Ingólfsdóttir, A., Walukiewicz, I. (eds.) ICALP 2008, Part II. LNCS, vol. 5126, pp. 579–591. Springer, Heidelberg (2008)

[GM84] Goldwasser, S., Micali, S.: Probabilistic encryption. Jour. of Computer and System Science 28(2), 270–299 (1984)

[GMR85] Goldwasser, S., Micali, S., Rackoff, C.: The knowledge complexity of interactive proof-systems. In: STOC, pp. 291–304 (1985)

[GMW86] Goldreich, O., Micali, S., Wigderson, A.: Proofs that yield nothing but their validity and a methodology of cryptographic protocol design. In: FOCS, pp. 174–187 (1986)

[GPSW06] Goyal, V., Pandey, O., Sahai, A., Waters, B.: Attribute-based encryption for fine-grained access control of encrypted data. In: ACM Conference on Computer and Communications Security, pp. 89–98 (2006)

[GPV08] Gentry, C., Peikert, C., Vaikuntanathan, V.: Trapdoors for hard lattices and new cryptographic constructions. In: STOC, pp. 197–206 (2008)

[GS02] Gentry, C., Silverberg, A.: Hierarchical ID-based cryptography. In: Zheng, Y. (ed.) ASIACRYPT 2002. LNCS, vol. 2501, pp. 548–566. Springer, Heidelberg (2002)

[GW09] Gentry, C., Waters, B.: Adaptive security in broadcast encryption systems (with short ciphertexts). In: Joux, A. (ed.) EUROCRYPT 2009. LNCS, vol. 5479, pp. 171–188. Springer, Heidelberg (2009)

[HL02] Horwitz, J., Lynn, B.: Toward hierarchical identity-based encryption. In: Knudsen, L.R. (ed.) EUROCRYPT 2002. LNCS, vol. 2332, pp. 466–481. Springer, Heidelberg (2002)

[KSW08] Katz, J., Sahai, A., Waters, B.: Predicate encryption supporting disjunctions, polynomial equations, and inner products. In: Smart, N.P. (ed.) EUROCRYPT 2008. LNCS, vol. 4965, pp. 146–162. Springer, Heidelberg (2008)

[LOS+10] Lewko, A., Okamoto, T., Sahai, A., Takashima, K., Waters, B.: Fully Secure Functional Encryption: Attribute-Based Encryption and (Hierarchical) Inner Product Encryption. In: Gilbert, H. (ed.) EUROCRYPT 2010. LNCS, vol. 6110, pp. 62–91. Springer, Heidelberg (2010)

[Nie02] Nielsen, J.B.: Separating Random Oracle Proofs from Complexity Theoretic Proofs: The Non-committing Encryption Case. In: Yung, M. (ed.) CRYPTO 2002. LNCS, vol. 2442, pp. 111–126. Springer, Heidelberg (2002)

[O'N10] O'Neill, A.: Definitional issues in functional encryption. Cryptology ePrint Archive, Report 2010/556 (2010), http://eprint.iacr.org/2010/556

[OSW07] Ostrovsky, R., Sahai, A., Waters, B.: Attribute-based encryption with non-monotonic access structures. In: ACM Conference on Computer and Communications Security, pp. 195–203 (2007)

[OT09] Okamoto, T., Takashima, K.: Hierarchical predicate encryption for inner-products. In: Matsui, M. (ed.) ASIACRYPT 2009. LNCS, vol. 5912, pp. 214–231. Springer, Heidelberg (2009)

[OT10] Okamoto, T., Takashima, K.: Fully secure functional encryption with general relations from the decisional linear assumption. In: Rabin, T. (ed.) CRYPTO 2010. LNCS, vol. 6223, pp. 191–208. Springer, Heidelberg (2010)

[SBC+07] Shi, E., Bethencourt, J., Chan, H.T.-H., Song, D.X., Perrig, A.: Multi-dimensional range query over encrypted data. In: IEEE Symposium on Security and Privacy, pp. 350–364 (2007)

[SF07] Sakai, R., Furukawa, J.: Identity-based broadcast encryption. Cryptology ePrint Archive, Report 2007/217 (2007), http://eprint.iacr.org/

[Sha84] Shamir, A.: Identity-based cryptosystems and signature schemes. In: Blakely, G.R., Chaum, D. (eds.) CRYPTO 1984. LNCS, vol. 196, pp. 47–53. Springer, Heidelberg (1985)

[SSW09] Shen, E., Shi, E., Waters, B.: Predicate privacy in encryption systems. In: Reingold, O. (ed.) TCC 2009. LNCS, vol. 5444, pp. 457–473. Springer, Heidelberg (2009)

[SW05] Sahai, A., Waters, B.: Fuzzy identity-based encryption. In: Cramer, R. (ed.) EUROCRYPT 2005. LNCS, vol. 3494, pp. 457–473. Springer, Heidelberg (2005)

[SW08] Shi, E., Waters, B.: Delegating capabilities in predicate encryption systems. In: Aceto, L., Damgård, I., Goldberg, L.A., Halldórsson, M.M., Ingólfsdóttir, A., Walukiewicz, I. (eds.) ICALP 2008, Part II. LNCS, vol. 5126, pp. 560–578. Springer, Heidelberg (2008)

[Wat05] Waters, B.: Efficient identity-based encryption without random oracles. In: Cramer, R. (ed.) EUROCRYPT 2005. LNCS, vol. 3494, pp. 114–127. Springer, Heidelberg (2005)

[Wat08] Waters, B.: Functional encryption:beyond public key cryptography. Power Point Presentation (2008), `http://userweb.cs.utexas.edu/ bwaters/presentations/files/functional.ppt`

[Wat11] Waters, B.: Ciphertext-policy attribute-based encryption: An expressive, efficient, and provably secure realization. In: Ishai, Y. (ed.) PKC 2011. LNCS, vol. 6597. Springer, Heidelberg (2011)

Concurrent Non-Malleable Zero Knowledge with Adaptive Inputs

Huijia Lin* and Rafael Pass**

Cornell University
{huijia,rafael}@cs.cornell.edu

Abstract. Concurrent non-malleable zero-knowledge ($\mathcal{CNMZK}$) considers the concurrent execution of zero-knowledge protocols in a setting where the attacker can simultaneously corrupt multiple provers and verifiers. We provide the first construction of a $\mathcal{CNMZK}$ protocol that, without any trusted set-up, remains secure even if the attacker may adaptively select the statements to receive proofs of; previous works only handle scenarios where the statements are fixed at the beginning of the execution, or chosen adaptively from a restricted set of statements.

1 Introduction

Zero-knowledge ($\mathcal{ZK}$) interactive proofs [GMR89] are fundamental constructs that allow the Prover to convince the Verifier of the validity of a mathematical statement $x \in L$, while providing *zero additional knowledge* to the Verifier. *Concurrent* $\mathcal{ZK}$, first introduced and achieved by Dwork, Naor and Sahai [DNS04], considers the execution of zero-knowledge protocols in an asynchronous and concurrent setting. In this model, an adversary acts as verifiers in many concurrent executions of the zero-knowledge protocol, and launches a coordinated attack on multiple independent provers to gain knowledge. *Non-malleable* $\mathcal{ZK}$, first introduced and achieved by Dolev, Dwork and Naor [DDN00], also considers the concurrent execution of zero-knowledge protocols, but in a different manner. In this model, an adversary concurrently participates in only two executions, but plays different roles in the two executions; in the first execution (called the left execution), it acts as a verifier, whereas in the second execution (called the right execution) it acts as a prover. The notion of *Concurrent Non-malleable* $\mathcal{ZK}$ ($\mathcal{CNMZK}$) considers both of the above attacks; the adversary may participate in an unbounded number of concurrent executions, playing the role of a prover in some, and the role of a verifier in others. Despite the generality of such an attacks scenario, this notion of security seems most appropriate for modeling the execution of cryptographic protocols in open networks, such as the Internet. Barak, Prabhakaran and Sahai (BPS) [BPS06] provided the the first $\mathcal{CNMZK}$

* Lin is supported by a Microsoft Research Fellowship.

** Pass is supported in part by a Microsoft New Faculty Fellowship, NSF CAREER Award CCF-0746990, AFOSR Award F56-8414, BSF Grant 2006317 and I3P grant 2006CS-001-0000001-02.

Y. Ishai (Ed.): TCC 2011, LNCS 6597, pp. 274–292, 2011.

argument for $\mathcal{NP}$ in the plain model (i.e., without any set-up assumptions); see also the more efficient instantiation of Ostrovsky, Pandey and Visconti [OPV10]. More recently, Lin, Pass, Tseng and Venkitasubramaniam (LPTV) [LPTV10] provided a somewhat different approach to constructing $\mathcal{CNMZK}$ protocols, improving the round-complexity of the BPS construction, as well as providing a construction of a $\mathcal{CNMZK}$ proof.

Adaptive Inputs Selection. All the above-mentioned feasibility results for $\mathcal{CNMZK}$, however, consider a quite restricted form of input selection: More precisely, whereas the attacker is allowed to adaptively select the statements it gives proofs of (on the right), the statements to receive proofs of (on the left) are assumed to be fixed before the execution begins.

Indeed, there is a sense in which this is necessary: as argued by Lindell [Lin03], if we consider a scenario where the left statement are chosen adaptively by an "environment" (think of this as some other arbitrary protocol running in the network), then the notion of $\mathcal{CNMZK}$ collapses down to the notion of *Universally Composable $\mathcal{ZK}$* [Can01], which is known to be unachievable without set-up [CKL03].

We here focus on the simpler case of just "self-composition": that is, we only consider the security of the $\mathcal{ZK}$ protocols (and thus we do not allow them to interact with other protocols in the network; this is similar to the original setting studied in the context of Concurrent $\mathcal{ZK}$). Yet, we want to capture a notion of security where also the statements in the left executions are adaptively chosen. The natural way to do this is to (just as in the definition of security of signature schemes [GMR89]) allow the attacker to adaptively select the statements it wants to hear proofs of on the left (as well as the statements it gives proofs of on the right); additionally we must restrict the attacker to only ask to hear proofs of statements that are true (or else we can never expect the conversation with the provers to be $\mathcal{ZK}$—if the statement is false, then the prover on the left will not be able to provide a proof, which thus reveals information). Once we consider such adaptive instance selection, we also need to specify where the *witnesses* for the left interaction come from; to make the definition as general as possible, we consider an arbitrary (potentially unbounded) *witnesses selecting* machine that receives as input the views of all parties (i.e., the honest prover, the honest verifiers, and the adversary) and outputs a witness w for any statement x requested by the adversary.

We call a $\mathcal{ZK}$ protocol that is secure in this setting a *$\mathcal{CNMZK}$ with Adaptive Input Selection*, or for short *Adaptive $\mathcal{CNMZK}$ ($\mathcal{ACNMZK}$)*. More precisely, a $\mathcal{ZK}$ protocol is $\mathcal{ACNMZK}$ if for every adversary A, there exists a computationally efficient simulator-extractor that can simulate both the left and the right interactions for A, while outputting a witness for every statement proved by the adversary in the right interactions. Our main result is the construction of an $\mathcal{ACNMZK}$ proof:

Theorem 1. *Assume the existence of collision-resistant hash functions. Then there exists a $\omega(\log^2 n)$-round concurrent non-malleable zero-knowledge proof with adaptive input selection (and with a black-box simulator) for all of $\mathcal{NP}$.*

Weaker Notions of Adaptive Input Selection. Our definition of adaptive input selection is strong in the sense that we allow the adversary to select *any* instance x (as long as it is true), and somehow "magically" it receives a proof of this statement. Often, it suffices to restrict to adversaries that only are allowed to request statements x for which there is an *efficient* way to recover a witness if having some auxiliary information. We may consider two ways of formalizing this.

- We may restrict the witness selecting machine to be a computationally bounded non-uniform algorithm that upon receiving a statement x (and potentially also the view of the adversary), *but not the view of the honest provers and verifiers*, outputs a witness w. This models a scenario where the adversary is restricted to only requesting proofs of statements x for which a witness can be efficiently computed using a "super witness". This is a natural extension of the fixed input scenario (again we have a witness—namely the "super witness"—that is fixed before the interaction) and indeed the result of [BPS06, LPTV10] handle also such a notion of adaptive input selection.
- A less restrictive method is to simply restrict the witness selecting machine to be a computationally bounded non-uniform algorithm (which still receives the views of all parties). In particular, this allows us to model a scenario where the adversary may request a proof about earlier proofs (e.g., a proof that the prover has behaved honestly in an earlier proof)—in such a scenario, the honest prover may be able to efficiently find a witness, but there might not exists an efficient algorithm without having access to the prover's random tape. As far as we know, none of the earlier results handle even such a restricted notion of adaptive input selection.

Other Related Work. We mention that there are several works constructing $\mathcal{CNMZK}$ protocols in various trusted set-up models. For instance, previous works [SCO$^+$01, CF01, DN02]) provide constructions of Universally Composable $\mathcal{ZK}$ in the Common Reference String (CRS) model; these protocol are thus also $\mathcal{ACNMZK}$.

An interesting recent work by Yao, Yung and Zhao [YYZ09] provide a construction of a $\mathcal{CNMZK}$ protocol in the Bare Public Key Model; their protocol is not UC secure but satisfies a notion CNMZK with adaptive inputs selection (for both the left and the right interaction); our definition of $\mathcal{ACNMZK}$ in the plain model is heavily inspired by their work.

Techniques. Our protocol is a close variant of the LPTV protocol; let us start by reviewing it. The protocol uses two main components. The first component is the notion of *concurrently extractable commitments* (CECom) introduced by Micciancio, Ong, Sahai, and Vadhan [MOSV06]. Informally, values committed to using a CECom can be extracted by a rewinding simulator even in the concurrent setting. In our protocol (as in most concurrent $\mathcal{ZK}$ protocols), the verifier commits to a random trapdoor using CECom, so that our $\mathcal{ZK}$ simulator may extract this trapdoor to perform simulation. The second component is the notion of *robust non-malleable commitments* (an extension due to Lin and Pass [LP09] of the notion of non-malleable commitments as defined by Dolev, Dwork, and Naor [DDN00]); roughly speaking these are non-malleable commitment schemes with an additional robustness property that makes them convenient to compose with other protocols.

The high-level idea behind the LPTV protocol (just as in the protocol of [BPS06]) is to start off with a *preamble phase* where the verifier uses a CECom to commit to a trapdoor; next in a *commit phase*, the prover commits to a witness of the proof statement using both a CECom and robust non-malleable commitments; and finally during a *proof phase*, the prover proves using a (stand-alone) $\mathcal{ZK}$ protocol that it has either committed to a valid witness, or a valid trapdoor in the commit phase. To prove security, LPTV provides a simulator that uses rewindings to extract out trapdoors (from the CECom in the preamble phase) to simulate the commit and proof phases of the left interactions, and uses rewindings again to extract the witnesses committed to by the adversary (from the CECom in the commit phase) on the right. The crux of the proof is then to show that even during the simulation, when the simulator commits to trapdoors (instead of real witnesses) in left interactions, the adversary still cannot commit to a trapdoor in right interactions, so that the values extracted out from the right interactions must be real witnesses. Very roughly speaking, this follows from the security guarantees of robust non-malleable commitments.

When considering adaptive input selection (for the left executions) a problem arises. First, proving indistinguishability of the simulation becomes problematic: in fact, getting a concurrent $\mathcal{ZK}$ protocol with adaptive input selection is already non-trivial (we call it Adaptive Concurrent Zero-Knowledge ($\mathcal{ACZK}$)); our core technical contribution is to provide a solution to this problem. The reason for this is that proving indistinguishability of the simulation requires performing a hybrid argument, where we switch the witness used in the left interactions from the trapdoors (used by the simulator) to the real witness (used by the prover). More precisely, we consider a hybrid H_i, where the first i left interactions are simulated using the trapdoors, and the later ones use the real witnesses. The problem is that the real witnesses might not be efficiently recoverable since the statements are chosen adaptively by the adversary (it is computed by a computationally unbounded witness-selecting machine); so the hybrid is not efficiently computable!

Our idea for circumventing this problem can be described as follows:

- First, we switch the order of the hybrids. We consider hybrids H_i where the first i left interactions are emulated using real witness and the later ones are simulated using trapdoors. The reason for doing this is that we can now non-uniformly fix the real witnesses of the first i left interactions by hard-coding the "prefix" of hybrid H_i before the ith left interaction; and then the remaining execution can be efficiently emulated using the real witnesses.
- But now the obstacle is that arguing indistinguishability of H_i and H_{i+1} becomes problematic. To show indistinguishability we need to show that simulating the ith left interaction using a real witness or trapdoor is indistinguishable (other interactions are simulated identically in the two hybrids).

It seems that this should just follow from the hiding and $\mathcal{ZK}$ property of the commit and proof phases of the left interaction. However, the problem is that (when trying to extract the trapdoors of the latter left interactions), we might be rewindings the ith left interaction. Our way around this problem is to add more CECom to the preamble phase; the idea is to show that there exists some alternative simulator, that generates a statistically close distribution, but is able to avoid rewinding the messages in the commit and proof phases of the left interaction that we want to violate indistinguishability of.

To also complete the proof of non-malleability, a second (very related problem) arises: namely, we need to argue that the witness committed to by the adversary on the right are valid even in simulation; this is usually done through a hybrid argument as well and relies on the robust non-malleability of the commitment scheme used in the commit phase (instead of the hiding and $\mathcal{ZK}$ properties). When doing this, we again run into the same problem as when showing indistinguishability of the simulation. Here the issue is that we need to ensure that the robust non-malleability property holds even under rewindings. We use the same idea to overcome this problem: as long as there are sufficiently many CECom in the preamble phase, we can describe an alternative simulator that produces a statistically close distribution without rewinding these commitments that we want to violate robust non-malleability of.

Overview. Section 2 contains the basic notations and definitions of $\mathcal{ACNMZK}$ and other primitives. In Section 3, we present our main result, a $\omega(\log^2 n)$-round $\mathcal{ACNMZK}$ proof system for all of $\mathcal{NP}$, from collision resistant hash functions. In Section 4.2, we first focus on showing the $\mathcal{ACZK}$ property of the protocol, which contains the main technical content of this paper; then in Section 5 we sketch how to extend this proof to also show the $\mathcal{ACNMZK}$ property.

2 Preliminaries

Let N denote the set of all positive integers. For any integer $n \in N$, let $[n]$ denote the set $\{1, 2, \ldots, n\}$, and let $\{0,1\}^n$ denote the set of n-bit strings, and ε the empty string. We assume familiarity with interactive Turing machines, interactive protocols, statistical/computational indistinguishability, notions of interactive proofs, zero-knowledge, (strong) witness-indistinguishability, and notions of statistically binding/hiding commitments. (See [Gol01] for formal definitions.)

2.1 Adaptive Concurrent Non-Malleable Zero-Knowledge

Our definition of adaptive concurrent non-malleable zero-knowledge is very similar to that of concurrent non-malleable zero-knowledge from [BPS06] (which in turn closely follows the definition of simulation extractability of [PR05]), with the only difference that now the adversary is allowed to adaptively select the statements it receives proofs to, subject to that they are true statements.

Let $\langle P, V\rangle$ be an interactive proof for a language $L \in \mathcal{NP}$ with witness relation R_L, and let n be the security parameter. Consider a man-in-the-middle adversary

A that participates in many left and right interactions in which $m = m(n)$ proofs take place. In the left interactions, the adversary A verifies the validity of statements $x_1, \ldots, x_m$ by interacting with an honest prover P, using identities $\mathsf{id}_1, \ldots, \mathsf{id}_m$. In the right interactions, A proves the validity of statements $\tilde{x}_1, \ldots, \tilde{x}_m$ to an honest verifier V, using identities $\tilde{\mathsf{id}}_1, \ldots, \tilde{\mathsf{id}}_m$. Prior to the interactions, all parties in the system receives as common input the security parameter in unary 1^n, and A receives as auxiliary input $z \in \{0,1\}^*$. Furthermore, at the beginning of each left (respectively right) interaction, the adversary adaptively selects the statement x_i (respectively $\tilde{x}_i$) and the identity id_i (respectively $\tilde{\mathsf{id}}_i$), with the only restriction that all the statements $x_1, \ldots, x_m$ chosen in the left interactions have to be true. Additionally in each left interaction, the prover P receives as local input a witness $w_i \in R_L(x_i)$, chosen adaptively by a *witness-selecting machine* M. More specifically, M is a (randomized) Turing machine that runs in *exponential* time, and whenever the adversary chooses a statement x_i for a left interaction, M on inputs the statement x_i and the current view of all parties (including the adversary, provers, and receivers), picks a witness $w_i \in R_L(x_i)$ as the private input of the prover P. Let $\mathsf{view}_{A,M}(n,z)$ denote a random variable that describes the view of A in the above experiment. Loosely speaking, an interactive proof is adaptive concurrent non-malleable zero-knowledge ($\mathcal{ACNMZK}$) if for all man-in-the-middle adversary A, there exists a probabilistic polynomial time machine (called the simulator-extractor) that can simulate both the left and the right interactions for A, while outputting a witness for every statement proved by the adversary in the right interactions.

Definition 1. *An interactive proof (P, V) for a language L with witness relation R_L is said to be* adaptive concurrent non-malleable zero-knowledge *if for every polynomial m, and every probabilistic polynomial-time man-in-the-middle adversary A that participates in at most $m = m(n)$ concurrent executions, there exists a probabilistic polynomial time machine S, such that, for every input-selecting machine M:*

1. *The following ensembles are computationally indistinguishable over $n \in N$*
   - $\{\mathsf{view}_{A,M}(n,z)\}_{n \in N, z \in \{0,1\}^*}$
   - $\{S_1(1^n, z)\}_{n \in N, z \in \{0,1\}^*}$

   where $S_1(1^n, z)$ denotes the first output of $S(1^n, z)$.
2. *Let $z \in \{0,1\}^*$ and $(\mathsf{view}, \boldsymbol{w})$ denote the output of $S(1^n, z)$. Let $\tilde{x}_1, \ldots, \tilde{x}_m$ be the statements of the right-interactions in* view*, and let $\mathsf{id}_1, \ldots, \mathsf{id}_m$ and $\tilde{\mathsf{id}}_1, \ldots, \tilde{\mathsf{id}}_m$ be the identities of the left-interactions and right-interactions in* view*. Then for every $i \in [m]$, if the ith right-interaction is accepting and $\tilde{\mathsf{id}}_i \neq \mathsf{id}_j$, $\boldsymbol{w}$ contains a witness w_i such that $R_L(\tilde{x}_i, w_i) = 1$.*

We also consider concurrent $\mathcal{ZK}$ with adaptive input selection. We say that an interactive proof (P, V) is adaptive concurrent $\mathcal{ZK}$ ($\mathcal{ACZK}$) if it satisfies the above definition with respect to adversaries that only receive proofs (and do not give proofs).

Remark 1. As mentioned before, the security proof in [BPS06, LPTV10] can be extended to show that their constructions of $\mathcal{CNMZK}$ protocols satisfy a notion of $\mathcal{CNMZK}$ with "weak" adaptive input selection, where the adversary can only choose to hear proofs of statements for which a witness can be computed efficiently *without knowing the random coins of the honest provers and verifiers.* Formally, the witness-selecting machine is restricted to be computationally bounded (i.e., a non-uniform $\mathcal{PPT}$) and only receive as input a statement x and the view of the adversary (but not the views of the left provers and right receivers.)

Remark 2. Universal Composability (UC) [Can01] considers a more generalized form of adaptive input selection, where both the statements and witnesses are chosen adaptively by a separate entity called the "environment", which may communicate with the adversary in an arbitrary way. In contrast, our definition of $\mathcal{ACNMZK}$ only allows the witnesses to be selected by a separate entity, whereas the statements are chosen directly by the adversary. It has been shown that UC $\mathcal{ZK}$ is unachievable without set-up [CKL03]. We mention that our construction actually satisfies a slight strengthening of the above definition of $\mathcal{ACNMZK}$, where the statements are adaptively chosen by a *stateless non-uniform* $\mathcal{PPT}$ machine that both the adversary and the simulator have oracle accesses to. Such a notion bring us closer to the definition of UC $\mathcal{ZK}$—in essence, the difference is that in UC $\mathcal{ZK}$ the statement selecting machine is not necessarily stateless; we defer the details to the full version.

Non-Malleable Commitment Schemes. We recall the definition of non-malleability from [LPV08] (which builds upon the definition of [DDN00, PR05]). Let $\langle C, R\rangle$ be a tag-based statistically binding commitment scheme, and let $n \in N$ be a security parameter. Consider a man-in-the-middle adversary A that, on auxiliary inputs n and z, participates in one left and one right interaction simultaneously. In the left interaction, the man-in-the-middle adversary A interacts with C, receiving a commitment to value v, using identity id of its choice. In the right interaction A interacts with R attempting to commit to a related value $\tilde{v}$, again using identity $\tilde{\mathsf{id}}$ of its choice. If the right commitment is invalid, or undefined, its value is set to $\perp$. Furthermore, if $\tilde{\mathsf{id}} = \mathsf{id}$, $\tilde{v}$ is also set to $\perp$—i.e., a commitment where the adversary copies the identity of the left interaction is considered invalid. Let $\mathsf{nmc}^A_{\langle C,R\rangle}(v, z)$ denote a random variable that describes the value $\tilde{v}$ and the view of A, in the above experiment.

Definition 2. *A statistically binding commitment scheme $\langle C, R\rangle$ is said to be* non-malleable (with respect to itself) *if for every polynomial $p(\cdot)$, and every probabilistic polynomial-time man-in-the-middle adversary A, the following ensembles are computationally indistinguishable.*

$$\left\{\mathsf{nmc}^A_{\langle C,R\rangle}(v, z)\right\}_{n\in N, v\in\{0,1\}^n, v'\in\{0,1\}^n, z\in\{0,1\}^*}$$

$$\left\{\mathsf{nmc}^A_{\langle C,R\rangle}(v', z)\right\}_{n\in N, v\in\{0,1\}^n, v'\in\{0,1\}^n, z\in\{0,1\}^*}$$

Non-Malleable Commitment Robust w.r.t. k-Round Protocols. The notion of non-malleability w.r.t. arbitrary k-round protocols is introduced in [LP09]. Unlike traditional definitions of non-malleability, which only consider man-in-the middle adversaries that participate in two (or more) executions of the *same* protocol, non-malleability w.r.t. arbitrary protocols considers a class of adversaries that can participate in a left interaction of any arbitrary protocol. Below we recall the definition. Consider a one-many man-in-the-middle adversary A that participates in one left interaction—communicating with a machine B—and one right interaction—acting as a commiter using the commitment scheme $\langle C, R\rangle$. As in the standard definition of non-malleability, A can adaptively choose the identity in the right interaction. We denote by $\mathsf{nmc}^{B,A}_{\langle C,R\rangle}(y, z)$ the random variable consisting of the view of $A(z)$ in a man-in-the-middle execution when communicating with $B(y)$ on the left and an honest receiver on the right, combined with the value $A(z)$ commits to on the right. Intuitively, we say that $\langle C, R\rangle$ is non-malleable w.r.t. B if $\mathsf{nmc}^{B,A}_{\langle C,R\rangle}(y_1, z)$ and $\mathsf{nmc}^{B,A}_{\langle C,R\rangle}(y_2, z)$ are indistinguishable, whenever interactions with $B(y_1)$ and $B(y_2)$ cannot be distinguished.

Definition 3. *Let B be a probabilistic polynomial time machine. We say the statistically binding commitment scheme $\langle C, R\rangle$ is* non-malleable w.r.t. B, *if for every probabilistic polynomial-time man-in-the-middle adversary A, and every two sequences $\{y^1_n\}_{n\in N}$ and $\{y^2_n\}_{n\in N}$ such that, for all probabilistic polynomial-time machine $\tilde{A}$, it holds that*

$$\left\{\langle B(y^1_n), \tilde{A}(z)\rangle(1^n)\right\}_{n\in N, z\in\{0,1\}^*} \approx \left\{\langle B(y^2_n), \tilde{A}(z)\rangle(1^n)\right\}_{n\in N, z\in\{0,1\}^*}$$

where $\langle B(y), \tilde{A}(z)\rangle(1^n)$ denotes the view of $\tilde{A}$ in interaction with B on common input 1^n, and private inputs z and y respectively, then it holds that:

$$\left\{nmc^{B,A}_{\langle C,R\rangle}(y^1_n, z)\right\}_{n\in N, z\in\{0,1\}^*} \approx \left\{nmc^{B,A}_{\langle C,R\rangle}(y^2_n, z)\right\}_{n\in N, z\in\{0,1\}^*}$$

We say that $\langle C, R\rangle$ is non-malleable w.r.t. k-round protocols if $\langle C, R\rangle$ is non-malleable w.r.t. any $\mathcal{PPT}$ machine B that interacts with the man-in-the-middle adversary in k rounds. Below, we focus on commitment schemes that are non-malleable w.r.t. itself and arbitrary $\ell(n)$-round protocols, where ℓ is a super-logarithmic function. We say that such a commitment scheme is robust w.r.t. $\ell(n)$-round protocols. The following result was shown in [LPV08].

Lemma 1 ([LPV08]). *Let $\ell(n)$ be a super-logarithmic function. Then there exists a $O(\ell(n))$-round statistically binding commitment scheme that is robust w.r.t. $\ell(n)$-round protocols, assuming that one-way functions exist.*

Concurrently Extractable Commitment Schemes. Micciancio, Ong, Sahai and Vadhan introduce and construct *concurrently extractable commitment schemes*, CECom, in [MOSV06]. The commitment scheme is an abstraction of the preamble stage of the concurrent zero-knowledge protocol of [PRS02]. Informally, values committed by CECom can be extracted by a rewinding extractor (e.g., the zero-knowledge simulator of [KP01, PRS02, PTV08]), even in the concurrent setting. In this work, we use the same construction as in [PRS02, MOSV06], but are unable to employ their analysis.

3 An $\mathcal{ACNMZK}$ Proof

In this section we construct an adaptive concurrent non-malleable zero-knowledge proof based on collision-resistant hash-functions. The construction is almost identical to the $\mathcal{CNMZK}$ proof system in [LPTV10], except that, the verifier is asked to provide more CECom commitments to its trapdoor at the beginning of the protocol, which, in the proof, facilitates the simulator-extractor to extract the trapdoor while strategically avoiding rewinding certain messages.

Let $\ell(n)$ be any super logarithmic function. Our adaptive concurrent non-malleable zero-knowledge protocol, ACNMZKProof, employs several commitment protocols. Let Com_{sh} be a 2-round statistically *hiding* commitment (based on collision-resistant hash-functions), Com_{sb} be a 2-round statistically *binding* commitment (based on one-way functions), and NMCom be an $O(\ell(n))$-round statistically binding commitment scheme that is robust w.r.t. $\ell(n)$-round protocols (based on one-way functions).

Our protocol also employs $\ell(n)$-round statistically hiding (respectively statistically binding) concurrently-extractable commitment schemes, CECom_{sh} (respectively CECom_{sb}). These schemes are essentially instantiations of the PRS preamble [PRS02], and can be constructed given Com_{sh} and Com_{sb}. Below we repeat their definitions.

To commit a n-bit string v, the commiter chooses $n \times \ell(n)$ pairs of random n-bit strings $(\alpha^0_{i,j}, \alpha^1_{i,j})$, $i \in [n], j \in [\ell(n)]$, such that $\alpha^0_{i,j} \oplus \alpha^1_{i,j} = v$ for every i and j. The sender then commits to v and each of the $2n\ell(n)$ strings in parallel using Com_{sh}. This is followed by $\ell(n)$ rounds of interactions. In the jth interaction, the receiver sends a random n-bit challenge $b_j = b_{1,j} \dots b_{n,j}$, and the commiter decommits the commitments of $\alpha^{b_{1,j}}_{1,j}, \dots, \alpha^{b_{n,j}}_{n,j}$ according to the challenge.

A valid decommitment of CECom_{sh} requires the commiter to decommit all initial commitments under scheme Com_{sh} (i.e., reveal the randomness of the commitments), and that the decommitted values satisfy $\alpha^0_{i,j} \oplus \alpha^1_{i,j} = v$ for every i and j.

A $\ell(n)$-round statistically binding concurrently-extractable commitment scheme, CECom_{sh}, is defined analogously as CECom_{sh} with the initial commitment Com_{sh} replaced by Com_{sb}. Additionally, we say a transcript of CECom_{sh} is *valid* if there exists a valid decommitment.

We now describe ACNMZKProof, our adaptive concurrent non-malleable zero-knowledge protocol. Protocol ACNMZKProof for a language $L \in \mathcal{NP}$ proceeds in six stages given a security parameter n, a common input statement $x \in \{0,1\}^n$, an identity id, and a private input $w \in R_L(x)$ to the Prover.

Stage 1: The Verifier chooses a random string $r \in \{0,1\}^n$ and commits to r using $k(n)+1$ invocations of CECom_{sh}, where $k(n)$ is the number of rounds in Stage 2-6 of the protocol; r is called the "fake witness".

Stage 2: The Prover commits to the witness w using CECom_{sb}.

Stage 3: The Prover commits to the witness w using NMCom with identity id.

Stage 4: The Prover commits to the witness w using NMCom with identity id, again.

Stage 5: The Verifier decommits the Stage 1 commitment to value r'.

Stage 6: The Prover using a $\omega(1)$-round $\mathcal{ZK}$ proof (e.g., [Blu86]), proves that the commitments in Stages 2, 3 and 4 all commit to the same value $\tilde{w}$ (with identity id), and that either $\tilde{w} \in R_L(x)$ or $\tilde{w} = r'$.

A formal description of the protocol can be found in Figure 1.

Protocol ACNMZKProof

Common Input: an instance x of a language L with witness relation R_L, an identifier id, and a security parameter n.

Auxiliary Input for Prover: a witness w, such that $(x, w) \in R_L(x)$.

Stage 1:

V uniformly chooses $r \in \{0,1\}^n$ (the "fake witness").

V commits to r using $k+1$ invocations of the protocol CECom_{sh}, where k is the number of rounds in Stage 2-6 of the protocol. Let $\mathcal{T}_1$ be the commitment transcript.

Stage 2:

P commits to w using protocol CECom_{sb}. Let $\mathcal{T}_2$ be the commitment transcript.

Stage 3:

P commits to w using protocol NMCom and identity id. Let $\mathcal{T}_3$ be the commitment transcript.

Stage 4:

P commits to w using protocol NMCom and identity id. Let $\mathcal{T}_4$ be the commitment transcript.

Stage 5:

V decommits $\mathcal{T}_1$ to value r; P aborts if no valid decommitment is given.

Stage 6:

P $\leftrightarrow$ V: a $\omega(1)$-round $\mathcal{ZK}$ proof [Blu86] of the statement: There exists $\tilde{w}$ such that

- $\tilde{w}$ is a valid decommitment of $\mathcal{T}_2$,
- *and* $\tilde{w}$ is a valid decommitment of $\mathcal{T}_3$ and $\mathcal{T}_4$ under identity id,
- *and* $\tilde{w} \in R_L(x)$ or $\tilde{w} = r$.

Fig. 1. An Adaptive Concurrent Non-Malleable $\mathcal{ZK}$ Proof for $\mathcal{NP}$

On Round Complexity: Since the protocol NMCom has $O(\ell(n))$ rounds, we have that $k(n) = O(\ell(n))$. Therefore, the round complexity of the protocol ACNMZKProof is $O(\ell^2(n)) = \omega(\log^2 n)$.

The above protocol is an extension of the Goldreich-Kahan protocol [GK96]. Completeness and Soundness follows using stand techniques; since the protocol is essentially the same as the $\mathcal{CNMZK}$ protocol in [LPTV10] (except that Stage 1 now contains many CECom's), we refer the reader to [LPTV10] for more details.

4 Proof of Security

The definition of $\mathcal{ACNMZK}$ requires a simulator-extractor S that is able to simulate the view of a man-in-the-middle adversary A (including both left and right interactions), while simultaneously extracting the witnesses to statements proved in the right interactions. We describe the construction of our simulator in Section 4.1, show that it is a correct $\mathcal{ACZK}$ simulator in Section 4.2, and extend this proof to show the $\mathcal{ACNMZK}$ property in Section 5.

4.1 Our Simulator-Extractor

Our simulator-extractor, S, is almost identical to the simulator extractor of the $\mathcal{CNMZK}$ protocol in [LPTV10], except that now, given more CECom_{sh}'s in Stage 1 of the protocol, S tries to extract a "fake" witness from every CECom_{sh} from the adversary in the left interactions, and aborts if the extraction fails for any of the commitment or the extracted value does not equal to the value that the adversary decommitment to later. Roughly speaking, S follows this strategy:

Simulating the view of the right interactions. S simply follows the honest verifier strategy.

Simulating the view of the left interactions. In each protocol execution, S first extracts the "fake witness" r from the $k(n)+1$ CECom_{sh}'s committed by A in Stage 1, then commits to r in Stage 2, 3, and 4, and finally simulates the $\mathcal{ZK}$ proof using r as a witness in Stage 6.

Extracting the witnesses. In each right interaction that completes successfully, S extracts a witness w from CECom_{sb} committed by A in Stage 2 of the protocol.

Thus, the main task of S is to extract the values committed by A, using CECom, in Stage 1 and 2 of the protocol. This is done by rewinding A during each CECom. To that end, we employ the "lazy KP" simulator of [PTV08], an oblivious simulator that is nearly identical to the Killian-Petrank (KP) simulator [KP01]. We also follow the analysis of [PTV08], which is in turn based on the analysis of [PRS02].

On a very high-level, S attempts to simulate the view of A (with "fake witnesses") in one continuous, straight-line manner (so as to not skew the output distribution); this is aided by numerous auxiliary rewinds that allows S to extract the "fake witnesses" in time. As implied by our simulation strategy, the view of A generated by S depends on the extracted "fake witnesses", but is otherwise independent of the interaction in auxiliary rewinds. (The simulator S is essentially identical to the simulator of the $\mathcal{CNMZK}$ protocol in [LPTV10]; we refer the reader to [LPTV10] for a more detailed description.)

It is useful to know that S may abort in two manners. At the end of a CECom, if S is unable to extract the committed value (the rewinds were unhelpful), S outputs $\bot_{ext}$. Or, in Stage 5 of a left interaction, if A decommits its Stage 1 CECom_{sh}'s to a value that is different from any of the $k(n)+1$ extracted values, S outputs $\bot_{bind}$. Conversely if S does not abort, then it must have extracted the committed value from every Stage 1 CECom_{sh} that it has encountered, and A must decommit to the extracted values (if A decommits at all). The following claim bounds the abort probability of S.

Claim 2. *S outputs $\bot_{ext}$ and $\bot_{bind}$ with negligible probability.*

The proof is identical to the proof of Claim 2 in [LPTV10], which in turn follow directly from the analysis of [PTV08] in the setting of concurrent $\mathcal{ZK}$; we refer the reader to [LPTV10] for a formal proof.

4.2 Proof of $\mathcal{ACZK}$

We first show that S is a valid $\mathcal{ACZK}$ simulator for the protocol $\mathsf{ACNMZKProof}$, that is, the view generated by S is indistinguishable from the real view of A.

Lemma 3. *For every witness-selecting machine M, the following ensembles are computationally indistinguishable over $n \in \mathsf{N}$:*

$$\{S_1(1^n, z)\}_{n\in \mathsf{N}, z\in\{0,1\}^*}$$
$$\{\mathsf{view}_{A,M}(1^n, z)\}_{n\in \mathsf{N}, z\in\{0,1\}^*}$$

To show Lemma 3, we introduce a series of hybrid simulators; the same hybrid simulators will also be helpful later in showing the $\mathcal{ACNMZK}$ property in Section 5. Hybrids hyb^i, $0 \le i \le m+1$ proceed in three steps.

Real Execution Phase: Run the honest man-in-the-middle execution with A until the ith left interaction starts: in left iteration $j < i$, run the witness-selecting machine M (on input the statement x_j of this interaction, and the current view of the adversary, prover and verifier) to compute a valid witnesses w_j and execute the honest prover strategy. Note that the Real Execution Phase may take exponential time. Let $\mathcal{V}_A$ be the view of A, and $\mathcal{V}_P$, $\mathcal{V}_V$ the view of the prover and verifier produced in this phase.

Simulation Phase: Feed A with $\mathcal{V}_A$. Run the following simulation strategy with A to complete the partial execution defined by $(\mathcal{V}_A, \mathcal{V}_P, \mathcal{V}_V)$.
- For every right interaction, emulate the interaction by following the honest verifier strategy from $\mathcal{V}_V$.
- For left interaction $j < i$, emulate the interaction by following the honest prover strategy from $\mathcal{V}_P$.

- For left interaction $j \geq i$, simulate the interaction using a "fake" witness, as S does.

Formally, this simulation strategy can be implemented as follows: construct another machine A' that internally incorporates A and simulates the first $i-1$ left and all right interactions for A honestly from $\mathcal{V}_P$ and $\mathcal{V}_V$, and forwards the rest $m-i+1$ left interactions externally. Then simply run S on A' and outputs the embedded view of A in the view of A' produced by S.

Output Phase: Output $\bot_{ext}$ or $\bot_{bind}$ if S returns $\bot_{ext}$ or $\bot_{bind}$; otherwise, output the view $\mathcal{V}$ of A embedded in the view of A' produced by S.

We also define hybrids hyb^i_+ that proceed identically to hyb^i except that, in the Simulation Phase, the ith left interaction is simulated using a real witness (rather than the "fake" witness). This can be done as the Real Execution Phase runs till the ith left interaction starts, and can also compute the real witness of the ith interaction. Note that these hybrids $\{\mathsf{hyb}^i\}$, $\{\mathsf{hyb}^i_+\}$ are only concerned with producing a view of A, and do not extract the witnesses of the right interactions.

By construction, hyb^i and hyb^i_+ abort only when S aborts. Hence by Claim 2, we have,

Claim 4. *For all i, hyb^i and hyb^i_+ output $\bot$ with negligible probability.*

By Claim 4, the output of hyb^1 is statistically close to the output of S running with A in its entirety. (They only differ when S aborts due to trying to extract witnesses of the right interactions from the CECom_{sb}'s committed by A.) The output of hyb^{m+1}, on the other hand, is identical to the real view of A. Therefore Lemma 3 directly follows from the next two claims:

Lemma 5. *The outputs of hyb^i_+ and hyb^{i+1} are statistically close.*

Proof. Ignoring the fact that hyb^i_+ and hyb^{i+1} may abort, their outputs are identical. This is because hyb^i_+ differs from hyb^{i+1} only in that when generating the output view, from the beginning of the ith left interaction until the beginning of the $i+1^{\text{st}}$ left interactions, hyb^i_+ employs *rewinds*. However, these rewinds do not extract any new "fake witnesses" for use in the output view, and do not skew the output distribution because the rewinding schedule (including which rewind determines the output view) is oblivious. Since both machines abort at most with negligible probability by Claim 4, their outputs are statistically close.

Lemma 6. *The outputs of hyb^i and hyb^i_+ are computationally indistinguishable.*

Proof. Assume for contradiction that there exists an adversary A and a polynomial p, such that, for infinitely many $n \in N$, hyb^i and hyb^i_+ are distinguishable with probability $1/p(n)$. Towards reaching a contradiction, note that hyb^i and hyb^i_+ differ only in how the ith left interaction is simulated (fake or real witness) in the rewindings. We thus want to violate the computational hiding property of Stage 2-4 of the protocol, or the strongly witness-indistinguishable property (implied by the $\mathcal{ZK}$ property) of Stage 6. However, two problems arise: (1) the

Real Execution Phase of the two hybrids takes exponential steps, and (2) Stage 2-6 of the ith left interaction maybe *rewound* by the simulator S. Fix a $n \in N$ for which our hypothesis holds. To overcome the first problem, by our hypothesis, there must exist an execution of the Real Execution Phase—defined by the views of the adversary $\mathcal{V}_A$, the left prover $\mathcal{V}_P$ and the right verifier $\mathcal{V}_V$ produced in this phase—such that, conditioned on $(\mathcal{V}_A, \mathcal{V}_P, \mathcal{V}_V)$ occurring in the two hybrids, hyb^i and hyb^i_+ are still distinguishable with probability $1/p(n)$. Given $(\mathcal{V}_A, \mathcal{V}_P, \mathcal{V}_V)$, the rest of the hybrids (i.e., the Simulation Phase and Output Phase) can be generated efficiently.

Now it only remains to handle the second problem, that is, Stage 2-6 of the ith left interaction may be rewound in the Simulation Phase. We consider another two hybrids $\tilde{\mathsf{hyb}}^i$ and $\tilde{\mathsf{hyb}}^i_+$, which proceed identically to hyb^i and hyb^i_+ respectively, except that, in the Simulation Phase, they employ the following alternative simulation strategy that avoids rewinding Stage 2-6 of the ith left interactions.

The Alternative Simulation Strategy of $\tilde{\mathsf{hyb}}^i$**:** The goal of this simulation strategy is to complete the partial execution $(\mathcal{V}_A, \mathcal{V}_P, \mathcal{V}_V)$ produced by the Real Execution Phase, without rewinding Stage 2-6 of the ith left interaction. Let $\{m_1, \ldots, m_t\}$ for $t = k(n)/2$, be the messages that A sends in Stage 2-6 of the ith left interaction; and a_i the reply to m_i from the left prover. Then the execution of A continuing from $\mathcal{V}_A$ is "equivalent" to the sequential execution of the following $t+1$ machines $A_1, \ldots, A_{t+1}$.

Machine A_i on input a partial view $\mathcal{V}_{i-1}$ of A up until the message m_{i-1} is sent and the reply a_{i-1} ($\mathcal{V}_0 = \mathcal{V}_A$ and $a_0 = \varepsilon$), continues the execution of A from $\mathcal{V}_{i-1}$, by feeding $\mathcal{V}_{i-1}$ and a_{i-1} to A, and forwarding every message from A externally; finally, it aborts when A terminates or sends the message m_i, and output the newly generated view $\mathcal{V}_i$ of A.

The alternative simulation strategy, instead of producing a simulated view of A "in one shot", produces the view "progressively" by simulating the view of $A_1, \ldots, A_{t+1}$ in sequence. Furthermore, the simulation strategy remembers all the "fake witnesses" it has extracted so far, and to simulate the view of A_i, it can use the "fake" witnesses extracted when simulating the views of A_j's with $j < i$. More precisely, let $\mathcal{S}^i$ ($\mathcal{S}^0 = \emptyset$) denote the set of "fake witnesses" extracted after simulating the views of the first i machines $A_1, \ldots, A_i$, and $(\mathcal{V}^j_A, \mathcal{V}^j_P, \mathcal{V}^j_V)$ ($(\mathcal{V}^0_A, \mathcal{V}^0_P, \mathcal{V}^0_V) = (\mathcal{V}_A, \mathcal{V}_P, \mathcal{V}_V)$) the partial execution produced after simulating the first i machines. In step $j \in [t+1]$,

1. Simulate the view of A_j continuing from $(\mathcal{V}^{j-1}_A, \mathcal{V}^{j-1}_P, \mathcal{V}^{j-1}_V)$ as in hyb^i—that is, emulate the first $i-1$ left and all the right interactions honestly from $\mathcal{V}^{j-1}_P$ and $\mathcal{V}^{j-1}_V$, and simulate the rest $m-i+1$ left interactions using "fake" witnesses—except that now the "fake" witnesses can be obtained through extracting from some CECom_{sh}'s in this step, or in previous steps, found in $\mathcal{S}$. (Output $\perp_{ext}$ if no such fake witness is available, and $\perp_{bind}$, if A_i decommits to a value different from any of the "fake" witnesses extracted.)

2. Set $\mathcal{V}_A^j$ to the view of A embedded in the simulated view of A_i (set $\mathcal{V}_P^j$ and $\mathcal{V}_V^j$ appropriately as well); add all the "fake" witnesses extracted in this step to $\mathcal{S}$.

Finally, $\tilde{\mathsf{hyb}}^i$ outputs $\mathcal{V} = \mathcal{V}_A^{t+1}$.

We remark that in step j, the only message that A_j receives belonging to Stage 2-6 of the ith left interaction is a_{j-1}. This is because A in A_j starts its execution from $\mathcal{V}_A^{j-1}$, after messages m_1 to m_{j-1} are sent, and is cutoff immediately after m_j is sent. Therefore, during the simulation with A_i, in every rewinding, A never sends m_1 to m_{j-1} again, and never receives a reply to m_j (as every time it does send m_j, it is cutoff immediately). Hence the only message it receive is a_{j-1}. Therefore, overall, the alternative simulation strategy never rewinds Stage 2-6 of the ith left interaction.

The Alternative Simulation Strategy of $\tilde{\mathsf{hyb}}_+^i$: Define $\tilde{\mathsf{hyb}}_+^i$ analogously for hyb_+^i. $\tilde{\mathsf{hyb}}_+^i$ proceeds identically to $\tilde{\mathsf{hyb}}^i$, except that in the simulation with A_j's, messages in Stage 2-6 of the ith left interaction are emulated using the real witness (as in hyb_+^i). As $\tilde{\mathsf{hyb}}^i$, $\tilde{\mathsf{hyb}}_+^i$ never rewinds Stage 2-6 of the ith left interaction.

Claim 7. *For all i, $\tilde{\mathsf{hyb}}^i$ and $\tilde{\mathsf{hyb}}_+^i$ output $\perp$ with negligible probability.*

Proof. It essentially follows from Claim 4 that the probabilities that $\tilde{\mathsf{hyb}}^i$ and $\tilde{\mathsf{hyb}}$ outputs $\perp_{bind}$ are negligible.

On the other hand, $\tilde{\mathsf{hyb}}^i$ (respectively $\tilde{\mathsf{hyb}}_+^i$) outputs $\perp_{ext}$ only if it fails to extract a "fake" witness for some left interaction $j \geq i$ (respectively $j > i$). Fix one such j. Since left interaction j starts completely after $\mathcal{V}_A$ (the view generated in the Real Execution Phase), the execution of this interaction occurs completely inside machines $A_1, \ldots, A_{t+1}$, where $t = k/2$. Then since the number of CECom_{sh}'s in Stage 1 of the left interaction is $k+1 > t+1$, there exists a machine $A_{j'}$, such that, during its execution, a complete CECom_{sh} from A is sent. Then in Step j' of $\tilde{\mathsf{hyb}}^i$ (respectively $\tilde{\mathsf{hyb}}_+^i$), the alternative simulation strategy must try to extract a "fake" witness from this CECom_{sh}, and by Claim 4, it succeeds except with negligible probability. Therefore, by union bound, the probability that $\tilde{\mathsf{hyb}}^i$ (respectively $\tilde{\mathsf{hyb}}_+^i$) outputs $\perp_{ext}$ is negligible.

Furthermore, ignoring the fact that $\tilde{\mathsf{hyb}}^i$ and hyb^i (resp., $\tilde{\mathsf{hyb}}_+^i$ and hyb_+^i) may abort, their outputs are identical, since the views of A in $\tilde{\mathsf{hyb}}^i$ and hyb^i are simulated identically. (This is because that the simulated view of A depends only on the value of the "fake" witnesses extracted, and is otherwise oblivious of the extraction strategy. Following from the same proof as in Claim 2, we have that for every left interaction in which the adversary successfully decommits the Stage 1 commitment (in Stage 5), the "fake" witnesses extracted in the two hybrids are identical, except from negligible probability. For the rest left interactions, the extracted "fake" witnesses are never used in the simulation.) Therefore,

Claim 8. *For all i, it holds that the outputs of $\tilde{\mathsf{hyb}}^i$ and hyb^i are statistically close, and the outputs of $\tilde{\mathsf{hyb}}^i_+$ and hyb^i_+ are statistically close.*

Combining Claim 8 with our hypothesis, we have that conditioned on $(\mathcal{V}_A, \mathcal{V}_P, \mathcal{V}_V)$ occurring in the two hybrids, $\tilde{\mathsf{hyb}}^i$ and $\tilde{\mathsf{hyb}}^i_+$ are distinguishable with probability at least $1/2p(n)$. Note that continuing from $(\mathcal{V}_A, \mathcal{V}_P, \mathcal{V}_V)$ the rest of the two hybrids can be efficiently generated, and the only difference between the two hybrids lies in how Stage 2-6 of the ith left interaction are simulated (using a fake or a real witness), which are *never rewound* in the two hybrids. Then it follows directly from the computational hiding property of Stage 2-4, and the strongly witness-indistinguishable property (implied by the $\mathcal{ZK}$ property) of Stage 6 that conditioned on $(\mathcal{V}_A, \mathcal{V}_P, \mathcal{V}_V)$, $\tilde{\mathsf{hyb}}^i$ and $\tilde{\mathsf{hyb}}^i_+$ are indistinguishable. This gives a contradiction.

5 Proof of $\mathcal{ACNMZK}$

As shown in the last section, the simulator constructed in Section 4.1 is a correct $\mathcal{ACZK}$ simulator; that is, the first output of S (i.e., $\mathsf{view} = S_1(1^n, z)$) is computationally indistinguishable from the real view of the adversary. To further show that S is also a correct $\mathcal{ACNMZK}$ simulator-extractor, it remains to show that the second output of S contains the valid $\mathcal{NP}$ witnesses of the statements proved in the right interactions (in view).

By construction, the witnesses that S outputs are just values it extracts out from the CECom_{sb}'s in Stage 2 of the right interactions. Therefore, if A always commits to valid witnesses using CECom_{sb} in the right interactions, by Claim 2 the simulator S would extract the valid witnesses except with negligible probability. Therefore, the following lemma establishes the correctness of the output witnesses:

Lemma 9. *For every $\mathcal{PPT}$ adversary A, there exists a negligible function ν, such that for every $n \in N$ and $z \in \{0,1\}^*$, the probability that A fails to commit to a valid witness in Stage 2 of a right interaction that is accepting and uses a different identity from all left interactions in $\mathsf{view} = S_1(1^n, z)$, is less than $\nu(n)$.*

Proof. Assume for contradiction that there exists a man-in-the-middle adversary A that participates in $m = m(n)$ left and right interactions, and a polynomial function p, such that for infinitely many $n \in N$ and $z \in \{0,1\}^*$, A *cheats* in an outcome of $S_1(1^n, z)$ with probability $1/p(n)$; by cheating, we mean that A fails to commit to a valid witness in Stage 2 of any right interaction that is accepting and uses a different identity from all the left interactions. (Note that A is not considered cheating if the simulator fails to output a view of A).

Consider again the series of hybrids, hyb^i and hyb^i_+, defined in section 4.2. Since the output of hyb^1 is statistically close to the output of S, by our hypothesis, the probability that A cheats in hyb^1 is non-negligible. On the other hand, in hyb^{m+1}, it follows from the soundness of Stage 6 that, except with negligible probability, in every accepting right interaction, A commits to either a real or a "fake" witness; it further follows from the statistically hiding property of Stage

1 and the (stand-alone) extractability of Stage 2 that, except with negligible probability, A never commits to a "fake" witness in any accepting right interactions. Hence, by union bound, except with negligible probability, A never cheats in hyb^{m+1}. It follows from Claim 6 that the probabilities of A cheating in hyb^i_+ and hyb^{i+1} differ by at most a negligible amount. Therefore, for infinitely many n, there must exist an $i = i(n)$, such that, the probabilities of A cheating in hyb^i and hyb^i_+ differ by at least a polynomial amount. Since the total number of right interactions is bounded by a polynomial, this implies that the probabilities that A cheats in a *randomly chosen* right interaction in the two hybrids differ by a polynomial amount.

Notice that the hybrids hyb^i and hyb^i_+ proceed identically up until the ith left interaction starts. After that, the only difference between the two experiments lies in how the ith left interaction is simulated (using either the fake or real witnesses). Towards reaching a contradiction, we want to claim that, by the non-malleability and $\ell(n)$-robustness of NMCom, the value A commits to in a randomly chosen right interaction is "computationally independent" from how Stage 2-6 of the ith left interaction are simulated. However, (as in proof of Lemma 5) two problems arise: one is that the Real Execution Phase of the two hybrids can not be generated efficiently, and the other is that both Stage 2-6 of the ith left and the randomly chosen right interactions might be rewound by S. We solve the two problem in the same way as in proof of Lemma 5: to overcome the first problem, we fix one execution of the Real Execution Phase $(\mathcal{V}_A, \mathcal{V}_P, \mathcal{V}_V)$ such that conditioned on it occurring, the two hybrids are still distinguishable with high probability; to overcome the second problem, we again consider two alternative hybrids $\hat{\mathsf{hyb}}^i$ and $\hat{\mathsf{hyb}}^i_+$, which proceed identically to hyb^i and hyb^i_+ respectively, except that, in the Simulation Phase, they employ an alternative simulation strategy that avoids rewinding Stage 2-6 of the ith left and the randomly picked right interactions. More precisely, $\hat{\mathsf{hyb}}^i$ and $\hat{\mathsf{hyb}}^i_+$ proceed almost identically to $\tilde{\mathsf{hyb}}^i$ and $\tilde{\mathsf{hyb}}^i_+$ in the proof of Lemma 5, except that now it "chops" up the execution of A into $k(n)+1$ phases $A_1, \ldots, A_{k+1}$, according to messages in Stage 2-6 of the ith left and the randomly picked right interactions, and simulates the views of $A_1, \ldots, A_{k+1}$ sequentially. It follows using the same argument that the outputs of $\hat{\mathsf{hyb}}^i$ and hyb^i, as well as that of $\hat{\mathsf{hyb}}^i_+$ and hyb^i_+, are statistically close.

Therefore by our hypothesis, the probabilities that A cheats in a *randomly chosen* right interaction in $\hat{\mathsf{hyb}}^i$ and $\hat{\mathsf{hyb}}^i_+$ differ by a polynomial amount. However, the only difference between the two hybrids lies in how Stage 2-6 of the ith left interaction are simulated (using a fake or a real witness), *and* the Stage 2-6 of the ith left and the randomly chosen right interactions *are never rewound* in the two hybrids. Then it follows using the same proof of Lemma 7 in [LPTV10] that, essentially by the non-malleability and $\ell(n)$-robustness of NMCom that the probability that A commits to a "fake" witness in Stage 2 of the randomly chosen right interaction differ by at most a negligible amount, which gives a contradiction.

Acknowledgement. We thank Vipul Goyal and the anonymous referees of TCC 2011 for many helpful comments and inspiring suggestions.

References

[BCC88] Brassard, G., Chaum, D., Crépeau, C.: Minimum disclosure proofs of knowledge. J. Comput. Syst. Sci. 37(2), 156–189 (1988)

[Blu86] Blum, M.: How to prove a theorem so no one else can claim it. In: Proc. of the International Congress of Mathematicians, pp. 1444–1451 (1986)

[BPS06] Barak, B., Prabhakaran, M., Sahai, A.: Concurrent non-malleable zero knowledge. In: FOCS, pp. 345–354 (2006)

[Can01] Canetti, R.: Universally composable security: A new paradigm for cryptographic protocols. In: FOCS 2001: Proceedings of the 42nd IEEE Symposium on Foundations of Computer Science, Washington, DC, USA, p. 136. IEEE Computer Society, Los Alamitos (2001)

[CF01] Canetti, R., Fischlin, M.: Universally Composable Commitments. In: Kilian, J. (ed.) CRYPTO 2001. LNCS, vol. 2139, pp. 19–40. Springer, Heidelberg (2001)

[CKL03] Canetti, R., Kushilevitz, E., Lindell, Y.: On the limitations of universally composable two-party computation without set-up assumptions. In: Biham, E. (ed.) EUROCRYPT 2003. LNCS, vol. 2656, pp. 68–86. Springer, Heidelberg (2003)

[DDN00] Dolev, D., Dwork, C., Naor, M.: Nonmalleable cryptography. SIAM Journal on Computing 30(2), 391–437 (2000)

[DN02] Damgård, I., Nielsen, J.B.: Perfect hiding and perfect binding universally composable commitment schemes with constant expansion factor. In: Yung, M. (ed.) CRYPTO 2002. LNCS, vol. 2442, pp. 581–596. Springer, Heidelberg (2002)

[DNS04] Dwork, C., Naor, M., Sahai, A.: Concurrent zero-knowledge. J. ACM 51(6), 851–898 (2004)

[GK96] Goldreich, O., Kahan, A.: How to construct constant-round zero-knowledge proof systems for NP. Journal of Cryptology 9(3), 167–190 (1996)

[GMR89] Goldwasser, S., Micali, S., Rackoff, C.: The knowledge complexity of interactive proof systems. SIAM Journal on Computing 18(1), 186–208 (1989)

[GMW91] Goldreich, O., Micali, S., Wigderson, A.: Proofs that yield nothing but their validity or all languages in NP have zero-knowledge proof systems. J. ACM 38(3), 690–728 (1991)

[Gol01] Goldreich, O.: Foundations of Cryptography — Basic Tools. Cambridge University Press, Cambridge (2001)

[KP01] Kilian, J., Petrank, E.: Concurrent and resettable zero-knowledge in poly-loalgorithm rounds. In: STOC 2001, pp. 560–569 (2001)

[Lin03] Lindell, Y.: Bounded-concurrent secure two-party computation without setup assumptions. In: STOC 2003, pp. 683–692 (2003)

[LP09] Lin, H., Pass, R.: Non-malleability amplification. In: STOC 2009, pp. 189–198 (2009)

[LPTV10] Lin, H., Pass, R., Tseng, W.-L.D., Venkitasubramaniam, M.: Concurrent non-malleable zero knowledge proofs. In: Rabin, T. (ed.) CRYPTO 2010. LNCS, vol. 6223, pp. 429–446. Springer, Heidelberg (2010)

[LPV08] Lin, H., Pass, R., Venkitasubramaniam, M.: Concurrent Non-malleable Commitments from Any One-Way Function. In: Canetti, R. (ed.) TCC 2008. LNCS, vol. 4948, pp. 571–588. Springer, Heidelberg (2008)

[MOSV06] Micciancio, D., Ong, S.J., Sahai, A., Vadhan, S.: Concurrent Zero Knowledge Without Complexity Assumptions. In: Halevi, S., Rabin, T. (eds.) TCC 2006. LNCS, vol. 3876, pp. 1–20. Springer, Heidelberg (2006)

[OPV10] Ostrovsky, R., Pandey, O., Visconti, I.: Efficiency preserving transformations for concurrent non-malleable zero knowledge. In: Micciancio, D. (ed.) TCC 2010. LNCS, vol. 5978, pp. 535–552. Springer, Heidelberg (2010)

[PR05] Pass, R., Rosen, A.: New and improved constructions of non-malleable cryptographic protocols. In: STOC 2005, pp. 533–542 (2005)

[PRS02] Prabhakaran, M., Rosen, A., Sahai, A.: Concurrent zero knowledge with logarithmic round-complexity. In: FOCS 2002, pp. 366–375 (2002)

[PTV08] Pass, R., Tseng, W.-L.D., Venkitasubramaniam, M.: Concurrent zero knowledge: Simplifications and generalizations(2008) (manuscript), `http://hdl.handle.net/1813/10772`

[SCO+01] De Santis, A., Di Crescenzo, G., Ostrovsky, R., Persiano, G., Sahai, A.: Robust non-interactive zero knowledge. In: Kilian, J. (ed.) CRYPTO 2001. LNCS, vol. 2139, pp. 566–598. Springer, Heidelberg (2001)

[YYZ09] Yao, A.C.-C., Yung, M., Zhao, Y.: Adaptive concurrent non-malleability with bare public-keys. CoRR, abs/0910.3282 (2009)

Round-Optimal Password-Based Authenticated Key Exchange

Jonathan Katz[1,⋆] and Vinod Vaikuntanathan[2,⋆⋆]

[1] University of Maryland, USA
jkatz@cs.umd.edu
[2] Microsoft Research
vinodv@alum.mit.edu

Abstract. We show a general framework for constructing password-based authenticated key exchange protocols with *optimal* round complexity — one message per party, sent simultaneously — in the standard model, assuming a common reference string. When our framework is instantiated using bilinear-map cryptosystems, the resulting protocol is also (reasonably) efficient. Somewhat surprisingly, our framework can be adapted to give protocols in the standard model that are *universally composable* while still using only one (simultaneous) round.

1 Password-Based Authenticated Key Exchange

Protocols for *authenticated key exchange* enable two parties to generate a shared, cryptographically strong key while communicating over an insecure network under the complete control of an adversary. Such protocols are among the most widely used and fundamental cryptographic primitives; indeed, agreement on a shared key is necessary before "higher-level" tasks such as encryption and message authentication become possible.

Parties must share *some* information in order for authenticated key exchange to be possible. It is well known that shared cryptographic keys — either in the form of public keys or a long, uniformly random symmetric key — suffice, and several protocols in this model, building on the classic Diffie-Hellman protocol [16] (which protects only against an eavesdropping adversary and provides no authentication at all) are known; see, e.g., [7, 4].

Password-based protocols allow users to "bootstrap" even a *very weak* (e.g., short) shared secret into a (much longer) cryptographic key. The canonical application here is authentication using *passwords*, though protocols developed in this context can be useful even when the shared secret has high min-entropy (but is not uniform) [9]. The security guaranteed by password-based protocols (roughly speaking) is that if the password is chosen uniformly[1] from a dictionary

---

⋆ Work done in part while visiting IBM. Research supported by NSF grant #0627306 and NSF CAREER award #0447075.

⋆⋆ Work done while at IBM.

[1] Although the usual presentation of PAK assumes a uniform password, known protocols work with passwords chosen from any (efficiently sampleable) distribution.

Y. Ishai (Ed.): TCC 2011, LNCS 6597, pp. 293–310, 2011.

of size D then an adversary who initiates Q "on-line" attacks — i.e., who actively interferes in Q sessions — has "advantage" at most Q/D. (This is inherent, as an adversary can always carry out Q impersonation attempts and succeed with this probability.) In particular, "off-line" dictionary attacks where an adversary enumerates passwords from the dictionary of potential passwords, and tries to match observed protocol executions to each one, are of no use.

Early work [20, 24] considered a "hybrid" setting where users share public keys in addition to a password. In the setting where *only* a password is shared, Bellovin and Merritt [6] proposed the first protocols for password-based authenticated key exchange (PAK) with heuristic arguments for their security. Several years later, provably secure PAK protocols were constructed [3,10,31] in the random oracle/ideal cipher models, and many improvements and generalizations of these protocols are known. In contrast, only a handful of PAK protocols are known in the so-called "standard model" (i.e., without random oracles):

- **General assumptions:** Goldreich and Lindell [19] gave the first PAK protocol in the standard model. Subsequent work of Barak et al. [2] shows a general feasibility result for computation over unauthenticated networks which implies a solution for PAK as a special case. These approaches gives the only PAK protocols for the plain model where there is no setup. (Nguyen and Vadhan [33] show efficiency improvements to the Goldreich-Lindell protocol, but achieve a weaker notion of security.) These approaches are impractical in terms of communication, computation, and round complexity. Moreover, they do not tolerate concurrent executions by the same party (unless additional setup is assumed). A recent protocol of Goyal et al. [21] addresses the issue of concurrent executions, but is still far from practical.
- **Efficient protocols:** Katz, Ostrovsky, and Yung [28] demonstrated the first *efficient* PAK protocol with a proof of security based on standard assumptions; extensions and improvements of their protocol were given in [18,13,27, 17,30]. Different constructions of efficient PAK protocols are given in [26,22]. In contrast to the works mentioned earlier, these approaches are secure even under concurrent executions by the same party. On the other hand, they require a *common reference string* (CRS). In practice, however, a CRS does not appear to be a serious drawback in the context of PAK where the CRS can be hard-coded into an implementation of the protocol. We note also that reliance on a CRS (or some other setup) is inherent for achieving *universally composable* PAK [13].

Round/message complexity of existing protocols. We distinguish between *rounds* and *messages*. Differing somewhat from the usual convention in the two-party setting (but matching the usual convention in the multi-party setting), we let a round consist of one message sent by each party simultaneously; note that in a one-round protocol each honest party's message (if any) cannot depend on the other party's message. We stress, however, that even for one-round protocols the adversary is always assumed to be *rushing*; i.e., the adversary may wait to receive an honest party's first-round message before sending its own.

Determining the optimal round complexity of key-exchange protocols is of both theoretical and practical interest, and has been studied in various settings. The original Diffie-Hellman protocol [16], which provides security against a passive eavesdropper, can be run in one round; one-round authenticated key exchange based on shared public/symmetric keys is also possible [25, 34]. One-round protocols for PAK are also known (e.g., [3]) in the random oracle model. All prior PAK protocols based on standard assumptions, though, require three or more rounds. We remark that the protocols in [26, 22] achieve *explicit* authentication in three rounds (whereas the protocols of [28,18,17,30] achieve only *implicit* authentication in three rounds, and require an additional round for explicit authentication), but the round complexity of these protocols cannot be reduced even if only implicit authentication is desired.

1.1 Our Results

We show a new framework for constructing *one-round* PAK protocols in the standard model (assuming a CRS), where each party may send their message *simultaneously.* (Once again, we stress that our security model allows for a "rushing" adversary who waits to see the message sent by a party before sending its response.) Our protocols achieve implicit authentication but can be extended to give explicit authentication using one additional round; it is not hard to see that explicit authentication is impossible in one round without stronger setup assumptions (e.g., a global clock).

Our framework relies on non-interactive zero-knowledge proofs (NIZK) and so, in general, may be computationally inefficient. When instantiating our framework using bilinear maps, however, we obtain a reasonably efficient solution (e.g., communicating a constant number of group elements).

Somewhat surprisingly, we can extend our framework to give a universally composable PAK protocol [12] *without increasing the round complexity at all* (and still without relying on random oracles). In contrast, the work of [13] shows a method (used also by [22]) for obtaining universal composability that requires additional messages/rounds. Abdalla et al. [1] show a universally composable PAK protocol, proven secure in the random oracle model, that requires three rounds. To the best of our knowledge, no prior universally composable protocol (whether in the random oracle model or not) can be run in only one round.

1.2 Our Techniques

At a basic level, we rely on smooth projective hash functions [14], as used in [18] (and implicitly in [28]); see Section 2.2 for a definition. The basic structure of previous protocols [28, 18], omitting many important details, is as follows:

First round: The client sends an encryption C of the password pw.
Second round: The server sends an encryption C' of pw, and a projected key $s' = \alpha(k', C, pw)$.
Third round: The client sends a projected key $s = \alpha(k, C', pw)$.

The client computes the session key as $H_k(C', pw) \cdot H_{s'}(C, pw, r)$, and the server computes the session key as $H_s(C', pw, r') \cdot H_{k'}(C, pw)$. (Here, r, r' is the randomness used to compute C, C', respectively.) Properties of the smooth projective hash function ensure that these are equal.

Two difficulties must be overcome in order to collapse a protocol of the above form to one round:

- In the smooth projective hash functions used in prior work, the "projection function" α was *adaptive*, and depended on both the hash key k and the element being hashed (i.e., (C, pw) in the above example). This leads to protocols requiring three rounds just to ensure correctness.

  Here we show a construction of CCA-secure encryption schemes with associated smooth projective hash functions whose projection function is *non-adaptive*, and depends only on the hash key k. This allows us to obtain the *functionality* of PAK in a single round, by having the client send $(\alpha(k), C)$ and the server send $(\alpha(k'), C')$ simultaneously.
- The above addresses correctness, but says nothing about security. The technical difficulty here is that an honestly generated client message $\mathsf{msg} = (s, C)$ might be forwarded by an adversary to *multiple* server instances (and vice versa), and it is required that the session keys computed in all these instances look random and independent to the adversary. (This issue does not arise in prior work because, roughly speaking, messages are *bound* to a single session by virtue of a signature verification key sent in the first round [28, 18] or a MAC derived from the shared session key [17]. Neither approach is viable if we want the entire protocol to take place in a single round.)

  Due to the above difficulty, the proof of security is the most technically challenging part of our work. Our proof relies on a technical lemma related to re-using both the hash keys and the inputs to the smooth projective hash function, and may be of independent interest.

Additional ideas are needed to obtain a *universally composable* protocol without increasing the number of rounds. We refer the reader to Section 5.1 for an overview of the techniques used there.

1.3 Outline of the Paper

In Section 2 we present a standard definition of security for PAK due to Bellare et al. [3]. We also review there the notion of smooth projective hashing, and prove a technical lemma regarding its usage. In Section 3 we describe our basic framework for constructing one-round PAK protocols, and prove security of this approach according to the definition of [3]. We discuss in Section 4 two instantiations of our framework: one based on the decisional Diffie-Hellman assumption, and a second, more efficient instantiation based on bilinear maps. In Section 5 we describe an extension of our framework that yields one-round, universally composable password-based authenticated key-exchange protocols.

2 Definitions and Background

Throughout, we denote the security parameter by n.

2.1 Password-Based Authenticated Key Exchange

Here we present a definition of security for PAK due to Bellare, Pointcheval, and Rogaway [3], based on prior work of [4, 5]. The text here is taken almost verbatim from [28].

Participants, passwords, and initialization. Prior to any execution of the protocol there is an initialization phase during which public parameters and a CRS are established. We assume a fixed set User of protocol participants (also called principals or users). For every distinct $U, U' \in \mathsf{User}$, users U and U' share a password $pw_{U,U'}$. We assume that each $pw_{U,U'}$ is chosen independently and uniformly from the set $[D] \stackrel{\text{def}}{=} \{1, \ldots, D\}$ for some integer D. (Our proof of security extends to more general cases, and we implicitly consider arbitrary password distributions in the setting of universal composability.)

Execution of the protocol. In the real world, a protocol determines how principals behave in response to input from their environment. In the formal model, these inputs are provided by the adversary. Each principal can execute the protocol multiple times (possibly concurrently) with different partners; this is modeled by allowing each principal to have an unlimited number of *instances* with which to execute the protocol. We denote instance i of user U as Π_U^i. Each instance may be used only once. The adversary is given oracle access to these different instances; furthermore, each instance maintains (local) state which is updated during the course of the experiment. In particular, each instance Π_U^i is associated with the following variables:

- sid_U^i, pid_U^i, and sk_U^i denote the *session id*, *partner id*, and *session key* for an instance, respectively. The session id is simply a way to keep track of different executions; we let sid_U^i be the (ordered) concatenation of all messages sent and received by Π_U^i. The partner id denotes the user with whom Π_U^i believes it is interacting. (Note that pid_U^i can never equal U.)
- acc_U^i and term_U^i are boolean variables denoting whether a given instance has accepted or terminated, respectively.

The adversary's interaction with the principals (more specifically, with the various instances) is modeled via access to *oracles* that we describe now:

- $\mathsf{Send}(U, i, \mathsf{msg})$ — This sends message msg to instance Π_U^i. This instance runs according to the protocol specification, updating state as appropriate. The message output by Π_U^i is given to the adversary.

  The adversary can "prompt" instance Π_U^i to initiate the protocol with partner U' by querying $\mathsf{Send}(U, i, U')$. In response to this query, instance Π_U^i outputs the first message of the protocol.

- $\mathsf{Execute}(U, i, U', j)$ — If Π_U^i and $\Pi_{U'}^j$ have not yet been used, this oracle executes the protocol between these instances and gives the transcript of this execution to the adversary. This oracle call represents passive eavesdropping of a protocol execution.
- $\mathsf{Reveal}(U, i)$ — This outputs the session key sk_U^i, modeling leakage of session keys due to, e.g., improper erasure of session keys after use, compromise of a host computer, or cryptanalysis.
- $\mathsf{Test}(U, i)$ — This oracle does not model any real-world capability of the adversary, but is instead used to define security. A random bit b is chosen; if $b = 1$ the adversary is given sk_U^i, and if $b = 0$ the adversary is given a session key chosen uniformly from the appropriate space.

Partnering. Let $U, U' \in \mathsf{User}$. Instances Π_U^i and $\Pi_{U'}^j$ are *partnered* if: (1) $\mathsf{sid}_U^i = \mathsf{sid}_{U'}^j \neq \text{NULL}$; and (2) $\mathsf{pid}_U^i = U'$ and $\mathsf{pid}_{U'}^j = U$.

Correctness. To be viable, a key-exchange protocol must satisfy the following notion of correctness: if Π_U^i and $\Pi_{U'}^j$ are partnered then $\mathsf{acc}_U^i = \mathsf{acc}_{U'}^j = \text{TRUE}$ and $\mathsf{sk}_U^i = \mathsf{sk}_{U'}^j$, i.e., they both accept and conclude with the same session key.

Advantage of the adversary. Informally, the adversary succeeds if it can guess the bit b used by the Test oracle. To formally define the adversary's success, we first define a notion of *freshness*. An instance Π_U^i is *fresh* unless one of the following is true at the conclusion of the experiment: (1) at some point, the adversary queried $\mathsf{Reveal}(U, i)$; or (2) at some point, the adversary queried $\mathsf{Reveal}(U', j)$, where $\Pi_{U'}^j$ and Π_U^i are partnered. We allow the adversary to succeed only if its Test query is made to a fresh instance; this is necessary for any reasonable definition of security.

An adversary $\mathcal{A}$ *succeeds* if it makes a single query $\mathsf{Test}(U, i)$ to a fresh instance Π_U^i, and outputs a bit b' with $b' = b$ (recall that b is the bit chosen by the Test oracle). We denote this event by Succ. The *advantage* of $\mathcal{A}$ in attacking protocol Π is given by $\mathsf{Adv}_{\mathcal{A},\Pi}(k) \stackrel{\text{def}}{=} 2 \cdot \Pr[\mathsf{Succ}] - 1$, where the probability is taken over the random coins used by the adversary and the random coins used during the course of the experiment (including the initialization phase).

It remains to define a secure protocol. A probabilistic polynomial-time (PPT) adversary can always succeed with probability 1 by trying all passwords one-by-one; this is possible since the size of the password dictionary is small. Informally, a protocol is secure if this is the best an adversary can do. Formally, an instance Π_U^i represents an *on-line attack* if both the following are true at the time of the Test query: (1) at some point, the adversary queried $\mathsf{Send}(U, i, *)$; and (2) at some point, the adversary queried $\mathsf{Reveal}(U, i)$ or $\mathsf{Test}(U, i)$. The number of on-line attacks represents a bound on the number of passwords the adversary could have tested in an on-line fashion.

Definition 1. *Protocol Π is a* secure protocol for password-based authenticated key exchange *if, for all dictionary sizes D and for all* PPT *adversaries $\mathcal{A}$ making at most $Q(n)$ on-line attacks, it holds that* $\mathsf{Adv}_{\mathcal{A},\Pi}(n) \leq Q(n)/D + \mathsf{negl}(n)$.

2.2 Smooth Projective Hash Functions

We provide a self-contained definitional treatment of smooth projective hash functions. These were introduced by Cramer and Shoup [14], and our discussion here is based on that of Gennaro and Lindell [18]. Rather than aiming for utmost generality, we tailor the definitions to our application.

Hard subset membership problems. Fix some integer D. Let $(\mathsf{Gen}, \mathsf{Enc}, \mathsf{Dec})$ be a CCA-secure labeled encryption scheme. For a given public key pk, we let C_{pk} denote the set of pairs of valid labels and valid ciphertexts with respect to pk, and require that this set be efficiently recognizable. For a given public key pk, define sets X and $\{L_{pw}\}_{pw\in[D]}$ as follows:

1. $X \stackrel{\text{def}}{=} \{(\mathsf{label}, C, pw)\}$, where $(\mathsf{label}, C) \in C_{pk}$ and $pw \in \{1, \ldots, D\}$.
2. $L_{pw} \stackrel{\text{def}}{=} \{(\mathsf{label}, \mathsf{Enc}_{pk}(\mathsf{label}, pw), pw)\}$, where $\mathsf{label} \in \{0,1\}^*$.

Let $L = \bigcup_{pw=1}^{D} L_i$, and note that $L \subset X$. It follows from CCA security of $(\mathsf{Gen}, \mathsf{Enc}, \mathsf{Dec})$ that the following is negligible for any polynomial-time $\mathcal{A}$:

$$\left| \Pr\left[\begin{array}{c} (pk, sk) \leftarrow \mathsf{Gen}(1^n); \\ (\mathsf{label}, pw) \leftarrow \mathcal{A}^{\mathsf{Dec}_{sk}(\cdot,\cdot)}(pk); \\ C \leftarrow \mathsf{Enc}_{pk}(\mathsf{label}, pw) \end{array} : \mathcal{A}^{\mathsf{Dec}_{sk}(\cdot,\cdot)}(C) = 1 \right] - \Pr\left[\begin{array}{c} (pk, sk) \leftarrow \mathsf{Gen}(1^n); \\ (\mathsf{label}, pw) \leftarrow \mathcal{A}^{\mathsf{Dec}_{sk}(\cdot,\cdot)}(pk); \\ C \leftarrow \mathsf{Enc}_{pk}(\mathsf{label}, 0) \end{array} : \mathcal{A}^{\mathsf{Dec}_{sk}(\cdot,\cdot)}(C) = 1 \right] \right|, \quad (1)$$

where $\mathcal{A}$ is disallowed from querying (label, C) to its decryption oracle.

Smooth projective hash functions. Fix pk and sets $X, \{L_i\}$ as above. A *smooth projective hash function* $\mathcal{H} = \{H_k\}_{k\in K}$ is a keyed function mapping elements in X to elements in some group G, along with a *projection function* $\alpha : K \to S$. Informally, if $x \in L$ then the value of $H_k(x)$ is uniquely determined by $s = \alpha(k)$ and x, whereas if $x \in X \setminus L$ then the value of $H_k(x)$ is statistically close to uniform given $\alpha(k)$ and x (assuming k was chosen uniformly in K). A smooth projective hash function is formally defined by a sampling algorithm that, given pk, outputs $(K, G, \mathcal{H} = \{H_k : X \to G\}_{k\in K}, S, \alpha : K \to S)$ such that:

1. There are efficient algorithms for (1) sampling a uniform $k \in K$, (2) computing $H_k(x)$ for $k \in K$ and $x \in X$, and (3) computing $\alpha(k)$ for $k \in K$.
2. For all $(\mathsf{label}, C, pw) \in L$, the value of $H_k(\mathsf{label}, C, pw)$ is uniquely determined by $\alpha(k)$. Moreover, there is an efficient algorithm that takes as input $s = \alpha(k)$ and $(\mathsf{label}, C, pw, r)$ for which $C = \mathsf{Enc}_{pk}(\mathsf{label}, pw; r)$, and outputs $H_k(\mathsf{label}, C, pw)$. (In other words, when $(\mathsf{label}, C, pw) \in L$ then $H_k(\mathsf{label}, C, pw)$ can be computed in two ways: either using k itself, or using $\alpha(k)$ and the randomness used to generate C.)
3. For any (even unbounded) function $f : S \to X \setminus L$, the following distributions have statistical difference negligible in n:

$$\{k \leftarrow K; s := \alpha(k) : (s, H_k(f(s)))\} \text{ and } \{k \leftarrow K; s := \alpha(k); g \leftarrow G : (s, g)\}$$

We stress that in the above we modify the definition from [18] in two ways: first, α is *non-adaptive*, and depends on k only (rather than both k and x); second, we require the above to hold even for *adaptive* choice of $f(s) \notin L$. Intuitively, the first modification helps us compress the number of rounds to one, whereas the second is necessary for proving security.

A technical lemma. We now prove a technical lemma regarding smooth projective hash functions. Somewhat informally, Gennaro and Lindell [18] showed that, for randomly generated pk and any label, pw, the distribution

$$\left\{k \leftarrow K; s := \alpha(k); C \leftarrow \mathsf{Enc}_{pk}(\mathsf{label}, pw) : \big(s, C, H_k(\mathsf{label}, C, pw)\big)\right\}$$

is computationally indistinguishable from the distribution

$$\{k \leftarrow K; s := \alpha(k); C \leftarrow \mathsf{Enc}_{pk}(\mathsf{label}, pw); g \leftarrow G : (s, C, g)\}.$$

(Note this holds even though $H_k(\mathsf{label}, C, pw)$ is uniquely determined by s and C) Here we show that this continues to hold even if hash keys and ciphertexts are *re-used* multiple times. That is, at a high level (ignoring labels and technical details), we show that the distribution

$$\left\{\begin{array}{c} k_1, \ldots, k_\ell \leftarrow K; \forall i : s_i := \alpha(k_i); \\ C_1, \ldots, C_\ell \leftarrow \mathsf{Enc}_{pk}(pw) \end{array} : \left(\{s_i\}, \{C_i\}, \{H_{k_i}(C_j, pw)\}_{i,j=1}^{\ell}\right)\right\}$$

is computationally indistinguishable from the distribution

$$\left\{\begin{array}{l} k_1, \ldots, k_\ell \leftarrow K; \forall i : s_i := \alpha(k_i); \\ C_1, \ldots, C_\ell \leftarrow \mathsf{Enc}_{pk}(pw); g_{i,j} \leftarrow G \end{array} : \left(\{s_i\}, \{C_i\}, \{g_{i,j}\}_{i,j=1}^{\ell}\right)\right\}.$$

Formally, fix a function $\ell = \ell(n)$, let $\mathcal{A}$ be an adversary, and let $b \in \{0, 1\}$. Consider the following experiment Expt_b:

1. Compute $(pk, sk) \leftarrow \mathsf{Gen}(1^n)$ and let $(K, G, \mathcal{H} = \{H_k : X \rightarrow G\}_{k \in K}, S, \alpha : K \rightarrow S)$ be a smooth projective hash function for pk. Give pk to $\mathcal{A}$.
2. Sample $k_1, \ldots, k_\ell \leftarrow K$, and let $s_i := \alpha(k_i)$ for all i. Give $s_1, \ldots, s_\ell$ to $\mathcal{A}$.
3. $\mathcal{A}$ may adaptively query a (modified) encryption oracle that takes as input (label, pw) and outputs a ciphertext $C \leftarrow \mathsf{Enc}_{pk}(\mathsf{label}, pw)$ along with
   (a) If $b = 0$, the values $H_{k_i}(\mathsf{label}, C, pw)$ for $i = 1$ to ℓ.
   (b) If $b = 1$, random values $g_1, \ldots, g_\ell \leftarrow G$.
4. $\mathcal{A}$ can also query a decryption oracle $\mathsf{Dec}_{sk}(\cdot, \cdot)$ at any point, except that it may not query any pair (label, C) where C was obtained from the encryption oracle on query (label, pw).
5. At the end of the experiment, $\mathcal{A}$ outputs a bit b'. We say $\mathcal{A}$ *succeeds* if $b' = b$.

A proof of the following appears in the full version of this work [29]:

Lemma 1. *Let* $(\mathsf{Gen}, \mathsf{Enc}, \mathsf{Dec})$ *be a CCA-secure labeled encryption scheme. For any polynomial* ℓ *and polynomial-time* $\mathcal{A}$, *we have* $\Pr[\mathcal{A}$ *succeeds*$] \leq \frac{1}{2} + \mathsf{negl}(n)$.

Public parameters: pk

User U		User U'
$k \leftarrow K;\ s := \alpha(k)$		$k' \leftarrow K;\ s' := \alpha(k')$
$\mathsf{label} := (U, U', s)$		$\mathsf{label}' := (U', U, s')$
$C \leftarrow \mathsf{Enc}_{pk}(\mathsf{label}, pw)$	$\xrightarrow{\quad s, C \quad}$	$C' \leftarrow \mathsf{Enc}_{pk}(\mathsf{label}', pw)$
	$\xleftarrow{\quad s', C' \quad}$	
$\mathsf{label}' := (U', U, s')$		$\mathsf{label} := (U, U', s)$
$sk_U := H_k(\mathsf{label}', C', pw) \cdot H_{k'}(\mathsf{label}, C, pw)$		$sk_{U'} := H_k(\mathsf{label}', C', pw) \cdot H_{k'}(\mathsf{label}, C, pw)$

Fig. 1. A one-round protocol for password-based authenticated key exchange

3 A Framework for One-Round PAK Protocols

Our protocol uses a chosen ciphertext-secure (CCA-secure) labeled public-key encryption scheme $(\mathsf{Gen}, \mathsf{Enc}, \mathsf{Dec})$, and a smooth projective hash function as described in Section 2.2.

Public parameters. The public parameters consist of a public key pk generated by $\mathsf{Gen}(1^n)$. No one need know or store the associated secret key. (For the specific instantiations given in Section 4, a public key can be derived from a common random string.) Let $(K, G, \mathcal{H} = \{H_k : X \to G\}_{k \in K}, S, \alpha : K \to S)$ be a smooth projective hash function for pk.

Protocol execution. Consider an execution of the protocol between users U and $U' \neq U$ holding a shared password pw. Our protocol is symmetric, and so we describe the execution from the point of view of U; see also Figure 1.

First, U chooses random hash key $k \leftarrow K$ and computes $s := \alpha(k)$. It then sets $\mathsf{label} := (U, U', s)$ and computes the ciphertext $C \leftarrow \mathsf{Enc}_{pk}(\mathsf{label}, pw)$. It sends the message (s, C).

Upon receiving the message (s', C'), user U does the following. If C' is not a valid ciphertext or $s' \notin S$, then U simply rejects. Otherwise, U sets $\mathsf{label}' := (U', U, s')$ and computes

$$\mathsf{sk}_U := H_k(\mathsf{label}', C', pw) \cdot H_{k'}(\mathsf{label}, C, pw).$$

U computes $H_k(\mathsf{label}', C', pw)$ using k, and can compute $H_{k'}(\mathsf{label}, C, pw)$ using $s' = \alpha(k')$ and the randomness it used to generate C. Correctness follows immediately from the definition of smooth projective hashing.

Theorem 1. *If* $(\mathsf{Gen}, \mathsf{Enc}, \mathsf{Dec})$ *is a CCA-secure labeled encryption scheme and* $(K, G, \mathcal{H} = \{H_k : X \to G\}_{k \in K}, S, \alpha : K \to S)$ *is a smooth projective hash* *function, then the protocol in Figure 1 is a secure protocol for password-based authenticated key exchange.*

The proof is in the full version of this work [29].

4 Instantiating the Building Blocks

We now discuss two possible instantiations of the building blocks required by the protocol of the previous section. Our first instantiation is based on the decisional Diffie-Hellman (DDH) assumption and (generic) simulation-sound non-interactive zero-knowledge (NIZK) proofs. (It could be based on the quadratic residuosity assumption or the Paillier assumption as well, much like in [18]. We omit further details.) Our second, more efficient construction is based on the decisional linear assumption [8] in groups with a bilinear map.

4.1 A Construction Based on the DDH Assumption

We first describe an encryption scheme and then the associated smooth projective hash function.

A CCA-secure encryption scheme. We construct a CCA-secure encryption scheme by applying the Naor-Yung/Sahai paradigm [32, 35] to the El Gamal encryption scheme. Briefly, the public key defines a group $\mathbb{G}$ of prime order p along with generators $g_1, h_1, g_2, h_2 \in \mathbb{G}$. The public key also contains a common random string crs for a (one-time) simulation-sound NIZK proof system [35].

Fixing $\mathbb{G}$, let $\mathsf{ElGamal}_{g,h}(m)$ denote an El Gamal encryption of $m \in \mathbb{G}$ with respect to (g, h); namely, $\mathsf{ElGamal}_{g,h}(m) \to (g^r,\ h^r \cdot m)$, where $r \in \mathbb{Z}_p$ is chosen uniformly at random. To encrypt a message $m \in \mathbb{G}$ in our CCA-secure scheme, the sender outputs the ciphertext $(\mathsf{ElGamal}_{g_1,h_1}(m),\ \mathsf{ElGamal}_{g_2,h_2}(m),\ \pi)$, where π is a simulation-sound NIZK proof that the same m is encrypted in both cases. Labels can be incorporated by including the label in the proof π; we omit the standard details.

Decryption of the ciphertext $(c_1, d_1, c_2, d_2, \pi)$ rejects if $c_1, d_1, c_2, d_2 \notin \mathbb{G}$ or if the proof π is invalid. (Note that the space of valid label/ciphertext pairs is efficiently recognizable without the secret key.) If the ciphertext is valid, then one of the two component ciphertexts is decrypted and the resulting message is output. The results of [35] show that this yields a CCA-secure (labeled) encryption scheme based on the DDH assumption and simulation-sound NIZK.

A smooth projective hash function. Fix a group $\mathbb{G}$ and a public key $pk = (g_1, h_1, g_2, h_2, \mathsf{crs})$ as above, and define sets X and $\{L_i\}$ as in Section 2.2. Define a smooth projective hash function as follows. The set of keys K consists of all four-tuples of elements in $\mathbb{Z}_p$. Given a valid label/ciphertext pair $(\mathsf{label}, C = (c_1, d_1, c_2, d_2, \pi))$ and key $k = (x_1, y_1, x_2, y_2)$, the hash function is defined as:

$$H_{(x_1,y_1,x_2,y_2)}\left(\mathsf{label}, (c_1, d_1, c_2, d_2, \pi), pw\right) = c_1^{x_1} \cdot (d_1/pw)^{y_1} \cdot c_2^{x_2} \cdot (d_2/pw)^{y_2} .$$

(Thus, the range of H is the group $\mathbb{G}$.) The projection function α is defined as:

$$\alpha(x_1, y_1, x_2, y_2) = (g_1^{x_1} \cdot h_1^{y_1},\ g_2^{x_2} \cdot h_2^{y_2})\,.$$

A proof of the following is given in the full version of this work [29].

Lemma 2. *$(K, \mathbb{G}, \mathcal{H} = \{H_k\}_{k \in K}, S, \alpha)$ as defined above is a smooth projective hash function for the hard subset membership problem $(X, \{L_i\})$.*

4.2 A Construction Based on the Decisional Linear Assumption

We now present a more efficient construction based on bilinear maps. The efficiency advantage is obtained by using a *specific* simulation-sound NIZK proof system, built using techniques adapted from [23,11]. Our construction here relies on the *decisional linear assumption* as introduced by Boneh et al. [8]; we refer the reader there for a precise statement of the assumption.

A CPA-secure encryption scheme. We start by describing a semantically secure encryption scheme, due to Boneh et al. [8], based on the decisional linear assumption; we then convert this into a CCA-secure encryption scheme via the same paradigm as above, but using an efficient simulation-sound NIZK proof system. The bilinear map itself is used only in the construction of the simulation-sound NIZK.

Fix groups $\mathbb{G}, \mathbb{G}_T$ of prime order p, and a bilinear map $e : \mathbb{G} \times \mathbb{G} \to \mathbb{G}_T$. The public key is $pk = (f, g, h) \in \mathbb{G}^3$, and the secret key is (α, β) such that $f = h^{1/\alpha}$ and $g = h^{1/\beta}$. A message $m \in \mathbb{G}$ is encrypted by choosing random $r, s \in \mathbb{Z}_p$ and computing the ciphertext $(f^r, g^s, h^{r+s} \cdot m)$. Given a ciphertext (c_1, c_2, c_3), we can recover m as $c_3 / c_1^{\alpha} c_2^{\beta}$.

A simulation-sound NIZK proof of plaintext equality. We can construct a (one-time) simulation-sound NIZK proof of plaintext equality for the encryption scheme described above using the techniques of [23, 11]. Details of the construction (which, while not entirely straightforward, are not the focus of this work) are given in Appendix A.

A CCA-secure encryption scheme. We obtain a CCA-secure encryption scheme by using the Naor-Yung/Sahai paradigm, as described previously. (The following discussion relies on the results of Appendix A.) The public key consists of group elements (f_1, g_1, f_2, g_2, h) used for encryption, in addition to any group elements needed for the CRS of the simulation-sound NIZK proof. Encryption of m, as described in Appendix A, is done by choosing $r_1, s_1, r_2, s_2 \in \mathbb{Z}_p$ and computing the ciphertext

$$(f_1^{r_1},\ g_1^{s_1},\ h^{r_1+s_1} \cdot m,\ f_2^{r_2},\ g_2^{s_2},\ h^{r_2+s_2} \cdot m,\ \pi),$$

where π denotes a simulation-sound NIZK proof that the same m was encrypted both times. (Once again, the space of valid label/ciphertext pairs is efficiently recognizable without the secret key.) It follows from [32, 35] that this yields a

CCA-secure scheme under the decisional linear assumption. Ciphertexts consist of 66 group elements altogether.

A smooth projective hash function. Fix $\mathbb{G}, \mathbb{G}_T$, and a public-key $pk = (f_1, g_1, f_2, g_2, h)$ as above, and define sets X and $\{L_i\}$ as in Section 2.2. We define a smooth projective hash function as follows. The set of keys K is the set of six-tuples of elements in $\mathbb{Z}_p$. Given a valid label/ciphertext pair $(\mathsf{label}, C = (c_1, d_1, e_1, c_2, d_2, e_2, \pi))$ and a key $k = (x_1, y_1, z_1, x_2, y_2, z_2) \in \mathbb{Z}_p^6$, the hash function is defined as

$$H_{(x_1,y_1,z_1,x_2,y_2,z_2)}(\mathsf{label}, C, pw) = c_1^{x_1} \cdot d_1^{y_1} \cdot (e_1/pw)^{z_1} \cdot c_2^{x_2} \cdot d_2^{y_2} \cdot (e_2/pw)^{z_2}.$$

(The range of H is $\mathbb{G}$ itself.) The projection function $\alpha : K \to \mathbb{G}^4$ is defined as:

$$\alpha(x_1, y_1, z_1, x_2, y_2, z_2) = (f_1^{x_1} h^{z_1},\ g_1^{y_1} h^{z_1},\ f_2^{x_2} h^{z_2},\ g_2^{y_2} h^{z_2}).$$

In Appendix B we show:

Lemma 3. *$(K, \mathbb{G}, \mathcal{H} = \{H_k\}_{k \in K}, S, \alpha)$ as defined above is a smooth projective hash function for the hard subset membership problem $(X, \{L_i\})$.*

5 A One-Round, Universally-Composable PAK Protocol

Canetti et al. [13] gave a definition of security for password-based authenticated key exchange in the universal composability (UC) framework [12]. Their definition guarantees a strong, simulation-based notion of security that, in particular, guarantees that security is maintained even when multiple protocols are run concurrently in an arbitrary network. For the specific case of password-based key exchange, the definition also has the advantage of *automatically* handling arbitrary (efficiently sampleable) distributions on passwords, and even correlations between passwords of different users. We refer to [13] for a more complete discussion, and a description of the password-based key-exchange functionality $\mathcal{F}_{\mathsf{pwKE}}$. We let $\hat{\mathcal{F}}_{\mathsf{pwKE}}$ denote the multi-session extension of $\mathcal{F}_{\mathsf{pwKE}}$.

5.1 Overview of the Construction

We do not know how to prove that the protocol from Section 3 is universally composable. The main difficulty is that the definition of PAK in the UC framework requires simulation *even if the adversary guesses the correct password.* (In contrast, in the proof of security in Section 3 we simply "give up" in case this ever occurs.) To see the problem more clearly, consider what happens in the UC setting when the simulator sends the first message of the protocol to the adversary, before the simulator knows the correct password. The simulator must send *some* ciphertext C as part of the first message, and this "commits" the simulator to some password pw. When the adversary sends the reply, the simulator can extract the adversary's "password guess" pw' and submit this guess to the ideal functionality. If this turns out to be the correct password, however, the simulator is stuck: it needs to compute a session key that matches the session key the adversary would compute, but the simulator is (information-theoretically!) unable to do so because it sent an incorrect ciphertext in the first message.

In prior work [13], the issue above was resolved by having one party send a "pre-commitment" to the password, and then running a regular PAK protocol along with a proof that the password being used in the protocol is the same as the password to which it "pre-committed". (The proof is set up in such a way that the simulator can equivocate this proof, but the adversary cannot.) This requires at least one additional round.

We take a different approach that does not affect the round complexity at all. Roughly, we modify the protocol from Figure 1 by having each party include as part of its message an encryption C_1 of its hash key k, along with a proof that C_1 encrypts a value k for which $\alpha(k) = s$. Now, even if the simulator is wrong in its guess of the password it will still be able to compute a session key by extracting this hash key from the adversary's message. A full description of the protocol is given in the following section.

While we do not describe in detail any instantiation of the components, we remark that it should be possible to use the same techniques as in Appendix A to construct (reasonably) efficient realizations of the necessary components using bilinear maps. We leave this for future work.

5.2 Description of the Protocol

In addition to the building blocks used in Section 3, here we also rely on an unbounded simulation-sound [15] NIZK proof system $(\mathsf{CRSGen}, \mathcal{P}, \mathcal{V})$ for a language L^* defined below.

Public parameters. The public parameters consist of two public keys pk_1, pk_2 generated by $\mathsf{Gen}(1^n)$ and a common random string crs for the simulation-sound NIZK proof system. Let $(K, G, \mathcal{H} = \{H_k : X \to G\}_{k \in K}, S, \alpha : K \to S)$ be a smooth projective hash function for pk_2.

Protocol execution. Consider an execution of the protocol between users U and $U' \neq U$ holding a shared password pw and a common session identifier ssid. (The ssid is an artifact of the UC framework, and it is guaranteed that (1) parties communicating with each other begin holding matching ssids, and (2) each ssid is used only once. Existence of these ssids is not essential to our proof of security, though it does make the proof somewhat simpler.) Our protocol is symmetric, and so we describe the execution from the point of view of U; see Figure 2.

First, U chooses a random hash key $k \leftarrow K$ and computes $s := \alpha(k)$. It then computes an encryption of k, namely $C_1 \leftarrow \mathsf{Enc}_{pk_1}(k)$. Define a language L^* as follows.

$$L^* \stackrel{\text{def}}{=} \{(s, C_1) : \exists k \in K \text{ and } \omega \text{ s.t } s = \alpha(k) \text{ and } C_1 = \mathsf{Enc}_{pk_1}(k; \omega)\}.$$

U computes an NIZK proof π that $(C_1, s) \in L^*$, using crs. It then sets $\mathsf{label} := (\mathsf{ssid}, U, U', s, C_1, \pi)$ and computes the ciphertext $C_2 \leftarrow \mathsf{Enc}_{pk_2}(\mathsf{label}, pw)$. The message it sends is (s, C_1, π, C_2).

Public Parameters: $(pk_1, pk_2, \mathsf{crs})$

User U		User U'
$k \leftarrow K;\ s := \alpha(k)$		$k' \leftarrow K;\ s' := \alpha(k')$
$C_1 \leftarrow \mathsf{Enc}_{pk_1}(k)$		$C'_1 \leftarrow \mathsf{Enc}_{pk_1}(k')$
$\pi := \mathcal{P}_{\mathsf{crs}}((s, C_1) \in L^*)$		$\pi' := \mathcal{P}_{\mathsf{crs}}((s', C'_1) \in L^*)$
$\mathsf{label} := (\mathsf{ssid}, U, U', s, C_1, \pi)$		$\mathsf{label}' := (\mathsf{ssid}, U', U, s', C'_1, \pi')$
$C_2 \leftarrow \mathsf{Enc}_{pk_2}(\mathsf{label}, pw)$	$\xrightarrow{s, C_1, \pi, C_2}$	$C'_2 \leftarrow \mathsf{Enc}_{pk_2}(\mathsf{label}', pw)$
	$\xleftarrow{s', C'_1, \pi', C'_2}$	
$\mathsf{label}' := (\mathsf{ssid}, U', U, s', C'_1, \pi')$		$\mathsf{label} := (\mathsf{ssid}, U, U', s, C_1, \pi)$
$sk_U := H_k(\mathsf{label}', C'_2, pw)$		$sk_{U'} := H_k(\mathsf{label}', C'_2, pw)$
$\cdot H_{k'}(\mathsf{label}, C_2, pw)$		$\cdot H_{k'}(\mathsf{label}, C_2, pw)$

Fig. 2. A universally composable protocol for password-based authenticated key exchange

Upon receiving the message (s', C'_1, π', C'_2), user U does the following. If the message is invalid (i.e., if verification of π' fails, or C'_2 is not a valid ciphertext, or $s' \notin S$), then U simply rejects. Otherwise, U sets $\mathsf{label}' := (\mathsf{ssid}, U', U, s', C'_1, \pi')$ and computes $\mathsf{sk}_U := H_k(\mathsf{label}', C'_2, pw) \cdot H_{k'}(\mathsf{label}, C_2, pw)$. Note U can compute $H_k(\mathsf{label}', C'_2, pw)$ since it knows k, and can compute $H_{k'}(\mathsf{label}, C_2, pw)$ using $s' = \alpha(k')$ and the randomness used to generate C_2. Correctness follows from the definition of smooth projective hashing. A proof of the following is given the full version of this work [29].

Theorem 2. *If* $(\mathsf{Gen}, \mathsf{Enc}, \mathsf{Dec})$ *is a CCA-secure public-key encryption scheme,* $(\mathsf{CRSGen}, \mathcal{P}, \mathcal{V})$ *is an unbounded simulation-sound NIZK proof system, and furthermore* $(K, G, \mathcal{H} = \{H_k : X \to G\}_{k \in K}, S, \alpha : K \to S)$ *is a smooth projective hash family, then the protocol in Figure 2 securely realizes* $\hat{\mathcal{F}}_{\mathsf{pwKE}}$ *in the* $\mathcal{F}_{\mathsf{crs}}$-*hybrid model.*

References

1. Abdalla, M., Catalano, D., Chevalier, C., Pointcheval, D.: Efficient two-party password-based key exchange protocols in the UC framework. In: Malkin, T.G. (ed.) CT-RSA 2008. LNCS, vol. 4964, pp. 335–351. Springer, Heidelberg (2008)
2. Barak, B., Canetti, R., Lindell, Y., Pass, R., Rabin, T.: Secure computation without authentication. In: Shoup, V. (ed.) CRYPTO 2005. LNCS, vol. 3621, pp. 361–377. Springer, Heidelberg (2005)

3. Bellare, M., Pointcheval, D., Rogaway, P.: Authenticated key exchange secure against dictionary attacks. In: Preneel, B. (ed.) EUROCRYPT 2000. LNCS, vol. 1807, pp. 139–155. Springer, Heidelberg (2000)
4. Bellare, M., Rogaway, P.: Entity authentication and key distribution. In: Stinson, D.R. (ed.) CRYPTO 1993. LNCS, vol. 773, pp. 232–249. Springer, Heidelberg (1994)
5. Bellare, M., Rogaway, P.: Provably secure session key distribution: The three party case. In: 27th Annual ACM Symposium on Theory of Computing (STOC), pp. 57–66. ACM Press, New York (1995)
6. Bellovin, S.M., Merritt, M.: Encrypted key exchange: Password-based protocols secure against dictionary attacks. In: IEEE Symposium on Security & Privacy, pp. 72–84. IEEE, Los Alamitos (1992)
7. Bird, R., Gopal, I., Herzberg, A., Janson, P., Kutten, S., Molva, R., Yung, M.: Systematic design of two-party authentication protocols. IEEE J. on Selected Areas in Communications 11(5), 679–693 (1993)
8. Boneh, D., Boyen, X., Shacham, H.: Short group signatures. In: Franklin, M. (ed.) CRYPTO 2004. LNCS, vol. 3152, pp. 41–55. Springer, Heidelberg (2004)
9. Boyen, X., Dodis, Y., Katz, J., Ostrovsky, R., Smith, A.: Secure remote authentication using biometric data. In: Cramer, R. (ed.) EUROCRYPT 2005. LNCS, vol. 3494, pp. 147–163. Springer, Heidelberg (2005)
10. Boyko, V., MacKenzie, P.D., Patel, S.: Provably secure password-authenticated key exchange using diffie-hellman. In: Preneel, B. (ed.) EUROCRYPT 2000. LNCS, vol. 1807, pp. 156–171. Springer, Heidelberg (2000)
11. Camenisch, J., Chandran, N., Shoup, V.: A public key encryption scheme secure against key dependent chosen plaintext and adaptive chosen ciphertext attacks. In: Joux, A. (ed.) EUROCRYPT 2009. LNCS, vol. 5479, pp. 351–368. Springer, Heidelberg (2009), `http://eprint.iacr.org/2008/375`
12. Canetti, R.: Universally composable security: A new paradigm for cryptographic protocols. In: 42nd Annual Symposium on Foundations of Computer Science (FOCS), pp. 136–145. IEEE, Los Alamitos (2001)
13. Canetti, R., Halevi, S., Katz, J., Lindell, Y., MacKenzie, P.D.: Universally composable password-based key exchange. In: Cramer, R. (ed.) EUROCRYPT 2005. LNCS, vol. 3494, pp. 404–421. Springer, Heidelberg (2005)
14. Cramer, R., Shoup, V.: Universal hash proofs and a paradigm for adaptive chosen ciphertext secure public-key encryption. In: Knudsen, L.R. (ed.) EUROCRYPT 2002. LNCS, vol. 2332, pp. 45–64. Springer, Heidelberg (2002)
15. De Santis, A., Di Crescenzo, G., Ostrovsky, R., Persiano, G., Sahai, A.: Robust non-interactive zero knowledge. In: Kilian, J. (ed.) CRYPTO 2001. LNCS, vol. 2139, pp. 566–598. Springer, Heidelberg (2001)
16. Diffie, W., Hellman, M.E.: New directions in cryptography. IEEE Trans. Information Theory 22(6), 644–654 (1976)
17. Gennaro, R.: Faster and shorter password-authenticated key exchange. In: Canetti, R. (ed.) TCC 2008. LNCS, vol. 4948, pp. 589–606. Springer, Heidelberg (2008)
18. Gennaro, R., Lindell, Y.: A framework for password-based authenticated key exchange. ACM Trans. Information and System Security 9(2), 181–234 (2006)
19. Goldreich, O., Lindell, Y.: Session-key generation using human passwords only. Journal of Cryptology 19(3), 241–340 (2006)
20. Gong, L., Lomas, T.M.A., Needham, R.M., Saltzer, J.H.: Protecting poorly chosen secrets from guessing attacks. IEEE J. Selected Areas in Communications 11(5), 648–656 (1993)

21. Goyal, V., Jain, A., Ostrovsky, R.: Password-authenticated session-key generation on the internet in the plain model. In: Rabin, T. (ed.) CRYPTO 2010. LNCS, vol. 6223, pp. 277–294. Springer, Heidelberg (2010)
22. Groce, A., Katz, J.: A new framework for efficient password-based authenticated key exchange. In: 17th ACM Conf. on Computer and Communications Security (CCCS), pp. 516–525. ACM Press, New York (2010)
23. Groth, J., Sahai, A.: Efficient non-interactive proof systems for bilinear groups. In: Smart, N.P. (ed.) EUROCRYPT 2008. LNCS, vol. 4965, pp. 415–432. Springer, Heidelberg (2008)
24. Halevi, S., Krawczyk, H.: Public-key cryptography and password protocols. ACM Trans. Information and System Security 2(3), 230–268 (1999)
25. Jeong, I.R., Katz, J., Lee, D.-H.: One-round protocols for two-party authenticated key exchange. In: Jakobsson, M., Yung, M., Zhou, J. (eds.) ACNS 2004. LNCS, vol. 3089, pp. 220–232. Springer, Heidelberg (2004)
26. Jiang, S., Gong, G.: Password based key exchange with mutual authentication. In: Handschuh, H., Hasan, M.A. (eds.) SAC 2004. LNCS, vol. 3357, pp. 267–279. Springer, Heidelberg (2004)
27. Katz, J., MacKenzie, P.D., Taban, G., Gligor, V.D.: Two-server password-only authenticated key exchange. In: Ioannidis, J., Keromytis, A.D., Yung, M. (eds.) ACNS 2005. LNCS, vol. 3531, pp. 1–16. Springer, Heidelberg (2005)
28. Katz, J., Ostrovsky, R., Yung, M.: Efficient password-authenticated key exchange using human-memorable passwords. In: Pfitzmann, B. (ed.) EUROCRYPT 2001. LNCS, vol. 2045, pp. 475–494. Springer, Heidelberg (2001)
29. Katz, J., Vaikuntanathan, V.: Round-optimal password-based authenticated key exchange, http://eprint.iacr.org/2010/368
30. Katz, J., Vaikuntanathan, V.: Smooth projective hashing and password-based authenticated key exchange from lattices. In: Matsui, M. (ed.) ASIACRYPT 2009. LNCS, vol. 5912, pp. 636–652. Springer, Heidelberg (2009)
31. MacKenzie, P.D., Patel, S., Swaminathan, R.: Password-authenticated key exchange based on RSA. In: Okamoto, T. (ed.) ASIACRYPT 2000. LNCS, vol. 1976, pp. 599–613. Springer, Heidelberg (2000)
32. Naor, M., Yung, M.: Universal one-way hash functions and their cryptographic applications. In: 21st Annual ACM Symposium on Theory of Computing (STOC), pp. 33–43. ACM Press, New York (1989)
33. Nguyen, M.-H., Vadhan, S.: Simpler session-key generation from short random passwords. Journal of Cryptology 21(1), 52–96 (2008)
34. Okamoto, T.: Authenticated key exchange and key encapsulation in the standard model. In: Kurosawa, K. (ed.) ASIACRYPT 2007. LNCS, vol. 4833, pp. 474–484. Springer, Heidelberg (2007)
35. Sahai, A.: Non-malleable non-interactive zero knowledge and adaptive chosen-ciphertext security. In: 40th Annual Symposium on Foundations of Computer Science (FOCS), pp. 543–553. IEEE, Los Alamitos (1999)

A A Simulation-Sound NIZK Proof of Plaintext Equality

Fix groups $\mathbb{G}, \mathbb{G}_T$ of prime order p, and a bilinear map $e : \mathbb{G} \times \mathbb{G} \to \mathbb{G}_T$ as in Section 4.2. Fix also two public keys $pk_1 = (f_1, g_1, h)$ and $pk_2 = (f_2, g_2, h)$. We encrypt a message m with respect to pk_1 by choosing random r, s and computing the ciphertext $(f_1^r, g_1^s, h^{r+s} \cdot m)$. We encrypt a message m with respect to pk_2 by choosing random $r, s \in \mathbb{Z}_p$ and computing the ciphertext $(f_2^r, g_2^s, h^{r+s} \cdot m)$. We stress that the public keys use the same value h.

We first describe a (potentially malleable) NIZK proof of plaintext equality. That is, given two ciphertexts (F_1, G_1, H_1) and (F_2, G_2, H_2) encrypted with respect to pk_1, pk_2, respectively, we describe a proof that these ciphertexts encrypt the same message. The observation is that plaintext equality is equivalent to the existence of $r_1, s_1, r_2, s_2 \in \mathbb{Z}_p$ such that:

$$F_1 = f_1^{r_1} \tag{2}$$
$$G_1 = g_1^{s_1} \tag{3}$$
$$F_2 = f_2^{r_2} \tag{4}$$
$$G_2 = g_2^{s_2} \tag{5}$$
$$H_1/H_2 = h^{r_1+s_1-r_2-s_2}. \tag{6}$$

As shown in [23] (see also [11, Section 4.4] for a self-contained description), NIZK proofs of satisfiability (with a CRS) can be constructed for a system of equations as above; since, in our case, we have 5 (linear) equations in 4 variables, proofs contain 22 group elements[2].

Camenisch et al. [11] show a construction of an *unbounded* simulation-sound NIZK. For our purposes, a simpler construction that is *one-time* simulation sound [35] suffices. Let (Gen, Sign, Vrfy) be a one-time signature scheme, where for simplicity we assume verification keys are elements of $\mathbb{G}$ (this can always be achieved using an extra step of hashing). To make the above (one-time) simulation-sound, we add group elements (f, g, h, F, G, H) to the CRS. Roughly, proofs of plaintext equality now contain:

1. A fresh signature verification key vk.
2. A proof that *either* there exists a satisfying assignment to Equations (2)–(6), *or* that the given tuple (f, g, h, F, G, H) is an encryption of vk. I.e., there exist r, s such that:
$$F = f^r, \qquad G = g^s, \qquad H/vk = h^{r+s}. \tag{7}$$
3. A signature σ (with respect to vk) on the proof from the previous step.

Noting that Equation (7) describes a system of 3 (linear) equations in 2 variables, and using the techniques from [11, Appendix A.2], an NIZK proof as required in step 2 can be done using 58 group elements, for a total of 60 group elements for the entire simulation-sound NIZK proof (assuming signatures are one group element for simplicity). See also footnote 2.

B Proof of Lemma 3

Sampling a uniform $k \in K$, computing $H_k(x)$ given $k \in K$ and $x \in X$ and computing $\alpha(k)$ for $k \in K$ are all easy.

[2] Our calculations here are based on the decisional linear assumption (the 2-linear assumption in the terminology of [11]). If we are willing to use the 1-linear assumption, the efficiency of our proofs can be improved.

We show that if $(\mathsf{label}, C, pw) \in L$, then $H_k(\mathsf{label}, C, pw)$ can be computed efficiently given $\alpha(k)$ and the randomness that was used to generate C. Since $(\mathsf{label}, C, pw) \in L$, we have that $C = (f_1^{r_1},\, g_1^{s_1},\, h^{r_1+s_1} f_1^{pw},\, f_2^{r_2},\, g_2^{s_2},\, h^{r_2+s_2} f_2^{pw})$ for some $r_1, s_1, r_2, s_2 \in \mathbb{Z}_p$. For $k = (x_1, y_1, z_1, x_2, y_2, z_2)$ we have

$$\begin{aligned} H_k(\mathsf{label}, C, pw) &= c_1^{x_1} d_1^{y_1} \cdot (e_1/f_1^{pw})^{z_1} \cdot c_2^{x_2} \cdot d_2^{y_2} \cdot (e_2/f_2^{pw})^{z_2} \\ &= (f_1^{x_1} h^{z_1})^{r_1} \cdot (g_1^{y_1} h^{z_1})^{s_1} \cdot (f_2^{x_2} h^{z_2})^{r_2} \cdot (g_2^{y_2} h^{z_2})^{s_2} . \end{aligned}$$

This can be computed easily given r_1, s_1, r_2, s_2, and

$$\alpha(k) \stackrel{\text{def}}{=} (f_1^{x_1} h^{z_1},\, g_1^{y_1} h^{z_1},\, f_2^{x_2} h^{z_2},\, g_2^{y_2} h^{z_2}) .$$

Next, we show that if $(\mathsf{label}, C, pw) \in X \setminus L$, then the value of $H_k(\mathsf{label}, C, pw)$ is uniform conditioned on $\alpha(k)$. (This holds even if (label, C, pw) are chosen adaptively depending on $\alpha(k)$.) Fix any $\alpha(k) = (S_1, S_2, S_3, S_4)$. Letting $\alpha_i = \log_h f_i$ and $\beta_i = \log_h g_i$, this value of $\alpha(k)$ constrains $k = (x_1, y_1, z_1, x_2, y_2, z_2)$ to satisfy

$$\begin{pmatrix} \alpha_1 & 0 & 1 & 0 & 0 & 0 \\ 0 & \beta_1 & 1 & 0 & 0 & 0 \\ 0 & 0 & 0 & \alpha_2 & 0 & 1 \\ 0 & 0 & 0 & 0 & \beta_2 & 1 \end{pmatrix} \begin{pmatrix} x_1 \\ y_1 \\ z_1 \\ x_2 \\ y_2 \\ z_2 \end{pmatrix} = \begin{pmatrix} \gamma_1 \\ \gamma_2 \\ \gamma_3 \\ \gamma_4 \end{pmatrix}, \tag{8}$$

where $\gamma_i = \log_h S_i$. For any $(\mathsf{label}, C, pw) \in X \setminus L$, we can write

$$C = (f_1^{r_1},\, g_1^{s_1},\, h^{r_1+s_1} f_1^{pw'},\, f_2^{r_2},\, g_2^{s_2},\, h^{r_2+s_2} f_2^{pw'},\, \pi)$$

for some $pw' \neq pw$. (We assume for simplicity that the same pw' is encrypted twice; since π is valid, this is the case with all but negligible probability.) We then have

$$\begin{aligned} & H_k(\mathsf{label}, C, pw) \\ &= f_1^{r_1 x_1} \cdot g_1^{s_1 y_1} \cdot h^{(r_1+s_1) z_1} \cdot \left(f_1^{\Delta}\right)^{z_1} \cdot f_2^{r_2 x_2} \cdot g_2^{s_2 y_2} \cdot h^{(r_2+s_2) z_2} \cdot \left(f_2^{\Delta}\right)^{z_2} \\ &= S_1^{r_1} S_2^{s_1} S_3^{r_2} S_4^{s_2} \cdot (f_1^{z_1} f_2^{z_2})^{\Delta}, \end{aligned} \tag{9}$$

where $\Delta = pw' - pw \neq 0$. For any $g \in \mathbb{G}$, we have $f_1^{z_1} f_2^{z_2} = g$ iff

$$\alpha_1 \cdot z_1 + \alpha_2 \cdot z_2 = \log_h g. \tag{10}$$

Since the system of equations given by (8) and (10) is under-defined, the probability that $f_1^{z_1} f_2^{z_2} = g$ is exactly $1/|\mathbb{G}|$ even conditioned on the value $\alpha(k)$. Looking at Equation (9), and noting that $S_1^{r_1} S_2^{s_1} S_3^{r_2} S_4^{s_2}$ is determined by $\alpha(k)$ and C, we conclude that the distribution of $H_k(\mathsf{label}, C, pw)$ is uniform in $\mathbb{G}$.

Bringing People of Different Beliefs Together to Do UC

Sanjam Garg[1], Vipul Goyal[2], Abhishek Jain[1], and Amit Sahai[1]

[1] UCLA
{sanjamg,abhishek,sahai}@cs.ucla.edu
[2] Microsoft Research, India
vipul@microsoft.com

Abstract. Known constructions of UC secure protocols are based on the premise that different parties collectively agree on some trusted setup. In this paper, we consider the following two intriguing questions: Is it possible to achieve UC if the parties do not want to put all their trust in *one* entity (or more generally, in one setup)? What if the parties have a difference of opinion about what they are willing to trust? The first question has been studied in only a limited way, while the second has never been considered before.

In this paper, we initiate a systematic study to answer the above questions. We consider a scenario with multiple setup instances where each party in the system has some individual *belief* (setup assumption in terms of the given setups). The belief of a party corresponds to what it is willing to trust and its security is guaranteed given that its belief "holds." The question considered is: "Given some setups and the (possibly) different beliefs of all the parties, when can UC security be achieved?" We present a general condition on the setups and the beliefs of all the parties under which UC security is possible. Surprisingly, we show that when parties have different beliefs, UC security can be achieved with a more limited "trust" than what is necessary in the traditional setting (where all parties have a common belief).

1 Introduction

Suppose Alice and Bob want to execute a UC-secure [4] protocol. They know that they will need to rely on some trust assumptions [6,7] in order to achieve UC security. Unfortunately, Alice and Bob have different beliefs about who is trustworthy: Suppose that Microsoft, Intel, Google, and Yahoo have all published common reference strings (CRS). Alice believes that either both Microsoft and Intel are trustworthy, or both Google and Yahoo are trustworthy. Bob, however, has a different view: Bob believes that either both Microsoft and Google are trustworthy, or both Intel and Yahoo are trustworthy. This seems like a horrible situation. Indeed, even if both Alice and Bob shared Alice's trust belief (or if they both shared Bob's trust belief), most UC-secure protocols would be impossible [13]. We show, surprisingly, that nevertheless in the situation above with

Y. Ishai (Ed.): TCC 2011, LNCS 6597, pp. 311–328, 2011.

asymmetric beliefs, Alice and Bob can execute a protocol that will guarantee UC security for *all parties whose trust beliefs turn out to be valid.* This paper is a systematic study of this problem – when can we guarantee UC security (in this sense) even when different parties have different trust beliefs.

Background. The last decade has seen a push towards obtaining secure computation protocols in the demanding network setting where there might be multiple concurrent protocol executions. The framework of universal composability (UC) was introduced by Canetti [4] to capture the security requirements in such a setting. However unfortunately, soon after the introduction of the UC framework, impossibility results were shown ruling out the existence of UC secure protocol for most functionalities of interest [6,7]. These results were further generalized [18,2] to rule out the existence of secure protocol even in various less demanding settings. These impossibility results refer to the "plain model" where the participating parties do not trust any external entity and they have no prior communication among themselves, etc.

To overcome these deep impossibility results and obtain secure protocols in the modern network setting, a number of different "setup assumptions" were introduced. A few examples follow. Canetti and Fischlin [6] and Canetti, Lindell, Ostrovsky and Sahai [8] consider the model where a trusted party publishes a "common reference string" (CRS). Canetti, Pass and Shelat [9] generalized these results by considering reference strings coming from an unknown distribution. Barak, Canetti, Nielsen and Pass [1] introduced the so called "registered public key" model; a variant of this model was considered by Canetti, Dodis, Pass and Walfish [5]. Katz [14] (and subsequently [10,11,19]) studied a model where the parties exchange tamper proof hardware tokens with each other. Under all of these settings, general positive results for all PPT computable functionalities have been shown in the UC framework. In addition, such positive results can also be obtained in the setting where a majority of the participants are assumed to be honest [3,4,15][1].

Now that there are a number of different options in what one might "assume about the world" to obtain UC protocols, the parties (or the system designer) would be forced to choose one. In some situation, such a choice would be natural from the environment where the protocols would be run. However in many other scenarios, it would be unclear which setup assumption is the right one: should the parties trust a CRS published by an authority or should they each register a public key with an authority? Should they assume that a majority of them are honest or should they not trust anyone and instead rely on tamper proof hardware tokens? Making such a choice can be non-trivial since a wrong choice could lead to a complete compromise of the system security.

In light of the above, we can consider the following two lines of thought. First, given the importance of making a correct choice, it is not unreasonable to imagine

[1] We slightly abuse the terminology and consider the honest majority setting as just another setup. Our study is targeted at assumptions which allow one to obtain UC protocols but can go wrong leading to a security compromise. Honest majority is one such assumption.

that in certain scenarios, the parties in the system cannot agree on a common choice, i.e., it is plausible that each party may have a different choice about what setup to use. For example, one party may want to use a CRS published by Microsoft while another may want to register keys with Google. Yet another party might want to place more trust in tamper-proofness of certain hardware tokens and not so much in any entity, and so on. In essence, each party may have a different trust assumption (referred to as the "belief" of the party) about the setups available. This gives rise to the following question:

Is it possible to construct UC protocols when parties have different beliefs about setups?

An orthogonal line of thought is that since a wrong choice could lead to a compromise of the system security, it might be desirable to diversify the risks involved and avoid a single point of failure. In other words, how about basing the protocol on a combination of setups rather than a single one? As an example, the parties might have access to a published CRS as well as have registered public keys with an authority. The protocol should retain its security even if one of these setup assumptions "breaks down" (for example, if the published CRS turned out to be adversarially chosen) but the other turned out to be "honestly chosen." Going a little further, there might be n instances of various setups and the protocol security should hold as long as (e.g.) either one of the first two or a majority of the rest were honestly done. More generally, we can consider the following question:

Is it possible to construct UC protocols using multiple setups when the parties share an arbitrary belief about the setups?

In this paper, we answer both the above questions in the affirmative. We remark that a general perception following the prior work on UC security is that a *common* trusted setup is necessary for achieving UC. As our results show, this is in fact *not* necessary. We further note that one would expect UC to be harder to achieve when parties may have asymmetric beliefs. On the contrary, we show that the level of trust needed to obtain UC can be significantly weakened in such a setting.

Related Work. There have been two previous works in the direction of basing UC secure protocols based on multiple setups. However the works have been much narrower in scope. Groth and Ostrovsky [13] consider the question of basing cryptography on multiple CRS. They showed how to construct UC protocols under the assumption that a majority of the given reference strings were honestly generated. Subsequently, Goyal and Katz [12] considered basing UC secure protocol under a combination of a CRS and the honest majority assumption.

These questions can be viewed as two special cases of our general question (in particular, the case of common belief). Our work subsumes these works and provides answers to a host of other such interesting questions.

Our Results. In this paper, we initiate a systematic study of the question of basing UC protocols on multiple different setups in a setting where each party has its own belief (setup assumption in terms of the given setups) about these setups. An informal problem statement is as follows. Consider a system with multiple setups and parties. Each party has a belief expressed as an arbitrary monotonic formula expressed in disjunctive normal form in terms of the setups, e.g. $(A \wedge B) \vee (C \wedge D)$ where A, B, C and D are some setups. We interpret this belief as– "Either setup A and setup B hold, or setup C and setup D hold." We ask the following question: "Given these setups and the different beliefs of all the parties, when can UC security be achieved?" In answering this question, we give a very general condition on the setups and the different beliefs of all the parties under which UC security is possible. This result is presented in Section 5.

Towards the goal of answering the above question, we first look at a simpler scenario in which all the parties have a "common belief," i.e., all the parties share the same belief about the setups in the system. We give a very general condition on the setups and the common belief of all the parties under which UC security is possible. This result is presented in Section 4.

As we discuss later in Section 2, a setup may be "corruptible" in multiple ways; as such, parties might be willing to put trust in the "extent" of corruption of a particular setup. Furthermore, different parties might be willing to put different levels of trust in the same setup. The results of Section 4 and 5 handle this case to some extent, but this scenario is handled in its full generality in Appendix B. We advise the reader to look at some of the interesting examples presented there.

In order to argue about security of a system with arbitrary setups, we abstract formal properties that essentially capture what "extra powers" a setup provides to a simulator over an adversary. These abstract properties allow us to categorize setups and argue UC security of protocols that can be constructed from them. We note that all known setups fit these definitions very well. We further note that our results generalize the previously known tight results of Groth and Ostrovsky [13] and Goyal and Katz [12]. Finally, we leave it as an open problem to study the tightness of our results in the general case.

Overview of Main Ideas. In past, UC protocols for different setups have been designed with very different techniques. Here, we wish to design a single protocol that simultaneously uses a combination of a number of different setups. Our starting point is the recent work of Lin, Pass and Venkitasubramaniam [16] which puts forward a unified framework for designing UC secure protocols. In the UC framework, to obtain positive results, the simulator is required to obtain some "extra power" over the adversary. Lin, Pass and Venkitasubramaniam observe that a general technique for constructing UC secure protocols is to have the simulator obtain a "trapdoor string" which is hard to compute for the adversary. This is formalized in the form of (two party) UC-puzzle protocols that enable the simulator to obtain such a trapdoor string (but prevent the adversary from doing so). Such a trapdoor string is already available to the simulator in the CRS model and hence designing a UC-puzzle is trivial in that case. However, even in case of other setup assumptions, Lin, Pass and Venkitasubramaniam show that generally it is possible to easily design such a UC-puzzle at the end of which the simulator obtains a trapdoor string (but the adversary does not).

Once we have a unified construction for UC protocols under different setup assumptions, we come to our main question: how do we fruitfully use multiple setups in a single protocol? Different setups might compose very differently with each others. A priori, it might seem that each pair of setups may compose in a unique way. Hence, considering the general question that we wish to study, it seems unclear how to proceed at all.

A key conceptual contribution of our work is a classification of the setup assumptions used to construct UC protocols. We observe that almost all setups can be classified among three types. To understand these different types, recall that in this work, we are concerned with unreliable setup assumptions that may actually turn out to be false. Coming back to the framework of Lin et al [16], our simulator would obtain a trapdoor string which would be hard to compute for the adversary (this is of course if the setup was "honest"). However what happens if the setup assumption was actually false (i.e., setup was "malicious")? We could have any one of the following three cases: (I) the adversary is able to obtain the trapdoor string (associated with the UC-puzzle) but not the simulator, (II) none of them are able to obtain the trapdoor string, and, (III) both the adversary and the simulator are able to obtain the trapdoor string[2]. Intuitively, the first case corresponds to *complete* corruption of the setup, while the other two cases correspond to *partial* corruption of the setup.

We are able to show that the above classification of setups solely decides the composability properties of different setups. In general, Type II and Type III setups have better composability properties than the Type I ones. For instance, in the special case where we have multiple instances of the same setup, a majority of them should be "honest" if the setup is of Type I. However if it is either of Type II or Type III, it is sufficient to have a *single* "honest" setup.

Going further, we note that following the work of Lin et al [16], the task of constructing UC secure protocols from any setup assumption reduces to the task of constructing a UC-puzzle (in the hybrid model of the corresponding setup). Then, in a scenario where all the parties share a common belief about the setups in the system, the task of constructing UC protocols reduces to task of constructing a UC-puzzle in the hybrid model of the multiple setups in the system. But what of the scenario where the parties have different beliefs about the setups? In this case, we show how to construct a *family* of UC-puzzles that in turn can be used to construct a family of "concurrent simulation-sound" zero knowledge protocols with "UC-simulation" property. For reasons discussed later in section 3 and 5, this is sufficient to construct UC protocols in such a setting.

Organization. We start by describing our model in Section 2. In Section 2.1, we recall the notion of UC-puzzles [16] and give their classification into various

[2] The fourth case is uninteresting as we argue later.

types. In Section 2.2, we give a classification of setups into types. We then present our positive result for the case where parties share a common belief in Section 4. Next, in Section 5, we present our positive result for the case where parties have different beliefs. Finally, in Appendix A and B, we extend our results to cover some more involved settings.

2 Our Model

Traditionally, protocols that utilize a single instance of a setup have been constructed, and proven to be universally composable as long as the setup is "honest" (i.e., the setup assumption holds). We, however, consider settings where a protocol may utilize more than one setup; in such a setting, one or more setups may in fact be "corrupted" (i.e., the setup assumption corresponding to a setup no longer holds because of possible control of the setup by an adversary). We wish to investigate when UC-security can be realized in such settings. To this end, we first consider an augmented modeling for setups (that could either be "honest" or "corrupted").

Modeling Setup Failure. Typically, a setup is modeled as a "trusted" ideal functionality that interacts with the parties in the system. Let $\mathcal{G}$ denote such an ideal functionality. In order to account for the scenario that a setup could in fact be corrupted by an adversary, we augment this model by considering another ideal functionality $m\mathcal{G}$ that represents a "malicious" version of $\mathcal{G}$. We refer to $m\mathcal{G}$ as a *failure mode* of $\mathcal{G}$. Then, we will model a setup as a pair $(\mathcal{G}, m\mathcal{G})$ of ideal functionalities, where $\mathcal{G}$ represents to the honest version of the setup while $m\mathcal{G}$ represents its failure mode.

For example, consider the common reference string (CRS) setup [6,8]. In previous works, the CRS setup is modeled as a trusted ideal functionality that samples a CRS from a specified distribution. A party in the system can query the ideal functionality, who in response, will return the CRS to the party. In our setting, we will model the CRS setup as a pair $(\mathcal{G}_{CRS}, m\mathcal{G}_{CRS})$ of ideal functionalities. Here $\mathcal{G}_{CRS}$ corresponds to the case where the CRS is generated honestly by the ideal functionality, such that no adversary can obtain any "trapdoor" information for the CRS. On the other hand, $m\mathcal{G}_{CRS}$ corresponds to the case where the functionality returns an adversarially chosen CRS; in particular, an adversary may be able to obtain some "trapdoor" information for the CRS.

It should be implicit that we only consider setups that are "sufficient" to realize UC-secure protocols. That is, we assume that given any setup $(\mathcal{G}, m\mathcal{G})$, it is possible to construct UC-secure protocols in the $\mathcal{G}$-hybrid model. We further assume that constructing UC-secure protocols is impossible in the $m\mathcal{G}$-hybrid model.

Multiple Failure Modes. The above modeling of setups is not quite complete, in that it is too restrictive to imagine that a setup may only have a single failure mode. Specifically, one could imagine a setup failing in multiple ways depending upon how it is corrupted by an adversary. For instance, let us consider the tamper-proof hardware token setup of Katz [14]. In the model of Katz, it is assumed that (a) parties can exchange tamper-proof hardware tokens with each other (b) a token "creator" cannot send messages to the token after giving the token to another party. This is modeled in the form of a wrapper functionality that takes a program code as an input from a party and then uses that code to "emulate" a token that interacts with the intended receiver. In this case, one can consider different possible corruptions of the wrapper functionality (each corresponding to a different failure mode $m\mathcal{G}$) where either or both of the above assumptions fail. In general, for a given honest version $\mathcal{G}$ of a setup, there may be multiple failure modes $m\mathcal{G}_1, m\mathcal{G}_2, \ldots, m\mathcal{G}_k$.

In the sequel, for simplicity of exposition, we will first restrict ourselves to the case where a setup only has a single failure mode. The results presented in Section 4 and 5 are obtained under this restriction. Later in Appendix B, we discuss how to extend our modeling to incorporate multiple failure modes and then explain how our results can be extended to this case.

Setup Types. Now recall that one of the main goals of this paper is to provide a way to construct UC secure protocols that utilize multiple different setups. A priori, it might seem that different setups may compose very differently with each other *depending upon their specific properties*, and that in the worst case, each pair of setups might compose in a unique way. However, we show that this is not the case; specifically, we give a classification of setups into different "Types," and then show that the Type of any setup solely decides the composability properties of that setup. We then study composition of setup types and give positive results for the feasibility of constructing UC-secure protocols in the presence of multiple setups (of possibly different Types).

Our classification of a setup into Types is based on the feasibility of constructing a specific primitive called "UC-puzzle" (in the hybrid model of the setup). We first recall the notion of UC puzzles in Section 2.1. Later, in Section 2.2, we give a classification of setups into Types.

2.1 UC Puzzles and Their Classification

Lin et al [16] introduced the notion of UC-puzzles, where, informally speaking, a UC-puzzle is a two-party protocol in a $\mathcal{G}$-hybrid model (where $\mathcal{G}$ is a trusted ideal functionality)[3] such that no real world adversary can complete the puzzle and also obtain a "trapdoor," while there exists a simulator that can "simulate" a puzzle execution (with the "correct" distribution) and also obtain a trapdoor. Looking ahead, our positive results rely crucially on UC-puzzles; therefore, we discuss them below in detail.

Recall that in our setting, setups are "corruptible." For instance, in the above example, the setup functionality in the system may in fact be $m\mathcal{G}$ (instead of $\mathcal{G}$) which is controlled by an adversary. Clearly, there may be no guarantee that

[3] We note that in the original definition of Lin et al [16], the existence of a an ideal functionality $\mathcal{G}$ is not compulsory. However, in such cases, one can imagine $\mathcal{G}$ to be an empty functionality.

the aforementioned properties of a UC-puzzle still hold in this case; as such, the original definition of [16] does not suffice for our purposes. To this end, we extend the original definition of UC-puzzles to account for such a scenario. We give an informal definition of a UC-puzzle as follows. Part of the definition below is taken almost verbatim from [16].

UC-puzzle. Let $(\mathcal{G}, m\mathcal{G})$ be a setup. A UC-puzzle is a pair $(\langle S, R\rangle, \mathcal{R})$, where $\langle S, R\rangle$ is a protocol between two parties—a sender S, and a receiver R—in the $\mathcal{F}$-hybrid model (where $\mathcal{F}$ is either $\mathcal{G}$ or $m\mathcal{G}$), and $\mathcal{R} \subseteq \{0,1\}^* \times \{0,1\}^*$ is an associated PPT computable relation.

I. If $\mathcal{F}$ is the "honest" ideal functionality $\mathcal{G}$, i.e., the puzzle is in the $\mathcal{G}$-hybrid model, then it must satisfy the following two properties.

Soundness. No PPT adversarial receiver R^* after an execution with an honest sender S can find (except with negligible probability) a trapdoor $\sigma \in \mathcal{R}(\texttt{trans})$, where $\texttt{trans}$ is the transcript of the puzzle execution.

Statistical Simulatability. Let $\mathcal{A}$ be a real world adversary (in an environment $\mathcal{Z}$) that participates as a sender in multiple concurrent executions of a UC-puzzle. Then, for every such $\mathcal{A}$, there exists a simulator $\mathcal{S}$ interacting only with $\mathcal{Z}$ such that no (possibly unbounded) $\mathcal{Z}$ can distinguish between an execution with $\mathcal{A}$ from an execution with $\mathcal{S}$, except with negligible probability. Further, for every completed puzzle execution, except with negligible probability, $\mathcal{S}$ outputs a trapdoor $\sigma \in \mathcal{R}(\texttt{trans})$, where $\texttt{trans}$ is the transcript of that puzzle execution.

II. Otherwise, if $\mathcal{F}$ is the "malicious" functionality $m\mathcal{G}$, i.e., the puzzle is in the $m\mathcal{G}$-hybrid model, then we consider three sub-cases and define different "types" for UC-puzzles. In order to describe these cases, we first define two properties that can be seen as "strong" negations of the two aforementioned properties of UC-puzzles[4].

Unsound. A UC-puzzle $(\langle S, R\rangle, \mathcal{R})$ is said to be Unsound in the $m\mathcal{G}$-hybrid model if there exists a PPT adversarial receiver R^* that can find (except with negligible probability) a trapdoor $\sigma \in \mathcal{R}(\texttt{trans})$, for a puzzle execution (with a transcript $\texttt{trans}$) with an honest sender S.

Unsimulatable. Further, we say that the UC-puzzle is Unsimulatable in the $m\mathcal{G}$-hybrid model if there exists an adversarial sender $\mathcal{A}$ such that no simulator $\mathcal{S}$ can obtain (except with negligible probability) a trapdoor $\sigma \in \mathcal{R}(\texttt{trans})$ for a completed puzzle execution (where $\texttt{trans}$ is the transcript of the puzzle execution).

[4] An intuitive explanation of the two new properties can be seen as follows. Informally speaking, we wish to say that a UC-puzzle in the $m\mathcal{G}$-hybrid model is Unsound if an adversary in the $m\mathcal{G}$-hybrid model enjoys the same power as any simulator in the $\mathcal{G}$-hybrid model. Similarly, a UC-puzzle in the $m\mathcal{G}$-hybrid model is Unsimulatable if a simulator in the $m\mathcal{G}$-hybrid model enjoys no extra power as compared to an adversary in the $\mathcal{G}$-hybrid model.

We now consider the following cases.

Type I. We say that a UC-puzzle is of Type I if it is Unsound and Unsimulatable in the $m\mathcal{G}$-hybrid model.

Type II. We say that a UC-puzzle is of Type II if it satisfies Soundness but is Unsimulatable in the $m\mathcal{G}$-hybrid model.

Type III. We say that a UC-puzzle is of Type III if it is Unsound but satisfies Statistical Simulatability in the $m\mathcal{G}$-hybrid model.

Finally, we can consider the case where both Soundness and Statistical Simulatability are satisfied in the $m\mathcal{G}$-hybrid model. We discard this case for the following reasons. Note that this case implies that we can construct UC-puzzles even in the $m\mathcal{G}$-hybrid model. Then, from the result of [16], it would follow that we can construct UC-secure protocols even in the $m\mathcal{G}$-hybrid model. Informally speaking, this means that the setup was not corrupted in any "interesting" way.

This completes our definition of UC-puzzles and their classification into Types.

2.2 Classification of Setups

Having defined different Types of UC-puzzles, we are now ready to define the Type of a setup based on the feasibility of constructing UC-puzzles in the hybrid model of that setup.

Definition 1 (Setup Types). *A setup $(\mathcal{G}, m\mathcal{G})$ is said to be of Type X if it is possible to construct a UC-puzzle of Type X in the $\mathcal{F}$-hybrid model, where $\mathcal{F}$ is either $\mathcal{G}$ or $m\mathcal{G}$, and $X \in \{I, II, III\}$.*

Setups with multiple Types. Note that it may be possible to construct multiple UC-puzzles of different Types in the hybrid model of a setup $(\mathcal{G}, m\mathcal{G})$; as such, the above definition allows a setup $(\mathcal{G}, m\mathcal{G})$ to have multiple Types. For simplicity of exposition, in the sequel, we will first assume that each setup has a unique Type. The results presented in Section 4 and 5 are obtained under this restriction. Later, in Appendix A, we give an example of a custom setup that has multiple types, and then explain how to extend our results to incorporate setups with multiple Types.

Classification of known setups. We now briefly discuss some known setups such as CRS [CF01, CLOS02, CPS07], tamper-proof hardware [Kat07], and key registration [BCNP04]. We first note that the only "natural" failure mode for the CRS setup corresponds to "complete corruption," where the CRS is chosen by the adversary. Then, it is not difficult to see that the CRS setup is of Type I. In contrast, the key registration setup and the hardware-token setup naturally allow "partial corruption". Let us first consider the key registration setup. Recall that in the key registration setup, it is assumed that parties can register their public keys in such a way that: (a) the public keys of the honest parties are "safe" (in the sense that the secret keys were chosen at random and kept secret from the adversary), and (b) the public keys of the corrupted parties are 'well-formed" (in the sense that the functionality has seen the corresponding secret keys). Then, an adversary may be able to corrupt the setup in multiple ways (each corresponding to a different failure mode) such that either or both of these assumptions are violated. The first failure mode corresponds to the complete corruption of the setup (i.e., both the above assumptions fail); in this case, the setup is of Type I. The second failure mode corresponds to the case where the public keys of corrupt parties may not be "well-formed"; however, the secret keys of the honest parties are still "safe." In this case, the setup is of Type II. Finally, in the third failure mode, the secret keys of honest parties may not be "safe"; however, the public keys of corrupted parties are still "well-formed." In this case, the setup is of Type III. In a similar way, one can consider different failure modes for the hardware token setup, each leading to a different Type. We refer the reader to the full version of the paper for details.

3 UC Security via UC-Puzzles

As observed by Lin et al [16], it is implicit from prior works [8,17,21,20] that the task of constructing UC-secure protocols for any well-formed[5] functionality reduces to the task of constructing a "concurrent simulation-sound" zero knowledge protocol (ssZK) with "UC simulation" property[6]. Very informally, these properties can be described as follows (the text is taken almost verbatim from [16]):

UC simulation: For every PPT adversary $\mathcal{A}$ receiving "honest" proofs of statements x using witness w, where (x, w) are chosen by the environment $\mathcal{Z}$, there exists a simulator $\mathcal{S}$ (that only gets statements x as input) such that no $\mathcal{Z}$ can distinguish (except with negligible probability) whether it is interacting with $\mathcal{A}$ or $\mathcal{S}$.

Concurrent simulation-soundness: An adversary who receives an unbounded number of concurrent *simulated* proofs, of statements chosen by $\mathcal{Z}$, cannot prove any false statements (except with negligible probability).

Lin et al [16] gave a unified framework for constructing UC-secure protocols from known setup assumptions like CRS [6,8], tamper-proof hardware tokens [14], key registration [1], etc. Specifically, Lin et al [16] gave a construction for an ssZK protocol from a UC-puzzle in a $\mathcal{G}$-hybrid model (where $\mathcal{G}$ is a "trusted" ideal functionality that may correspond to, for instance, a CRS setup), and a strongly non-malleable witness indistinguishable (SNMWI) argument of knowledge (see [16] for details).

We note that following the work of [16], the task of constructing UC secure protocols from any setup assumption reduces to the task of constructing a UC-puzzle (in the hybrid model of the corresponding setup)[7]. Then, looking ahead, the positive results in this paper crucially rely on the framework of [16].

[5] We refer the reader to [8] for a definition of "well-formed" functionalities.

[6] Formally, this can be modeled as implementing a specific "zero knowledge proof of membership" functionality.

[7] Note that [16] gave a construction of an SNMWI protocol based on one-way functions.

4 Common Belief about the Setups

We first consider the setting where all the parties in the system share a *common*, though arbitrary, belief about the setups present in the system. For instance, consider an example where there are three CRS setups in the system. In this case, all the parties may share a common belief that either (a) the first CRS is honestly generated, or (b) both the second and the third CRSs are honestly generated. Ideally, given any function f, we would like to construct a protocol Π (in the hybrid model of the three CRSs) that securely realizes f when either of the above two cases is actually true. In this section, we investigate the possibility of constructing such protocols. In particular, we will give a condition that takes into account the setups present in the system and the common belief shared by the parties; we then show that constructing secure protocols (where security is defined in the above sense) is possible if our condition is satisfied. We first introduce some notation and definitions.

Belief Set. In the above example, we can express the common belief of the parties in the form of a DNF formula written as CRS_1 OR (CRS_2 AND CRS_3), where CRS_i denotes the ith CRS. The DNF formula, in turn, can be represented as a set $\Sigma = \{T_1, T_2\}$ where T_1 is a set that consists of CRS_1, and T_2 is a set that consists of the CRS_2 and CRS_3.

We generalize this in the following manner. Let U be the set of all the setups present in the system. Then, we will represent the common belief of the parties as a set $\Sigma = \{T_1, \ldots, T_k\}$ (where k is an arbitrary, possibly exponential, value), where each member T_i is a subset of U. The common belief of the parties is expressed as follows: $\exists T_i \in \Sigma$ such each setup in T_i holds[8]. We will refer to this set Σ as the *belief set.*

Σ-secure protocols. We would like to construct secure computation protocols that realize UC security if the common belief of the parties about the setups in the system is actually true. We formalize this in following definition of *Σ-secure* protocols.

Definition 2 (Σ-secure protocols). *Let there be n setups in the system denoted by $(\mathcal{G}_1, m\mathcal{G}_1), \ldots, (\mathcal{G}_n, m\mathcal{G}_n)$. Let $\mathcal{F}_1, \ldots, \mathcal{F}_n$ be n ideal functionalities, where $\forall i, \mathcal{F}_i$ is either $\mathcal{G}_i$ or $m\mathcal{G}_i$. Let $\Sigma = \{T_1, \ldots, T_k\}$ be a belief set over the n setups that represents the common belief of the parties. We say that a protocol Π Σ-securely realizes a functionality f in the $(\mathcal{F}_1, \ldots, \mathcal{F}_n)$-hybrid model if Π UC-securely realizes f in the $(\mathcal{F}_1, \ldots, \mathcal{F}_n)$-hybrid model whenever $\exists T_i \in \Sigma$ such that each setup in T_i holds (i.e., $\mathcal{F}_j$ is the ideal functionality $\mathcal{G}_j$, for each setup $(\mathcal{G}_j, m\mathcal{G}_j) \in T_i$).*

Remark. We note that our security definition does not immediately comply with the traditional UC framework where setups are "incorruptible". To this

[8] Here, and throughout the text, it should be implicit that whenever we say a setup X holds, what we actually mean is that the *setup assumption* corresponding to setup X holds.

end, we consider a simple modification to the UC framework where the adversary is allowed to choose (before the start of the protocol) the setups that she wishes to corrupt. Of course, this information is not known to the protocol designer, since in that case, achieving UC security is easy – the parties could simply ignore the corrupt setups and use only the honest ones.

UC-compatible Belief Sets. Before we discuss our positive result on constructing Σ-secure protocols, we first define the notion of a *UC-compatible* belief set, that is central to our result.

Definition 3 (UC-compatible belief set). *A belief set $\Sigma = \{T_1, T_2 \ldots T_k\}$, where $\forall i \in [k]$, $T_i \neq \emptyset$, is said to be UC-compatible if $\forall i, j \in [k]$, at least one of the following conditions hold:*

- $T_i \cap T_j \neq \emptyset$.
- *Both T_i and T_j contain at least one (not necessarily the same) Type II setup.*
- *Both T_i and T_j contain at least one (not necessarily the same) Type III setup.*
- *Either T_i or T_j contains at least one Type II setup as well as at least one Type III setup.*

We are now ready to state our main result for the common belief case.

Theorem 1 (Main result for common belief case). *Let there be n setups in the system denoted by $(\mathcal{G}_1, m\mathcal{G}_1), \ldots, (\mathcal{G}_n, m\mathcal{G}_n)$. Let $\mathcal{F}_1, \ldots, \mathcal{F}_n$ be n ideal functionalities, where $\forall i, \mathcal{F}_i$ is either $\mathcal{G}_i$ or $m\mathcal{G}_i$. Let Σ be a belief set over the n setups that represents the common belief of the parties. If Σ is UC-compatible, then for every well-formed functionality f, there exists a non-trivial protocol Π that Σ-securely realizes f in the $(\mathcal{F}_1, \ldots, \mathcal{F}_n)$-hybrid model.*

Proof (Sketch). As noted in Section 3, it follows from the work of [16] that the task of constructing UC secure protocols from any setup assumption reduces to the task of constructing a UC-puzzle (in the hybrid model of the corresponding setup). Then, we note that in order to prove Theorem 1, it suffices to construct a puzzle $\langle S, R \rangle$ with an associated relation $\mathcal{R}$ such that $(\langle S, R \rangle, \mathcal{R})$ is a UC-puzzle in the $(\mathcal{F}_1, \ldots, \mathcal{F}_n)$-hybrid model if the belief set Σ is UC-compatible. We now briefly explain the puzzle construction.

Consider the n setups in the system. Recall that if a setup is of type t_i, then there exists a UC-puzzle (in the hybrid model of that setup) of type t_i. Then, our new puzzle protocol $\langle S, R \rangle$ is simply a sequential composition of the n puzzle protocols obtained from the n setups. Defining the associated relation $\mathcal{R}$ (and hence the trapdoor) is more tricky. Let $\Sigma = \{T_1, \ldots, T_k\}$, where Σ is the belief set. Recall that as per definition 2, our (final) protocol should be UC-secure whenever there exists $T_i \in \Sigma$ such that each setup in T_i holds (in this case, we say that set T_i is good). To this end, we define k trapdoors σ_i, one corresponding to each set T_i; here, the main requirement is that if a set T_i is good, then (a) the simulator can obtain the trapdoor σ_i, but (b) no adversary can obtain any of the k trapdoors. We make the following observations: (a) the simulator can obtain the trapdoor for the UC-puzzle corresponding to each setup present in a good set (but the adversary cannot), (b) the simulator can obtain the trapdoor for a Type III UC-puzzle even if the corresponding setup does *not* hold (hence, these trapdoors are for "free"). In light of these observations, we define σ_i to contain the trapdoor for the UC-puzzle corresponding to each setup present in T_i; additionally, σ_i contains the trapdoor for each Type III UC-puzzle.

Now suppose that $\exists i \in [k]$ such that T_i is good. Further, (in the worst case) suppose that each $T_j \neq T_i$ is such that none of the setups in T_j holds (we will call such a set T_j to be bad). By definition, the simulator can obtain the trapdoor σ_i. Further, no adversary can obtain the trapdoor σ_i. In order to argue that no adversary can obtain any of the remaining $k-1$ trapdoors (corresponding to the bad sets), we make use of the fact that Σ is UC-compatible. That is, since Σ is UC-compatible, for each bad set T_j, at least one of the four conditions (c.f. Definition 3) must hold. The proof follows by a case analysis. Here, in addition to the earlier observations, we use the fact that no adversary can obtain a trapdoor for a Type II UC-puzzle (even if the corresponding setup does *not* hold). Due to lack of space, we defer the details to the full version.

5 Different Beliefs about the Setups

In this section, we consider the setting where the parties in the system have different and independent beliefs about the setups in the system. Note that in such a scenario, depending upon the *reality* of how all the setups are implemented (for e.g., a third party that publishes a CRS may or may not be honest), the beliefs of some parties about the setups may turn out to be true, while that of other parties may turn out to be false. Then let us consider what would be an appropriate security definition for secure computation protocols in such a setting. Ignoring for a moment whether the definition is actually realizable, note that an acceptable security definition must provide standard security guarantees for at least those parties whose beliefs about the setups turn out to be true. But what of the parties whose belief about the setups turns out to be false? Note that a party would not expect the protocol to be secure if its belief about the setups turns out to be false. In light of this observation, below we consider a security definition that provides security for only those honest parties whose belief about the setups turns out to be true; any other party is considered to be adversarial, and therefore, no security is provided for such a party. As we show later, our definition is actually realizable when the beliefs of the parties about the setups satisfy a specific property (see below for more details). We now give more details.

Let $P_1, \ldots, P_m$ denote the parties in the system. We will use the notion of a belief set (as defined in Section 4) to represent the independent belief of each party about the setups in the system. Specifically, let $\Sigma_i = \{T_{i,1}, \ldots, T_{i,k_i}\}$ denote the belief set of P_i.

$(\Sigma_1, \ldots, \Sigma_m)$-secure protocols. As mentioned above, we would like to construct protocols that realize UC security with the (natural) condition that security is provided only for those (honest) parties whose beliefs about the setups turn out to be true. To formally capture the fact that no security provided for a specific set of parties, we consider a minor modification in the standard model of UC security [4]. Specifically, let $U = \{P_1, \ldots, P_m\}$ denote the set of parties in the system. Let $H \subseteq U$ be a set of parties for whom we wish to provide standard security guarantees. We stress that H is not an a-priori fixed set of parties. Specifically, in our setting, H is determined once the adversary decides which setups in the system it wishes to corrupt. Then, in the modified UC framework, at the beginning of the protocol, the adversary $\mathcal{A}$ is required to corrupt each party $P_i \in U \setminus H$. The adversary can then further corrupt parties in H depending upon the corruption model (static or adaptive). We wish to provide security for the *honest* parties in H. We are now ready to define $(\Sigma_1, \ldots, \Sigma_m)$-*secure protocols.*

Definition 4 ($(\Sigma_1, \ldots, \Sigma_m)$-secure protocols). *Let there be n setups in the system denoted by $(\mathcal{G}_1, m\mathcal{G}_1), \ldots, (\mathcal{G}_n, m\mathcal{G}_n)$. Let $\mathcal{F}_1, \ldots, \mathcal{F}_n$ be n functionalities, where $\forall i, \mathcal{F}_i$ is either $\mathcal{G}_i$ or $m\mathcal{G}_i$. Let $U = \{P_1, \ldots, P_m\}$ denote the set of parties in the system. For every i, let $\Sigma_i = \{T_{i,1}, \ldots, T_{i,k_i}\}$ be a belief set over the n setups that represents the independent belief of P_i. We say that a protocol Π $(\Sigma_1, \ldots, \Sigma_m)$-securely realizes a functionality f in the $(\mathcal{F}_1, \ldots, \mathcal{F}_n)$-hybrid model if $\forall H \subseteq \{P_1, \ldots, P_m\}$, Π UC-securely realizes f in the $(\mathcal{F}_1, \ldots, \mathcal{F}_n)$-hybrid model against all adversaries that initially corrupt all parties in $U \setminus H$, when $\forall P_i \in H$, $\exists j \in [k_i]$ such that each setup in $T_{i,j} \in \Sigma_i$ holds (i.e., $\mathcal{F}_\ell = \mathcal{G}_\ell$ for each setup $(\mathcal{G}_\ell, m\mathcal{G}_\ell) \in T_{i,j}$).*

UC-compatibility for Collection of Belief Sets. The notion of UC-compatibility (as defined in Section 4) is central to our results. Here, we extend this notion to a *collection* of belief sets.

Definition 5 (UC-compatible collection of belief sets). *A collection of belief sets $\Sigma_1, \ldots, \Sigma_m$ where $\Sigma_i = \{T_{i,1}, \ldots, T_{i,k_i}\}$ and $T_{i,\ell} \neq \emptyset \ \forall i \in [m], \ell \in [k_i]$, is said to be UC-compatible if $\forall i, j \in [m]$ and $\forall \ell \in [k_i], \hat{\ell} \in [k_j]$, at least one of the following conditions hold:*

- $T_{i,\ell} \cap T_{j,\hat{\ell}} \neq \emptyset$.
- *Both $T_{i,\ell}$ and $T_{j,\hat{\ell}}$ contain at least one (not necessarily the same) Type II setup.*
- *Both $T_{i,\ell}$ and $T_{j,\hat{\ell}}$ contain at least one (not necessarily the same) Type III setup.*
- *Either $T_{i,\ell}$ or $T_{j,\hat{\ell}}$ contains at least one Type II setup as well as at least one Type III setup.*

Remark. Note that it is possible that given two belief sets Σ_1, Σ_2, neither of them is UC-compatible but the collection $\{\Sigma_1, \Sigma_2\}$ is UC-compatible.

We are now ready to state our main result for the different beliefs case.

Theorem 2 (Main result for different beliefs case). *Let there be n setups in the system denoted by $(\mathcal{G}_1, m\mathcal{G}_1), \ldots, (\mathcal{G}_n, m\mathcal{G}_n)$. Let $\mathcal{F}_1, \ldots, \mathcal{F}_n$ be n ideal functionalities, where $\forall i, \mathcal{F}_i$ is either $\mathcal{G}_i$ or $m\mathcal{G}_i$. Let $P_1, \ldots, P_m$ be m parties. For every i, let Σ_i be a belief set over the n setups that represents the independent* *belief of P_i. If the collection of belief sets $\Sigma_1, \ldots, \Sigma_m$ is UC-compatible, then for every well-formed functionality f, there exists a non-trivial protocol Π that $(\Sigma_1, \ldots, \Sigma_m)$-securely realizes f in the $(\mathcal{F}_1, \ldots, \mathcal{F}_n)$-hybrid model.*

Proof (Idea). As noted earlier in Section 3, the task of constructing UC-secure protocols for any well-formed functionality reduces to the task of constructing an ssZK protocol. Intuitively, this is because given a functionality f, we can start with a semi-honest secure computation protocol Π for f, and then "compile" Π with an ssZK protocol to obtain a UC-secure protocol against active adversaries. Furthermore, following the result of [16], given any setup assumption (such as CRS), the above task is further reduced to the task of constructing a UC-puzzle in the hybrid model of the corresponding setup.

Now recall that in light of the above, we proved our positive result in the common belief case by simply constructing a UC-puzzle if the belief set (that represents the common belief of the parties) is UC-compatible. In the different beliefs case, however, instead of constructing a single UC-puzzle, we will construct a *family* of UC-puzzles if the collection of belief sets $\Sigma_1, \ldots, \Sigma_m$ (where Σ_i represents the belief of party P_i) is UC-compatible. Specifically, for each pair of parties P_i and P_j, we will construct two different UC-puzzles, (a) one where P_i (resp., P_j) acts as the sender (resp., receiver) and (b) the other where the roles of P_i and P_j are reversed. Then, given such a family of UC-puzzles, we can construct a family of ssZK protocols where the protocols in the family are concurrent simulation-sound with respect to each other. Specifically, for each pair of parties P_i and P_j, we can construct two different ssZK protocols, (a) one where P_i (resp., P_j) acts as the prover (resp., verifier), and (b) the other, where the roles of P_i and P_j are reversed. Finally, in order to construct a UC-secure protocol for any well-formed functionality f, we can start with a semi-honest protocol Π for f, and then "compile" Π with the above family of ssZK protocols in the following manner. Whenever a party P_i sends a protocol message to P_j, it proves that it has "behaved honestly so far in the protocol" by running an execution of the "appropriate" ssZK protocol (i.e., where P_i and P_j play the roles of the prover and verifier respectively) from the above family.

Due to lack of space, the details of the proof are deferred to the full version.

References

1. Barak, B., Canetti, R., Nielsen, J., Pass, R.: Universally composable protocols with relaxed set-up assumptions. In: FOCS (2004)
2. Barak, B., Prabhakaran, M., Sahai, A.: Concurrent non-malleable zero knowledge. In: FOCS (2006)
3. Ben-Or, M., Goldwasser, S., Wigderson, A.: Completeness theorems for non-cryptographic fault-tolerant distributed computation. In: STOC (1988)
4. Canetti, R.: Universally composable security: A new paradigm for cryptographic protocols. In: FOCS (2001)
5. Canetti, R., Dodis, Y., Pass, R., Walfish, S.: Universally composable security with global setup. In: Vadhan, S.P. (ed.) TCC 2007. LNCS, vol. 4392, pp. 61–85. Springer, Heidelberg (2007)

6. Canetti, R., Fischlin, M.: Universally composable commitments. In: Kilian, J. (ed.) CRYPTO 2001. LNCS, vol. 2139, pp. 19–40. Springer, Heidelberg (2001)
7. Canetti, R., Kushilevitz, E., Lindell, Y.: On the limitations of universally composable two-party computation without set-up assumptions. J. Cryptology 19 (2006)
8. Canetti, R., Lindell, Y., Ostrovsky, R., Sahai, A.: Universally composable two-party and multi-party secure computation. In: STOC (2002)
9. Canetti, R., Pass, R., Shelat, A.: Cryptography from sunspots: How to use an imperfect reference string. In: FOCS (2007)
10. Chandran, N., Goyal, V., Sahai, A.: New constructions for UC secure computation using tamper-proof hardware. In: Smart, N.P. (ed.) EUROCRYPT 2008. LNCS, vol. 4965, pp. 545–562. Springer, Heidelberg (2008)
11. Damgård, I., Nielsen, J.B., Wichs, D.: Isolated proofs of knowledge and isolated zero knowledge. In: Smart, N.P. (ed.) EUROCRYPT 2008. LNCS, vol. 4965, pp. 509–526. Springer, Heidelberg (2008)
12. Goyal, V., Katz, J.: Universally composable multi-party computation with an unreliable common reference string. In: Canetti, R. (ed.) TCC 2008. LNCS, vol. 4948, pp. 142–154. Springer, Heidelberg (2008)
13. Groth, J., Ostrovsky, R.: Cryptography in the multi-string model. In: Menezes, A. (ed.) CRYPTO 2007. LNCS, vol. 4622, pp. 323–341. Springer, Heidelberg (2007)
14. Katz, J.: Universally composable multi-party computation using tamper-proof hardware. In: Naor, M. (ed.) EUROCRYPT 2007. LNCS, vol. 4515, pp. 115–128. Springer, Heidelberg (2007)
15. Kushilevitz, E., Lindell, Y., Rabin, T.: Information-theoretically secure protocols and security under composition. In: STOC (2006)
16. Lin, H., Pass, R., Venkitasubramaniam, M.: A unified framework for concurrent security: universal composability from stand-alone non-malleability. In: STOC (2009)
17. Lindell, Y.: Bounded-concurrent secure two-party computation without setup assumptions. In: STOC (2003)
18. Lindell, Y.: Lower bounds for concurrent self composition. In: Naor, M. (ed.) TCC 2004. LNCS, vol. 2951, pp. 203–222. Springer, Heidelberg (2004)
19. Moran, T., Segev, G.: David and goliath commitments: UC computation for asymmetric parties using tamper-proof hardware. In: Smart, N.P. (ed.) EUROCRYPT 2008. LNCS, vol. 4965, pp. 527–544. Springer, Heidelberg (2008)
20. Pass, R.: Bounded-concurrent secure multi-party computation with a dishonest majority. In: STOC (2004)
21. Pass, R., Rosen, A.: Bounded-concurrent secure two-party computation in a constant number of rounds. In: FOCS (2003)

A Setups with Multiple Types

In the preceding text, for simplicity of exposition, we first restricted ourselves to the setting where each setup has a unique type. In this section, we discuss how our results can be extended to incorporate setups with multiple Types. Due to lack of space, we only provide an informal treatment of the results here. We refer the reader to the full version for more details.

Extending our results of Section 4 and 5. For simplicity of exposition, we will only consider the case where a setup is of all three types – Type I, Type II and Type III (other cases can be handled in a similar manner). The main idea that allows to handle setups with multiple types is that we can think of a setup $(\mathcal{G}, m\mathcal{G})$ as three separate setups of unique type. If $(\mathcal{G}, m\mathcal{G})$ "holds", then each of the three setups must "hold," while if it is "corrupt", then each of the three setups are "corrupt." Then, roughly speaking, simply replacing $(\mathcal{G}, m\mathcal{G})$ with these three setups of unique types allows us to directly use the results of Section 4 and Section 5.

More specifically, recall that a belief set (as defined in Section 4) is expressed in terms of setups with unique types. If, instead, a belief set has a setup with multiple types, then we can modify the belief set and express it in terms of setups with unique types by following the above trick. Once we reduce all the setups in the system into setups of unique types and the belief set of parties are expressed in terms of these setups, then we can directly apply Theorem 1 to obtain a possibility result in the common belief case. In case parties have different belief sets as in Section 5, then we will need to express belief sets of all parties in terms of setups of unique types. Then we can directly apply Theorem 2 to obtain a possibility result for the distinct belief case.

B Setups with Multiple Failure Modes

In the preceding text, for simplicity of exposition, we first restricted ourselves to the setting where a setup has only a single failure mode. We now briefly discuss how to handle setups with multiple failure modes. Due to lack of space, our discussion will be informal and brief. We refer the reader to the full version for details.

Example. We first discuss a motivating example to highlight the importance of handling setups with multiple failure modes. Consider a system with three instances of the key registration setup. Recall that a key registration setup is based on the following two assumptions: (1) the secret keys of honest parties are "safe", and (2) the public keys of corrupted parties are "well-formed." Then, the parties in the system may have the following common belief: either

- the first key registration functionality is "honest", but in case it fails, then either (1) or (2) still holds (i.e., it never fails completely), or
- the second and third key registration functionalities are "honest", but in case the second functionality fails, then (1) still holds, while if the third functionality fails, then (2) still holds.

We stress that the above example is very "natural", and that in a real-world scenario, the parties may be willing to put trust into the "extent" of corruption of a setup. (In some scenarios, this may be due to some physical constraints imposed upon an adversary because of how the setup is done.) It is interesting to note that UC is indeed possible in the above example. We now briefly explain how to extend our model and our earlier results to accommodate setups with multiple failure modes.

Extending our model to accommodate multiple failure modes. Consider a setup modeled by an ideal functionality $\mathcal{G}$ with failure modes $m\mathcal{G}_1, m\mathcal{G}_2, \ldots,$

$m\mathcal{G}_\ell$. For lack of a better terminology, we will refer to each pair $(\mathcal{G}, m\mathcal{G}_i)$ (that was originally referred to as a setup) as a *setup mode*. Then, we can define Types for a setup mode in the same manner as we defined Types as in Definition 1. A setup mode $(\mathcal{G}, m\mathcal{G})$ is said to be of Type X if it is possible to construct a UC-puzzle of Type X in $(\mathcal{G}, m\mathcal{G})$ hybrid model. Note that the above definition allows a setup mode to have multiple Types. For simplicity of exposition, in this subsection, we will restrict ourselves to the case where a setup mode has only a single Type. We stress that this restriction can be easily removed by using techniques from Appendix A.

Extending our results of Section 4 and Section 5. Due you lack of space, we informally explain how our results in the common belief case can be extended. We refer the reader to the full version of the paper for details on the different beliefs case.

We note that the key issue that arises because of multiplicity of failure modes is in the definition of UC-compatible belief set (c.f. Definition 3). We start by giving a more general definition of UC-compatible belief set (see below) which takes into account the multiplicity of failure modes for setups. We then briefly argue that given this more general definition of UC-compatible belief set, Theorem 1 is still applicable.

Definition 6 (Generalization of Definition 3). *A belief set $\Sigma = \{T_1, T_2 \ldots T_k\}$, where $\forall i \in [k]$, $T_i \neq \emptyset$, is said to be UC-compatible if $\forall i, j \in [k]$, at least one of the following conditions hold:*

- *There exists a setup $\mathcal{G}$ such that both T_i and T_j contain a setup mode $(\mathcal{G}, \star)$.*
- *Both T_i and T_j contain at least one (not necessarily the same) Type II setup mode.*
- *Both T_i and T_j contain at least one (not necessarily the same) Type III setup mode.*
- *Either T_i or T_j contains at least one Type II setup mode as well as at least one Type III setup mode.*

Definition 6 differs from Definition 3 in two ways: (a) The first condition of Definition 6 ignores the failure mode of the setups when taking intersection. However, we note that this is not really a fundamental difference because the failure modes of the setups were irrelevant in the first condition of Definition 3 as well. This is because we earlier restricted setups to exhibit only a single failure mode. Intuitively, this condition captures the case when the sets T_i and T_j share a setup and that setup *holds*. Hence, the failure mode is of no consequences. (b) The remaining three conditions in Definition 6 differ from the corresponding conditions in Definition 3 in the usage of setup modes instead of setups. We note, however, that these conditions in Definition 3 (resp., Definition 6) rely only on the types of the setup (resp., setup mode), and disregard the actual setups themselves. Hence, intuitively, the fact the same setup is leading to different types is of no consequences. Due to the above reasons, the proof of Theorem 1 easily extends to the general case with multiple failure modes.

Secure Two-Party Computation via Cut-and-Choose Oblivious Transfer

Yehuda Lindell* and Benny Pinkas**

Dept. of Computer Science, Bar-Ilan University, Israel

Abstract. Protocols for secure two-party computation enable a pair of parties to compute a function of their inputs while preserving security properties such as privacy, correctness and independence of inputs. Recently, a number of protocols have been proposed for the efficient construction of two-party computation secure in the presence of malicious adversaries (where security is proven under the standard simulation-based ideal/real model paradigm for defining security). In this paper, we present a protocol for this task that follows the methodology of using cut-and-choose to boost Yao's protocol to be secure in the presence of malicious adversaries. Relying on specific assumptions (DDH), we construct a protocol that is significantly more efficient and far simpler than the protocol of Lindell and Pinkas (Eurocrypt 2007) that follows the same methodology. We provide an exact, concrete analysis of the efficiency of our scheme and demonstrate that (at least for not very small circuits) our protocol is more efficient than any other known today.

1 Introduction

1.1 Background

Protocols for secure two-party computation enable a pair of parties P_1 and P_2 with private inputs x and y, respectively, to compute a function f of their inputs while preserving a number of security properties. The most central of these properties are *privacy* (meaning that the parties learn the output $f(x, y)$ but nothing else), *correctness* (meaning that the output received is indeed $f(x, y)$ and not something else), and *independence of inputs* (meaning that neither party can choose its input as a function of the other party's input). The standard way of formalizing these security properties is to compare the output of a real protocol execution to an "ideal execution" in which the parties send their inputs to an incorruptible trusted party who computes the output for the parties. Informally speaking, a protocol is then secure if no real adversary attacking the real protocol can do more harm than an ideal adversary (or simulator) who interacts in the ideal model [13,14,26,2,3]. An important parameter when considering this

* `lindell@cs.biu.ac.il`.Research supported by the European Research Council as part of the ERC project LAST, and by the ISRAEL SCIENCE FOUNDATION (781/07).

** `benny@pinkas.net` Research supported by the European Research Council as part of the ERC project SFEROT and by the EU project CACE.

Y. Ishai (Ed.): TCC 2011, LNCS 6597, pp. 329–346, 2011.

problem relates to the power of the adversary. The two most studied models are the *semi-honest model* (where the adversary follows the protocol specification exactly but tries to learn more than it should by inspecting the protocol transcript) and the *malicious model* (where the adversary can follow any arbitrary polynomial-time strategy).

In the 1980s powerful feasibility results were proven, showing that *any* probabilistic polynomial-time functionality can be securely computed in the presence of semi-honest adversaries [33] and in the presence of malicious adversaries [13]. These results showed that it is possible to achieve such secure protocols, but did not demonstrate how to do so efficiently (where by efficiency we mean a protocol that can be implemented and run in practice). To be more exact, the protocol of [33] for semi-honest adversaries is efficient. However, achieving security efficiently for the case of malicious adversaries is far more difficult. In fact, until recently, no efficient general protocols were known at all, where a general protocol is one that can be used for computing *any* functionality.

This situation has changed in the past few years, possibly due to increasing interest from outside the cryptographic community in secure protocols that are efficient enough to be used in practice. The result has been that a number of secure two-party protocols were presented that are secure in the presence of malicious adversaries, where security is rigorously proven according to the aforementioned ideal/real model paradigm [20,24,28,18]. Interestingly, these protocols all take novel, different approaches and so the secure-protocol skyline is more diverse than before, providing the potential for taking the protocols a step closer to very high efficiency. These protocols are discussed in more detail in Section 1.3.

We remark that the protocol of [24] has been implemented for the non-trivial problem of securely computing the AES block cipher (pseudorandom function), where one party's input is a secret key and the other party's input is a value to be "encrypted" [31]. A Boolean circuit for computing this function was designed with approximately 33,000 gates, and the protocol of [24] was implemented for this circuit. Experiments showed that the running-time of the protocol was between 18 and 40 minutes, depending on the assumptions taken on the primitives used to implement the protocol. Although this is quite a long time, for some applications it can be reasonable. In addition, it demonstrates that it is *possible* to securely compute functions with large circuits, and motivates the search for finding even more efficient protocols that can widen the applicability of such computations in real-world settings.

1.2 Our Results

In this paper, we follow the construction paradigm of [24] and significantly simplify and improve the efficiency of their construction. The approach of [24] is to carry out a basic cut-and-choose on the garbled circuit construction of Yao [33] (we assume familiarity with Yao's protocol). That is, party P_1 constructs s copies of a garbled circuit and sends them to P_2, who then asks P_1 to open half of them in order to verify that they are correctly constructed. If all of the opened circuits are indeed correct, then it is guaranteed that a *majority* of the unopened half are also correct, except with probability that is negligible in a statistical security parameter s. Thus, P_1 and P_2 evaluate the remaining $s/2$ circuits, and P_2 takes the output that appears in most of the evaluated circuits. As discussed in [24], P_2 cannot abort in the case that not all of the $s/2$ circuits evaluate to the same value, even though in such a case it knows that P_1 is cheating. The reason for this is that P_1 may construct a circuit that computes f in the case that P_2's first bit equals 0, and otherwise it outputs random garbage. Now, with probability 1/2 this faulty circuit is not opened and so is one of the circuits to be evaluated. In this case, if P_2 would abort when it saw random garbage then P_1 would know that P_2's first input bit equals 1. For this reason, P_2 takes the majority output and ignores minority values without aborting.

Although intuitively appealing, the cut-and-choose approach introduces a number of difficulties which significantly affect the efficiency of the protocol of [24]. First, since the parties need to evaluate $s/2$ circuits rather than one, there needs to be a mechanism to ensure that they use the same input in all evaluations (the solution for this for P_2's inputs is easy, but for P_1's inputs turns out to be hard). The mechanism used in [24] required constructing and sending $s^2\ell$ commitments. In the implementation by [31], they set $s = 160$ and $\ell = 128$. Thus, the overhead due to these consistency proofs is the computation and transmission of 3,276,800 commitments! Another problem that arises in the protocol of [24] is that a malicious P_1 can input an incorrect key into one of the oblivious transfers used for P_2 to obtain the keys associated with its input wires in the garbled circuit. For example, it can set all the keys associated with 0 for P_2's first input bit to be garbage, thereby making it impossible for P_2 to decrypt any circuit if its first input bit indeed equals 0. In contrast, P_1 can make all of the other keys be correct. In this case, P_1 is able to learn P_2's first input bit, merely by whether P_2 obtains an output or not. The important observation is that the checks on the garbled circuit carried out by P_2 do not detect this because there is a separation between the cut-and-choose checks and the oblivious transfer. The solution to this problem in [24] requires making the circuit larger and significantly increasing the size of the inputs by replacing each input bit with the exclusive-or of multiple random input bits. Finally, the analysis of [24] yields an error of $2^{-s/17}$. Thus, in order to obtain an error level of 2^{-40} the parties need to exchange 680 circuits. We remark that it has been conjectured in [31] that the true error level of the protocol is $2^{-s/4}$; however, this has not been proven.

Our protocol. We solve the aforementioned problems in a way that is far simpler and far more efficient than in [24]. In addition, we reduce the error probability to $2^{-0.311s}$ and thus for an error of 2^{-40} it suffices to send only 128 circuits. This is an important improvement because the experiments of [31] demonstrate that the bottleneck in efficiency is not the exponentiations, but rather the number of circuits and the commitments for proving consistency. Thus, in our protocol we moderately increase the number of exponentiations, while reducing the number of circuits, completely removing the commitments, and also removing the need to increase the size of the inputs. We remark that the price for these improvements is that our protocol relies heavily on the decisional Diffie-Hellman (DDH) assumption, while the protocol of [24] used general assumptions only. We now proceed to describe our two main techniques:

1. Our solution for ensuring consistency of P_1's inputs is to have P_1 determine the keys associated with its own input bits via a Diffie-Hellman pseudorandom synthesizer [27]. That is, P_1 chooses values $g^{a_1^0}, g^{a_1^1}, \ldots, g^{a_\ell^0}, g^{a_\ell^1}$ and $g^{r_1}, \ldots, g^{r_s}$ and then sets the keys associated with its ith input bit in the jth circuit to be $g^{a_i^0 \cdot r_j}, g^{a_i^1 \cdot r_j}$. Given all of the $\{g^{a_i^0}, g^{a_i^1}, g^{r_j}\}$ values and any subset of keys of P_1's input generated in this way, the remaining keys associated with its input are pseudorandom by the DDH assumption. Furthermore, it is possible for P_1 to efficiently prove that it is using the same input in all circuits when the keys have this nice structure.
2. As we have described, the reason that the inputs and circuits were needed to be made larger in [24] is due to the fact that the cut-and-choose circuit checks were separated from the oblivious transfer. In order to solve this problem, we introduce a new primitive called *cut-and-choose oblivious transfer*. This is an ordinary oblivious transfer with the sender inputting many pairs $(x_1^0, x_1^1), \ldots, (x_s^0, x_s^1)$, and the receiver inputting bits $\sigma_1, \ldots, \sigma_s$. However, the receiver also inputs a set $\mathcal{J} \subset [s]$ of size exactly $s/2$. Then, the receiver obtains $x_i^{\sigma_i}$ for every i (as in a regular oblivious transfer) along with both values (x_j^0, x_j^1) for every $j \in \mathcal{J}$. The use of this primitive in our protocol intertwines the oblivious transfer and the circuit checks and solves the aforementioned problem. We also show how to implement this primitive in a highly efficient way, under the DDH assumption. We believe that this primitive is of independent interest, and could be useful in many cut-and-choose scenarios.

Efficiency analysis. Our entire protocol, including all subprotocols, is explicitly written and analyzed in a concrete and exact way for efficiency. Considerable effort has been made to optimize the constructions and reduce the constants throughout. We believe that this is of great importance when the focus of a result is *efficiency*. See Section 1.3 for a summary of the exact complexity of our protocol, and Section 3 for a complete analysis, with optimizations in Section 3.3.

Variants. Another advantage of our protocol over that of [24] is that we obtain a universally composable [4] variant that is only slightly less efficient than the stand-alone version. This is because our simulator only rewinds during zero-knowledge protocols. These protocols are also Σ protocols and so can be efficiently transformed into universally composable zero-knowledge. As with our basic protocol, we provide an explicit description of this transformation and analyze its exact efficiency. Finally, we also show how our protocol yields a more efficient construction for security in the presence of covert adversaries [1], when high values of the deterrent factor ϵ are desired.

1.3 Comparison to Other Protocols

We provide an analysis of the efficiency of recent protocols for secure two-party computation. Each protocol takes a different approach, and thus the approaches may yield more efficient instantiations in the future. Nevertheless, as we will show, our protocol is significantly more efficient than the current best instantiations of the other approaches (at least, for not very small circuits).

- **Committed input method (Jarecki-Shmatikov [20]):** The secure two-party protocol of [20] works by constructing a single circuit and proving that it is correct. The novelty of this protocol is that this can be done with only a constant number of (large modulus) exponentiations per gate of the circuit. Thus, for circuits that are relatively small, this can be very efficient. However, an exact count gives that approximately 720 exponentiations are required per gate. Thus, even for small circuits, this protocol is not yet practical. For large circuits like AES with 33,000 gates, the number of exponentiations is very large ($23,760,000$ for AES), and is not realistic. (The authors comment that if efficient batch proofs can be found for the languages they require then this can be significantly improved. However, to the best of our knowledge, no such improvements have yet been made. Furthermore, for a large circuit we believe it unlikely that this method will yield a highly efficient protocol).
- **LEGO (Nielsen-Orlandi [28]):** The LEGO protocol [28] follows the cut-and-choose methodology in a completely different way. Specifically, the circuit constructor first sends the receiver many gates, and the receiver checks that they are correctly constructed by asking for some to be opened. After this stage, the parties interact in a way that enables the gates to be securely soldered (like Lego blocks) into a correct circuit. Since it is not guaranteed that all of the gates are correct, but just a vast majority, a fault tolerant circuit of size $O(s \cdot |C|/\log|C|)$ is constructed, where s is a statistical security parameter. The error as a function of s is 2^{-s} and the constant inside the "O" notation for the number of exponentiations is 32 [29]. Thus, for an error of 2^{-40} we have that the overall number of exponentiations carried out by the parties is $1280 \cdot |C|/\log|C|$. For large circuits, like that of AES, this is unlikely to be practical. (For example, for the AES circuit with 33,000 gates we have that the parties need to carry out $2,816,000$ exponentiations. Observe that due to the size of the circuit, the $\log|C|$ factor is significant in making the protocol more efficient than [20], as predicted in [28]. This protocol also relies on the DDH assumption. It is worthy to note that exponentiations in this protocol are in a regular "Diffie-Hellman" group and so Elliptic curves can be used, in contrast to [20] who work in $\mathbb{Z}_N^*$).
- **Virtual multiparty method (Ishai et al. [18,19]):** This method works by having the parties simulate a virtual multiparty protocol with an honest majority. The cost of the protocol essentially consists of the cost of running a semi-honest protocol for computing the multiplication of additive shares, for every multiplication carried out by a party in a multiparty protocol with honest majority. Thus, the actual efficiency of the protocol depends heavily on the multiparty protocol to be simulated, and the semi-honest protocols used for simulating the multiparty protocol. An asymptotic analysis demonstrates that this method may be competitive. However, no concrete analysis has been carried out, and it is currently an open question whether or not it

is possible to instantiate this protocol in a way that will be competitive with other known protocols.

- **Cut-and-choose on circuits (Lindell-Pinkas [24]):** Since this protocol has been discussed at length above, we just briefly recall that the complexity of the protocol is $O(\ell)$ oblivious transfers for input-length ℓ (where the constant inside here is not small because of the need to increase the number of P_2's inputs), and the construction and computation of s garbled circuits and of $s^2\ell$ commitments. In addition, the proven error of the protocol is $2^{-s/17}$ and its conjectured error is $2^{-s/4}$. The actual value has a significant impact on the efficiency.

In contrast to the above, the complexity of our protocol is as follows. The parties need to compute $15s\ell + 39\ell + 10s + 6$ exponentiations, where ℓ is the input length and s is a statistical security parameter discussed below. We further show that with optimizations the $15s\ell$ component can be brought down to just $\mathbf{5.66}s\ell$ **full exponentiations**, and if preprocessing can be used then only $s\ell/2$ full exponentiations need to be computed after the inputs become known. In addition, the protocol requires the exchange of $\mathbf{7}s\ell + \mathbf{22}\ell + \mathbf{7}s + \mathbf{5}$ **group elements**, and has **12 rounds of communication**. Finally, there are $\mathbf{6.5}|C|s$ **symmetric encryptions** for constructing and decrypting the garbled circuits and $\mathbf{4}|C|s$ ciphertexts sent for transmitting these circuits. An important factor here is the value of s needed. The error of our protocol is $\mathbf{2^{-0.311s}}$ and so for an error of 2^{-40} it suffices to set $s = \mathbf{128}$. (The overhead of computing an AES circuit, after preprocessing, with $|C| = 33,000$, and $s = \ell = 128$, is therefore about $93,000$ exponentiations, $27,500,000$ symmetric encryptions, and communicating 28.6 Mbytes, where about 95% of the communication is spent on sending the garbled circuits). Finally, we stress also that all of our exponentiations are of the basic Diffie-Hellman type and so can be implemented over Elliptic curves, which is much cheaper than RSA-type operations.

Full version. In this extended abstract we do not have space to present the proofs of security of our protocols. A full version of this paper appears in the *Cryptology ePrint Archive* (report 2010/284).

2 Cut-and-Choose Oblivious Transfer

2.1 The Functionality and Construction Overview

Our protocol for secure two-party computation uses a new primitive that we call *cut-and-choose oblivious transfer*. Loosely speaking, a cut-and-choose OT is a batch oblivious transfer protocol (meaning an oblivious transfer for multiple pairs of inputs) with the additional property that the receiver can choose a subset of the pairs (of a predetermined size) for which it learns *both* values. This is a very natural primitive which has clear applications for protocols that are based on cut-and-choose, as is our protocol here for general two-party computation.

The cut-and-choose OT functionality, denoted $\mathcal{F}_{\mathbf{ccot}}$, with parameter s, is formally defined in Figure 1, together with a variant functionality that we will need, which considers the case that R is forced to use the *same* choice σ in every transfer. This variant is denoted $\mathcal{F}^{\boldsymbol{S}}_{\mathbf{ccot}}$.

FIGURE 1 (The cut-and-choose OT functionalities)

The cut-and-choose OT functionality $\mathcal{F}_{\mathbf{ccot}}$:

- **Inputs:**
  - S inputs a vector of pairs $\overline{x} = \{(x^i_0, x^i_1)\}^s_{i=1}$
  - R inputs $\sigma_1, \ldots, \sigma_s \in \{0,1\}$ and a set of indices $\mathcal{J} \subset [s]$ of size exactly $s/2$.
- **Output:** If $\mathcal{J}$ is not of size $s/2$ then S and R receive $\perp$ as output. Otherwise,
  - For every $j \in \mathcal{J}$ the receiver R obtains the pair (x^j_0, x^j_1).
  - For every $j \notin \mathcal{J}$ the receiver R obtains $x^j_{\sigma_j}$.

The single-choice cut-and-choose OT functionality $\mathcal{F}^{\boldsymbol{S}}_{\mathbf{ccot}}$:

- **Inputs:** The same as above, but with R having only a single input bit σ.
- **Output:** As above, but with R obtaining the value x^j_σ for every $j \notin \mathcal{J}$.

In order to motivate the usefulness of this functionality, we describe its use in our protocol. Oblivious transfer is used in Yao's protocol so that the party computing the garbled circuit (call it P_2) can obtain the keys (garbled values) on the wires corresponding with its input while keeping its input secret. It is crucial that P_2 obtain only a single key for each wire, since this is what ensures that it can only obtain a single output. When applying cut-and-choose, many circuits are constructed and then half of them are opened, where opening means that P_2 receives all of the input keys to the circuit, enabling it to decrypt all garbled gates and check that they were correctly constructed. By using cut-and-choose OT, P_2 receives all of its keys in the circuits to be opened directly, in contrast to having P_1 send them separately after the indices of the circuits to be opened are sent from P_2 to P_1. The advantage of this approach is that P_1 cannot use *different* keys in the OT and when opening the circuit. See Section 3.1 for discussion on why this is important.

In cut-and-choose on Yao's protocol, one oblivious transfer is needed for every bit of P_2's input (equivalently, every wire on the circuit), and P_2 should receive the keys associated with this fixed bit in all of the circuits. In order to ensure that P_2 uses the same input in all circuits, we devised a *single-choice* variant of cut-and-choose OT. In the full version, we separately present the basic variant since it is of independent interest and may be useful in other applications.

2.2 Constructing a Single-Choice Cut-and-Choose OT Protocol

The starting point for our construction of cut-and-choose OT is the universally composable protocol of Peikert et al. [30]; we refer only to the instantiation of their protocol based on the DDH assumption because this is the most efficient. However, our protocol can use any of their instantiations. The protocol of [30] is cast in the common reference string (CRS) model, where the CRS is a tuple (g_0, g_1, h_0, h_1) where g_0 is a generator of a group of order q (in which DDH is assumed to be hard), $g_1 = (g_0)^y$ for some random y, and it holds that $h_0 = (g_0)^a$ and $h_1 = (g_1)^b$ where $a \neq b$. We first observe that it is possible for the receiver to choose this tuple itself, as long as it proves that it indeed fulfills the property that $a \neq b$. Furthermore, this can be proven very efficiently by setting $b = a + 1$; in this case, the proof that $b = a + 1$ is equivalent to proving that $(g_0, g_1, h_0, \frac{h_1}{g_1})$ is a Diffie-Hellman tuple (note that the security of [30] is based only on $a \neq b$ and not on these values being independent of each other). We thus obtain a highly efficient version of the protocol of [30] in the stand-alone model.

Next, observe that the protocol of [30] has the property that if (g_0, g_1, h_0, h_1) *is* a Diffie-Hellman tuple (i.e., if $a = b$) then it is possible for the receiver to learn both values (of course, in a real execution this cannot happen because the receiver proves that $a \neq b$). This property is utilized by [30] to prove universal composability; in their case the simulator can choose the CRS so that $a = b$ and then learn both inputs of the sender, something that is needed for proving simulation-based security. However, in our case, we *want* the receiver to be able to sometimes learn both inputs of the sender. We can therefore utilize this exact property and have the receiver choose s pairs $\{(h_0^j, h_1^j)\}_{j=1}^s$ such that for $s/2$ of the pairs (h_0^j, h_1^j) it holds that $a \neq b$ (ensuring that it learns only one input) and for $s/2$ of the pairs (h_0^j, h_1^j) it holds that $a = b$ (enabling it to learn both inputs by actually running the UC simulator). This therefore provides the exact cut-and-choose property in the OT that is needed. Of course, the receiver must also prove that it behaved in this way. Specifically, it proves in zero-knowledge that $s/2$ out of s pairs are such that $a \neq b$. This proof too can be computed at low cost using the technique of Cramer et al. [7]; see the full version of the paper for a full description and efficiency analysis of the zero-knowledge protocol.

In the oblivious transfer protocol, the receiver with an input σ chooses a random r and sends $(g_\sigma)^r$ and $(h_\sigma^1)^r, \ldots, (h_\sigma^s)^r$ to the sender, using the $g_0, g_1, (h_0^j, h_1^j)$ values sent previously. In order to ensure that the single-choice property holds, the receiver R must prove that it used the same σ in every computation of $(h_\sigma^j)^r$. The protocol has been carefully designed so that the way that R chooses the values enables this proof to be carried out efficiently. Specifically, the required zero-knowledge proof is that there exists an $r \in \mathbb{Z}_q$ such that either $g' = (g_0)^r$ and $h_j = (h_0^j)^r$ for every $1 \leq j \leq s$, or $g' = (g_1)^r$ and $h_j = (h_1^j)^r$ for every $1 \leq j \leq s$, which is just a proof that one of two sets of tuples are all of the Diffie-Hellman type; see the protocol specification below and the full version for details. The cost of this proof is $s + 18$ exponentiations and the exchange of 10 group elements.

PROTOCOL 2 (Single-Choice Cut-and-Choose Oblivious Transfer)

- **Inputs:** *The sender's input is a vector of s pairs (x_0^j, x_1^j) and the receiver's input is a single bit $\sigma \in \{0, 1\}$ and a set $\mathcal{J} \subset [s]$ of size exactly $s/2$.*

- **Auxiliary input:** *Both parties hold a security parameter* 1^n *and* $(\mathbb{G}, q, g_0)$, *where* $\mathbb{G}$ *is an efficient representation of a group of order* q *with a generator* g_0, *and* q *is of length* n.
- **Setup phase:**
  1. *R chooses a random* $y \leftarrow \mathbb{Z}_q$ *and sets* $g_1 = (g_0)^y$.
  2. *For every* $j \in \mathcal{J}$, *R chooses a random* $\alpha_j \leftarrow \mathbb{Z}_q$ *and computes* $h_0^j = g_0^{\alpha_j}$ *and* $h_1^j = g_1^{\alpha_j}$.
  3. *For every* $j \notin \mathcal{J}$, *R chooses random* $\alpha_j \leftarrow \mathbb{Z}_q$ *and computes* $h_0^j = g_0^{\alpha_j}$ *and* $h_1^j = g_1^{\alpha_j+1}$.
  4. *R sends* $(g_1, h_0^1, h_1^1, \ldots, h_0^s, h_1^s)$ *to S*
  5. *R proves using a zero-knowledge proof of knowledge to S that* $s/2$ *of the tuples* $(g_0, g_1, h_0^j, \frac{h_1^j}{g_1})$ *are DH tuples (and through this, that the tuples* (g_0, g_1, h_0^j, h_1^j) *are* not *DH tuples). If S rejects the proof then it outputs* $\perp$ *and halts.*
- **Transfer phase (for every** j**):**
  1. *The receiver chooses a random value* $r \leftarrow \mathbb{Z}_q$ *and computes* $g' = (g_\sigma)^r$. *Then, for every* j, *it computes* $h_j = (h_\sigma^j)^r$. *It sends* $(g', h_1, \ldots, h_s)$ *to the sender.*
  2. *The receiver proves in zero knowledge that either all* $\{(g_0, h_0^j, g', h_j)\}_{j=1}^s$ *or all* $\{(g_1, h_1^j, g', h_j)\}_{j=1}^s$ *are Diffie-Hellman tuples.*
  3. *The sender operates in the following way:*
     - *Define the function* $RAND(w, x, y, z) = (u, v)$, *where* $u = (w)^s \cdot (y)^t$ *and* $v = (x)^s \cdot (z)^t$, *and the values* $s, t \leftarrow \mathbb{Z}_q$ *are random.*
     - *S computes the pairs* $(u_0^j, v_0^j) = RAND(g_0, g_j, h_0^j, h_j)$ *and* $(u_1^j, v_1^j) = RAND(g_1, g_j, h_1^j, h_j)$.
     - *S sends the receiver the values* (u_0^j, w_0^j) *where* $w_0^j = v_0^j \cdot x_0^j$, *and* (u_1^j, w_1^j) *where* $w_1^j = v_1^j \cdot x_1^j$, *for every* j.
- **Output:**
  1. *For every* $j \in \{1, \ldots, s\}$, *the receiver computes* $x_\sigma^j = w_\sigma^j / (u_\sigma^j)^r$.
  2. *For every* $j \in \mathcal{J}$, *the receiver also computes* $x_{1-\sigma}^j = w_{1-\sigma}^j / (u_{1-\sigma}^j)^{r \cdot z}$ *where* $z = y^{-1}$ *mod* q *if* $\sigma = 0$, *and* $z = y$ *if* $\sigma = 1$.

The fact that the output obtained is correct follows from the correctness (and simulation strategy) of the protocol of [30]. We prove the following:

Proposition 3. *If the Decisional Diffie-Hellman assumption holds in the group* $\mathbb{G}$, *then Protocol 2 securely realizes the* $\mathcal{F}_{\mathrm{ccot}}^S$ *functionality in the presence of malicious adversaries.*

The proof of Proposition 3 is based on the following ideas. Suppose first that the adversary controls the receiver; we briefly describe an ideal-model simulator "interacting" with the adversary. The simulator receives from the adversary the values (g_0, g_1) and $(h_0^1, h_1^1, \ldots, h_s^0, h_s^1)$ of the setup phase, and then runs the extractor of the zero-knowledge proof of knowledge to learn the witness set. The extracted witnesses identify exactly the set $\mathcal{J}$ of $s/2$ indices for which (g_0, g_1, h_0^j, h_1^j) is a Diffie-Hellman tuple, and in addition provides the simulator with the set of values α_j for all $j \notin \mathcal{J}$. Next, in the transfer phase, given g' and h_j, the simulator can extract the value of σ by simply checking if $(g')^{\alpha_j} = h_j$; alternatively, it can run the extractor for the zero-knowledge proof of the transfer phase. The simulator then sends the set $\mathcal{J}$ and bit σ to the trusted party and obtains the corresponding outputs. Now, for every $j \in \mathcal{J}$, the simulator received both values x_0^j and x_1^j and so can compute (u_0^j, w_0^j) and (u_1^j, w_1^j) just like the honest sender. In contrast, for every $j \notin \mathcal{J}$, the simulator received only x_σ^j. In this case, it can still compute (u_σ^j, w_σ^j) like the honest sender. Regarding $(u_{1-\sigma}^j, w_{1-\sigma}^j)$, these can just be taken as uniformly distributed values in $\mathbb{G}$, by the property of the $RAND$ transformation when applied to non-Diffie-Hellman tuples.

Next consider the case that the adversary controls the sender. In this case, the simulator chooses the (h_0^j, h_1^j) values such that (g_0, g_1, h_0^j, h_1^j) is a Diffie-Hellman tuple for *all* j. The simulator then "cheats" in the zero-knowledge proof by generating a *simulated* zero-knowledge proof with the adversary. Given that now all (g_0, g_1, h_0^j, h_1^j) are Diffie-Hellman tuples, the simulator can extract both inputs (x_0^j, x_1^j) for *every* $j = 1, \ldots, s$, using the same computation as an honest receiver would for $j \in \mathcal{J}$. Finally, the simulator sends the values $\{(x_0^j, x_1^j)\}_{j=1}^s$ to the trusted party, outputs whatever the adversary outputs, and halts. The fact that this is indistinguishable from a real execution follows directly from the indistinguishability of simulated zero-knowledge proofs from real proofs, and from the DDH assumption.

Exact efficiency. In the full version, we present an exact analysis of this protocol, including the cost of all zero-knowledge proofs. The result is that the protocol requires **$20.5s + 24$ exponentiations**, the exchange of **$11s + 15$ group elements**, and **6 rounds of communication**.

2.3 Batch Single-Choice Cut-and-Choose OT

In our protocol we need to carry out cut-and-choose oblivious transfers for all wires in the circuit (where for each wire the input used is P_2's input bit). However, the subset of indices for which the receiver obtains both pairs must be the same in all transfers. This is due to the fact that this determines which of the circuits are checked and which are evaluated, and it is crucial that in all of the evaluated circuits P_2 only receives one value on every wire. We call a functionality that achieves this "batch single-choice cut-and-choose OT" and denote it $\mathcal{F}_{\rm ccot}^{S,B}$.

In order to realize this functionality it suffices to run the setup phase of Protocol 2 once (this ensures that the set $\mathcal{J}$ is the same in all executions). Then, the transfer phase of the single-choice protocol is run ℓ times *in parallel* (with the single choice in the ith execution being some σ_i). We remark that parallel composition holds here because the simulation only rewinds in the transfer phase for the zero-knowledge protocol, which is zero-knowledge under parallel composition. We have the following:

Proposition 4. *Assuming that the Decisional Diffie-Hellman assumption holds in* $\mathbb{G}$*, the above-described protocol securely realizes* $\mathcal{F}_{\rm ccot}^{S,B}$ *in the presence of malicious adversaries.*

Exact efficiency. The setup phase here remains the same, and including the zero-knowledge proof it costs $9s+5$ exponentiations and the exchange of $5s+5$ group elements. The transfer phase is repeated ℓ times, where each transfer incurs a cost of $11.5s+19$ exponentiations and the exchange of $6s+10$ group elements. We conclude that there are $\mathbf{11.5s\ell+19\ell+9s+5}$ **exponentiations**, $\mathbf{6s\ell+10\ell+5s+5}$ **group elements** sent and **6 rounds of communication**. In Section 3.3 we observe that $10.5s\ell$ of the exponentiations are "fixed-base" and thus the overall effective cost is about $\mathbf{4.5s\ell}$ **exponentiations**.

3 The Protocol for Secure Two-Party Computation

3.1 Protocol Description

Before describing the protocol in detail, we first present an intuitive explanation of the different steps, and their purpose:

Step 1: P_1 constructs s copies of a Yao garbled circuit for computing the function. The keys (garbled values) on the wires of the s copies of the circuit are all random, except for the keys corresponding to P_1's input wires, which are chosen in a special way. Namely, P_1 chooses random values $a_1^0, a_1^1, \ldots, a_\ell^0, a_\ell^1$ (where the length of P_1's input is ℓ) and $r_1, \ldots, r_s$, and sets the keys on the wire associated with its ith input in the jth circuit to be $g^{a_i^0 \cdot r_j}$ and $g^{a_i^1 \cdot r_j}$. Note that the $2\ell+s$ values $g^{a_1^0}, g^{a_1^1}, \ldots, g^{a_\ell^0}, g^{a_\ell^1}, g^{r_1}, \ldots, g^{r_s}$ constitute commitments to all $2\ell s$ keys[1]. (The keys are actually a pseudorandom synthesizer [27], and therefore if some of the keys are revealed, the remaining keys remain pseudorandom).

Step 2: The parties execute batch single-choice cut-and-choose OT. P_1 inputs the key-pairs for all wires associated with P_2's input, and P_2 inputs its input and a random set $\mathcal{J} \subset [s]$ of size $s/2$. The result is that P_2 learns all the keys on the wires associated with its own input for $s/2$ of the circuits as indexed by $\mathcal{J}$ (called check circuits), and in addition learns the keys corresponding to its actual input in these wires in the remaining circuits (called evaluation circuits).

Step 3: P_1 sends P_2 the garbled circuits, and the values $g^{a_1^0}, g^{a_1^1}, \ldots, g^{a_\ell^0}, g^{a_\ell^1}$, $g^{r_1}, \ldots, g^{r_s}$ which are commitments to all the keys on the wires associated with P_1's input. Observe that at this stage P_1 is fully committed to all s circuits, but does not yet know which circuits are to be opened.

---

[1] The actual symmetric keys used are derived from the $g^{a_i^0 \cdot r_j}, g^{a_i^1 \cdot r_j}$ values using a randomness extractor; a universal hash function suffices for this [6,16]. The only subtlety is that P_1 must be fully committed to the garbled circuits, including these symmetric keys, before it knows which circuits are to be checked. However, randomness extractors are not $1-1$ functions. This is solved by having P_1 send the seed for the extractor before Step 4 below. Observe that the $\{g^{a_i^0}, g^{a_i^1}, g^{r_j}\}$ values and the seed for the extractor fully determine the symmetric keys, as required.

Step 4: P_2 reveals to P_1 its choice of check circuits and proves that was indeed its choice by sending, for each check circuit, *both* values on the wire associated with P_2's first input bit. Note that P_2 can know both these values only for circuits that are check circuits.

Step 5: To completely decrypt the check circuits in order to check that they were correctly constructed, P_2 also needs to obtain all the keys on the wires associated with P_1's input. Therefore, if the jth circuit is a check circuit, then P_1 sends r_j to P_2. Given all of the $g^{a_i^0}, g^{a_i^1}$ values and r_j, party P_2 can compute all of the keys $g^{a_i^0 \cdot r_j}, g^{a_i^1 \cdot r_j}$ in the jth circuit by itself (and P_1 cannot change the values). Furthermore, this reveals nothing about the keys in the evaluation circuits.

Step 6: Given all of the keys on all of the input wires, P_2 checks the validity of the $s/2$ check circuits. This ensures that P_2 will catch P_1 with high probability if many of the garbled circuits generated by P_1 do not compute the correct function. Thus, unless P_2 detects cheating, it is assured that a majority of the evaluation circuits are correct.

Step 7: All that remains is for P_1 to send P_2 the keys associated with its actual input, and then P_2 will be able to compute the evaluation circuits. This raises a problem as to how P_2 can be sure that P_1 sends keys that correspond to the same input in all circuits. This brings us to the way that P_1 chose these keys (via the Diffie-Hellman pseudorandom synthesizer). Specifically, for every wire i and evaluation-circuit j, party P_1 sends P_2 the value $g^{a_i^{x_i} \cdot r_j}$ where x_i is the ith bit of P_1's input. P_1 then proves in zero-knowledge that the same $a_i^{x_i}$ exponent appears in all of the values sent. Essentially, this is a proof that the values constitute an "extended" Diffie-Hellman tuple and thus this statement can be proven very efficiently.

Step 8: Finally, given the keys associated with P_1's inputs and its own inputs, P_2 evaluates the evaluation circuits and obtains their output values. Recall, however, that the checks above only guarantee that a majority of the circuits are correct, and not that all of them are. Therefore, P_2 outputs the value that is output from the majority of the evaluation circuits. We stress that if P_2 sees different outputs in different circuits, and thus knows for certain that P_1 has tried to cheat, it must ignore this observation and output the majority value (or otherwise it might leak information to P_1, as in the example described in Section 1.2).

We remark on one type of attack discussed in [21,24]. The concern there was that P_1 would use correct keys for all of P_2's input bits when opening the check circuit, but would use incorrect keys in some of the oblivious transfers. This is problematic because if P_1 input incorrect keys for the zero value of P_2's first input bit, and correct keys for all other values, then P_2 would not detect any misbehavior if its first input bit equals 1. However, if its first input bit equals 0 then it would have to abort (because it would not be able to decrypt any of the evaluation circuits). This results in P_1 learning P_2's first input bit with probability 1. In order to solve this problem in [24] it was necessary to split P_2's input bits into random shares, thereby increasing the size of the input to the circuit and the size of the circuit itself. In contrast, this attack does not arise here at all because P_2 obtains all of the keys associated with its input bits in the cut-and-choose oblivious transfer, and the values are not sent separately for check and evaluation circuits. Thus, if P_1 attempts a similar attack here for a small number of circuits then it will not be the majority and so does not matter, and if it does so for a large number of circuits then it will be caught with overwhelming probability. We now proceed to the full protocol description.

PROTOCOL 5 (Computing $f(x, y)$)

Inputs: *P_1 has input $x \in \{0,1\}^\ell$ and P_2 has input $y \in \{0,1\}^\ell$.*

Auxiliary input: *a statistical security parameter s, the description of a circuit C such that $C(x, y) = f(x, y)$, and $(\mathbb{G}, q, g)$ where $\mathbb{G}$ is a cyclic group with generator g and* prime *order q, and q is of length n.*

The protocol:

1. Input key choice and circuit preparation:
   (a) *P_1 chooses random values $a_1^0, a_1^1, \ldots, a_\ell^0, a_\ell^1 \in_R \mathbb{Z}_q$ and $r_1, \ldots, r_s \in_R \mathbb{Z}_q$.*
   (b) *Let $w_1, \ldots, w_\ell$ be the input wires corresponding to P_1's input in C, and denote by $w_{i,j}$ the instance of wire w_i in the jth garbled circuit, and by $k_{i,j}^b$ the key associated with bit b on wire $w_{i,j}$. Then, P_1 sets the keys for its input wires to: $k_{i,j}^0 = H(g^{a_i^0 \cdot r_j})$ and $k_{i,j}^1 = H(g^{a_i^1 \cdot r_j})$, where H is a suitable randomness extractor [6,16]; see also [10].*
   (c) *P_1 constructs s independent copies of a garbled circuit of C, denoted $GC_1, \ldots, GC_s$, using random keys except for wires $w_1, \ldots, w_\ell$ for which the keys are as above.*
2. Oblivious transfers: *P_1 and P_2 run batch single-choice cut-and-choose oblivious transfer with parameters ℓ (the number of parallel executions) and s (the number of pairs in each execution):*
   (a) *P_1 defines vectors $\boldsymbol{z}_1, \ldots \boldsymbol{z}_\ell$ so that $\boldsymbol{z}_i$ contains the s pairs of random symmetric keys associated with P_2's ith input bit y_i in all garbled circuits $GC_1, \ldots, GC_s$.*
   (b) *P_2 inputs a random subset $\mathcal{J} \subset [s]$ of size exactly $s/2$ and bits $\sigma_1, \ldots, \sigma_\ell \in \{0,1\}$, where $\sigma_i = y_i$ for every i.*
   (c) *P_2 receives all the keys associated with its input wires in all circuits GC_j for $j \in \mathcal{J}$, and receives the keys associated with its input y on its input wires in all other circuits.*
3. Send circuits and commitments: *P_1 sends P_2 the garbled circuits (i.e., the gate and output tables), the "seed" for the randomness extractor H, and the following "commitment" to the garbled values associated with P_1's input wires: $\left\{(i, 0, g^{a_i^0}), (i, 1, g^{a_i^1})\right\}_{i=1}^{\ell}$ and $\left\{(j, g^{r_j})\right\}_{j=1}^{s}$.*
4. Send cut-and-choose challenge: *P_2 sends P_1 the set $\mathcal{J}$ along with the* pair *of keys associated with its first input bit y_1 in every circuit GC_j for $j \in \mathcal{J}$. If the values received by P_1 are incorrect, it outputs $\perp$ and aborts. Circuits GC_j for $j \in \mathcal{J}$ are called* check-circuits, *and for $j \notin \mathcal{J}$ are called* evaluation-circuits.

5. Send all input garbled values in check-circuits: *For every* check-circuit GC_j*, party* P_1 *sends the value* r_j *to* P_2*, and* P_2 *checks that these are consistent with the pairs* $\{(j, g^{r_j})\}_{j \in \mathcal{J}}$ *received in Step 3. If not,* P_2 *aborts outputting* $\perp$.
6. Correctness of check circuits: *For every* $j \in \mathcal{J}$*,* P_2 *uses the* $g^{a_i^0}, g^{a_i^1}$ *values it received in Step 3, and the* r_j *values it received in Step 5, to compute the values* $k^0_{i,j} = H(g^{a_i^0 \cdot r_j}), k^1_{i,j} = H(g^{a_i^1 \cdot r_j})$ *associated with* P_1*'s input in* GC_j*. In addition it sets the garbled values associated with its own input in* GC_j *to be as obtained in the cut-and-choose OT. Given all the garbled values for all input wires in* GC_j*, party* P_2 *decrypts the circuit and verifies that it is a garbled version of* C.
7. P_1 sends its garbled input values in the evaluation-circuits:
   (a) *P_1 sends the keys associated with its inputs in the evaluation circuits: For every* $j \notin \mathcal{J}$ *and every wire* $i = 1, \ldots, \ell$*, party* P_1 *sends the value* $k'_{i,j} = g^{a_i^{x_i} \cdot r_j}$*;* P_2 *sets* $k_{i,j} = H(k'_{i,j})$.
   (b) *P_1 proves that all input values are consistent: For every input wire* $i = 1, \ldots, \ell$*, party* P_1 *proves* in parallel *that there exists a value* $\sigma_i \in \{0, 1\}$ *such that for every* $j \notin J$*,* $k'_{i,j} = g^{a_i^{\sigma_i} \cdot r_j}$*. If any of the proofs fail, then* P_2 *aborts and outputs* $\perp$.
8. Circuit evaluation: *P_2 uses the keys associated with* P_1*'s input obtained in Step 7a and the keys associated with its own input obtained in Step 2c to evaluate the evaluation circuits* GC_j *for every* $j \notin \mathcal{J}$*. If a circuit decryption fails, then* P_2 *sets the output of that circuit to be* $\perp$*. Party* P_2 *takes the output that appears in most circuits, and outputs it.*

3.2 Properties

The security of the protocol is expressed in the following theorem, which is proved in the full version of the paper:

Theorem 6. *Assume that the decisional Diffie-Hellman assumption is hard in* $\mathbb{G}$*, that the protocol used in Step 2 securely computes the batch single-choice cut-and-choose oblivious transfer functionality, that the protocol used in Step 7b is a zero-knowledge proof of knowledge, and that the symmetric encryption scheme used to generate the garbled circuits is secure. Then, Protocol 5 securely computes the function* f *in the presence of malicious adversaries.*

We remark here on one aspect of the proof that is crucial to the concrete efficiency of the protocol. Party P_1 can successfully cheat if it manages to pass the cut-and-choose test with a majority of the evaluation circuits being incorrect. To do this, at least $s/4$ circuits must be incorrect. In the proof of security we show that the probability that at least this many circuits are incorrect without P_2 catching P_1 is approximately $2^{-0.311s}$ where the approximation is due to Stirling's formula. Based on this, it suffices to use 128 garbled circuits in order to obtain an error of 2^{-40}. (We also compared the exact bound to this approximation on the concrete value of $s = 128$, to verify that the approximation is good for s of this size).

Exact efficiency. An exact analysis of the protocol yields that there are **$15s\ell + 39\ell + 10s + 5$ exponentiations** (of which $11.5s\ell$ are for Step 2, performing the OTs), **$7s\ell + 22\ell + 7s + 5$ group elements** sent and **12 rounds of communication**. In addition, there are **$6.5|C|s$ symmetric encryptions**, of which $4|C|s$ encryptions for constructing all s garbled circuits, and $2|C|s$ encryptions for P_2 to check $s/2$ of them. Finally, there are **$4|C|s$ ciphertexts** sent for transmitting these circuits. The overhead of the protocol can be improved by different optimizations, as shown in Section 3.3 below.

3.3 Optimizations

Fixed-base exponentiations. Exponentiations are commonly computed by repeated squaring, which for a group of order q of length m bits requires on average $1.5m$ multiplications for a full exponentiation. If multiple exponentiations of the same base are computed, then the repeated binary powers of the base can be computed once for all exponentiations, reducing the amortized overhead of an exponentiation on average to $0.5m$ multiplications. All but $s\ell$ of the exponentiations in our protocol are fixed-based, and thus taking this into account the effective overhead of the exponentiations is equivalent to that of just **$5.66s\ell$ full exponentiations.**

Reducing the computation of P_2 in Step 6. In Step 6 of Protocol 5, P_2 performs $s\ell$ exponentiations in order to compute the garbled values associated with P_1's input in the check circuits. Namely, given the $(i, 0, g^{a_i^0}), (i, 1, g^{a_i^1})$ tuples and r_j for every $j \in \mathcal{J}$, party P_2 computes $g^{a_i^0 \cdot r_j}, g^{a_i^1 \cdot r_j}$ for all $i = 1 \ldots \ell$ and $j \in \mathcal{J}$. This step costs $s\ell$ exponentiations (2ℓ exponentiations for each of the $s/2$ check circuits). We can reduce this to about a quarter by having P_1 send the $g^{a_i^0 \cdot r_j}, g^{a_i^1 \cdot r_j}$ values to P_2 and prove that they are correct (not in zero-knowledge).

The protocol is modified by changing Step 6 as follows (recall that P_2 already has all of the $(i, 0, g^{a_i^0}), (i, 1, g^{a_i^1})$ tuples and r_j values):

1. P_1 sends P_2 all of the values $k'^0_{i,j} = g^{a_i^0 \cdot r_j}$ and $k'^1_{i,j} = g^{a_i^1 \cdot r_j}$ for $i = 1, \ldots, \ell$ and $j \in \mathcal{J}$.
2. P_2 chooses random values $\gamma_i^0, \gamma_i^1 \in [1, 2^L]$ for $i = 1, \ldots, \ell$.
3. For every $j \in \mathcal{J}$, party P_2 computes the values $\alpha_j = \left(\prod_{i=1}^{\ell} (g^{a_i^0})^{\gamma_i^0} \cdot (g^{a_i^1})^{\gamma_i^1}\right)^{r_j}$ and $\beta_j = \prod_{i=1}^{\ell} (k'^0_{i,j})^{\gamma_i^0} \cdot (k'^1_{i,j})^{\gamma_i^1}$ Note that computing α_j requires only a single *full* exponentiation since the value $(g^{a_i^0})^{\gamma_i^0} \cdot (g^{a_i^1})^{\gamma_i^1}$ can be computed once for all j.
4. P_2 accepts P_1's input if and only if $\alpha_j = \beta_j$ for all $j \in \mathcal{J}$.

Claim 7. *The probability that P_2 accepts if there exists an $i \in \{1, \ldots, \ell\}$ and $j \in \mathcal{J}$ such that $k'^0_{i,j} \neq g^{a_i^0 \cdot r_j}$ or $k'^1_{i,j} \neq g^{a_i^1 \cdot r_j}$ is at most $\frac{s}{2} \cdot 2^{-L}$.*

Preprocessing. The bulk of the exponentiations performed in the protocol can be precomputed. Step 1 of the protocol, where P_1 computes its input keys, can clearly be computed before P_1 receives its inputs. Step 2 executes the oblivious transfers. It can be slightly changed to be run before P_2 receives its inputs: P_2 can execute this step with random inputs $\sigma_1, \ldots, \sigma_\ell$. Then, when it receives its input bits $y_1, \ldots, y_\ell$, it sends to P_1 a string of correction bits $y_1 \oplus \sigma_1, \ldots, y_\ell \oplus \sigma_\ell$. P_1 exchanges the roles of the two keys of input wires of P_2 for which it receives a correction bit with the value 1. (The security proof can be easily adapted for this variant of the protocol). Given this change, both Steps 1 and 2 can be precomputed. These steps account for $13.5s\ell$ of the $15s\ell$ exponentiations of the protocol, where the remaining $1.5s\ell$ exponentiations are fixed base. This means that if preprocessing is used, then after receiving their inputs the parties need to effectively compute only $\boldsymbol{s\ell/2}$ **full exponentiations**.

4 Universal Composability and Covert Adversaries

4.1 Universally Composable Two-Party Computation

The simulators in the proof of Theorem 6 carry out no rewinding, and likewise the intermediate simulators used to prove the reductions. Thus, if the protocols used to compute the batch cut-and-choose oblivious transfer functionality and the zero-knowledge proof of knowledge of Step 7b are universally composable, then so is Protocol 5. In order to obtain this property, we simply need to apply a transformation from Sigma protocols to universally-composable zero knowledge, which can be achieved efficiently using universally-composable commitments. Details of how this can be achieved are given in the full version of the paper. We have the following:

Theorem 8. *Assume that the decision Diffie-Hellman assumption holds. Then, for every efficiently computable two-party function f with inputs of length ℓ, there exists a universally composable protocol that securely computes f in the commitment-hybrid model in the presence of malicious adversaries, with 8 rounds of computation and $O(s\ell + s^2)$ exponentiations.*

4.2 Covert Security

In the model of security in the presence of covert adversaries [1], the requirement is that any cheating by an adversary will be caught with some probability ϵ. The value of ϵ taken depends on the application, the ramifications to an adversary being caught, the value to an adversary of successfully cheating (if not caught) and so on. The analysis of our protocol shows that for *every* value of s (even if s is very small) the probability that an adversary can cheat without being caught is at most $2^{-\frac{s}{4}+1}$. This immediately yields a protocol that is secure in the presence of covert adversaries, as stated in the following theorem.

Theorem 9. *Assume that the decisional Diffie-Hellman assumption is hard in $\mathbb{G}$, that the protocol used in Step 2 securely computes the batch single-choice cut-and-choose oblivious transfer functionality, that the protocol used in Step 7b is a zero-knowledge proof of knowledge, and that the symmetric encryption scheme* *used to generate the garbled circuit is secure. Then, for any integer $s > 4$, Protocol 5 securely computes the function f in the presence of covert adversaries with ϵ-deterrent (under the strong explicit cheat formulation), for $\epsilon = 1 - 2^{-\frac{s}{4}+1}$.*

We stress that our protocol is significantly more efficient than the protocols of [1] and [15] when values of ϵ that are greater than 1/2 are desired. For example, in order to obtain an ϵ-deterrent of 0.98, the protocol of [1] requires using 50 garbled circuits. However, taking $s = 50$ in our protocol here yields an ϵ-deterrent of $1 - 2^{-11.5}$ which is much much larger.

Acknowledgements

We thank Bo Zhang for pointing out an error in the single-choice cut-and-choose oblivious transfer protocol in an earlier version of this work.

References

1. Aumann, Y., Lindell, Y.: Security Against Covert Adversaries: Efficient Protocols for Realistic Adversaries. Journal of Cryptology 23(2), 281–343 (2010)
2. Beaver, D.: Foundations of Secure Interactive Computing. In: Feigenbaum, J. (ed.) CRYPTO 1991. LNCS, vol. 576, pp. 377–391. Springer, Heidelberg (1992)
3. Canetti, R.: Security and Composition of Multi-party Cryptographic Protocols. Journal of Cryptology 13(1), 143–202 (2000)
4. Canetti, R.: Universally Composable Security: A New Paradigm for Cryptographic Protocols. In: 42nd FOCS, pp. 136–145 (2001), Full version `http://eprint.iacr.org/2000/067`
5. Canetti, R., Fischlin, M.: Universally Composable Commitments. In: Kilian, J. (ed.) CRYPTO 2001. LNCS, vol. 2139, pp. 19–40. Springer, Heidelberg (2001)
6. Carter, L., Wegman, M.N.: Universal Classes of Hash Functions. JCSS 18(2), 143–154 (1979)
7. Cramer, R., Damgård, I., Schoenmakers, B.: Proof of partial knowledge and simplified design of witness hiding protocols. In: Desmedt, Y.G. (ed.) CRYPTO 1994. LNCS, vol. 839, pp. 174–187. Springer, Heidelberg (1994)
8. Damgård, I.: On Σ Protocols, `http://www.daimi.au.dk/~ivan/Sigma.pdf`
9. Dodis, Y., Shoup, V., Walfish, S.: Efficient Constructions of Composable Commitments and Zero-Knowledge Proofs. In: Wagner, D. (ed.) CRYPTO 2008. LNCS, vol. 5157, pp. 515–535. Springer, Heidelberg (2008)
10. Dodis, Y., Gennaro, R., Håstad, J., Krawczyk, H., Rabin, T.: Randomness Extraction and Key Derivation Using the CBC, Cascade and HMAC Modes. In: Franklin, M. (ed.) CRYPTO 2004. LNCS, vol. 3152, pp. 494–510. Springer, Heidelberg (2004)
11. Garay, J., MacKenzie, P., Yang, K.: Strengthening Zero-Knowledge Protocols Using Signatures. In: Biham, E. (ed.) EUROCRYPT 2003. LNCS, vol. 2656, pp. 177–194. Springer, Heidelberg (2003)
12. Goldreich, O.: Foundations of Cryptography: Volume 2 – Basic Applications. Cambridge University Press, Cambridge (2004)
13. Goldreich, O., Micali, S., Wigderson, A.: How to Play any Mental Game – A Completeness Theorem for Protocols with Honest Majority. In: 19th STOC, pp. 218–229 (1987), For details see [12, Chapter 7]

14. Goldwasser, S., Levin, L.: Fair Computation of General Functions in Presence of Immoral Majority. In: Menezes, A., Vanstone, S.A. (eds.) CRYPTO 1990. LNCS, vol. 537, pp. 77–93. Springer, Heidelberg (1991)
15. Goyal, V., Mohassel, P., Smith, A.: Efficient Two Party and Multi Party Computation Against Covert Adversaries. In: Smart, N.P. (ed.) EUROCRYPT 2008. LNCS, vol. 4965, pp. 289–306. Springer, Heidelberg (2008)
16. Hastad, J., Impagliazzo, R., Levin, L., Luby, M.: Construction of a Pseudo-random Generator from any One-way Function. SIAM Journal on Computing 28(4), 1364–1396 (1999)
17. Hazay, C., Nissim, K.: Efficient Set Operations in the Presence of Malicious Adversaries. In: Nguyen, P.Q., Pointcheval, D. (eds.) PKC 2010. LNCS, vol. 6056, pp. 312–331. Springer, Heidelberg (2010)
18. Ishai, Y., Prabhakaran, M., Sahai, A.: Founding Cryptography on Oblivious Transfer – Efficiently. In: Wagner, D. (ed.) CRYPTO 2008. LNCS, vol. 5157, pp. 572–591. Springer, Heidelberg (2008)
19. Ishai, Y., Prabhakaran, M., Sahai, A.: Secure Arithmetic Computation with No Honest Majority. In: Reingold, O. (ed.) TCC 2009. LNCS, vol. 5444, pp. 294–314. Springer, Heidelberg (2009)
20. Jarecki, S., Shmatikov, V.: Efficient Two-Party Secure Computation on Committed Inputs. In: Naor, M. (ed.) EUROCRYPT 2007. LNCS, vol. 4515, pp. 97–114. Springer, Heidelberg (2007)
21. Kiraz, M., Schoenmakers, B.: A Protocol Issue for the Malicious Case of Yao's Garbled Circuit Construction. In: Proceedings of 27th Symposium on Information Theory in the Benelux, pp. 283–290 (2006)
22. Lindell, Y.: Parallel Coin-Tossing and Constant-Round Secure Two-Party Computation. Journal of Cryptology 16(3), 143–184 (2003)
23. Lindell, Y., Pinkas, B.: A Proof of Yao's Protocol for Secure Two-Party Computation. The Journal of Cryptology 22(2), 161–188 (2009)
24. Lindell, Y., Pinkas, B.: An Efficient Protocol for Secure Two-Party Computation in the Presence of Malicious Adversaries. In: Naor, M. (ed.) EUROCRYPT 2007. LNCS, vol. 4515, pp. 52–78. Springer, Heidelberg (2007)
25. MacKenzie, P., Yang, K.: On Simulation-Sound Trapdoor Commitments. In: Cachin, C., Camenisch, J.L. (eds.) EUROCRYPT 2004. LNCS, vol. 3027, pp. 382–400. Springer, Heidelberg (2004)
26. Micali, S., Rogaway, P.: Secure Computation. In: Feigenbaum, J. (ed.) CRYPTO 1991. LNCS, vol. 576, pp. 392–404. Springer, Heidelberg (1992)
27. Naor, M., Reingold, O.: Synthesizers and Their Application to the Parallel Construction of Psuedo-Random Functions. In: 36th FOCS, pp. 170–181 (1995)
28. Nielsen, J.B., Orlandi, C.: LEGO for Two-Party Secure Computation. In: Reingold, O. (ed.) TCC 2009. LNCS, vol. 5444, pp. 368–386. Springer, Heidelberg (2009)
29. Orlandi, C.: Personal communication (2010)
30. Peikert, C., Vaikuntanathan, V., Waters, B.: A Framework for Efficient and Composable Oblivious Transfer. In: Wagner, D. (ed.) CRYPTO 2008. LNCS, vol. 5157, pp. 554–571. Springer, Heidelberg (2008)
31. Pinkas, B., Schneider, T., Smart, N.P., Williams, S.C.: Secure Two-Party Computation Is Practical. In: Matsui, M. (ed.) ASIACRYPT 2009. LNCS, vol. 5912, pp. 250–267. Springer, Heidelberg (2009)
32. Shamir, A.: How to Share a Secret. Communications of the ACM 22(11), 612–613 (1979)
33. Yao, A.C.: How to Generate and Exchange Secrets. In: 27th FOCS, pp. 162–167 (1986)

Practical Adaptive Oblivious Transfer from Simple Assumptions

Matthew Green* and Susan Hohenberger**

Johns Hopkins University
{mgreen,susan}@cs.jhu.edu

Abstract. In an adaptive oblivious transfer (OT) protocol, a sender commits to a database of messages and then repeatedly interacts with a receiver in such a way that the receiver obtains one message per interaction of his choice (and nothing more) while the sender learns nothing about any of the choices. Recently, there has been significant effort to design practical adaptive OT schemes and to use these protocols as a building block for larger database applications. To be well suited for these applications, the underlying OT protocol should: (1) support an efficient initialization phase where *one* commitment can support an arbitrary number of receivers who are guaranteed of having the same view of the database, (2) execute transfers in time independent of the size of the database, and (3) satisfy a strong notion of security under a simple assumption in the standard model.

We present the first adaptive OT protocol simultaneously satisfying these requirements. The sole complexity assumption required is that given (g, g^a, g^b, g^c, Q), where g generates a bilinear group of prime order p and a, b, c are selected randomly from $\mathbb{Z}_p$, it is hard to decide if $Q = g^{abc}$. All prior protocols in the standard model either do not meet our efficiency requirements or require dynamic "q-based" assumptions.

Our construction makes an important change to the established "assisted decryption" technique for designing adaptive OT. As in prior works, the sender commits to a database of n messages by publishing an encryption of each message and a signature on each encryption. Then, each transfer phase can be executed in time *independent* of n as the receiver blinds one of the encryptions and proves knowledge of the blinding factors and a signature on this encryption, after which the sender helps the receiver decrypt the chosen ciphertext. One of the main obstacles to designing an adaptive OT scheme from a simple assumption is realizing a suitable signature for this purpose (i.e., enabling signatures on group elements in a manner that later allows for efficient proofs.) We make the observation that a secure signature scheme is not necessary for this paradigm, provided that signatures can only be forged in certain ways.

* Supported by NSF CNS-1010928. This publication was also made possible by Grant Number HHS 90TR0003/01. Its contents are solely the responsibility of the authors and do not necessarily represent the official views of the HHS.

** Supported by NSF CAREER CNS-1053886, a Microsoft New Faculty Fellowship and a Google Research Award.

Y. Ishai (Ed.): TCC 2011, LNCS 6597, pp. 347–363, 2011.

We then show how to efficiently integrate an insecure signature into a secure adaptive OT construction.

1 Introduction

Oblivious transfer OT [35,39] is a two-party protocol, where a Sender with messages $M_1, \ldots, M_N$ and a Receiver with indices $\sigma_1, \ldots, \sigma_k \in [1, N]$ interact in such a way that at the end the Receiver obtains $M_{\sigma_1}, \ldots, M_{\sigma_k}$ without learning anything about the other messages and the Sender does not learn anything about the choices $\sigma_1, \ldots, \sigma_k$. In the *adaptive* OT setting [33], the Receiver may obtain $M_{\sigma_{i-1}}$ before deciding on σ_i [33].

Our Goals. Adaptive OT is an interesting primitive. Like non-adaptive OT, it is a key building block for secure multi-party computation [40,19,28]. More practically, it captures the way an oblivious medical, financial or patent database would be accessed. Recently, there has been a focus on designing practical, privacy-preserving databases with access controls [15,8] or pricing mechanisms [36] based on adaptive OT. Unfortunately, researchers trying to design more-complex systems on top of current adaptive OT protocols do not have any ideal choices. For a database with N messages supporting U Receivers with security parameter λ, such a protocol must be:

1. Extremely efficient, even when N, the database size, is large. In particular, the cost to transfer one message to one Receiver should depend only on the security parameter and not on N. I.e., a Receiver should not have to do work proportional to the size of the database to download one file. (This rules out a number of naive approaches as discussed below.)
2. Furthermore, since few databases serve only one user, it should be possible to extend the protocol to the case where there are *many* Receivers, each of whom receives a consistent view of the database. In particular, the ideal situation, which we achieve in this work, is to have a *non-interactive* initialization phase, where the Sender can do $O(\lambda N)$ work to form a commitment that can then be used for an arbitrary number of receivers. Several prior works (e.g., [10,22,27,36]) support a relatively efficient initialization phase with $O(\lambda(N+U))$ total work. By adding a CRS and making some modifications, this can likely be reduced to $O(\lambda N)$ (although the complexity assumptions will still be an issue.) What one wishes to avoid, however, is an initialization phase that requires $O(\lambda N U)$ total work. I.e., the sender should not have to set up a *unique* database containing *all* of its files for *each* of its users. (This also rules out some basic approaches.)
3. Finally, since this protocol is designed to be a building block of larger applications, it is critical that it be a solid one. In particular, it should satisfy a strong notion of security (i.e., full-simulatability or UC) under a mild complexity assumption in the standard model. Unfortunately, while

sufficiently practical protocols exist, they either require random oracles [10,22], dynamic[1] assumptions [10,22,27,36] or interactive assumptions [38].

Thus, a new construction based on new techniques is needed.

From Non-Adaptive to Adaptive OT for Single Receivers. Since it is known how to build non-adaptive OT protocols based on simple assumptions [21,32,34] such as Decisional Diffie-Hellman and Quadratic Residuosity, it is natural to ask why constructing adaptive protocols has proven so difficult. Given any fully-simulatable 1-out-of-N non-adaptive OT protocol, one can build a fully-simulatable k-out-of-N adaptive OT protocol for a *single* Receiver by sequentially executing k instances of the non-adaptive protocol and, before each execution, having the sender prove in zero-knowledge that the sequence of N messages used in execution i is the same as the sequence of N messages used in execution $i-1$ [10]. Unfortunately, for security parameter λ, this protocol requires a total of $O(Nk\lambda)$ work to transfer k messages for (only) *one* Receiver and is thus impractical for any application involving large databases.

Thus, when Camenisch, Neven and shelat [10] began to reinvestigate this problem in 2007, they stressed that the real challenge was to build an OT scheme where the sender makes an initial commitment to the database (which is assumed to be broadcast to all receivers), and then the two parties only exchange a *constant number* of group elements per transfer.

Our Contributions. We present an efficient, adaptive oblivious transfer protocol which is fully-simulatable under a simple, static assumption. The sole complexity assumption required is that given (g, g^a, g^b, g^c, Q), where g generates a bilinear group of prime order p and a, b, c are selected randomly from $\mathbb{Z}_p$, it is hard to decide if $Q = g^{abc}$. This assumption called *Decisional 3-Party Diffie-Hellman* has been used in prior works [31,5,25]. Our protocol is practical, although more costly than the very efficient Camenisch et al. protocol [10] by a constant factor. The database commitment in our scheme requires roughly $(9 + 7N)$ group elements, whereas the commitment in [10] required roughly $(3 + 2N)$ group elements. By including the mild Decision Linear assumption [4], we can efficiently make this initialization phase *non-interactive* as we discuss in Section 3.

Our construction introduces a twist on the *assisted decryption* approach to OT design, where the underlying signatures need not be existentially unforgeable provided that certain forgeries are not permitted. As we discuss, these techniques may be useful in simplifying the complexity assumptions in schemes beyond OT such as F-signatures and anonymous credentials [1].

---

[1] These are also called *parametric* or *q-based* assumptions. An example is q-Strong Diffie-Hellman [3] (q-SDH): given $(g, g^x, g^{x^2}, g^{x^3}, \ldots, g^{x^q})$, where g generates a group of prime order p and x is a random value in $\mathbb{Z}_p$, it is hard to compute $(g^{1/(x+c)}, c)$ for any $c \in \mathbb{Z}_p^*$. Typically, when q-SDH is used as the foundation of an adaptive OT scheme, q must dynamically adjust to the number of files in the database. Thus, the assumption required actually changes based on how the protocol is used.

Protocol	Initialization Cost	Transfer Cost	Assumption	Security Defn
Folklore	$\cdot$	$O(\lambda N)$	general assumptions	Full Sim
KN [29]	$O(\lambda(N+U))$	$O(\lambda N)$	Decisional nth Residuosity/DDH	Full Sim
NP [33]	$\cdot$	$O(\lambda \lg(N))$	DDH + OT_1^2	Half Sim
KNP [30]	$O(\lambda NU)$	$O(\lambda)$	DDH	Full Sim*
CNS [10]	$O(\lambda(N+U))$	$O(\lambda)$	q-Power DDH + q-Strong DH	Full Sim
GH [22]	$O(\lambda(N+U))$	$O(\lambda)$	Decision Linear + q-Hidden LRSW	UC
JL [27]	$O(\lambda(N+U))$	$O(\lambda)$	Comp. Dec. Residuosity + q-DDHI	Full Sim
RKP [36]	$O(\lambda(N+U))$	$O(\lambda)$	DLIN + q-Hidden SDH + q-TDH	UC
§2.1	$O(\lambda(N+U))$	$O(\lambda)$	Decision 3-Party DH	Full Sim
§3	$O(\lambda N)$	$O(\lambda)$	Decision 3-Party DH + DLIN	Full Sim

Fig. 1. Survey of adaptive k-out-of-N Oblivious Transfer protocols secure in the standard model. Let λ be the security parameter, N the size of the database and U the number of receivers. The horizontal lines separate the schemes into efficiency categories, which improve as one scans down the table. While the least efficient categories can be realized using assumptions such as DDH, all prior attempts to achieve practicality have required a dynamic q-based complexity assumption. A $*$ denotes the construction meets a strictly weaker notion than the standard used in the other works.

Intuition behind our $\mathsf{OT}_{k\times 1}^N$ *Construction.* As with most previous $\mathsf{OT}_{k\times 1}^N$ constructions, our construction uses a technique known as *assisted decryption*. For $i = 1$ to N, the Sender commits to his database by encrypting each message as $C_i = \text{Enc}(M_i)$, and publishes a public key and ciphertexts $(pk, C_1, \ldots, C_N)$. The Receiver then checks that each ciphertext is well-formed. To obtain a message, the Sender and Receiver engage in a *blind* decryption protocol, i.e., an interactive protocol in which the Sender does not view the ciphertext he decrypts, but where the Receiver is convinced that decryption was done correctly.

The difficulty here is to prevent the Receiver from abusing the decryption protocol, e.g., by requesting decryptions of ciphertexts which were either not produced by the Sender or have been mauled. The folklore solution is to have the Receiver provide a proof that his request corresponds to $C_1 \vee C_2 \vee \ldots \vee C_N$. Of course, the cost of each transfer is now dependent on the total database size and thus this solution is no (asymptotically) better than the trivial solution mentioned above.

In Eurocrypt 2007, Camenisch, Neven and shelat [10] were the first to propose a method for executing "assisted decryption" efficiently. The sender signed each ciphertext value. The receiver was required to prove knowledge of a corresponding signature before the sender would assist him in decrypting a ciphertext. This clever approach reduced the $O(N\lambda)$ work per transfer required above, to only $O(\lambda)$ work, where λ is a security parameter.

More specifically, Camenisch, Neven and shelat [10] used a deterministic encryption scheme and a signature with a particular structure: for $pk = (g, g^x, H = e(g,h))$ and $sk = h$, let $C_i = \left(g^{\frac{1}{x+i}}, M_i \cdot e(g,h)^{\frac{1}{x+i}}\right)$. Recall that $g^{1/(x+i)}$ is a weak Boneh-Boyen signature [3] on i under g^x, and here only a polynomial number of "messages" (1 to N) are signed. While this scheme supports an elegant and efficient blind decryption protocol, it also requires strong q-based assumptions for both the indistinguishability of the ciphertexts as well as the unforgeability of the weak Boneh-Boyen signature. It is based on the q-Strong Diffie-Hellman and the q-Power Decisional Diffie-Hellman assumptions. The latter assumption states that given $(g, g^x, g^{x^2}, \ldots, g^{x^q}, H)$, where $g \in \mathbb{G}$ and $H \in \mathbb{G}_T$, it is hard to distinguish the vector of elements $(H^x, H^{x^2}, \ldots, H^{x^q})$ from a vector of random elements in $\mathbb{G}_T$. In essence, the rigid structure of this (and all prior) constructions appear to require a similarly structured complexity assumption, which grows with the database size.

To move past this, we will "loosen" the structure of the ciphertext and signature enough to break the dependence on a structured assumption, but not so much as to ruin our ability to perform efficient proofs. Finding this balance proved highly non-trivial.

We now turn to how our construction works. We will encrypt using the Boneh-Boyen IBE [2], which has a public key $pk = (g, g_1 = g^a, g_2, h)$ and encrypts M as $(g^r, (g_1^i h)^r, e(g_1, g_2)^r M)$ for identity i and randomness $r \in \mathbb{Z}_p$. Then we will sign r. To do this, we need a standard model signature scheme from a simple assumption (which is itself somewhat rare.) We choose the *stateful* signatures of Hohenberger-Waters [26], which has a public key $pk = (g, g^b, u, v, d, w, z, h)$ and signs M as $(\sigma_1, \sigma_2, s, i)$ for state i and randomness $s, t \in \mathbb{Z}_p$, where $\sigma_1 = g^t$, $\sigma_2 = (u^M v^s d)^b (w^{\lceil \lg(i) \rceil} z^i h)^t$.

Attempt 1. Now, consider the construction obtained by combining the BB IBE, secure under Decisional Bilinear Diffie-Hellman, with the HW signature, secure under the Computational Diffie-Hellman assumption. Here we will encrypt the ith message using identity i (in the BB IBE) and state i (in the HW signature), with an extra u^r term to allow the Receiver to verify well-formedness:

$$g^r, \quad (g_1^i h)^r, \quad e(g_1, g_2)^r M, \quad g^t, \quad (u^r v^s d)^b (w^{\lceil \lg(i) \rceil} z^i h)^t, \quad u^r, \quad s$$

The Receiver can verify the well-formedness of the ith ciphertext $(c_1, \ldots, c_7)$ by checking that $e((g_1^i h), c_1) = e(g, c_2)$, $e(g, c_6) = e(c_1, u)$ and

$$e(g, c_5) = e(c_6 v^{c_7} d, g^b) e(w^{\lceil \lg(i) \rceil} z^i h, c_4).$$

It is important that the Receiver can verify the well-formedness of the ciphertext-signature pair, so that the simulator can properly extract the messages from a cheating Sender during the proof of security. It is a nice additional feature that our verification is public and non-interactive.

Attempt 2. However, the above construction still has a lot of problems. Recall that we want the Receiver to ask for a blind decryption of a given ciphertext by (somehow) sending in blinded portions of the ciphertext, proving that these portions are linked to r and proving that he knows a signature on r. Unfortunately, efficiently proving knowledge of the HW signature is problematic due to the $\lceil \lg(i) \rceil$ exponent. We could do this using a range proof [13,9,6,7], however, this would require that we introduce stronger assumptions such as Strong RSA or q-Strong Diffie-Hellman. We could instead do a bit-by-bit proof, but this would severely hurt our efficiency. Instead, our solution is to drop this term entirely from the HW signature to obtain the ciphertext:

$$g^r, \quad (g_1^i h)^r, \quad e(g_1, g_2)^r M, \quad g^t, \quad (u^r v^s d)^b (z^i h)^t, \quad u^r, \quad s$$

One major issue is that dropping this term breaks the unforgeability of the signature scheme. Indeed, it is now possible for anyone to efficiently compute a signature on any index over a certain polynomial threshold as set in the proof of security. However, we specifically chose to encrypt with the Boneh-Boyen IBE for this purpose. We will set our parameters so that an adversary is free to forge signatures with states of $N+1$ and higher, where N is the size of our database. The key idea is that asking for decryptions on *different identities* will not help a malicious Receiver obtain information about the database messages; indeed, we could even hand him the secret key for those identities. This makes our proof much more efficient, however, there is still a large problem.

Attempt 3. To argue, in the proof of security, that no malicious Receiver can forge signatures on a state $i \in [1, N]$, we must *extract* this signature and its forgery message from the proof of knowledge. However, we cannot extract the "message" r from a cheating Receiver, because an honest Receiver will not know the randomness used in the ciphertexts created by the Sender. The most we can ask a Receiver to prove knowledge of is the signature on r comprised of (c_4, c_5, c_6, c_7) and the value g^r. Thus, we cannot extract from the Receiver a valid forgery of the HW signatures.

Moreover, we need a stronger security guarantee than HW signatures gave us (i.e., existential unforgeability under adaptive chosen message attack [20].) We need that: it is not only the case that an adversary cannot produce a pair (m, σ) for a new m; now the adversary cannot even produce the pair (g^m, σ) for a new m, where σ is a signature on m. Do such powerful signatures exist?

Indeed, this security notion was formalized as F-signatures by Belenkiy, Chase, Kohlweiss and Lysyanskaya [1], where they also required q-based complexity assumptions for their construction. Fortunately, we are able to show that the HW signatures (and our mangled version of them without the $w^{\lceil \lg(i) \rceil}$ term) remain F-unforgeable for $F(m) = g^m$ under a simple static assumption. (See [26] or the full version of this work [23] for the full details on HW; the mangled version is proven as part of the OT system in Section 2.2.) We tie both this version of the signature scheme and the Boneh-Boyen IBE together under a single assumption: given (g, g^a, g^b, g^c), it is hard to decide if $Q = g^{abc}$.

Comparison to Prior Work. Let us briefly compare our approach to prior works; see Figure 1 for more. As we mention above, Camenisch, Neven and shelat [10] gave the first efficient, fully-simulatable construction for adaptive (and non-adaptive) OT. It is secure in the standard model under the q-Strong Diffie-Hellman and the q-Power Decisional Diffie-Hellman assumptions.

They also provided a scheme in the random oracle model from unique blind signatures.

Green and Hohenberger [21] provided an adaptive OT construction in the random oracle model based on the Decisional Bilinear Diffie-Hellman assumption, namely, that given (g, g^a, g^b, g^c, Q), it is hard to decide if $Q = e(g, g)^{abc}$. In their construction, the Sender encrypted each message i under identity i using a IBE system. Then they provided a blind key extraction protocol, where the Receiver could blindly obtain a secret key for any identity of her choice.

In the assisted decryption setting, Green and Hohenberger [22] took an approach similar to [10] to achieve UC security. It was based on the Decision Linear and q-Hidden LRSW assumptions, in the asymmetric setting. The latter assumption implies that DDH must hold in both $\mathbb{G}_1$ and $\mathbb{G}_2$.

Jarecki and Liu [27] took an alternative view: for $pk = g^x$, let $C_i = M_i \cdot g^{1/(x+i)}$. Recall that $g^{1/(x+i)}$ is also the Dodis-Yampolskiy pseudorandom function on input i [18]. This essentially simplifies the Camenisch et al. construction and allows a fully-simulatable scheme based on the Composite Decisional Residuosity and q-Decisional Diffie-Hellman Inversion assumptions. The blind decryption protocol involves obliviously evaluating the PRF on input i, which requires some costly zero knowledge proofs. However, this protocol is interesting as the only efficient and fully-simulatable protocol that does not require bilinear groups.

Rial, Kohlweiss and Preneel [36] presented a *priced* version of UC-secure adaptive OT using the assisted decryption approach. In priced OT, the obliviousness property must hold, even though the items being sold may have unique prices. The scheme is secure in the standard model under the Decision Linear, q-Triple Diffie-Hellman, and q-Hidden Strong Diffie-Hellman assumptions.

Unfortunately, all of these constructions have a rigid structure and seem to require a structured complexity assumption. We show that this structure can be enforced, not on the message itself, but rather through the *identity* of the encryption and the *state* of the signature. This provides us with enough glue to keep the security of the scheme together without overdoing it.

Recently, Kurosawa and Nojima [29] and Chen, Chou and Hou [14] gave adaptive OT constructions which purported to improve the underlying complexity assumptions of the schemes above, but which actually resorted to $O(N\lambda)$ transfer cost. It was already known how to achieve this level of (in)efficiency from *general* assumptions, including those of [29,14], by following the folklore method for building adaptive OT from any non-adaptive OT system, as described in [17,10] and the opening of our introduction. Moreover, [14] is set in the random oracle model.

Very recently[2], Kurosawa, Nojima and Phong [30] gave an adaptive OT construction from DDH with $O(\lambda)$ transfers. However, their work has several technical issues. First, their construction does not satisfy the standard full simulation definition used in [10,21,22,27,36] and this work. In [30], if a receiver ever requests the same file twice (say, she downloads a patent one day, deletes it, then

[2] The work of [30] appeared after the initial posting of this work [23].

downloads it again a month later), then this can be detected by the sender. This is at odds with the full simulation definition where the adversarial sender is only told by the ideal functionality that a file has been requested and thus cannot detect a repeated download. Second, it is not obvious how to modify their construction to satisfy the full simulation notion. One approach might be to make the receiver stateful and store every file she ever requests. This has the obvious drawback of requiring permanent storage of the decrypted messages, which may not be practical and is not a requirement in other works. Moreover, subtle technical issues arise as to what the receiver sends during a repeated query round. Third, their construction requires a very expensive initialization procedure where the sender must transmit, then receive back and store $O(N\lambda)$ bits for *each* receiver. In contrast, all prior practical work [10,21,22,27,36] and our results only require that the sender publish and store *one* $O(N\lambda)$ bit database for *all* receivers.

Thus, we build on this body of prior work to present the first efficient scheme satisfying the standard notion of full simulation from a simple assumption in the standard model.

2 Technical Preliminaries

Bilinear Groups. Let BMsetup be an algorithm that, on input 1^κ, outputs the parameters for a bilinear mapping as $\gamma = (p, g, \mathbb{G}, \mathbb{G}_T, e)$, where g generates $\mathbb{G}$, the groups $\mathbb{G}, \mathbb{G}_T$ have prime order $p \in \Theta(2^\kappa)$, and $e : \mathbb{G} \times \mathbb{G} \rightarrow \mathbb{G}_T$. Two algebraic properties required are that: (1) if g generates $\mathbb{G}$, then $e(g, g) \neq 1$ and (2) for all $a, b \in \mathbb{Z}_p$, it holds that $e(g^a, g^b) = e(g, g)^{ab}$.

Assumption 1 (Decisional 3-Party Diffie-Hellman (3DDH) [31,5,25]) *Let $\mathbb{G}$ be a group of prime order $p \in \Theta(2^\lambda)$. For all p.p.t. adversaries $\mathcal{A}$, the following probability is 1/2 plus an amount negligible in λ:*

$$\Pr[g, z_0 \leftarrow \mathbb{G}; a, b, c \leftarrow \mathbb{Z}_p; z_1 \leftarrow g^{abc}; d \leftarrow \{0,1\}; d' \leftarrow \mathcal{A}(g, g^a, g^b, g^c, z_d) : d = d'].$$

Proofs of Knowledge. We use known zero-knowledge and witness indistinguishable techniques for proving statements about discrete logarithms and their natural extensions to proving statements about bilinear groups, such as (1) proof of knowledge of a discrete logarithm modulo a prime [37] and (2) proof of the disjunction or conjunction of any two statements [16]. These are typically interactive, 4-round protocols. We discuss further implementation details in the full version [23].

When referring to the proofs above, we will use the notation of Camenisch and Stadler [11]. For instance, $ZKPoK\{(x, h) : y = g^x \wedge H = e(y, h) \wedge (1 \leq x \leq n)\}$ denotes a zero-knowledge proof of knowledge of an integer x and a group element $h \in \mathbb{G}$ such that $y = g^x$ and $H = e(y, h)$ holds and $1 \leq x \leq n$. All values not enclosed in ()'s are assumed to be known to the verifier.

2.1 The Construction

Our $\mathsf{OT}^N_{k\times 1}$ protocol appears in Figure 2. This protocol follows the *assisted (or blind) decryption* paradigm pioneered by [10,22,27]. The Sender begins the OT protocol by encrypting each message in the database and publishing these values to the Receiver. The Receiver then checks that each ciphertext is well-formed. For each of k transfers, the two parties co-operatively execute a protocol following which (1) the Receiver obtains the decryption of at most one ciphertext, while (2) the Sender learns nothing about *which* ciphertext was decrypted. We require that the interactive decryption protocol run in time independent of the size of the database.

The encryption scheme that we use is a novel combination of the Boneh-Boyen IBE scheme [2] and a (insecure) version of the Hohenberger-Waters signatures [26]. We present methods for proving knowledge of such signatures and obtaining a blind decryption. Of course, in an adaptive OT scheme, the difficulty is always in getting all elements of the fully-simulatable proof of security to work out. There are many subtle details in basing the security for any database of size N under the one simple assumption that given (g, g^a, g^b, g^c), it is hard to decide if $Q = g^{abc}$.

Ciphertext Structure. In Figure 2, we reference a $\mathsf{VerifyCiphertext}$ algorithm for verifying the well-formedness of a ciphertext. Let us explain that now. The Sender's public parameters pk include $\gamma = (p, g, \mathbb{G}, \mathbb{G}_T, e)$ and generators $(g_1, g_2, h, g_3, g_4, u, v, d) \in \mathbb{G}^8$. For message $M \in \mathbb{G}_T$, identity $j \in \mathbb{Z}_p$, and random values $r, s, t \in \mathbb{Z}_p$ we can express a ciphertext as:

$$C = \Big(g^r,\ (g_1^j h)^r,\ M \cdot e(g_1, g_2)^r,\ g^t,\ (u^r v^s d)^b (g_3^j h)^t,\ u^r,\ s\Big)$$

Given only pk, j, the $\mathsf{VerifyCiphertext}$ function validates that the ciphertext has this structure. We define it as follows.

$\mathsf{VerifyCiphertext}(pk, C, j)$. Parse C as $(c_1, \ldots, c_7)$ and pk to obtain $g, g_1, h, g_3, g_4, u, v, d$. This routine outputs 1 if and only if the following equalities hold:

$$\begin{aligned} e(g_1^j h, c_1) &= e(g, c_2) \ \wedge \\ e(g, c_6) &= e(c_1, u) \ \wedge \\ e(g, c_5) &= e(g_4, c_6 v^{c_7} d)\, e(c_4, g_3^j h) \end{aligned}$$

This function will always output 1 when input a properly-formed ciphertext.

2.2 Security Analysis

We now show that the $\mathsf{OT}^N_{k\times 1}$ protocol above is sender-secure and receiver-secure in the full-simulation model under the Decisional 3-Party Diffie-Hellman assumption (3DDH). We will address Sender and Receiver security separately.

A note on the PoK protocols. For generality, our security proofs use the terms $\epsilon_{ZK}, \epsilon_{WI}$ to indicate the maximal advantage that every p.p.t. distinguisher has in distinguishing simulated ZKPoKs from real ones (*resp.* WI proofs on different witnesses). We additionally use ϵ_{Ext} to indicate the maximum probability that

$\mathsf{S_I}(M_1,\ldots,M_N)$	$\mathsf{R_I}()$
1. Select $\gamma = (p, g, \mathbb{G}, \mathbb{G}_T, e) \leftarrow \mathsf{BMsetup}(1^\kappa)$ and $a, b \xleftarrow{\$} \mathbb{Z}_p$, choose $g_2, g_3, h, u, v, d \xleftarrow{\$} \mathbb{G}$ and set $g_1 \leftarrow g^a, g_4 \leftarrow g^b$. Let $pk = (\gamma, g_1, g_2, g_3, g_4, h, u, v, d)$ and $sk = (a, b)$.	
2. For $j = 1$ to N, select $r_j, s_j, t_j \xleftarrow{\$} \mathbb{Z}_p$ and set: $C_j \leftarrow [g^{r_j},\ (g_1^j h)^{r_j},\ M_j e(g_1, g_2)^{r_j},\ g^{t_j},\ (u^{r_j} v^{s_j} d)^b (g_3^j h)^{t_j},\ u^{r_j},\ s_j]$.	
3. Send $(pk, C_1, \ldots, C_N)$ to Receiver.	
4. Conduct $ZKPoK\{(a) : g_1 = g^a\}$.	
	5. Verify pk and the proof[a]. Check for $j = 1$ to N: $\mathsf{VerifyCiphertext}(pk, C_j, j) = 1$. If any check fails, output $\perp$.
Output $S_0 = (pk, sk)$.	Output $R_0 = (pk, C_1, \ldots, C_N)$.
$\mathsf{S_T}(S_{i-1})$	$\mathsf{R_T}(R_{i-1}, \sigma_i)$
	1. Parse C_{σ_i} as $(c_1, \ldots, c_7)$, select $x, y \xleftarrow{\$} \mathbb{Z}_p$ and compute $v_1 \leftarrow g^x c_1$.
	2. Send v_1 to Sender, and conduct: $WIPoK\{(\sigma_i, x, c_2, c_4, c_5, c_6, c_7) : e(v_1/g^x, (g_1^{\sigma_i} h)) = e(c_2, g) \wedge e(c_6, g) = e(v_1/g^x, u) \wedge e(c_5, g) = e(c_6 v^{c_7} d, g_4) e(c_4, g_3^{\sigma_i} h)\}$.
3. Set $R \leftarrow e(v_1, g_2^a)$.	
4. Send R to Receiver and conduct: $ZKPoK\{(a) : R = e(v_1, g_2^a) \wedge g_1 = g^a\}$.	
	5. If the proof does not verify, output $\perp$. Else output $M'_{\sigma_i} \leftarrow \frac{c_3 \cdot e(g_1, g_2)^x}{R}$.
Output $S_i = S_{i-1}$.	Output $R_i = (R_{i-1}, M'_{\sigma_i})$.

[a] By verify pk, we mean check that γ contains parameters for a bilinear map, where p is prime and g generates $\mathbb{G}$ with order p. Also, verify that the remaining pk elements are members of $\mathbb{G}$.

Fig. 2. Our adaptive $\mathsf{OT}^N_{k\times 1}$ protocol. VerifyCiphertext is described above.

the extractor for a PoK fails (soundness). We propose to use four-round Schnorr proofs which are zero-knowledge/WI ($\epsilon_{WI} = \epsilon_{ZK} = 0$) and computationally sound under the Discrete Logarithm assumption (which is naturally implied by 3DDH). Our security proofs employ the knowledge extractors for these proofs-of-knowledge, which we will define as $\mathsf{E}_1, \mathsf{E}_2, \mathsf{E}_3$[3].

[3] These correspond respectively to the proofs $ZKPoK\{(a) : g_1 = g^a\}$, $WIPoK\{(\sigma_i, x, y, z, c_4, c_5, c_6, c_7) : \ldots\}$, and $ZKPoK\{(a) : R = e(v_1, g_2^a) \wedge g_1 = g^a\}$.

SENDER SECURITY. Given a (possibly cheating) real-world receiver $\hat{\mathsf{R}}$, we show how to construct an ideal-world receiver $\hat{\mathsf{R}}'$ such that all p.p.t. distinguishers have at most negligible advantage in distinguishing the distribution of an honest real-world sender S interacting with $\hat{\mathsf{R}}$ ($\mathbf{Real}_{\mathsf{S},\hat{\mathsf{R}}}$) from that of $\hat{\mathsf{R}}'$ interacting with the honest ideal-world sender S' ($\mathbf{Ideal}_{\mathsf{S}',\hat{\mathsf{R}}'}$). Let us now describe the operation of $\hat{\mathsf{R}}'$, which runs $\hat{\mathsf{R}}$ internally, interacting with it in the role of the Sender:

1. To begin, $\hat{\mathsf{R}}'$ selects a random collection of messages $\bar{M}_1, \ldots, \bar{M}_N \stackrel{\$}{\leftarrow} \mathbb{G}_T$ and follows the S_I algorithm (from Figure 2) with these as input up to the point where it obtains $(pk, C_1, \ldots, C_N)$.
2. It sends $(pk, C_1, \ldots, C_N)$ to $\hat{\mathsf{R}}$ and then *simulates* the interactive proof $ZKPoK\{(a) : g_1 = g^a\}$. (Even though $\hat{\mathsf{R}}'$ knows $sk = a$, it ignores this value and simulate this proof step.)
3. For each of k transfers initiated by $\hat{\mathsf{R}}$,
   (a) $\hat{\mathsf{R}}'$ verifies the received WIPoK and uses the knowledge extractor E_2 to obtain the values $\sigma_i, x, y, c_1, c_2, c_3, c_4$ from it. $\hat{\mathsf{R}}'$ aborts and outputs error when E_2 fails.
   (b) When $\sigma_i \in [1, N]$, $\hat{\mathsf{R}}'$ queries the trusted party T to obtain M_{σ_i}, parses C_{σ_i} as $(c_1, \ldots, c_7)$ and responds with $R = \frac{c_3 e(g_1, g_2)^x}{M_{\sigma_i}}$ (if T returns $\perp$, $\hat{\mathsf{R}}'$ aborts the transfer). When $\sigma_i \notin [1, N]$, $\hat{\mathsf{R}}'$ follows the normal protocol. In both cases, $\hat{\mathsf{R}}'$ simulates $ZKPoK\{(a) : R = e(v_1, g_2^a) \wedge g_1 = g^a\}$.
4. $\hat{\mathsf{R}}'$ uses $\hat{\mathsf{R}}$'s output as its own.

Theorem 2. *Let ϵ_{ZK} be the maximum advantage with which any p.p.t. algorithm distinguishes a simulated ZKPoK, and ϵ_{Ext} be the maximum probability that the extractor E_2 fails (with ϵ_{ZK} and ϵ_{Ext} both negligible in κ). If all p.p.t. algorithms have negligible advantage $\leq \epsilon$ at solving the 3DDH problem, then:*

$$\mathbf{Pr}\left[D(\mathbf{Real}_{\mathsf{S},\hat{\mathsf{R}}}(N, k, M_1, \ldots, M_N, \Sigma)) = 1\right] - \mathbf{Pr}\left[D(\mathbf{Ideal}_{\mathsf{S}',\hat{\mathsf{R}}'}(N, k, M_1, \ldots, M_N, \Sigma)) = 1\right] \leq (k+1)\epsilon_{ZK} + k\epsilon_{Ext} + N\epsilon\left(1 + \frac{p}{p-1}\right).$$

A proof of Theorem 2 is sketched in Appendix A.1 and detailed in [23].

RECEIVER SECURITY. For any real-world cheating sender $\hat{\mathsf{S}}$ we can construct an ideal-world sender $\hat{\mathsf{S}}'$ such that all p.p.t. distinguishers have negligible advantage at distinguishing the distribution of the real and ideal experiments. Let us now describe the operation of $\hat{\mathsf{S}}'$, which runs $\hat{\mathsf{S}}$ internally, interacting with it in the role of the Receiver.

1. To begin, $\hat{\mathsf{S}}'$ runs the R_I algorithm, with the following modification: when $\hat{\mathsf{S}}$ proves knowledge of a, $\hat{\mathsf{S}}'$ uses the knowledge extractor E_1 to extract a, outputting error if the extractor fails. Otherwise, it has obtained the values $(pk, C_1, \ldots, C_N)$.

2. For $i = 1$ to N, $\hat{\mathsf{S}}'$ decrypts each of $\hat{\mathsf{S}}$'s ciphertexts $C_1, \ldots, C_N$ using the value a as a decryption key,[4] and sends the resulting $M_1^*, \ldots, M_N^*$ to the trusted party T.
3. Whenever T indicates to $\hat{\mathsf{S}}'$ that a transfer has been initiated, $\hat{\mathsf{S}}'$ runs the transfer protocol with $\hat{\mathsf{S}}$ on the fixed index 1. If the transfer succeeds, $\hat{\mathsf{S}}'$ returns the bit 1 (indicating success) to T, or 0 otherwise.
4. $\hat{\mathsf{S}}'$ uses $\hat{\mathsf{S}}$'s output as its own.

Theorem 3. *Let ϵ_{WI} be the maximum advantage that any p.p.t. algorithm has at distinguishing a WIPoK, and let ϵ_{Ext} be the maximum probability that the extractor E_1 fails. Then $\forall$ p.p.t. D:*

$$\mathbf{Pr}\left[D(\mathbf{Real}_{\hat{\mathsf{S}},\mathsf{R}}(N,k,M_1,\ldots,M_N,\Sigma)) = 1\right] - \mathbf{Pr}\left[D(\mathbf{Ideal}_{\hat{\mathsf{S}}',\mathsf{R}'}(N,k,M_1,\ldots,M_N,\Sigma)) = 1\right] \leq (k+1)\epsilon_{Ext} + k\epsilon_{WI}.$$

A proof of Theorem 3 is sketched in Appendix A.2 and detailed in [23].

3 Efficiently Supporting Multiple Receivers

Adaptive Oblivious Transfer ($\mathsf{OT}_{k\times 1}^N$) is traditionally defined as a protocol between a Sender and a single Receiver. However, the way it is typically used in practical works such as Coull et al. [15] and Camenisch et al. [8] is that $U \geq 1$ distinct Receivers all interact with a single Sender.

Extending the full simulation definition to cover this explicitly is rather straightforward. We do so in the full version [23]. The main technical concern is that every Receiver should have the same view of the database. That is, if two Receivers make a request on index i and neither transfer resulted in an error, then those Receivers must have obtained the same message. In [23] we explain why our construction would satisfy such a notion – namely, that all Receivers share the same database and a Receiver does not accept a message unless the Sender can prove that it correctly corresponds to this database. For simplicity, we assume secure channels for the transfer phase.

Eliminating the $O(\lambda U)$ term. Interestingly, we can also improve the efficiency of our initialization protocol when multiple Receivers are present. In the current protocol of Figure 2, the Sender must conduct the proof of knowledge $ZKPoK\{(a) : g_1 = g^a\}$ with each Receiver. This can be accomplished using a very inexpensive interactive four-round proof in the standard model.

Fortunately even this minimal per-user initialization can be eliminated by assuming a Common Reference String shared by the Sender and all Receivers and using an NIZKPoK to broadcast this proof to all Receivers. To instantiate this proof, we suggest the efficient proof system of Groth and Sahai [24],

[4] Parse C_i as $(c_1, \ldots, c_7)$ and decrypt as $M_i^* = c_3/e(c_1, g_2^a)$. As noted in Section 3, one can modify the protocol so that the Sender conducts a PoK of the value g_2^a.

which permits proofs of pairing-based statements under the Decision Linear assumption [4]. One wrinkle in this approach is that our proof of Receiver security assumes that the simulator can extract the trapdoor $a \in \mathbb{Z}_p$ from the ZKPoK, in order to decrypt the ciphertext vector $C_1, \ldots, C_N$. However, the knowledge extractor for the Groth-Sahai proof system is limited in that it can only extract elements of the bilinear image group $\mathbb{G}$. Fortunately for our purposes, the value $g_2^a \in \mathbb{G}$ can be used as an alternative trapdoor (see Section 2.2 for details). Thus when using the Groth-Sahai system we must re-write the proof as $NIZKPoK\{(g_2^a) : e(g_1, g_2) = e(g_2^a, g)\}$[5].

4 Conclusions and Open Problems

We presented the first efficient, adaptive oblivious transfer protocol which is fully-simulatable under simple, static assumptions. Our protocol is practical and can be used as a building block in larger database applications, such as [15,36,8], as a step to reducing the overall assumptions on the system.

We leave open several interesting problems. First, we use standard zero-knowledge proof and extraction techniques which require rewinding, and thus, our scheme is not UC-secure. A natural question is whether one can obtain UC-security by replacing our interactive proofs with the non-interactive Groth-Sahai proofs [24]. Unfortunately, this is not an easy substitution. Our security proofs use techniques from the Boneh-Boyen cryptosystem that depend fundamentally on the ability to extract *integers* from the Receiver's proof of knowledge during the Transfer phase. The Groth-Sahai proof system is only F-extractable, meaning that one can obtain only group elements from the extractor (even when the proof is over integer witnesses). One can easily substitute a bit-by-bit proof of each integer, but we would hope to identify a more practical approach.

It would be interesting to know if the observations about and manipulations of the Hohenberger-Waters signatures [26] identified in this work could be extended to applications such as anonymous credentials and electronic cash, where most efficient constructions still require random oracles or strong complexity assumptions. One of the main difficulties is that many interesting protocols require an F-signature together with an efficient range proof (i.e., method for proving in zero-knowledge that a committed value lies within a public range.) Currently, the only efficient techniques for doing the latter require either the Strong RSA assumption [13,9,6] or (more recently) the q-Strong Diffie-Hellman assumption [7,12]. (Here q need only be linked to a security parameter, e.g., $q = 256$.) It would be interesting if range proofs under weaker assumptions could be devised.

---

[5] As mentioned by Groth and Sahai, statements of this form must be slightly re-written to enable full zero knowledge. The equivalent statement is $ZKPoK\{(g_2^a, g_1') : e(g_2^a, g)e(g_1', g_2^{-1}) = 1 \wedge e(g_1', g) = e(g_1, g)\}$.

References

1. Belenkiy, M., Chase, M., Kohlweiss, M., Lysyanskaya, A.: P-signatures and Noninteractive Anonymous Credentials. In: Canetti, R. (ed.) TCC 2008. LNCS, vol. 4948, pp. 356–374. Springer, Heidelberg (2008)
2. Boneh, D., Boyen, X.: Efficient Selective-ID Secure Identity-Based Encryption Without Random Oracles. In: Cachin, C., Camenisch, J.L. (eds.) EUROCRYPT 2004. LNCS, vol. 3027, pp. 223–238. Springer, Heidelberg (2004)
3. Boneh, D., Boyen, X.: Short Signatures Without Random Oracles. In: Cachin, C., Camenisch, J.L. (eds.) EUROCRYPT 2004. LNCS, vol. 3027, pp. 56–73. Springer, Heidelberg (2004)
4. Boneh, D., Boyen, X., Shacham, H.: Short Group Signatures. In: Franklin, M. (ed.) CRYPTO 2004. LNCS, vol. 3152, pp. 41–55. Springer, Heidelberg (2004)
5. Boneh, D., Sahai, A., Waters, B.: Fully Collusion Resistant Traitor Tracing with Short Ciphertexts and Private Keys. In: Vaudenay, S. (ed.) EUROCRYPT 2006. LNCS, vol. 4004, pp. 573–592. Springer, Heidelberg (2006)
6. Boudot, F.: Efficient Proofs that a Committed Number Lies in an Interval. In: Preneel, B. (ed.) EUROCRYPT 2000. LNCS, vol. 1807, pp. 431–444. Springer, Heidelberg (2000)
7. Camenisch, J., Chaabouni, R., Shelat, A.: Efficient Protocols for Set Membership and Range Proofs. In: Pieprzyk, J. (ed.) ASIACRYPT 2008. LNCS, vol. 5350, pp. 234–252. Springer, Heidelberg (2008)
8. Camenisch, J., Dubovitskaya, M., Neven, G.: Oblivious transfer with access controls. In: ACM CCS 2009, pp. 131–140 (2009)
9. Camenisch, J., Michels, M.: Proving in Zero-Knowledge that a Number n is the Product of Two Safe Primes. In: Stern, J. (ed.) EUROCRYPT 1999. LNCS, vol. 1592, pp. 107–122. Springer, Heidelberg (1999)
10. Camenisch, J., Neven, G., Shelat, A.: Simulatable Adaptive Oblivious Transfer. In: Naor, M. (ed.) EUROCRYPT 2007. LNCS, vol. 4515, pp. 573–590. Springer, Heidelberg (2007)
11. Camenisch, J., Stadler, M.: Efficient Group Signature Schemes for Large Groups. In: Kaliski Jr., B.S. (ed.) CRYPTO 1997. LNCS, vol. 1294, pp. 410–424. Springer, Heidelberg (1997)
12. R. Chaabouni, H. Lipmaa, A. Shelat. Additive combinatorics and discrete logarithm based range protocols (2009), Cryptology ePrint Archive: 2009/469 `http://eprint.iacr.org/2009/469.pdf`
13. Chan, A., Frankel, Y., Tsiounis, Y.: Easy Come - Easy Go Divisible Cash. In: Nyberg, K. (ed.) EUROCRYPT 1998. LNCS, vol. 1403, pp. 561–575. Springer, Heidelberg (1998)
14. Chen, Y., Chou, J.-S., Hou, X.-W.: A novel k-out-of-n oblivious transfer protocols based on bilinear pairings (2010), Cryptology ePrint Archive: Report 2010/027
15. Coull, S., Green, M., Hohenberger, S.: Controlling Access to an Oblivious Database Using Stateful Anonymous Credentials. In: Jarecki, S., Tsudik, G. (eds.) PKC 2009. LNCS, vol. 5443, pp. 501–520. Springer, Heidelberg (2009)
16. Cramer, R., Damgård, I., Schoenmakers, B.: Proof of Partial Knowledge and Simplified Design of Witness Hiding Protocols. In: Desmedt, Y.G. (ed.) CRYPTO 1994. LNCS, vol. 839, pp. 174–187. Springer, Heidelberg (1994)
17. Dodis, Y., Halevi, S., Rabin, T.: A Cryptographic Solution to a Game Theoretic Problem. In: Bellare, M. (ed.) CRYPTO 2000. LNCS, vol. 1880, pp. 112–130. Springer, Heidelberg (2000)

18. Dodis, Y., Yampolskiy, A.: A Verifiable Random Function with Short Proofs and Keys. In: Vaudenay, S. (ed.) PKC 2005. LNCS, vol. 3386, pp. 416–431. Springer, Heidelberg (2005)
19. Goldreich, O., Micali, S., Wigderson, A.: How to play any mental game or a completeness theorem for protocols with honest majority. In: STOC 1987, pp. 218–229 (1987)
20. Goldwasser, S., Micali, S., Rivest, R.L.: A digital signature scheme secure against adaptive chosen-message attacks. SIAM J. Computing 17(2) (1988)
21. Green, M., Hohenberger, S.: Blind Identity-Based Encryption and Simulatable Oblivious Transfer. In: Kurosawa, K. (ed.) ASIACRYPT 2007. LNCS, vol. 4833, pp. 265–282. Springer, Heidelberg (2007)
22. Green, M., Hohenberger, S.: Universally Composable Adaptive Oblivious Transfer. In: Pieprzyk, J. (ed.) ASIACRYPT 2008. LNCS, vol. 5350, pp. 179–197. Springer, Heidelberg (2008)
23. Green, M., Hohenberger, S.: Practical adaptive oblivious transfer from simple assumptions (2010), The full version of this work appears in the Cryptology ePrint Archive: Report 2010/109
24. Groth, J., Sahai, A.: Efficient Non-interactive Proof Systems for Bilinear Groups. In: Smart, N.P. (ed.) EUROCRYPT 2008. LNCS, vol. 4965, pp. 415–432. Springer, Heidelberg (2008)
25. Hohenberger, S., Rothblum, G.N., Shelat, A., Vaikuntanathan, V.: Securely Obfuscating Re-encryption. In: Vadhan, S.P. (ed.) TCC 2007. LNCS, vol. 4392, pp. 233–252. Springer, Heidelberg (2007)
26. Hohenberger, S., Waters, B.: Realizing Hash-and-Sign Signatures under Standard Assumptions. In: Joux, A. (ed.) EUROCRYPT 2009. LNCS, vol. 5479, pp. 333–350. Springer, Heidelberg (2009)
27. Jarecki, S., Liu, X.: Efficient Oblivious Pseudorandom Function with Applications to Adaptive OT and Secure Computation of Set Intersection. In: Reingold, O. (ed.) TCC 2009. LNCS, vol. 5444, pp. 577–594. Springer, Heidelberg (2009)
28. Kilian, J.: Founding cryptography on oblivious transfer. In: STOC 1988, pp. 20–31 (1988)
29. Kurosawa, K., Nojima, R.: Simple Adaptive Oblivious Transfer without Random Oracle. In: Matsui, M. (ed.) ASIACRYPT 2009. LNCS, vol. 5912, pp. 334–346. Springer, Heidelberg (2009)
30. Kurosawa, K., Nojima, R., Phong, L.T.: Efficiency-Improved Fully Simulatable Adaptive OT under the DDH Assumption. In: Garay, J.A., De Prisco, R. (eds.) SCN 2010. LNCS, vol. 6280, pp. 172–181. Springer, Heidelberg (2010)
31. Laguillaumie, F., Paillier, P., Vergnaud, D.: Universally Convertible Directed Signatures. In: Roy, B. (ed.) ASIACRYPT 2005. LNCS, vol. 3788, pp. 682–701. Springer, Heidelberg (2005)
32. Lindell, Y.: Efficient Fully-Simulatable Oblivious Transfer. In: Malkin, T.G. (ed.) CT-RSA 2008. LNCS, vol. 4964, pp. 52–70. Springer, Heidelberg (2008)
33. Naor, M., Pinkas, B.: Oblivious Transfer with Adaptive Queries. In: Wiener, M. (ed.) CRYPTO 1999. LNCS, vol. 1666, pp. 573–590. Springer, Heidelberg (1999)
34. Peikert, C., Vaikuntanathan, V., Waters, B.: A Framework for Efficient and Composable Oblivious Transfer. In: Wagner, D. (ed.) CRYPTO 2008. LNCS, vol. 5157, pp. 554–571. Springer, Heidelberg (2008)
35. Rabin, M.: How to exchange secrets by oblivious transfer. Technical Report TR-81, Aiken Computation Laboratory, Harvard University (1981)

36. Rial, A., Kohlweiss, M., Preneel, B.: Universally Composable Adaptive Priced Oblivious Transfer. In: Shacham, H., Waters, B. (eds.) Pairing 2009. LNCS, vol. 5671, pp. 231–247. Springer, Heidelberg (2009)
37. Schnorr, C.-P.: Efficient signature generation for smart cards. Journal of Cryptology 4(3), 239–252 (1991)
38. Sun, H.-M., Chen, Y., Chou, J.-S.: An efficient secure oblivious transfer (2009), Cryptology ePrint Archive: Report 2009/521
39. Wiesner, S.: Conjugate coding. SIGACT News 15, 78–88 (1983)
40. Yao, A.: How to generate and exchange secrets. In: FOCS, pp. 162–167 (1986)

A Proof Sketches of Sender and Receiver Security

Complete proofs of sender and receiver security appear in the full version [23].

A.1 Proof Sketch of Sender Security (Theorem 2)

Proof. We will begin with $\mathbf{Real}_{\mathsf{S},\hat{\mathsf{R}}}$, then modify the distribution via a series of hybrid games until we arrive at a distribution identical to that of $\mathbf{Ideal}_{\mathsf{S}',\hat{\mathsf{R}}'}$. Let us define these hybrids as follows:

Game 0. The real-world experiment conducted between S and $\hat{\mathsf{R}}$ ($\mathbf{Real}_{\mathsf{S},\hat{\mathsf{R}}}$).
Game 1. This game modifies **Game 0** as follows: (1) each of S's ZKPoK executions is replaced with a *simulated* proof of the same statement,[6] and (2) the knowledge extractor E_2 is used to obtain the values $(\sigma_i, x, y, z, \bar{c}_4, \bar{c}_5, \bar{c}_6, \bar{c}_7)$ from each of $\hat{\mathsf{R}}$'s transfer queries. Whenever the extractor fails, S terminates the experiment and outputs the distinguished symbol error.
Game 2. This game modifies **Game 1** such that, whenever the extracted value $\sigma_i \in [1, N]$, S's response R is computed using the following approach: parse $C_{\sigma_i} = (c_1, \ldots, c_7)$ and set $R = \frac{c_3 e(g_1, g_2)^x}{M_{\sigma_i}}$. When $\sigma_i \notin [1, N]$, the response is computed using the normal protocol.
Game 3. This game modifies **Game 2** by replacing the input to S_I with a dummy vector of random messages $\bar{M}_1, \ldots, \bar{M}_N \in \mathbb{G}_T$. However when S computes a response value using the technique of **Game 2**, the response is based on the original message vector $M_1, \ldots, M_N$. We claim that the distribution of this game is equivalent to that of $\mathbf{Ideal}_{\mathsf{S}',\hat{\mathsf{R}}'}$.

Let us now consider the following Lemmas. For notational convenience, define:

$$\mathbf{Adv}\left[\,\mathbf{Game\ i}\,\right] = \mathbf{Pr}\left[\,D(\mathbf{Game\ i}) = 1\,\right] - \mathbf{Pr}\left[\,D(\mathbf{Game\ 0}) = 1\,\right].$$

Lemma 1. *If all p.p.t. algorithms D distinguish a simulated ZKPoK with advantage at most ϵ_{ZK} and the extractor E_2 fails with probability at most ϵ_{Ext}, then* $\mathbf{Adv}\left[\,\mathbf{Game\ 1}\,\right] \leq (k+1)\cdot\epsilon_{ZK} + k\cdot\epsilon_{Ext}$.

[6] This includes at most $k+1$ PoK executions, including the initial $ZKPoK\{(a) : g_1 = g^a\}$ and each subsequent proof $ZKPoK\{(a) : R = e(v_1, g_2^a) \wedge g_1 = g^a\}$.

Lemma 2. *If no p.p.t. algorithm has advantage* $> \epsilon$ *in solving the 3DDH problem, then*

$$\mathbf{Adv}\left[\,\mathbf{Game\ 2}\,\right] - \mathbf{Adv}\left[\,\mathbf{Game\ 1}\,\right] \leq \frac{Np}{p-1} \cdot \epsilon.$$

Lemma 3. *If no p.p.t adversary has advantage* $> \epsilon$ *at solving the 3DDH problem, then*

$$\mathbf{Adv}\left[\,\mathbf{Game\ 3}\,\right] - \mathbf{Adv}\left[\,\mathbf{Game\ 2}\,\right] \leq N \cdot \epsilon.$$

Proof of the above lemmas is in [23]. By summing over hybrids **Game 0** to **Game 3**, we obtain $\mathbf{Adv}\left[\,\mathbf{Game\ 3}\,\right] \leq (k+1)\epsilon_{ZK} + k\epsilon_{Ext} + N\epsilon(1 + \frac{p}{p-1})$. For the Schnorr proofs we use, $\epsilon_{ZK} = 0$. This concludes the proof of Sender security.

A.2 Proof Sketch of Receiver Security (Theorem 3)

Proof. We again arrive at the ideal-world sender via a series of hybrid games:

Game 0. The real-world experiment conducted between $\hat{\mathsf{S}}$ and R ($\mathbf{Real}_{\hat{\mathsf{S}},\mathsf{R}}$).
Game 1. A modification of **Game 0** in which R applies the knowledge extractor E_1 to $\hat{\mathsf{S}}$'s proof ZKPoK$\{a : g_1 = g^a\}$. If this extraction fails, R aborts and outputs $\perp$. Further, for transfers $i = 1$ through k, R uses the knowledge extractor E_3 on $\hat{\mathsf{S}}$'s proof $ZKPoK\{(a) : R = e(v_1, g_2^a) \wedge g_1 = g^a\}$ to extract the values a, aborting if the extractor fails (or returns inconsistent values).
Game 2. For transfer $i = 1$ to k, modify R's request such that $\sigma_i = 1$. The distribution of this game is identical to that of $\mathbf{Ideal}_{\hat{\mathsf{S}}',\mathsf{R}'}$.

Lemma 4. *If the extractor* E_1 *and* E_3 *fail with probability at most* ϵ_{Ext}*, then* $\mathbf{Adv}\left[\,\mathbf{Game\ 1}\,\right] \leq (k+1)\epsilon_{Ext}$.

Lemma 5. *If the Receiver's WIPoK is distinguishable with maximum advantage* ϵ_{WI}*, then*

$$\mathbf{Adv}\left[\,\mathbf{Game\ 2}\,\right] - \mathbf{Adv}\left[\,\mathbf{Game\ 1}\,\right] = \leq k \cdot \epsilon_{WI}.$$

Proof of the above lemmas is in [23]. Summing the differences, we have

$$\mathbf{Adv}\left[\,\mathbf{Game\ 2}\,\right] - \mathbf{Adv}\left[\,\mathbf{Game\ 0}\,\right] = (k+1)\epsilon_{Ext} + k\epsilon_{WI}.$$

For the Schnorr proofs we use, $\epsilon_{WI} = 0$. This concludes the proof of Receiver security.

Completeness Theorems with Constructive Proofs for Finite Deterministic 2-Party Functions

Daniel Kraschewski and Jörn Müller-Quade

Institute of Cryptography and Security, Faculty of Informatics,
Karlsruhe Institute of Technology, Germany
{kraschewski,mueller-quade}@kit.edu

Abstract. In this paper we present simple but comprehensive combinatorial criteria for completeness of finite deterministic 2-party functions with respect to information-theoretic security. We give a general protocol construction for efficient and statistically secure reduction of oblivious transfer to any finite deterministic 2-party function that fulfills our criteria. For the resulting protocols we prove universal composability. Our results are tight in the sense that our criteria still are necessary for any finite deterministic 2-party function to allow for implementation of oblivious transfer with statistical privacy and correctness.

We unify and generalize results of Joe Kilian (1991, 2000) in two ways. Firstly, we show that his completeness criteria also hold in the UC framework. Secondly, what is our main contribution, our criteria also cover a wide class of primitives that are not subject of previous criteria. We show that there are non-trivial examples of finite deterministic 2-party functions that are neither symmetric nor asymmetric and therefore have not been covered by existing completeness criteria so far.

As a corollary of our work, every finite deterministic 2-party function is either complete or can be considered equivalent to a non-complete symmetric 2-party function—this assertion holds true with respect to active adversaries as well as passive adversaries. Thereby known results on non-complete symmetric 2-party functions are strengthened.

Keywords: oblivious transfer, complete primitives, information-theoretic security, universal composability, secure function evaluation.

1 Introduction

Oblivious transfer in the sense of a trusted erasure channel (Rabin-OT) was introduced in [27] and later in [4] proven to be equivalent to $\binom{2}{1}$-OT, where a receiver Bob may learn only one of two bits sent by Alice. Oblivious transfer turned out to be complete in the sense that every secure multiparty computation can be implemented using OT [14,10,7,13]. Thereby, enduring interest in OT arised in cryptography and for numerous primitives it has been investigated, whether they allow for implementation of OT. In our work we exhaustively treat this question

Y. Ishai (Ed.): TCC 2011, LNCS 6597, pp. 364–381, 2011.

for the class of "finite deterministic 2-party functions", sometimes also referred to as "crypto gates". Such primitives are characterized by some finite alphabetes $\Upsilon_{\mathrm{A}}, \Upsilon_{\mathrm{B}}, \Omega_{\mathrm{A}}, \Omega_{\mathrm{B}}$ and some mappings $f_{\mathrm{A}} \in \Omega_{\mathrm{A}}^{\Upsilon_{\mathrm{A}} \times \Upsilon_{\mathrm{B}}}$, $f_{\mathrm{B}} \in \Omega_{\mathrm{B}}^{\Upsilon_{\mathrm{A}} \times \Upsilon_{\mathrm{B}}}$, such that on input $x \in \Upsilon_{\mathrm{A}}$ from Alice and $y \in \Upsilon_{\mathrm{B}}$ from Bob the primitive outputs $f_{\mathrm{A}}(x, y)$ to Alice and $f_{\mathrm{B}}(x, y)$ to Bob.

1.1 Related Work

In the literature one finds OT protocols for the bounded-classical-storage model [2] and the bounded-quantum-storage model [8] as well as noisy classical [6,30] and quantum channels [24,25], the latter taking commitments for granted. Further, there are reductions of $\binom{2}{1}$-OT to weaker OT versions that leak additional information [5,9,29] and to Rabin-OT [4]. OT-combiners implement OT from granted sets of OTs with faulty members [26,11]. For reversing the direction of $\binom{2}{1}$-OT a protocol is known with optimal number of OT calls [28]. Relative to complexity assumptions all-or-nothing laws have been shown [1,12,23], i.e. all non-trivial primitives are complete. Our work has several, nowadays folklore reduction techniques in common with all the aforementioned literature.

We unify and generalize the results of [15,16], where completeness criteria for symmetric (i.e. both parties receive the same output) and asymmetric (i.e. only one party learns the function output) 2-party functions were provided with respect to information-theoretic security. We import a main argument for the necessity of our criteria from [15]. Our sufficiency proof is independent from [15,16], since our results are more general and we use a very strict notion of security.

There are also results regarding whether various symmetric 2-party functions can be reduced to each other [22] and what can be implemented from scratch when there is only a passive adversary [21,20]. A corollary of our work extends all these results to non-symmetric primitives; some results of [20] already build on an early manuscript of our work [18].

1.2 Our Contribution

We expose a wide class of complete finite deterministic 2-party functions that are essentially neither symmetric nor asymmetric and hence are not subject of statistical completeness criteria in the literature so far. Further, by surprisingly simple combinatorial criteria to the respective function tables we give a precise characterization of *all* finite deterministic 2-party functions that allow for statistically secure implementation of OT. We provide an efficient and universally composable protocol scheme for OT from any finite deterministic 2-party function fulfilling our criteria. Our results are tight, as the necessity of our criteria still holds when only correctness and privacy of the implemented OT are required.

As a remarkable corollary of our work all non-complete finite deterministic 2-party functions turn out symmetric. This strengthens several known results for non-complete symmetric 2-party functions [21,22,20].

2 Presentation of Our Results

In this section we briefly present our results. Thereto, we first refer to the security notion that we use (Sec. 2.1), then introduce and motivate the notations needed for formulation of our results (Sec. 2.2) and, last but not least, state our completeness criteria in form of a Classification Theorem (Sec. 2.3).

2.1 Notion of Security

We prove our classification results in the UC framework [3] with static corruption and statistical security, i.e. the adversarial entities $\mathcal{A}, \mathcal{S}$ and the environment $\mathcal{Z}$ are computationally unbounded. Nonetheless, in our case the running time of a simulator $\mathcal{S}$ will always be polynomial in the running time of the according adversary $\mathcal{A}$. Since we implement $\binom{2}{1}$-OT from given 2-party functions, in the real model there always is a hybrid functionality that provides access to the latter (see Fig. 1). Since $\binom{2}{1}$-OT can be considered a special 2-party function that on input $(b_0, b_1) \in \{0,1\}^2$ from Alice and $c \in \{0,1\}$ from Bob outputs b_c to Bob and a special "nothing" symbol $\perp$ to Alice, we omit an explicit definition of the ideal functionality $\mathcal{F}_{\mathrm{OT}}$.

Functionality: $\mathcal{F}_{\mathrm{SFE}}^{(F)}$

Let F be characterized by the output functions $f_{\mathrm{A}} : \Upsilon_{\mathrm{A}} \times \Upsilon_{\mathrm{B}} \to \Omega_{\mathrm{A}}$ and $f_{\mathrm{B}} : \Upsilon_{\mathrm{A}} \times \Upsilon_{\mathrm{B}} \to \Omega_{\mathrm{B}}$, where $\Upsilon_{\mathrm{A}}, \Omega_{\mathrm{A}}$ are Alice's input and output alphabet and $\Upsilon_{\mathrm{B}}, \Omega_{\mathrm{B}}$ are Bob's input and output alphabet.

- Upon receiving input (x, i) from Alice, verify that $(x, i) \in \Upsilon_{\mathrm{A}} \times \mathbb{N}$ and that there is no recorded tuple $(\tilde{x}, i, \texttt{Alice})$; else ignore that input. Next, record $(x, i, \texttt{Alice})$ and send $(\texttt{processing}, \texttt{Alice}, i)$ to the adversary.
- Upon receiving input (y, i) from Bob, verify that $(y, i) \in \Upsilon_{\mathrm{B}} \times \mathbb{N}$ and that there is no recorded tuple $(\tilde{y}, i, \texttt{Bob})$; else ignore that input. Next, record $(y, i, \texttt{Bob})$ and send $(\texttt{processing}, \texttt{Bob}, i)$ to the adversary.
- Upon receiving a message $(\texttt{Delivery}, \texttt{Alice}, i)$ from the adversary, verify that there are recorded tuples $(x, i, \texttt{Alice})$ and $(y, i, \texttt{Bob})$ and the former is not marked; else ignore that input. Next, mark the recorded tuple $(x, i, \texttt{Alice})$, compute $a \leftarrow f_{\mathrm{A}}(x, y)$ and output (a, i) to Alice.
- Upon receiving a message $(\texttt{Delivery}, \texttt{Bob}, i)$ from the adversary, verify that there are recorded tuples $(x, i, \texttt{Alice})$ and $(y, i, \texttt{Bob})$ and the latter is not marked; else ignore that input. Next, mark the recorded tuple $(y, i, \texttt{Bob})$, compute $b \leftarrow f_{\mathrm{B}}(x, y)$ and output (b, i) to Bob.

When a party is corrupted, the adversary is granted unrestricted access to the channel between $\mathcal{F}_{\mathrm{SFE}}^{(F)}$ and the corrupted party, including the ability of deleting and/or forging arbitrary messages.

Fig. 1. The ideal functionality for secure evaluation of a 2-party function F. Adapted and simplified version of the Secure Function Evaluation functionality in [3]. Note that via the parameter i only the same multi-session ability is achieved as in [3] by multiple session IDs.

	0	1	2
0	0/0	0/0	0/0
1	0/0	1/1	0/1
2	0/0	0/1	1/1

	0	1	2
0	0/0	1/1	0/1
1	0/0	0/0	0/0
2	$\top$/0	$\top$/1	$\bot$/1

Fig. 2. Function tables of two 2-party functions that are consistent renamings of each other (Alice's inputs label the rows, Bob's inputs label the columns; outputs are denoted a/b, meaning that Alice learns a and Bob learns b). We just interchanged the first two rows and applied an injective function to Alice's outputs in the third row; i.e. $\sigma_{\mathrm{A}}(0) = 1$, $\sigma_{\mathrm{A}}(1) = 0$, $\rho_{\mathrm{A}}(2, 0) = (2, \top)$, $\rho_{\mathrm{A}}(2, 1) = (2, \bot)$, everything else just is mapped to itself.

2.2 Basic Concepts

A finite deterministic 2-party function can be characterized by its input and output alphabets and output functions (q.v. Fig. 1). By $\mathfrak{F}_{\mathsf{fin,det}}$ we denote the set of all tuples $(\Upsilon_{\mathrm{A}}, \Upsilon_{\mathrm{B}}, \Omega_{\mathrm{A}}, \Omega_{\mathrm{B}}, f_{\mathrm{A}}, f_{\mathrm{B}})$, where $\Upsilon_{\mathrm{A}}, \Upsilon_{\mathrm{B}}, \Omega_{\mathrm{A}}, \Omega_{\mathrm{B}}$ are non-empty finite alphabets and $f_{\mathrm{A}}, f_{\mathrm{B}}$ are mappings from $\Upsilon_{\mathrm{A}} \times \Upsilon_{\mathrm{B}}$ to Ω_{A} and from $\Upsilon_{\mathrm{A}} \times \Upsilon_{\mathrm{B}}$ to Ω_{B} respectively. For convenience we will not always differentiate pedantically between the mathematical object $F \in \mathfrak{F}_{\mathsf{fin,det}}$ and the corresponding primitive $\mathcal{F}_{\mathrm{SFE}}^{(F)}$, but from the context should be always clear what is meant.

Our notion of $\mathfrak{F}_{\mathsf{fin,det}}$ turns out a bit too detailed, since Alice and Bob can always relabel their input-output tuples of a given 2-party function without any side effects. There is no need for distinguishing between some $F \in \mathfrak{F}_{\mathsf{fin,det}}$ and any relabelled version of F. By the following definition we can abstract from those irrelevant details (q.v. Fig. 2).

Definition 1 (Consistent renamings). *Let* $F := (\Upsilon_{\mathrm{A}}, \Upsilon_{\mathrm{B}}, \Omega_{\mathrm{A}}, \Omega_{\mathrm{B}}, f_{\mathrm{A}}, f_{\mathrm{B}}) \in \mathfrak{F}_{\mathsf{fin,det}}$ *and* $F' := (\Upsilon'_{\mathrm{A}}, \Upsilon'_{\mathrm{B}}, \Omega'_{\mathrm{A}}, \Omega'_{\mathrm{B}}, f'_{\mathrm{A}}, f'_{\mathrm{B}}) \in \mathfrak{F}_{\mathsf{fin,det}}$. *Then* F *and* F' *are* consistent renamings *of each other, if there exist some injective mappings* $\rho_{\mathrm{A}} : \Upsilon_{\mathrm{A}} \times \Omega_{\mathrm{A}} \to \Upsilon'_{\mathrm{A}} \times \Omega'_{\mathrm{A}}$ *and* $\rho_{\mathrm{B}} : \Upsilon_{\mathrm{B}} \times \Omega_{\mathrm{B}} \to \Upsilon'_{\mathrm{B}} \times \Omega'_{\mathrm{B}}$ *and some bijective mappings* $\sigma_{\mathrm{A}} : \Upsilon_{\mathrm{A}} \to \Upsilon'_{\mathrm{A}}$ *and* $\sigma_{\mathrm{B}} : \Upsilon_{\mathrm{B}} \to \Upsilon'_{\mathrm{B}}$*, such that for all* $x \in \Upsilon_{\mathrm{A}}$*,* $y \in \Upsilon_{\mathrm{B}}$ *it holds:*

$$\rho_{\mathrm{A}}\big(x, f_{\mathrm{A}}(x, y)\big) = \big(\sigma_{\mathrm{A}}(x), f'_{\mathrm{A}}(\sigma_{\mathrm{A}}(x), \sigma_{\mathrm{B}}(y))\big)$$
$$\rho_{\mathrm{B}}\big(y, f_{\mathrm{B}}(x, y)\big) = \big(\sigma_{\mathrm{B}}(y), f'_{\mathrm{B}}(\sigma_{\mathrm{A}}(x), \sigma_{\mathrm{B}}(y))\big)$$

Moreover, there may exist input symbols that are kind of "redundant" in the sense that an actively corrupted party can always input some corresponding "dominating" input symbols and at the same time perfectly simulate honest behaviour. This concept plays an important role in our proofs and results. We formally grasp it by the following definition.

Definition 2 (Redundancy). *Given* $F = (\Upsilon_{\mathrm{A}}, \Upsilon_{\mathrm{B}}, \Omega_{\mathrm{A}}, \Omega_{\mathrm{B}}, f_{\mathrm{A}}, f_{\mathrm{B}}) \in \mathfrak{F}_{\mathsf{fin,det}}$*, an input symbol* $y' \in \Upsilon_{\mathrm{B}}$ *is* redundant*, if there exists some corresponding* dominating *input symbol* $y \in \Upsilon_{\mathrm{B}} \setminus \{y'\}$*, such that the following two conditions hold:*

1. *For all* $x \in \Upsilon_{\mathrm{A}}$ *we have that* $f_{\mathrm{A}}(x, y) = f_{\mathrm{A}}(x, y')$*, i.e. from her own output Alice does never learn whether Bob did input* y *or* y'*.*

	0	1
0	0/0	0/0
1	0/0	0/1

	0	1
0	0/0	0/0
1	0/0	1/0

	0	1
0	0/0	0/0
1	0/0	1/1

Fig. 3. Function tables of the three different types of OT-cores (up to consistent renaming)

2. *For all $x, x' \in \Upsilon_{\mathrm{A}}$ with $f_{\mathrm{B}}(x,y') \neq f_{\mathrm{B}}(x',y')$ we have that $f_{\mathrm{B}}(x,y) \neq f_{\mathrm{B}}(x',y)$, i.e. by inputting y instead of y' Bob gets exactly the same or strictly more information.*

For input symbols $x \in \Upsilon_{\mathrm{A}}$ redundancy is defined analogously. If neither Υ_{A} nor Υ_{B} contains any redundant input symbols, F is called redundancy-free.

W.l.o.g. actively corrupted parties always use dominating input symbols instead of the corresponding redundant ones. Also, there is no need to constrain what honest parties may learn. Therefore, in presence of an active adversary we can consider any 2-party functions equivalent when they only differ in some redundant input symbols.

Definition 3 (Equivalence). *Let $F := (\Upsilon_{\mathrm{A}}, \Upsilon_{\mathrm{B}}, \Omega_{\mathrm{A}}, \Omega_{\mathrm{B}}, f_{\mathrm{A}}, f_{\mathrm{B}}) \in \mathfrak{F}_{\mathsf{fin,det}}$ and $F' := (\Upsilon'_{\mathrm{A}}, \Upsilon'_{\mathrm{B}}, \Omega'_{\mathrm{A}}, \Omega'_{\mathrm{B}}, f'_{\mathrm{A}}, f'_{\mathrm{B}}) \in \mathfrak{F}_{\mathsf{fin,det}}$. Then F and F' are called* equivalent, *if they can be transformed into consistent renamings of each other by successive*[1] *removal of redundant input symbols from $\Upsilon_{\mathrm{A}}, \Upsilon_{\mathrm{B}}$, $\Upsilon'_{\mathrm{A}}, \Upsilon'_{\mathrm{B}}$ and according adjustment of $f_{\mathrm{A}}, f_{\mathrm{B}}, f'_{\mathrm{A}}, f'_{\mathrm{B}}$. Let $[F]$ denote the resulting equivalence class.*

Given $F \in \mathfrak{F}_{\mathsf{fin,det}}$, one can show quite easily that all redundancy-free $\bar{F}, \bar{F}' \in [F]$ are consistent renamings of each other, i.e. the redundancy-free version of F is unique up to consistent renaming.

2.3 The Classification Theorem

With the concepts from Sec. 2.2 we can now formulate our completeness criteria.

Definition 4 (Symmetric 2-party functions). *Let $F' \in \mathfrak{F}_{\mathsf{fin,det}}$. If F' is a consistent renaming of some $F = (\Upsilon_{\mathrm{A}}, \Upsilon_{\mathrm{B}}, \Omega_{\mathrm{A}}, \Omega_{\mathrm{B}}, f_{\mathrm{A}}, f_{\mathrm{B}}) \in \mathfrak{F}_{\mathsf{fin,det}}$ with $\Omega_{\mathrm{A}} = \Omega_{\mathrm{B}}$ and $f_{\mathrm{A}} = f_{\mathrm{B}}$, then F' is called* symmetric.

Definition 5 (OT-cores). *Let $F := (\Upsilon_{\mathrm{A}}, \Upsilon_{\mathrm{B}}, \Omega_{\mathrm{A}}, \Omega_{\mathrm{B}}, f_{\mathrm{A}}, f_{\mathrm{B}}) \in \mathfrak{F}_{\mathsf{fin,det}}$. Then a quadruple $(x, x', y, y') \in \Upsilon_{\mathrm{A}}^2 \times \Upsilon_{\mathrm{B}}^2$ is an* OT-core *of F, if the following three conditions are met (q.v. Fig. 3):*

1. *We have that $f_{\mathrm{A}}(x,y) = f_{\mathrm{A}}(x,y')$.*

[1] Note that a step-by-step removal of one symbol at a time is crucial here. There may exist distinct input symbols that dominate each other but must not be removed both.

	1
0	0/0
1	0/1

	0	1
1	0/0	1/0

	0	1
0	0/0	0/0
1	0/0	1/1

Fig. 4. Redundancy-free versions of the three different types of OT-cores (cf. Fig. 3), when there are no other input symbols around

2. *We have that* $f_{\mathrm{B}}(x, y) = f_{\mathrm{B}}(x', y)$.
3. *We have that* $f_{\mathrm{A}}(x', y) \neq f_{\mathrm{A}}(x', y')$ *or* $f_{\mathrm{B}}(x, y') \neq f_{\mathrm{B}}(x', y')$ *(or both).*

Theorem 1 (Classification theorem). *For each* $F \in \mathfrak{F}_{\mathsf{fin,det}}$ *it holds:*

1. *For the* $\mathcal{F}_{\mathrm{SFE}}^{(F)}$*-hybrid model there exists an OT protocol that is statistically secure against passive adversaries, iff* F *has an OT-core.*
2. *If for the* $\mathcal{F}_{\mathrm{SFE}}^{(F)}$*-hybrid model there does not exist any OT protocol that is statistically secure against passive adversaries, then* F *is symmetric.*
3. *For the* $\mathcal{F}_{\mathrm{SFE}}^{(F)}$*-hybrid model there exists an OT protocol that is statistically secure against active adversaries, iff the redundancy-free version of* F *has an OT-core.*
4. *If for the* $\mathcal{F}_{\mathrm{SFE}}^{(F)}$*-hybrid model there does not exist any OT protocol that is statistically secure against active adversaries, then the redundancy-free version of* F *is symmetric.*

Note that, when there is an active adversary, only the third function in Fig. 3 is complete on its own. The redundancy free versions of the other two functions just collapse to simple binary channels (q.v. Fig. 4). This collapsing can be prevented by additional input symbols. In Fig. 5 one can see, how OT-cores can be complemented to redundancy-free 2-party functions of minimum size.

For *symmetric* and *asymmetric* 2-party functions our completeness criteria coincide with the criteria from [15,16]. More concretely, we can directly translate the completeness criteria of [15,16] to our notations as follows.

Completeness criteria of [15]: A *symmetric* 2-party function F is complete, iff it has an OT-core. This holds true, regardless whether the adversary is active or passive.

Completeness criteria of [16]: Given an active adversary, an *asymmetric* 2-party function F' (with Bob being the receiver) is complete, iff for every input symbol $y \in \Upsilon_{\mathrm{B}}$ there exists some other input symbol $y' \in \Upsilon_{\mathrm{B}}$ that is not dominated by y; in other words, F' is complete, iff its redundancy-free version is non-trivial in the sense that both input alphabets have cardinality 2 or more. Given only a passive adversary, an *asymmetric* 2-party function F' is complete, iff it has an OT-core.

However, our criteria are much more comprehensive than that of [15,16], since ours also cover 2-party functions that are neither *symmetric* nor *asymmetric*. An illustrating example is the third function in Fig. 5, which is complete but not subject of the criteria in [15,16].

	0	1
0	0/0	0/0
1	0/0	0/1
2	0/1	0/0

	0	1
0	0/0	0/0
1	0/0	0/1
2	0/1	0/2
3	0/2	0/2

	0	1
0	0/0	0/0
1	0/0	0/1
2	0/1	1/2

	0	1
0	0/0	0/0
1	0/0	1/1

Fig. 5. Function tables of the four minimal complete 2-party functions. Up to consistent renaming and interchanging the roles of Alice and Bob every function table of a complete 2-party function $F \in \mathfrak{F}_{\mathsf{fin,det}}$ contains at least one of these examples as a submatrix.

3 How to Prove the Classification Theorem

In this section we give an intuitive exposition of how we prove our Classification Theorem. Due to space limitations we can only sketch the main ideas; for formal proofs we refer to the full version [19].

A fundamental tool in our proof strategy is the connection between presence of OT-cores and the question whether a 2-party function is symmetric.

Lemma 1 (Symmetrization lemma). *Each $F \in \mathfrak{F}_{\mathsf{fin,det}}$ that does not have any OT-core is symmetric (in the sense of Definition 4).*

One way to prove this lemma can be sketched as follows. Given a 2-party function $F := (\Upsilon_{\mathrm{A}}, \Upsilon_{\mathrm{B}}, \Omega_{\mathrm{A}}, \Omega_{\mathrm{B}}, f_{\mathrm{A}}, f_{\mathrm{B}}) \in \mathfrak{F}_{\mathsf{fin,det}}$, we can define an equivalence relation on $(\Upsilon_{\mathrm{A}} \times \Omega_{\mathrm{A}}) \cup (\Upsilon_{\mathrm{B}} \times \Omega_{\mathrm{B}})$ induced as follows:

$$(x, a) \sim (y, b) \quad :\Leftarrow \quad f_{\mathrm{A}}(x, y) = a \ \wedge \ f_{\mathrm{B}}(x, y) = b$$

Let the according equivalence classes be denoted by $[x, a]$ or $[y, b]$. For all $x, x' \in \Upsilon_{\mathrm{A}}$, $a, a' \in \Omega_{\mathrm{A}}$ some simple induction yields the following implication (else F would have an OT-core):

$$(x, a) \sim (x', a') \quad \Rightarrow \quad \{y \in \Upsilon_{\mathrm{B}} \mid f_{\mathrm{A}}(x, y) = a\} = \{y \in \Upsilon_{\mathrm{B}} \mid f_{\mathrm{A}}(x', y) = a'\}$$

Thereby, we cannot find any $x \in \Upsilon_{\mathrm{A}}$, $a, a' \in \Omega_{\mathrm{A}}$ with $a \neq a'$ and $(x, a) \sim (x, a')$; the analog holds for $y \in \Upsilon_{\mathrm{A}}$, $b, b' \in \Omega_{\mathrm{A}}$. Hence, via the mappings $\rho_{\mathrm{A}} : (x, a) \mapsto \big(x, [x, a]\big)$ and $\rho_{\mathrm{B}} : (y, b) \mapsto \big(y, [y, b]\big)$ we get a consistent renaming of F and this consistent renaming is obviously symmetric.

By the Symmetrization Lemma and some results in the literature we can already argue for the assertions 1 and 2 of our Classification Theorem. On the one hand, when F has no OT-core, F can be considered symmetric by our Symmetrization Lemma. However, in [15] it has been shown that no reduction of OT to a symmetric 2-party function without OT-core can yield correctness and privacy at the same time, even if there is only a passive adversary—Alice can always exactly determine Bob's information about her inputs to the underlying 2-party function and vice versa.

	0	1	2	3
0	0/0	0/0	0/0	0/0
1	0/0	1/0	0/1	1/1
2	0/1	1/1	1/2	0/2

Fig. 6. A complete 2-party function that needs some carefully chosen, non-symmetric input distribution

On the other hand, when F has an OT-core and there is only a passive adversary, we can trivially implement one of the 2-party functions in Fig. 3. However, each of them can be transformed into a non-trivial noisy channel (shown to be complete in [6]) by the following protocol with expected 4 function calls. Alice first inputs a random bit b and then the inverse $\neg b$; Bob inputs independent random bits in both steps. The protocol is restarted until nowhere output "1" occurs. Afterwards Alice uses the last value of b as a one-time pad, which Bob knows with probability $\frac{2}{3}$.

Once assertion 1 of the Classification Theorem is shown, assertion 2 follows by the Symmetrization Lemma. Analogously assertion 4 follows from assertion 3, so all we have to do is proving assertion 3. One direction, the necessity of OT-cores, already follows from the passive case. Proving sufficiency for the active case is much more challenging and can be seen as our main contribution.

Our overall strategy for reducing OT in presence of an active adversary to a finite deterministic 2-party function having an OT-core proceeds in two steps. First, Alice and Bob generate some amount of correlated data by repeated invocation of the 2-party function with randomized input. Within a subsequent test step each party has to partially unveil its data, so that significant cheating can be detected. Then, on top of the remaining data an invocation of OT is built. In Sec. 3.1 we examine what input distributions are adequate and how the test step has to be performed. In Sec. 3.2 we construct a protocol for OT from such correlated data and we examine its security.

3.1 Secure Generation of Correlated Data

We start our examination with some negative example (see Fig. 6), which shows that choosing an adequate input distribution is not trivial. In the first place, the example in Fig. 6 shows that letting Alice and Bob use uniformly random input is not necessarily secure. In our example there would be an undetectable cheating strategy[2] for a corrupted Bob: He picks a uniformly random input symbol from $\{2,3\}$ instead of $\{0,1,2,3\}$ and after each invocation of the 2-party function with probability $\frac{1}{2}$ locally relabels his input-output tuple by $(2,0) \mapsto (0,0)$, $(2,1) \mapsto (0,0)$, $(2,2) \mapsto (1,1)$, $(3,0) \mapsto (1,0)$, $(3,1) \mapsto (1,0)$, $(3,2) \mapsto (0,1)$. Thereby he can perfectly simulate honest behaviour, but at the same time does learn all of Alice's inputs to the 2-party function.

[2] Note that such an undetectable cheating strategy cannot exist for symmetric 2-party functions, as there Alice will notice any change in Bob's output distribution.

We circumvent this problem by more asymmetric input distributions: We pick an OT-core and let the corresponding input symbols be input with relatively high probability, while all other input symbols have a relatively low probability and are only needed for the test step. However, the example in Fig. 6 also shows that we must choose the OT-core carefully. E.g. the OT-core in the upper left corner would be a bad choice, since the abovementioned cheating strategy can be adjusted to every protocol that assigns equal probability to Bob's input symbols "0" and "1". Still, significant cheating is possible for any input distribution with high probability for "0" and "1", as inputting "0" and "1" each once can be perfectly simulated by inputting "2" and "3" each once.

Actually, a main part of our work consists in proving that there always exists a "good" OT-core, if only the redundancy-free version of the considered 2-party function has any OT-core at all. In Fig. 6 one "good" OT-core corresponds to inputs $\{0,1\}$ from Alice and $\{1,2\}$ from Bob: By occasionally inputting "2" Alice can check that Bob does not too often use other input symbols than $\{1,2\}$ (on input "2" she must not get output "0" too often) and that he does input "1" and "2" each with the right frequency (on input "1" she must get output "1" and "0" with according frequency), while Bob also sees Alice's actual input distribution (it is close to Bob's output distribution on input "2"). However, as the first function in Fig. 5 shows, in general it will not suffice that the participants only pay attention to their own input-output distributions. Since in this example Alice's output always is "0", only by unveiling some random subset of his input-output tuples Bob can prove that he did use a prescribed input distribution; e.g. he will be caught cheating with high probability when he claims to have input "0" sufficiently often but can never distinguish whether Alice did input "0" or "2". Again, for a meaningful test it is necessary that Alice uses her complete input alphabet.

These examples motivate that always all input symbols should be used with some non-zero probability. In the following we first sketch our protocol for generation of correlated data, then we introduce some algebraic structure that abstractly represents how a corrupted party can deviate from the protocol; finally we argue that there always is an OT-core that is "robust" against all such cheating strategies.

Our protocol for generating correlated data basically proceeds as follows:

1. **Invocation of F:** Alice and Bob call the underlying 2-party function F with randomized input for k times (k being the security parameter) and record their respective input-output tuples. A protocol parameter assigns what probability mass functions are to be used.
2. **Control A:** Alice challenges Bob on some polynomial subset of the recorded data, where he has to reveal his input-output tuples. Alice aborts the protocol if Bob obviously lies (i.e. his announcement is inconsistent with Alice's recorded input-output tuples) or his input distribution appears faulty. The test set is then removed from the recorded data.
3. **Control B:** This step equals the previous one with interchanged roles of Alice and Bob.

4. **Output:** Both parties announce where they have used input symbols that were only for test purposes. All corresponding elements are removed from the recorded input-output tuples by both parties. When too much of the recorded data has been deleted, the protocol is aborted; else each party outputs its remaining string of recorded input-output tuples.

We call this scheme *offline protocol*, since after the protocol step **Invocation of** F never again access to F is needed.

At this point we want to emphasize that although offline protocols are not completely symmetric in Alice and Bob, all of our arguments are. This convenient circumstance is predicated on the fact that a corrupted party only can get some polynomially small advantage by adversarial choice of the test set in protocol step **Control A** or **Control B** respectively. Our protocol in Sec. 3.2 for reduction of OT to correlated data is robust against such polynomially small advantages.

Now we define and investigate a class of functions $\eta : \Upsilon_{\mathrm{A}} \times \Upsilon_{\mathrm{B}}^2 \to \mathbb{R}_{\geq 0}$ that characterize how a corrupted Bob may cheat in an offline protocol. For symmetry reasons our results will directly carry over to the case that Alice is corrupted. Our intuition is that $\eta(x, y, y')$ quantifies the relative frequency of events in protocol step **Control A**, where F was invoked with input (x, y), but Bob successfully claims that he did input y'. We call such functions *cheating situations*. For convenience we use the notation $\eta(X, Y, Y') := \sum_{x \in X, y \in Y, y' \in Y'} \eta(x, y, y')$ for any $X \subseteq \Upsilon_{\mathrm{A}}$, $Y, Y' \subseteq \Upsilon_{\mathrm{B}}$. Also for convenience, we speak of a situation $(x, y)_F$ when we mean that F was called with input x from Alice and input y from Bob. We have the following six conditions to cheating situations:

1. It holds that $\eta(\Upsilon_{\mathrm{A}}, \Upsilon_{\mathrm{B}}, \Upsilon_{\mathrm{B}}) = 1$.
2. For all $x \in \Upsilon_{\mathrm{A}}$ it holds that $\eta(x, \Upsilon_{\mathrm{B}}, \Upsilon_{\mathrm{B}}) > 0$, i.e. Alice did use her complete input alphabet.
3. For all $x \in \Upsilon_{\mathrm{A}}$, $y \in \Upsilon_{\mathrm{B}}$ it holds that $\eta(x, y, \Upsilon_{\mathrm{B}}) = \eta(x, \Upsilon_{\mathrm{B}}, \Upsilon_{\mathrm{B}}) \cdot \eta(\Upsilon_{\mathrm{A}}, y, \Upsilon_{\mathrm{B}})$, i.e. Bob's actual input distribution is independent of Alice's input distribution.
4. For all $x \in \Upsilon_{\mathrm{A}}$, $y' \in \Upsilon_{\mathrm{B}}$ it holds that $\eta(x, \Upsilon_{\mathrm{B}}, y') = \eta(x, \Upsilon_{\mathrm{B}}, \Upsilon_{\mathrm{B}}) \cdot \eta(\Upsilon_{\mathrm{A}}, \Upsilon_{\mathrm{B}}, y')$, i.e. Bob's claimed input distribution appears independent of Alice's input distribution.
5. (a) For all $x \in \Upsilon_{\mathrm{A}}$, $y, y' \in \Upsilon_{\mathrm{B}}$ with $f_{\mathrm{A}}(x, y) \neq f_{\mathrm{A}}(x, y')$ it holds that $\eta(x, y, y') = 0$; else in the test step **Control A** Bob would be caught cheating immediately.
   (b) For all $x, x' \in \Upsilon_{\mathrm{A}}$, $y, y' \in \Upsilon_{\mathrm{B}}$ that fulfill $f_{\mathrm{B}}(x, y) = f_{\mathrm{B}}(x', y)$ and $f_{\mathrm{B}}(x, y') \neq f_{\mathrm{B}}(x', y')$, it holds that $\eta(x, y, y') = \eta(x', y, y') = 0$; else Bob would run an overwhelming risk of being caught cheating, since he cannot distinguish between situations $(x, y)_F$ and $(x', y)_F$ but must perfectly distinguish between these situations in the test step **Control A**.

Given some 2-party function $F \in \mathfrak{F}_{\mathsf{fin,det}}$, the set $\mathfrak{N}_F$ of all according cheating sitations has a very handy algebraic structure. On the one hand, cheating situations can be considered independent of (honest) Alice's input distribution, since they can canonically be rescaled to every input distribution that has non-zero probability for all $x \in \Upsilon_{\mathrm{A}}$. On the other hand, when we fix Alice's input distribution, i.e. for all $x \in \Upsilon_{\mathrm{A}}$ the $\eta(x, \Upsilon_{\mathrm{B}}, \Upsilon_{\mathrm{B}})$ are fixed, then our six conditions can be subsumed by a linear equation system, i.e. the set of all remaining cheating situations is a convex and bounded polytope in the linear space $\mathbb{R}^{\Upsilon_{\mathrm{A}} \times \Upsilon_{\mathrm{B}}^2}$.

Also the abovementioned conditions 5a and 5b play a fundamental role in our proofs. Therefore we sum them up by an extra notation. Given $F = (\Upsilon_{\mathrm{A}}, \Upsilon_{\mathrm{B}}, \Omega_{\mathrm{A}}, \Omega_{\mathrm{B}}, f_{\mathrm{A}}, f_{\mathrm{B}}) \in \mathfrak{F}_{\mathsf{fin,det}}$ and $x \in \Upsilon_{\mathrm{A}}$, $y, y' \in \Upsilon_{\mathrm{B}}$, let $(x, y) \overset{F}{\rightsquigarrow} (x, y')$ denote that the following two conditions are fulfilled:

- It holds that $f_{\mathrm{A}}(x, y) = f_{\mathrm{A}}(x, y')$.
- For all $\tilde{x} \in \Upsilon_{\mathrm{A}}$ with $f_{\mathrm{B}}(x, y) = f_{\mathrm{B}}(\tilde{x}, y)$ it holds that $f_{\mathrm{B}}(x, y') = f_{\mathrm{B}}(\tilde{x}, y')$.

The intuition behind that notation is that Bob can claim a situation $(x, y)_F$ to be a situation $(x, y')_F$, iff $(x, y) \overset{F}{\rightsquigarrow} (x, y')$. At least he cannot do so too often, if $(x, y) \overset{F}{\not\rightsquigarrow} (x, y')$. For all cheating situations η and all $x \in \Upsilon_{\mathrm{A}}$, $y, y' \in \Upsilon_{\mathrm{B}}$ with $(x, y) \overset{F}{\not\rightsquigarrow} (x, y')$ it holds that $\eta(x, y, y') = 0$.

Note that the "$\overset{F}{\rightsquigarrow}$"-relation links cheating situations to redundancy matters, since an input symbol $y' \in \Upsilon_{\mathrm{B}}$ is redundant, iff there exists some $y \in \Upsilon_{\mathrm{B}} \setminus \{y'\}$ with $(x, y) \overset{F}{\rightsquigarrow} (x, y')$ for all $x \in \Upsilon_{\mathrm{A}}$. In other words, the "$\overset{F}{\rightsquigarrow}$"-relation describes some kind of "local redundancy".

Given that Alice is uncorrupted, for every non-aborted run of an offline protocol there exists with overwhelming probability some cheating situation η, such that up to some polynomially small error the mappings $(x, y) \mapsto \eta(x, \Upsilon_{\mathrm{B}}, y)$ and $(x, y) \mapsto \eta(x, y, \Upsilon_{\mathrm{B}})$ describe the prescribed and the actual joint input distribution to the underlying 2-party function respectively. Thus we have to look for some kind of "robust" OT-cores $(\tilde{x}, \tilde{x}', \tilde{y}, \tilde{y}')$, so that there does not exist any essentially non-trivial cheating situation η with $\eta(\Upsilon_{\mathrm{A}}, \Upsilon_{\mathrm{B}}, \{\tilde{y}, \tilde{y}'\}) = 1$. More concretely, we will show that whenever a redundancy-free 2-party function $F \in \mathfrak{F}_{\mathsf{fin,det}}$ has any OT-core $(\tilde{x}, \tilde{x}', \tilde{y}, \tilde{y}')$, then F also has an OT-core $(\tilde{x}, \tilde{x}', \bar{y}, \bar{y}')$, such that for every cheating situation η with $\eta(\Upsilon_{\mathrm{A}}, \Upsilon_{\mathrm{B}}, \{\bar{y}, \bar{y}'\}) = 1$ it holds that $\eta(\Upsilon_{\mathrm{A}}, \Upsilon_{\mathrm{B}}, y) = \eta(\Upsilon_{\mathrm{A}}, y, \Upsilon_{\mathrm{B}})$ for all $y \in \Upsilon_{\mathrm{B}}$, i.e. Bob practically cannot lie about his actual input distribution when he is demanded to use no other input symbols than $\bar{y}, \bar{y}'$.

Note that Alice's input symbols $\tilde{x}, \tilde{x}'$ have remained the same; hence in a second step we can analogously find an OT-core $(\bar{x}, \bar{x}', \bar{y}, \bar{y}')$ that is also "robust" against all relevant cheating attempts of Alice and stays "robust" against a possibly malicious Bob.

Given an OT-core $(\tilde{x}, \tilde{x}', \tilde{y}, \tilde{y}')$ of a redundancy-free 2-party function $F \in \mathfrak{F}_{\mathsf{fin,det}}$, we can find an OT-core with the desired "robustness" by just picking some $\bar{y}, \bar{y}' \in \Upsilon_{\mathrm{B}}$, such that $(\tilde{x}, \tilde{x}', \bar{y}, \bar{y}')$ is an OT-core and the following set has minimum size (q.v. Fig. 7):

$$\Phi(\bar{y}, \bar{y}') \;:=\; \left\{ y \in \Upsilon_{\mathrm{B}} \;\middle|\; \forall x \in \Upsilon_{\mathrm{A}} : (x, y) \overset{F}{\rightsquigarrow} (x, \bar{y}) \vee (x, y) \overset{F}{\rightsquigarrow} (x, \bar{y}') \right\}$$

Intuitively spoken, within an offline protocol that assigns high input probability only to $\bar{y}, \bar{y}'$ Bob cannot use any input symbol $y \in \Upsilon_{\mathrm{B}} \setminus \Phi(\bar{y}, \bar{y}')$ too often; at

	0	1	2	3	4	5
0	0/0	0/0	0/0	0/0	0/0	1/*
1	0/0	1/0	1/0	0/0	0/1	*/*
2	0/1	0/1	0/1	0/1	0/2	*/*
3	0/1	0/1	0/1	0/2	0/2	*/*
4	0/2	0/1	0/1	0/2	0/2	*/*
5	0/3	0/2	0/2	0/3	0/3	*/*
6	0/3	0/2	0/3	0/3	0/3	*/*
7	0/3	0/3	0/3	0/3	0/3	*/*

Fig. 7. Example for illustration of the construction of Φ and Y, Y'. From the first two rows one can infer that $(0,1,0,1)$ is an OT-core and $\Phi(0,1) \subseteq \{0,1,2,3,4\}$, regardless of the wildcards "$*$". The other six rows just make the function redundancy-free, but still allow that $\Phi(0,1) \supseteq \{0,1,2,3,4\}$. Thereby, for the OT-core in the upper left corner we have that $\Phi(0,1) = \{0,1,2,3,4\}$ and $Y = \{0,3\}$ and $Y' = \{1,2,4\}$. However, we would not pick this OT-core but $(0,1,0,4)$ or $(0,1,3,4)$ instead, since $\Phi(0,4) = \Phi(3,4) = \{0,3,4\} \subsetneq \Phi(0,1)$, as Alice can distinguish between $\{0,3,4\}$ and $\{1,2\}$ by her output in the second row. Note that analogously $\Phi(1,2) = \{1,2\}$, but $(0,1,1,2)$ is not an OT-core.

least for some specific $x \in \Upsilon_{\mathrm{A}}$ he practically cannot claim a situation $(x,y)_F$ to be $(x,\bar{y})_F$ or $(x,\bar{y}')_F$ without being caught cheating. In general it will not necessarily hold that $\Phi(\bar{y},\bar{y}') = \{\bar{y},\bar{y}'\}$, nonetheless we can show now that the chosen OT-core $(\tilde{x},\tilde{x}',\bar{y},\bar{y}')$ is "robust" in the abovementioned sense. So, let some arbitrary cheating situation η with $\eta(\Upsilon_{\mathrm{A}},\Upsilon_{\mathrm{B}},\{\bar{y},\bar{y}'\}) = 1$ be given. By the following eight steps we show that $\eta(\Upsilon_{\mathrm{A}},\Upsilon_{\mathrm{B}},y) = \eta(\Upsilon_{\mathrm{A}},y,\Upsilon_{\mathrm{B}})$ for all $y \in \Upsilon_{\mathrm{B}}$.

1. Since the "$\overset{F}{\rightsquigarrow}$"-relation is transitive, we observe that $\Phi(y,y') \subseteq \Phi(\bar{y},\bar{y}')$ for all $y,y' \in \Phi(\bar{y},\bar{y}')$.
2. We want to exploit the minimality of $\Phi(\bar{y},\bar{y}')$, but it yields that $\left|\Phi(\bar{y},\bar{y}')\right| \leq \left|\Phi(y,y')\right|$ only in case that $(\tilde{x},\tilde{x}',y,y')$ is an OT-core. However, note that $f_{\mathrm{A}}(\tilde{x},\bar{y}) = f_{\mathrm{A}}(\tilde{x},\bar{y}')$, since $(\tilde{x},\tilde{x}',\bar{y},\bar{y}')$ is an OT-core. Furthermore, for all $y \in \Phi(\bar{y},\bar{y}')$ by definition of Φ we especially have that $(\tilde{x},y) \overset{F}{\rightsquigarrow} (\tilde{x},\bar{y})$ or $(\tilde{x},y) \overset{F}{\rightsquigarrow} (\tilde{x},\bar{y}')$, what in turn implies that $f_{\mathrm{A}}(\tilde{x},y) = f_{\mathrm{A}}(\tilde{x},\bar{y})$ or $f_{\mathrm{A}}(\tilde{x},y) = f_{\mathrm{A}}(\tilde{x},\bar{y}')$. Putting things together, we can conclude that $f_{\mathrm{A}}(\tilde{x},y) = f_{\mathrm{A}}(\tilde{x},y')$ for all $y,y' \in \Phi(\bar{y},\bar{y}')$. Therefore, by the following construction we can split $\Phi(\bar{y},\bar{y}')$ into disjoint subsets Y, Y', such that $(\tilde{x},\tilde{x}',y,y')$ actually is an OT-core for all $y \in Y$, $y' \in Y'$. We define (q.v. Fig. 7):

$$Y := \left\{y \in \Phi(\bar{y},\bar{y}') \;\middle|\; f_{\mathrm{A}}(\tilde{x}',\bar{y}) = f_{\mathrm{A}}(\tilde{x}',y) \;\wedge\; f_{\mathrm{B}}(\tilde{x},y) = f_{\mathrm{B}}(\tilde{x}',y)\right\}$$
$$Y' := \left\{y' \in \Phi(\bar{y},\bar{y}') \;\middle|\; f_{\mathrm{A}}(\tilde{x}',\bar{y}) \neq f_{\mathrm{A}}(\tilde{x}',y') \;\vee\; f_{\mathrm{B}}(\tilde{x},y') \neq f_{\mathrm{B}}(\tilde{x}',y')\right\}$$

Now, by the minimality of $\Phi(\bar{y},\bar{y}')$ and our observation in step 1 it follows that $\Phi(\bar{y},\bar{y}') = \Phi(y,y')$ for all $y \in Y$, $y' \in Y'$.

3. Now, for each $(x,\hat{y}) \in \Upsilon_A \times \Phi(\bar{y},\bar{y}')$ at least one of the following assertions must hold true:

$$\forall y \in Y : (x,\hat{y}) \overset{F}{\rightsquigarrow} (x,y) \qquad\qquad \forall y' \in Y' : (x,\hat{y}) \overset{F}{\rightsquigarrow} (x,y')$$

Otherwise we had some $x \in \Upsilon_A$, $\hat{y} \in \Phi(\bar{y},\bar{y}')$, $y \in Y$, $y' \in Y'$, such that $(x,\hat{y}) \overset{F}{\not\rightsquigarrow} (x,y)$ and $(x,\hat{y}) \overset{F}{\not\rightsquigarrow} (x,y')$ and thereby $\hat{y} \notin \Phi(y,y')$, what would be a contradiction to $\hat{y} \in \Phi(\bar{y},\bar{y}') = \Phi(y,y')$ (cf. the final sentence of step 2).

4. For every $\hat{y} \in \Phi(\bar{y},\bar{y}') \setminus \{\bar{y}\}$ we find some $x \in \Upsilon_A$, such that $(x,\hat{y}) \overset{F}{\not\rightsquigarrow} (x,\bar{y})$ and $\forall y' \in Y' : (x,y') \overset{F}{\not\rightsquigarrow} (x,\bar{y})$.

   This follows from our observation in step 3, F being redundancy-free and the transitivity of the "$\overset{F}{\rightsquigarrow}$"-relation. Since F is redundancy-free, we find some $x \in \Upsilon_A$, such that $(x,\hat{y}) \overset{F}{\not\rightsquigarrow} (x,\bar{y})$. This not only is one part of the claim above, but it also yields by step 3 that $(x,\hat{y}) \overset{F}{\rightsquigarrow} (x,y')$ for all $y' \in Y'$, since $\bar{y} \in Y$ by construction of Y. Now, if we could find any $y' \in Y'$ with $(x,y') \overset{F}{\rightsquigarrow} (x,\bar{y})$, in contradiction to our choice of x this would imply that $(x,\hat{y}) \overset{F}{\rightsquigarrow} (x,\bar{y})$, due to the transitivity of the "$\overset{F}{\rightsquigarrow}$"-relation.

5. For all $\hat{y} \in \Phi(\bar{y},\bar{y}') \setminus \{\bar{y}\}$ we have that $\eta(\Upsilon_A, Y\setminus\{\hat{y}\}, \Upsilon_B) \geq \eta(\Upsilon_A, \Upsilon_B, \bar{y})$, i.e. Bob's claimed input frequency of $\bar{y}$ cannot be greater than his actual overall input frequency of symbols in $Y\setminus\{\hat{y}\}$.

   Otherwise we could find some $\hat{y} \in \Phi(\bar{y},\bar{y}') \setminus \{\bar{y}\}$, such that $\eta(x,\Upsilon_B,\bar{y}) > \eta(x, Y\setminus\{\hat{y}\}, \Upsilon_B)$ for all $x \in \Upsilon_A$ (cf. the conditions 3 and 4 to cheating situations). However, by step 4 we can choose x such that Bob cannot claim any situation $(x,y')_F$ with $y' \in Y' \cup \{\hat{y}\}$ to be a situation $(x,\bar{y})_F$; the same holds for $y' \in \Upsilon_B \setminus \Phi(\bar{y},\bar{y}')$ by definition of Φ. He may do so only for situations $(x,y')_F$ with $y' \in Y\setminus\{\hat{y}\}$, but these are too few, as $\eta(x,\Upsilon_B,\bar{y}) > \eta(x, Y\setminus\{\hat{y}\}, \Upsilon_B)$.

6. We observe that $\eta(\Upsilon_A, \Upsilon_B\setminus\Phi(\bar{y},\bar{y}'), \Upsilon_B) = 0$, since $\eta(\Upsilon_A, \Upsilon_B, \{\bar{y},\bar{y}'\}) = 1$ by assumption, i.e. $\eta(\Upsilon_A, \Upsilon_B, \Upsilon_B\setminus\{\bar{y},\bar{y}'\}) = 0$, and $\eta(\Upsilon_A, \Upsilon_B\setminus\Phi(\bar{y},\bar{y}'), \{\bar{y},\bar{y}'\}) = 0$ by construction of Φ.

7. For every $\hat{y}' \in \Phi(\bar{y},\bar{y}')\setminus\{\bar{y}'\}$ we have that $\eta(\Upsilon_A, Y\cup\{\hat{y}'\}, \Upsilon_B) \leq \eta(\Upsilon_A, \Upsilon_B, \bar{y})$, i.e. Bob's claimed input frequency of $\bar{y}$ cannot be less than his actual overall input frequency of symbols in $Y \cup \{\hat{y}'\}$.

   Since the assertion of step 3 is symmetric in Y and Y', analogously to step 4 for every $\hat{y}' \in \Phi(\bar{y},\bar{y}')\setminus\{\bar{y}'\}$ we find some $x \in \Upsilon_A$, such that $\forall y \in Y \cup \{\hat{y}'\}$: $(x,y) \overset{F}{\not\rightsquigarrow} (x,\bar{y}')$. We can use that to prove the analog of step 5: For every $\hat{y}' \in \Phi(\bar{y},\bar{y}')\setminus\{\bar{y}'\}$ we have that $\eta(\Upsilon_A, Y'\setminus\{\hat{y}'\}, \Upsilon_B) \geq \eta(\Upsilon_A, \Upsilon_B, \bar{y}')$. Moreover, we have that $\eta(\Upsilon_A, \Phi(\bar{y},\bar{y}'), \Upsilon_B) = 1$ by step 6 and that $\eta(\Upsilon_A, \Upsilon_B, \{\bar{y},\bar{y}'\}) = 1$ by assumption. Conclusively, for all $\hat{y}' \in \Phi(\bar{y},\bar{y}')\setminus\{\bar{y}'\}$ we get that $\eta(\Upsilon_A, \Upsilon_B, \bar{y}) = 1 - \eta(\Upsilon_A, \Upsilon_B, \bar{y}') \geq 1 - \eta(\Upsilon_A, Y'\setminus\{\hat{y}'\}, \Upsilon_B) = \eta(\Upsilon_A, Y\cup\{\hat{y}'\}, \Upsilon_B)$.

8. By combination of step 5 and step 7, for all $\hat{y}, \hat{y}' \in \Phi(\bar{y},\bar{y}')$ with $\hat{y}' \neq \bar{y}'$ and $\hat{y} \neq \bar{y}$ we can now conclude that $\eta(\Upsilon_A, Y\cup\{\hat{y}'\}, \Upsilon_B) \leq \eta(\Upsilon_A, Y\setminus\{\hat{y}\}, \Upsilon_B)$. This can be exploited as follows. On the one hand, we can choose $\hat{y} = \bar{y}'$, i.e. $Y\setminus\{\hat{y}\} = Y$, whereby for all $\hat{y}' \in Y'\setminus\{\bar{y}'\}$ it follows that $\eta\big(\Upsilon_A, \hat{y}', \Upsilon_B\big) \leq 0$, i.e. $\eta\big(\Upsilon_A, Y'\setminus\{\bar{y}'\}, \Upsilon_B\big) = 0$. On the other hand, we can choose $\hat{y}' = \bar{y}$, i.e.

$Y \cup \{\hat{y}'\} = Y$, whereby for all $\hat{y} \in Y \backslash \{\bar{y}\}$ it follows that $\eta(\Upsilon_{\rm A}, \hat{y}, \Upsilon_{\rm B}) \leq 0$, i.e. $\eta(\Upsilon_{\rm A}, Y \backslash \{\bar{y}\}, \Upsilon_{\rm B}) = 0$. Conclusively, using that $\eta(\Upsilon_{\rm A}, \Upsilon_{\rm B} \backslash \Phi(\bar{y}, \bar{y}'), \Upsilon_{\rm B}) = 0$ by step 6, we get that $\eta(\Upsilon_{\rm A}, \Upsilon_{\rm B} \backslash \{\bar{y}, \bar{y}'\}, \Upsilon_{\rm B}) = 0$, i.e. $\eta(\Upsilon_{\rm A}, \{\bar{y}, \bar{y}'\}, \Upsilon_{\rm B}) = 1$. Now, since $\eta(\Upsilon_{\rm A}, \Upsilon_{\rm B}, \{\bar{y}, \bar{y}'\}) = 1$ by assumption and neither $\bar{y}$ nor $\bar{y}'$ is redundant, one can infer rather straightforwardly that $\eta(\Upsilon_{\rm A}, \Upsilon_{\rm B}, y) = \eta(\Upsilon_{\rm A}, y, \Upsilon_{\rm B})$ for all $y \in \Upsilon_{\rm B}$, as claimed.

3.2 Reduction of OT to Correlated Data

We now sketch a protocol that implements OT from the correlated data produced by an appropriate offline protocol. Within this sketch we also informally argue for the protocol's security. Given a redundancy-free 2-party function F that has some OT-core $(\tilde{x}, \tilde{x}', \tilde{y}, \tilde{y}')$, the protocol proceeds as follows:

0. W.l.o.g. we may assume that the OT-core $(\tilde{x}, \tilde{x}', \tilde{y}, \tilde{y}')$ is of the first or last type in Fig. 3; else we interchange the roles of Alice and Bob. W.l.o.g. we also assume that Alice's and Bob's actual input and output symbols coincide with that of Fig. 3, i.e. $\tilde{x} = \tilde{y} = 0$ and so on. Furthermore, w.l.o.g. we assume that $(\tilde{x}, \tilde{x}', \tilde{y}, \tilde{y}')$ is a "robust" OT-core, whose existence we have shown in Section 3.1.
1. Alice and Bob execute an offline protocol (as sketched in Section 3.1), where the probability mass functions $n_{\rm A}$ and $n_{\rm B}$ that stand for Alice's and Bob's prescribed input distribution respectively, are such that $n_{\rm A}(0) \approx \frac{1}{3}$ and $n_{\rm A}(1) \approx \frac{2}{3}$ and $n_{\rm B}(0) \approx n_{\rm B}(1) \approx \frac{1}{2}$. Note that in general these will not be the exact input probabilities, as for meaningful tests in the protocol steps **Control A** and **Control B** we still need all other inputs to be used with some polynomial frequency. However, for growing security parameter the relative frequency of the other inputs may polynomially converge to zero. Further note that even if a party is corrupted, its actual input distribution in non-aborted protocol runs must be polynomially close to honest behaviour, since $(\tilde{x}, \tilde{x}', \tilde{y}, \tilde{y}')$ was chosen to be a "robust" OT-core.
2. We want to handle all possible types of OT-cores analogously, therefore we let Alice announce where she got output "1". All corresponding input-output tuples are deleted from the recorded data by both parties. When Alice tries to delete too little, Bob aborts the protocol. He also aborts the protocol when he has to delete some input-output tuple other than $(1, f_{\rm B}(1,1))$. Since Alice cannot distinguish between situations $(0,0)_F$ and $(0,1)_F$, this forces her to play honestly up to some polynomially small fraction of the recorded data.
3. Now most of the remaining input-output tuples belong to situations $(0,0)_F$, $(0,1)_F$, $(1,0)_F$. Since all according outputs are "0", it suffices that Alice and Bob henceforth only keep track of their recorded input strings. Note that at this stage about one quarter of the remaining recorded data belongs to situations $(0,0)_F$, one quarter to $(0,1)_F$ and one half to $(1,0)_F$.
4. Alice deletes some elements from her recorded input string, such that afterwards the string is balanced (i.e. it contains the same number of "0"s

and "1"s). She annonces the corresponding indices to Bob, who deletes the according elements from his recorded data. If Alice tries to delete too much, Bob aborts the protocol.

5. Alice randomly permutes her recorded input string, such that afterwards each element at an odd possition is different from its subsequent element. She announces the permutation to Bob, who permutes his input string accordingly. Thereby their input strings become strings of pairs (each starting at an odd position), such that a pair "01" or "10" on Bob's side indicates the respective inverted pair "10" or "01" on Alice's side and a pair "00" on Bob's side gives him no information about the pair on Alice's side. If Bob finds a pair "11" (starting at an odd position), he aborts the protocol. Note that about half of Bob's pairs are "00", one quarter is "01" and one quarter is "10".

   Further note that primarily there is only one way Alice may get some additional information about where Bob has "00"-pairs: She chooses the permutation adversarially, so that some "11"-pairs are produced on her side. However, since her input string is roughly balanced since the beginning of step 3, she must produce roughly as much "00"-pairs as "11"-pairs on her side and for each "00"-pair she is caught cheating by Bob with probability $\frac{1}{2}$. So even a corrupted Alice may know at most polynomially few positions where Bob has "00"-pairs.

6. Since Bob now can reconstruct about half of Alice's input string and Alice has only few information about where exactly Bob can do that, we can treat the recorded data like the result of Rabin-OT calls and adapt standard reduction techniques[3]. To that effect we rename Alices input string into a string of half length over the alphabet $\{0,1\}$ and accordingly for Bob over the alphabet $\{0,1,\perp\}$; in particular the renaming is "01"$\mapsto$"0", "10"$\mapsto$"1" on Alice's side and "10"$\mapsto$"0", "01"$\mapsto$"1", "00"$\mapsto$"$\perp$" on Bob's side. When a party cheated, we can represent that by a special symbol "$\top$" in that party's string. However, the symbol "$\top$" may occur only with some polynomial relative frequency, say less than $k^{-\gamma}$. Let $\kappa := \lceil k^{1-\gamma} \rceil$.
7. Now, let $b_0, b_1 \in \{0,1\}$ be Alice's $\binom{2}{1}$-OT input and let $c \in \{0,1\}$ be Bob's choice bit. Alice chooses two random bit strings $\tilde{b}_0, \tilde{b}_1 \in \{0,1\}^\kappa$ with $\bigoplus_{j=1}^{\kappa} \tilde{b}_0[j] = b_0$ and $\tilde{b}_0[j] \oplus \tilde{b}_1[j] = b_0 \oplus b_1$ for $j = 1, \ldots, \kappa$. Bob chooses a random bit string $\tilde{c} \in \{0,1\}^\kappa$ with $\bigoplus_{j=1}^{\kappa} \tilde{c}[j] = c$.
8. Alice and Bob respectively partition their recorded input strings into κ consecutive substrings of equal length l with l as large as possible; remaining elements are just discarded. Let $\tilde{s}_{\mathrm{A}}^{(j)}$ denote Alices j-th substring and $\tilde{s}_{\mathrm{B}}^{(j)}$ Bob's j-th substring. Note that by our choice of κ at least one of the $\tilde{s}_{\mathrm{A}}^{(j)}$ does not contain the symbol "$\top$". Further note that for each $\tilde{s}_{\mathrm{B}}^{(j)}$ about half of the

[3] Note that due to a subtle issue we cannot directly apply the results of [5,9,29] for reduction of OT to weak OT; e.g. in our case a corrupted Alice can choose to learn some prefix of Bob's string. In contrast, weak OT does not allow the adversary to influence when exactly additional information is leaked.

contained elements equal "⊥", because of the permutation at the beginning of step 3.

For $j = 1, \ldots, \kappa$ now the following subprotocol is executed:

(a) Bob chooses some disjoint random sets $K_0^{(j)}, K_1^{(j)} \subseteq \{1, \ldots, l\}$ of equal cardinality $\lceil \frac{l}{3} \rceil$, such that no element of $\tilde{s}_{\mathrm{B}}^{(j)}$ indexed by $K_{\tilde{c}[j]}^{(j)}$ is "⊥". He announces $(K_0^{(j)}, K_1^{(j)})$ to Alice. Note that Alice does not get any information about at least one of the $\tilde{c}[j]$, since the corresponding $\tilde{s}_{\mathrm{A}}^{(j)}$ does not contain the symbol "⊤". Hence she stays ignorant of Bob's choice bit c.

(b) For $i = 0, 1$ Alice uses the XOR of the elements in $\tilde{s}_{\mathrm{A}}^{(j)}$ indexed by $K_i^{(j)}$ as a one-time pad for $\tilde{b}_i[j]$. She sends the according cyphertexts to Bob, who learns $\tilde{b}_{\tilde{c}[j]}[j]$ by reconstructing the needed one-time pad from $\tilde{s}_{\mathrm{B}}^{(j)}$. Note that for each j Bob cannot get some information about both bits $\tilde{b}_0[j], \tilde{b}_1[j]$ at the same time, since more than one third of the elements in $\tilde{s}_{\mathrm{B}}^{(j)}$ equals "⊥". Hence he may learn at most one of Alice's $\binom{2}{1}$-OT inputs b_0, b_1.

9. Alice outputs the nothing symbol "⊥" and Bob computes and outputs $b_c = \bigoplus_{j=1}^{\kappa} \tilde{b}_{\tilde{c}[j]}[j]$. Correctness of Bob's output can be shown by induction on the Hamming weight of $\tilde{c}$.

We conclude our work with some remarks about how one can prove universal composability of this protocol, i.e. that it is simulatable in the ideal model (q.v. Section 2.1). Access to the underlying 2-party function F is in the ideal model only simulated, so the simulator can compute all the $\tilde{s}_{\mathrm{A}}^{(j)}$ or $\tilde{s}_{\mathrm{B}}^{(j)}$ respectively and hence extract the OT input of a corrupted Alice or Bob. Moreover, when Bob is corrupted, the simulator can fake a real protocol run that matches the ideal Alice's inputs b_0, b_1 as follows: Just before step 8b is entered the κ-th time, the simulator inputs the extracted choice bit c into the ideal functionality $\mathcal{F}_{\mathrm{OT}}$, thus learning b_c, and then revises $\tilde{b}_0[\kappa]$ and $\tilde{b}_1[\kappa]$ accordingly.

4 Conclusion

In this paper we showed that there is a wide class of primitives that have not been covered by existing completeness criteria, namely all 2-party functions that are essentially neither symmetric nor asymmetric. We solved this open problem by presenting simple but comprehensive criteria that combinatorially classify all complete deterministic 2-party functions with finite input and output alphabets. We proved constructively that our criteria are sufficient in the UC framework, which is the most restrictive common notion of security we know. Our criteria also turn out necessary even with respect to very weak notions of security. Therefore we consider them valid for virtually all reasonable security notions.

A remarkable corollary of our work is that every non-complete deterministic 2-party function with finite input and output alphabets is essentially symmetric. Thereby we extended the results of [21,22,20] to non-symmetric 2-party functions. The questions treated there become trivial for complete primitives and we have shown that every essentially non-symmetric 2-party function actually is complete.

Acknowledgements. We want to thank Mike Rosulek and the anonymous reviewers of TCC 2011 for helpful comments.

References

1. Beimel, A., Malkin, T., Micali, S.: The All-or-Nothing Nature of Two-Party Secure Computation. In: Wiener, M.J. (ed.) CRYPTO 1999. LNCS, vol. 1666, pp. 80–97. Springer, Heidelberg (1999)
2. Cachin, C., Crépeau, C., Marcil, J.: Oblivious transfer with a memory-bounded receiver. In: Proceedings of FOCS 2001, pp. 493–502 (1998)
3. Canetti, R.: Universally composable security: A new paradigm for cryptographic protocols. In: Proceedings of FOCS 2001, pp. 136–145 (2001), revised version online `http://eprint.iacr.org/2000/067`
4. Crépeau, C.: Equivalence between Two Flavours of Oblivious Transfers. In: Pomerance, C. (ed.) CRYPTO 1987. LNCS, vol. 293, pp. 350–354. Springer, Heidelberg (1988)
5. Crépeau, C., Kilian, J.: Weakening security assumptions and oblivious transfer (abstract). In: Goldwasser, S. (ed.) CRYPTO 1988. LNCS, vol. 403, pp. 2–7. Springer, Heidelberg (1990)
6. Crépeau, C., Morozov, K., Wolf, S.: Efficient Unconditional Oblivious Transfer from Almost Any Noisy Channel. In: Blundo, C., Cimato, S. (eds.) SCN 2004. LNCS, vol. 3352, pp. 47–59. Springer, Heidelberg (2005)
7. Crépeau, C., van de Graaf, J., Tapp, A.: Committed Oblivious Transfer and Private Multi-party Computation. In: Coppersmith, D. (ed.) CRYPTO 1995. LNCS, vol. 963, pp. 110–123. Springer, Heidelberg (1995)
8. Damgård, I., Fehr, S., Renner, R., Salvail, L., Schaffner, C.: A Tight High-Order Entropic Quantum Uncertainty Relation with Applications. In: Menezes, A. (ed.) CRYPTO 2007. LNCS, vol. 4622, pp. 360–378. Springer, Heidelberg (2007)
9. Damgård, I., Kilian, J., Salvail, L.: On the (Im)possibility of Basing Oblivious Transfer and Bit Commitment on Weakened Security Assumptions. In: Stern, J. (ed.) EUROCRYPT 1999. LNCS, vol. 1592, pp. 56–73. Springer, Heidelberg (1999)
10. Goldwasser, S., Levin, L.A.: Fair Computation of General Functions in Presence of Immoral Majority. In: Menezes, A., Vanstone, S.A. (eds.) CRYPTO 1990. LNCS, vol. 537, pp. 77–93. Springer, Heidelberg (1991)
11. Harnik, D., Ishai, Y., Kushilevitz, E., Nielsen, J.B.: OT-Combiners via Secure Computation. In: Canetti, R. (ed.) TCC 2008. LNCS, vol. 4948, pp. 393–411. Springer, Heidelberg (2008)
12. Harnik, D., Naor, M., Reingold, O., Rosen, A.: Completeness in two-party secure computation: A computational view. Journal of Cryptology 19(4), 521–552 (2006)
13. Ishai, Y., Prabhakaran, M., Sahai, A.: Founding Cryptography on Oblivious Transfer – Efficiently. In: Wagner, D. (ed.) CRYPTO 2008. LNCS, vol. 5157, pp. 572–591. Springer, Heidelberg (2008)
14. Kilian, J.: Founding cryptography on oblivious transfer. In: Proceedings of STOC 1988, pp. 20–31. ACM, New York (1988)
15. Kilian, J.: A general completeness theorem for two-party games. In: Proceedings of STOC 1991, pp. 553–560. ACM, New York (1991)

16. Kilian, J.: More general completeness theorems for secure two-party computation. In: Proceedings of STOC 2000, pp. 316–324. ACM, New York (2000)
17. Kraschewski, D.: Vollständigkeitskriterien von kryptographischen Primitiven. Diploma thesis, Institut für Algorithmen und Kognitive Systeme, Universität Karlsruhe (2006)
18. Kraschewski, D., Müller-Quade, J.: Completeness theorems with constructive proofs for symmetric, asymmetric and general 2-party-functions. Unpublished manuscript of the present work with different and more complicated proof techniques, based on the first author's diploma thesis [17] (2008)
19. Kraschewski, D., Müller-Quade, J.: Completeness theorems with constructive proofs for finite deterministic 2-party functions (full version). Cryptology ePrint Archive, Report 2010/654 (2010), Full version of the present work, online available at `http://eprint.iacr.org/2010/654`
20. Künzler, R., Müller-Quade, J., Raub, D.: Secure Computability of Functions in the IT Setting with Dishonest Majority and Applications to Long-Term Security. In: Reingold, O. (ed.) TCC 2009. LNCS, vol. 5444, pp. 238–255. Springer, Heidelberg (2009)
21. Kushilevitz, E.: Privacy and communication complexity. SIAM Journal on Discrete Mathematics 5(2), 273–284 (1992)
22. Maji, H.K., Prabhakaran, M., Rosulek, M.: Complexity of Multi-party Computation Problems: The Case of 2-Party Symmetric Secure Function Evaluation. In: Reingold, O. (ed.) TCC 2009. LNCS, vol. 5444, pp. 256–273. Springer, Heidelberg (2009)
23. Maji, H.K., Prabhakaran, M., Rosulek, M.: A Zero-One Law for Cryptographic Complexity with Respect to Computational UC Security. In: Rabin, T. (ed.) CRYPTO 2010. LNCS, vol. 6223, pp. 595–612. Springer, Heidelberg (2010)
24. Mayers, D.: On the Security of the Quantum Oblivious Transfer and Key Distribution Protocols. In: Coppersmith, D. (ed.) CRYPTO 1995. LNCS, vol. 963, pp. 124–135. Springer, Heidelberg (1995)
25. Mayers, D.: Quantum Key Distribution and String Oblivious Transfer in Noisy Channels. In: Koblitz, N. (ed.) CRYPTO 1996. LNCS, vol. 1109, pp. 343–357. Springer, Heidelberg (1996)
26. Meier, R., Przydatek, B., Wullschleger, J.: Robuster Combiners for Oblivious Transfer. In: Vadhan, S.P. (ed.) TCC 2007. LNCS, vol. 4392, pp. 404–418. Springer, Heidelberg (2007)
27. Rabin, M.O.: How to exchange secrets by oblivious transfer. Technical report, Aiken Computation Laboratory, Harvard University (1981)
28. Wolf, S., Wullschleger, J.: Oblivious Transfer Is Symmetric. In: Vaudenay, S. (ed.) EUROCRYPT 2006. LNCS, vol. 4004, pp. 222–232. Springer, Heidelberg (2006)
29. Wullschleger, J.: Oblivious-transfer amplification. In: Naor, M. (ed.) EUROCRYPT 2007. LNCS, vol. 4515, pp. 555–572. Springer, Heidelberg (2007)
30. Wullschleger, J.: Oblivious Transfer from Weak Noisy Channels. In: Reingold, O. (ed.) TCC 2009. LNCS, vol. 5444, pp. 332–349. Springer, Heidelberg (2009)

A Zero-One Law for Secure Multi-party Computation with Ternary Outputs

Gunnar Kreitz

KTH – Royal Institute of Technology
gkreitz@kth.se

Abstract. There are protocols to privately evaluate any function in the passive (honest-but-curious) setting assuming that the honest nodes are in majority. For some specific functions, protocols are known which remain secure even without an honest majority. The seminal work by Chor and Kushilevitz [7] gave a complete characterization of Boolean functions, showing that each Boolean function either requires an honest majority, or is such that it can be privately evaluated regardless of the number of colluding nodes.

The problem of discovering the threshold for secure evaluation of more general functions remains an open problem. Towards a resolution, we provide a complete characterization of the security threshold for functions with three different outputs. Surprisingly, the zero-one law for Boolean functions extends to $\mathbb{Z}_3$, meaning that each function with range $\mathbb{Z}_3$ either requires honest majority or tolerates up to n colluding nodes.

1 Introduction

Multi-party secure function evaluation (SFE) is a cornerstone of modern cryptography, and has been extensively studied since it was introduced by Yao [15]. In this work we consider the joint evaluation by n parties of a public n-ary function f in such a way that no collusion of parties learns anything more than what they do by knowing their own inputs and seeing the output. We consider the *symmetric* case where all participants receive the same output.

Several models of adversaries occur in the SFE literature. A first distinction is whether the adversary has limited computational power (*computational security*) or not (*information-theoretic security*). A second important distinction is whether the parties corrupted by the adversary must still follow the protocol (*passive*) or not (*active*). In the present work, we are concerned with information-theoretic security and all adversaries considered are passive. We assume that the parties communicate over a complete network with *private channels*, meaning that the adversary cannot see messages sent between two honest parties.

Another important limitation put upon the adversary is which parties she can corrupt. The most common adversary is allowed to corrupt up to a threshold $t \leq n$ participants for some t which is typically a function of n. We say that a function for which there is a protocol tolerating up to t corruptions is t-private. In this paper, we will only consider threshold adversaries. More general adversarial

Y. Ishai (Ed.): TCC 2011, LNCS 6597, pp. 382–399, 2011.

models have also been studied, both in terms of a more general specification of the parties the adversary can corrupt by Hirt and Maurer [10] and considering a mix active and passive adversarial corruptions by Beerliová-Trubíniová *et al.* [2].

There exist protocols to securely evaluate any function $\lfloor (n-1)/2 \rfloor$-privately in our setting by Ben-Or, Goldwasser, and Wigderson [3], and Chaum, Crépeau, and Damgård [4]. For some functions, in particular Boolean disjunction, this has been proved to be an upper bound meaning that there are no protocols to evaluate them which remain secure against more than $\lfloor (n-1)/2 \rfloor$ colluding parties. For other functions, in particular summation over a finite Abelian group, there are n-private protocols. This raises the question of determining the privacy threshold of functions.

Chor and Kushilevitz [7] completely answered the question for Boolean functions. They proved a zero-one law showing that each Boolean function is either $\lfloor (n-1)/2 \rfloor$-private (and not $\lceil n/2 \rceil$-private) or n-private. Their work presents a proof that a function containing an OR-like substructure (an *embedded* OR) is $\lfloor (n-1)/2 \rfloor$-private and that all Boolean functions without such a substructure can be computed by a single Boolean summation.

Proving that a function f cannot be t-privately computed is often done by a partition argument, reducing to the two-party case. In these proofs, the parties are partitioned into two parts of size $\leq t$ and we think of f as a two-party function with each party supplying all inputs for one set of the partition. If the two-party function is not 1-private, then f is not t-private. Chor and Ishai [6] analyzed partition arguments and gave a generalization partitioning the parties into $k > 2$ sets which increases the power of the framework. However, in this paper, we will only need partitioning arguments with two sets.

Chor, Geréb-Graus, and Kushilevitz [5] showed that for every t, $\lceil n/2 \rceil \leq t \leq n-2$ there exists a function such that it is t-private but not $(t+1)$-private. We remark that the functions they construct in their proofs have very large ranges which grow exponentially with t.

The privacy of symmetric[1] functions with Boolean arguments has been studied by Chor and Shani [9]. For such functions, they prove a necessary condition on the preimages of outputs for the function to be $\lceil n/2 \rceil$-private. They also define a class called dense symmetric functions where this necessary condition is also sufficient for n-privacy. Thus, they also prove a zero-one law where for a class of functions, where each function in the class is either n-private or not $\lceil n/2 \rceil$-private.

For two-party computation, a complete characterization of the 1-private functions was made independently by Beaver [1] and Kushilevitz [14]. They both show that a function f is 1-private if and only if it is decomposable, and for decomposable functions, there is a straightforward 1-private protocol. One of our protocols, Protocol 3, can be viewed as a generalization of the protocol for decomposable functions to the multi-party case.

[1] Here, symmetric means the standard notion of a symmetric function, not the SFE-specific notion that all parties receive the same output.

Künzler, Müller-Quade, and Raub [13] give a combinatorial classification of functions computable in several different adversarial models, including the information-theoretic passive model which we work with in this paper. However, in this setting, they consider the broadcast model of communication which gives different results from private channels. For instance, summation is not n-private in the broadcast channel model.

1.1 Our Contribution

In this work, we extend the zero-one law of Boolean privacy to functions with three outputs. For notational convenience, we talk about functions with range $\mathbb{Z}_3$, but we would like to emphasize our results do not depend on any algebraic structure over the range of the function. More formally, we prove the following statement:

Theorem 1 (Main theorem). *For every n-argument function $f : A_1 \times \ldots \times A_n \to \mathbb{Z}_3$, f is either n-private, or it is $\lfloor (n-1)/2 \rfloor$-private and not $\lceil n/2 \rceil$-private.*

The core part of our proof is a structure lemma (Lemma 8) showing that every function f with range $\mathbb{Z}_3$ must have at least one of three properties (which we define more formally later):

- f has an embedded OR
- f is a permuted sum
- f is collapsible.

We provide protocols for n-privately evaluating those functions of the two latter types which do not contain an embedded OR.

Our definition of an embedded OR is a generalization of the one commonly found in the literature, but the presence of one implies that there is no protocol which can securely evaluate f and tolerate more than t colluding parties for some t (but potentially for a $t > \lceil n/2 \rceil$).

Finally, we prove (Theorem 22) that the existence of an embedded OR (in our generalized sense) also implies the existence of a "small" embedded OR, giving $t = \lceil n/2 \rceil$. By combining this result with our structure lemma and the result from [7] that a function with an embedded OR of size at most $\lceil n/2 \rceil$ cannot be $\lceil n/2 \rceil$-privately computed, our main theorem follows. We state the proof more formally in Section 6.

We remark that while our statements are true for $n = 2$, there are complete classifications [1,14] for the 2-party case which are simpler than ours (for $n = 2$, our protocols reduce to decomposition) and not limited to functions with range $\mathbb{Z}_3$. Our contribution lies in the case when $n \geq 3$.

The proof of our theorems are significantly more involved than the analogous proofs for Boolean functions. In several of our proofs we need to apply a fairly extensive case analysis.

Our result answers in part a question raised by Chor and Ishai [6] by showing that partition reductions (with only two sets) are universal for proving non-privacy of functions mapping to $\mathbb{Z}_3$.

2 Notation and Preliminary Theorems

We use boldface letters to refer to vectors, like: $\boldsymbol{x}$, $\boldsymbol{y}$. We work with functions with range $\mathbb{Z}_3$, and use the three Greek letters α, β, and γ to denote the three different outputs of the function. We take as convention that the three represent distinct outputs (so $\alpha \neq \beta \neq \gamma$). Sometimes we need to discuss an output as being not α, which we denote by $\not\alpha$.

In the proceeding discussion, we often need to discuss the behavior of a subfunction when keeping some subset of its arguments fixed. To simplify this discussion, we introduce some notation. For disjoint $S_1, S_2, S_3 \subseteq [n]$ we define

$$\begin{aligned} f^{\boldsymbol{a}}_{\{S_1\}}(\boldsymbol{x}) &\stackrel{\text{def}}{=} f(\{x_i\}_{i\in S_1}, \{a_i\}_{i\in S_1^C}) \\ f^{\boldsymbol{a}}_{\{S_1,S_2\}}(\boldsymbol{x},\boldsymbol{y}) &\stackrel{\text{def}}{=} f(\{x_i\}_{i\in S_1}, \{y_i\}_{i\in S_2}, \{a_i\}_{i\in(S_1\cup S_2)^C}) \\ f^{\boldsymbol{a}}_{\{S_1,S_2,S_3\}}(\boldsymbol{x},\boldsymbol{y},\boldsymbol{z}) &\stackrel{\text{def}}{=} f(\{x_i\}_{i\in S_1}, \{y_i\}_{i\in S_2}, \{z_i\}_{i\in S_3}, \{a_i\}_{i\in(S_1\cup S_2\cup S_3)^C}) . \end{aligned}$$

We sometimes consider singleton sets S_1, S_2, S_3 and then denote them simply by their only element, with some abuse of notation. That is,

$$\begin{aligned} f^{\boldsymbol{a}}_{\{i\}}(x) &\stackrel{\text{def}}{=} f(a_1,\ldots,a_{i-1},x,a_{i+1},\ldots,a_n) \\ f^{\boldsymbol{a}}_{\{i,j\}}(x,y) &\stackrel{\text{def}}{=} f(a_1,\ldots,a_{i-1},x,a_{i+1},\ldots,a_{j-1},y,a_{j+1},\ldots,a_n) , \end{aligned}$$

and analogously for $f^{\boldsymbol{a}}_{\{i,j,k\}}(x,y,z)$ and $f^{\boldsymbol{a}}_{\{i,j,k,l\}}(x,y,z,w)$.

We need to describe details of functions' behaviors, and adopt a geometric viewpoint. In the proofs, we speak of inputs as being neighbors and of rows, diagonals, and rectangles and induced rectangles in the function table. By neighbors we mean points at Hamming distance 1. By a row, we mean the values taken by the function fixing all but one values, i.e. the values $f^{\boldsymbol{a}}_{\{i\}}(x)$ for all $x \in A_1$ with a fixed i and $\boldsymbol{a}$ which are clear from the context. By a rectangle, we mean the values $f^{\boldsymbol{e}}_{\{S_1,S_2\}}(\boldsymbol{a},\boldsymbol{c}), f^{\boldsymbol{e}}_{\{S_1,S_2\}}(\boldsymbol{a},\boldsymbol{d}), f^{\boldsymbol{e}}_{\{S_1,S_2\}}(\boldsymbol{b},\boldsymbol{c}), f^{\boldsymbol{e}}_{\{S_1,S_2\}}(\boldsymbol{b},\boldsymbol{d})$. Note that a rectangle by this definition is a high-dimensional structure. By induced rectangle, we mean a rectangle as before but where $|S_1| = |S_2| = 1$, thus looking like a rectangle in the function table. We only use the concept of a diagonal of a 2×2 induced rectangle. For fixed inputs $\boldsymbol{a}$ and dimensions i, j we say that $f^{\boldsymbol{a}}_{\{i,j\}}(x_1,y_1), f^{\boldsymbol{a}}_{\{i,j\}}(x_2,y_2)$ is a diagonal for $x_1 \neq x_2$ and $y_1 \neq y_2$.

Definition 1 (Redundant inputs). *For an n-argument function f, we say that inputs $x, y, x \neq y$ are* redundant *for player k if for all $\boldsymbol{a}$ it holds that $f^{\boldsymbol{a}}_{\{k\}}(x) = f^{\boldsymbol{a}}_{\{k\}}(y)$.*

Definition 2 (Normalized function). *An n-argument function f with no redundant inputs for any player is said to be* normalized.

We take as convention that all functions are normalized. This assumption is without loss of generality as a function can easily be normalized by for each set of redundant inputs removing all but one. A protocol for evaluating the normalized function can be used to evaluate the original function as well by performing the same procedure.

To prove Theorem 1, we make use of a theorem by Chor and Kushilevitz [7] which states that there is no 1-private protocol for a 2-party computation of disjunction. Through standard simulation techniques, this gives impossibility results for multi-party protocols of functions containing an OR-like substructure. This is commonly referred to as an embedded OR, or a corner. We formally define an embedded OR and then restate their result. For a two-party function, the definition is straightforward:

Definition 3 (Embedded OR (2 parties)). *We say that a two-argument function f contains an* embedded OR *if there exists inputs x_1, x_2, y_1, y_2 ($x_1 \neq x_2, y_1 \neq y_2$) such that $f(x_1, y_1) = f(x_1, y_2) = f(x_2, y_1) \neq f(x_2, y_2)$.*

However, when considering the n-party case, the definition of an embedded OR becomes slightly more complex. In particular, we need our definition to capture the size of the collusion required to realize an embedded OR, as that size also limits the impossibility result that follows from the existence of such an embedded OR. To this end, we define an embedded OR as having a degree k. We remark that Kilian *et al.* [11] define an embedded OR as one of degree 1. Much of the previous literature has mostly been concerned with Boolean functions, and then, the existence of an embedded OR (of any degree) implies the existence of one of degree 1, as proved in [11]. However, for functions with larger ranges, the situation is more complex, as shown by our Theorem 22.

Definition 4 (Embedded OR (n parties, induced, generalized), corner-free). *We say that an n-argument function f contains an* embedded OR *of degree k if there exists disjoint subsets $S_1, S_2 \subset [n]$ where $|S_1|, |S_2| \leq k$, and values $\boldsymbol{a}$ such that the two-argument function $f'(\boldsymbol{x}, \boldsymbol{y}) = f^{\boldsymbol{a}}_{\{S_1,S_2\}}(\boldsymbol{x}, \boldsymbol{y})$ contains an embedded* OR. *We refer to an embedded* OR *of degree 1 as an* induced *embedded* OR, *and one of degree greater than 1 as a* generalized *embedded* OR. *A function without an embedded* OR *(of any degree) is said to be* corner-free.

With the definitions in place, we are ready to restate a result by Chor and Kushilevitz [7]. The result we need was not presented as a separate lemma in their paper, but instead follows as a corollary from two of their lemmas which we restate in simplified form.

Lemma 2 (Partition lemma, [7]). *Let $f : A_1 \times \ldots \times A_n \to R$ be $\lceil n/2 \rceil$-private. Then for every subset S_1 of size $\lceil n/2 \rceil$, the two-argument function $f'(\boldsymbol{x}, \boldsymbol{y}) = f_{\{S_1, S_1^C\}}(\boldsymbol{x}, \boldsymbol{y})$ is 1-private.*

Lemma 3 (Corners lemma, [7]). *A two-argument function is not 1-private if it contains an embedded* OR.

Corollary 4. *A function containing an embedded* OR *of degree at most $\lceil n/2 \rceil$ is not $\lceil n/2 \rceil$-private.*

We also make use of [7, Theorem 4] which states that a corner-free Boolean function can be expressed as a Boolean sum:

Theorem 5 ([7]). *For a corner-free Boolean function f there are functions f_i such that $f(x_1, \ldots, x_n) = \sum_{i=1}^{n} f_i(x_i)$ where the sum is computed modulo 2.*

We formally restate the theorem from [11] showing that a generalized embedded OR in a Boolean function implies an induced embedded OR. In our terminology:

Theorem 6 ([11]). *A Boolean function f containing an embedded* OR *contains an embedded* OR *of degree 1.*

We show functions and subfunctions which depend on up to 4 arguments. To be able to draw them, we show 2-dimensional projections separated by lines with vertical lines indicating a 3rd dimension and horizontal lines indicating a 4th dimension. We present sample function in Figure 1, showing a function which contains an embedded OR of degree 2 but does not contain an embedded OR of degree 1. The highlighted embedded OR occurs with the subsets $S_1 = \{P_1, P_3\}$ and $S_2 = \{P_2, P_4\}$ with inputs $(2, 1)$ and $(1, 2)$ for S_1 and $(1, 1)$ and $(2, 2)$ for S_2. As the function is drawn, the coalition S_1 in the embedded OR controls the horizontal position, and S_2 controls the vertical position.

0	**1**	**1**	2
1	0	2	1
2	0	0	1
0	**2**	**1**	0

Fig. 1. An example function containing an embedded OR of degree 2 (highlighted)

We use the following lemma which we believe is well-known. A proof is included in the full version of this paper [12].

Lemma 7. *If an n-argument function $f : A_1 \times \ldots \times A_n \to G$, where G is an Abelian group, has the property that for every pair of dimensions j, k and inputs $x_1, x_2, y_1, y_2, \boldsymbol{a}$ the following equality holds:*

$$f^{\boldsymbol{a}}_{\{j,k\}}(x_1, y_1) + f^{\boldsymbol{a}}_{\{j,k\}}(x_2, y_2) = f^{\boldsymbol{a}}_{\{j,k\}}(x_1, y_2) + f^{\boldsymbol{a}}_{\{j,k\}}(x_2, y_1), \qquad (1)$$

then f can be rewritten as $f(x_1, \ldots, x_n) = \sum_{i=1}^{n} f_i(x_i)$.

3 A Structure Lemma

The main step towards proving Theorem 1 is the establishment of a structure lemma for functions with range $\mathbb{Z}_3$. Thus, we turn toward some global properties of functions (as opposed to the comparatively local property of the existence of an embedded OR). The first such property captures the case when we can split the range of a function into two parts, and compute a Boolean sum to discover which part the output lies in. If we can then proceed with further such subdivisions until we arrive at a single possible output, this immediately gives a protocol to compute f. We prove that this further subdivision is always possible for corner-free f with range $\mathbb{Z}_3$ in Lemma 21. We remark that this is a further generalization of the multi-party decomposability defined in [13], which in turn was a generalization of 2-party decomposability defined in [14]. We show a collapsible function and the generalized decomposition of it in Figure 2.

Definition 5 (Collapsible). *We say that a function $f : A_1 \times \dots A_n \to R$ is* collapsible *if there is a subset $R', \emptyset \subset R' \subset R$ such that the Boolean function*

$$f'(\boldsymbol{x}) = \begin{cases} 1 & \textit{if } f(\boldsymbol{x}) \in R' \\ 0 & \textit{otherwise} \end{cases}$$

does not contain an embedded OR *and can thus be n-privately computed. We refer to f' as being collapsed.*

For a collapsible function f with range $\mathbb{Z}_3$ if f is collapsible we can choose R' with two elements α, β and say that f is collapsible by collapsing α and β.

0	1	2	2	2	0
1	0	2	2	2	1
2	2	0	1	0	2

(a) Collapsible f

1	1	0	0	0	1
1	1	0	0	0	1
0	0	1	1	1	0

(b) f collapsed

Fig. 2. An example collapsible function and the collapsed function

Summation in a finite Abelian group is a function which is known to be n-private [8]. In a summation, the effect of one party's input can be thought of as applying a permutation to the sum of the other parties' inputs. We generalize this by defining a permuted sum where we give one of the parties a special role and let her input select an arbitrary permutation to be applied to the sum of the other parties' inputs. All functions which are sums, i.e. can be rewritten as $\sum_{i=1}^{n} f_i(x_i)$, are also permuted sums. In our applications, the sum may be a Boolean sum or over $\mathbb{Z}_3$. We show two example functions which are permuted sums in Figure 3

Definition 6 (Permuted sum). *We say that a function is a* permuted sum *if it can be written as $\pi_{x_i}(\sum_{j \neq i} f_j(x_j))$ where π_x is a permutation. We refer to party i as the* permuter.

With these definitions, we are now ready to state and prove our structure lemma:

Lemma 8 (Structure lemma). *For every normalized n-argument function $f : A_1 \times \dots \times A_n \to \mathbb{Z}_3$, at least one of the following holds:*

- *f has an embedded* OR
- *f is a permuted sum*
- *f is collapsible*

0	0	1	1	2	2
1	2	0	2	0	1

(a) f

0	1	2	0	2	1	1	0	2
1	2	0	2	1	0	0	2	1
2	0	1	1	0	2	2	1	0

(b) g

Fig. 3. Two example permuted sums. In f, party 2 (selecting column) is the permuter selecting one of the 6 permutations. The function $g = \pi_{x_3}(x_1 + x_2)$ where π_1 is the identity permutation, $\pi_2 = (12)$ and $\pi_3 = (01)$.

We present protocols for n-privately evaluating permuted sums (Protocol 2) and collapsible functions (Protocol 3) which do not contain an embedded OR. In Theorem 22 we show that if f contains an embedded OR, it also contains a small embedded OR. This, together with Corollary 4 concludes the proof of our Theorem 1.

To prove the structure lemma, we perform a case-analysis based on a property of f we call a link:

Definition 7 (Link, link-free). *We say that an n-argument function has a* link (over output α) *in dimension k if there exists inputs $x, y, x \neq y$, and $\boldsymbol{a}$ such that $\alpha = f^{\boldsymbol{a}}_{\{k\}}(x) = f^{\boldsymbol{a}}_{\{k\}}(y)$. We say that f has* links in c dimensions *if there are precisely c distinct k such that f has a link in dimension k. We say that a function is* link-free *if it has no links.*

Lemma 9. *In a corner-free n-argument function $f : A_1 \times \ldots \times A_n \to \mathbb{Z}_3$, if there are links between inputs x and y in dimension k over two distinct outputs, then x and y are redundant for player k.*

Proof. Let f have links over α and β between inputs x and y in dimension k. That is, there exists values $\boldsymbol{a}, \boldsymbol{b}$ such that $f^{\boldsymbol{a}}_{\{k\}}(x) = f^{\boldsymbol{a}}_{\{k\}}(y) = \alpha$, and $f^{\boldsymbol{b}}_{\{k\}}(x) = f^{\boldsymbol{b}}_{\{k\}}(y) = \beta$. Suppose that for some $\boldsymbol{c}$ we have $f^{\boldsymbol{c}}_{\{k\}}(x) \neq f^{\boldsymbol{c}}_{\{k\}}(y)$. Then one of $f^{\boldsymbol{c}}_{\{k\}}(x)$ and $f^{\boldsymbol{c}}_{\{k\}}(y)$ equals α or β. If one of them is α then f has an embedded OR with $S_1 = \{k\}$, $S_2 = \{k\}^C$ using inputs (x, y) and $(\boldsymbol{a}, \boldsymbol{c})$. If one is β then f has an embedded OR with $S_1 = \{k\}$, $S_2 = \{k\}^C$ using inputs (x, y) and $(\boldsymbol{b}, \boldsymbol{c})$. □

Looking at the proof of Lemma 9 we begin to see the importance of the small range of the function to the analysis. It also highlights the added complexities compared to the Boolean case, as for a Boolean function any link implies that two inputs are redundant. From the lemma and its proof follow two corollaries about normalized functions with range $\mathbb{Z}_3$:

Corollary 10. *For a normalized n-argument function $f : A_1 \times \ldots \times A_n \to \mathbb{Z}_3$ with a link over α in dimension k for inputs x, y, for all $\boldsymbol{a}$, we have that $f^{\boldsymbol{a}}_{\{k\}}(x)$ uniquely determines $f^{\boldsymbol{a}}_{\{k\}}(y)$. More specifically, the possible combinations of values are $(\alpha, \alpha); (\beta, \gamma); (\gamma, \beta)$.*

Proof. Follows from the proof of Lemma 9. □

Corollary 11. *A normalized* n*-argument function* $f : A_1 \times \ldots \times A_n \to \mathbb{Z}_3$ *cannot have links over* α *in dimension* k *for inputs* x, y*, and* x, z*.*

Proof. By Corollary 10 the value at x determines the value at both y and z and hence inputs y and z are redundant. □

Analogously to an embedded OR, we introduce notation for the various 2×2 substructures in a function. Apart from the embedded OR, two of them feature prominently in our proofs. Firstly, a 2×2 substructure with one output occurring on the diagonal, and the two other values occurring once each on the opposite diagonal is called Aff_3. Secondly, a 2×2 substructure where one output is on one diagonal, and another is on the other is referred to as an XOR. For the XOR, we also define the type of an XOR as the pair (without order) of outputs in the XOR. All the substructures which can occur (up to symmetries) are depicted in Figure 4. A 2×2 substructure where only one output occurs is called constant, and if we want to emphasize that it is the output α which occurs, we write (α)-constant.

$\alpha\ \alpha$	$\alpha\ \alpha$	$\alpha\ \alpha$	$\alpha\ \alpha$	$\alpha\ \beta$	$\alpha\ \beta$
$\alpha\ \alpha$	$\alpha\ \beta$	$\beta\ \beta$	$\beta\ \gamma$	$\beta\ \alpha$	$\gamma\ \alpha$
(a) Constant	(b) OR	(c) 2-link	(d) Link	(e) XOR	(f) Aff_3

Fig. 4. The six 2×2 substructures

Definition 8 (Type of an XOR). *If an* XOR *consists of outputs* α *and* β *we say that it is an* XOR *of type* (α, β)*, denoted* (α, β)-XOR*. The order of elements is not important, so for functions to* $\mathbb{Z}_3$ *there are three possible types of* XOR*:* (α, β)*,* (α, γ)*,* (β, γ)*.*

Our name Aff_3 comes from the fact that it can be expressed as an affine function modulo 3, analogously to the fact that XOR can be expressed as a sum modulo 2. We do not need that it is affine, but we make use of the fact that a function where all subfunctions are of the form Aff_3 can be written as a sum on the form $\sum_{i=1}^{n} f_i(x_i)$ with summation in $\mathbb{Z}_3$.

Lemma 12. *An* n*-argument corner-free function* $f : A_1 \times \ldots \times A_n \to \mathbb{Z}_3$ *such that all* 2×2 *subfunctions are of the form* Aff_3 *can be expressed as* $\sum_{i=1}^{n} f_i(x_i)$ *with summation in* $\mathbb{Z}_3$*.*

Proof. By Lemma 7 we need to verify that (1) holds for all 2×2 subfunctions, which are all of the form Aff_3. For all ways of assigning 0, 1 and 2 (distinctly) to α, β, γ we have that $2\alpha \equiv \beta + \gamma \pmod 3$. As $2 \equiv -1 \pmod 3$ this is equivalent to $\alpha + \beta + \gamma \equiv 0 \pmod 3$. □

In our proof of Lemma 8 we consider the substructures occurring in f. We begin by establishing three preliminary lemmas. The lemmas come into play primarily in cases when f contains links in few dimensions (none or one), and if f has an XOR spanned by dimensions i, j and f is link-free in those dimensions, then $|A_i| = |A_j| = 2$, giving some intuition for the condition on the size of the two inputs in the lemmas. We highlight the proof idea for each of the lemmas and give full proofs in the appendix.

Lemma 13. *Let f be an n-argument corner-free function $f : A_1 \times \ldots \times A_n \to \mathbb{Z}_3$ with i, j such that $|A_i| = |A_j| = 2$ such that for all $\boldsymbol{a}$, $f^{\boldsymbol{a}}_{\{i,j\}}$ is an* XOR. *If all three types of* XOR*'s occur then there is a dimension k such that the input in dimension k determines the type of* XOR.

Proof (Idea). We show that if no such k exists then f contains an embedded OR. Full proof in Section A.1. □

Lemma 14. *Let f be an n-argument corner-free function $f : A_1 \times \ldots \times A_n \to \mathbb{Z}_3$ with i, j such that $|A_i| = |A_j| = 2$ and an output α such that for all $\boldsymbol{a}$ precisely one diagonal of $f^{\boldsymbol{a}}_{\{i,j\}}$ has two α's. Then f is collapsible.*

Proof (Idea). If f is not collapsible then the collapsed function contains an embedded OR. We show that this implies an embedded OR in f as well. Full proof in Section A.2. □

Lemma 15. *An n-argument corner-free function $f : A_1 \times \ldots \times A_n \to \mathbb{Z}_3$ with i, j such that $|A_i| = |A_j| = 2$ and such that for some $\boldsymbol{a}$, $f^{\boldsymbol{a}}_{\{i,j\}}$ is an* Aff_3 *and for some $\boldsymbol{b}$, $f^{\boldsymbol{b}}_{\{i,j\}}$ is an* XOR *is collapsible.*

Proof (Idea). We prove that f fulfills the conditions of Lemma 14. Full proof in Section A.3. □

Our proof of Lemma 8 proceeds in three separate lemmas, depending on whether the function f is link-free (Lemma 16), has links in one dimension (Lemma 17), or if it has links in two or more dimensions (Lemma 18). As the proofs are long and consist mainly of case analysis, we simply state the lemmas here. The proof of the first lemma is given in the appendix, and the two others in the full version of this paper [12].

Lemma 16. *Every n-argument link-free, corner-free function $f : A_1 \times \ldots \times A_n \to \mathbb{Z}_3$ is collapsible or a permuted sum.*

Proof (Idea). Case analysis showing we can apply one of Lemma 13, Lemma 14 and Lemma 15. Full proof in Section A.4 □

Lemma 17. *Every n-argument function $f : A_1 \times \ldots \times A_n \to \mathbb{Z}_3$ with links in 1 dimension and without an embedded* OR *is collapsible or a permuted sum.*

Proof (Idea). Case analysis showing we can apply one of Lemma 13, Lemma 14 and Lemma 15. Proof in the full version of this paper [12]. □

Lemma 18. *Every n-argument function $f : A_1 \times \ldots \times A_n \to \mathbb{Z}_3$ with links in 2 or more dimensions and without an embedded* OR *is collapsible.*

Proof (Idea). We show that all links must be over the same output. This gives some implications for the substructures of f which we use to show f must be collapsible. Proof in the full version of this paper [12]. □

4 Protocols

With the structure lemma established, we can now turn to the question of n-private protocols for collapsible functions and permuted sums. From the definitions of the two classes, we have two natural and easy protocols. The main problem we need to address in this section is proving the existence of a protocol for collapsible functions. For a function which is collapsible by collapsing β and γ it is clear from the definition that we can n-privately evaluate if the output is α or if it is one of β and γ. The key issue is to prove that we can then proceed with a second step where we can n-privately evaluate whether the output is β or if it is γ.

The construction of this second step relies on the passive model of adversaries and the knowledge that the output of the function is not α. Thus, in our second step we compute a sum which may have different outputs at points where the original function had α's. Such a construction is inherently insecure with active adversaries, as they may switch inputs between the first step of the decomposition and the second and would then learn some information about the other parties' inputs.

In both of our protocols we use a subprotocol by Chor and Kushilevitz [8] for n-private summation over any finite Abelian group. For completeness, we include a description of their protocol as Protocol 1. When used in our protocol for a permuted sum, the summation is either Boolean or in $\mathbb{Z}_3$ depending on the function f (but not on the inputs).

Protocol 1 (Summation [8]). The protocol for summation where party P_i participates with input x_i proceeds as follows:

1. In round $1 \le i \le n-2$, party P_i sums all its received messages, $w_i = \sum_{j=1}^{i-1} z_{j,i}$. Then, it chooses random group elements $z_{i,i+1}, z_{i,i+2}, \ldots, z_{i,n-1}$. Finally, it computes $z_{i,n}$ such that $x_i + w_i = \sum_{j=i+1}^{n} z_{i,j}$ and sends $z_{i,j}$ to P_j $(j > i)$.
2. In round $n-1$, party P_{n-1} computes $z_{n-1,n} = x_{n-1} + \sum_{j=1}^{n-2} z_{j,n-1}$ and sends $z_{n-1,n}$ to P_n.
3. In round n, party P_n computes the sum s as $s = x_n + \sum_{j=1}^{n-1} z_{j,n}$.

All sums are computed over some fixed finite Abelian group.

Protocol 2 (Permuted sum). The protocol for evaluating a permuted sum f, where party P_i (without loss of generality we assume the permuter is party n) participates with input x_i proceeds as follows:

1. Use Protocol 1 to privately compute $s = \sum_{j=1}^{n-1} f_j(x_j)$ such that only the permuter learns s.
2. The permuter computes the output as $\pi_{x_n}(s)$ and sends it to the other parties.

The sum is computed modulo 2, or 3, depending on f.

Protocol 3 (Collapsible). The protocol for evaluating a function f collapsible with partition $R' = \{\gamma\}$, where party P_i participates with input x_i proceeds as follows:

1. Use Protocol 1 to compute $s = \sum_{i=1}^{n} f_i(x_i) \pmod 2$, with f_i such that $s = 1$ iff $f(\boldsymbol{x}) = \gamma$
2. If $s = 0$, compute $s' = \sum_{i=1}^{n} g_i(x_i) \pmod 4$, with g_i such that $f(\boldsymbol{x}) = \alpha$ implies $s' = 0$, and $f(\boldsymbol{x}) = \beta$ implies $s' = 2$.

The correctness of Protocol 2 follows immediately from the definition of a permuted sum. In Protocol 3, since f is collapsible, the functions f_i exist by the definition of a collapsible function. However, the existence of appropriate g_i is not as straightforward. We prove, constructively, in Lemma 21 that they always exist for corner-free collapsible functions with range $\mathbb{Z}_3$. We stress that the choice of g_i does not depend on the input $\boldsymbol{x}$, but only on the function f.

The privacy of both these protocols is straightforward, and we only sketch the arguments.

Theorem 19. *Protocol 2 is n-private.*

Proof. The subprotocol used for summation was proven to be n-private in [8]. Due to the structure of the function, we see that the permuter, P_n, learns the sum s from $f(\boldsymbol{x})$ and x_n, since $s = \pi_{x_n}^{-1}(f(\boldsymbol{x}))$. □

Theorem 20. *Protocol 3 is n-private.*

Proof. The subprotocol used for summation was proven to be n-private in [8]. When the output is γ then, by the privacy of the summation sub-protocol, the protocol is private. Furthermore, when the output is one of α, β, then the privacy of the composed protocol also follow directly from the privacy of the subprotocols. The first sum only reveals that the output is one of α, β, and then, the condition on g_i is sufficient to guarantee that the sum s' reveals nothing but whether the output is α or β, as with a passive adversary we are guaranteed that s' is either 0 or 2. □

While the privacy is straightforward, the proof that there are functions g_i as required by Protocol 3 is rather involved and we simply state the lemma here and give the proof in [12]. One may intuitively expect that such functions could simply be Boolean, but it turns out that for some f we do need the full range of $\mathbb{Z}_4$.

Lemma 21. *Protocol 3 can evaluate all corner-free, collapsible functions with range $\mathbb{Z}_3$.*

Proof (Idea). We construct a function g such that $f(\boldsymbol{x}) = \alpha \implies g(\boldsymbol{x}) = 0$ and $f(\boldsymbol{x}) = \beta \implies g(\boldsymbol{x}) = 2$. By case analysis on the induced rectangles in g, we show that g satisfies the conditions of Lemma 7 and hence there are g_i as required by Protocol 3. Proof in the full version of this paper [12]. □

5 An Embedded OR Implies a Small Embedded OR

Previously, we have often assumed that functions are free of embedded OR's of *any* degree (i.e., that they are corner-free). However, to be able to apply Corollary 4 we need to show that a sufficiently small embedded OR exists.

For Boolean functions f, if f has an embedded OR of any degree, then it also has an embedded OR of degree 1, as proved in [11], explaining the zero-one nature of Boolean privacy.

It turns out that for functions with range $\mathbb{Z}_3$, similarly to the Boolean case, the presence of a large embedded OR implies that the function also contains a small one. We state the theorem here and give the proof in [12].

Theorem 22. *Every n-argument function $f : A_1 \times \ldots A_n \to \mathbb{Z}_3$ that has an embedded* OR *of any degree has an embedded* OR *of degree at most* 3*. Furthermore, every* 4*-argument function $f : A_1 \times A_2 \times A_3 \times A_4 \to \mathbb{Z}_3$ that has an embedded* OR*, also has one of degree at most* 2*.*

Proof (Idea). The basic idea is similar to that used in the proof of Theorem 6. However, while the boolean case is fairly straightforward, our proof results in a fairly extensive case analysis. Proof in the full version of this paper [12]. □

6 Proof of the Main Theorem

We now conclude by re-stating our main theorem and presenting the proof.

Theorem 23 (Main theorem). *For every n-argument function $f : A_1 \times \ldots \times A_n \to \mathbb{Z}_3$, f is either n-private, or it is $\lfloor (n-1)/2 \rfloor$-private and not $\lceil n/2 \rceil$-private.*

Proof. If f is corner-free, then by Lemma 8 it is a permuted sum, collapsible, or both. Thus, it can be n-privately computed by Protocol 2 or Protocol 3.

If f is not corner-free, then by Theorem 22 it contains an embedded OR of degree at most $\lceil n/2 \rceil$. Thus, by Corollary 4, f is not $\lceil n/2 \rceil$-private. □

Acknowledgements

I would like to thank Johan Håstad for his many helpful insights, Dominik Raub for introducing me to this problem and for the interesting discussions, and Per Austrin for Protocol 3.

References

1. Beaver, D.: Perfect privacy for two-party protocols. In: Feigenbaum, J., Merritt, M. (eds.) Proceedings of DIMACS Workshop on Distributed Computing and Cryptology, vol. 2, pp. 65–77. American Mathematical Society, Providence (1989)
2. Beerliová-Trubíniová, Z., Fitzi, M., Hirt, M., Maurer, U.M., Zikas, V.: MPC vs. SFE: Perfect security in a unified corruption model. In: Canetti, R. (ed.) TCC 2008. LNCS, vol. 4948, pp. 231–250. Springer, Heidelberg (2008)

3. Ben-Or, M., Goldwasser, S., Wigderson, A.: Completeness theorems for non-cryptographic fault-tolerant distributed computation (extended abstract). In: STOC, pp. 1–10. ACM, New York (1988)
4. Chaum, D., Crépeau, C., Damgård, I.: Multiparty unconditionally secure protocols (extended abstract). In: STOC, pp. 11–19. ACM, New York (1988)
5. Chor, B., Geréb-Graus, M., Kushilevitz, E.: On the structure of the privacy hierarchy. J. Cryptology 7(1), 53–60 (1994)
6. Chor, B., Ishai, Y.: On privacy and partition arguments. Inf. Comput. 167(1), 2–9 (2001)
7. Chor, B., Kushilevitz, E.: A zero-one law for boolean privacy. SIAM J. Discrete Math. 4(1), 36–47 (1991)
8. Chor, B., Kushilevitz, E.: A communication-privacy tradeoff for modular addition. Inf. Process. Lett. 45(4), 205–210 (1993)
9. Chor, B., Shani, N.: The privacy of dense symmetric functions. Computational Complexity 5(1), 43–59 (1995)
10. Hirt, M., Maurer, U.M.: Player simulation and general adversary structures in perfect multiparty computation. J. Cryptology 13(1), 31–60 (2000)
11. Kilian, J., Kushilevitz, E., Micali, S., Ostrovsky, R.: Reducibility and completeness in private computations. SIAM J. Comput. 29(4), 1189–1208 (2000)
12. Kreitz, G.: A zero-one law for secure multi-party computation with ternary outputs (full version), Cryptology ePrint Archive, Report 2011/002 (2011), `http://www.eprint.iacr.org/`
13. Künzler, R., Müller-Quade, J., Raub, D.: Secure Computability of Functions in the IT Setting with Dishonest Majority and Applications to Long-Term Security. In: Reingold, O. (ed.) TCC 2009. LNCS, vol. 5444, pp. 238–255. Springer, Heidelberg (2009)
14. Kushilevitz, E.: Privacy and communication complexity. SIAM J. Discrete Math. 5(2), 273–284 (1992)
15. Yao, A.C.-C.: Protocols for secure computations (extended abstract). In: FOCS, pp. 160–164. IEEE, Los Alamitos (1982)

A Proofs

A.1 Proof of Lemma 13

Lemma 24. *Let f be an n-argument corner-free function $f : A_1 \times \ldots \times A_n \to \mathbb{Z}_3$ with i, j such that $|A_i| = |A_j| = 2$ such that for all $\boldsymbol{a}$, $f^{\boldsymbol{a}}_{\{i,j\}}$ is an* XOR. *If all three types of* XOR*'s occur then there is a dimension k such that the input in dimension k determines the type of* XOR.

Proof. We assume that there is an (α, β)-XOR and an (α, γ)-XOR at Hamming distance 1. We denote the dimension by which they differ by k and relabel the inputs in dimension k such that $f^{\boldsymbol{a}}_{\{i,j,k\}}(\cdot, \cdot, 1)$ is an (α, β)-XOR and $f^{\boldsymbol{a}}_{\{i,j,k\}}(\cdot, \cdot, 2)$ is an (α, γ)-XOR.

We proceed to show that there is no $\boldsymbol{b}$ such that $f^{\boldsymbol{b}}_{\{i,j,k\}}(\cdot, \cdot, 1)$ or $f^{\boldsymbol{b}}_{\{i,j,k\}}(\cdot, \cdot, 2)$ is a (β, γ)-XOR. Assume to the contrary that there is a $\boldsymbol{b}$ such that $f^{\boldsymbol{b}}_{\{i,j,k\}}(\cdot, \cdot, 1)$ is a (β, γ)-OR (the case if it occurs at input 2 in dimension k is analogous). We

α	β	α	γ
β	α	γ	α
γ	β		
β	γ		

Fig. 5. Illustration of a contradiction in the proof of Lemma 13

illustrate this case in Figure 5 where we for simplicity show $\boldsymbol{b}$ as differing from $\boldsymbol{a}$ in only one dimension, which is not something we assume in the proof.

What values can the function take at $f^{\boldsymbol{b}}_{\{i,j,k\}}(1,1,2)$? We claim that any output at that position would violate the assumption that f is corner-free. In Figure 5 we can see why this is true for a simple function (writing α, β or γ anywhere in the missing 2×2 field can be verified to result in an embedded OR). For brevity, we only discuss the case when $f^{\boldsymbol{b}}_{\{i,j,k\}}(1,1,2) = \alpha$ here (the other cases are almost identical and the core idea is captured by Figure 5). Proofs of all three cases are given in the full version of this paper [12].

Assume $f^{\boldsymbol{b}}_{\{i,j,k\}}(1,1,2) = \alpha$. Then we can find y (equal to 1 or 2) such that $f^{\boldsymbol{a}}_{\{i,j,k\}}(1,y,2) = \alpha$ as $f^{\boldsymbol{a}}_{\{i,j,k\}}(\cdot,\cdot,2)$ is an (α,γ)-XOR. We can also find x such that $f^{\boldsymbol{a}}_{\{i,j,k\}}(x,y,1) = \alpha$ as $f^{\boldsymbol{a}}_{\{i,j,k\}}(\cdot,\cdot,1)$ is an (α,β)-XOR. However, as $f^{\boldsymbol{b}}_{\{i,j,k\}}(\cdot,\cdot,1)$ is a (β,γ)-XOR we are guaranteed that $f^{\boldsymbol{b}}_{\{i,j,k\}}(x,1,1) \neq \alpha$. Thus, f contains an embedded OR with $S_1 = \{i,k\}$ and $S_2 = S_1^C$ using $(1,2);(x,1)$ on S_1 and $y;1$ or j and $\boldsymbol{a};\boldsymbol{b}$ on the rest of S_2.

We now conclude that there is no $\boldsymbol{b}$ such that $f^{\boldsymbol{b}}_{\{i,j,k\}}(\cdot,\cdot,1)$ or $f^{\boldsymbol{b}}_{\{i,j,k\}}(\cdot,\cdot,2)$ is a (β,γ)-XOR. As f has all types of XOR's, there must still be a (β,γ)-XOR in the function. Thus, we see that $|A_k| \geq 3$, and we can assume there is a $\boldsymbol{b}$ such that $f^{\boldsymbol{b}}_{\{i,j,k\}}(\cdot,\cdot,3)$ is a (β,γ)-XOR.

We claim that $f^{\boldsymbol{a}}_{\{i,j,k\}}(\cdot,\cdot,3)$ must be a (β,γ)-XOR. To see this we observe that if it was another type of XOR, then by the same proof that showed that there is no (β,γ)-XOR for $k = 1,2$ we could have shown that there was not (β,γ)-XOR for $k = 3$, but we know that $f^{\boldsymbol{b}}_{\{i,j,k\}}(\cdot,\cdot,3)$ is a (β,γ)-XOR. We now see that for a given x_k, all XOR's must be of the same type as that at $f^{\boldsymbol{a}}_{\{i,j,x_k\}}(\cdot,\cdot,k)$ which concludes our proof. □

A.2 Proof of Lemma 14

Lemma 25. *Let f be an n-argument corner-free function $f : A_1 \times \ldots \times A_n \to \mathbb{Z}_3$ with i,j such that $|A_i| = |A_j| = 2$ and an output α such that for all $\boldsymbol{a}$ precisely one diagonal of $f^{\boldsymbol{a}}_{\{i,j\}}$ has two α's. Then f is collapsible.*

Proof. We claim that f is collapsible by collapsing β and γ. To prove this we show that the collapsed function

$$g(\boldsymbol{x}) = \begin{cases} 1 & \text{if } f(\boldsymbol{x}) \in \{\beta,\gamma\} \\ 0 & \text{if } f(\boldsymbol{x}) = \alpha \end{cases}$$

does not contain an embedded OR of degree 1. Then by Theorem 6 we have that g is corner-free and Theorem 5 implies that the collapsed function can be written as a Boolean sum.

We begin by observing that as each 2×2 plane spanned by dimensions i, j contains exactly one diagonal with α's, each such 2×2 plane contains two α's, as if it had three α's it would be an embedded OR and if it had four α's both diagonals would have two α's. We further make the observation that a pair of neighboring outputs in dimension i (and analogously in j) are such that exactly one of them is α. More formally, for all $\boldsymbol{c}$ if $f^{\boldsymbol{c}}_{\{i\}}(1) = \alpha$ then $f^{\boldsymbol{c}}_{\{i\}}(2) \neq \alpha$ and if $f^{\boldsymbol{c}}_{\{i\}}(1) \neq \alpha$ then $f^{\boldsymbol{c}}_{\{i\}}(2) = \alpha$.

As g is Boolean, by Theorem 6 we know that if g has an embedded OR (of any degree), it also has an embedded OR of degree 1. We assume by contradiction that there is an embedded OR of degree 1 in g. We reorder inputs and dimensions such that the embedded OR is spanned by dimensions $1, 2$ using inputs $(1, 2); (1, 2)$, with other inputs as $\boldsymbol{a}$. We say that $g^{\boldsymbol{a}}_{\{1,2\}}$ is an embedded OR with slight abuse of notation (as $|A_1|$ or $|A_2|$ could be greater than 2). We see that the embedded OR cannot have three 1's as g takes the value 0 where f takes the value α, so an embedded OR with three 0's corresponds to an embedded OR with three α in f, which is corner-free. Thus, the embedded OR must have three 1's.

From our observation we know that each 2×2 plane in g spanned by i, j has two 0's, so there cannot be an OR in g with three 1's spanned by dimensions i, j. Thus, at least one of i and j must be different from both 1 and 2. We assume $i \neq 1, 2$ and reorder inputs such that the embedded OR occurs when $x_i = 1$. Let $\boldsymbol{b}$ be $\boldsymbol{a}$ with the value at x_i removed.

We now consider what values occur at $f^{\boldsymbol{b}}_{\{1,2,i\}}(\cdot, \cdot, 2)$. We know that of the four outputs of $f^{\boldsymbol{b}}_{\{1,2,i\}}(\cdot, \cdot, 1)$ one is α and three are different from α. But by our observation, this implies that of the four outputs of $f^{\boldsymbol{b}}_{\{1,2,i\}}(\cdot, \cdot, 2)$ three are α and one is different from α. This concludes our proof as it shows that an embedded OR in g implies an embedded OR in f which we assumed to be corner-free. □

A.3 Proof of Lemma 15

Lemma 26. *An n-argument corner-free function $f : A_1 \times \ldots \times A_n \to \mathbb{Z}_3$ with i, j such that $|A_i| = |A_j| = 2$ and such that for some $\boldsymbol{a}$, $f^{\boldsymbol{a}}_{\{i,j\}}$ is an* Aff$_3$ *and for some $\boldsymbol{b}$, $f^{\boldsymbol{b}}_{\{i,j\}}$ is an* XOR *is collapsible.*

Proof. Let the output that appears twice in $f^{\boldsymbol{a}}_{\{i,j\}}$ be α, and reorder inputs such that $f^{\boldsymbol{a}}_{\{i,j\}}(1, 1) = f^{\boldsymbol{a}}_{\{i,j\}}(2, 2) = \alpha$.

We now claim that $f^{\boldsymbol{b}}_{\{i,j\}}(1, 2) = f^{\boldsymbol{b}}_{\{i,j\}}(2, 1) = \alpha$. As $f^{\boldsymbol{b}}_{\{i,j\}}$ is an XOR we know that $f^{\boldsymbol{b}}_{\{i,j\}}(1, 2) = f^{\boldsymbol{b}}_{\{i,j\}}(2, 1)$. If $f^{\boldsymbol{b}}_{\{i,j\}}(1, 2) = f^{\boldsymbol{b}}_{\{i,j\}}(2, 1) \in \{\beta, \gamma\}$ then there is an embedded OR with $S_1 = \{i, j\}$ and $S_2 = S_1^C$ as using inputs $(1, 2); (2, 1)$ on S_1 and $\boldsymbol{a}; \boldsymbol{b}$ on S_2. We assume the other diagonal of the XOR consists of β's, i.e. $f^{\boldsymbol{b}}_{\{i,j\}}(1, 1) = f^{\boldsymbol{b}}_{\{i,j\}}(2, 2) = \beta$, and that $f^{\boldsymbol{a}}_{\{i,j\}}(1, 2) = \beta$, $f^{\boldsymbol{a}}_{\{i,j\}}(2, 1) = \gamma$. This is without loss of generality as we can relabel outputs and switch the roles of parties 1 and 2.

$$\begin{array}{cc} \boldsymbol{\alpha}\ \beta & \beta\ \boldsymbol{\alpha} \\ \gamma\ \alpha & \alpha\ \boldsymbol{\beta} \\ \text{(a) } f^{\boldsymbol{a}}_{\{i,j\}} & \text{(b) } f^{\boldsymbol{b}}_{\{i,j\}} \end{array}$$

Fig. 6. An XOR and Aff_3 in f. The outputs involved in proving that f has no links in dimension i are highlighted.

We claim that the function f cannot have any links in dimensions i or j. To see this for dimension i, we see that $f^{\boldsymbol{a}}_{\{i,j\}}(1,1) = \alpha$ and $f^{\boldsymbol{a}}_{\{i,j\}}(2,1) = \gamma$ but also $f^{\boldsymbol{b}}_{\{i,j\}}(1,2) = \alpha$ and $f^{\boldsymbol{b}}_{\{i,j\}}(2,2) = \beta$. Thus, the value of $f^{\boldsymbol{c}}_{\{i\}}(2)$ is not a function of the value of $f^{\boldsymbol{c}}_{\{i\}}(1)$ for all $\boldsymbol{c}$ and the contrapositive form of Corollary 10 gives that f cannot have a link between inputs 1 and 2 in dimension i. Similarly for dimension j, we have that $f^{\boldsymbol{a}}_{\{i,j\}}(2,2) = \alpha$ and $f^{\boldsymbol{a}}_{\{i,j\}}(2,1) = \gamma$, but also $f^{\boldsymbol{b}}_{\{i,j\}}(1,2) = \alpha$ and $f^{\boldsymbol{b}}_{\{i,j\}}(1,1) = \beta$. This demonstrates that $f^{\boldsymbol{c}}_{\{j\}}(1)$ is not a function of $f^{\boldsymbol{c}}_{\{j\}}(2)$ for all $\boldsymbol{c}$, and by the contrapositive form of Corollary 10, there is no link between inputs 2 and 1 in dimension j.

We proceed by proving that for all $\boldsymbol{c}$ precisely one of the two diagonals of $f^{\boldsymbol{c}}_{\{i,j\}}$ contains two α's. What are the possible values for $(f^{\boldsymbol{c}}_{\{i,j\}}(1,2), f^{\boldsymbol{c}}_{\{i,j\}}(2,1))$? We proved (when $\boldsymbol{c} = \boldsymbol{b}$ but we made no use of any properties of $\boldsymbol{b}$) that they cannot be (β,β) or (γ,γ). Furthermore, as $f^{\boldsymbol{b}}_{\{i,j\}}(1,2) = f^{\boldsymbol{b}}_{\{i,j\}}(2,1) = \alpha$ it cannot be that precisely one of the values is α, as then f would have an embedded OR with $S_1 = \{i,j\}$ and $S_2 = S_1^C$ using inputs $(1,2);(2,1)$ on S_1 and $\boldsymbol{b};\boldsymbol{c}$ on S_2. Thus the only remaining possibilities are (α,α); (β,γ); (γ,β). As f has no links in dimension i or j we see that in the first case neither $f^{\boldsymbol{c}}_{\{i,j\}}(1,1)$ nor $f^{\boldsymbol{c}}_{\{i,j\}}(2,2)$ can equal α. In the two latter cases we have again by the link-freeness in dimensions i and j that $f^{\boldsymbol{c}}_{\{i,j\}}(1,1) = f^{\boldsymbol{c}}_{\{i,j\}}(2,2) = \alpha$. By Lemma 14 we have that f is collapsible as claimed. □

A.4 Proof of Lemma 16

Lemma 27. *Every n-argument link-free, corner-free function $f : A_1 \times \ldots \times A_n \to \mathbb{Z}_3$ is collapsible or a permuted sum.*

Proof. For a link-free and corner-free function, only two types of induced rectangles are possible: XOR and Aff_3. If f contains an XOR spanned by dimensions i and j, then $|A_i| = |A_j| = 2$ since f is link-free.

We proceed with a case analysis. If f does not contain an XOR, then we select the first case. Otherwise, we pick an arbitrary XOR occurring in f and fix the dimensions i and j spanning it, and denote by $\boldsymbol{a}$ a set of inputs such that $f^{\boldsymbol{a}}_{\{i,j\}}$ is an (α,β)-XOR (if f has an XOR, we can relabel outputs such that there is an (α,β)-XOR). When we have fixed dimensions i, j we select among the four last cases of our proof based *only* on the 2×2-planes spanned by dimensions i, j.

Case 1: Only Aff_3 (all dimensions). If all induced rectangles of f are of the form Aff_3, then f satisfies the condition of Lemma 12 and is a (permuted) sum.

Case 2: Both XOR and Aff_3 (spanned by i, j). By Lemma 15, f is collapsible.

Case 3: Only XOR, one type of XOR (spanned by i, j). If only one type of XOR's occur, then f is a Boolean corner-free function and by Theorem 5 we have that f is a sum, and thus also a permuted sum.

Case 4: Only XOR, two types of XOR (spanned by i, j). We assume that f contains XOR's of types (α, β) and (α, γ). Then α occurs on exactly one diagonal of all 2×2 planes spanned by dimensions i, j and by Lemma 14 f is collapsible by collapsing β and γ.

Case 5: Only XOR, three types of XOR (spanned by i, j). By Lemma 13 we see that there must be a dimension k such that the input in dimension k determines the type of the XOR. Reorder inputs such that for input 1 in dimension k the 2×2-planes spanned by i and j are (α, β)-XOR's. We let $\boldsymbol{a} = (1)$ and $S_1 = \{k\}^C$ and see that $f^{\boldsymbol{a}}_{\{S_1\}}$ is a Boolean corner-free function. Thus, Theorem 5 implies that $f^{\boldsymbol{a}}_{\{S_1\}}(x_1, \ldots, x_{k-1}, x_{k+1}, \ldots, x_n) = \sum_{i \neq k} f_i(x_i)$ with the sum computed modulo 2.

We claim that f is a permuted sum with P_k as the permuter and the sum computed modulo 2. To see this, we prove that for all $x_k \in A_k$ and for all inputs $\boldsymbol{b}$ we have $f^{\boldsymbol{b}}_{\{k\}}(x_k) = \pi_{x_k}\{f^{\boldsymbol{b}}_{\{k\}}(1)\}$. As f is link-free, we have that $f^{\boldsymbol{b}}_{\{k\}}(x_k) \neq f^{\boldsymbol{b}}_{\{k\}}(1)$. If x_k is such that the 2×2-planes spanned by dimensions 1 and 2 when the input in dimension k is x_k are (α, β)-XOR then this means that $f^{\boldsymbol{b}}_{\{k\}}(1) = \alpha \implies f^{\boldsymbol{b}}_{\{k\}}(x_k) = \beta$ and $f^{\boldsymbol{b}}_{\{k\}}(1) = \beta \implies f^{\boldsymbol{b}}_{\{k\}}(x_k) = \alpha$. Similarly if the XOR's are (α, γ)-XOR's $f^{\boldsymbol{b}}_{\{k\}}(1) = \alpha \implies f^{\boldsymbol{b}}_{\{k\}}(x_k) = \gamma$, and as the 2×2-planes are XOR's we have $f^{\boldsymbol{b}}_{\{k\}}(1) \neq \alpha \implies f^{\boldsymbol{b}}_{\{k\}}(x_k) = \alpha$. The case for x_k with (β, γ)-XOR's is analogous, concluding the proof. □

PCPs and the Hardness of Generating Private Synthetic Data*

Jonathan Ullman** and Salil Vadhan***

School of Engineering and Applied Sciences &
Center for Research on Computation and Society
Harvard University, Cambridge, MA
{jullman,salil}@seas.harvard.edu

Abstract. Assuming the existence of one-way functions, we show that there is no polynomial-time, differentially private algorithm $\mathcal{A}$ that takes a database $D \in (\{0,1\}^d)^n$ and outputs a "synthetic database" $\widehat{D}$ all of whose two-way marginals are approximately equal to those of D. (A two-way marginal is the fraction of database rows $x \in \{0,1\}^d$ with a given pair of values in a given pair of columns). This answers a question of Barak et al. (PODS '07), who gave an algorithm running in time $\mathrm{poly}(n, 2^d)$.

Our proof combines a construction of hard-to-sanitize databases based on digital signatures (by Dwork et al., STOC '09) with encodings based on probabilistically checkable proofs.

We also present both negative and positive results for generating "relaxed" synthetic data, where the fraction of rows in D satisfying a predicate c are estimated by applying c to each row of $\widehat{D}$ and aggregating the results in some way.

Keywords: privacy, digital signatures, inapproximability, constraint satisfaction problems, probabilistically checkable proofs.

1 Introduction

There are many settings in which it is desirable to share information about a database that contains sensitive information about individuals. For example, doctors may want to share information about health records with medical researchers, the federal government may want to release census data for public information, and a company like Netflix may want to provide its movie rental database for a public competition to develop a better recommendation system. However, it is important to do this in way that preserves the "privacy" of the individuals whose records are in the database. This privacy problem has been studied by statisticians and the database security community for a number of years (cf., [1,8,15]), and recently the theoretical computer science community has developed an appealing new approach to the problem, known as *differential privacy*. (See the surveys [10,9]).

---

* A full version of this paper appears on ECCC [28].

** `http://seas.harvard.edu/~jullman`. Supported by NSF grant CNS-0831289.

*** `http://seas.harvard.edu/~salil`. Supported by NSF grant CNS-0831289.

Y. Ishai (Ed.): TCC 2011, LNCS 6597, pp. 400–416, 2011.

Differential Privacy. A randomized algorithm $\mathcal{A}$ is defined to be *differentially private* [11] if for every two databases $D = (x_1, \ldots, x_n)$, $D' = (x'_1, \ldots, x'_n)$ that differ on exactly one row, the distributions $\mathcal{A}(D)$ and $\mathcal{A}(D')$ are "close" to each other. Formally, we require that $\mathcal{A}(D)$ and $\mathcal{A}(D')$ assign the same probability mass to every event, up to a multiplicative factor of $e^\epsilon \approx 1 + \epsilon$, where ϵ is typically taken to be a small constant. (In addition to this multiplicative factor, it is often allowed to also let the probabilities to differ by a negligible additive term). This captures the idea that no individual's data has a significant influence on the output of $\mathcal{A}$ (provided that data about an individual is confined to one or a few rows of the database). Differential privacy has several nice properties lacking in previous notions, such as being agnostic to the adversary's prior information and degrading smoothly under composition.

With this model of privacy, the goal becomes to design algorithms $\mathcal{A}$ that simultaneously meet the above privacy guarantee and give "useful" information about the database. For example, we may have a true query function c in which we're interested, and the goal is to design $\mathcal{A}$ that is differentially private (with ϵ as small as possible) and estimates c well (e.g. the error $|\mathcal{A}(D) - c(D)|$ is small with high probability). For example, if $c(D)$ is the fraction of database rows that satisfy some property — a *counting query* — then it is known that we can take $\mathcal{A}(D)$ to equal $c(D)$ plus random Laplacian noise with standard deviation $O(1/(\epsilon n))$, where n is the number of rows in the database and ϵ is the measure of differential privacy [5]. A sequence of works [7,13,5,11] has provided a very good understanding of differential privacy in an interactive model in which real-valued queries c are made and answered one at a time. The amount of noise that one needs when responding to a query c should be based on the sensitivity of c, as well as the total number of queries answered so far.

However, for many applications, it would be more attractive to do a noninteractive data release, where we compute and release a single, differentially private "summary" of the database that enables others to determine accurate answers to a large class of queries. What form should this summary take? The most appealing form would be a *synthetic database*, which is a new database $\widehat{D} = \mathcal{A}(D)$ whose rows are "fake", but come from the same universe as those of D and are guaranteed to share many statistics with those of D (up to some accuracy). Some advantages of synthetic data are that it can be easily understood by humans, and statistical software can be run directly on it without modification. For example, these considerations led the German Institute for Employment Research to adopt synthetic databases for the release of employment statistics [25].

Previous Results on Synthetic Data. The first result on producing differentially private synthetic data came in the work of Barak et al. [3]. Given a database D consisting of n rows from $\{0,1\}^d$, they show how to construct a differentially private synthetic database $\widehat{D}$, also of n rows from $\{0,1\}^d$, in which the full "contingency table," consisting of all conjunctive counting queries, is approximately preserved. That is, for every conjunction $c(x_1, \ldots, x_n) = x_{i_1} \wedge x_{i_2} \wedge \cdots \wedge x_{i_k}$ for $i_1, \ldots, i_k \in [d]$, the fraction of rows in $\widehat{D}$ that satisfy c equals the fraction of rows in D that satisfy c up to an additive error of $2^{O(d)}/n$. The running time of their algorithm is $\mathrm{poly}(n, 2^d)$, which is feasible for small values of d. They pose as an open problem whether the running time of their algorithm can be improved for the case where we only want to preserve the *k-way marginals* for small k (e.g. $k = 2$). These are the counting queries corresponding to conjunctions of up to k literals. Indeed, there are only $O(d)^k$ such conjunctions, and we can produce differentially private estimates for all the corresponding counting queries in time $\mathrm{poly}(n, d^k)$ by just adding noise $O(d)^k/n$ to each one. Moreover, a version of the Barak et al. algorithm [3] can ensure that even these noisy answers are consistent with a real database[1].

A more general and dramatic illustration of the potential expressiveness of synthetic data came in the work of Blum, Ligett, and Roth [6]. They show that for every class $\mathcal{C} = \{c : \{0,1\}^d \to \{0,1\}\}$ of predicates, there is a differentially private algorithm A that produces a synthetic database $\hat{D} = \mathcal{A}(D)$ such that all counting queries corresponding to predicates in $\mathcal{C}$ are preserved to within an accuracy of $\tilde{O}((d \log(|\mathcal{C}|)/n)^{1/3})$, with high probability. In particular, with $n = \mathrm{poly}(d)$, the synthetic data can provide simultaneous accuracy for an *exponential-sized* family of queries (e.g. $|\mathcal{C}| = 2^d$). Unfortunately, the running time of the BLR mechanism is also exponential in d.

Dwork et al. [12] gave evidence that the large running time of the BLR mechanism is inherent. Specifically, assuming the existence of one-way functions, they exhibit an efficiently computable family $\mathcal{C}$ of predicates (e.g. consisting of circuits of size d^2) for which it is infeasible to produce a differentially private synthetic database preserving the counting queries corresponding to $\mathcal{C}$ (for databases of any $n = \mathrm{poly}(d)$ number of rows). For non-synthetic data, they show a close connection between the infeasibility of producing a differentially private summarization and the existence of efficient "traitor-tracing schemes." However, these results leave open the possibility that for *natural* families of counting queries (e.g. those corresponding to conjunctions), producing a differentially private synthetic database (or non-synthetic summarization) can be done efficiently. Indeed, one may have gained optimism by analogy with the early days of computational learning theory, where one-way functions were used to show hardness of learning arbitrary efficiently computable concepts in computational learning theory but natural subclasses (like conjunctions) were found to be learnable [29].

Our Results. We prove that it is infeasible to produce synthetic databases preserving even very simple counting queries, such as 2-way marginals:

Theorem 1. *Assuming the existence of one-way functions, there is a constant $\gamma > 0$ such that for every polynomial p, there is no polynomial-time, differentially private algorithm A that takes a database $D \in (\{0,1\}^d)^{p(d)}$ and produces a synthetic database $\hat{D} \in (\{0,1\}^d)^*$ such that $|c(D) - c(\hat{D})| \le \gamma$ for all 2-way marginals c.*

(Recall that a 2-way marginal $c(D)$ computes the fraction of database rows satisfying a conjunction of two literals, i.e. the fraction of rows $x_i \in \{0,1\}^d$ such that $x_i(j) = b$ and $x_i(j') = b'$ for some columns $j, j' \in [d]$ and values $b, b' \in \{0,1\}$). In fact, our impossibility result extends from conjunctions of 2 literals to any family of constant arity predicates that contains a function depending on at least two variables, such as parities of 3 literals.

[1] Technically, this "real database" may assign fractional weight to some rows.

As mentioned earlier, all 2-way marginals *can* be easily summarized with non-synthetic data (by just adding noise to each of the $(2d)^2$ values). Thus, our result shows that requiring a synthetic database may severely constrain what sorts of differentially private data releases are possible. (Dwork et al. [12] also showed that there exists a poly(d)-sized family of counting queries that are hard to summarize with synthetic data, thereby separating synthetic data from non-synthetic data. Our contribution is to show that such a separation holds for a very simple and natural family of predicates, namely 2-way marginals).

This separation between synthetic data and non-synthetic data seems analogous to the separations between proper and improper learning in computational learning theory [24,16], where it is infeasible to learn certain concept classes if the output hypothesis is constrained to come from the same representation class as the concept, but it becomes feasible if we allow the output hypothesis to come from a different representation class. This gives hope for designing efficient, differentially private algorithms that take a database and produce a compact summary of it that is not synthetic data but somehow can be used to accurately answer exponentially many questions about the original database (e.g. all marginals). The negative results of [12] on non-synthetic data (assuming the existence of efficient traitor-tracing schemes) do not say anything about natural classes of counting queries, such as marginals.

To bypass the complexity barrier stated in Theorem 1, it may not be necessary to introduce exotic data representations; some mild generalizations of synthetic data may suffice. For example, several recent algorithms [6,27,14] produce several synthetic databases, with the guarantee that the *median* answer over these databases is approximately accurate. More generally, we can consider summarizations of a database D that that consist of a collection $\hat{D}$ of rows from the same universe as the original database, and where we estimate $c(D)$ by applying the predicate c to each row of $\hat{D}$ and then aggregating the results via some aggregation function f. With standard synthetic data, f is simply the average, but we may instead allow f to take a median of averages, or apply an affine shift to the average. For such *relaxed synthetic data*, we prove the following results:

- There is a constant k such that counting queries corresponding to k-juntas (functions depending on at most k variables) cannot be accurately and privately summarized as relaxed synthetic data with a median-of-averages aggregator, or with a symmetric and monotone aggregator (that is independent of the predicate c being queried).
- For every constant k, counting queries corresponding to k-juntas *can* be accurately and privately summarized as relaxed synthetic data with an aggregator that applies an affine shift to the average (where the shift does depend on the predicate being queried).

Techniques. Our proof of Theorem 1 and our other negative results are obtained by combining the hard-to-sanitize databases of Dwork et al. [12] with PCP reductions. They construct a database consisting of valid message-signature pairs (m_i, σ_i) under a digital signature scheme, and argue that any differentially private sanitizer that preserves accuracy for the counting query associated with the signature verification predicate can be used to forge valid signatures. We replace each message-signature pair (m_i, σ_i) with a PCP encoding π_i that proves that (m_i, σ_i) satisfies the signature verification algorithm. We then argue that if accuracy is preserved for a large fraction of the (constant arity) constraints of the PCP verifier, then we can "decode" the PCP either to violate privacy (by recovering one of the original message-signature pairs) or to forge a signature (by producing a new message-signature pair).

We remark that error-correcting codes were already used in [12] for the purpose of producing a fixed polynomial-sized set of counting queries that can be used for all verification keys. Our observation is that by using *PCP* encodings, we can reduce not only the number of counting queries in consideration, but also their computational complexity.

Our proof has some unusual features among PCP-based hardness results:

- As far as we know, this is the first time that PCPs have been used in conjunction with cryptographic assumptions for a hardness result. (They have been used together for positive results regarding computationally sound proof systems [21,22,4]). It would be interesting to see if such a combination could be useful in, say, computational learning theory (where PCPs have been used for hardness of "proper" learning [2,17] and cryptographic assumptions for hardness of representation-independent learning [29,20]).
- While PCP-based inapproximability results are usually stated as Karp reductions, we actually need them to be *Levin* reductions — capturing that they are reductions between search problems, and not just decision problems. (Previously, this property has been used in the same results on computationally sound proofs mentioned above).

2 Preliminaries

2.1 Sanitizers

Let a *database* $D \in (\{0,1\}^d)^n$ be a matrix of n rows, $x_1, \ldots, x_n$, corresponding to people, each of which contains d binary attributes. A *sanitizer* $\mathcal{A} : (\{0,1\}^d)^n \to \mathcal{R}$ takes a database and outputs some data structure in $\mathcal{R}$. In the case where $\mathcal{R} = (\{0,1\}^d)^{\hat{n}}$ (an $\hat{n}$-row database) we say that $\mathcal{A}$ outputs a *synthetic database*.

We would like such sanitizers to be both *private* and *accurate*. In particular, the notion of privacy we are interested in is as follows

Definition 2 (Differential Privacy). *[11] A sanitizer $\mathcal{A} : (\{0,1\}^d)^n \to \mathcal{R}$ is (ϵ, δ)-*differentially private *if for every two databases $D_1, D_2 \in (\{0,1\}^d)^n$ that differ on exactly one row, and every subset $S \subseteq \mathcal{R}$*

$$\Pr[\mathcal{A}(D_1) \in S] \leq e^{\epsilon} \Pr[\mathcal{A}(D_2) \in S] + \delta$$

*In the case where $\delta = 0$ we say that $\mathcal{A}$ is ϵ-*differentially private.

Since a sanitizer that always outputs 0 satisfies Definition 2, we also need to define what it means for a database to be accurate. In this paper we consider accuracy with respect to counting queries. Consider a set $\mathcal{C}$ consisting of boolean predicates $c : \{0,1\}^d \to \{0,1\}$, which we call a *concept class*. Then each predicate c induces a *counting query* that on database $D = (x_1, \ldots, x_n) \in (\{0,1\}^d)^n$ returns

$$c(D) = \frac{1}{n} \sum_{i=1}^{n} c(x_i)$$

If the output of $\mathcal{A}$ is a synthetic database $\widehat{D} \in (\{0,1\}^d)^*$, then $c(\mathcal{A}(D))$ is simply the fraction of rows of $\widehat{D}$ that satisfy the predicate c. However, if $\mathcal{A}$ outputs a data structure that is not a synthetic database, then we require that there is an *evaluator* function $\mathcal{E} : \mathcal{R} \times \mathcal{C} \to \mathbb{R}$ that estimates $c(D)$ from the output of $\mathcal{A}(D)$ and the description of c. For example, $\mathcal{A}$ may output a vector $Z = (c(D) + Z_c)_{c \in \mathcal{C}}$ where Z_c is a random variable for each $c \in \mathcal{C}$, and $\mathcal{E}(Z, c)$ is the c-th component of $Z \in \mathcal{R} = \mathbb{R}^{|\mathcal{C}|}$. Abusing notation, we will write $c(\mathcal{A}(D))$ as shorthand for $\mathcal{E}(\mathcal{A}(D), c)$.

We will say that $\mathcal{A}$ is "accurate" for the concept class $\mathcal{C}$ if the estimates $c(\mathcal{A}(D))$ are close to the fractional counts $c(D)$. Formally

Definition 3 (Accuracy). *An output Z of sanitizer $\mathcal{A}(D)$ is α-*accurate *for a concept class $\mathcal{C}$ if*

$$\forall c \in \mathcal{C}, |c(Z) - c(D)| \leq \alpha.$$

*A sanitizer $\mathcal{A}$ is (α, β)-*accurate *for a concept class $\mathcal{C}$ if for every database D,*

$$\Pr_{\mathcal{A}'s \text{ coins}} [\forall c \in \mathcal{C}, \ |c(\mathcal{A}(D)) - c(D)| \leq \alpha] \geq 1 - \beta$$

In this paper we say $f(n) = \text{negl}(n)$ if $f(n) = o(n^{-c})$ for every $c > 0$ and say that $f(n)$ is *negligible*. We use $|s|$ to denote the length of the string s, and $s_1 \| s_2$ to denote the concatenation of s_1 and s_2.

2.2 Hardness of Sanitizing

Differential privacy is a very strong notion of privacy, so it is common to look for hardness results that also apply to weaker notions of privacy. These hardness results show that every sanitizer must be "blatantly non-private" in some sense. In this paper our notion of blatant non-privacy roughly states that there exists an efficient adversary who can find a row of the original database using only the output from any efficient sanitizer. Such definitions are also referred to as "row non-privacy." We define hardness-of-sanitization with respect to a particular concept class, and want to exhibit a distribution on databases for which it would be infeasible for any efficient sanitizer to give accurate output without revealing a row of the database. Specifically, following [12], we define the following notions

Definition 4 (Database Distribution Ensemble). *Let $\mathcal{D} = \mathcal{D}_d$ be an ensemble of distributions on d-column databases with $n+1$ rows $D \in (\{0,1\}^d)^{n+1}$. Let $(D, D', i) \leftarrow_R \tilde{\mathcal{D}}$ denote the experiment in which we choose $D_0 \leftarrow_R \mathcal{D}$ and $i \in [n]$ uniformly at random, and set D to be the first n rows of D_0 and D' to be D with the i-th row replaced by the $(n+1)$-st row of D_0.*

Definition 5 (Hard-to-sanitize Distribution). *Let $\mathcal{C}$ be a concept class, $\alpha \in [0, 1]$ be a parameter, and $\mathcal{D} = \mathcal{D}_d$ be a database distribution ensemble.*

The distribution $\mathcal{D}$ is $(\alpha, \mathcal{C})$-hard-to-sanitize *if there exists an efficient adversary $\mathcal{T}$ such that for any alleged polynomial-time sanitizer $\mathcal{A}$ the following conditions hold:*

1. *Whenever $\mathcal{A}(D)$ is α-accurate, then $\mathcal{T}(\mathcal{A}(D))$ outputs a row of D:*

$$\Pr_{\substack{(D,D',i) \leftarrow_R \tilde{\mathcal{D}} \\ \mathcal{A}'s \text{ and } \mathcal{T}'s \text{ coins}}} [(\mathcal{A}(D) \text{ is } \alpha\text{-accurate for } \mathcal{C}) \wedge (\mathcal{T}(\mathcal{A}(D)) \cap D = \emptyset)] \leq \text{negl}(d).$$

2. *For every efficient sanitizer $\mathcal{A}$, $\mathcal{T}$ cannot extract x_i from the database D':*

$$\Pr_{\substack{(D,D',i) \leftarrow_R \tilde{\mathcal{D}} \\ \mathcal{A}'s \text{ and } \mathcal{T}'s \text{ coins}}} [\mathcal{T}(\mathcal{A}(D')) = x_i] \leq \text{negl}(d)$$

   where x_i is the i-th row of D.

In [12], it was shown that every distribution that is $(\alpha, \mathcal{C})$-hard-to-sanitize in the sense of Definition 5, is also hard to sanitize while achieving even weak differential privacy.

Claim 6. *[12] If a distribution ensemble $\mathcal{D} = \mathcal{D}_d$ on $n(d)$-row databases is $(\alpha, \mathcal{C})$-hard-to-sanitize, then for every constant $a > 0$ and every $\beta = \beta(d) \leq 1 - 1/\text{poly}(d)$, no efficient sanitizer that is (α, β)-accurate with respect to $\mathcal{C}$ can achieve $(a \log(n), (1-8\beta)/2n^{1+a})$-differential privacy.*

In particular, for all constants $\epsilon, \beta > 0$, no polynomial-time sanitizer can achieve (α, β)-accurateness and $(\epsilon, \text{negl}(n))$-differential privacy.

We could use a weaker definition of hard-to-sanitize distributions, which would still suffice to rule out differential privacy, that only requires that for every efficient $\mathcal{A}$, there exists an adversary $\mathcal{T}_{\mathcal{A}}$ that almost always extracts a row of D from every α-accurate output of $\mathcal{A}(D)$. In our definition we require that there exists a fixed adversary $\mathcal{T}$ that almost always extracts a row of D from every α-accurate output of any efficient $\mathcal{A}$. Reversing the quantifiers in this fashion only makes our negative results stronger.

In this paper we are concerned with sanitizers that output synthetic databases, so we will relax Definition 5 by restricting the quantification over sanitizers to only those sanitizers that output synthetic data.

Definition 7 (Hard-to-sanitize Distribution as Synthetic Data). *A database distribution ensemble $\mathcal{D}$ is* $(\alpha, \mathcal{C})$-hard-to-sanitize as synthetic data *if the conditions of Definition 5 hold for every sanitizer $\mathcal{A}$ that outputs a synthetic database.*

3 Relationship with Hardness of Approximation

The objective of a privacy-preserving sanitizer is to reveal some properties of the underlying database without giving away enough information to reconstruct that database. This requirement has different implications for sanitizers that produce synthetic databases and those with arbitrary output.

The SuLQ framework of [5] is a well-studied, efficient technique for achieving (ϵ, δ)-differential privacy, with non-synthetic output. To get accurate, private output for a family of counting queries with predicates in $\mathcal{C}$, we can release a vector of noisy counts $(c(D) + Z_c)_{c\in\mathcal{C}}$ where the random variables $(Z_c)_{c\in\mathcal{C}}$ are drawn independently from a distribution suitable for preserving privacy. (e.g. a Laplace distribution with standard deviation $O(|\mathcal{C}|/\epsilon n)$).

Consider the case of an n-row database D that contains satisfying assignments to a 3CNF formula φ, and suppose our concept class includes all disjunctions on three literals (or, equivalently, all conjunctions on three literals). Then the technique above releases a set of noisy counts that describes a database in which every clause of φ is satisfied by most of the rows of D. However, sanitizers that output accurate synthetic databases are required to produce a database that consists of rows that satisfy most of the clauses of φ.

Because of the noise added to the output, the requirement of a synthetic database does not strictly force the sanitizer to find a satisfying assignment for the given 3CNF. However, it is known to be NP-hard to find even approximate satisfying assignments for many constraint satisfaction problems. In our main result, Theorem 14, we will show that there exists a distribution over databases that is hard-to-sanitize with respect to synthetic data for any concept class that is sufficient to express a hard-to-approximate constraint satisfaction problem.

3.1 Hard to Approximate CSPs

We define a *constraint satisfaction problem* to be the following.

Definition 8 (Constraint Satisfaction Problem (CSP)). *For a function $q = q(d) \leq d$, a* family of $q(d)$-CSPs, *denoted $\Gamma = (\Gamma_d)_{d\in\mathbb{N}}$, is a sequence of sets Γ_d of boolean predicates on $q(d)$ variables. If $q(d)$ and Γ_d do not depend on d then we refer to Γ as a* fixed family of q-CSPs.

For every $d \geq q(d)$, let $\mathcal{C}_\Gamma^{(d)}$ be the class consisting of all predicates $c : \{0,1\}^d \to \mathbb{R}$ of the form $c(u_1, \ldots, u_d) = \gamma(u_{i_1}, \ldots, u_{i_{q(d)}})$ for some $\gamma \in \Gamma_d$ and $i_1, \ldots, i_{q(d)} \in [d]$. We call $\mathcal{C}_\Gamma = \cup_{d=0}^{\infty} \mathcal{C}_\Gamma^{(d)}$ the class of constraints of *Γ. Finally, we say a multiset $\varphi \subseteq \mathcal{C}_\Gamma^{(d)}$ is a* d-variable instance of *$\mathcal{C}_\Gamma$ and each $\varphi_i \in \varphi$ is a* constraint of *φ.*

We say that an assignment x satisfies *the constraint φ_i if $\varphi_i(u) = 1$. For $\varphi = \{\varphi_1, \ldots, \varphi_m\}$, define*

$$\mathrm{val}(\varphi, u) = \frac{\sum_{i=1}^m \varphi_i(u)}{m} \qquad \textit{and} \qquad \mathrm{val}(\varphi) = \max_{u\in\{0,1\}^d} \mathrm{val}(\varphi, u).$$

Our hardness results will apply to concept classes $\mathcal{C}_\Gamma^{(d)}$ for CSP families Γ with certain additional properties. Specifically we define,

Definition 9 (Nice CSP). *A family $\Gamma = (\Gamma_d)_{d\in\mathbb{N}}$ of $q(d)$-CSPs* nice *if*

1. *$q(d) = d^{1-\Omega(1)}$,*
2. *for every $d \in \mathbb{N}$, Γ_d contains a non-constant predicate $\varphi^* : \{0,1\}^{q(d)} \to \{0,1\}$. Moreover, φ^* and two assignments $u_0^*, u_1^* \in \{0,1\}^{q(d)}$ such that $\varphi^*(u_0) = 0$ and $\varphi^*(u_1) = 1$ can be found in time* $\mathrm{poly}(d)$.

We note that any fixed family of q-CSP that contains a non-constant predicate is a nice CSP. Indeed, these CSPs (e.g. conjunctions of 2 literals) are the main application of interest for our results. However it will sometimes be useful to work with generalizations to nice CSPs with predicates of non-constant arity.

For our hardness result, we will need to consider a strong notion of hard constraint satisfaction problems, which is related to probabilistically checkable proofs. First we recall the standard notion of hardness of approximation under Karp reductions. (stated for additive, rather than multiplicative approximation error)

Definition 10 (inapproximability under Karp reductions). *For functions* $\alpha, \gamma : \mathbb{N} \to [0,1]$. *A family of CSPs* $\Gamma = (\Gamma_d)_{d \in \mathbb{N}}$ *is* (α, γ)-hard-to-approximate under Karp reductions *if there exists a polynomial-time computable function* R *such that for every circuit* C *with input size* $\overline{d}$, *if we set* $\varphi_C = R(C) \subseteq \mathcal{C}_\Gamma^{(d)}$ *for some* $d = \mathrm{poly}(\overline{d})$, *then*

1. *if* C *is satisfiable, then* $\mathrm{val}(\varphi_C) \geq \gamma(d)$, *and*
2. *if* C *is unsatisfiable, then* $\mathrm{val}(\varphi_C) < \gamma(d) - \alpha(d)$.

For our hardness result, we will need a stronger notion of inapproximability, which says that we can efficiently transform satisfying assignments of C into solutions to φ_C of high value, and vice-versa.

Definition 11 (inapproximability under Levin reductions). *For functions* $\alpha, \gamma : \mathbb{N} \to [0,1]$. *A family of CSPs* $\Gamma = (\Gamma_d)_{d \in \mathbb{N}}$ *is* (α, γ)-hard-to-approximate under Levin reductions *if there exist polynomial-time computable functions* R, *Enc*, *Dec such that for every circuit* C *with input of size* $\overline{d}$ *if we set* $\varphi_C = R(C) \subseteq \mathcal{C}_\Gamma^{(d)}$ *for some* $d = \mathrm{poly}(\overline{d})$, *then*

1. *for every* $u \in \{0,1\}^{\overline{d}}$ *such that* $C(u) = 1$, $\mathrm{val}(\varphi_C, Enc(u, C)) \geq \gamma(d)$,
2. *and for every* $\pi \in \{0,1\}^d$ *such that* $\mathrm{val}(\varphi_C, \pi) \geq \gamma(d) - \alpha(d)$, $C(Dec(\pi, C)) = 1$,
3. *and for every* $u \in \{0,1\}^{\overline{d}}$, $Dec(Enc(u, C)) = u$

When we do not wish to specify the value γ *we will simply say that the family* Γ *is* α-hard-to-approximate under Levin reductions *to indicate that there exists such a* $\gamma \in (\alpha, 1]$. *If we drop the requirement that* R *is efficiently computable, then we say that* Γ *is* (α, γ)-hard-to-approximate under inefficient Levin reductions.

The notation Enc, Dec reflects the fact that we think of the set of assignments π such that $\mathrm{val}(\varphi_C, \pi) \geq \gamma$ as a sort of error-correcting code on the satisfying assignments to C. Any π' with value close to γ can be decoded to a valid satisfying assignment.

Levin reductions are a stronger notion of reduction than Karp reductions. To see this, let Γ be α-hard-to-approximate under Levin reductions, and let R, Enc, Dec be the functions described in Definition 11. We now argue that for every circuit C, the formula $\varphi_C = R(C)$ satisfies conditions 1 and 2 of Definition 10. Specifically, if there exists an assignment $u \in \{0,1\}^{\overline{d}}$ that satisfies C, then $Enc(u, C)$ satisfies at least a γ fraction of the constraints of φ_C. Conversely if any assignment $\pi \in \{0,1\}^d$ satisfies at least a $\gamma - \alpha$ fraction of the constraints of φ_C, then $Dec(\pi, C)$ is a satisfying assignment of C.

Variants of the PCP Theorem can be used to show that essentially every class of CSP is hard-to-approximate in this sense. We restrict to CSP's that are closed under complement as it suffices for our application.

Theorem 12 (variant of PCP Theorem). *For every fixed family of CSPs Γ that is closed under negation and contains a function that depends on at least two variables, there is a constant $\alpha = \alpha(\Gamma) > 0$ such that Γ is α-hard to approximate under Levin reductions.*

It seems likely that optimized PCP/inapproximability results (like [19]) are also Levin reductions, which would yield fairly large values for α for natural CSPs (e.g. $\alpha = 1/8 - \epsilon$ if Γ contains all conjunctions of 3-literals, because then $\mathcal{C}_\Gamma$ contains MAX 3-SAT).

For some of our results we will need CSPs that are very hard to approximate (under possibly inefficient reductions), which we can obtain by "sequential repetition" of constant-error PCPs.

Theorem 13 (variant of PCP Theorem with subconstant error). *There is a constant C such that for every $\epsilon = \epsilon(d) > 2^{-\mathrm{poly}(d)}$, the constraint family $\Gamma = (\Gamma_d)_{d \in \mathbb{N}}$ of $k(d)$-clause 3-CNF formulas is $(1 - \epsilon(d), 1)$-hard-to-approximate under inefficient Levin reductions, for $k(d) = C \log(1/\epsilon(d))$.*

Further discussion of these theorems can be found in the full version of this paper.

4 Hard-to-Sanitize Distributions from Hard CSPs

In this section we prove that to efficiently produce a synthetic database that is accurate for the constraints of a CSP that is hard-to-approximate under Levin reductions, we must pay constant error in the worst case. Following [12], we start with a digital signature scheme, and a database of valid message-signature pairs. There is a verifying circuit C_{vk} and valid message-signature pairs are satisfying assignments to that circuit. Now we encode each row of database using the function Enc, described in Definition 11, that maps satisfying assignments to C_{vk} to assignments of the CSP instance $\varphi_{C_{vk}} = R(C_{vk})$ with value at least γ. Then, any assignment to the CSP instance that satisfies a $\gamma - \alpha$ fraction of clauses can be decoded to a valid message-signature pair. The database of encoded message-signature pairs is what we will use as our hard-to-sanitize distribution.

4.1 Main Hardness Result

We are now ready to state and prove our hardness result. Let $\Gamma = (\Gamma_d)_{d \in \mathbb{N}}$ be a family of $q(d)$-CSPs and let $\mathcal{C}_\Gamma = \cup_{d=1}^{\infty} \mathcal{C}_\Gamma^{(d)}$ be the class of constraints of Γ, which was constructed in Definition 8. We now state our hardness result.

Theorem 14. *Let $\Gamma = (\Gamma_d)_{d \in \mathbb{N}}$ be a family of nice $q(d)$-CSPs such that $\Gamma_d \cup \neg\Gamma_d$ is $\alpha(d)$-hard-to-approximate under (possibly inefficient) Levin reductions for $\alpha = \alpha(d) \in (0, 1/2)$. Assuming the existence of one-way functions, for every polynomial $n(d)$, there exists a distribution ensemble $\mathcal{D} = \mathcal{D}_d$ on $n(d)$-row databases that is $(\alpha(d), \mathcal{C}_\Gamma^{(d)})$-hard-to-sanitize as synthetic data.*

Proof. Let $\Pi = (Gen, Sign, Ver)$ be a digital signature scheme where it is even hard to produce a new signature for a previously signed message[2]. and let Γ be a family of CSPs that is α-hard-to-approximate under Levin reductions. Let R, Enc, Dec be the polynomial-time functions and $\gamma = \gamma(d) \in (\alpha, 1]$ be the parameter from Definition 11. Let $\kappa = d^\tau$ for a constant $\tau > 0$ to be defined later.

Let $n = n(d) = \text{poly}(d)$. We define the database distribution ensemble $\mathcal{D} = \mathcal{D}_d$ to generate $n+1$ random message-signature pairs and then encode them as PCP witnesses with respect to the signature-verification algorithm. We also encode the verification key for the signature scheme using the non-constant constraint $\varphi^* : \{0,1\}^{q(d)} \to \{0,1\}$ in Γ_d and the assignments $u_0^*, u_1^* \in \{0,1\}^{q(d)}$ such that $\varphi^*(u_0^*) = 0$ and $\varphi^*(u_1^*) = 1$, as described in the definition of nice CSPs (Definition 9).

Database Distribution Ensemble $\mathcal{D} = \mathcal{D}_d$:

$(sk, vk) \leftarrow_{\text{R}} Gen(1^\kappa)$, let $vk = vk_1 vk_2 \ldots vk_\ell$, where $\ell = |vk| = \text{poly}(\kappa)$
$(m_1, \ldots, m_{n+1}) \leftarrow_{\text{R}} (\{0,1\}^\kappa)^{n+1}$
for $i = 1$ to $n+1$ **do**
  $x_i := Enc(m_i \| Sign_{sk}(m_i), C_{vk}) \| u^*_{vk_1} \| u^*_{vk_2} \| \cdots \| u^*_{vk_\ell}$, padded with zeros to be of length exactly d
end for
return $D_0 := (x_1, \ldots, x_{n+1})$

Recall that $s_1 \| s_2$ denotes the concatenation of the strings s_1 and s_2. Note that the length of x_i before padding is $\text{poly}(\kappa) + q(d)\text{poly}(\kappa) \leq d^{1-\Omega(1)}\text{poly}(d^\tau)$, so we can choose the constant $\tau > 0$ to be small enough that the length of x before padding is at most d and the above is well defined.

Every valid pair $(m, Sign_{sk}(m))$ is a satisfying assignment of the circuit C_{vk}, hence every row of D_0 constructed in this way will satisfy at least a γ fraction of the clauses of the formula $\varphi_{C_{vk}} = R(C_{vk})$. Additionally, for every bit of the verification key, there is a block of $q(d)$ bits in each row that contains either a satisfying assignment or a non-satisfying assignment of φ^*, depending on whether that bit of the key is 1 or 0. Specifically, let $L = |Enc(m_i \| Sign_{sk}(m_i))|$ in the construction of D_0 and for $j = 1, 2, \ldots, \ell$, let $\varphi_j^*(x) = \varphi^*(x_{L+(j-1)q+1}, x_{L+(j-1)q+2}, \ldots, x_{L+jq})$. Then, by construction, $\varphi_j^*(D_0) = vk_j$, the j-th bit of the verification key. Note that $\varphi_j^* \in \mathcal{C}_\Gamma^{(d)}$ for $j = 1, 2, \ldots, \ell$, by our construction of $\mathcal{C}_\Gamma^{(d)}$ (Definition 8).

We now prove the following two lemmas that will establish $\mathcal{D}$ is hard-to-sanitize:

Lemma 15. *There exists a polynomial-time adversary $\mathcal{T}$ such that for every polynomial-time sanitizer $\mathcal{A}$,*

$$\Pr_{\substack{(D,D',i) \leftarrow_R \tilde{\mathcal{D}} \\ \mathcal{A}'s \text{ and } \mathcal{T}'s \text{ coins}}} \left[(\mathcal{A}(D) \text{ is } \alpha\text{-accurate for } \mathcal{C}_\Gamma^{(d)}) \wedge (\mathcal{T}(\mathcal{A}(D)) \cap D = \emptyset) \right] \leq \text{negl}(d) \tag{1}$$

[2] These digital signature schemes are defined formally in the full version of this paper. In [18] it is shown how to modify known constructions [23,26] to obtain a such a digital signature scheme from any one-way function.

Proof. Our privacy adversary tries to find a row of the original database by trying to PCP-decode each row of the "sanitized" database and then re-encoding it. In order to do so, the adversary needs to know the verification key used in the construction of the database, which it can discover from the answers to the queries φ_j^*, defined above. Formally, we define the privacy adversary by means of a subroutine that tries to learn the verification key and then PCP-decode each row of the input database:

Subroutine $\mathcal{K}(\widehat{D})$:

Let d be the dimension of rows in $\widehat{D}$, $\kappa = d^\tau$, $\ell = |vk| = \text{poly}(\kappa)$.
for $j = 1$ to ℓ **do**
  $\widehat{vk_j} = \left[\varphi_j^*(\widehat{D}) \text{ rounded to } \{0,1\}\right]$
end for
return $\widehat{vk_1}\|\widehat{vk_2}\|\ldots\|\widehat{vk_\ell}$

Subroutine $\mathcal{T}_0(\widehat{D})$:

Let $\hat{n}$ be the number of rows in $\widehat{D}$, $\widehat{vk} = \mathcal{K}(\widehat{D})$
for $i = 1$ to $\hat{n}$ **do**
  if $C_{\widehat{vk}}(Dec(\hat{x}_i, C_{\widehat{vk}})) = 1$ **then**
    return $Dec(\hat{x}_i, C_{\widehat{vk}})$
  end if
end for
return $\perp$

Privacy Adversary $\mathcal{T}(\widehat{D})$:

Let $\widehat{vk} = \mathcal{K}(\widehat{D})$.
return $Enc(\mathcal{T}_0(\widehat{D}), C_{\widehat{vk}})$

Let $\mathcal{A}$ be a polynomial-time sanitizer, we will show that Inequality (1) holds.

Claim 16. *If $\widehat{D} = \mathcal{A}(D)$ is α-accurate for $\mathcal{C}_\Gamma^{(d)}$, then $\mathcal{T}_0(\widehat{D})$ outputs a pair (m, σ) s.t. $C_{vk}(m, \sigma) = 1$.*

Proof. First we argue that if $\widehat{D}$ is α-accurate for $\mathcal{C}_\Gamma^{(d)}$ for $\alpha < 1/2$, then $\mathcal{K}(\widehat{D}) = vk$, where vk is the verification key used in the construction of D_0. By construction, $\varphi_j^*(D) = vk_j$. If $vk_j = 0$ and $\widehat{D}$ is α-accurate for D then $\varphi_j^*(\widehat{D}) \leq \alpha < 1/2$, and $\widehat{vk_j} = vk_j$. Similarly, if $vk_j = 1$ then $\varphi_j^*(\widehat{D}) \geq 1 - \alpha > 1/2$, and $\widehat{vk_j} = vk_j$. Thus, for the rest of the proof we will be justified in substituting vk for $\widehat{vk}$.

Next we show that if $\widehat{D}$ is α-accurate, then $\mathcal{T}_0(\widehat{D}) \neq \perp$. It is sufficient to show there exists $\hat{x}_i \in \widehat{D}$ such that $\text{val}(\varphi_{C_{vk}}, x_i) \geq \gamma - \alpha$, which implies $C_{vk}(Dec(\hat{x}_i, C_{vk})) = 1$.

Since every $(m_i, Sign_{sk}(m_i))$ pair is a satisfying assignment to C_{vk}, the definition of *Enc* (Definition 11) implies that each row x_i of D has $\text{val}(\varphi_{C_{vk}}, x_i) \geq \gamma$. Thus if $\varphi_{C_{vk}} = \{\varphi_1, \ldots, \varphi_m\}$, then

$$\frac{1}{m}\sum_{j=1}^{m}\varphi_j(D) = \frac{1}{m}\sum_{j=1}^{m}\left(\frac{1}{n}\sum_{i=1}^{n}\varphi_j(x_i)\right) = \frac{1}{n}\sum_{i=1}^{n}\text{val}(\varphi_{C_{vk}}, x_i) \geq \gamma.$$

Since $\widehat{D}$ is α-accurate for $\mathcal{C}_{\Gamma}^{(d)}$, and for every constraint φ_j, either $\varphi_j \in \Gamma$ or $\neg\varphi_j \in \Gamma$, then for every constraint $\varphi_j \in \varphi_{C_{vk}}$, we have $\varphi_j(\widehat{D}) \geq \varphi_j(D) - \alpha$. Thus

$$\frac{1}{\hat{n}} \sum_{i=1}^{\hat{n}} \mathrm{val}(\varphi_{C_{vk}}, \hat{x}_i) = \frac{1}{m} \sum_{j=1}^{m} \varphi_j(\widehat{D}) \geq \frac{1}{m} \sum_{j=1}^{m} \varphi_j(D) - \alpha \geq \gamma - \alpha.$$

So for at least one row $\hat{x} \in \widehat{D}$ it must be the case that $\mathrm{val}(\varphi_{C_{vk}}, \hat{x}) \geq \gamma - \alpha$. The definition of *Dec* (Definition 11) implies $C_{vk}(Dec(\hat{x}, C_{vk})) = 1$.

Now notice that if $\mathcal{T}_0(\mathcal{A}(D))$ outputs a valid message-signature pair but $\mathcal{T}(\mathcal{A}(D)) \cap D = \emptyset$, then this means $\mathcal{T}_0(\mathcal{A}(D))$ is forging a new signature not among those used to generate D, violating the security of the digital signature scheme. Formally, we construct a signature forger as follows:

Forger $\mathcal{F}(vk)$ with oracle access to $Sign_{sk}$:

Use the oracle $Sign_{sk}$ to generate an n-row database D just as in the definition of $\mathcal{D}_d$ (consisting of PCP encodings of valid message-signature pairs and an encoding of vk).
Let $\widehat{D} := \mathcal{A}(D)$
return $\hat{x}^* := \mathcal{T}_0(\widehat{D})$

Notice that running $\mathcal{F}$ in a chosen-message attack is equivalent to running $\mathcal{T}$ in the experiment of inequality (1), except that $\mathcal{F}$ does not re-encode the output of $\mathcal{T}_0(\mathcal{A}(D))$. By the super-security of the signature scheme, if the $\hat{x}^*$ output by $\mathcal{F}$ is a valid message-signature pair (as holds if $\mathcal{A}(D)$ is α-accurate for $\mathcal{C}_{\Gamma}^{(d)}$, by Claim 16), then it must be one of the message-signature pairs used to construct D (except with probability $\mathrm{negl}(\kappa) = \mathrm{negl}(d)$). This implies that $\mathcal{T}(\mathcal{A}(D)) = Enc(\hat{x}^*, C_{vk}) \in D$ (except with negligible probability). Thus, we have

$$\Pr_{\substack{(D,D',i) \leftarrow_R \tilde{\mathcal{D}} \\ \mathcal{A}'s \text{ coins}}} [\mathcal{A}(D) \text{ is } \alpha\text{-accurate for } \mathcal{C}_{\Gamma}^{(d)} \Rightarrow \mathcal{T}(\mathcal{A}(D)) \in D] \geq 1 - \mathrm{negl}(d),$$

which is equivalent to the statement of the lemma.

Lemma 17

$$\Pr_{\substack{(D,D',i) \leftarrow_R \tilde{\mathcal{D}} \\ \mathcal{A}'s \text{ and } \mathcal{T}'s \text{ coins}}} [\mathcal{T}(\mathcal{A}(D')) = x_i] \leq \mathrm{negl}(d)$$

Proof. Since the messages m_i used in D_0 are drawn independently, D' contains no information about the message m_i, thus no adversary can, on input $\mathcal{A}(D')$ output the target row x_i except with probability $2^{-\kappa} = \mathrm{negl}(d)$.

These two claims suffice to establish that $\mathcal{D}$ is $(\alpha, \mathcal{C}_{\Gamma})$-hard-to-sanitize as synthetic data.

Theorem 1 in the introduction follows by combining Theorems 12 and 14.

5 Relaxed Synthetic Data

The proof of Theorem 14 requires that the sanitizer output a synthetic database. In this section we present similar hardness results for sanitizers that produce other forms of output, as long as they still produce a collection of elements from $\{0,1\}^d$, that are interpreted as the data of (possibly "fake") individuals. More specifically, we consider sanitizers that output a database $\widehat{D} \in (\{0,1\}^d)^{\hat{n}}$ but are evaluated by applying a predicate c to each row and then applying a function f to the resulting bits and the predicate c. For example, when the sanitizer outputs a synthetic database, we have $f(b_1,\ldots,b_{\hat{n}},c) = (1/\hat{n})\sum_{i=1}^{\hat{n}} b_i$, which is just the fraction of rows that get labeled with a 1 by the predicate c (independent of c).

We now give a formal definition of *relaxed synthetic data*

Definition 18 (Relaxed Synthetic Data). *A sanitizer $\mathcal{A} : (\{0,1\}^d)^n \to (\{0,1\}^d)^{\hat{n}}$ with evaluator $\mathcal{E}$ outputs* relaxed synthetic data *for a family of predicates $\mathcal{C}$ if there exists $f : \{0,1\}^{\hat{n}} \times \mathcal{C} \to [0,1]$ such that for every $c \in \mathcal{C}$*

$$\mathcal{E}(\widehat{D},c) = f(c(\hat{x}_1), c(\hat{x}_2), \ldots, c(\hat{x}_{\hat{n}}), c),$$

and f is monotone[3] *in the first $\hat{n}$ inputs.*

This relaxed notion of synthetic data is of interest because many natural approaches to sanitizing yield outputs of this type. In particular, several previous sanitization algorithms [6,27,14] produce a *set* of synthetic databases and answer a query by taking a median over the answers given by the individual databases. We view such databases as a single synthetic database but require that f have a special form. Unfortunately, the sanitizers of [14] and [27] run in time exponential in the dimension of the data, d, and the results of the next subsection show this limitation is inherent even for simple concept classes.

We now give an informal description of our hardness results for different forms of relaxed synthetic data. Our proofs use the same construction of hard-to-sanitize databases as Theorem 14 with a modified analysis and parameters to show that the output must still contain a PCP-decodable row. Formal statements and proofs of all of the following statements can be found in the full version of this paper.

- We say that a sanitizer outputs *medians of synthetic data* if it satisfies Definition 18 with

$$\mathcal{E}(\hat{x}_1,\ldots,\hat{x}_{\hat{n}},c) = \text{median}\left\{\frac{1}{|S_1|}\sum_{i\in S_1} c(\hat{x}_i), \ldots, \frac{1}{|S_\ell|}\sum_{i\in S_\ell} c(\hat{x}_i)\right\}$$

  for some partition $[\hat{n}] = S_1 \cup S_2 \cdots \cup S_\ell$. We rule out efficient sanitizers with medians of synthetic data for CSPs that are hard-to-approximate under Levin reductions within a multiplicative factor larger than 2. By Theorem 13, these CSPs include k-clause 3-CNF formulas for some constant k.

[3] Given two vectors $a = (a_1,\ldots,a_n)$ and $b = (b_1,\ldots,b_n)$ we say $b \succeq a$ iff $b_i \geq a_i$ for every $i \in [n]$. We say a function $f : \{0,1\}^n \to [0,1]$ is *monotone* if $b \succeq a \Longrightarrow f(b) \geq f(a)$.

- We say that a sanitizer outputs *symmetric relaxed synthetic data* if it satisfies Definition 18 with $\mathcal{E}(\hat{x}_1, \dots, \hat{x}_{\hat{n}}, c) = g((1/\hat{n}) \sum_{i=1}^{\hat{n}} c(\hat{x}_i))$ for a monotone function $g : [0, 1] \to [0, 1]$. These evaluators are symmetric both in that g does not depend on the predicate and that g only depends on the fraction of rows that satisfy the predicate. We rule out efficient sanitizers with symmetric relaxed synthetic data for CSPs that are hard-to-approximate under Levin reductions within an additive factor larger than $1/2$. By Theorem 13, these CSPs include k-clause CNF formulas, for some constant k.
- We show that no efficient sanitizer can produce accurate relaxed synthetic data for a sequence of CSPs that is $(1 - \mathrm{negl}(d))$-hard-to-approximate under inefficient Levin reductions. By Theorem 13, these CSPs include 3-CNF formulas of $\omega(\log d)$ clauses.

5.1 Positive Results for Relaxed Synthetic Data

We also show that there exists an efficient sanitizer for the family of all constant-arity predicates. As an intermediate step, we also show there exists an efficient sanitizer as symmetric relaxed synthetic data for any family of parity predicates. Our results show that relaxed synthetic data allows for more efficient sanitization than standard synthetic data, since Theorem 14 rules out an accurate, efficient sanitizer with standard synthetic data, even for 3-literal parity predicates. Our result for parities also shows that our hardness result for symmetric relaxed synthetic data is tight with respect to the required hardness of approximation, since the class of 3-literal parity predicates is $(1/2 - \epsilon)$-hard-to-approximate [19].

A function $f : \{0, 1\}^d \to \{0, 1\}$ is a *k-junta* if it depends on at most k variables. Let $\mathcal{J}_{d,k}$ be the set of all k-juntas on d variables.

Theorem 19. *There exists an ϵ-differentially private sanitizer that runs in time* poly(n, d) *and produces relaxed synthetic data and is (α, β)-accurate for $\mathcal{J}_{d,k}$ when*

$$n \geq \frac{C\binom{d}{\leq k} \log\left(\binom{d}{\leq k}/\beta\right)}{\alpha\epsilon}$$

for a sufficiently large constant C, where $\binom{d}{\leq k} = \sum_{i=0}^{k} \binom{d}{i}$.

To prove Theorem 19, we start with a sanitizer for *parity predicates*. A function $\chi : \{0, 1\}^d \to \{-1, 1\}$ is a *parity predicate*[4] if there exists a vector $s \in \{0, 1\}^d$ s.t. $\chi(x) = \chi_s(x) = (-1)^{\langle x, s \rangle}$.

Theorem 20. *Let $\mathcal{P}$ be a family of parity predicates such that $\chi_{0^d} \notin \mathcal{P}$. There exists an ϵ-differentially private sanitizer $\mathcal{A}(D, \mathcal{P})$ that runs in time* poly(n, d) *and produces symmetric relaxed synthetic data that is (α, β)-accurate for $\mathcal{P}$ when*

$$n \geq \frac{2|\mathcal{P}| \log(2|\mathcal{P}|/\beta)}{\alpha\epsilon}.$$

[4] In the preliminaries we define a predicate to be a $\{0, 1\}$-valued function but our definition naturally generalizes to $\{-1, 1\}$-valued functions. For $c : \{0, 1\}^d \to \{-1, 1\}$ and database $D = (x_1, \dots, x_n) \in (\{0, 1\}^d)^n$, we define $c(D) = \frac{1}{n} \sum_{i=1}^{n} c(x_i)$.

Without relaxed synthetic data, Theorems 19 and 20 can be achieved by simply releasing a vector of noisy answers to the queries [11]. Our sanitizer begins with this vector of noisy answers and constructs relaxed synthetic data from those answers. Our technique is similar to that of Barak et. al. [3], which also begins with noisy answers and constructs a (standard) synthetic database that gives approximately the same answers. However, they construct their synthetic database by solving a linear program over a set of 2^d variables, each of which represents the frequency of one of the possible rows $x \in \{0,1\}^d$. Thus their sanitizer runs in time exponential in d.

Our sanitizer also starts with a vector of noisy answers to parity queries and *efficiently* constructs symmetric relaxed synthetic data that gives answers to each query that are close to the initial noisy answers after applying a *fixed linear scaling*. To construct each row of the synthetic database, $\widehat{D}$, we select a random parity query in $\chi \in \mathcal{P}$ and then sample a row $x \in \{0,1\}^d$ such that the expectation of $\chi(x)$ is equal to the initial noisy estimate of $\chi(D)$; it can be shown that for every $\chi' \neq \chi$ (except χ_{0^d}), the expectation of $\chi'(x)$ is zero. Thus we can estimate the value of $\chi(D)$ by taking $\chi(\widehat{D})$ and multiplying by $|\mathcal{P}|$. We then show that if we apply our sanitizer to the family $\mathcal{P}_{d,k}$ containing all parity predicates on d variables that depend on at most k variables, the result is also accurate for the family $\mathcal{J}_{d,k}$ of k-juntas after applying an affine shift that depends on the average value of the junta of interest.

A complete discussion of these results is deferred to the full version of this paper.

Acknowledgments

We thank Boaz Barak, Irit Dinur, Cynthia Dwork, Vitaly Feldman, Oded Goldreich, Johan Håstad, Valentine Kabanets, Dana Moshkovitz, Anup Rao, Guy Rothblum, and Les Valiant for helpful conversations. We also thank the anonymous reviewers for helpful comments.

References

1. Adam, N.R., Wortmann, J.: Security-control methods for statistical databases: A comparative study. ACM Computing Surveys 21, 515–556 (1989)
2. Alekhnovich, M., Braverman, M., Feldman, V., Klivans, A.R., Pitassi, T.: The complexity of properly learning simple concept classes. J. Comput. Syst. Sci. 74, 16–34 (2008)
3. Barak, B., Chaudhuri, K., Dwork, C., Kale, S., McSherry, F., Talwar, K.: Privacy, accuracy, and consistency too: A holistic solution to contingency table release. In: Proceedings of the 26th Symposium on Principles of Database Systems, pp. 273–282 (2007)
4. Barak, B., Goldreich, O.: Universal arguments and their applications. SIAM J. Comput. 38, 1661–1694 (2008)
5. Blum, A., Dwork, C., McSherry, F., Nissim, K.: Practical privacy: The SuLQ framework. In: Proceedings of the 24th ACM SIGMOD-SIGACT-SIGART Symposium on Principles of Database Systems (June 2005)
6. Blum, A., Ligett, K., Roth, A.: A learning theory approach to non-interactive database privacy. In: Proceedings of the 40th ACM SIGACT Symposium on Thoery of Computing (2008)
7. Dinur, I., Nissim, K.: Revealing information while preserving privacy. In: Proceedings of the Twenty-Second ACM SIGACT-SIGMOD-SIGART Symposium on Principles of Database Systems, pp. 202–210 (2003)

8. Duncan, G.: Confidentiality and statistical disclosure limitation. In: International Encyclopedia of the Social and Behavioral Sciences. Elsevier, Amsterdam (2001)
9. Dwork, C.: A firm foundation for private data analysis. Communications of the ACM (to appear)
10. Dwork, C.: Differential privacy. In: Bugliesi, M., Preneel, B., Sassone, V., Wegener, I. (eds.) ICALP 2006, Part II. LNCS, vol. 4052, pp. 1–12. Springer, Heidelberg (2006)
11. Dwork, C., McSherry, F., Nissim, K., Smith, A.: Calibrating noise to sensitivity in private data analysis. In: Halevi, S., Rabin, T. (eds.) TCC 2006. LNCS, vol. 3876, pp. 265–284. Springer, Heidelberg (2006)
12. Dwork, C., Naor, M., Reingold, O., Rothblum, G., Vadhan, S.: When and how can privacy-preserving data release be done efficiently? In: Proceedings of the 2009 International ACM Symposium on Theory of Computing (STOC) (2009)
13. Dwork, C., Nissim, K.: Privacy-preserving datamining on vertically partitioned databases. In: Franklin, M. (ed.) CRYPTO 2004. LNCS, vol. 3152, pp. 528–544. Springer, Heidelberg (2004)
14. Dwork, C., Rothblum, G., Vadhan, S.P.: Boosting and differential privacy. In: Proceedings of FOCS 2010 (2010)
15. Evfimievski, A., Grandison, T.: Privacy Preserving Data Mining (a short survey). In: Encyclopedia of Database Technologies and Applications. Information Science Reference (2006)
16. Feldman, V.: Hardness of proper learning. In: The Encyclopedia of Algorithms. Springer, Heidelberg (2008)
17. Feldman, V.: Hardness of approximate two-level logic minimization and PAC learning with membership queries. Journal of Computer and System Sciences 75(1), 13–26 (2009), `http://dx.doi.org/10.1016/j.jcss.2008.07.007`
18. Goldreich, O.: Foundations of Cryptography, vol. 2. Cambridge University Press, Cambridge (2004)
19. Håstad, J.: Some optimal inapproximability results. J. ACM. 48, 798–859 (2001)
20. Kearns, M.J., Valiant, L.G.: Cryptographic limitations on learning boolean formulae and finite automata. J. ACM. 41, 67–95 (1994)
21. Kilian, J.: A note on efficient zero-knowledge proofs and arguments (extended abstract). In: STOC (1992)
22. Micali, S.: Computationally sound proofs. SIAM J. Comput. 30, 1253–1298 (2000)
23. Naor, M., Yung, M.: Universal one-way hash functions and their cryptographic applications. In: STOC, pp. 33–43 (1989)
24. Pitt, L., Valiant, L.G.: Computational limitations on learning from examples. J. ACM 35, 965–984 (1988)
25. Reiter, J.P., Drechsler, J.: Releasing multiply-imputed synthetic data generated in two stages to protect confidentiality. Iab discussion paper, Intitut für Arbeitsmarkt und Berufsforschung (IAB), Nürnberg, Institute for Employment Research, Nuremberg, Germany (2007), `http://ideas.repec.org/p/iab/iabdpa/200720.html`
26. Rompel, J.: One-way functions are necessary and sufficient for secure signatures. In: STOC, pp. 387–394 (1990)
27. Roth, A., Roughgarden, T.: Interactive privacy via the median mechanism. In: STOC 2010 (2010)
28. Ullman, J., Vadhan, S.P.: PCPs and the hardness of generating synthetic data. Electronic Colloquium on Computational Complexity (ECCC) 17, 17 (2010)
29. Valiant, L.G.: A theory of the learnable. Communications of the ACM 27(11), 1134–1142 (1984)

Limits of Computational Differential Privacy in the Client/Server Setting

Adam Groce, Jonathan Katz, and Arkady Yerukhimovich

Dept. of Computer Science
University of Maryland
{agroce,jkatz,arkady}@cs.umd.edu

Abstract. Differential privacy is a well established definition guaranteeing that queries to a database do not reveal "too much" information about specific individuals who have contributed to the database. The standard definition of differential privacy is information theoretic in nature, but it is natural to consider *computational* relaxations and to explore what can be achieved with respect to such notions. Mironov et al. (Crypto 2009) and McGregor et al. (FOCS 2010) recently introduced and studied several variants of computational differential privacy, and show that in the *two-party* setting (where data is split between two parties) these relaxations can offer significant advantages.

Left open by prior work was the extent, if any, to which computational differential privacy can help in the usual *client/server* setting where the entire database resides at the server, and the client poses queries on this data. We show, for queries with output in $\mathbb{R}^n$ (for constant n) and with respect to a large class of utilities, that any computationally private mechanism can be converted to a statistically private mechanism that is equally efficient and achieves roughly the same utility.

1 Introduction

A statistical database holds data representing some population. It is often desirable to allow clients to query this database to learn properties of the underlying population. However, it is also important to protect the privacy of the individual users whose data is contained in the database. This conflict between utility and privacy has motivated a significant amount of research in recent years, and several definitions of privacy as well as techniques for achieving these definitions have appeared in the literature.

The foundational definition of privacy in this setting is that of *differential privacy* [6,5,3]. Very coarsely, this definition can be viewed as limiting the amount of information the answer to some query reveals about any particular user in the database. The standard definition of differential privacy is very strong, requiring unconditional privacy guarantees against computationally unbounded adversaries. Despite this fact, there has been a good amount of success in designing differentially private mechanisms for many types of queries and in various settings [1,5,12,2,9].

Y. Ishai (Ed.): TCC 2011, LNCS 6597, pp. 417–431, 2011.

Recently, Mironov et al. [11] introduced various notions of *computational* differential privacy and explored relations between them. There are several reasons to consider such relaxations of differential privacy. In practice a computational notion of security suffices, yet the stringent notion of (statistical) differential privacy rules out some mechanisms that are intuitively secure: e.g., a differentially private mechanism implemented using pseudorandom noise in place of truly random noise, or a differentially private mechanism implemented using secure multi-party computation [4,11]. One might hope that by considering a relaxed definition we can circumvent limitations or impossibility results that arise in the information-theoretic setting, in the same way that computationally secure notions of encryption allow bypassing known bounds for perfectly secure encryption. Recent results [11,10] show that this is the case in the *two-party* setting where the database is partitioned between two parties who wish to evaluate some query over their joint data. Specifically, McGregor et al. [10] show a strong separation between the accuracy that can be obtained when using differential privacy as opposed to using *computational* differential privacy.

McGregor et al. [10], however, leave open the analogous question in the more widely studied *client/server* setting where a server holds the entire database on which a client may pose queries. Indeed, they explicitly remark [10, Section 1]:

> *[Our] strong separation between (information-theoretic) differential privacy and computational differential privacy ... stands in sharp contrast with the client-server setting where... there are not even candidates for a separation.*

It is this question we address in this paper.

1.1 Summary of Our Results

There are (at least) two notions of computational privacy that can be considered: IND-CDP and SIM-CDP. These notions are introduced in [11], where it is shown that any SIM-CDP mechanism is also IND-CDP (the other direction is not known); thus, SIM-CDP is a possibly stronger definition. (Mironov et al. also define the notion of $\text{SIM}_{\forall\exists}$-CDPbut this notion is equivalent to IND-CDP.) We review these definitions in Section 2.

There are two measures one could hope to improve upon when moving from the setting of (statistical) differential privacy to the setting of computational differential privacy: the best possible *utility* (or *accuracy*) that can be achieved, and the *efficiency* of implementing a mechanism that achieves some level of utility. With respect to the definitions given by Mironov et al., it is not hard to see that the best achievable utility cannot be improved as long as the utility is an efficiently computable function of the database and the output of the mechanism. (This is an immediate consequence of the SIM-CDP and $\text{SIM}_{\forall\exists}$-CDP definitions, since otherwise the utility function itself serves as a distinguisher.) The interesting question is therefore to look for improvements in the efficiency, e.g., to show that the best possible utility *for polynomial-time mechanisms* is better in the computational case, or even to show a polynomial factor improvement in the efficiency in moving from one case to the other. Unfortunately, we show two negative results indicating that such improvements are unlikely in certain natural settings:

1. Our first result concerns *black-box* constructions of computationally secure mechanisms from a wide range of cryptographic primitives including trapdoor permutations, collision-resistant hash functions, and/or random oracles. Roughly, we show that for any black-box construction of a computationally private mechanism there exists a corresponding *statistically* private mechanism that performs just as well in terms of both efficiency and utility (with respect to any utility measure).
2. Our main results rules out improvements by *arbitrary* mechanisms, for a specific (but large) class of queries and utility measures. That is, for queries with output in $\mathbb{R}^n$ (for constant n) and a natural class of utilities, we show that *any* computationally private mechanism can be converted to a statistically private mechanism that is roughly as efficient and achieves almost the same utility.

Each result applies to both the IND-CDP and SIM-CDP definitions.

We believe our results represent an important step in understanding the benefits and limitations of computational notions of privacy. Although we show negative results, they may point toward specific situations where computational differential privacy gives some advantage. We leave it as an open question to find utility measures or query classes with respect to which computational differential privacy *can* help in the client/server setting, or to extend our impossibility results to show that no such improvements can be hoped for.

Limitations of our results. There are several types of queries to which our results do not apply. The most important are queries with outputs that cannot naturally be thought of as tuples of real numbers. This includes, e.g., queries that return classifiers (as in [9]), graphs, or synthetic databases.

Our results also do not apply, in general, to queries that return output in $\mathbb{R}^n$ for "large" n (i.e., n that grows with the security parameter k). In particular, this means that our results are somewhat limited when it comes to analyzing differential privacy of multiple queries. (Note that n queries with outputs in $\mathbb{R}$ can be viewed as a single query with output in $\mathbb{R}^n$.) Our results do apply to any *constant* number of queries. In addition, using composition properties of differential privacy, our results apply to the case where arbitrarily many queries are answered, and all queries are answered independently (i.e., the server maintains no state). However, in some cases it is known that answering many queries at the same time can be done with better privacy than would be achieved by answering each query independently; in such cases our results do not apply.

Our results also hold only for restricted classes of utility functions. For example, they do not apply when there is no polynomial bound on the error.

2 (Computational) Differential Privacy

We begin by reviewing definitions for the various notions of differential privacy that will be discussed in this paper. All have roughly the same underlying intuition, but the technical differences are crucial. We begin by defining "adjacent" databases.

Definition 1. *Two databases $D, D' \in \mathcal{D}$ are* adjacent *if they differ in at most 1 entry.*

Differential privacy guarantees that the results of queries on two adjacent databases cannot be distinguished very well. This is a very strong privacy guarantee, that in particular ensures that the presence or absence of any one user in the database cannot affect the results very much.

One way to formalize this notion is to require that no set of answers can be significantly more likely to result from D than from D'. Formalizing this yields the by-now-standard notion of (statistical) differential privacy:

Definition 2. *A randomized function $f : \mathcal{D} \to \mathcal{R}$ is* ϵ-differentially private *(ϵ-DP) if for all adjacent databases $D, D' \in \mathcal{D}$ and all subsets $S \subset \mathcal{R}$:*

$$\Pr[f(D) \in S] \leq e^{\epsilon} \times \Pr[f(D') \in S].$$

This is the strongest definition of differential privacy. It can, in fact, be criticized as too strong. For example, consider a set of responses that are possible outputs when querying D but impossible when querying D'. The existence of such responses violates differential privacy, even if the probability of outputting one of these responses is small. To allow for this sort of situation one can consider a slightly weaker notion of differential privacy, called (ϵ, δ)-differential privacy, that allows a small additive factor in the inequality [4].

Definition 3. *A randomized function $f : \mathcal{D} \to \mathcal{R}$ is* (ϵ, δ)-differentially private *((ϵ, δ)-DP) if for all adjacent databases $D, D' \in \mathcal{D}$ and all subsets $S \subset \mathcal{R}$:*

$$\Pr[f(D) \in S] \leq e^{\epsilon} \times \Pr[f(D') \in S] + \delta.$$

It is worth noting that while for ϵ-DP it is sufficient to require the inequality in the definition to hold pointwise, for (ϵ, δ)-differential privacy it is important to explicitly consider all subsets S.

We say a family of mechanisms $\{f_k\}$ is *efficient* if the running time of $f_k(D)$ is at most $\mathsf{poly}(|D|, k)$. A family $\{f_k\}$ is *uniform* if there is a Turing machine f such that $f(k, D) = f_k(D)$. It is reasonable even in the information-theoretic setting to consider a family of mechanisms $\{f_k\}$ indexed by a security parameter k, and to require that δ become negligible in k.

Definition 4. *Let ϵ be an arbitrary function. A family of randomized functions $\{f_k\}_{k \in \mathbb{N}}$ is $(\epsilon, \mathsf{negl})$-DP if there exists a negligible function δ such that each f_k is $(\epsilon(k), \delta(k))$-DP.*

The above definitions are all information-theoretic in nature, but it is natural to consider computational variants. Mironov et al. [11] propose two definitions of computational differential privacy, SIM-CDP and IND-CDP. Roughly, one can view IND-CDP as an "indistinguishability-based" relaxation whereas SIM-CDP is a "simulation-based" notion. SIM-CDP is at least as strong as IND-CDP [11], but the converse is not known. All the definitions can be presented for either uniform or non-uniform adversaries; for consistency with [11], we give non-uniform definitions here. While we state our results for the case of non-uniform adversaries, our results all carry over to the uniform setting as well.

IND-CDP provides perhaps the most natural relaxation of differential privacy.

Definition 5 (IND-CDP). *Let ϵ be an arbitrary function. A family of functions $\{f_k\}_{k\in\mathbb{N}}$ is ϵ-IND-CDP if for every non-uniform polynomial-time $\mathcal{A}$ and every sequence $\{(D_k, D'_k)\}_{k\in\mathbb{N}}$ of (ordered pairs of) polynomial-size, adjacent databases, there is a negligible function* negl *such that*

$$\Pr[\mathcal{A}(f_k(D_k)) = 1] \leq e^{\epsilon(k)} \times \Pr[\mathcal{A}(f_k(D'_k)) = 1] + \mathsf{negl}(k).$$

The notion of SIM-CDP requires that there be a statistically private mechanism that is indistinguishable from the mechanism under consideration.

Definition 6 (SIM-CDP). *Let ϵ be an arbitrary function. A family of functions $\{f_k\}_{k\in\mathbb{N}}$ is ϵ-SIM-CDP if there exists a family of functions $\{F_k\}_{k\in\mathbb{N}}$ that is $(\epsilon, \mathsf{negl})$-DP and is computationally indistinguishable from $\{f_k\}$. The latter means there is a negligible function* negl *such that for any non-uniform polynomial-time $\mathcal{A}$ and any database D:*

$$\left|\Pr[\mathcal{A}(f_k(D)) = 1] - \Pr[\mathcal{A}(F_k(D)) = 1]\right| \leq \mathsf{negl}(k).$$

In [11] it is required that $\{F_k\}_{k\in\mathbb{N}}$ be ϵ-DP (rather than $(\epsilon, \mathsf{negl})$-DP). Thus our definition is slightly weaker, which makes our impossibility results stronger.

We also recall the notion of $\text{SIM}_{\forall\exists}$-CDP, which weakens SIM-CDP by reversing the order of quantifiers in the definition: here, the statistically private mechanism F is allowed to be different for each pair of databases (D, D'). Crucially for our purposes, this definition is known to be equivalent to IND-CDP [11].

Definition 7 ($\text{SIM}_{\forall\exists}$-CDP). *Let ϵ be an arbitrary function. A family of functions $\{f_k\}_{k\in\mathbb{N}}$ is ϵ-$\text{SIM}_{\forall\exists}$-CDP if for all sequences of (unordered pairs of) adjacent databases $\{\{D_k, D'_k\}\}_{k\in\mathbb{N}}$ there is a family of functions $\{F_k\}_{k\in\mathbb{N}}$ such that:*

1. *$\{F_k\}$ is ϵ-DP on $\{\{D_k, D'_k\}\}_{k\in\mathbb{N}}$; i.e., for all subsets $S \subset \mathcal{R}$ we have*

$$\Pr[F_k(D_k) \in S] \leq e^{\epsilon(k)} \times \Pr[F_k(D'_k) \in S].$$

2. *$f_k(D_k)$ and $f_k(D'_k)$ are indistinguishable from $F_k(D_k)$ and $F_k(D'_k)$ respectively. Formally, for any non-uniform, polynomial-time adversary $\mathcal{A}$*

$$\left|\Pr[\mathcal{A}(f_k(D_k)) = 1] - \Pr[\mathcal{A}(F_k(D_k)) = 1]\right| \leq \mathsf{negl}(k),$$

and similarly for D'_k.

Thus far we have only discussed privacy but have not mentioned *utility*. In general, we assume a utility measure U that takes as input a database D and the output of some mechanism $f(D)$ and returns a real number. In Section 4 we consider a specific class of utilities.

3 Limitations on Black-Box Constructions

Here we show that black-box constructions (of a very general sort) cannot help in the setting of computational differential privacy. (We refer the reader to [13] for further discussion and definitional treatment of black-box constructions.) For concreteness, in the technical discussion we focus on black-box constructions from one-way functions, but at the end of the section we discuss generalizations of the result.

Roughly, a fully black-box construction of an ϵ-IND-CDP mechanism from a one-way function is a family of polynomial-time oracle machines $\{f_k^{(\cdot)}\}_{k\in\mathbb{N}}$ such that for every $\mathcal{A}$ and every $\mathcal{O}$ that is one-way against $\mathcal{A}$ it holds that $\{f_k^{\mathcal{O}}\}_{k\in\mathbb{N}}$ is ϵ-IND-CDP against $\mathcal{A}$. It would make sense also to impose a utility condition on the construction (which could be viewed as a correctness requirement on the constructions), but we do not do so here.

Theorem 1. *If there exists a fully black-box construction $\{f_k\}_{k\in\mathbb{N}}$ of an ϵ-IND-CDP mechanism from one-way functions, then there exists an $(\epsilon, \mathsf{negl})$-DP family $\{f'_k\}_{k\in\mathbb{N}}$ that is roughly as efficient and such that, for all databases D and utility measures U,*

$$\left|\mathbf{E}\left[U(D, f_k^{\mathcal{O}}(D))\right] - \mathbf{E}\left[U(D, f'_k(D))\right]\right| \leq \mathsf{negl}(k),$$

where the expectations are both taken over the randomness of the mechanism, and the expectation on the left is additionally taken over random choice of a function $\mathcal{O}$.

Proof. The key idea behind the proof is as follows: a random function is one-way with overwhelming probability [8,7]; thus, the mechanism $f_k^{\mathcal{O}}$ with $\mathcal{O}$ chosen at random is also ϵ-IND-CDP. Since the construction is fully black-box (and hence relativizing), one-wayness of $\mathcal{O}$ (and hence indistinguishability of the mechanism) holds even for an unbounded adversary as long as the adversary makes only polynomially many queries to $\mathcal{O}$. We construct f'_k by having it simply run f_k as a subroutine, simulating a random function $\mathcal{O}$ on behalf of f_k. This idea is motivated by analogous techniques used in [7].

Let Func denote the set of length-preserving functions from $\{0,1\}^*$ to $\{0,1\}^*$, and let f'_k be as just described. Then for any adjacent databases D, D' and any (unbounded) $\mathcal{A}$:

$$\Pr[\mathcal{A}(f'_k(D)) = 1] = \Pr_{\mathcal{O}\leftarrow\mathsf{Func}}[\mathcal{A}(f_k^{\mathcal{O}}(D)) = 1]$$

and

$$\Pr[\mathcal{A}(f'_k(D')) = 1] = \Pr_{\mathcal{O}\leftarrow\mathsf{Func}}[\mathcal{A}(f_k^{\mathcal{O}}(D')) = 1].$$

Letting OWF denote the event that $\mathcal{O}$ is one-way, we have

$$\begin{aligned}\Pr[\mathcal{A}(f'_k(D)) = 1] &\le \Pr\left[\mathcal{A}(f_k^{\mathcal{O}}(D)) = 1 \mid \mathsf{OWF}\right] + \mathsf{negl}(k) \\ &\le e^{\epsilon(k)} \times \Pr\left[\mathcal{A}(f_k^{\mathcal{O}}(D')) = 1 \mid \mathsf{OWF}\right] + \mathsf{negl}'(k) \\ &\le e^{\epsilon(k)} \times \Pr[\mathcal{A}(f'_k(D')) = 1] + \mathsf{negl}''(k).\end{aligned}$$

The second inequality holds since $\{f_k\}$ is a fully black-box construction of an ϵ-IND-CDP mechanism from one-way functions. (Note that, above, $\mathcal{A}$ is not given access to $\mathcal{O}$ at all.) But the condition that

$$\Pr[\mathcal{A}(f'_k(D)) = 1] \le e^{\epsilon(k)} \times \Pr[\mathcal{A}(f'_k(D')) = 1] + \mathsf{negl}''(k)$$

for an unbounded $\mathcal{A}$ is equivalent to $(\epsilon, \mathsf{negl})$-differential privacy.

The claim regarding the utility of $\{f'_k\}$ follows by a similar argument. (Note that we do not require that U be efficiently computable.)

Note that the above proof holds not just for constructions based on one-way functions, but for any black-box construction from a primitive P that can be instantiated with a random object. This includes, e.g., ideal ciphers, collision-resistant hash functions, and trapdoor permutations [7].

4 Limitations for Computational Differential Privacy

In the previous section we ruled out black-box constructions from general assumptions, but with regard to arbitrary measures of utility and arbitrary mechanisms. Here, we focus on *arbitrary* mechanisms with output in $\mathbb{R}^n$ (for constant n), and a large, but specific, class of efficiently computable utilities. Specifically, we look at utilities defined by (a generalization of) the L_p norm.

Definition 8 (L_p-norm). *The L_p-norm of a vector $\mathbf{x} \in \mathbb{R}^n$, denoted $||\mathbf{x}||_p$, is defined as*

$$||\mathbf{x}||_p \stackrel{\text{def}}{=} (|x_1|^p + |x_2|^p + \ldots + |x_n|^p)^{1/p}$$

for $p \in \mathbb{N}^+$, where x_i is the ith coordinate of $\mathbf{x}$. (We do not deal with the L_0 norm in this paper.) We also allow $p = \infty$, where

$$||\mathbf{x}||_\infty \stackrel{\text{def}}{=} \max(|x_1|, |x_2|, \ldots, |x_n|).$$

A natural notion of utility would be to look at the average distance (in some L_p norm) from the true answer to the output of the mechanism. We broaden this to include things like mean-squared error that are commonly used in statistics.

Definition 9 (Average (p, v)-error). *Let $f_k : \mathcal{D} \to \mathbb{R}^n$ be a mechanism for answering a query $q : \mathcal{D} \to \mathbb{R}^n$. The* average (p, v)-error *(also called the vth* moment of the L_p error*) of this mechanism ($p > 0, v \ge 1$) on database D is*

$$\sigma_{p,v}(q, D, f_k) \stackrel{\text{def}}{=} \mathbf{E}\left[||f_k(D) - q(D)||_p^v\right].$$

We often refer to the above as "error" rather than "utility"; lower error values are good, whereas lower utility values are bad. We remark that we can handle utility measures beyond the above, as long as they satisfy a technical requirement that follows from our proof. Since we do not currently have any clean way to state this requirement, we do not discuss it further.

Given a mechanism $\{f_k : \mathcal{D} \to \mathbb{R}^n\}_{k\in\mathbb{N}}$ for answering a query $q : \mathcal{D} \to \mathbb{R}^n$, we say *the average* (p, v)*-error of* $\{f_k\}$ *is polynomially bounded* if there is a polynomial err such that, for all D and k, we have

$$\sigma_{p,v}(q, D, f_k) \leq \mathsf{err}(k).$$

Theorem 2, below, shows that nothing can be gained by using computational differential privacy rather than statistical differential privacy, as long as we consider mechanisms whose error is polynomially bounded. Before giving the formal theorem statement and proof in the following section, we give an intuitive explanation here.

Let f_k be a polynomial-time ϵ-SIM-CDP mechanism for answering some query $q : \mathcal{D} \to \mathbb{R}^n$, where we assume that f_k also has output in $\mathbb{R}^n$ (and n is independent of k). Let $p > 0, v \geq 1$ be arbitrary, and assume the average (p, v)-error of f_k is polynomially bounded with error bound err. We claim there is an $(\epsilon, \mathsf{negl})$-DP mechanism $\hat{f}_k$ with essentially the same running time[1] as f_k, and such that $\sigma_{p,v}(q, D, \hat{f}_k) < \mathsf{err}(k) + \mathsf{negl}(k)$.

Let $\{F_k\}$ be a mechanism that is $(\epsilon, \mathsf{negl})$-DP and indistinguishable from $\{f_k\}$. Such a mechanism is guaranteed to exist by definition of SIM-CDP. Note that $\{F_k\}$ may be much less efficient than $\{f_k\}$, and may not even be implementable in polynomial time. On the other hand, F_k and f_k must induce distributions over $\mathbb{R}^n$ that are, in some sense, very close. Intuitively, in any "box" in $\mathbb{R}^n$ of noticeable size, the probabilities with which the outputs of F_k or f_k lie in that cell must be roughly equal; if not, the difference in probabilities could be used to distinguish F_k and f_k (since membership in the box can be efficiently tested).

We derive $\hat{f}_k$ by adding a small amount of uniform noise to the output of f_k. Carefully setting the amount of noise to be sufficiently small, we can bound the error introduced in moving from f_k to $\hat{f}_k$. To analyze privacy of the resulting mechanism, we look at the mechanism $\hat{F}_k$ where a small amount of uniform noise is added to F_k. For any particular value x, the probability with which $\hat{f}_k$ (resp., $\hat{F}_k$) outputs x is proportional to the probability that f_k (resp., F_k) outputs a value within a box centered at x. This box is sufficiently big so that $\hat{F}_k$ and $\hat{f}_k$ have similar probabilities of outputting any particular value.

While $\hat{F}_k$ and $\hat{f}_k$ have similar probabilities of outputting any particular value, these small differences could, in principle, compound and become unacceptably large when summed over all values in some set $S \subset \mathbb{R}^n$. To show that such differences do not grow too large, we use the fact that f_k has polynomially bounded error. This allows us to break our analysis into two parts: one focusing on a region S_c "close" to the correct answer $q(D)$, and the other focusing on $S_f = S \setminus S_c$. We show that

[1] Specifically, $\hat{f}_k$ runs f_k and adds a random number to its output.

$$\left|\Pr[\hat{f}_k(D) \in S_c] - \Pr[\hat{F}_k(D) \in S_c]\right|$$

is small, using the argument discussed above; we also show that

$$\max\{\Pr[\hat{f}_k(D) \in S_f], \Pr[\hat{F}_k(D) \in S_f]\}$$

is small by the polynomial bound on the error. Combined, this shows that for every S, the difference

$$\left|\Pr[\hat{f}_k(D) \in S] - \Pr[\hat{F}_k(D) \in S]\right|$$

is small, as required. Since F_k, and hence $\hat{F}_k$, is *statistically* differentially private, this means that $\hat{f}_k$ is also.

Formal details are given in the following section.

4.1 Statement and Proof of the Main Result

We first present a proof that applies to the (stronger) SIM-CDP definition. We then outline the changes needed to prove the result for the case of IND-CDP.

Theorem 2. *Fix $p > 0, v \geq 1$. Let $\{f_k : \mathcal{D} \to \mathbb{R}^n\}$ be an efficient ϵ-SIM-CDP mechanism whose average (p, v)-error is polynomially bounded by* err. *Then there is an efficient $(\epsilon, \mathsf{negl})$-DP mechanism $\{\hat{f}_k\}$ with $\sigma_{p,v}(q, D, \hat{f}_k) < \mathsf{err}(k) + \mathsf{negl}(k)$.*

Moreover, $\hat{f}_k$ has essentially the same running time as f_k; specifically, $\hat{f}_k$ only adds uniform noise to f_k.

Proof. Let $\{F_k\}$ be an $(\epsilon, \mathsf{negl})$-DP family of mechanisms that is indistinguishable from $\{f_k\}$. Let negl_1 be a negligible function such that for any non-uniform polynomial-time $\mathcal{A}$ and any database D,

$$\left|\Pr[\mathcal{A}(f_k(D)) = 1] - \Pr[\mathcal{A}(F_k(D)) = 1]\right| \leq \mathsf{negl}_1(k).$$

(Such a function exists by definition of SIM-CDP.)

Since $\{f_k\}$ is efficient, its output must have some polynomial length. We assume that f_k (and hence F_k) give output in fixed-point notation with k bits of precision. Formally, let $\mathbb{R}_k$ be the set

$$\mathbb{R}_k = \{x \in \mathbb{R} \mid \exists j \in \mathbb{Z} : x = j \cdot 2^{-k}\};$$

then we assume that f_k gives output in $\mathbb{R}_k^n$. (More generally, the proof given here works when the precision is any polynomial in k. Moreover, fixed-point notation is not essential; in particular, the proof can be modified for the case when the output of f_k is given in floating-point notation.) For $x \in \mathbb{R}$ and $k \in \mathbb{N}$, define $\lceil x \rceil_k \stackrel{\text{def}}{=} \lceil x \cdot 2^k \rceil \cdot 2^{-k}$ to be the value x "rounded up" so that it lies in $\mathbb{R}_k$.

A set $\mathcal{B} \subset \mathbb{R}^n$ is a *box* if it a Cartesian product of closed intervals in $\mathbb{R}$. Abusing notation, we call a sequence $\{\mathcal{B}_k\}$ of boxes $\mathcal{B}_k \subset \mathbb{R}_k^n$ a box as well. The following is an immediate consequence of the SIM-CDP definition (recall the definition requires indistinguishability against non-uniform adversaries):

Lemma 1. *For any box $\{\mathcal{B}_k\}$ and any database D:*

$$|\Pr[f_k(D) \in \mathcal{B}_k] - \Pr[F_k(D) \in \mathcal{B}_k]| \leq \mathsf{negl}_1(k).$$

We next define two mechanisms $\{\hat{F}_k\}$ and $\{\hat{f}_k\}$ that are "noisy" versions of $F(D)$ and $f(D)$, respectively. Because we are dealing with discrete rather than continuous values, the definition is more complicated than simply adding uniform noise in some range.

Set $c(k) = \left\lceil \sqrt[4n]{\mathsf{negl}_1(k)} \right\rceil_k$ For $\mathbf{x} \in \mathbb{R}_k^n$, let $\mathcal{B}_{c,k}(\mathbf{x})$ denote the box with radius $c(k)$ (in the L_∞ norm) centered at $\mathbf{x}$; that is,

$$\mathcal{B}_{c,k}(\mathbf{x}) = \{\mathbf{y} \in \mathbb{R}_k^n \ : \ ||\mathbf{y} - \mathbf{x}||_\infty \leq c(k)\}.$$

Mechanism $\{\hat{f}_k\}$ is defined as follows: $\hat{f}_k(D)$ computes $f_k(D)$, and then outputs a uniform value in $\mathcal{B}_{c,k}(f(D))$. (This is equivalent to adding uniform, independent, discretized noise from $[-c(k), c(k)]$ to each coordinate of $f(D)$.) Mechanism $\{\hat{F}_k\}$ is defined to be the analogous mechanism that adds noise to F instead of f.

$B_{c,k}(\mathbf{x})$ contains $\left(c(k) \cdot 2^{k+1} + 1\right)^n$ points and thus, for any D and $\mathbf{x} \in \mathbb{R}_k^n$:

$$\Pr[\hat{F}_k(D) = \mathbf{x}] = \left(c(k) \cdot 2^{k+1} + 1\right)^{-n} \cdot \Pr[F_k(D) \in B_{c,k}(\mathbf{x})]$$

and

$$\Pr[\hat{f}_k(D) = \mathbf{x}] = \left(c(k) \cdot 2^{k+1} + 1\right)^{-n} \cdot \Pr[f_k(D) \in B_{c,k}(\mathbf{x})].$$

Taking $\mathcal{B}_k = \mathcal{B}_{c,k}(\mathbf{x}_k)$ (for an arbitrary sequence $\{\mathbf{x}_k\}$ with $\mathbf{x}_k \in \mathbb{R}_k^n$) in Lemma 1, we obtain:

$$\begin{aligned} &\left|\Pr[\hat{F}_k(D) = \mathbf{x}_k] - \Pr[\hat{f}_k(D) = \mathbf{x}_k]\right| \\ &= \left(c(k) \cdot 2^{k+1} + 1\right)^{-n} \cdot \left|\Pr[F_k(D) \in B_{c,k}(\mathbf{x}_k)] - \Pr[f_k(D) \in B_{c,k}(\mathbf{x}_k)]\right| \\ &\leq \left(c(k) \cdot 2^{k+1} + 1\right)^{-n} \cdot \mathsf{negl}_1(k). \end{aligned} \quad (1)$$

The above holds for an arbitrary database D, and so it also holds for any adjacent database D'.

$\hat{F}_k$ applies post-processing to the output of F_k, so $\{\hat{F}_k\}$ is also $(\epsilon, \mathsf{negl})$-DP. Let negl_2 be a negligible function such that for all sets S and adjacent databases D and D' it holds that

$$\Pr[\hat{F}_k(D) \in S] \leq e^{\epsilon(k)} \times \Pr[\hat{F}_k(D') \in S] + \mathsf{negl}_2(k). \quad (2)$$

Our goal is to prove that $\hat{f}_k(D)$ is *statistically* close to $\hat{F}_k(D)$, for any D, which will then imply the theorem. We have already shown (cf. Equation (1)) that the distributions of $\hat{f}_k(D)$ and $\hat{F}_k(D)$ are *pointwise* negligibly close. We need to show that this is true also for arbitrary subsets. To do this, we first use the polynomial error bound on f_k to argue that f_k (and hence $\hat{f}_k$) must put relatively low weight on outputs that are far from the correct output. Formally:

Lemma 2. *There is a polynomial b such that, for any D, we have*

$$\sigma_{p,v}(q, D, \hat{f}_k) \leq \mathsf{err}(k) + c(k) \cdot b(k).$$

The lemma follows from the observation that, for any fixed output $y = f_k(D)$, the output $\hat{y} = \hat{f}_k(D)$ satisfies

$$||\hat{y} - q(D)||_p \leq ||y - q(D)||_p + n \cdot c(k).$$

The proof of the lemma is tedious, and so we defer it to Appendix A.

Fix an arbitrary D. We now show that with high probability the output of $\hat{f}_k(D)$ is close to the true answer $q(D)$. Set $z(k) = \left\lceil \frac{1}{\sqrt[4n]{\mathsf{negl}_1(k)}} \right\rceil_k$, and define

$$\mathsf{Close}_k \stackrel{\text{def}}{=} \{\mathbf{x} \in \mathbb{R}_k^d : ||\mathbf{x} - q(D)||_p^v \leq z(k)\};$$

i.e., these are the points close to $q(D)$. Let $\mathsf{Far}_k \stackrel{\text{def}}{=} \mathbb{R}_k^n \setminus \mathsf{Close}_k$. Because the average error of $\hat{f}_k$ is at most $\mathsf{err}(k) + b(k) \cdot c(k)$, we have

$$\Pr[\hat{f}_k(D) \in \mathsf{Far}_k] \leq \left(\mathsf{err}(k) + b(k) \cdot c(k)\right) / z(k). \tag{3}$$

Indistinguishability of $\{f_k\}$ and $\{F_k\}$, and the manner in which $\{\hat{f}_k\}$ and $\{\hat{F}_k\}$ are constructed, implies that $\{\hat{f}_k\}$ and $\{\hat{F}_k\}$ are indistinguishable as well. As in the proof of Lemma 1, this means that

$$\left|\Pr[\hat{f}_k(D) \in \mathsf{Far}_k] - \Pr[\hat{F}_k(D) \in \mathsf{Far}_k]\right| \leq \mathsf{negl}_1(k).$$

Combining this with Equation (3) yields

$$\Pr[\hat{F}_k(D) \in \mathsf{Far}_k] \leq \left(\mathsf{err}(k) + b(k) \cdot c(k)\right) / z(k) + \mathsf{negl}_1(k).$$

We now use the above results to relate the probabilities that $\hat{F}_k(D)$ or $\hat{f}_k(D)$ lie within some arbitrary set. The number of points in Close_k is bounded from above by $(z(k) \cdot 2^{k+1} + 1)^n$, since its size is largest (for fixed $z(k)$) when $p = \infty$ and $v = 1$. For any $S_k \subset \mathbb{R}_k^n$, we can thus lower-bound $\Pr[\hat{F}_k(D) \in S_k]$ via

$$\begin{aligned} \Pr[\hat{F}_k(D) \in S_k] &= \sum_{\mathbf{x} \in S_k} \Pr[\hat{F}_k(D) = \mathbf{x}] \\ &\geq \sum_{\mathbf{x} \in S_k \cap \mathsf{Close}_k} \Pr[\hat{F}_k(D) = \mathbf{x}] \\ &\geq \sum_{\mathbf{x} \in S_k \cap \mathsf{Close}_k} \left(\Pr[\hat{f}_k(D) = \mathbf{x}] - \left(c(k) \cdot 2^{k+1} + 1\right)^{-n} \cdot \mathsf{negl}_1(k)\right), \end{aligned}$$

using Equation (1), which bounds the difference in probabilities between $\hat{f}_k$ and $\hat{F}_k$ pointwise. Continuing, we have

$$\begin{aligned}
&\Pr[\hat{F}_k(D) \in S_k] \\
&\geq \Pr[\hat{f}_k(D) \in S_k \cap \mathsf{Close}_k] - \left(z(k) \cdot 2^{k+1} + 1\right)^n \cdot \left(c(k) \cdot 2^{k+1} + 1\right)^{-n} \cdot \mathsf{negl}_1(k) \\
&\geq \Pr[\hat{f}_k(D) \in S_k \cap \mathsf{Close}_k] - \left(\frac{z(k)+1}{c(k)}\right)^n \cdot \mathsf{negl}_1(k) \\
&\quad + \left(\Pr[\hat{f}_k(D) \in S_k \cap \mathsf{Far}_k] - \left(\mathsf{err}(k) + b(k) \cdot c(k)\right)/z(k)\right) \\
&\geq \Pr[\hat{f}_k(D) \in S_k] - \left(\frac{z(k)+1}{c(k)}\right)^n \cdot \mathsf{negl}_1(k) - \left(\mathsf{err}(k) + b(k) \cdot c(k)\right)/z(k). \qquad (4)
\end{aligned}$$

Similarly, we can upper-bound $\Pr[\hat{F}_k(D) \in S_k]$ via

$$\begin{aligned}
&\Pr[\hat{F}_k(D) \in S_k] \\
&\leq \sum_{\mathbf{x} \in S_k \cap \mathsf{Close}_k} \Pr[\hat{F}_k(D) = \mathbf{x}] + \Pr[\hat{F}_k(D') \in \mathsf{Far}_k] \\
&\leq \sum_{\mathbf{x} \in S_k \cap \mathsf{Close}_k} \left(\Pr[\hat{f}_k(D) = \mathbf{x}] + \left(c(k) \cdot 2^{k+1} + 1\right)^{-n} \cdot \mathsf{negl}_1(k)\right) \\
&\quad + \Pr[\hat{F}_k(D) \in \mathsf{Far}_k] \\
&\leq \Pr[\hat{f}_k(D) \in S_k] + \left(\frac{z(k)+1}{c(k)}\right)^n \cdot \mathsf{negl}_1(k) \\
&\quad + \left(\mathsf{err}(k) + b(k) \cdot c(k)\right)/z(k) + \mathsf{negl}_1(k). \qquad (5)
\end{aligned}$$

Equations (4) and (5) hold for an arbitrary database D, and thus also hold for any adjacent database D'. Substituting into Equation (2) and simplifying, we obtain

$$\begin{aligned}
&\Pr[\hat{f}_k(D) \in S_k] \\
&\leq e^{\epsilon(k)} \times \Pr[\hat{f}_k(D') \in S_k] \\
&\quad + \left(e^{\epsilon(k)} + 1\right) \times \left(\left(\frac{z(k)+1}{c(k)}\right)^n \mathsf{negl}_1(k) + \left(\mathsf{err}(k) + b(k) \cdot c(k)\right)/z(k)\right) \\
&\quad + e^{\epsilon(k)} \cdot \mathsf{negl}_1(k) + \mathsf{negl}_2(k).
\end{aligned}$$

We show that the additive terms are all negligible. Note first that

$$\begin{aligned}
\left(\frac{z(k)+1}{c(k)}\right)^n \cdot \mathsf{negl}_1(k) &\leq \left(\frac{\frac{1}{\sqrt[4n]{\mathsf{negl}_1(k)}} + 2}{\sqrt[4n]{\mathsf{negl}_1(k)}}\right)^n \cdot \mathsf{negl}_1(k) \\
&\leq \left(\frac{3}{\sqrt[2n]{\mathsf{negl}_1(k)}}\right)^n \mathsf{negl}_1(k) \\
&\leq 3^n \cdot \sqrt{\mathsf{negl}_1(k)},
\end{aligned}$$

which is negligible in k (recall n is constant). To bound $\left(\mathsf{err}(k)+b(k)\cdot c(k)\right)/z(k)$, take k large enough so that $b(k)\cdot c(k)\leq \mathsf{err}(k)$ (this is always possible, since c is negligible while err and b are polynomial). We then have

$$\frac{\mathsf{err}(k)+b(k)\cdot c(k)}{z(k)}\leq 2\cdot\mathsf{err}(k)\cdot\sqrt[4n]{\mathsf{negl}_1(k)},$$

which is negligible. We conclude that $\{\hat{f}_k\}$ is (ϵ,negl)-DP.

The case of IND-CDP. A result analogous to the above holds also for the case of IND-CDP. This follows fairly easily using the equivalent formulation of IND-CDP in terms of $\text{SIM}_{\forall\exists}$-CDP. The difference between SIM-CDP and $\text{SIM}_{\forall\exists}$-CDP is with respect to the order of quantifiers, but this has no real effect on our proof. Note, in particular, that our construction of $\{\hat{f}_k\}$ does not depend, either explicitly or implicitly, on $\{F_k\}$.

Acknowledgments

We thank the referees for their detailed comments which allowed us to simplify parts of our proof. The second author thanks Adam Smith for suggesting the problem, and for several interesting discussions on the topic of computational differential privacy.

References

1. Blum, A., Dwork, C., McSherry, F., Nissim, K.: Practical privacy: The SuLQ framework. In: 24th ACM Symposium on Principles of Database Systems (PODS), pp. 128–138. ACM Press, New York (2005)
2. Blum, A., Ligett, K., Roth, A.: A learning theory approach to non-interactive database privacy. In: 40th Annual ACM Symposium on Theory of Computing (STOC), pp. 609–618. ACM Press, New York (2008)
3. Dwork, C.: Differential privacy. In: Bugliesi, M., Preneel, B., Sassone, V., Wegener, I. (eds.) ICALP 2006, Part II. LNCS, vol. 4052, pp. 1–12. Springer, Heidelberg (2006)
4. Dwork, C., Kenthapadi, K., McSherry, F., Mironov, I., Naor, M.: Our data, ourselves: Privacy via distributed noise generation. In: Vaudenay, S. (ed.) EUROCRYPT 2006. LNCS, vol. 4004, pp. 486–503. Springer, Heidelberg (2006)
5. Dwork, C., McSherry, F., Nissim, K., Smith, A.: Calibrating noise to sensitivity in private data analysis. In: Halevi, S., Rabin, T. (eds.) TCC 2006. LNCS, vol. 3876, pp. 265–284. Springer, Heidelberg (2006)
6. Dwork, C., Nissim, K.: Privacy-preserving datamining on vertically partitioned databases. In: Franklin, M. (ed.) CRYPTO 2004. LNCS, vol. 3152, pp. 528–544. Springer, Heidelberg (2004)
7. Gennaro, R., Gertner, Y., Katz, J., Trevisan, L.: Bounds on the efficiency of generic cryptographic constructions. SIAM Journal on Computing 35(1), 217–246 (2005)
8. Impagliazzo, R., Rudich, S.: Limits on the provable consequences of one-way permutations. In: 21st Annual ACM Symposium on Theory of Computing (STOC), pp. 44–61. ACM Press, New York (1989)

9. Kasiviswanathan, S.P., Lee, H.K., Nissim, K., Raskhodnikova, S., Smith, A.: What can we learn privately? In: 49th Annual Symposium on Foundations of Computer Science (FOCS), pp. 531–540. IEEE, Los Alamitos (2008)
10. McGregor, A., Mironov, I., Pitassi, T., Reingold, O., Talwar, K., Vadhan, S.P.: The limits of two-party differential privacy. In: 51st Annual Symposium on Foundations of Computer Science (FOCS), pp. 81–90. IEEE, Los Alamitos (2010)
11. Mironov, I., Pandey, O., Reingold, O., Vadhan, S.: Computational differential privacy. In: Halevi, S. (ed.) CRYPTO 2009. LNCS, vol. 5677, pp. 126–142. Springer, Heidelberg (2009)
12. Nissim, K., Raskhodnikova, S., Smith, A.: Smooth sensitivity and sampling in private data analysis. In: 39th Annual ACM Symposium on Theory of Computing (STOC), pp. 75–84. ACM Press, New York (2007)
13. Reingold, O., Trevisan, L., Vadhan, S.P.: Notions of reducibility between cryptographic primitives. In: Naor, M. (ed.) TCC 2004. LNCS, vol. 2951, pp. 1–20. Springer, Heidelberg (2004)

A Proof of Lemma 2

Let Y_k be the set of possible distances between two points in $\mathbb{R}_k^n$; i.e.,

$$Y_k \stackrel{\text{def}}{=} \{y \in \mathbb{R} \mid y = ||\mathbf{x}_1 - \mathbf{x}_2||_p \text{ for some } \mathbf{x}_1, \mathbf{x}_2 \in \mathbb{R}_k^n\}.$$

Let $p_{y,k} \stackrel{\text{def}}{=} \Pr\left[y - 2^{-k} < ||f_k(D) - q(D)||_p \leq y\right]$. Then, by the assumption of our theorem,

$$\sigma_{p,v}(q, D, f_k) \leq \sum_{y \in Y_k} p_{y,k} \cdot y^v \leq \mathsf{err}(k).$$

We can upper-bound $\sigma_{p,v}(q, D, \hat{f}_k)$ by assuming that the noise added by $\hat{f}_k$ moves the output further away from the correct answer $q(D)$. In the worst case (when $p = 1$), this increases the distance between the output and $q(D)$ by at most $c'(k) \stackrel{\text{def}}{=} n \cdot c(k)$. Therefore,

$$\sigma_{p,v}(q, D, \hat{f}_k) \leq \sum_{y \in Y_k} p_{y,k} \cdot (y + c'(k))^v.$$

Using Taylor's theorem, $(y + c'(k))^v \leq y^v + v \cdot (y + c'(k))^{v-1} \cdot c'(k)$. Thus, for k sufficiently large it holds that

$$\begin{aligned}
\sigma_{p,v}(q, D, \hat{f}_k) &\leq \sum_{y \in Y_k} p_{y,k} \cdot \left(y^v + v \cdot (y + c'(k))^{v-1} \cdot c'(k)\right) \\
&\leq \mathsf{err}(k) + \sum_{y \in Y_k} p_{y,k} \cdot \left(v \cdot (y + c'(k))^{v-1} \cdot c'(k)\right) \\
&\leq \mathsf{err}(k) + v \cdot c'(k) \cdot \sum_{y \in Y_k} p_{y,k} \cdot (y + n)^{v-1},
\end{aligned}$$

using for the last inequality the fact that $c'(k) \leq n$ for k large enough.

If $y \leq n$ then $(y+n)^{v-1} \leq (2n)^{v-1}$, while if $y \geq n$ then $(y+n)^{v-1} \leq (2y)^{v-1}$. As a result, we can bound the expression above as

$$\begin{aligned}
&\sigma_{p,v}(q, D, \hat{f}_k) \\
&\leq \mathsf{err}(k) + v \cdot c'(k) \cdot \sum_{y \in Y_k} p_{y,k} \cdot 2^{v-1} \cdot (n^{v-1} + y^{v-1}) \\
&\leq \mathsf{err}(k) + v \cdot c'(k) \cdot \left(\sum_{y \in Y_k} p_{y,k} \cdot 2^{v-1} n^{v-1} + \sum_{y \in Y_k} p_{y,k} \cdot 2^{v-1} y^{v-1} \right) \\
&\leq \mathsf{err}(k) + v \cdot c'(k) \cdot \left(2^{v-1} n^{v-1} + 2^{v-1} \sum_{y \in Y_k} p_{y,k} \cdot y^{v-1} \right).
\end{aligned}$$

Since $y > 0$, we have $y^{v-1} \leq y^v + 1$. Then:

$$\begin{aligned}
\sigma_{p,v}(q, D, \hat{f}_k) &\leq \mathsf{err}(k) + v \cdot c'(k) \cdot \left(2^{v-1} n^{v-1} + 2^{v-1} \sum_{y \in Y_k} p_{y,k} \cdot (y^v + 1) \right) \\
&\leq \mathsf{err}(k) + v \cdot c'(k) \cdot \left(2^{v-1} n^{v-1} + 2^{v-1} \cdot (\mathsf{err}(k) + 1) \right) \\
&\leq \mathsf{err}(k) + c(k) \cdot \left(2^{v-1} v \cdot n^v + 2^{v-1} v \cdot n \cdot (\mathsf{err}(k) + 1) \right).
\end{aligned}$$

Since err is polynomial and n, v are constants, this completes the proof.

Towards Privacy for Social Networks: A Zero-Knowledge Based Definition of Privacy

Johannes Gehrke, Edward Lui, and Rafael Pass*

Cornell University
{johannes,luied,rafael}@cs.cornell.edu

Abstract. We put forward a zero-knowledge based definition of privacy. Our notion is strictly stronger than the notion of differential privacy and is particularly attractive when modeling privacy in social networks. We furthermore demonstrate that it can be meaningfully achieved for tasks such as computing averages, fractions, histograms, and a variety of graph parameters and properties, such as average degree and distance to connectivity. Our results are obtained by establishing a connection between zero-knowledge privacy and sample complexity, and by leveraging recent sublinear time algorithms.

1 Introduction

Data privacy is a fundamental problem in today's information age. Enormous amounts of data are collected by government agencies, search engines, social networking systems, hospitals, financial institutions, and other organizations, and are stored in databases. There are huge social benefits in analyzing this data; however, it is important that sensitive information about individuals who have contributed to the data is not leaked to users analyzing the data. Thus, one of the main goals is to release statistical information about the population who have contributed to the data without breaching their individual privacy.

Many privacy definitions and schemes have been proposed in the past (see [4] and [11] for surveys). However, many of them have been shown to be insufficient by describing realistic attacks on such schemes (e.g., see [19]). The notion of *differential privacy* [8,7], however, has remained strong and resilient to these attacks. Differential privacy requires that when one person's data is added or removed from the database, the output of the database access mechanism changes very little so that the output before and after the change are "ϵ-close" (where a specific notion of closeness of distributions is used). This notion has quickly become the standard notion of privacy, and mechanisms for releasing a variety of functions (including histogram queries, principal component analysis, learning, and many more (see [6] for a recent survey)) have been developed.

---

* Pass is supported in part by a Microsoft New Faculty Fellowship, NSF CAREER Award CCF-0746990, AFOSR Award F56-8414, BSF Grant 2006317 and I3P grant 2006CS-001-0000001-02.

Y. Ishai (Ed.): TCC 2011, LNCS 6597, pp. 432–449, 2011.

As we shall argue, however, although differential privacy provides a strong privacy guarantee, there are realistic social network settings where these guarantees might not be strong enough. Roughly speaking, the notion of differential privacy can be rephrased as requiring that whatever an adversary learns about an individual could have been recovered about the individual had the adversary known every other individual in the database (see the appendix of [8] for a formalization of this statement). Such a privacy guarantee is not sufficiently strong in the setting of social networks where an individual's *friends* are strongly correlated with the individual; in essence, "if I know your friends, I know you". (Indeed, a recent study [17] indicates that an individual's sexual orientation can be accurately predicted just by looking at the person's Facebook friends.) We now give a concrete example to illustrate how a differentially private mechanism can violate the privacy of individuals in a social network setting.

Example 1 (Democrats vs. Republicans). Consider a social network of n people that are grouped into cliques of size 200. In each clique, either at least 80% of the people are Democrats, or at least 80% are Republicans. However, assume that the number of Democrats overall is roughly the same as the number of Republicans. Now, consider a mechanism that computes the proportion (in $[0,1]$) of Democrats in each clique and adds just enough Laplacian noise to satisfy ϵ-differential privacy for a small ϵ, say $\epsilon = 0.1$. For example, to achieve ϵ-differential privacy, it suffices to add $Lap(\frac{1}{200\epsilon})$ noise[1] to each clique independently, since if a single person changes his or her political preference, the proportion for the person's clique changes by $\frac{1}{200}$ (see Proposition 1 in [8]).

Since the mechanism satisfies ϵ-differential privacy for a small ϵ, one may think that it is safe to release such information without violating the privacy of any particular person. That is, the released data should not allow us to guess correctly with probability significantly greater than $\frac{1}{2}$ whether a particular person is a Democrat or a Republican. However, this is not the case. With $\epsilon = 0.1$, $Lap(\frac{1}{200\epsilon})$ is a small amount of noise, so with high probability, the data released will tell us the main political preference for any particular clique. An adversary that knows which clique a person is in will be able to correctly guess the political preference of that person with probability close to 80%.

Remark 1. In the above example, we assume that the graph structure is known and that the adversary can identify what clique an individual is in. Such information is commonly available: Graph structures of (anonymized) social networks are often released; these may include a predefined or natural clustering of the people (nodes) into cliques. Furthermore, an adversary may often also figure out the identity of various nodes in the graph (see [1,16]); in fact, by participating in the social network before the anonymized graph is published, an adversary can even target specific individuals of his or her choice (see [1]).

Differential privacy says that the output of the mechanism does not depend much on any particular individual's data in the database. Thus, in the above example, a person has little reason not to truthfully report his political preference.

[1] $Lap(\lambda)$ is the Laplace distribution with probability density function $f_\lambda(x) = \frac{1}{2\lambda}e^{\frac{|x|}{\lambda}}$.

However, this does not necessarily imply that the mechanism does not violate the person's privacy. In situations where a social network provides auxiliary information about an individual, that person's privacy can be violated even if he decides to not have his information included!

It is already known that differential privacy may not provide a strong enough privacy guarantee when an adversary has specific auxiliary information about an individual. For example, it was pointed out in [7] that if an adversary knows the auxiliary information "person A is two inches shorter than the average American woman", and if a differentially private mechanism accurately releases the average height of American women, then the adversary learns person A's height (which is assumed to be sensitive information in this example). In this example, the adversary has very specific auxiliary information about an individual that is usually hard to obtain. However, in the Democrats vs. Republicans example, the auxiliary information (the graph and clique structure) about individuals is more general and more easily accessible. Since social network settings contain large amounts of auxiliary information and correlation between individuals, differential privacy is usually not strong enough in such settings.

One may argue that there are versions of differential privacy that protect the privacy of groups of individuals, and that the mechanism in the Democrats vs. Republicans example does not satisfy these stronger definitions of privacy. While this is true, the main point here is that differential privacy will not protect the privacy of an individual, even though the definition is designed for individual privacy. Furthermore, even if we had used a differentially private mechanism that ensures privacy for groups of size 200 (i.e., the size of each clique), it might still be possible to deduce information about an individual by looking at the *friends of the friends* of the individual; this includes a significantly larger number of individuals[2].

1.1 Towards a Zero-Knowledge Definition of Privacy

In 1977, Dalenius [5] stated a privacy goal for statistical databases: anything about an individual that can be learned from the database can also be learned without access to the database. This would be a very desirable notion of privacy. Unfortunately, Dwork and Naor [7,9] demonstrated a general impossibility result showing that a formalization of Dalenius's goal along the lines of semantic security for cryptosystems cannot be achieved, assuming that the database gives any non-trivial utility.

Our aim is to provide a privacy definition along the lines of Dalenius, and more precisely, relying on the notion of zero-knowledge from cryptography. In this context, the traditional notion of zero-knowledge says that an adversary gains essentially "zero additional knowledge" by accessing the mechanism. More precisely, whatever an adversary can compute by accessing the mechanism can essentially also be computed without accessing the mechanism. A mechanism

[2] The number of "friends of friends" is usually larger than the square of the number of friends (see [23]).

satisfying this property would be private but utterly useless, since the mechanism provides essentially no information. The whole point of releasing data is to provide utility; thus, this extreme notion of zero-knowledge, which we now call "complete zero-knowledge", is not very applicable in this setting.

Intuitively, we want the mechanism to not release any additional information beyond some *"aggregate information"* that is considered acceptable to release. To capture this requirement, we use the notion of a "simulator" from zero-knowledge, and we require that a simulator with the acceptable aggregate information can essentially compute whatever an adversary can compute by accessing the mechanism. Our zero-knowledge privacy definition is thus stated relative to some class of algorithms providing acceptable aggregate information.

Aggregate Information. The question is how to define appropriate classes of aggregate information. We focus on the case where the aggregate information is any information that can be obtained from k random samples/rows (each of which corresponds to one individual's data) of the database, where the data of the person the adversary wants to attack has been concealed. The value of k can be carefully chosen so that the aggregate information obtained does not allow one to infer (much) information about the concealed data. The simulator is given this aggregate information and has to compute what the adversary essentially computes, even though the adversary has access to the mechanism. This ensures that the mechanism does not release any additional information beyond this "k random sample" aggregate information given to the simulator.

Differential privacy can be described using our zero-knowledge privacy definition by considering simulators that are given aggregate information consisting of the data of all but one individual in the database; this is the same as aggregate information consisting of "k random samples" with $k = n$, where n is the number of rows in the database (recall that the data of the individual the adversary wants to attack is concealed), which we formally prove later. For k less than n, such as $k = \log n$ or $k = \sqrt{n}$, we obtain notions of privacy that are stronger than differential privacy. For example, we later show that the mechanism in the Democrats vs. Republicans example does not satisfy our zero-knowledge privacy definition when $k = o(n)$ and n is sufficiently large.

We may also consider more general models of aggregate information that are specific to graphs representing social networks; in this context we focus on random samples with some exploration of the neighborhood of each sample.

1.2 Our Results

We consider two different settings for releasing information. In the first setting, we consider statistical (row) databases in a setting where an adversary might have auxiliary information, such as from a social network, and we focus on releasing traditional statistics (e.g., averages, fractions, histograms, etc.) from a database. As explained earlier, differential privacy may not be strong enough in such a setting, so we use our zero-knowledge privacy definition instead. In the second setting, we consider graphs with personal data that represent social networks, and we focus on releasing information directly related to a social network, such as properties of the graph structure.

Setting #1. Computing functions on databases with zero-knowledge privacy: In this setting, we focus on computing functions mapping databases to $\mathbb{R}^m$. Our main result is a characterization of the functions that can be released with zero-knowledge privacy in terms of their *sample complexity*—i.e., how accurate the function can be approximated using random samples from the input database. More precisely, functions with low sample complexity can be computed accurately by a zero-knowledge private mechanism, and vice versa. It is already known that functions with low sample complexity can be computed with differential privacy (see [8]), but here we show that the stronger notion of zero-knowledge privacy can be achieved. In this result, the zero-knowledge private mechanism we construct simply adds Laplacian noise appropriately calibrated to the sample complexity of the function.

Many common queries on statistical databases have low sample complexity, including averages, sum queries, and coarse histogram queries. (In general, it would seem that any "meaningful" query function for statistical databases should have relatively low sample complexity if we think of the rows of the database as random samples from some large underlying population). As a corollary of our characterization we get zero-knowledge private mechanisms for all these functions providing decent utility guarantees. These results can be found in Section 3.

Setting #2. Releasing graph structure information with zero-knowledge privacy: In this setting, we consider a graph representing a social network, and we focus on privately releasing information about the structure of the graph. We use our zero-knowledge privacy definition, since the released information can be combined with auxiliary information such as an adversary's knowledge and/or previously released data (e.g., graph structure information) to breach the privacy of individuals.

The connection between sample complexity and zero-knowledge privacy highlights an interesting connection between *sublinear time algorithms* and privacy. As it turns out, many of the recently developed sublinear algorithms on graphs proceed by picking random samples (and next performing some local exploration); we are able to leverage these algorithms to privately release graph structure information, such as average degree and distance to properties such as connectivity and cycle-freeness. We discuss these results in Section 4.

2 Zero-Knowledge Privacy

2.1 Definitions

Let $\mathcal{D}$ be the class of all databases whose rows are tuples from some relation/universe X. For convenience, we will assume that X contains a tuple $\perp$, which can be used to conceal the true value of a row. Given a database D, let

$|D|$ denote the number of rows in D. For any integer n, let $[n]$ denote the set $\{1, \ldots, n\}$. For any database $D \in \mathcal{D}$, any integer $i \in [|D|]$, and any $v \in X$, let (D_{-i}, v) denote the database D with row i replaced by the tuple v.

In this paper, mechanisms, adversaries, and simulators are simply randomized algorithms that play certain roles in our definitions. Let San be a mechanism that operates on databases in $\mathcal{D}$. For any database $D \in \mathcal{D}$, any adversary A, and any $z \in \{0,1\}^*$, let $Out_A(A(z) \leftrightarrow San(D))$ denote the random variable representing the output of A on input z after interacting with the mechanism San operating on the database D. Note that San can be interactive or non-interactive. If San is non-interactive, then $San(D)$ sends information (e.g., a sanitized database) to A and then halts immediately; the adversary A then tries to breach the privacy of some individual in the database D.

Let agg be any class of randomized algorithms that provide aggregate information to simulators, as described in Section 1.1. We refer to agg as a *model of aggregate information.*

Definition 1. *We say that San is* **ϵ-zero-knowledge private with respect to** agg *if there exists a $T \in agg$ such that for every adversary A, there exists a simulator S such that for every database $D \in X^n$, every $z \in \{0,1\}^*$, every integer $i \in [n]$, and every $W \subseteq \{0,1\}^*$, the following hold:*

- $\Pr[Out_A(A(z) \leftrightarrow San(D)) \in W] \leq e^{\epsilon} \cdot \Pr[S(z, T(D_{-i}, \perp), i, n) \in W]$
- $\Pr[S(z, T(D_{-i}, \perp), i, n) \in W] \leq e^{\epsilon} \cdot \Pr[Out_A(A(z) \leftrightarrow San(D)) \in W]$

The probabilities are over the random coins of San and A, and T and S, respectively.

Intuitively, the above definition says that whatever an adversary can compute by accessing the mechanism can essentially also be computed without accessing the mechanism but with certain aggregate information (specified by agg). The adversary in the latter scenario is represented by the simulator S. The definition requires that the adversary's output distribution is close to that of the simulator. This ensures that the mechanism essentially does not release any additional information beyond what is allowed by agg. When the algorithm T provides aggregate information to the simulator S, the data of individual i is concealed so that the aggregate information does not depend directly on individual i's data. However, in the setting of social networks, the aggregate information may still depend on people's data that are correlated with individual i in reality, such as the data of individual i's friends. Thus, the role played by agg is very important in the context of social networks.

To measure the closeness of the adversary's output and the simulator's output, we use the same closeness measure as in differential privacy (as opposed to, say, statistical difference) for the same reasons. As explained in [8], consider a mechanism that outputs the contents of a randomly chosen row. Suppose agg is defined so that it includes the algorithm that simply outputs its input $(D_{-i}, \perp)$ to the simulator (which is the case of differential privacy; see Section 1.1 and 2.2). Then, a simulator can also choose a random row and then simulate the adversary with the chosen row sent to the simulated adversary. The real adversary's output will be very close to the simulator's output in statistical difference ($1/n$ to be precise); however, it is clear that the mechanism always leaks private information about some individual.

Remark 2. Our ϵ-zero-knowledge privacy definition can be easily extended to $(\epsilon, \delta(\cdot))$-zero-knowledge privacy, where we also allow an additive error of $\delta(n)$ on the RHS of the inequalities. We can further extend our definition to $(c, \epsilon, \delta(\cdot))$-zero-knowledge privacy to protect the privacy of any group of c individuals simultaneously. To obtain this more general definition, we would change "$i \in [n]$" to "$I \subseteq [n]$ with $|I| \leq c$", and "$S(z, (D_{-i}, \perp), i, n)$" to "$S(z, (D_{-I}, \perp), I, n)$". We use this more general definition when we consider group privacy.

Remark 3. In our zero-knowledge privacy definition, we consider computationally unbounded simulators. We can also consider PPT simulators by requiring that the mechanism San and the adversary A are PPT algorithms, and agg is a class of PPT algorithms. All of these algorithms would be PPT in n, the size of the database. With minor modifications, the results of this paper would still hold in this case.

The choice of agg determines the type and amount of aggregate information given to the simulator, and should be decided based on the context in which the zero-knowledge privacy definition is used. The aggregate information should not depend much on data that is highly correlated with the data of a single person, since such aggregate information may be used to breach the privacy of that person. For example, in the context of social networks, such aggregate information should not depend much on any person and the people closely connected to that person, such as his or her friends. By choosing agg carefully, we ensure that the mechanism essentially does not release any additional information beyond what is considered acceptable. We first consider the model of aggregate information where T in the definition of zero-knowledge privacy chooses $k(n)$ random samples. Let $k : \mathbb{N} \to \mathbb{N}$ be any function.

- $RS(k(\cdot)) = k(\cdot)$ random samples: the class of algorithms T such that on input a database $D \in X^n$, T chooses $k(n)$ random samples (rows) from D uniformly without replacement, and then performs any computation on these samples without reading any of the other rows of D. Note that with such samples, T can emulate choosing $k(n)$ random samples with replacement, or a combination of without replacement and with replacement.

$k(n)$ should be carefully chosen so that the aggregate information obtained does not allow one to infer (much) information about the concealed data. For $k(n) = 0$, the simulator is given no aggregate information at all, which is the case of complete zero-knowledge. For $k(n) = n$, the simulator is given all the rows of the original database except for the target individual i, which is the case of differential privacy (as we prove later). For $k(n)$ strictly in between 0 and n, we obtain notions of privacy that are stronger than differential privacy. For example, one can consider $k(n) = o(n)$, such as $k(n) = \log n$ or $k(n) = \sqrt{n}$.

In the setting of a social network, $k(n)$ can be chosen so that when $k(n)$ random samples are chosen from $(D_{-i}, \perp)$, with very high probability, for (almost) all individuals j, very few of the $k(n)$ chosen samples will be in individual j's local neighborhood in the social network graph. This way, the aggregate information released by the mechanism depends very little on data that is highly correlated with the data of a single individual. The choice of $k(n)$ would depend on various properties of the graph structure, such as clustering coefficient, edge density, and degree distribution. The choice of $k(n)$ would also depend on the amount of correlation between the data of adjacent or close vertices (individuals) in the graph, and the type of information released by the mechanism. In this model of aggregate information, vertices (individuals) in the graph with more adjacent vertices (e.g., representing friends) may have less privacy than those with fewer adjacent vertices. However, this is often the case in social networks, where having more links/connections to other people may result in less privacy.

In the remainder of this section, we focus primarily on the $RS(k(\cdot))$ model of aggregate information. In Section 4, we consider other models of aggregate information that take more into consideration the graph structure of a social network. Note that zero-knowledge privacy does not necessarily guarantee that the privacy of every individual is completely protected. Zero-knowledge privacy is defined with respect to a model of aggregate information, and such aggregate information may still leak some sensitive information about an individual in certain scenarios.

Composition: Just as for differentially private mechanisms, mechanisms that are ϵ-zero-knowledge private with respect to $RS(k(\cdot))$ also compose nicely.

Proposition 1. *Suppose San_1 is ϵ_1-zero-knowledge private with respect to $RS(k_1(\cdot))$ and San_2 is ϵ_2-zero-knowledge private with respect to $RS(k_2(\cdot))$. Then, the mechanism obtained by composing San_1 with San_2 is $(\epsilon_1+\epsilon_2)$-zero-knowledge private with respect to $RS((k_1+k_2)(\cdot))$.*

See the full version of this paper ([12]) for the proof.

Graceful Degradation for Group Privacy: A nice feature of differential privacy is that ϵ-differential privacy implies $(c, c\epsilon)$-differential privacy for groups of size c (see [7] and the appendix in [8]). However, the $c\epsilon$ appears in the exponent of e in the definition of $(c, c\epsilon)$-differential privacy, so the degradation is exponential in c. Thus, the group privacy guarantee implied by ϵ-differential privacy is not very meaningful unless the group size c is small. We do not have a group privacy guarantee for pure ϵ-zero-knowledge privacy; however, we do have a group privacy guarantee for $(\epsilon, \delta(\cdot))$-zero-knowledge privacy with respect to $RS(k(\cdot))$ that does not degrade at all for ϵ, and only degrades linearly for $\delta(\cdot)$ with increasing group size.

Proposition 2. *Suppose San is $(\epsilon, \delta(\cdot))$-zero-knowledge private with respect to $RS(k(\cdot))$. Then, for every $c \geq 1$, San is also $(c, \epsilon, \delta_c(\cdot))$-zero-knowledge private with respect to $RS(k(\cdot))$, where $\delta_c(n) = \delta(n) + e^{\epsilon}(c-1) \cdot \frac{k(n)}{n}$.*

See the full version of this paper for the proof. Intuitively, for $k(n)$ sufficiently smaller than n, $(\epsilon, \delta(\cdot))$-zero-knowledge privacy with respect to $RS(k(\cdot))$ actually implies some notion of group privacy, since the algorithm T (in the privacy definition) chooses each row with probability $k(n)/n$. Thus, T chooses any row of a fixed group of c rows with probability at most $ck(n)/n$. If this probability is very small, then the output of T and thus the simulator S does not depend much on any group of c rows.

2.2 Differential Privacy vs. Zero-Knowledge Privacy

In this section, we compare differential privacy to our zero-knowledge privacy definition. We first state the definition of differential privacy in a form similar to our zero-knowledge privacy definition in order to more easily compare the two. For any pair of databases $D_1, D_2 \in X^n$, let $H(D_1, D_2)$ denote the number of rows in which D_1 and D_2 differ, comparing row-wise.

Definition 2. *We say that San is* **ϵ-differentially private** *if for every adversary A, every $z \in \{0,1\}^*$, every pair of databases $D_1, D_2 \in X^n$ with $H(D_1, D_2) \leq 1$, and every $W \subseteq \{0,1\}^*$, we have*

$$\Pr[Out_A(A(z) \leftrightarrow San(D_1)) \in W] \leq e^{\epsilon} \cdot \Pr[Out_A(A(z) \leftrightarrow San(D_2)) \in W],$$

where the probabilities are over the random coins of San and A. For (c, ϵ)-
differential privacy *(for groups of size c), the "$H(D_1, D_2) \leq 1$" is changed to "$H(D_1, D_2) \leq c$".*

Proposition 3. *Suppose San is ϵ-zero-knowledge private with respect to any class agg. Then, San is 2ϵ-differentially private.*

Proposition 4. *Suppose San is ϵ-differentially private. Then, San is ϵ-zero-knowledge private with respect to $RS(n)$.*

See the full version of this paper for the proof of Propositions 3 and 4.

Remark 4. If we consider PPT simulators in the definition of zero-knowledge privacy instead of computationally unbounded simulators, then we require *San* in Proposition 4 to be PPT as well.

Combining Propositions 3 and 4, we see that our zero-knowledge privacy definition includes differential privacy as a special case (up to a factor of 2 for ϵ).

2.3 Revisiting the Democrats vs. Republicans Example

Recall the Democrats vs. Republicans example in the introduction. The mechanism in the example is ϵ-differentially private for some small ϵ, even though the privacy of individuals is clearly violated. However, the mechanism is not zero-knowledge private in general. Suppose that the people's political preferences are stored in a database $D \in X^n$.

Proposition 5. *Fix $\epsilon > 0$, $c \geq 1$, and any function $k(\cdot)$ such that $k(n) = o(n)$. Let San be a mechanism that on input $D \in X^n$ computes the proportion of Democrats in each clique and adds $Lap(\frac{c}{200\epsilon})$ noise to each proportion independently. Then, San is (c, ϵ)-differentially private, but for every sufficiently large n, San is not ϵ'-zero-knowledge private with respect to $RS(k(\cdot))$ for any constant $\epsilon' > 0$.*

See the full version of this paper for the proof. Intuitively, the last part of the proposition holds because for sufficiently large n, with high probability there exists some clique such that an adversary having only $k(n) = o(n)$ random samples would not have any samples in that clique. Thus, with high probability, there exists some clique that the adversary knows nothing about. Therefore, the adversary does gain knowledge by accessing the mechanism, which gives some information about every clique since the amount of noise added to each clique is constant.

Remark 5. In the Democrats vs. Republicans example, even if San adds $Lap(\frac{1}{\epsilon})$ noise to achieve $(200, \epsilon)$-differential privacy so that the privacy of each clique (and thus each person) is protected, the mechanism would still fail to be ϵ'-zero-knowledge private with respect to $RS(k(\cdot))$ for any constant $\epsilon' > 0$ when n is sufficiently large (see Proposition 5). Thus, zero-knowledge privacy with respect to $RS(k(\cdot))$ with $k(n) = o(n)$ seems to provide an unnecessarily strong privacy guarantee in this particular example. However, this is mainly because the clique size is fixed and known to be 200, and we have assumed that the only correlation between people's political preferences that exists is within a clique. In a more realistic social network, there would be cliques of various sizes, and the correlation between people's data would be more complicated. For example, an adversary knowing your friends' friends may still be able to infer a lot of information about you.

3 Characterizing Zero-Knowledge Privacy

In this section, we focus on constructing zero-knowledge private mechanisms that compute a function mapping databases in X^n to $\mathbb{R}^m$, and we characterize the set of functions that can be computed with zero-knowledge privacy. These are precisely the functions with low sample complexity, i.e., can be approximated (accurately) using only limited information from the database, such as k random samples.

We quantify the error in approximating a function $g : X^n \to \mathbb{R}^m$ using L_1 distance. Let the L_1-sensitivity of g be defined by $\Delta(g) = \max\{||g(D') - g(D'')||_1 : D', D'' \in X^n \text{ s.t. } H(D', D'') \leq 1\}$. Let $\mathcal{C}$ be any class of randomized algorithms.

Definition 3. *A function $g : X^n \to \mathbb{R}^m$ is said to have* (δ, β)-**sample complexity** *with respect to $\mathcal{C}$ if there exists an algorithm $T \in \mathcal{C}$ such that for every input $D \in X^n$, we have $T(D) \in \mathbb{R}^m$ and*

$$\Pr[||T(D) - g(D)||_1 \leq \delta] \geq 1 - \beta.$$

T is said to be a (δ, β)-**sampler** *for g with respect to $\mathcal{C}$.*

Remark 6. If we consider PPT simulators in the definition of zero-knowledge privacy instead of computationally unbounded simulators, then we would require here that $\mathcal{C}$ is a class of PPT algorithms (PPT in n, the size of the database). Thus, in the definition of (δ, β)-sample complexity, we would consider a family of functions (one for each value of n) that can be computed in PPT, and the sampler T would be PPT in n.

It was shown in [8] that functions with low sample complexity with respect to $RS(k(\cdot))$ have low sensitivity as well.

Lemma 1 ([8]). *Suppose $g : X^n \to \mathbb{R}^m$ has (δ, β)-sample complexity with respect to $RS(k(\cdot))$ for some $\beta < \frac{1-k(n)/n}{2}$. Then, $\Delta(g) \leq 2\delta$.*

As mentioned in [8], the converse of the above lemma is not true, i.e., not all functions with low sensitivity have low sample complexity (see [8] for an example). This should be no surprise, since functions with low sensitivity have accurate differentially private mechanisms, while functions with low sample complexity have accurate zero-knowledge private mechanisms. We already know that zero-knowledge privacy is stronger than differential privacy, as illustrated by the Democrats vs. Republicans example.

We now state how the sample complexity of a function is related to the amount of noise a mechanism needs to add to the function value in order to achieve a certain level of zero-knowledge privacy.

Proposition 6. *Suppose $g : X^n \to [a, b]^m$ has (δ, β)-sample complexity with respect to some $\mathcal{C}$. Then, the mechanism $San(D) = g(D) + (X_1, \ldots, X_m)$, where $X_j \sim Lap(\lambda)$ for $j = 1, \ldots, m$ independently, is* $\ln((1-\beta)e^{\frac{\Delta(g)+\delta}{\lambda}} + \beta e^{\frac{(b-a)m}{\lambda}})$*-zero-knowledge private with respect to $\mathcal{C}$.*

The intuition is that the sampling error gets blurred by the noise added.

Proof. Let T be a (δ, β)-sampler for g with respect to $\mathcal{C}$. Let A be any adversary. Let S be a simulator that, on input $(z, T(D_{-i}, \perp), i, n)$, first checks whether $T(D_{-i}, \perp)$ is in $[a, b]^m$; if not, S projects $T(D_{-i}, \perp)$ onto the set $[a, b]^m$ (with respect to L_1 distance) so that the accuracy of $T(D_{-i}, \perp)$ is improved and $||g(D) - T(D_{-i}, \perp)||_1 \leq (b-a)m$ always holds, which we use later. From here on, $T(D_{-i}, \perp)$ is treated as a random variable that reflects the possible modification S may perform. The simulator S computes $T(D_{-i}, \perp) + (X_1, \ldots, X_m)$, which we will denote using the random variable $S'(z, T(D_{-i}, \perp), i, n)$. S then simulates the computation of $A(z)$ with $S'(z, T(D_{-i}, \perp), i, n)$ sent to A as a message, and outputs whatever A outputs.

Let $D \in X^n$, $z \in \{0,1\}^*$, $i \in [n]$. Fix $x \in T(D_{-i}, \perp)$ and $s \in \mathbb{R}^m$. Then, we have

$$
\begin{aligned}
&\max\left\{\frac{f_\lambda(s-g(D))}{f_\lambda(s-x)}, \frac{f_\lambda(s-x)}{f_\lambda(s-g(D))}\right\} \\
&= \max\left\{e^{(\frac{1}{\lambda}\cdot(||s-x||_1-||s-g(D)||_1))}, e^{(\frac{1}{\lambda}\cdot(||s-g(D)||_1-||s-x||_1))}\right\} \\
&\le e^{(\frac{1}{\lambda}\cdot||g(D)-x||_1)} \le e^{(\frac{1}{\lambda}\cdot(||g(D)-g(D_{-i},\perp)||_1+||g(D_{-i},\perp)-x||_1))} \\
&\le e^{(\frac{1}{\lambda}\cdot(\Delta(g)+||g(D_{-i},\perp)-x||_1))}. \qquad (1)
\end{aligned}
$$

Since $||g(D)-x||_1 \le (b-a)m$ always holds, we also have

$$
\max\left\{\frac{f_\lambda(s-g(D))}{f_\lambda(s-x)}, \frac{f_\lambda(s-x)}{f_\lambda(s-g(D))}\right\} \le e^{(\frac{1}{\lambda}\cdot||g(D)-x||_1)} \le e^{\frac{(b-a)m}{\lambda}}. \qquad (2)
$$

Since T is a (δ,β)-sampler for g, we have $\Pr[||g(D_{-i},\perp)-T(D_{-i},\perp)||_1 \le \delta] \ge 1-\beta$. Thus, using (1) and (2) above, we have

$$
\ln\left(\frac{\sum_{x\in T(D_{-i},\perp)} f_\lambda(s-x)\cdot \Pr[T(D_{-i},\perp)=x]}{f_\lambda(s-g(D))}\right) \le \ln((1-\beta)e^{\frac{\Delta(g)+\delta}{\lambda}}+\beta e^{\frac{(b-a)m}{\lambda}}).
$$

Now, using (1) and (2) again, we also have

$$
\begin{aligned}
&\ln\left(\frac{f_\lambda(s-g(D))}{\sum_{x\in T(D_{-i},\perp)} f_\lambda(s-x)\cdot \Pr[T(D_{-i},\perp)=x]}\right) \\
&= -\ln\left(\frac{\sum_{x\in T(D_{-i},\perp)} f_\lambda(s-x)\cdot \Pr[T(D_{-i},\perp)=x]}{f_\lambda(s-g(D))}\right) \\
&\le -\ln((1-\beta)e^{-\frac{\Delta(g)+\delta}{\lambda}}+\beta e^{-\frac{(b-a)m}{\lambda}}) = \ln(((1-\beta)e^{-\frac{\Delta(g)+\delta}{\lambda}}+\beta e^{-\frac{(b-a)m}{\lambda}})^{-1}) \\
&\le \ln((1-\beta)e^{\frac{\Delta(g)+\delta}{\lambda}}+\beta e^{\frac{(b-a)m}{\lambda}}),
\end{aligned}
$$

where the last inequality follows from the fact that the function $f(x)=x^{-1}$ is convex for $x>0$. Then, for every $s\in\mathbb{R}^n$, we have

$$
\begin{aligned}
&\left|\ln\left(\frac{\Pr[San(D)=s]}{\Pr[S'(z,T(D_{-i},\perp),i,n)=s]}\right)\right| \\
&= \left|\ln\left(\frac{f_\lambda(s-g(D))}{\sum_{x\in T(D_{-i},\perp)} f_\lambda(s-x)\cdot \Pr[T(D_{-i},\perp)=x]}\right)\right| \\
&\le \ln((1-\beta)e^{\frac{\Delta(g)+\delta}{\lambda}}+\beta e^{\frac{(b-a)m}{\lambda}}).
\end{aligned}
$$

Thus, for every $W\subseteq\{0,1\}^*$, we have $\left|\ln\left(\frac{\Pr[Out_A(A(z)\leftrightarrow San(D))\in W]}{\Pr[S(z,T(D_{-i},\perp),i,n)\in W]}\right)\right| \le \ln((1-\beta)e^{\frac{\Delta(g)+\delta}{\lambda}}+\beta e^{\frac{(b-a)m}{\lambda}})$. □

Corollary 1. *Suppose* $g: X^n \to [a,b]^m$ *has* (δ,β)*-sample complexity with respect to* $RS(k(\cdot))$ *for some* $\beta < \frac{1-k(n)/n}{2}$*. Then, the mechanism* $San(D) = g(D)+(X_1,\ldots,X_m)$*, where* $X_j \sim Lap(\lambda)$ *for* $j=1,\ldots,m$ *independently, is* $\ln((1-\beta)e^{\frac{3\delta}{\lambda}}+\beta e^{\frac{(b-a)m}{\lambda}})$*-zero-knowledge private with respect to* $RS(k(\cdot))$*.*

Proof. This follows from combining Proposition 6 and Lemma 1.

Using Proposition 6, we can recover the basic mechanism in [8] that is ϵ-differentially private.

Corollary 2. *Let $g : X^n \to [a,b]^m$ and $\epsilon > 0$. A mechanism San for g that adds $Lap(\frac{\Delta(g)}{\epsilon})$ noise to $g(D)$ is ϵ-zero-knowledge private with respect to $RS(n)$.*

Proof. We note that every function $g : X^n \to \mathbb{R}^m$ has $(0,0)$-sample complexity with respect to $RS(n)$. The corollary follows by applying Proposition 6.

We now show how the zero-knowledge privacy and utility properties of a mechanism computing a function is related to the sample complexity of the function. A class of algorithms agg is said to be *closed under postprocessing* if for any $T \in agg$ and any algorithm M, the composition of M and T (i.e., the algorithm that first runs T and then runs M on the output of T) is also in agg. We note that $RS(k(\cdot))$ is closed under postprocessing.

Proposition 7. *Let agg be any class of algorithms that is closed under postprocessing, and suppose a function $g : X^n \to \mathbb{R}^m$ has a mechanism San such that the following hold:*

– *Utility:* $\Pr[||San(D) - g(D)||_1 \le \delta] \ge 1 - \beta$ *for every $D \in X^n$*
– *Privacy: San is ϵ-zero-knowledge private with respect to agg.*

Then, g has $(\delta, \frac{\beta+(e^\epsilon-1)}{e^\epsilon})$-sample complexity with respect to agg.

See the full version of this paper for the proof. The intuition is that the zero-knowledge privacy of San guarantees that San can be simulated by a simulator S that is given aggregate information provided by some algorithm $T \in agg$. Thus, an algorithm that runs T and then S will be able to approximate g with accuracy similar to that of San.

3.1 Some Simple Examples of Zero-Knowledge Private Mechanisms

Example 2 (**Averages**). Fix $n > 0$, $k = k(n)$. Let $avg : [0,1]^n \to [0,1]$ be defined by $avg(D) = \frac{\sum_{i=1}^n D_i}{n}$, and let $San(D) = avg(D) + Lap(\lambda)$, where $\lambda > 0$. Let T be an algorithm that, on input a database $D \in [0,1]^n$, chooses k random samples from D (uniformly), and then outputs the average of the k random samples. By Hoeffding's inequality, we have $\Pr[|T(D) - avg(D)| \le \delta] \ge 1 - 2e^{-2k\delta^2}$. Thus, avg has $(\delta, 2e^{-2k\delta^2})$-sample complexity with respect to $RS(k(\cdot))$. By Proposition 6, San is $\ln(e^{\frac{1}{\lambda}(\frac{1}{n}+\delta)} + 2e^{\frac{1}{\lambda}-2k\delta^2})$-zero-knowledge private with respect to $RS(k(\cdot))$.

Let $\epsilon \in (0,1]$. We choose $\delta = \frac{1}{k^{1/3}}$ and $\lambda = \frac{1}{\epsilon}(\frac{1}{n} + \delta) = \frac{1}{\epsilon}(\frac{1}{n} + \frac{1}{k^{1/3}})$ so that $\ln(e^{\frac{1}{\lambda}(\frac{1}{n}+\delta)} + 2e^{\frac{1}{\lambda}-2k\delta^2}) = \ln(e^\epsilon + 2e^{\frac{\epsilon}{1/n+k^{-1/3}}-2k^{1/3}}) \le \ln(e^\epsilon + 2e^{-k^{1/3}}) \le \epsilon + 2e^{-k^{1/3}}$. Thus, we have the following result:

– By adding $Lap(\frac{1}{\epsilon}(\frac{1}{n} + \frac{1}{k^{1/3}})) = Lap(O(\frac{1}{\epsilon k^{1/3}}))$ noise to $avg(D)$, San is $(\epsilon + 2e^{-k^{1/3}})$-zero-knowledge private with respect to $RS(k(\cdot))$.

Example 3 (**Fraction of rows satisfying some property** P**).** Let $P : X \to \{0,1\}$ be the predicate representing some property of a row. Let $g : X^n \to [0,1]$ be defined by $g(D) = \frac{\sum_{i=1}^{n} P(D_i)}{n}$, which is the fraction of rows satisfying property P. Since $g(D)$ can be viewed as the average of the numbers $\{P(D_i)\}_{i=1}^{n}$, we can get the same result as in the example for averages.

Example 4 (**Histograms).** We can easily construct a zero-knowledge private mechanism (with respect to $RS(k(\cdot))$) that computes a histogram with m bins by estimating each bin count separately using $k(n)/m$ random samples each and then applying Proposition 6. Alternatively, we can construct a mechanism by composing San_i for $i = 1, \ldots, m$, where San_i is any zero-knowledge private mechanism (with respect to $RS(\frac{1}{m}k(\cdot))$) for estimating the number of rows in the ith bin, and then applying our composition result (Propsition 1).

Example 5 (**Sample and DP-Sanitize).** Our example mechanism for computing averages comes from the general connection between sample complexity and zero-knowledge privacy (Proposition 6), which holds for any model of aggregate information. For computing averages, we can actually construct a mechanism with (usually) better utility by choosing $k(n)$ random samples without replacement from the input database $D \in X^n$ and then running a differentially private mechanism on the chosen samples. It is not hard to show that such a mechanism is zero-knowledge private with respect to $RS(k(\cdot))$. In general, this "sample and DP-sanitize" method works for query functions that can be approximated using random samples (e.g., averages, fractions, and histograms), and allows us to convert differentially private mechanisms to zero-knowledge private mechanisms with respect to $RS(k(\cdot))$. (See the full version of this paper for more details.)

3.2 Answering a Class of Queries Simultaneously

In the full version of this paper, we generalize the notion of sample complexity (with respect to $RS(k(\cdot))$ to classes of query functions and show a connection between differential privacy and zero-knowledge privacy for any class of query functions with low sample complexity. In particular, we show that for any class $\mathcal{Q}$ of query functions that can be approximated simultaneously using random samples, any differentially private mechanism that is useful for $\mathcal{Q}$ can be converted to a zero-knowledge private mechanism that is useful for $\mathcal{Q}$, similar to the "Sample and DP-sanitize" method. We also show that any class of fraction queries (functions that compute the fraction of rows satisfying some property P) with low VC dimension can be approximated simultaneously using random samples, so we can use the differentially private mechanisms in [2] and [10] to obtain zero-knowledge private mechanisms (with respect to $RS(k(\cdot))$ for any class of fraction queries with low VC dimension.

4 Zero-Knowledge Private Release of Graph Properties

In this section, we first generalize statistical (row) databases to graphs with personal data so that we can model a social network and privately release information that is dependent on the graph structure. We then discuss how to model privacy in a social network, and we construct a sample of zero-knowledge private mechanisms that release certain information about the graph structure of a social network.

We represent a social network using a graph whose vertices correspond to people (or other social entities) and whose edges correspond to social links between them, and a vertex can have certain personal data associated with it. There are various types of information about a social network one may want to release, such as information about the people's data, information about the structure of the social network, and/or information that is dependent on both. In general, we want to ensure privacy of each person's personal data as well as the person's links to other people (i.e., the list of people the person is linked to via edges).

To formally model privacy in social networks, let $\mathcal{G}_n$ be a class of graphs on n vertices where each vertex includes personal data. (When we refer to a graph $G \in \mathcal{G}_n$, the graph always includes the personal data of each vertex.) The graph structure is represented by an adjacency matrix, and each vertex's personal data is represented by a tuple in X. For the privacy of individuals, we use our zero-knowledge privacy definition with some minor modifications:

- ϵ-zero-knowledge privacy is defined as before except we change "database $D \in X^n$" to "graph $D \in \mathcal{G}_n$", and we define $(D_{-i}, \perp)$ to be the graph D except the personal data of vertex i is replaced by $\perp$, and all the edges incident to vertex i are removed (by setting the corresponding entries in the adjacency matrix to 0); thus $(D_{-i}, \perp)$ is essentially D with person i's personal data and links removed.

We now consider functions $g : \mathcal{G}_n \to \mathbb{R}^m$, and we redefine the L_1-sensitivity of g to be $\Delta(g) = \max\{||g(D') - g(D'')||_1 : D', D'' \in \mathcal{G}_n \text{ s.t. } (D'_{-i}, \perp) = (D''_{-i}, \perp) \text{ for some } i \in [n]\}$. We also redefine $RS(k(\cdot))$ so that the algorithms in $RS(k(\cdot))$ are given a graph $D \in \mathcal{G}_n$ and are allowed to choose $k(n)$ random vertices without replacement and read their personal data; however, the algorithms are not allowed to read the structure of the graph, i.e., the adjacency matrix. It is easy to verify that all our previous results still hold when we consider functions $g : \mathcal{G}_n \to \mathbb{R}^m$ on graphs and use the new definition of $\Delta(g)$ and $RS(k(\cdot))$.

Since a social network has more structure than a statistical database containing a list of values, we consider more general models of aggregate information that allow us to release more information about social networks:

- $RSE(k(\cdot), s) = k(\cdot)$ random samples with exploration: the class of algorithms T such that on input a graph $D \in \mathcal{G}_n$, T chooses $k(n)$ random vertices uniformly without replacement. For each chosen vertex v, T is allowed to explore the graph locally at v until s vertices (including the sampled vertex) have been visited. The data of any visited vertex can be read. (RSE stands for "random samples with exploration".)

- $RSN(k(\cdot), d) = k(\cdot)$ random samples with neighborhood: same as $RSE(k(\cdot), s)$ except that while exploring locally, instead of exploring until s vertices have been visited, T is allowed to explore up to a distance of d from the sampled vertex. (RSN stands for "random samples with neighborhood".)

Note that these models of aggregate information include $RS(k(\cdot))$ as a special case. We can also consider variants of these models where instead of allowing the data of any visited vertex to be read, only the data of the $k(n)$ randomly chosen vertices can be read. (The data of the "explored" vertices cannot be read.)

Remark 7. In the above models, vertices (people) in the graph with high degree may be visited with higher probability than those with low degree. Thus, the privacy of these people may be less protected. However, this is often the case in social networks, where people with very many friends will naturally have less privacy than those with few friends.

We now show how to combine Proposition 6 (the connection between sample complexity and zero-knowledge privacy) with recent sublinear time algorithms to privately release information about the graph structure of a social network. For simplicity, we assume that the degree of every vertex is bounded by some constant $d_{\max}$ (which is often the case in a social network anyway)[3].

Let $\mathcal{G}_n$ be the set of all graphs on n vertices where every vertex has degree at most $d_{\max}$. We assume that $d_{\max}$ is publicly known. Let $M = \frac{d_{\max} n}{2}$ be an upper bound on the number of edges of a graph in $\mathcal{G}_n$. For any graph $G \in \mathcal{G}$, the (relative) distance from G to the some property Π, denoted $dist(G, \Pi)$, is the least number of edges that need to be modified (added/removed) in G in order to make it satisfy property Π, divided by M.

Theorem 1. *Let Conn, Eul, and CycF be the property of being connected, Eulerian*[4]*, and cycle-free, respectively. Let $\bar{d}(G)$ denote the average degree of a vertex in G. Then, for the class of graphs $\mathcal{G}_n$, we have the following results:*

1. *The mechanism $San(G) = dist(G, Conn) + Lap(\frac{2/n+\delta}{\epsilon})$ is $\epsilon + e^{-(K-\epsilon/\delta)}$-zero-knowledge private with respect to $RSE(k(\cdot), s)$, where $k(n) = O(\frac{K}{(\delta d_{\max})^2})$ and $s = O(\frac{1}{\delta d_{\max}})$.*
2. *The mechanism $San(G) = dist(G, Eul) + Lap(\frac{4/n+\delta}{\epsilon})$ is $\epsilon + e^{-(K-\epsilon/\delta)}$-zero-knowledge private with respect to $RSE(k(\cdot), s)$, where $k(n) = O(\frac{K}{(\delta d_{\max})^2})$ and $s = O(\frac{1}{\delta d_{\max}})$.*
3. *The mechanism $San(G) = dist(G, CycF) + Lap(\frac{2/n+\delta}{\epsilon})$ is $\epsilon + e^{-(K-\epsilon/\delta)}$-zero-knowledge private with respect to $RSE(k(\cdot), s)$, where $k(n) = O(\frac{K}{\delta^2})$ and $s = O(\frac{1}{\delta d_{\max}})$.*

---

[3] Weaker results can still be established without this assumption.

[4] A graph G is Eulerian if there exists a path in G that traverses every edge of G exactly once.

4. *The mechanism* $San(G) = \bar{d}(G) + Lap(\frac{2d_{\max}/n + \delta L}{\epsilon})$ *is* $\epsilon + e^{-(K - \epsilon/\delta)}$*-zero-knowledge private with respect to* $RSN(k(\cdot), 2)$*, where* $k(n) = O(K\sqrt{n}\log^2 n \cdot \frac{1}{\delta^{9/2}}\log(\frac{1}{\delta}))$*. (Here, we further assume that every graph in* $\mathcal{G}$ *has no isolated vertices and the average degree of a vertex is bounded by* L*.)*

The results of the above theorem are obtained by combining Proposition 6 (the connection between sample complexity and zero-knowledge privacy) with sublinear time algorithms from [22] (for results 1, 2, and 3) and [15] (for result 4). Intuitively, the sublinear algorithms give bounds on the sample complexity of the functions ($dist(G, Conn)$, etc.) with respect to $RSE(k(\cdot), s)$ or $RSN(k(\cdot), d)$.

There are already many (non-private) sublinear time algorithms for computing information about graphs whose accuracy is proved formally (e.g., see [15,3,22,13,18,14,24]) or demonstrated empirically (e.g, see [21,20]). We leave for future work to investigate whether these (or other) sublinear algorithms can be used to get zero-knowledge private mechanisms.

Acknowledgements

We thank Cynthia Dwork, Ilya Mironov, and Omer Reingold for helpful discussions, and we also thank the anonymous reviewers for their helpful comments.

This material is based upon work supported by the National Science Foundation under Grants 0627680 and 1012593, by the New York State Foundation for Science, Technology, and Innovation under Agreement C050061, and by the iAd Project funded by the Research Council of Norway. Any opinions, findings and conclusions or recommendations expressed in this material are those of the author(s) and do not necessarily reflect the views of the sponsors.

References

1. Backstrom, L., Dwork, C., Kleinberg, J.: Wherefore art thou r3579x?: anonymized social networks, hidden patterns, and structural steganography. In: WWW 2007: Proc. of the 16th International Conference on World Wide Web, pp. 181–190 (2007)
2. Blum, A., Ligett, K., Roth, A.: A learning theory approach to non-interactive database privacy. In: STOC 2008: Proc. of the 40th Annual ACM Symposium on Theory of Computing, pp. 609–618 (2008)
3. Chazelle, B., Rubinfeld, R., Trevisan, L.: Approximating the minimum spanning tree weight in sublinear time. SIAM J. Comput. 34(6), 1370–1379 (2005)
4. Chen, B.C., Kifer, D., LeFevre, K., Machanavajjhala, A.: Privacy-preserving data publishing. Foundations and Trends in Databases 2(1-2), 1–167 (2009)
5. Dalenius, T.: Towards a methodology for statistical disclosure control. Statistik Tidskrift 15, 429–444 (1977)
6. Dwork, C.: The differential privacy frontier. In: Reingold, O. (ed.) TCC 2009. LNCS, vol. 5444, pp. 496–502. Springer, Heidelberg (2009)
7. Dwork, C.: Differential privacy. In: Bugliesi, M., Preneel, B., Sassone, V., Wegener, I. (eds.) ICALP 2006. LNCS, vol. 4052, pp. 1–12. Springer, Heidelberg (2006)

8. Dwork, C., McSherry, F., Nissim, K., Smith, A.: Calibrating noise to sensitivity in private data analysis. In: Halevi, S., Rabin, T. (eds.) TCC 2006. LNCS, vol. 3876, pp. 265–284. Springer, Heidelberg (2006)
9. Dwork, C., Naor, M.: On the difficulties of disclosure prevention in statistical databases or the case for differential privacy (2008)
10. Dwork, C., Rothblum, G., Vadhan, S.: Boosting and differential privacy. In: Proc. of the 51st Annual IEEE Symposium on Foundations of Computer Science (2010)
11. Fung, B.C.M., Wang, K., Chen, R., Yu, P.S.: Privacy-preserving data publishing: A survey of recent developments. ACM Comput. Surv. 42(4), 1–53 (2010)
12. Gehrke, J., Lui, E., Pass, R.: Towards privacy for social networks: A zero-knowledge based definition of privacy (2011) (manuscript)
13. Goldreich, O., Ron, D.: Property testing in bounded degree graphs. In: Proc. of the 29th annual ACM Symposium on Theory of Computing, pp. 406–415 (1997)
14. Goldreich, O., Ron, D.: A sublinear bipartiteness tester for bounded degree graphs. In: Proc. of the 30th Annual ACM Symposium on Theory of Computing, pp. 289–298 (1998)
15. Goldreich, O., Ron, D.: Approximating average parameters of graphs. Random Struct. Algorithms 32(4), 473–493 (2008)
16. Hay, M., Miklau, G., Jensen, D., Towsley, D., Weis, P.: Resisting structural re-identification in anonymized social networks. Proc. VLDB Endow. 1, 102–114 (2008)
17. Jernigan, C., Mistree, B.: Gaydar (2009), `http://www.telegraph.co.uk/technology/facebook/6213590/Gay-men-can-be-identified-by-their-Facebook-friends.html`
18. Kaufman, T., Krivelevich, M., Ron, D.: Tight bounds for testing bipartiteness in general graphs. SIAM J. Comput. 33(6), 1441–1483 (2004)
19. Kifer, D.: Attacks on privacy and definetti's theorem. In: SIGMOD Conference, pp. 127–138 (2009)
20. Krishnamurthy, V., Faloutsos, M., Chrobak, M., Lao, L., Cui, J.-H., Percus, A.G.: Reducing large internet topologies for faster simulations. In: Boutaba, R., Almeroth, K.C., Puigjaner, R., Shen, S., Black, J.P. (eds.) NETWORKING 2005. LNCS, vol. 3462, pp. 328–341. Springer, Heidelberg (2005)
21. Leskovec, J., Faloutsos, C.: Sampling from large graphs. In: KDD 2006: Proc. of the 12th ACM SIGKDD International Conference on Knowledge Discovery and Data Mining, pp. 631–636 (2006)
22. Marko, S., Ron, D.: Approximating the distance to properties in bounded-degree and general sparse graphs. ACM Trans. Algorithms 5(2), 1–28 (2009)
23. Newman, M.E.J.: Ego-centered networks and the ripple effect. Social Networks 25(1), 83–95 (2003)
24. Parnas, M., Ron, D.: Testing the diameter of graphs. Random Struct. Algorithms 20(2), 165–183 (2002)

On the Black-Box Complexity of Optimally-Fair Coin Tossing

Dana Dachman-Soled[1], Yehuda Lindell[2], Mohammad Mahmoody[3], and Tal Malkin[1]

[1] Columbia University
{dglasner,tal}@cs.columbia.edu
[2] Bar-Ilan University
lindell@macs.biu.ac.il
[3] Cornell University
mohammad@cs.cornell.edu

Abstract. A fair two-party coin tossing protocol is one in which both parties output the same bit that is almost uniformly distributed (i.e., it equals 0 and 1 with probability that is at most negligibly far from one half). It is well known that it is *impossible* to achieve fair coin tossing even in the presence of fail-stop adversaries (Cleve, FOCS 1986). In fact, Cleve showed that for every coin tossing protocol running for r rounds, an efficient fail-stop adversary can bias the output by $\Omega(1/r)$. Since this is the best possible, a protocol that limits the bias of any adversary to $O(1/r)$ is called *optimally-fair*. The only optimally-fair protocol that is known to exist relies on the existence of oblivious transfer, because it uses general secure computation (Moran, Naor and Segev, TCC 2009). However, it is possible to achieve a bias of $O(1/\sqrt{r})$ in r rounds relying only on the assumption that there exist one-way functions. In this paper we show that it is impossible to achieve optimally-fair coin tossing via a black-box construction from one-way functions for r that is less than $O(n/\log n)$, where n is the input/output length of the one-way function used. An important corollary of this is that it is impossible to construct an optimally-fair coin tossing protocol via a black-box construction from one-way functions whose round complexity is *independent* of the security parameter n determining the security of the one-way function being used. Informally speaking, the main ingredient of our proof is to eliminate the random-oracle from "secure" protocols with "low round-complexity" and simulate the protocol securely against semi-honest adversaries in the plain model. We believe our simulation lemma to be of broader interest.

Keywords: black-box separations, coin tossing, optimally-fair coin tossing, round-complexity, lower-bound.

1 Introduction

We study the fundamental problem of (two-party) coin tossing, where two mutually distrustful parties wish to generate a common random bit. Ideally, this

Y. Ishai (Ed.): TCC 2011, LNCS 6597, pp. 450–467, 2011.

bit should be almost completely unbiased (namely be equal to 1 with probability that is at most negligibly far from 1/2). Furthermore, by the definition of a secure coin tossing protocol, if the two parties follow the protocol then they must both output the same random bit. Unfortunately, however, as shown in a classic result by Cleve [C86], if one of the parties may deviate from the protocol (even if the deviation is only "fail-stop" meaning that the adversary merely aborts early), then secure coin tossing cannot be achieved. In fact, Cleve proved that for any coin tossing protocol running for r rounds there exists an efficient fail-stop adversary that can bias the resulting bit by at least $\Omega(1/r)$.

On the positive side, an early result by Blum [B82] uses one-way functions to construct a coin tossing protocol in a weaker model, where an unbiased output is achieved if both parties complete the protocol, but if a malicious party aborts early, the honest party does not output any bit. This protocol was used by Cleve [C86] to construct a coin tossing protocol that runs for r rounds and for which no efficient adversary can bias the output bit by more than $O(1/\sqrt{r})$ assuming that one-way functions exist[1].

This gap between the lower and upper bounds in [C86] remained open for more than two decades. Recently, it was closed by Moran et al. [MNS09], who constructed a protocol for coin tossing that matches the lower-bound of [C86]. Specifically, they constructed an $O(r)$-round protocol with the property that no adversary can bias the output by more than $O(1/r)$. Thus, they demonstrated that the $\Omega(1/r)$ lower-bound is tight. We call such a protocol *optimally-fair* because no protocol can achieve lower bias.

Interestingly, the protocol of [MNS09] uses general secure computation and thus requires the assumption that oblivious transfer exists (or any assumption implying it, like enhanced trapdoor permutations). In contrast, the coin tossing protocol of Blum [B82] and the protocol of [C86] achieving bias of $O(1/\sqrt{r})$ can be constructed from any one-way function. This disparity was observed by [MNS09] who state: *"A challenging problem is to either achieve the optimal bias based on seemingly weaker assumptions (e.g., one-way functions), or to demonstrate that oblivious transfer is in fact essential."*

In this paper we take a step toward answering this question, and show that one-way functions are not sufficient for achieving optimally-fair coin tossing via *black-box reductions* when the number of rounds r is $o(n/\log n)$ for security parameter n (i.e., the input/output length of the one-way function). We note that the protocols mentioned above of [C86, MNS09] are indeed black-box

Theorem 1 (Main Theorem, Informal). *Let Π be a black-box construction for two-party optimally-fair coin tossing based on one-way functions with input and output length n. Then the number of rounds r of interaction in Π is at least $r = \Omega(n/\log n)$.*

1 Essentially, this protocol works by running Blum's protocol r times sequentially and outputting the bit that appeared in most executions. (If one of the parties halts prematurely, then the other party takes locally chosen uniformly distributed bits as the output bits for the remaining Blum executions.)

In fact, we prove something even stronger: – **Stronger primitives.** The same result holds even if the primitive used in the construction is an exponentially-hard one-way function or an exponentially hard collision resistant hash function $h\colon \{0,1\}^n \mapsto \{0,1\}^{\theta(n)}$ (or in fact any primitive which can be derived in a black-box manner from a random oracle). The result holds also for more structured primitives such as one-way permutation. The latter extension is based on the simple observation that a random function and a random permutation can not be distinguished with "few" queries asked by the construction. We refer the reader for the full proof of these extensions to the full version of the paper.
– **Optimality of the bias.** The same result holds even when Π achieves any $o(1/\sqrt{r})$ bias (not only for optimally-fair protocols with a bias of $O(1/r)$).

Our main technical lemma in order to prove Theorem 1 is to show how to remove random oracles from certain secure protocols in the random oracle models which we believe to be of independent interest.

Lemma 1 (Simulation Lemma, Informal). *Let Π be a two-party protocol in the random oracle model in which the parties query a (random) oracle of input/output length n, ask a total of $m = \mathrm{poly}(n)$ queries and communicate for $o(n/\log n)$ rounds. Then there are two protocols: Π_E (the extended protocol) and Π_T (the threshold-simulation protocol) such that the following holds.* **(a)** *In Π_E the parties act as Π but the ask up to $2^{o(n)}$ extra queries from the oracle.* **(b)** *Π_T is performed in the plain model without the random oracle.* **(c)** *The joint views of the parties in Π_E and Π_T are λ-close for an arbitrary parameter $\lambda = 1/\mathrm{poly}(n)$.*

The high level structure of the proof of Theorem 1 is to use the simulation lemma and the result of [CI93] which breaks any coin-tossing protocol in the plain model with "few" rounds. See Section 1.1 for more details.

We also observe that our simulation lemma can be used to derive impossibility results in the context of secure two-party computation of non-trivial functions. Kushilevitz [K92] classified the finite functions that have perfectly secure two-party protocols against semi-honest adversaries and called them "decomposable functions". Maji, Prabhakaran and Rosulek [MPR09] extended this result to the regime of statistical security and showed that only decomposable functions can have (randomized) two-party protocols which are statistically secure against semi-honest parties. The latter result together with our simulation lemma imply that if a function is not decomposable, it can not have a black-box secure protocol based on one-way function (or based on the other primitives mentioned above) with $o(n/\log n)$ rounds of communication. The steps of the proof of this result are very similar to the case of coin-tossing described in Theorem 1. See Section 1.1 for more details and see the full version of the paper for the complete proof.

Discussion and Implications. Our lower-bound proves that either there is no black-box construction of optimally-fair coin tossing from any of the primitives mentioned in Theorem 2, or if there is any such construction it will suffer from an almost linear $\widetilde{\Omega}(n)$ lower-bound on its round-complexity (which arguably is the main efficiency measure) depending on the security parameter of the primitive used. Such a construction, where the number of rounds, and thus the bias, depends on the security parameter, seems counter-intuitive (yet see the comparison below with statistically hiding commitments which do have constructions with the number of rounds depending on the security parameter).

In particular, our negative result implies that the use of oblivious transfer (as an assumption stronger than one-way function) in the construction of [MNS09], achieving $O(1/r)$ bias for any r, is *inherent*. Moreover, the construction of [C86], using commitments (that can be constructed in a black-box way from one-way functions) and achieving $O(1/\sqrt{r})$ bias for any r, is actually *optimal* (as Theorem 2 holds for any $o(1/\sqrt{r})$ bias).

It is also interesting to contrast our lower bound with the original impossibility result of Cleve [C86]. One way to view the result of [C86] is as a proof that in order to achieve $O(1/r)$ bias any protocol must have at least $\Omega(r)$ rounds of interaction. Our lower bound then says that it is only possible to achieve $O(1/r)$ bias with r rounds when relying on one-way functions (or any of the primitives mentioned in Theorem 2) for $r = \Omega(n/\log n)$ which is very large. In particular, it is not possible to construct a protocol (using a black-box reduction) whose round efficiency depends *only* on the desired bias and is independent of the security parameter n used to determine the input length to the one-way function. This has the ramification that increasing the security parameter in order to obtain a stronger guarantee of invertibility of the one-way function (to get a more secure protocol) has an effect also on the round complexity of the protocol.

Black-Box Separations. One of the main goals of modern cryptography has been to identify the minimal assumptions necessary to construct secure cryptographic primitives. For example, [Y82, GM84, R90, HILL99, GGM86, LR88, IL89, NY89, N91] have shown that private key encryption, pseudorandom generators, pseudorandom functions and permutations, bit commitment, and digital signatures exist if and only if one-way functions exist. On the other hand, some cryptographic primitives such as public key encryption, oblivious transfer, and key agreement are not known to be equivalent to one way functions. Thus, it is natural to ask whether the existence of one-way functions implies these primitives. However, it seems unclear how to formalize such a question; since it is widely believed that both one-way functions and public key encryption exist, this would imply in a trivial logical sense that the existence of one-way functions implies the existence of public key encryption. Thus, we can only hope to rule out restricted types of constructions that are commonly used to prove implications in cryptography. Impagliazzo and Rudich [IR89] were the first to develop a technique to rule out the existence of an important class of reductions between primitives known as black-box reductions. Intuitively, this is a reduction where the primitive is treated as an oracle or a "black-box". There are actually several flavors of black-box reductions (fully black-box, semi black-box and weakly black-box [RTV04]). In our work, we only deal with fully black-box reduction, and so we will focus on this notion here. Informally, a fully black-box reduction from a primitive $\mathcal{Q}$ to a primitive $\mathcal{P}$ is a pair of *oracle* PPT Turing machines (G, S) such that the following two properties hold:

Correctness: For every implementation f of primitive $\mathcal{P}$, $g = G^f$ implements $\mathcal{Q}$.

Security: For every implementation f of primitive $\mathcal{P}$, and every adversary A, if A breaks G^f (as an implementation of $\mathcal{Q}$) then $S^{A,f}$ breaks f. (Thus, if f is "secure", then so is G^f.)

We remark that an *implementation* of a primitive is any specific scheme that meets the requirements of that primitive (e.g., an implementation of a public-key encryption scheme provides samplability of key pairs, encryption with the public-key, and decryption with the private key). Correctness thus states that when G is given oracle access to any valid implementation of $\mathcal{P}$, the result is a valid implementation of $\mathcal{Q}$. Furthermore, security states that any adversary breaking G^f yields an adversary breaking f. The reduction here is *fully* black-box in the sense that the adversary S breaking f uses A in a black-box manner.

Comparison to Similar Lower-Bounds on the Round-Complexity. The only similar lower-bound on the round-complexity of black-box constructions that we are aware of is the result of Haitner, Hoch, Reingold, and Segev [HHRS07] which deals with the round-efficiency of statistically hiding commitment schemes. Interestingly, our lower-bound is exactly the same as that of [HHRS07] which also is based on the security parameter of the one-way function used in the construction . It seems that the techniques used in [HHRS07] and our techniques explained below are quite different. This raises the question of whether there are more connections between the two results. For instance, is it possible to simplify any of these arguments using ideas from the other work? More importantly, this suggests the intriguing possibility that perhaps a positive solution for optimally-fair coin tossing from one-way functions can be achieved with $O(n/\log n)$ rounds, using the techniques which are used in constructing the positive results of $O(n/\log n)$-round statistically hiding commitments [NOVY98, HR07, HNO +09].

1.1 Our Technique

We recall a result of Cleve and Impagliazzo [CI93] which shows that for any coin tossing protocol with r rounds, there exists a *computationally unbounded* adversary who can achieve bias of at least $\Omega(1/\sqrt{r})$. Moreover, this adversary follows the protocol as specified, except that it may abort prematurely; as such the adversary is fail-stop. We show that a black-box construction of an $o(n/\log n)$-round coin tossing from own-way functions with input/output length n (or in fact any primitive which is implied by a random-function in a black-box way) will essentially suffer from the same attack of [CI93] and thus cannot guarantee any bias below $\Omega(1/\sqrt{r})$ through a black-box proof of security.

We start by assuming that there is a black-box construction Π of optimally-fair coin tossing from one-way function with $r = o(n/\log n)$ rounds. A random function is one-way with overwhelming probability, so informally speaking, if we feed the construction Π with a random function it should still be an optimally-fair coin tossing protocol. In fact, something stronger happens when a construction based on one-way function is fed with a random function: Such a construction will now be secure even against computationally *unbounded* adversaries who are allowed to ask $2^{o(n)}$ oracle queries to the random oracle. The reason for this is that if there were such an adversary, then the security reduction will imply that there is an adversary inverting a random function with $2^{o(n)}$ number of queries (see the proof of Theorem 2 for more details) which is not possible. We will take advantage of this stronger property to derive the contradiction by presenting a $2^{o(n)}$-query attack whenever the round complexity is $o(n/\log n)$. The idea of feeding a black-box construction with a random-function and enhancing its security, and then deriving contradiction by a simple counting argument (rather than refuting the relativizing reductions [IR89]—which is a much harder task) is also employed in previous works such as [GGKT05, BM07].

Our main technical step will be to show that the round-complexity of $o(n/\log n)$ for the black-box construction of coin tossing implies the existence of a $2^{o(n)}$-query adversary who is able to bias the output bit by $\omega(1/r)$. In fact we show how to achieve bias $\Omega(1/\sqrt{r}) = \omega(1/r)$. The existence of such an attack implies the result because by the security reduction the ability to bias the protocol yields an adversary inverting the one-way function. Our $2^{o(n)}$-query attacker runs the protocol (of the corresponding party) honestly except that it gathers more information about the random oracle along the execution of the protocol by asking $\text{poly}(n,r)^r$ (which is $2^{o(n)}$ for $r = o(n/\log n)$) more queries and achieves bias of $\Omega(1/\sqrt{r})$ by deciding to stop at some point during the protocol.

We shall emphasize that the reason that we can *not* directly use the attack of [CI93] in the presence of a random oracle is that, even conditioned on the transcript of the interaction, the random oracle builds *dependencies* between the views of Alice and Bob. However the attack of [CI93] essentially uses the fact that conditioned on the transcript the views of Alice and Bob are independent in a plain protocol (where no random oracle is used). Thus we need to find a way to "kill" this dependency to be able to use their attack.

Our $2^{o(n)}$-query attacker uses special properties of an attack given by Barak and Mahmoody [BM09] to break any key-agreement protocol with an optimal number of queries to the random oracle. The attacker of [BM09]—which here we call the "independence learning algorithm", or the simply the learning algorithm for short—gets as input a threshold parameter ε which controls its efficiency and accuracy at the same time. Roughly speaking if Alice and Bob ask m oracle queries in their execution, it will lead to $O(m/\varepsilon)$ queries asked by the learner and the error of $m\varepsilon$. This learning algorithm can be described more naturally as an *online* algorithm which learns certain oracle queries during the interaction between Alice and Bob (despite the fact that passive adversaries can always wait till the end of the interaction). Our attacker uses this learning algorithm internally and feeds it with *different* values for the threshold parameter ε for each round; the parameter ε taken grows exponentially with the round numbers. Due to the heavy use of the threshold parameter of the learning algorithm in our attack, we call it the "threshold attacker" TA. Note that since the learning algorithm only requires the knowledge of the public transcripts, both Alice and Bob can run the learning algorithm in any two-party protocol (e.g., a coin tossing protocol rather than a key-agreement protocol). Thus our threshold attacker TA, which is in fact executed by either Alice or Bob, can also run the learning algorithm during the coin tossing protocol.

The Threshold Attacker—More Details. For an arbitrary two-party protocol Π in the random oracle model (or any other oracle model) we can think of "curious" parties who run the protocol honestly but will ask more oracle queries along their execution of the protocol[2]. We use the terminology of [GIMS10] and call such a game a *curious extension* of the original protocol Π. To get the threshold attacker, Alice or Bob (whoever is performing the attack) will need to play a curious extension of the original protocol by asking up to $2^{o(n)}$ oracle queries. Here we will only deal with an extension based on the learning algorithm of [BM09]. That is, the attacking party runs the learning algorithm along the honest execution of the original coin-tossing protocol and decides to abort prematurely. We let the parties take turn in simulating the learning algorithm in the following way: Whenever Alice (or Bob) is sending a message w_i, they attach to it the set of query/answer pairs that the learning algorithm would learn after w_i is sent across the channel. For brevity we call this specific curious extension in which both Alice and Bob run the learning algorithm along the original game (and attach the learner's view of each round to their messages) simply "the extended execution" of the original protocol (without referring to the learning algorithm explicitly). We show how our threshold attacker can perform their attack in the extended execution.

We prove that the extended protocol has the interesting property that now Alice and Bob can in fact "simulate" the random oracle on their own (using their private randomness) in a way that their views are statistically close to those in the execution of the original extended game in the random oracle model. To perform the simulation, Alice and Bob will answer their queries to the random oracle using fresh randomness unless they have asked this query at some point before (and thus chose the answer already) or that they are told by the other party what the answer to this query should be (through the extra messages simulating the learner's view).

To prove that the above simple simulation is indeed a statistically-close simulation of the extension game we need to show that (unless with small probability) there is no inconsistencies between the oracle answers chosen by Alice and Bob for their oracle queries. Here we crucially use the fact that the learning algorithm provides enough information along the game so that Alice and Bob will always choose consistent oracle answers for their queries. Suppose that Alice is sending a message w_i and is also attaching a list of $k \approx m/\varepsilon_i$ simulated learning queries to the message w_i where ε_i is the learner's threshold used in round i by Alice and m is the total number of queries in the original protocol. For any query q among these k queries which are being asked by Alice from the random oracle (and thus

[2] This is slightly different from the semi-honest parties who run the protocol honestly without asking more oracle queries and only later analyze their view of the interaction.

being simulated) for the first time, we want that q is *not* among Bob's "private" queries which was simulated at some point before (yet is not announced through the learner's simulation). The learner's algorithm has the property that if Bob uses threshold ε_{i-1} to simulate the learner in the previous round $i-1$ then any such query q has chance of at most ε_{i-1} to be a "private" query of Bob. Therefore, by a union bound, the probability that any of these k queries cause an inconsistency is at most $\approx k\varepsilon_{i-1} = m\varepsilon_{i-1}/\varepsilon_i$. By taking $\varepsilon_{i-1} \ll \varepsilon_i/m$, we can control the probability of such event to be arbitrary small. This clarifies why we end up using exponentially smaller thresholds for smaller rounds.

Finally, since we could simulate the extended execution through a plain protocol, we can use the *inefficient* attack of [CI93], which can be applied to any plain protocol and apply it to the simulation of the extension game. Since the extended execution and its simulation are statistically close experiments, we conclude that almost the same bias would be achieved by the attacker in the extension execution with only $2^{o(n)}$ queries and so we are done.

A Parallel Work. The threshold simulation technique was discovered independently in a parallel work by Maji and Prabhakaran [MP10] in the context of using random oracle for the aim of achieving statistically secure protocols.

2 Definitions and Useful Lemmas

Definition 1 (coin tossing from one-way function). *For (interactive) oracle algorithms A, B we call $\Pi = (A, B)$ a black-box construction of coin tossing with bias at most δ based on exponentially-hard one-way functions with security parameter n, if the following properties hold:*

- *A and B have their own private randomness R_A, R_B. They take as input 1^n and run in time* $\text{poly}(n)$ *and interact for $r(n) = \text{poly}(n)$ number of rounds.*
- **Completeness:** *For any function $f\colon \{0,1\}^n \mapsto \{0,1\}^n$, when A and B are given oracle access to f, then at the end of the protocol A's output a and B's output b are such that $a = b$ and b is considered the output of the protocol. Also if during the protocol A (resp., B) receives the special message $\perp$ (denoting that the other party has stopped playing in the protocol) then A (resp., B) outputs a bit a (resp b) on their own which is considered as the output of the protocol.*
- **Security (against bias δ):** *There is an oracle algorithm S running in time $2^{o(n)}$ with the following property. For any $f\colon \{0,1\}^n \mapsto \{0,1\}^n$ given as oracle, if $\widehat{A}$ (resp., $\widehat{B}$) is a malicious interactive algorithm interacting with B (resp., A) which makes the output bit b to be $\delta(n)$-biased, then $S^{f,\widehat{A}}$ (given oracle access to f and $\widehat{A}$) breaks the security of f (as an exponentially-hard one-way function).*

We denote by $(a|b) \leftarrow \langle \widehat{A}, B\rangle$ (resp. $(a|b) \leftarrow \langle A, \widehat{B}\rangle$) the joint output of $\widehat{A}$ and B (resp. A and $\widehat{B}$) generated by an interaction of $\widehat{A}$ and B (resp. A and $\widehat{B}$).

The proof of the following two lemmas can be verified by inspection.

Lemma 2 (Inverting Random Functions). *Let A be a* computationally unbounded *oracle algorithm given oracle access to a random function $f\colon \{0,1\}^n \mapsto \{0,1\}^n$ (the randomness of f is chosen after A is fixed). Then if A asks at most $2^{\alpha n}$ queries from f, the probability that A can successfully invert a given input $y = f(U_n)$ (to any preimage of y) is at most $2 \cdot 2^{(\alpha-1)n} + 2^{-n}$ which is negligible for any constant $\alpha < 1$.*

Lemma 3 (Inverting Random Functions with a Fixed Subdomain). *Let $S \subset \{0,1\}^n$ be of size $|S| \leq 2^{\beta n}$ for $\beta < 1$, and let $f_S\colon S \mapsto \{0,1\}^n$ be a fixed function. Let F be the set of all functions $f\colon \{0,1\}^n \mapsto \{0,1\}^n$ which are equal to f_S over S. Now, let A be a* computationally unbounded *oracle algorithm which can depend on f_S and is given oracle access to a random function $f \xleftarrow{R} F$ (the randomness of f is chosen after A is fixed). Then if A asks at most $2^{\alpha n}$ queries from f, the probability that A can successfully invert a given input $y = f(U_n)$ (to any preimage of y) is at most $2 \cdot (2^{(\alpha-1)n} + 2^{(\beta-1)n}) + 2^{-n}$ which is negligible for any constants $\alpha < 1$ and $\beta < 1$.*

3 Simulation Lemma

In this section, we present a general lemma that holds for any two-party protocol in the random oracle model. This lemma will be useful for proving our result on coin tossing, but also has applications to general two-party computation as we describe below.

Lemma 4 (Simulation Lemma). *Let Π be a two-party protocol between Alice and Bob in the random oracle model where they ask at most m oracle queries and interact for r rounds. Then there exist protocols Π_T and Π_E called the λ-threshold simulation and λ-extended execution of Π such that the views of Alice and Bob (as a jointly distributed random variable) in Π_T and Π_E are λ-close. Moreover, the following properties hold:*

- *Π_T makes no oracle queries.*
- *For $\lambda = 1/\operatorname{poly}(n)$, $r = o(n/\log n)$ and $m = \operatorname{poly}(n)$, Π_E makes at most $2^{o(n)}$ queries.*
- *Let $W^{\Pi} = [w_1^{\Pi}, \ldots, w_i^{\Pi}]$ be the sequence of messages sent between Alice and Bob so far in an execution of protocol Π relative to oracle f with random tapes R_A, R_B respectively. For $\lambda = 1/\operatorname{poly}(n)$, $r = o(n/\log n)$ and $m = \operatorname{poly}(n)$, both Alice and/or Bob can make at most $2^{o(n)}$ queries and produce the transcript $W^{\Pi_E} = [w_1^{\Pi_E}, \ldots, w_i^{\Pi_E}]$ that is generated by an execution of the protocol Π_E relative to oracle f with random tapes R_A, R_B.*

The above lemma implies the following corollary:

Corollary 1. *Let $p = 1/\operatorname{poly}(n)$ and let Q be some two-party cryptographic task such that for every implementation Π_{plain} in the plain model with $r =$*

$o(n/\log n)$ rounds, there is a computationally-unbounded, semi-honest adversary which breaks the security of Π_{plain} with probability p. Let Π be a black-box construction of Q with r rounds based on exponentially-hard one-way functions with security parameter n (i.e. the input/output length of f). Then $r = \Omega(n/\log n)$.

The corollary follows from Lemma 4 due to the following: Assume such a construction Π exists with $r = o(n/\log n)$ rounds. Now consider Π_T, the λ-threshold simulation of Π. Since Π_T also has $r = o(n/\log n)$ rounds and does not make calls to the oracle, we have by hypothesis that there is an unbounded attacker $\widehat{A}$ (resp. $\widehat{B}$) which breaks the security of Π_T with probability $p = 1/\mathrm{poly}(n)$. Now, for $\lambda \leq p/2 = 1/\mathrm{poly}(n)$, we have that the views of Alice and Bob (as a jointly distributed random variable) in Π_T and in the λ-extended exection, Π_E, are λ-close. Moreover, given the transcript generated by Π, Alice (resp. Bob) can make at most $2^{o(n)}$ queries and produce the corresponding transcript of Π_E. Thus, there is a threshold attacker TA which plays the part of Alice (resp. Bob) in Π, makes at most $2^{o(n)}$ queries to compute the messages of Π_E, runs $\widehat{A}$ (resp. $\widehat{B}$) internally while simulating the view of $\widehat{A}$ (resp. $\widehat{B}$) using the λ-close view produced by Π_E and finally outputs whatever $\widehat{A}$ (resp. $\widehat{B}$) outputs. So TA breaks the security of Π_E (and thus of Π) with probability $p/2$, where the probability is computed over the randomness of f. Having the threshold attacker TA the proof can be concluded as follows:

(a) Since the attacker TA breaks security with probability $p/2 = 1/\mathrm{poly}(n)$, by an averaging argument, for at least $p/4$ fraction of the functions $f\colon \{0,1\}^n \mapsto \{0,1\}^n$, the attacker TA^f breaks security with probability $p/4$. We call such function f, a good function. **(b)** Using the security reduction S, for all good functions f, S^{f,TA^f} inverts $y = f(U_n)$ with probability at least $2^{-o(n)}$. **(c)** We can combine the algorithms S and TA to get a single oracle algorithm T^f which inverts $f(U_n)$ with probability $2^{-o(n)}$ when f is a good function by asking only $2^{o(n)}$ queries to f. Which means that in this case T asks only $2^{o(n)}$ oracle queries and inverts a *random* f with probability at least $p/4 \cdot 2^{-o(n)} = 2^{-o(n)}$ (because f is a good function with probability at least $p/4$). The latter contradicts Lemma 2.

Before we prove Lemma 4, we review relevant previous work.

The Independence Learner of [BM09]. Here we describe the properties of the attacker of Barak and Mahmoody [BM09] presented in the context of breaking any key agreement protocol with optimal number of queries to the random oracle. Since the main property of the learning algorithm is that conditioned on the learner's information Alice and Bob's views are almost independent, we call this attack the independence learning algorithm.

Lemma 5 (The Independence Learner of [BM09]). *Let Σ be any two-party protocol in the random oracle model (with arbitrary number of rounds) between Alice and Bob in which Alice and Bob ask at most m queries from the random oracle H. Then there is a universal constant c and a (computationally unbounded)* independence learning *algorithm which is given a parameter ε (called the* threshold*) as input and has the following properties. For brevity we denote the independence learning algorithm by Eve.*

- *Eve only has access the public messages sent between Alice and Bob and can ask queries from the random oracle.*
- (cm/ε)**-Efficiency:** *Eve is deterministic and, over the randomness of the oracle and Alice and Bob's private randomness, the expected number of Eve queries from the oracle H is at most cm/ε.*
- *Eve asks its queries along the game. Namely, although Eve can wait till the end and then ask all of her queries, her description defines which queries to be asked right after each message is sent across the public channel. So the learning algorithm is divided into the same number of rounds as the protocol.*
- $(c\sqrt{m\varepsilon})$**-Security:** *Let $W = [w_1, \dots, w_i]$ be the sequence of messages sent between Alice and Bob so far, and let I be the list of oracle query/answer pairs that Eve has asked till the end of the i'th round, and let $\mathbf{AB} = (\mathbf{A}, \mathbf{B})$ be the joint distribution over the views of Alice and Bob only conditioned on (W, I). By $\mathbf{A}$ and $\mathbf{B}$ we refer to the projections of $\mathbf{AB}$ over its first or second components (referring to the view of either Alice or Bob only) as random variables. For a specific view $A \leftarrow \mathbf{A}$ for Alice, by $Q(A)$ we refer to the set of oracle queries that A contains. We also use the notation $Q(I)$ to refer to the queries denoted in I.*

  With probability at least $1 - c\sqrt{m\varepsilon}$ over the randomness of Alice, Bob, and the random oracle H the following holds at all moments during the protocol when Eve is done with her learning phase in that round: There are independent distributions $\widehat{\mathbf{A}}, \widehat{\mathbf{B}}$ such that:
  1. *The statistical distance between $\widehat{\mathbf{A}} \times \widehat{\mathbf{B}}$ and $\mathbf{AB}$ is at most $\Delta(\widehat{\mathbf{A}} \times \widehat{\mathbf{B}}, \mathbf{AB}) \leq c\sqrt{m\varepsilon}$.*
  2. *For every oracle query $q \notin Q(I)$, it holds that $\Pr[q \in Q(\widehat{\mathbf{A}}) \cup Q(\widehat{\mathbf{B}})] \leq \varepsilon$.*
- **Robustness.** *The learning algorithm is robust to the input parameter ε in the following sense. If the parameter ε changes in the interval $\varepsilon \in [\varepsilon_1, \varepsilon_2]$ arbitrarily during the learner's execution (even inside a learning phase of a specific round), it still preserves $O(cm/\varepsilon_1)$-efficiency and $(c\sqrt{m\varepsilon_2})$-security.*

Lemma 5 is implicit in [BM09], and we show how to derive it from the explicit results of [BM09] in the full version of the paper.

Given a protocol Π, we now describe the λ-extended execution, Π_E, and the λ-threshold simulation, Π_T, of Π that were mentioned in Lemma 4.

Definition 2 (Extended Execution). *Let Π be a two-party protocol between Alice and Bob in the random oracle model where they ask at most m oracle queries and interact for r rounds. The extended execution Π_E of Π gets as input a parameter λ and simulates the original protocol Π in the random oracle model as follows.*

- *Let $\varepsilon_r = \frac{1}{m} \cdot \left(\frac{\lambda}{9rc}\right)^2$ and for $j \in \{r, r-1, \dots, 2\}$ define $\varepsilon_{j-1} = \varepsilon_j \cdot \frac{\lambda^2}{90r^2m}$. Note that if r, λ, m are $\leq \mathrm{poly}(n)$, then $\varepsilon_r = 1/\mathrm{poly}(n)$ and $\varepsilon_1 = \mathrm{poly}(n)^{-r}$.*
- *Now imagine an Eve who runs the independence learner of Lemma 5 and uses ε_i as its learning parameter in the learning phase after the i'th round.*

– *In round i, the party who is sending the message w_i, also runs the i'ih round of the learning phase of Eve and attaches to w_i the list of all the query/answer pairs that are the result of this learning algorithm. Note that since Eve's algorithm is only depending on the messages being sent and her previous knowledge about the oracle, the parties are able to do this job.*

Definition 3 (Threshold Simulation). *Let Π be a two-party protocol between Alice and Bob in the random oracle model where they ask at most m oracle queries and interact for r rounds. A threshold simulation Π_T of Π gets as input a parameter λ and simulates the original protocol Π* plainly *as follows.*

– *The parameters ε_i for $i \in [r]$ are defined similar to the extended execution.*
– *In the i'th round the party who sends the i'th message tries to simulate the i'th round of the extended execution but* without *using a random oracle. The way the simulation is done is as follows: To compute the message w_i, suppose q is a query to be asked from the oracle. Now if q is in the set of queries learned by Eve so far or if q was asked previously by the same party, the same answer will be returned which was used before. But, if the query q is new, a fresh random answer will be used. The same is also done to answer any query that the learning algorithm Eve tries to learn.*

The following lemma explains why a threshold simulation is indeed a good simulation of the extended execution.

Lemma 6 (Properties of the Threshold Simulation). *Let Π be a two-party protocol between Alice and Bob in the random oracle model where they ask at most m oracle queries and let Π_T and Π_E be in order its λ-threshold simulation and λ-extended execution. Then the views of Alice and Bob (as a jointly distributed random variable) in Π_T and Π_E are λ-close.*

Proof. It is easy to see that the extended execution and the threshold simulation will be exactly the same games until the following happens: A party, say Alice sends a message w_i along with the simulation of Eve's i'th round, but one of these queries (which are asked in this round either for her own protocol or to simulate Eve) will hit one of Bob's "private" queries which are *not* announced through Eve's previous simulated query/answers. We show that this "bad" event happens with probability at most λ.

Note that by the robustness of the independence learner Eve and by the choice of the (largest) parameter $\varepsilon_r = \frac{1}{m} \cdot \left(\frac{\lambda}{9rc}\right)^2$, Eve's algorithm remains at least $c\sqrt{m\varepsilon} = \lambda/(9r)$ secure in round i. So, except with probability at most $r \cdot \lambda/(9r) = \lambda/9$ we can pretend (as a mental experiment) that at all moments the security requirement of the learning algorithm holds with probability 1 rather than $1 - c\sqrt{m\varepsilon}$. In the following we show that (up to the bad event mentioned above which happens with probability at most $\lambda/9$) the probability that an "inconsistency" happens in round i is at most $\lambda/(3r)$, and thus we will be done by a union bound. By inconsistency we mean that Alice announces (a different) answer for an oracle query that is privately asked by Bob already (or vice versa).

Suppose Alice is sending the message in the i'th round and suppose no inconsistency has happened so far. Let fix $W = [w_1, \ldots, w_{i-1}]$ to be the sequence of the messages sent till this moment and let I be the union Eve's simulated queries till the end of the $(i-1)$'th round. An inconsistency in round i can happen as follows: one of the queries asked by Alice (either to run her own protocol or to simulate Eve) hits one of Bob's private queries. We bound this probability conditioned on any fixed (W, I) over which the security property of the learner holds (as we said this property will hold with probability at least $1 - \lambda/9$).

As a mental experiment we can continue the game (after fixing (W, I)) by sampling from the random variable $(A, B) \leftarrow \mathbf{AB}$ for the views of Alice and Bob so far conditioned on (W, I) and then continue Alice's simulation. Let assume for a moment that we sample $(A, B) \leftarrow \widehat{\mathbf{A}} \times \widehat{\mathbf{B}}$ rather than from $\mathbf{AB}$. We bound the probability of any inconsistency in the former case to be $2\lambda/(9r)$, and since the distributions $\mathbf{AB}$ and $\widehat{\mathbf{A}} \times \widehat{\mathbf{B}}$ are $\lambda/(9r)$ close, it follows that the probability of any inconsistency in this round is bounded by $2 \cdot \lambda/(9r) + \lambda/(9r) = \lambda/(3r)$ which is small enough for us.

But now we use the security property of the independence learner. Note that when we get the sample $(A, B) \leftarrow \widehat{\mathbf{A}} \times \widehat{\mathbf{B}}$, A and B are sampled *independently*. So, we can sample A first, continue Alice's computation, and then sample $B \leftarrow \widehat{\mathbf{B}}$ at the end (and we will abort if the private queries collide). The number of queries that Alice will ask to run her own protocol is at most m. By the efficiency property of the learning algorithm applied to round i, the number of Eve's simulated queries in this round are, on average, at most cm/ε_i. By a Markov bound, this number is at most $\frac{cm}{\varepsilon_i} \cdot \frac{9r}{\lambda}$ with probability at least $1 - \lambda/(9r)$. So except with probability $\lambda/(9r)$ the total number of queries asked by Alice in this round is at most $m + 9cmr/(\varepsilon_j \lambda) < 10cmr/(\varepsilon_j \lambda)$. Note that the probability that any of these $10cmr/(\varepsilon_j \lambda)$ queries are among the private queries of a sample from $\widehat{\mathbf{B}}$ (sampled as Bob's view) is at most ε_{j-1}. So, by a union bound, the probability that at least one of these queries hits $\widehat{\mathbf{B}}$'s private queries is at most $\frac{10cmr}{\varepsilon_j \lambda} \cdot \varepsilon_{j-1} = \lambda/(9r)$ and this finishes the proof.

So, all left to do is to count how many queries are asked by our λ-extended execution Π_E and show that it is (say on average) at most $2^{o(n)}$. This is indeed the case because of the robustness and the efficiency properties of the learning algorithm. The smallest threshold used in our attack is $\varepsilon_1 = \text{poly}(n)^{-r}$ because $\lambda = 1/r$ and $r = \text{poly}(n), m = \text{poly}(n)$. Therefore our attacker asks at most $O(m/\varepsilon_1)$ number of queries on average which for $r = o(n/\log n)$ is at most $O(m/\varepsilon_1) = \text{poly}(n)^r = 2^{o(n)}$.

4 Proof of the Main Theorem

In this section we first prove our main theorem for the case of exponentially-hard one-way function as the primitive used. Extending the proof to stronger primitives implied by a random oracle is discussed at the end.

Theorem 2 (Main Theorem, Formal). *Let Π be a black-box construction for two-party coin tossing (between Alice and Bob) with bias at most $o(1/\sqrt{r})$ (where r is the number of rounds in Π) based on exponentially-hard one-way functions with security parameter n (i.e., the input/output length of f). Then $r = \Omega(n/\log n)$.*

Proof. For sake of contradiction let assume that such construction exists with $r = o(n/\log n)$ round complexity. The proof goes through the following steps. We first feed Alice and Bob's protocols in the construction Π with a *random* function $f: \{0,1\}^n \mapsto \{0,1\}^n$. We show that in that setting at least one of the parties can ask $n^{O(r)}$ queries to f and bias the output by at least $\Omega(1/\sqrt{r})$ by a fail-stop attack. The probability over which the bias is computed also includes the randomness of f. As in Section 3, we call this attacker the threshold attacker, TA. Having the threshold attacker TA the proof can be concluded as follows.

(a) Since the attacker TA achieves bias $\delta = \Omega(1/\sqrt{r})$ and since the bias is always $\delta < 1$, therefore by an averaging argument, for at least $\delta/2$ fraction of the functions $f: \{0,1\}^n \mapsto \{0,1\}^n$, the attacker TA^f achieves bias at least $\delta/2 = \Omega(1/\sqrt{r})$. We call such function f, a good function. **(b)** Using the security reduction S, for all good functions f, S^{f,TA^f} inverts $y = f(U_n)$ with probability at least $2^{-o(n)}$. **(c)** We can combine the algorithms S and TA to get a single oracle algorithm T^f which inverts $f(U_n)$ with probability $2^{-o(n)}$ when f is a good function by asking only $2^{o(n)}\,\mathrm{poly}(n)^r$ queries to f. For $r = o(n/\log n)$, it holds that $\mathrm{poly}(n)^r = 2^{o(n)}$, which means that in this case T asks only $2^{o(n)} \cdot 2^{o(n)} = 2^{o(n)}$ oracle queries and inverts a *random* f with probability at least $\frac{\delta}{2} \cdot 2^{-o(n)} = 2^{-o(n)}$ (because f is a good function with probability at least $\delta/2$). The latter contradicts Lemma 2.

In the following we first describe the results that we borrow or derive from previous work needed for our threshold attacker TA, and then will describe and prove the properties of TA.

The Fail Stop Attacker of [CI93]. Cleve and Impagliazzo [CI93] showed that when computationally unbounded parties participate in any coin tossing protocol, at least one of them can bias the output bit by following the protocol honestly and aborting at some point based on the information provided to them by their view.

Lemma 7 (The Attacker of [CI93]). *Let Σ be any two-party protocol for coin tossing between Alice and Bob with r rounds of interaction. Then either Alice or Bob can bias the output bit by $\Omega(1/\sqrt{r})$ in the fail-stop model through a computationally* unbounded *attack.*

4.1 Our Threshold Attacker

In this section we use the attack of Lemma 7 as well as the results of Section 3 to finish the proof of Theorem 2 by presenting our threshold attacker. We will do so first in a special case where the protocol Π is of a special form which we call *instant*. The case of instant constructions carries the main ideas of the proof. Later we prove Theorem 2 for constructions which are not necessarily instant.

Definition 4 (Instant Constructions). *A black-box construction of coin tossing is an* instant *construction if whenever a party aborts the protocol, the other party decides on the output bit* without *asking any additional queries to its oracle.*

We note that the protocol of Cleve [C86] which achieves bias at most $O(1/\sqrt{r})$ based on one-way function is in fact an instant construction.

Given an instant coin-tossing protocol Π, we apply Lemma 4 to obtain the λ-threshold simulation and λ-extended execution of Π, Π_T, Π_E. Since the threshold simulation, Π_T, is a plain protocol we can apply Lemma 7 to get an attack of bias $\Omega(1/\sqrt{r})$ by either Alice or Bob. Now if we take the simulation parameter λ to be at most $1/r = o(1/\sqrt{r})$, then the same exact attack will also give a bias of $\Omega(1/\sqrt{r}) - o(1/\sqrt{r}) = \Omega(1/\sqrt{r})$ in the extended execution. Here we crucially rely on the instant property because of the following: As soon as Alice or Bob (who is the attacker) stops continuing the game, the other party in the threshold simulation will decide on the final output bit by looking at their current view. But this last step will not be statistically close between the extended execution and the threshold execution if in the extended execution the deciding party chooses the output after asking more queries. In other words, if the party who announces the output bit (not the attacker) wants to ask more oracle queries to compute the output bit, there should be some simulated random answers chosen by the corresponding party in the threshold simulation to on behalf of these queries, but that step is not taken care of by Lemma 6 (because the aborted party is *not* given the learning algorithm's queries for the aborted round). By Lemma 4, our attacker asks at most $2^{o(n)}$ queries.

Before going over how to handle the non-instant constructions we clarify that extending Theorem 2 to stronger primitives such as exponentially-hard collision resistant hash function is immediate. All one has to do is to substitute the collision resistant hash functions $h\colon \{0,1\}^n \mapsto \{0,1\}^{n/2}$ used in the construction by a random function $f\colon \{0,1\}^n \mapsto \{0,1\}^{n/2}$ (which is in fact a $2^{\Omega(n)}$-secure hash function). To provide access to a family of hash functions one can use the random oracle over larger domains of input/output length $3n$ and use the first n bits of the input as the index to the hash family and simply throw away the last $\frac{5n}{2}$ bits of the output. The rest of the proof remains the same.

Handling Non-instant Constructions. It is instructing to recall that given a random oracle there is indeed a one-round protocol which is optimally-fair: Alice asks $H(0)$ (assuming that the random oracle is Boolean) and then sends $H(0)$ to Bob which is the final output bit. If Alice aborts and does not send $H(0)$, Bob will go ahead and ask $H(0)$ himself and takes that as the final output bid. It is clear that this trivial protocol is completely fair because $H(0)$ is an unbiased bit. Also note that the proof of the previous section handing the instant constructions works just as well for protocols which use a truly random oracle (rather than a one-way function) as their primitive used. So it should be of no surprise that the proof of the instant case does not immediately generalize to cover all the black-box constructions (the trivial coin-tossing protocol based on random oracle is clearly a non-instant protocol). To handle the non-instant constructions we inherently need to use the fact that the constructions we deal with are optimally-fair protocols given any *one-way* function as the primitive used. In the following we show how this stronger requirement of the construction gives us what we need in Theorem 2.

Making constructions almost instant. It is easy to see that any construction for coin tossing can be changed into an equivalent protocol which is "almost" an instant one. Namely, whenever a party A is sending a message m, it can also consider the possibility that the other party B will abort the game right after A sends his message. So, during the computation of m, A can go ahead and ask whatever query from the oracle which is needed to compute the final bit in case B aborts. This way, A will not need to ask any oracle queries in case B aborts in this round. By doing this change (which clearly does not affect the security of the protocol) the construction becomes "almost" instant. The point is that the receiver of the *first* message can not follow the change suggested here because they do not send any message before the first round. Therefore, in the following we only consider constructions which are "almost-instant" (i.e., the only moment that a party might violate the instant property is when the sender of the first message aborts the protocol, and the receiver might still need to ask oracle queries before deciding on the output.)

Handling almost-instant constructions. Suppose Π is an almost-instant construction. Suppose Π_E and Π_T be in order Π's extended execution and the threshold simulation games. The proof of Lemma 6 shows that if no party aborts the experiments Π_E and Π_T are λ-close. The discussion following the proof of Lemma 6 shows that if one of the parties runs the same fail-stop attack in Π_E and Π_T the experiments are *still* λ-close conditioned on the assumption that the round in which the abort happens is any round other than the first one. So, all we need to handle is the case in which the sender of the first message (which we assume to be Alice) aborts the game in the first round (after asking some oracle queries). In the following we focus on this specific cease.

Note that when aborted in the first round Bob can *not* simply simulate the extended execution by using fresh randomness to answer his oracle queries in order to decide the output bit. If he does so it might not be consistent with Alice's queries asked before aborting and thus it will not be a good simulation[3]. Despite this issues, if we are somehow magically guaranteed that when aborted in the first round, none of Bob's queries to compute the output bit collides with Alice's queries asked before, then we can still use fresh random answers to answer Bob's queries to compute the output bit.

Suppose after Alice computes her message but *right before* she sends this message we run the independence learning algorithm with parameter $\lambda/(10m)$.

[3] This will be more clear if one consider the trivial protocol mentioned above which uses a truly random oracle. If Alice aborts whenever $H(0) = 0$, and if Bob uses a fresh random answer whenever he gets aborted by Alice, then the final output will be equal to 1 with probability 3/4 which is clearly a huge bias!

This learning algorithm will announce a set of $O(10m^2/\lambda)$ queries and answers conditioned on which any other query has a chance of at most $\lambda/(10m)$ of being asked by Alice in her computation of the first message. Let the set S be the set of all these $O(10m^2/\lambda)$ queries and let $f(S)$ be their answers. By the security property of the learning algorithm, conditioned on S and $f(S)$, an aborted Bob will not ask any query out of S which collides with Alice's private queries out of S before aborting (unless with probability at most $O(\lambda)$).

The idea is to sample the set S and $f(S)$ once for all, and hardwire them into the random oracle and Alice and Bob's algorithms. This way, simulating Bob's queries with random answers after being aborted will not lead to any inconsistency with Alice's queries unless with probability at most $O(\lambda)$. But if we *fix* the answer of such queries that might hurt the protocol's fairness. At this point we use the fact that the construction is supposed to be fair given any one-way function (and not necessarily a random function). Any random oracle is one-way with overwhelming probability even if we fix a subdomain $S \subseteq \{0,1\}^n$, $|S| \leq \text{poly}(n)$ of its domain and this idea is formalized in Lemma 3. Namely, if we hardwire the random function over a subdomain $S \subseteq \{0,1\}^n$, $|S| \leq \text{poly}(n)$ we can still use the same exact proof as the case of instant constructions for Theorem 2 with the only difference that now we will use Lemma 3 rather than Lemma 2.

Acknowledgement

We thank the anonymous reviewers for numerous valuable comments.

References

[B82] Blum, M.: Coin flipping by telephone - a protocol for solving impossible problems. In: COMPCON, pp. 133–137 (1982)

[BM07] Barak, B., Mahmoody, M.: Lower bounds on signatures from symmetric primitives. In: FOCS: IEEE Symposium on Foundations of Computer Science (FOCS) (2007)

[BM09] Barak, B., Mahmoody-Ghidary, M.: Merkle puzzles are optimal — an $o(n^2)$-query attack on any key exchange from a random oracle. In: Halevi, S. (ed.) CRYPTO 2009. LNCS, vol. 5677, pp. 374–390. Springer, Heidelberg (2009)

[C86] Cleve, R.: Limits on the security of coin flips when half the processors are faulty (extended abstract). In: STOC, pp. 364–369 (1986)

[CI93] Cleve, R., Impagliazzo, R.: Martingales, collective coin flipping and discrete control processes (1993) (unpublished)

[GGKT05] Gennaro, R., Gertner, Y., Katz, J., Trevisan, L.: Bounds on the efficiency of generic cryptographic constructions. SICOMP: SIAM Journal on Computing 35 (2005)

[GGM86] Goldreich, O., Goldwasser, S., Micali, S.: How to construct random functions. J. ACM 33(4), 792–807 (1986)

[GIMS10] Goyal, V., Ishai, Y., Mahmoody, M., Sahai, A.: Interactive locking, zero-knowledge pCPs, and unconditional cryptography. In: Rabin, T. (ed.) CRYPTO 2010. LNCS, vol. 6223, pp. 173–190. Springer, Heidelberg (2010)

[GM84] Goldwasser, S., Micali, S.: Probabilistic encryption. J. Comput. Syst. Sci. 28(2), 270–299 (1984)

[HHRS07] Haitner, I., Hoch, J.J., Reingold, O., Segev, G.: Finding collisions in interactive protocols - a tight lower bound on the round complexity of statistically-hiding commitments. In: FOCS, pp. 669–679 (2007)

[HILL99] Håstad, J., Impagliazzo, R., Levin, L.A., Luby, M.: A pseudorandom generator from any one-way function. SIAM J. Comput. 28(4), 1364–1396 (1999)

[HNO⁺09] Haitner, I., Nguyen, M.-H., Ong, S.J., Reingold, O., Vadhan, S.: Statistically hiding commitments and statistical zero-knowledge arguments from any one-way function. SIAM Journal on Computing 39(3), 1153–1218 (2009)

[HR07] Haitner, I., Reingold, O.: A new interactive hashing theorem. In: IEEE Conference on Computational Complexity (CCC) (2007)

[IL89] Impagliazzo, R., Luby, M.: One-way functions are essential for complexity based cryptography (extended abstract). In: FOCS, pp. 230–235 (1989)

[IR89] Impagliazzo, R., Rudich, S.: Limits on the provable consequences of one-way permutations. In: STOC, pp. 44–61 (1989)

[K92] Kushilevitz, E.: Privacy and communication complexity. SIAM J. Discrete Math 5(2), 273–284 (1992)

[LR88] Luby, M., Rackoff, C.: How to construct pseudorandom permutations from pseudorandom functions. SIAM J. Comput. 17(2), 373–386 (1988)

[MNS09] Moran, T., Naor, M., Segev, G.: An optimally fair coin toss. In: Reingold, O. (ed.) TCC 2009. LNCS, vol. 5444, pp. 1–18. Springer, Heidelberg (2009)

[MP10] Maji, H., Prabhakaran, M.: Personal communication (2010)

[MPR09] Maji, H.K., Prabhakaran, M., Rosulek, M.: Complexity of multi-party computation problems: The case of 2-party symmetric secure function evaluation. In: Reingold, O. (ed.) TCC 2009. LNCS, vol. 5444, pp. 256–273. Springer, Heidelberg (2009)

[N91] Naor, M.: Bit commitment using pseudorandomness. J. Cryptology 4(2), 151–158 (1991)

[NOVY98] Naor, M., Ostrovsky, R., Venkatesan, R., Yung, M.: Perfect zero-knowledge arguments for NP using any one-way permutation. JCRYPTOL: Journal of Cryptology 11 (1998)

[NY89] Naor, M., Yung, M.: Universal one-way hash functions and their cryptographic applications. In: STOC, pp. 33–43 (1989)

[R90] Rompel, J.: One-way functions are necessary and sufficient for secure signatures. In: STOC, pp. 387–394 (1990)

[RTV04] Reingold, O., Trevisan, L., Vadhan, S.: Notions of reducibility between cryptographic primitives. In: Naor, M. (ed.) TCC 2004. LNCS, vol. 2951, pp. 1–20. Springer, Heidelberg (2004)

[Y82] Yao, A.C.-C.: Theory and applications of trapdoor functions. In: FOCS, pp. 80–91 (1982)

Tight Bounds for Classical and Quantum Coin Flipping

Esther Hänggi[1] and Jürg Wullschleger[2]

[1] Computer Science Department, ETH Zurich, Zürich, Switzerland
[2] DIRO, Université de Montréal, Quebec, Canada
McGill University, Quebec, Canada

Abstract. *Coin flipping* is a cryptographic primitive for which strictly better protocols exist if the players are not only allowed to exchange classical, but also quantum messages. During the past few years, several results have appeared which give a tight bound on the range of implementable unconditionally secure coin flips, both in the classical as well as in the quantum setting and for both weak as well as strong coin flipping. But the picture is still incomplete: in the quantum setting, all results consider only protocols with *perfect correctness*, and in the classical setting tight bounds for strong coin flipping are still missing.

We give a general definition of coin flipping which unifies the notion of strong and weak coin flipping (it contains both of them as special cases) and allows the honest players to abort with a certain probability. We give tight bounds on the achievable range of parameters both in the classical and in the quantum setting.

1 Introduction

Coin flipping (or coin tossing) as a cryptographic primitive has been introduced by Blum [5] and is one of the basic building blocks of *secure two-party computation* [21].

Coin flipping can be defined in several ways. The most common definition, sometimes called *strong coin flipping*, allows two honest players to receive a uniform random bit $c \in \{0, 1\}$, such that a dishonest player cannot *increase* the probability of any output. A dishonest player may, however, abort the protocol, in which case the honest player gets the erasure symbol Δ as output[1]. A weaker definition, called *weak coin flipping*, only requires that each party cannot increase the probability of their preferred value.

Without any additional assumptions, unconditionally secure weak coin flipping (and therefore also strong coin flipping) cannot be implemented by a classical protocol. This follows from a result by Hofheinz, Müller-Quade and Unruh [9],

[1] The dishonest player may abort after receiving the output bit, but before the honest player gets the output bit. This allows cases where the honest player gets, for example, 0 with probability 1/2 and Δ otherwise. There exists also a definition of coin flipping where a dishonest player does not have this unfair advantage, and the honest player must always get a uniformly random bit, no matter what the other player does. See [7,12,16].

Y. Ishai (Ed.): TCC 2011, LNCS 6597, pp. 468–485, 2011.

which implies that if two honest players always receive the same uniform bit, then there always exists one player that can force the bit to be his preferred value with certainty.

If the players can communicate using a quantum channel, unconditionally secure coin flipping is possible to some extent. The bounds of the possibilities have been investigated by a long line of research. Aharanov *et al.* [1] presented a strong coin flipping protocol where no quantum adversary can force the outcome to a certain value with probability larger than 0.914. This bound has been improved by Ambainis [2] and independently by Spekkens and Rudolph [18] to 0.75 (see also [8] for a different protocol). For weak coin flipping, Spekkens and Rudolph [19] presented a protocol where the dishonest player cannot force the outcome to its preferred value with probability larger than $1/\sqrt{2} \approx 0.707$. (Independently, Kerenidis and Nayak [10] showed a slightly weaker bound of 0.739). This bound has further been improved by Mochon, first to 0.692 [13] and finally to $1/2 + \varepsilon$ for any constant $\varepsilon > 0$ [15], therefore getting arbitrarily close to the theoretical optimum. For strong coin flipping, on the other hand, this is not possible, since it has been shown by Kitaev [11] (see [3] for a proof) that for any quantum protocol there is always a player able to force an outcome with probability at least $1/\sqrt{2}$. Chailloux and Kerenidis [6] showed that a bound of $1/\sqrt{2} + \varepsilon$ for any constant $\varepsilon > 0$ can be achieved, by combining two classical protocols with Mochon's result: They first showed that an *unbalanced* weak coin flip can be implemented using many instances of weak coin flips, and then that one instance of an unbalanced weak coin flip suffices to implement a strong coin flip with optimal achievable bias.

1.1 Limits of Previous Results

In all previous work on quantum coin flipping, honest players are required to output a *perfect* coin flip, i.e., the probability of both values has to be exactly 1/2, and the players must never disagree on the output or abort. However, in practice the players may very well be willing to allow a small probability of error even if both of them are honest. Furthermore, a (quantum) physical implementation of any protocol will always contain some noise and, therefore, also some probability to disagree or abort. This requirement is, therefore, overly strict and raises the question how much the cheating probability can be reduced when allowing an error of the honest players.

It is well-known that there exist numerous cryptographic tasks where allowing an (arbitrarily small) error can greatly improve the performance of the protocol. For example, as shown in [4], the amount of secure AND gates (or, alternatively, oblivious transfers) needed between two parties to test equality of two strings is only $O(\log 1/\varepsilon)$ for any small error $\varepsilon > 0$, while it is exponential in the length of the inputs in the perfect case. Considering reductions from oblivious transfer to different variants of oblivious transfer where the players can use quantum communication, it has recently been shown that introducing a small error can reduce the amount of oblivious transfer needed by an arbitrarily large factor [20].

It can easily be seen that *some* improvement on the achievable parameters must be possible also in the case of coin flipping: In any protocol, the honest players can simply flip the output bit with some probability. This increases the error, but decreases the bias. In the extreme case, the two players simply flip two independent coins and output this value. This prohibits any bias from the adversary, at the cost of making the players disagree with probability 1/2.

The only bounds on coin flipping we are aware of allowing for an error of the honest players have been given in the classical setting. An impossibility result for weak coin flipping has been given in [9]. Nguyen, Frison, Phan Huy and Massar presented in [17] a slightly more general bound and a protocol that achieves the bound in some cases.

1.2 Contribution

We introduce a general definition of coin flipping, characterized by 6 parameters, which we denote by

$$\mathrm{CF}(p_{00}, p_{11}, p_{0*}, p_{1*}, p_{*0}, p_{*1}) .$$

The value p_{ii} (where $i \in \{0,1\}$) is the probability that two honest players output i and the value p_{*i} (p_{i*}) is the maximal probability that the first (second) player can force the honest player to output i. With probability $1 - p_{00} - p_{11}$, two honest players will abort the protocol and output a dummy symbol[2]. This new definition has two main advantages:

- It generalizes both weak and strong coin flipping, but also allows for additional types of coin flips which are unbalanced or lay somewhere between weak and strong.
- It allows two honest players to abort with some probability.

We will first consider classical protocols (Section 3), and give tight bounds for all parameters. The impossibility result (Lemma 5) uses a similar proof technique as Theorem 7 in [9]. In combination with two protocols showing that this bound can be reached (Lemma 4), we obtain the following theorem.

Theorem 1. *Let* $p_{00}, p_{11}, p_{0*}, p_{1*}, p_{*0}, p_{*1} \in [0,1]$. *There exists a classical protocol that implements an unconditionally secure* $\mathrm{CF}(p_{00}, p_{11}, p_{0*}, p_{1*}, p_{*0}, p_{*1})$ *if and only if*

$$\begin{aligned} p_{00} &\leq p_{0*}p_{*0} , \\ p_{11} &\leq p_{1*}p_{*1} , \quad \textit{and} \\ p_{00} + p_{11} &\leq p_{0*}p_{*0} + p_{1*}p_{*1} - \max(0, p_{0*} + p_{1*} - 1)\max(0, p_{*0} + p_{*1} - 1) . \end{aligned}$$

[2] Similar to [9], we can require that two honest players always output the same values, since the players can always add a final round to check if they have the same value and output the dummy symbol if the values differ.

For weak coin flipping, i.e., $p_{*1} = 1$ and $p_{0*} = 1$, the bound of Theorem 1 simplifies to the condition that $p_{00} \leq p_{*0}$, $p_{11} \leq p_{1*}$, and

$$1 - p_{00} - p_{11} \geq (1 - p_{*0})(1 - p_{1*}) ,$$

which is the bound that is also implied by Theorem 7 in [9].

In Section 4, we consider the quantum case, and give tight bounds for all parameters. The quantum protocol (Lemma 10) bases on one of the protocols presented in [6], and is a classical protocol that uses an unbalanced quantum weak coin flip as a resource. The impossibility result follows from the proof of Kitaev's bound on quantum strong coin flipping (Lemma 11).

Theorem 2. *Let $p_{00}, p_{11}, p_{0*}, p_{1*}, p_{*0}, p_{*1} \in [0, 1]$. For any $\varepsilon > 0$, there exists a quantum protocol that implements an unconditionally secure* $\mathrm{CF}(p_{00}, p_{11}, p_{0*} + \varepsilon, p_{1*} + \varepsilon, p_{*0} + \varepsilon, p_{*1} + \varepsilon)$ *if*

$$\begin{aligned} p_{00} &\leq p_{0*} p_{*0} , \\ p_{11} &\leq p_{1*} p_{*1} , \quad \textit{and} \\ p_{00} + p_{11} &\leq 1 . \end{aligned}$$

If these bounds are not satisfied then there does not exist a quantum protocol for $\varepsilon = 0$.

Our results, therefore, give the exact trade-off between weak vs. strong coin flipping, between bias vs. abort-probability, and between classical vs. quantum coin flipping. (Some of these trade-offs are shown in Figures 1 and 2). They imply, in particular, that quantum protocols can achieve strictly better bounds

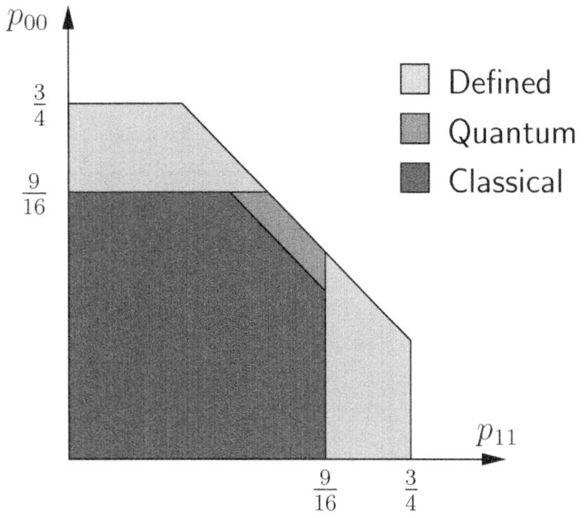

Fig. 1. For values $p_{0*} = p_{*0} = p_{1*} = p_{*1} = 3/4$, this figure shows the achievable values of p_{00} and p_{11} in the classical and the quantum setting. The light grey area is the set of all coin flips that can be defined. (See Definition 1).

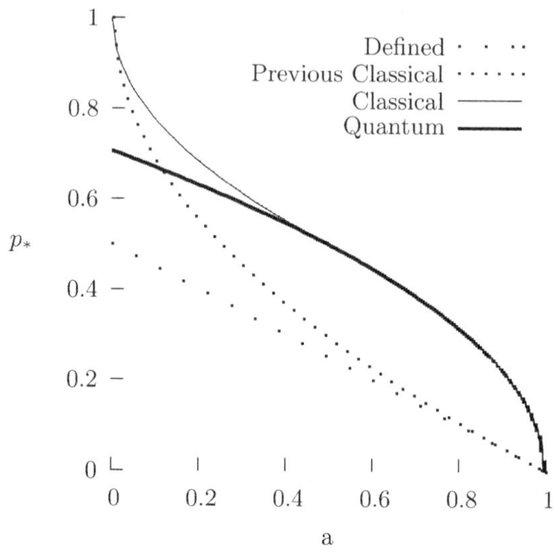

Fig. 2. This graph shows the optimal bounds for symmetric coin flipping of the form $\mathrm{CF}((1-a)/2,(1-a)/2,p_*,p_*,p_*,p_*)$. The value p_* is the maximal probability that any player can force the coin to be a certain value, and a is the abort probability. Therefore, the smaller p_* for a fixed value of a, the better is the protocol. The definition of coin flipping (Definition 1) implies that $p_* \geq (1-a)/2$. Hence, the theoretically optimal bound is $p_* = (1-a)/2$. In the quantum case, the optimal achievable bound is $p_* = \sqrt{(1-a)/2}$. In the classical case the optimal achievable bound is $p_* = 1 - \sqrt{a/2}$ for $a < 1/2$ and the same as the quantum bound for $a \geq 1/2$. The best previously known classical lower bounds from [9,17] was $p_* \geq 1 - \sqrt{a}$.

if $p_{0*} + p_{1*} > 1$ and $p_{*0} + p_{*1} > 1$. Outside this range classical protocols attain the same bounds as quantum protocols.

Since the optimal quantum protocol is a classical protocol using quantum weak coin flips as a resource, the possibility to do weak coin flipping, as shown by Mochon [15], can be seen as the crucial difference between the classical and the quantum case.

2 Preliminaries

In a classical protocol, the two players (Alice and Bob) are restricted to classical communication. Both players are given unlimited computing power and memory, and are able to locally sample random variables from any distribution. In a quantum protocol, the two players may exchange quantum messages. They have unlimited quantum memory and can perform any quantum computation on it. All operations are noiseless. At the beginning of the protocol, the players do not share any randomness or entanglement. While honest players have to follow the protocol, we do not make any assumption about the behaviour of the malicious players. We assume that the adversary is static, i.e., any malicious player is malicious from the beginning. Furthermore, we require that the protocol has a finite number of rounds.

Definition 1. *Let* $p_{00}, p_{11}, p_{0*}, p_{1*}, p_{*0}, p_{*1} \in [0,1]$*, such that* $p_{00} + p_{11} \leq 1$*,* $p_{00} \leq \min\{p_{0*}, p_{*0}\}$ *and* $p_{11} \leq \min\{p_{1*}, p_{*1}\}$ *holds*[3]*. A protocol implements a* $\mathrm{CF}(p_{00}, p_{11}, p_{0*}, p_{1*}, p_{*0}, p_{*1})$*, if the following conditions are satisfied:*

- *If both players are honest, then they output the same value* $i \in \{0,1\}$ *with probability* p_{ii} *and* Δ *with probability* $1 - p_{00} - p_{11}$*.*[2]
- *For any dishonest Alice, the probability that Bob outputs* 0 *is at most* p_{*0}*, and the probability that he outputs* 1 *is at most* p_{*1}*.*
- *For any dishonest Bob, the probability that Alice outputs* 0 *is at most* p_{0*}*, and the probability that she outputs* 1 *is at most* p_{1*}*.*

Definition 1 generalizes the notion of both weak and strong coin flips and encompasses, in fact, the different definitions given in the literature.

- A perfect weak coin flip is a
$$\mathrm{CF}\left(\frac{1}{2}, \frac{1}{2}, 1, \frac{1}{2}, \frac{1}{2}, 1\right) .$$
- A perfect strong coin flip is a
$$\mathrm{CF}\left(\frac{1}{2}, \frac{1}{2}, \frac{1}{2}, \frac{1}{2}, \frac{1}{2}, \frac{1}{2}\right) .$$
- The weak coin flip with error $\varepsilon > 0$ of [15] is a
$$\mathrm{CF}\left(\frac{1}{2}, \frac{1}{2}, 1, \frac{1}{2} + \varepsilon, \frac{1}{2} + \varepsilon, 1\right) .$$
- The unbalanced weak coin flip $\mathrm{WCF}(z, \varepsilon)$ of [6] is a
$$\mathrm{CF}\left(z, 1 - z, 1, 1 - z + \varepsilon, z + \varepsilon, 1\right) .$$
- The strong coin flip of [6] is a
$$\mathrm{CF}\left(\frac{1}{2}, \frac{1}{2}, \frac{1}{\sqrt{2}} + \varepsilon, \frac{1}{\sqrt{2}} + \varepsilon, \frac{1}{\sqrt{2}} + \varepsilon, \frac{1}{\sqrt{2}} + \varepsilon\right) .$$

Note that $\mathrm{CF}(p_{00}, p_{11}, p_{0*}, p_{1*}, p_{*0}, p_{*1})$ can also be defined as an ideal functionality that is equivalent to the above definition. Such a functionality would look like this: If there is any corrupted player, then the functionality first asks him to send a bit $b \in \{0,1\}$ that indicates which value he prefers. The functionality then

[3] The last two conditions are implied by the fact that a dishonest player can always behave honestly. Hence, he can always bias the coin to $i \in \{0,1\}$ with probability p_{ii}.

flips a coin $c \in \{0, 1, \Delta\}$, where the probabilities depend on b and on which player is corrupted. For example, if the first player is corrupted and $b = 0$, then $c = 0$ will be chosen with probability p_{*0}, $c = 1$ with probability $\min(p_{*1}, 1 - p_{*0})$ and Δ otherwise. The functionality then sends c to the adversary, and the adversary chooses whether he wants to abort the protocol or not. If he does not abort, the honest player receives c (which might already be Δ), and Δ otherwise. If none of the players are corrupted, the functionality chooses a value $c \in \{0, 1, \Delta\}$ which takes on $i \in \{0, 1\}$ with probability p_{ii} and sends c to the two players.

3 Classical Coin Flipping

3.1 Protocols

Protocol `CoinFlip1`:
Parameters: $p_{0*}, p_{1*}, p_{*0}, p_{*1} \in [0, 1]$, $p_{0*} + p_{1*} \leq 1$.

1. Alice flips a three-valued coin a such that the probability that $a = i$ is p_{i*} for $i = \{0, 1\}$, and $a = \Delta$ otherwise. She sends a to Bob.
2. If $a = \Delta$, Bob outputs $b = \Delta$. If $a \neq \Delta$, Bob flips a coin b such that $b = a$ with probability p_{*a} and $b = \Delta$ otherwise. Bob sends b to Alice and outputs b.
3. If $b = a$ Alice outputs b, otherwise Δ.

Lemma 1. *If either $p_{0*} + p_{1*} \leq 1$ or $p_{*0} + p_{*1} \leq 1$, then there exists a classical coin-flipping protocol with $p_{00} = p_{0*}p_{*0}$ and $p_{11} = p_{1*}p_{*1}$.*

Proof. If $p_{0*} + p_{1*} \leq 1$, they use Protocol `CoinFlip1`. (If $p_{*0} + p_{*1} \leq 1$, they exchange the role of Alice and Bob). By construction, a malicious Bob cannot bias Alice's output by more than p_{i*}, and a malicious Alice cannot bias Bob's output by more than p_{*i}. Honest players output the value 0 with probability $p_{0*}p_{*0}$ and 1 with probability $p_{1*}p_{*1}$. □

Protocol `CoinFlip2`:
Parameters: $p, x_0, x_1, y_0, y_1 \in [0, 1]$.

1. Alice flips a coin $a \in \{0, 1\}$ such that $a = 0$ with probability p and sends it to Bob.
2. Bob receives the coin a and flips a coin $b \in \{0, 1\}$ such that the probability that $b = a$ is x_a. He sends b to Alice. If $b = a$ he outputs b.
3. If $b = a$, then Alice outputs b. If $a \neq b$, then Alice flips a coin c, such that with probability y_b, $c = b$ and else $c = \Delta$. She sends c to Bob and outputs it.
4. If $c = b$ Bob outputs c, else Δ.

Lemma 2. *If* $p_{0*} + p_{1*} > 1$, $p_{*0} + p_{*1} > 1$, $p_{00} \leq p_{0*}p_{*0}$, $p_{11} \leq p_{1*}p_{*1}$, *and*

$$p_{00} + p_{11} = p_{0*}p_{*0} + p_{1*}p_{*1} - (p_{0*} + p_{1*} - 1)(p_{*0} + p_{*1} - 1) \tag{1}$$

then there exists a classical protocol implementing $\mathrm{CF}(p_{00}, p_{11}, p_{0*}, p_{1*}, p_{*0}, p_{*1})$.

Proof. We use Protocol `CoinFlip2` and choose the parameters

$$x_i := p_{*i}\,, \quad y_0 := \frac{p_{0*} - p}{1 - p}\,, \quad y_1 := \frac{p_{1*} + p - 1}{p}\,, \quad p := \frac{p_{00} - p_{0*} + p_{0*}p_{*1}}{p_{*0} + p_{*1} - 1}\,.$$

Note that if $p = 1$ then y_0 is undefined (and the same holds for y_1 if $p = 0$), but this does not cause any problem since in this case the parameter y_0 is never used in the protocol.

We now verify that these parameters are between 0 and 1. We have $y_0, y_1 \in [0, 1]$, if $p \in [1 - p_{1*}, p_{0*}]$. To see that p lies indeed in this interval, note that the upper bound follows from

$$p = \frac{p_{00} - p_{0*} + p_{0*}p_{*1}}{p_{*0} + p_{*1} - 1} \leq \frac{p_{0*}p_{*0} - p_{0*} + p_{0*}p_{*1}}{p_{*0} + p_{*1} - 1} = \frac{p_{0*}(p_{*0} + p_{*1} - 1)}{p_{*0} + p_{*1} - 1} = p_{0*}\,.$$

For the lower bound, note that

$$\begin{aligned} 1 - p &= \frac{p_{*0} + p_{*1} - 1}{p_{*0} + p_{*1} - 1} - \frac{p_{00} - p_{0*} + p_{0*}p_{*1}}{p_{*0} + p_{*1} - 1} \\ &= \frac{p_{*0} + p_{*1} - 1 - p_{00} + p_{0*} - p_{0*}p_{*1}}{p_{*0} + p_{*1} - 1} \\ &= \frac{p_{1*}p_{*0} - p_{1*} + p_{11}}{p_{*0} + p_{*1} - 1}\,, \end{aligned} \tag{2}$$

where we have used that

$$\begin{aligned} & p_{*0} + p_{*1} - 1 - p_{00} + p_{0*} - p_{0*}p_{*1} \\ & \overset{(1)}{=} p_{*0} + p_{*1} - 1 - (p_{0*}p_{*0} + p_{1*}p_{*1} - (p_{0*} + p_{1*} - 1)(p_{*0} + p_{*1} - 1) - p_{11}) \\ & \quad + p_{0*} - p_{0*}p_{*1} \\ & = p_{*0} + p_{*1} - 1 - p_{0*}p_{*0} - p_{1*}p_{*1} + p_{0*}p_{*0} + p_{0*}p_{*1} - p_{0*} + p_{1*}p_{*0} + p_{1*}p_{*1} \\ & \qquad - p_{1*} - p_{*0} - p_{*1} + 1 + p_{11} + p_{0*} - p_{0*}p_{*1} \\ & = p_{1*}p_{*0} - p_{1*} + p_{11}\,. \end{aligned}$$

Therefore

$$p = 1 - \frac{p_{11} - p_{1*} + p_{1*}p_{*0}}{p_{*0} + p_{*1} - 1} \geq 1 - \frac{p_{*1}p_{1*} - p_{1*} + p_{1*}p_{*0}}{p_{*0} + p_{*1} - 1} = 1 - p_{1*}\,.$$

It follows that $p, x_0, x_1, y_0, y_1 \in [0, 1]$.

If both players are honest, then the probability that they both output 0 is

$$\begin{aligned} px_0 + (1-p)(1-x_1)y_0 &= px_0 + (1-p)(1-x_1)\frac{p_{0*} - p}{1-p} \\ &= pp_{*0} + (1-p_{*1})(p_{0*} - p) \\ &= pp_{*0} - p(1-p_{*1}) + p_{0*}(1-p_{*1}) \\ &= \frac{p_{00} - p_{0*} + p_{0*}p_{*1}}{p_{*0} + p_{*1} - 1}(p_{*0} + p_{*1} - 1) + p_{0*}(1-p_{*1}) \\ &= p_{00} \ . \end{aligned}$$

The probability that they both output 1 is

$$\begin{aligned} p(1-x_0)y_1 + (1-p)x_1 &= p(1-p_{*0})\frac{p_{1*} + p - 1}{p} + (1-p)p_{*1} \\ &= (1-p_{*0})(p_{1*} + p - 1) + (1-p)p_{*1} \\ &= p_{1*}(1-p_{*0}) - (1-p)(1-p_{*0}) + (1-p)p_{*1} \\ &= p_{1*}(1-p_{*0}) + (1-p)(p_{*1} + p_{*0} - 1) \\ &\overset{(2)}{=} p_{1*}(1-p_{*0}) + \frac{p_{1*}p_{*0} - p_{1*} + p_{11}}{p_{*0} + p_{*1} - 1}(p_{*1} + p_{*0} - 1) \\ &= p_{11} \ . \end{aligned}$$

If Alice is malicious, she can bias Bob to output value i either by sending i as first message hoping that Bob does not change the value, which has probability $x_i = p_{*i}$; or by sending the value $1-i$ hoping that Bob changes the value, which occurs with probability $1 - x_{1-i} = 1 - p_{*1-i} \leq p_{*i}$. Hence, she succeeds with probability p_{*i}.

Bob can bias Alice to output value i by sending $b = i$ independently of what Alice had sent as first message. For $i = 0$, Alice will accept this value with probability

$$p + (1-p)y_0 = p + (1-p)\frac{p_{0*} - p}{1-p} = p_{0*}$$

and for $i = 1$ with probability

$$1 - p + py_1 = 1 - p + p\frac{p_{1*} + p - 1}{p} = p_{1*} \ . \tag{3}$$

□

In order to show that all values with $p_{00} + p_{11}$ below the bound given in (1) can be reached, we will need additionally the following lemma.

Lemma 3. *If there exists a protocol P that implements a coin flip* $\mathrm{CF}(p_{00}, p_{11}, p_{0*}, p_{1*}, p_{*0}, p_{*1})$*, then, for any $p'_{00} \leq p_{00}$ and $p'_{11} \leq p_{11}$, there exists a protocol P' that implements a coin flip* $\mathrm{CF}(p'_{00}, p'_{11}, p_{0*}, p_{1*}, p_{*0}, p_{*1})$.

Proof. P' is defined as follows: The players execute protocol P. If the output is $i \in \{0,1\}$, then Alice changes to Δ with probability $1 - p'_{ii}/p_{ii}$. If Alice changes to Δ, Bob also changes to Δ. Obviously, the cheating probabilities are still bounded by $p_{0*}, p_{1*}, p_{*0}, p_{*1}$, which implies that that protocol P' implements a $\mathrm{CF}(p'_{00}, p'_{11}, p_{0*}, p_{1*}, p_{*0}, p_{*1})$. □

Combining Lemmas 1, 2 and 3, we obtain Lemma 4.

Lemma 4. *Let* $p_{00}, p_{11}, p_{0*}, p_{1*}, p_{*0}, p_{*1} \in [0,1]$. *There exists a classical protocol that implements* $\mathrm{CF}(p_{00}, p_{11}, p_{0*}, p_{1*}, p_{*0}, p_{*1})$ *if*

$$\begin{aligned} p_{00} &\le p_{0*}p_{*0} , \\ p_{11} &\le p_{1*}p_{*1} , \textit{ and} \\ p_{00} + p_{11} &\le p_{0*}p_{*0} + p_{1*}p_{*1} - \max(0, p_{0*} + p_{1*} - 1)\max(0, p_{*0} + p_{*1} - 1) . \end{aligned}$$

Proof. If $p_{0*} + p_{1*} > 1$ and $p_{*0} + p_{*1} > 1$, then Lemmas 2 and 3 imply the bound. Otherwise, i.e., if either $p_{0*} + p_{1*} \le 1$ or $p_{*0} + p_{*1} \le 1$, then $\max(0, p_{0*} + p_{1*} - 1)\max(0, p_{*0} + p_{*1} - 1) = 0$ and the bound is implied by Lemmas 1 and 3. □

3.2 Impossibilities

The following lemma shows that the bounds obtained in Lemma 4 are optimal. The proof uses the same idea as the proof of Theorem 7 in [9].

Lemma 5. *Let the parameters* $p_{00}, p_{11}, p_{0*}, p_{1*}, p_{*0}, p_{*1}$ *be* $\in [0,1]$. *A coin flip* $\mathrm{CF}(p_{00}, p_{11}, p_{0*}, p_{1*}, p_{*0}, p_{*1})$ *can only be implemented by a classical protocol if*

$$\begin{aligned} p_{00} &\le p_{0*}p_{*0} , \\ p_{11} &\le p_{1*}p_{*1} , \textit{ and} \\ p_{00} + p_{11} &\le p_{0*}p_{*0} + p_{1*}p_{*1} - \max(0, p_{0*} + p_{1*} - 1)\max(0, p_{*0} + p_{*1} - 1) . \end{aligned}$$

Proof. We can assume that the output is a deterministic function of the transcript of the protocol. This can be enforced by adding an additional round at the end of the protocol where the two players tell each other what they are going to output. Since we do not require the protocol to be efficient, Lemma 7 in [9] implies that we can also assume that the honest parties maintain no internal state except for the list of previous messages.

For any partial transcript t of a protocol, we define p^t_{0*} as the maximum over all transcripts starting with t, i.e., the maximum probability with which Bob can force Alice to output 0, given the previous interaction has given t. In the same way, we define p^t_{1*}, p^t_{*0}, p^t_{*1}. We define p^t_{00} and p^t_{11} as the probabilities that the output of the honest players will be 00 and 11, respectively, given the previous interaction has given t. We will now do an induction over all transcripts, showing that for all t, we have

$$\begin{aligned} p^t_{00} &\le p^t_{0*}p^t_{*0} , \\ p^t_{11} &\le p^t_{1*}p^t_{*1} , \text{ and} \\ p^t_{00} + p^t_{11} &\le p^t_{0*}p^t_{*0} + p^t_{1*}p^t_{*1} - \max(0, p^t_{0*} + p^t_{1*} - 1)\max(0, p^t_{*0} + p^t_{*1} - 1) . \end{aligned}$$

For complete transcripts t, each honest player will output either 0, 1 or Δ with probability 1 and we always have $p^t_{0*} + p^t_{1*} - 1 \leq 0$ and $p^t_{*0} + p^t_{*1} - 1 \leq 0$. Therefore, we only need to check that $p^t_{00} \leq p^t_{0*} p^t_{*0}$ and $p^t_{11} \leq p^t_{1*} p^t_{*1}$. For $j \in \{0,1\}$, if $p^t_{jj} = 1$, then $p^t_{j*} = p^t_{*j} = 1$, so the condition is satisfied. In all the other cases we have $p^t_{jj} = 0$, in which case the condition is satisfied as well.

Let t now be a partial transcript, and let Alice be the next to send a message. Let M be the set of all possible transcripts after Alice has sent her message. For the induction step, we now assume that the statement holds for all transcript in M, and show that then it must also hold for t. Let r_i be the probability that an honest Alice will choose message $i \in M$. By definition, we have

$$p^t_{00} = \sum_{i \in M} r_i p^i_{00}, \quad p^t_{11} = \sum_{i \in M} r_i p^i_{11}, \quad p^t_{0*} = \sum_{i \in M} r_i p^i_{0*}, \quad p^t_{1*} = \sum_{i \in M} r_i p^i_{1*} ,$$

$$p^t_{*0} = \max_{i \in M} p^i_{*0}, \quad p^t_{*1} = \max_{i \in M} p^i_{*1} .$$

For $j \in \{0,1\}$ it holds that

$$p^t_{jj} = \sum_{i \in M} r_i p^i_{jj} \leq \sum_{i \in M} r_i p^i_{j*} p^i_{*j} \leq \sum_{i \in M} r_i p^i_{j*} p^t_{*j} = p^t_{j*} p^t_{*j} ,$$

which shows the induction step for the first two inequalities. To show the last inequality, let

$$f(a,b,c,d) := ac + bd - \max(0, a+b-1)\max(0, c+d-1) ,$$

where $a,b,c,d \in [0,1]$. If we fix the values c and d, we get the function $f_{c,d}(a,b) := f(a,b,c,d)$. It consists of two linear functions: If $a+b \leq 1$, we have

$$f_{c,d}(a,b) = ac + bd ,$$

and if $a+b \geq 1$ we have

$$f_{c,d}(a,b) = ac + bd - (a+b-1)\max(0, c+d-1) .$$

Note that these two linear functions are equal if $a+b=1$, and we have $(a+b-1)\max(0,c+d-1) \geq 0$ if $a+b \geq 1$. It follows that $f_{c,d}(a,b)$ is concave, meaning that for all $\gamma, a, b, a', b' \in [0,1]$, we have

$$\gamma f_{c,d}(a,b) + (1-\gamma) f_{c,d}(a',b') \leq f_{c,d}(\gamma a + (1-\gamma)a', \gamma b + (1-\gamma)b') . \tag{4}$$

Since for any $a+b \neq 1$ and $c+d \neq 1$

$$\frac{\partial}{\partial c} f(a,b,c,d) \geq 0 \qquad \text{and} \qquad \frac{\partial}{\partial d} f(a,b,c,d) \geq 0 , \tag{5}$$

we have $f(a,b,c',d) \geq f(a,b,c,d)$ for $c' \geq c$ and $f(a,b,c,d') \geq f(a,b,c,d)$ for $d' \geq d$. Hence,

$$\begin{aligned}
&p^t_{00} + p^t_{11} \\
&\quad = \sum_{i \in M} r_i (p^i_{00} + p^i_{11}) \\
&\quad \leq \sum_{i \in M} r_i \left(p^i_{0*} p^i_{*0} + p^i_{1*} p^i_{*1} - \max(0, p^i_{0*} + p^i_{1*} - 1) \max(0, p^i_{*0} + p^i_{*1} - 1)\right) \\
&\quad \leq \sum_{i \in M} r_i \left(p^i_{0*} p^t_{*0} + p^i_{1*} p^t_{*1} - \max(0, p^i_{0*} + p^i_{1*} - 1) \max(0, p^t_{*0} + p^t_{*1} - 1)\right) \\
&\quad \overset{(4)}{\leq} p^t_{0*} p^t_{*0} + p^t_{1*} p^t_{*1} - \max(0, p^t_{0*} + p^t_{1*} - 1) \max(0, p^t_{*0} + p^t_{*1} - 1) ,
\end{aligned}$$

and the inequalities also hold for t. The statement follows by induction. □

From Lemmas 4 and 5 we obtain Theorem 1.

4 Quantum Coin Flipping

4.1 Protocols

An *unbalanced weak coin flip with error* ε WCF(z, ε) is a CF$(z, 1-z, 1, 1-z+\varepsilon, z+\varepsilon, 1)$, i.e., a coin flip where Alice wins with probability z, Bob with probability $1-z$ and both cannot increase their probability to win by more than ε. (They may, however, decrease the probability to 0). Let WCF$(z) :=$ WCF$(z, 0)$.

It has been shown by Mochon [15] that weak coin flipping can be implemented with an arbitrarily small error.

Theorem 3 (Mochon [15]). *For any constant* $\varepsilon > 0$*, there exists a quantum protocol that implements* WCF$(1/2, \varepsilon)$.

In [14], Mochon showed that quantum coin-flipping protocols compose sequentially. Implicitly using this result, Chailloux and Kerenidis showed that an unbalanced weak coin flip can be implemented from many instances of (balanced) weak coin flips.

Proposition 1 (Chailloux, Kerenidis [6]). *For all* $z \in [0,1]$*, there exists a classical protocol that uses* k *instances of* WCF$(1/2, \varepsilon)$ *and implements* WCF$(x, 2\varepsilon)$*, for a value* $x \in [0,1]$ *with* $|x - z| \leq 2^{-k}$.

The following lemma shows that the parameter z can be slightly changed without increasing the error much.

Lemma 6. *For any* $1 > z' > z > 0$*, there exists a classical protocol that uses* 1 *instance of* WCF(z', ε) *and implements* WCF$(z, \varepsilon + z' - z)$.

Proof. The protocol first calls WCF(z', ε). If Alice wins, i.e., if the output is 0, then she changes the output bit to 1 with probability $1 - z/z'$, and sends the bit to Bob. Bob only accepts changes from 0 to 1, but not from 1 to 0. Alice can force the coin to be 0 with probability at most $z' + \varepsilon = z + (\varepsilon + z' - z)$. Let $x \in [0, 1 - z' + \varepsilon]$ be the probability with which a cheating Bob forces the protocol WCF(z', ε) to output 1. Alice will output 1 with probability

$$x + (1-x)\left(1 - \frac{z}{z'}\right) = 1 - \frac{z}{z'} + x \cdot \frac{z}{z'} \leq 1 - \frac{z}{z'} + (1 - z' + \varepsilon) \cdot \frac{z}{z'}$$
$$= 1 - z + \varepsilon \cdot \frac{z}{z'} \leq 1 - z + \varepsilon \,.$$ □

Note that for $z \in \{0,1\}$, the implementation of WCF$(z, 0)$ is trivial. Hence, Theorem 3, Proposition 1 and Lemma 6 imply together that WCF(z, ε) can be implemented for any $z \in [0,1]$ with an arbitrarily small error ε. To simplify the analysis of our protocols, we will assume that we have access to WCF(z) for any $z \in [0,1]$. The following lemma shows that when WCF(z) is replaced by WCF(z, ε), the bias of the output is increased by at most 2ε.

Lemma 7. *Let P be a protocol that implements* CF$(p_{00}, p_{11}, p_{0*}, p_{1*}, p_{*0}, p_{*1})$ *using one instance of* WCF(z). *If* WCF(z) *is replaced by* WCF(z, ε), *then P implements* CF$(p_{00}, p_{11}, p_{0*} + 2\varepsilon, p_{1*} + 2\varepsilon, p_{*0} + 2\varepsilon, p_{*1} + 2\varepsilon)$.

Proof. Let us compare two settings: one where the players execute P using one instance of WCF(z, ε), and the other where they use one instance of WCF(z). When both players are honest, the two settings are obviously identical. Let Alice be honest and Bob malicious. For each setting, we can define an event that occurs with probability at most ε, such that under the condition that the two events do not occur, WCF(z) and WCF(z, ε) and hence the whole protocol are identical. The probability that the two events do not occur is at least $1 - 2\varepsilon$ by the union bound. Therefore, the probabilities that the honest player outputs 0 (or 1) differ by at most 2ε. The statement follows. □

The following protocol is a generalization of the strong coin-flipping protocol S from [6]. It gives optimal bounds for the case where the honest players never abort, i.e., $p_{00} + p_{11} = 1$.

Protocol `QCoinFlip1`:
Parameters: $x, z_0, z_1, p_0, p_1 \in [0,1]$.

- Alice flips a coin $a \in \{0,1\}$ such that the probability that $a = 0$ is x and sends a to Bob.
- Alice and Bob execute WCF(z_a).
- If Alice wins, i.e., the outcome is 0, then both output a.
- If Bob wins, then he flips a coin b such that $b = a$ with probability p_a. Both output b.

Lemma 8. *Let* $p_{0*}, p_{1*}, p_{*0}, p_{*1} \in [0,1]$ *where* $p_{*0} + p_{*1} > 1$, $p_{0*} + p_{1*} > 1$ *and* $p_{*0}p_{0*} + p_{*1}p_{1*} = 1$. *Given access to one instance of* $\mathrm{WCF}(z)$, *we can implement a* $\mathrm{CF}(p_{00}, p_{11}, p_{0*}, p_{1*}, p_{*0}, p_{*1})$ *where* $p_{00} = p_{0*}p_{*0}$ *and* $p_{11} = p_{1*}p_{*1}$.

Proof. We execute Protocol `QCoinFlip1`, choosing the parameters

$$p_i := 1 - p_{*1-i}\,, \quad z_0 := \frac{p_{*0} + p_{*1} - 1}{p_{*1}}\,, \quad z_1 := \frac{p_{*0} + p_{*1} - 1}{p_{*0}}\,,$$
$$\text{and} \quad x := \frac{p_{0*}p_{*0} + p_{*1} - 1}{p_{*0} + p_{*1} - 1}\,.$$

Note that

$$1 - z_0 = \frac{1 - p_{*0}}{p_{*1}} \quad \text{and} \quad 1 - z_1 = \frac{1 - p_{*1}}{p_{*0}}\,.$$

Since $1 - p_{*0} < p_{*1}$ and $1 - p_{*1} < p_{*0}$, these values are between 0 and 1, and hence also z_0 and z_1 are between 0 and 1. From $p_{0*} \leq 1$ follows that $x \leq 1$, and from $p_{*0}p_{0*} + p_{*1} \geq p_{*0}p_{0*} + p_{*1}p_{1*} = 1$ that $x \geq 0$. Furthermore, we have

$$z_0 + (1 - z_0)p_0 = \frac{p_{*0} + p_{*1} - 1}{p_{*1}} + \frac{(1 - p_{*1})(1 - p_{*0})}{p_{*1}} = p_{*0} \tag{6}$$

and

$$z_1 + (1 - z_1)p_1 = \frac{p_{*0} + p_{*1} - 1}{p_{*0}} + \frac{(1 - p_{*1})(1 - p_{*0})}{p_{*0}} = p_{*1}\,.$$

Alice can bias Bob's coin to 0 with probability

$$\max\{z_0 + (1 - z_0)p_0; (1 - p_1)\} = p_{*0}$$

and to 1 with probability

$$\max\{z_1 + (1 - z_1)p_1; (1 - p_0)\} = p_{*1}\,.$$

The probability that Bob can bias Alice's coin to 0 is

$$\begin{aligned} x + (1 - x)(1 - z_1) &= (1 - z_1) + xz_1 \\ &= \frac{1 - p_{*1}}{p_{*0}} + \frac{p_{0*}p_{*0} + p_{*1} - 1}{p_{*0} + p_{*1} - 1} \cdot \frac{p_{*0} + p_{*1} - 1}{p_{*0}} \\ &= p_{0*} \end{aligned}$$

and the probability that he can bias it to 1 is

$$\begin{aligned} (1 - x) + x(1 - z_0) &= 1 - xz_0 \\ &\overset{(6)}{=} 1 - \frac{p_{0*}p_{*0} + p_{*1} - 1}{p_{*0} + p_{*1} - 1} \cdot \frac{p_{*0} + p_{*1} - 1}{p_{*1}} \\ &= 1 - \frac{p_{0*}p_{*0} + p_{*1} - 1}{p_{*1}} \\ &= \frac{1 - p_{0*}p_{*0}}{p_{*1}} \\ &= \frac{p_{1*}p_{*1}}{p_{*1}} = p_{1*}\,. \end{aligned}$$

Furthermore, two honest players output 0 with probability

$$\begin{aligned}
xz_0 + x(1-z_0)p_0 + (1-x)(1-z_1)(1-p_1) \\
&= x(z_0 + (1-z_0)p_0) + (1-x)\frac{1-p_{*1}}{p_{*0}}p_{*0} \\
&= xp_{*0} + (1-x)(1-p_{*1}) \\
&= 1 - p_{*1} + x(p_{*0} + p_{*1} - 1) \\
&= p_{0*}p_{*0} \\
&= p_{00}
\end{aligned}$$

and 1 with probability $1 - p_{00} = 1 - p_{0*}p_{*0} = p_{1*}p_{*1} = p_{11}$. □

The following protocol gives optimal bounds for the general case. It uses one instance of the above protocol, and lets Alice and Bob abort in some situations.

Protocol `QCoinFlip2`:
Parameters: Protocol P, $\varepsilon_0, \varepsilon_1 \in [0, \frac{1}{2}]$.

- Alice and Bob execute the coin-flipping protocol P.
- If Alice obtains 0, she changes to Δ with probability ε_0. If Bob obtains 1, he changes to Δ with probability ε_1. If either Alice or Bob has changed to Δ, they both output Δ, otherwise they output the value obtained from P.

Lemma 9. *Let $p_{0*}, p_{1*}, p_{*0}, p_{*1} \in [0,1]$ where $p_{*0} + p_{*1} > 1$, $p_{0*} + p_{1*} > 1$ and $p_{0*}p_{*0} + p_{*1}p_{*1} \leq 1$. Given access to $\mathrm{WCF}(z)$ for any $z \in [0,1]$, we can implement a $\mathrm{CF}(p_{00}, p_{11}, p_{0*}, p_{1*}, p_{*0}, p_{*1})$ where $p_{00} = p_{0*}p_{*0}$ and $p_{11} = p_{1*}p_{*1}$.*

Proof. From $p_{*0} + p_{*1} > 1$ and $p_{0*} + p_{1*} > 1$ follows that either $p_{*0} + p_{1*} > 1$ or $p_{0*} + p_{*1} > 1$. Without loss of generality, let us assume that $p_{*0} + p_{1*} > 1$.

Let

$$p'_{0*} := \min\left(1, \frac{1 - p_{1*}p_{*1}}{p_{*0}}\right) \quad \text{and} \quad p'_{*1} := \frac{1 - p'_{0*}p_{*0}}{p_{1*}} .$$

First, note that since $p_{0*} \leq \frac{1-p_{1*}p_{*1}}{p_{*0}}$ we have $p'_{0*} \geq p_{0*}$. Obviously, we also have $p'_{0*} \leq 1$. Since $p'_{0*} \leq \frac{1-p_{1*}p_{*1}}{p_{*0}}$, we have

$$p'_{*1} = \frac{1 - p'_{0*}p_{*0}}{p_{1*}} \geq \frac{1 - \frac{1-p_{1*}p_{*1}}{p_{*0}}p_{*0}}{p_{1*}} = \frac{p_{1*}p_{*1}}{p_{1*}} = p_{*1} .$$

In order to see that $p'_{*1} \leq 1$, we need to distinguish two cases. Since $p'_{0*} := \min\left(1, \frac{1-p_{1*}p_{*1}}{p_{*0}}\right)$, it holds that either $p'_{0*} = 1$ or $p'_{0*} = \frac{1-p_{1*}p_{*1}}{p_{*0}}$. In the first case,

$$p'_{*1} = \frac{1 - p_{*0}}{p_{1*}} < \frac{p_{1*}}{p_{1*}} = 1 ,$$

and the claim holds. In the second case,

$$p'_{*1} = \frac{1 - p'_{*0}p_{*0}}{p_{1*}} = \frac{1 - (1 - p_{1*}p_{*1})}{p_{1*}} = p_{*1} \leq 1 ,$$

and the claim also holds. Therefore, $p'_{*1} \leq 1$.

Since $p'_{0*}p_{*0} + p_{1*}p'_{*1} = 1$, according to Lemma 8, we can use protocol `QCoinFlip1` to implement a $\mathrm{CF}(p'_{00}, p'_{11}, p'_{0*}, p_{1*}, p_{*0}, p'_{*1})$, where $p'_{00} = p'_{0*}p_{*0}$ and $p'_{11} = p_{1*}p'_{*1}$. Using that protocol as protocol P, let Alice and Bob execute protocol `QCoinFlip2` with $\varepsilon_0 := 1 - p_{0*}/p'_{0*}$, and $\varepsilon_1 := 1 - p_{*1}/p'_{*1}$.

The probability that Bob can bias Alice to 0 is now $(1-\varepsilon_0)p'_{0*} = p_{0*}$, and the probability that Alice can bias Bob to 1 is now $(1 - \varepsilon_1)p'_{*1} = p_{*1}$. Furthermore, the probability that two honest players output both 0 is $(1 - \varepsilon_0)p'_{00} = (1 - \varepsilon_0)p'_{0*}p_{*0} = p_{0*}p_{*0}$ and the probability that they both output 1 is $(1 - \varepsilon_1)p'_{11} = (1 - \varepsilon_1)p_{1*}p'_{*1} = p_{1*}p_{*1}$. □

Lemma 10. *Let $p_{00}, p_{11}, p_{0*}, p_{1*}, p_{*0}, p_{*1} \in [0, 1]$ with*

$$p_{00} \leq p_{0*}p_{*0} ,$$
$$p_{11} \leq p_{1*}p_{*1} , \quad \textit{and}$$
$$p_{00} + p_{11} \leq 1 .$$

Then, for any constant $\varepsilon > 0$, there exists a quantum protocol that implements $\mathrm{CF}(p_{00}, p_{11}, p_{0*} + \varepsilon, p_{1*} + \varepsilon, p_{*0} + \varepsilon, p_{*1} + \varepsilon)$.

Proof. Let us first assume that $p_{*0} + p_{*1} > 1$ and $p_{0*} + p_{1*} > 1$. We reduce the value of p_{0*} to p_{00}/p_{*0} and the value of p_{1*} to p_{11}/p_{*1}, which ensures that $p_{0*}p_{*0} + p_{1*}p_{*1} \leq 1$. Now we can apply Lemma 9, together with Theorem 3, Proposition 1 and Lemmas 6, 7 and 3.

If the assumption does not hold then either $p_{*0} + p_{*1} \leq 1$ or $p_{0*} + p_{1*} \leq 1$. In this case, we can apply Lemmas 1 and 3. □

4.2 Impossibilities

In order to see that the bound obtained in Section 4.1 is tight, we can use the proof of Kitaev [11] (printed in [3]) showing that an adversary can always bias the outcome of a strong quantum coin-flipping protocol. In fact, Equations (36) - (38) in [3] imply that for any quantum coin-flipping protocol, it must hold that $p_{11} \leq p_{1*}p_{*1}$. In the same way, it can be proven that $p_{00} \leq p_{0*}p_{*0}$. We obtain the following lemma.

Lemma 11. *A $\mathrm{CF}(p_{00}, p_{11}, p_{0*}, p_{1*}, p_{*0}, p_{*1})$ can only be implemented by a quantum protocol if*

$$p_{00} \leq p_{0*}p_{*0} ,$$
$$p_{11} \leq p_{1*}p_{*1} , \quad \textit{and}$$
$$p_{00} + p_{11} \leq 1 .$$

Lemma 10 and 11 imply together Theorem 2.

5 Conclusions

We have shown tight bounds for a general definition of coin flipping, which give trade-offs between weak vs. strong coin flipping, between bias vs. abort probability, and between classical vs. quantum protocols.

Our result extends the work of [6], and shows that the whole advantage of the quantum setting lies in the ability to do weak coin flips (as shown by Mochon [15]). If weak coin flips are available in the classical setting, classical protocols can achieve the same bounds as quantum protocols.

For future work, it would be interesting to see if similar bounds hold for the definition of coin flipping without the possibility for the malicious player to abort.

Acknowledgments. We thank Thomas Holenstein, Stephanie Wehner and Severin Winkler for helpful discussions, and the anonymous referees for their helpful comments. This work was funded by the Swiss National Science Foundation (SNSF), an ETHIIRA grant of ETH's research commission, the U.K. EPSRC, grant EP/E04297X/1 and the Canada-France NSERC-ANR project FREQUENCY. Most of this work was done while JW was at the University of Bristol.

References

1. Aharonov, D., Ta-Shma, A., Vazirani, U.V., Yao, A.C.: Quantum bit escrow. In: STOC 2000: Proceedings of the 32nd Annual ACM Symposium on Theory of Computing, pp. 705–714 (2000)
2. Ambainis, A.: A new protocol and lower bounds for quantum coin flipping. In: STOC 2001: Proceedings of the 33rd Annual ACM Symposium on Theory of Computing, pp. 134–142 (2001)
3. Ambainis, A., Buhrman, H., Dodis, Y., Röhrig, H.: Multiparty quantum coin flipping. In: CCC 2004: Proceedings of the 19th Annual IEEE Conference on Computational Complexity, pp. 250–259 (2004)
4. Beimel, A., Malkin, T.: A quantitative approach to reductions in secure computation. In: Naor, M. (ed.) TCC 2004. LNCS, vol. 2951, pp. 238–257. Springer, Heidelberg (2004)
5. Blum, M.: Coin flipping by telephone a protocol for solving impossible problems. SIGACT News 15(1), 23–27 (1983)
6. Chailloux, A., Kerenidis, I.: Optimal quantum strong coin flipping. In: FOCS 2009: Proceedings of the 50th Annual IEEE Symposium on Foundations of Computer Science, pp. 527–533 (2009)
7. Cleve, R.: Limits on the security of coin flips when half the processors are faulty. In: STOC 1986: Proceedings of the 18th Annual ACM Symposium on Theory of Computing, pp. 364–369 (1986)
8. Colbeck, R.: An entanglement-based protocol for strong coin tossing with bias 1/4. Physics Letters A 362(5-6), 390–392 (2007)
9. Hofheinz, D., Müller-Quade, J., Unruh, D.: On the (Im-)Possibility of extending coin toss. In: Vaudenay, S. (ed.) EUROCRYPT 2006. LNCS, vol. 4004, pp. 504–521. Springer, Heidelberg (2006)

10. Kerenidis, I., Nayak, A.: Weak coin flipping with small bias. Information Processing Letters 89(3), 131–135 (2004)
11. Kitaev, A.: Quantum coin-flipping. In: QIP 2003 (2002), Slides `http://www.msri.org/publications/ln/msri/2002/qip/kitaev/1/index.html`
12. Lo, H.-K., Chau, H.F.: Why quantum bit commitment and ideal quantum coin tossing are impossible. Physica D 120(1-2), 177–187 (1998)
13. Mochon, C.: Quantum weak coin-flipping with bias of 0.192. In: FOCS 2004: Proceedings of the 45th Annual IEEE Symposium on Foundations of Computer Science, pp. 2–11 (2004)
14. Mochon, C.: Serial composition of quantum coin flipping and bounds on cheat detection for bit commitment. Physical Review A 70(3), 32312 (2004)
15. Mochon, C.: Quantum weak coin flipping with arbitrarily small bias (2007), `http://arxiv.org/abs/0711.4114`
16. Moran, T., Naor, M., Segev, G.: An optimally fair coin toss. In: Reingold, O. (ed.) TCC 2009. LNCS, vol. 5444, pp. 1–18. Springer, Heidelberg (2009)
17. Nguyen, A.T., Frison, J., Huy, K.P., Massar, S.: Experimental quantum tossing of a single coin. New Journal of Physics 10(8), 83037 (2008)
18. Spekkens, R.W., Rudolph, T.: Degrees of concealment and bindingness in quantum bit commitment protocols. Physical Review A 65(1), 12310 (2001)
19. Spekkens, R.W., Rudolph, T.: A quantum protocol for cheat-sensitive weak coin flipping. Physical Review Letters 89(22), 227901 (2002)
20. Winkler, S., Wullschleger, J.: On the efficiency of classical and quantum oblivious transfer reductions. In: Rabin, T. (ed.) CRYPTO 2010. LNCS, vol. 6223, pp. 707–723. Springer, Heidelberg (2010)
21. Yao, A.C.: Protocols for secure computations. In: FOCS 1982: Proceedings of the 23rd Annual IEEE Symposium on Foundations of Computer Science, pp. 160–164 (1982)

Exploring the Limits of Common Coins Using Frontier Analysis of Protocols

Hemanta K. Maji[1,⋆], Pichayoot Ouppaphan[1,⋆⋆], Manoj Prabhakaran[1,⋆], and Mike Rosulek[2]

[1] Department of Computer Science, University of Illinois, Urbana-Champaign
{hmaji2,pouppap2,mmp}@uiuc.edu
[2] Department of Computer Science, University of Montana
mikero@cs.umt.edu

Abstract. In 2-party secure computation, access to common, trusted randomness is a fundamental primitive. It is widely employed in the setting of computationally bounded players (under various complexity assumptions) to great advantage. In this work we seek to understand the power of trusted randomness, primarily in the computationally unbounded (or information theoretic) setting. We show that a source of common randomness does not add any additional power for secure evaluation of deterministic functions, even when one of the parties has arbitrary influence over the distribution of the common randomness. Further, common randomness helps only in a trivial sense for realizing randomized functions too (namely, it only allows for sampling from publicly fixed distributions), if UC security is required.

To obtain these impossibility results, we employ a recently developed protocol analysis technique, which we call the *frontier analysis*. This involves analyzing carefully defined "frontiers" in a weighted tree induced by the protocol's execution (or executions, with various inputs), and establishing various properties regarding one or more such frontiers. We demonstrate the versatility of this technique by employing carefully chosen frontiers to derive the different results. To analyze randomized functionalities we introduce a frontier argument that involves a geometric analysis of the space of probability distributions.

Finally, we relate our results to computational intractability questions. We give an equivalent formulation of the "cryptomania assumption" (that there is a semi-honest or standalone secure oblivious transfer protocol) in terms of UC-secure reduction among randomized functionalities. Also, we provide an *unconditional result* on the uselessness of common randomness, even in the computationally bounded setting.

Our results make significant progress towards understanding the exact power of shared randomness in cryptography. To the best of our knowledge, our results are the first to comprehensively characterize the power of large classes of randomized functionalities.

⋆ Partially supported by NSF grants CNS 07-47027 and CNS 07-16626.
⋆⋆ Supported by NSF grant CNS 0851957 for undergraduate research.

Y. Ishai (Ed.): TCC 2011, LNCS 6597, pp. 486–503, 2011.

1 Introduction

In this work, we consider a fundamental question: *How cryptographically useful is a trusted source of public coins?*

While there are several instances in cryptography where a common random string or a trusted source of public coins is very useful (e.g. [3,5]), we show severe limitations to its usefulness[1] in secure two-party computation, without — and sometimes even with — computational intractability assumptions. In contrast, it is well known that more general correlated private random variables can be extremely powerful [2]. Given that for semi-honest security common randomness is useless (as one of the parties could sample and broadcast it), it is not surprising that it should turn out to be not as powerful as general correlated random variables. However, despite its fundamental nature, the exact power of common randomness has not yet been characterized. Here, we provide tight characterizations of what can be achieved with a source of common randomness, in various settings of 2-party computation. We show:

- For two-party secure function evaluation (SFE) of *deterministic* functions, being given a source of common randomness is useless, irrespective of any computational complexity assumptions, when considering security in the standalone setting[2].
- Clearly a source of common randomness can be useful for realizing *randomized* functionalities. However, in the case of UC security, we show that a source of common coins can be useful only in a trivial sense (unless restricted to the computationally bounded setting, and intractability assumptions are employed). We show that any UC-secure protocol that uses common coins for evaluating a randomized function can be replaced by a protocol of the following simple form: one of the parties announces a probability distribution, based deterministically on its input, and then the two parties sample an outcome from this distribution using freshly sampled common coins. We call the resulting functionality a *publicly-selectable source.*
- We relate computational intractability assumptions to secure reductions among randomized functionalities, giving evidence that common randomness is useful only under strong computational assumptions. In particular we show that common randomness can be used to UC-securely realize a symmetric functionality with bi-directional influence (i.e., the output is influenced by both the parties' inputs) if and only if there exists a semi-honest secure protocol for oblivious transfer.

These results are actually proven for a class of sources more general than coin tossing, namely *selectable sources* – that let one of the parties (secretly) specify

[1] We say that a source of common randomness is useless in realizing some 2-party functionality $\mathcal{F}$ if either $\mathcal{F}$ could be realized without using the given source or $\mathcal{F}$ cannot be realized even given the source. Note that we consider only the feasibility question and not any efficiency issues.

[2] In the case of UC security, it follows from the results in [16] that a source of common randomness is useless except in Cryptomania, where it is a complete functionality.

which among a set of distributions should be used by the source. We highlight two aspects of these results:

Non-blackbox analysis of protocols. In deriving the impossibility results our analysis crucially relies on the communication and information structure of protocols. We build on the "frontier analysis" paradigm in [8,15,16], but significantly extend its power, among other things, to enable analyzing protocols for arbitrary randomized functionalities, and protocols using randomized functionalities.

These results (and hence proofs) are necessarily of a *non-relativizing* nature — if the protocol has access to another trusted functionality (more sophisticated than common randomness), the impossibility results no longer hold. Specifics about the common randomness functionality are (and must be) used in our proofs. Such low-level analysis of protocols, we believe, is crucial to understanding the power and complexity of multi-party computation primitives.

Understanding randomized functionalities. Secure evaluation of *randomized* functions has in general been a poorly understood area. In particular, to date it remains open to characterize which randomized functions can be securely realized even against computationally unbounded passive (honest-but-curious) adversaries — a problem that was solved for deterministic functions twenty years ago [1,13]. Much of the study of randomized functionalities has been focused on in-depth understanding of the simplest such functionality — namely generating shared fair coins (e.g., see [7,10,8,18] and references therein). Our results provide significant insight into *other* randomized functionalities as well, and their connections to computational intractability assumptions. In particular, our results involve two interesting classes of randomized functionalities, namely *selectable sources* and *publicly-selectable sources.*

1.1 Overview

Frontier analysis. The bulk of our results take the form of statements of cryptographic impossibility. That is, we show that a protocol for a given cryptographic task is impossible (or else implies a certain computational primitive like one-way functions). Such impossibility results have been a core challenge in cryptography. In this work, we present a powerful battery of techniques that we use to analyze 2-party protocols, which we broadly call "frontier analysis."

The basic outline of a frontier analysis is as follows. We first interpret a protocol as a tree of possible transcripts, with weights corresponding to the probability that the protocol assigns to each message, based on the parties' inputs. Within this tree, we identify "frontiers", which are simply a collection of nodes (partial transcripts) that form a cut and an independent set. Intuitively, these frontiers correspond to points in the protocol when some condition is satisfied for the first time, where the condition in question depends on the kind of analysis needed: for example, the first place the transcript leaks "significant" information about a party's input, or the first place that common coins have made a "significant" influence on the protocol's output.

Impossibility proofs using frontier analysis proceed by showing that frontiers of certain kind exist, often showing that multiple frontiers must be encountered in a specific order, and then showing that an adversary can effect an attack by exploiting the properties of these frontiers. The interested reader can find a high level discussion on frontier analysis as a tool for protocol analysis in the full version of this paper [14].

Common coins are not useful in SFE protocols. We show that against computationally unbounded adversaries (more precisely, against adversaries that can break one-way functions), any 2-party deterministic SFE (in which both parties receive the same output) functionality that can be securely realized given a trusted coin-tossing functionality can in fact be securely realized without it. This is most interesting for the standalone setting, because if one-way functions do exist then a standalone-secure coin-tossing protocols exist, so again access to a trusted coin-tossing functionality is redundant[3].

We start off by showing that there is no secure protocol for evaluating boolean XOR given a coin-tossing functionality. In many ways these functionalities have similar "complexity" (in particular, neither is complete, and both are trivial to realize against passive adversaries), so establishing a qualitative separation between them is interesting in itself. In a protocol for XOR, either party may be the first to reveal information about their input, and the two parties can even gradually reveal more and more information about their input in an interleaved fashion. We define a frontier corresponding to the first point at which some party has revealed "significant" information about its input. Then we define an attack that can be carried out when the protocol crosses this frontier. Since a large class of SFE functionalities can be used to securely realize XOR, the impossibility extends to these functionalities as well.

We then use the combinatorial characterizations of Symmetric Secure Function Evaluation (SSFE) functionalities (obtained using frontier analysis) from [15] to extend the result to arbitrary SSFE functionalities (instead of just XOR). Further, using an extension of a result in [11], we extend this to arbitrary SFE functionalities by associating a symmetric SFE with every general SFE that has a secure protocol using a source of common randomness.

For randomized SFE, common coins help only in a trivial sense. We show that common coins are useful in constructing UC-secure protocols for randomized SFE functionalities only for the class of publicly-selectable sources (Theorem 2). For this result, we exploit the versatility of the frontier analysis and also employ a geometric analysis of the space of effective probability distributions.

---

[3] A recent result in [16] gives a sharp result for the case of UC security: the coin-tossing functionality is useful in realizing further deterministic SFE functionalities if and only if there exists a semi-honest oblivious transfer protocol. However neither the result nor the approach in [16] extends to the standalone setting. Also, our result is applicable to not just symmetric functionalities and coin-tossing, but extends to general SFE functionalities and all selectable sources.

The frontier analysis is carried out for an SSFE functionality, and then the result is extended to general SFE functionality separately. For a randomized SSFE functionality, for each pair of inputs, the output is specified by a distribution (over a finite output alphabet). This distribution can be represented as a vector in d-dimensional real space where d is the size of the output alphabet. By considering all possible inputs, we obtain a set of points in this space as legitimate output distributions. But since the parties can choose their input according to any distribution they wish, the entire convex hull of these points is the set of legitimate output distributions. Note that the vertices of this polytope correspond to the output distributions for various specific input choices.

In analyzing a protocol for such a functionality, we define *two* very different frontiers: one intuitively captures the last point in the protocol where the parties' inputs have any noticeable influence over the output distribution. The other intuitively captures the first point where the common coins have had a non-trivial influence on the output distribution.

Defining these frontiers is a delicate task, but once they are defined, we can show that, for the protocol to be UC-secure, the two frontiers must be encountered in the order listed above. Thus there is always a point within the protocol where the parties' inputs have stopped influencing the output, yet the public coins have not yet started influencing the output in a non-trivial way. At this point, we can show that the output distribution is uniquely determined, and that the subsequent coins are simply used to sample from this publicly-chosen distribution.

Then, on each node in the first frontier the conditional output distribution is still within the polytope. On the other hand, since the input influence has ceased at this point, for any fixed input, its output distribution must be determined by this frontier: i.e., it must be a convex combination of the conditional output distributions at the nodes on the frontier. That is, the output distribution for this input is a convex combination of conditional output distributions which are all themselves within the polytope. Now, (without loss of generality, as it turns out) we can consider inputs whose output distributions are vertices of the polytope. Then, *for all nodes in the frontier* the conditional output distribution must coincide with the final distribution itself. Thus on reaching this frontier in the protocol, the output distribution is revealed (as a deterministic function of the inputs) and the rest of the protocol simply samples from this distribution.

Finally, we extend this result also to general SFE (instead of just symmetric SFE) functionalities, in the same way as for deterministic functionalities.

Selectable sources. Selectable sources are an interesting class of randomized functionalities with an intermediate level of complexity: they can be more complex than a (fixed) source of common randomness, yet they are simple enough that we can show that they are as useless as common randomness when it comes to securely realizing deterministic functionalities. The extension is observed by following the analysis for the case of the source of common randomness, and identifying the properties that it relies on. We do not know at this point whether these are exactly all the functionalities which are useless for realizing SFE functionalities, but based on our understanding so far, we conjecture that they are.

Connections to Computational Intractability. Finally, we relate our results to computational intractability questions. The attacks based on frontier analysis can often be extended to the computationally bounded setting, if one-way functions do not exist (as was pointed out in [16]). We show that this is indeed the case for our attacks. In fact, our first application of such an extension is to obtain an *unconditional* result about the uselessness of selectable sources in realizing deterministic secure function evaluation with standalone security. For this we use the fact that if one-way functions do not exist we can attack any given protocol, whereas if one-way functions do exist then we can realize any selectable source functionality (with standalone security) and then again they are useless.

We also generalize a result in [16] to the setting of randomized functionalities. There it was shown that if any non-trivial deterministic functionality is UC-securely realizable using access to common randomness, then there exists an oblivious transfer protocol secure against semi-honest adversaries. We generalize common randomness to any selectable source, and also generalize non-trivial deterministic functionalities to randomized SSFE functionalities with both parties' inputs having an influence on the output.

Related Results. Frontier analysis is possibly implicit in previous works on proving impossibility or lower bounds for protocols. For instance, the analysis in [8] very well fits our notion of what frontier analysis is. The analysis of protocols in [6,1,13] also have some elements of a frontier analysis, but of a rudimentary form which was sufficient for analysis of perfect security. In [15] frontier analysis was explicitly introduced and used to prove several protocol impossibility results and characterizations. [12] also presented similar results and used somewhat similar techniques (but relied on analyzing the protocol by rounds, instead of frontiers, and suffered limitations on the round complexity of the protocols for which the impossibility could be shown).

2 Preliminaries

We say that a function $\nu : \mathbb{N} \to \mathbb{R}$ is *negligible* if for every polynomial p, $\nu(k) < 1/p(k)$ for sufficiently large k. If $\mathcal{D}, \mathcal{D}'$ are discrete probability distributions with support S, we write $\mathrm{SD}(\mathcal{D}, \mathcal{D}')$ to denote the statistical distance of the distributions, defined as $\mathrm{SD}(\mathcal{D}, \mathcal{D}') = \frac{1}{2}\sum_{s\in S} |\mathcal{D}(s) - \mathcal{D}'(s)|$.

Security. We use standard conventions and terminology for the security of protocols for multi-party computation tasks. A protocol is secure if for every adversary in the real world (in which parties execute a protocol), there is an adversary, or *simulator*, in the ideal world (in which the task is carried out on behalf of the parties by a trusted third party called a *functionality*) that achieves the same effect. A *semi-honest* or *passive* adversary is one which is not allowed to deviate from the protocol. *Standalone* security is achieved if the simulator is allowed to rewind the adversary; *Universally composable (UC)* security [4] is achieved if the simulation is straight-line (i.e., never rewinds the adversary). In this work, we exclusively consider *static* adversaries, who do not adaptively corrupt honest parties during the execution of a protocol.

The *plain model* is a real world in which protocols only have access to a simple communication channel; a *hybrid model* is a real world in which protocols can additionally use a particular trusted functionality. While hybrid worlds are usually considered only for UC security, we also use the terminology in the setting of standalone security. We note that protocols for *non-reactive* functionalities (i.e., those which receive input from all parties, then give output, and then stop responding) do securely compose even in the standalone security setting.

2.1 Functionalities

We focus on classifying several important subclasses of functionalities.

Secure function evaluation (SFE). A 2-party secure function evaluation (SFE) functionality is specified by two functions $f_1 : X \times Y \to Z$ and $f_2 : X \times Y \to Z$, where X and Y are finite sets. The functionality waits for input $x \in X$ from Alice and $y \in Y$ from Bob, then delivers $f_1(x, y)$ and $f_2(x, y)$ to them, respectively. There is no fairness guarantee: if a party is corrupt, it can obtain its own output first and decide whether the output should be delivered to the other party.

If $f_1 = f_2$ are identical we call it a *symmetric* SFE (or SSFE) functionality. SSFE functionalities are the most fundamental, and have been studied since Yao first introduced the concept of multi-party computation [20]. We can specify an SSFE function by simply giving its function table, where the rows correspond to an input of Alice, and columns correspond to an input of Bob. For instance, the XOR functionality has function table $\begin{array}{|cc|}\hline 0 & 1 \\ 1 & 0 \\ \hline\end{array}$.

Randomized functionalities. A randomized SFE functionality is specified by functions $f_1, f_2 : X \times Y \times R \to Z$. The functionality takes inputs $x \in X$ from Alice, $y \in Y$ from Bob, uniformly samples $r \in R$ and outputs $f_1(x, y, r)$ and $f_2(x, y, r)$ to Alice and Bob, respectively. An important example is the common randomness functionality, denoted by $\mathcal{F}_{\mathsf{coin}}$ (with $X = Y = \{0\}$, $R = \{0, 1\}$, and $f_1(x, y, r) = f_2(x, y, r) = r$). Note that for a given pair of inputs, the outputs to Alice and Bob could be correlated as the same value r is used in both.

We identify two important subclasses of randomized SSFE functionalities:

Selectable sources: One in which one party's input does not affect the output. That is, functions which can be written as $f(x, y, r) = h(x, r)$ for some function h. Note that for different values of x, the function's output distribution may be arbitrary.

Publicly-selectable sources: Those functions which can be written as $f(x, y, r) = (g(x), h(g(x), r))$, for some functions g and h. In this case, the function's output distribution for different values of x must be either identical (when $g(x) = g(x')$) or have disjoint supports (when $g(x) \neq g(x')$, which is included in the function's output). Intuitively, the function's output determines the identity of the random distribution $h(g(x), \cdot)$ that was used.

In these two classes of functionalities, only one party can influence the output, so we say they have *uni-directional influence*. If there exists inputs x, x', x'' for Alice and y, y', y'' for Bob so that $f(x, y') \not\equiv f(x, y'')$, and $f(x', y) \not\equiv f(x'', y)$, then both parties can potentially influence the output, and we say that the functionality has *bi-directional influence*.

Isomorphism. $\mathcal{F}$ and $\mathcal{G}$ are isomorphic[4] if either functionality can be UC-securely realized using the other functionality by a protocol that is "local" in the following sense: to realize $\mathcal{F}$ given $\mathcal{G}$ (say), each party maps its input (possibly probabilistically) to inputs for the functionality $\mathcal{G}$, calls $\mathcal{G}$ once with that input and, based on their private input, the output obtained from $\mathcal{G}$, and possibly private random coins, locally computes the final output, without any other communication. It is easy to see that isomorphism is an equivalence relation.

Usefulness of a source. We say that a source of common randomness $\mathcal{G}$ is **useless** in realizing a 2-party functionality $\mathcal{F}$ if either $\mathcal{F}$ could be securely realized in the plain model (i.e., without using $\mathcal{G}$) or $\mathcal{F}$ cannot be securely realized even in the $\mathcal{G}$-hybrid model. Note that we consider only the feasibility question and not any efficiency issues.

2.2 Frontier Analysis

Protocols and transcript trees. We view a 2-party protocol as a weighted tree of possible transcripts. The leaves of the tree correspond to completed transcripts, on which both parties give output. The tree's internal nodes alternate between "Alice" and "Bob" nodes, corresponding to points in the protocol (identified by partial transcripts) at which Alice and Bob send messages, respectively. Given a party's private input and the transcript so far (i.e., a node in the tree), the protocol assigns probabilities to the outgoing edges (i.e., possible next messages). In some settings we also consider nodes corresponding to invocations of ideal functionalities (like $\mathcal{F}_{\mathsf{coin}}$), when appropriate. For these the protocol tree assigns probabilities to the outputs of the functionality (the corresponding "messages" included in the transcripts for these steps) according to the probabilities of parties' inputs and the functionality's internal randomness. An execution of the protocol corresponds to a traversal from root to leaf in the tree.

Probabilities and frontiers. We write $\Pr[v|x, y]$ for the probability that the protocol visits node v (equivalently, generates a transcript with v as a prefix) when executed honestly on inputs x and y. Suppose $\pi_A(x, vb)$ is the probability that when Alice executes the protocol honestly with input x and the transcript so far is v, her next message is b. Similarly, we define a probability π_B for Bob. Then (assuming Alice speaks first in the protocol):

$$\Pr[v|x, y] = \pi_A(x, v_1)\pi_B(y, v_1v_2)\cdots = \left[\prod_{i \text{ odd}} \pi_A(x, v_1 \cdots v_i)\right]\left[\prod_{i \text{ even}} \pi_B(y, v_1 \cdots v_i)\right]$$

[4] The definition given here is a generalization for randomized functionalities of the definition from [15].

If we define $\alpha(v,x)$ and $\beta(v,y)$ to be the two parenthesized quantities (equivalently, the product of weights from Alice nodes and Bob nodes in the transcript tree, respectively), then we have $\Pr[v|x,y] = \alpha(v,x)\beta(v,y)$. Thus, in a plain protocol, the two parties make independent contributions to the probability of each transcript. In fact, even if the protocol is allowed to use a selectable source, this property still holds (see Section 4). This property of protocols is crucially used in all frontier analysis in this work.

When S is a set of independent nodes in the transcript tree (prefix-free partial transcripts), we define $\Pr[S|x,y] = \sum_{v \in S} \Pr[v|x,y]$, as all the probabilities in the summation are for mutually exclusive events. If $\Pr[F|x,y] = 1$, then we call F a *frontier*. Equivalently, a frontier is a maximal independent set in the transcript tree. In general, a frontier represents a point in the protocol where a certain event happens, usually defined in terms of the probabilities α and β.

3 Handling General SFE Functionalities

Frontier analysis is most naturally applied to protocols realizing SSFE functionalities — that is, functionalities which give the same output to both parties. So we derive our results for such functionalities. However, we can then extend our characterizations to apply to SFE functionalities with unrestricted outputs using the following lemma (see the full version of this paper [14]):

Lemma 1. *Suppose $\mathcal{H}$ is a functionality that has a passive-secure protocol in the plain model. If $\mathcal{H}$ is useful in UC- or standalone-securely realizing a (possibly randomized) SFE functionality $\mathcal{F}$, then there exists a* **symmetric** *SFE functionality $\mathcal{F}^*$ such that $\mathcal{F}^*$ is isomorphic to $\mathcal{F}$, and $\mathcal{H}$ is useful in (respectively, UC- or standalone-) securely realizing $\mathcal{F}^*$.*

Here, being useful or not is in the sense of the definition given in Section 2.1.

Proving Lemma 1 essentially involves relating SSFE and SFE functionalities. As it turns out, relating symmetric and unrestricted functionalities is most convenient in the setting of passive security. In that setting, we associate with every SFE functionality $\mathcal{F}$ a symmetric functionality which is simply the maximal "common information" provided to the two parties by $\mathcal{F}$. (See proof in the full version [14] for a combinatorial description of this function) Following [11] it is not hard to show that if an SFE functionality $\mathcal{G}$ is not isomorphic to its (symmetric-output) common information functionality then $\mathcal{G}$ must be complete in the passive security setting.

To apply this result, however, we must be careful in relating passive security and active security. It is not necessarily the case that an actively secure protocol implies a passively secure protocol (since in the passive security setting, the security reduction must map passively corrupt adversaries to passively corrupt simulators). In [14] we show that every SFE functionality is isomorphic to a functionality that is "deviation-revealing" [19]. Such functionalities have the property that active-secure protocols imply passive-secure protocols. Using these two results, we are able to transition from active to passive security, and then argue about generalized vs. symmetric output.

4 Selectable Sources Are Useless for Deterministic SFE

In this section we will show that any selectable source is useless for securely realizing any deterministic SFE functionality against computationally unbounded adversaries. In particular this shows that $\mathcal{F}_{\mathsf{coin}}$ is useless for realizing any deterministic SFE functionality.

Theorem 1. *Suppose $\mathcal{F}$ is a 2-party deterministic SFE and $\mathcal{G}$ is a selectable source. Then $\mathcal{F}$ has a standalone-secure (resp. UC-secure) protocol in the $\mathcal{G}$-hybrid model against computationally unbounded adversaries if and only if $\mathcal{F}$ has a standalone-secure (resp. UC-secure) protocol in the plain model.*

To give an overview of our techniques, we present the result for the special case of $\mathcal{F} = \mathcal{F}_{\mathsf{xor}}$ and $\mathcal{G} = \mathcal{F}_{\mathsf{coin}}$. Then we describe the modifications necessary to consider arbitrary $\mathcal{F}$ and arbitrary selectable source $\mathcal{G}$, respectively.

The case of $\mathcal{F}_{\mathsf{xor}}$ and $\mathcal{F}_{\mathsf{coin}}$. This special case illustrates our new frontier-based attack. It is well-known that there is no standalone-secure (or UC-secure) protocol for $\mathcal{F}_{\mathsf{xor}}$ in the plain model (cf. the complete characterization of [12,15]). Also note that standalone security is a special case of UC security. Thus it suffices to show the following:

Lemma 2. *There is no standalone-secure protocol for $\mathcal{F}_{\mathsf{xor}}$ using $\mathcal{F}_{\mathsf{coin}}$, against computationally unbounded adversaries.*

Proof (Sketch). The main novelty in this proof (compared to the techniques in [15]) is the nature of the frontier we consider, in a semi-honest protocol. For semi-honest security, $\mathcal{F}_{\mathsf{xor}}$ does not have a canonical order for *what information* must be revealed by the two parties. This thwarts the analysis in [15], which depends on defining frontiers corresponding to what information is revealed in what order. Nevertheless we show that using a frontier parameterized by a threshold μ on (an appropriately defined notion of) *how much information* is revealed about a party's input, one can devise an attack on purported protocol for $\mathcal{F}_{\mathsf{xor}}$ in the $\mathcal{F}_{\mathsf{coin}}$-hybrid model.

For simplicity, first assume that we are given a protocol π for $\mathcal{F}_{\mathsf{xor}}$ in the *plain* model (i.e., let us ignore $\mathcal{F}_{\mathsf{coin}}$ for the moment). Let α and β be defined as in Section 2. Then for every node v in the transcript tree of π, define

$$\delta_A(v, x, x') = \frac{|\alpha(v,x) - \alpha(v,x')|}{\alpha(v,x) + \alpha(v,x')} \quad \text{and} \quad \delta_B(v, y, y') = \frac{|\beta(v,y) - \beta(v,y')|}{\beta(v,y) + \beta(v,y')}.$$

δ_A and δ_B are well-defined after we exclude any nodes that have $\alpha(v,x) = \alpha(v,x') = 0$ or $\beta(v,y) = \beta(v,y') = 0$. Intuitively, $\delta_A(v,x,x')$ and $\delta_B(v,y,y')$ measure how much the transcript reveals about the distinction between x and x', or y and y', respectively. A δ value of 0 means that the partial transcript v is independent of the choice between the two inputs; a value of 1 means that the transcript v is only consistent with one of the two inputs.

Then given a parameter μ, we define a frontier F as follows:

$$F = \left\{ v \;\middle|\; \begin{array}{c} \max\{\delta_A(v,0,1), \delta_B(v,0,1)\} \geq \mu \\ \text{and no proper prefix of } v \text{ also satisfies this condition} \end{array} \right\}$$

Intuitively, F is the first place at which one of the parties has revealed "significant" information about its input, where significance is measured by μ.

Now we sketch an attack based on this frontier. (The actual proof and calculations in [14] follow a slightly different argument, but using the same frontier). Suppose by symmetry that on an honest execution, the protocol assigns the majority of the weight on F to transcripts v satisfying $\delta_B(v,0,1) \geq \mu$. Then, intuitively, Alice can launch an attack as follows. She runs the protocol honestly (say, with input 0) until reaching F. Then at F, the transcript is correlated with Bob's input enough so that Alice can guess Bob's input with bias roughly $\mu/2$. On the other hand, since $\delta_A(v,0,1) < \mu$ with good probability at this point of the protocol, both values for Alice's input are somewhat likely explanations for the transcript seen so far. Therefore if Alice changes her input at this point (by sampling a state consistent with the current transcript and the new input), the outcome of the protocol will change with all but negligible probability, thanks to the correctness guarantee of the protocol. Thus, Alice can significantly correlate her *effective* input with Bob's, so that Bob's output is biased significantly towards 1 (when Bob picks his input at random). But this is a behavior that is not possible in an ideal-world interaction with $\mathcal{F}_{\mathsf{xor}}$, so it constitutes a violation of the security of π.

The only difference when attacking a protocol in the $\mathcal{F}_{\mathsf{coin}}$-hybrid model is that the common coins also influence the probabilities of partial transcripts. One may consider the probability of a partial transcript v (which includes outputs of $\mathcal{F}_{\mathsf{coin}}$) as a product of $\alpha(v,x)$, $\beta(v,y)$, and a contribution $\gamma(v)$ from the combined calls to $\mathcal{F}_{\mathsf{coin}}$. However, $\gamma(v)$ does not depend on x or y, so we can absorb its contribution into (arbitrarily) $\alpha(v,x)$ and the analysis remains valid[5].

Uselessness of $\mathcal{F}_{\mathsf{coin}}$ for any SFE $\mathcal{F}$. First we consider the case when $\mathcal{F}$ is a *symmetric* SFE functionality. We use the characterization of SSFE functionalities

[5] Note that α and β are defined only in terms of honest behavior by the parties, so that every call to $\mathcal{F}_{\mathsf{coin}}$ delivers its output to both parties in our analysis and associated attack. (Only corrupt parties can prevent output delivery in a functionality with no output fairness guarantee.) Thus our attacks neither rely on fairness nor crucially exploit unfairness in the source of common coins; the adversaries we construct will always choose to deliver the outputs of the setup functionality.

with standalone-secure protocols from [15] to show that if an SSFE functionality $\mathcal{F}$ has no standalone-secure protocol in the plain model, then either there is a standalone-secure protocol for $\mathcal{F}_{\mathsf{xor}}$ in the $\mathcal{F}$-hybrid model, or else there is a frontier-based attack that violates standalone security of every purported protocol for $\mathcal{F}$ in the plain model.

In the first case, Lemma 2 demonstrates that $\mathcal{F}$ can have no standalone-secure protocol in the $\mathcal{F}_{\mathsf{coin}}$-hybrid world. In the second case, we observe that the frontier-based attacks go through unaltered even if the protocols are allowed access to $\mathcal{F}_{\mathsf{coin}}$. This is because the frontier attack merely relies on the fact that in a protocol, given a transcript prefix v, the next message depends only on one of Alice and Bob's inputs. However, this is true even if the protocol has access to $\mathcal{F}_{\mathsf{coin}}$— the bits from $\mathcal{F}_{\mathsf{coin}}$ being independent of both parties' inputs.

This allows us to conclude that in either case, there can be no protocol for $\mathcal{F}$ in the $\mathcal{F}_{\mathsf{coin}}$-hybrid model, giving us the following lemma (see the full version [14] for more details).

Lemma 3. *If $\mathcal{F}$ is a 2-party deterministic SSFE that has no standalone-secure (resp. UC-secure) protocol against unbounded adversaries in the plain model, then $\mathcal{F}$ has no standalone-secure (resp. UC-secure) protocol in the $\mathcal{F}_{\mathsf{coin}}$-hybrid model.*

Replacing $\mathcal{G}$ with an arbitrary selectable source. Our analysis goes through with minimal modification when $\mathcal{F}_{\mathsf{coin}}$ is replaced by an arbitrary selectable source. Recall that in a selectable source functionality $\mathcal{G}$, only one party can influence the output at a time (depending on which "direction" $\mathcal{G}$ is used in). When $\mathcal{G}$ is used such that only Alice influences the output, the influence on the transcript's probability can be collected into the term $\alpha(v, x)$. Similarly, when only Bob can influence the output of $\mathcal{G}$, the influence can be collected into the term $\beta(v, y)$. Therefore, we can still write $\Pr[v|x, y] = \alpha(v, x)\beta(v, y)$ for appropriate α and β. Each invocation of $\mathcal{G}$ is an atomic event with respect to the frontiers and to the adversary's changes in behavior in our our attacks.

Extending to general SFE functionalities. Finally, we prove Theorem 1, using Lemma 1. Note that a selectable source has a passive secure protocol (Alice samples an output and gives it to Bob). Thus if there exists any SFE functionality $\mathcal{F}$ for which some selectable source is useful in (UC- or standalone-) securely realizing, then by Lemma 1 selectable source is useful in (UC- or standalone-) securely realizing some SSFE functionality as well, contradicting Lemma 3.

5 Coins Are Useless for Randomized SFE

In this section, we characterize the set of randomized SFE functionalities that can be reduced to $\mathcal{F}_{\mathsf{coin}}$.

Since $\mathcal{F}_{\mathsf{coin}}$ itself is not securely realizable (in the UC or standalone model) against computationally unbounded adversaries, common randomness clearly allow more functionalities to be securely realized. In particular common randomness can be used to generate a shared sample from a publicly agreed-upon distribution. However, we show that this is essentially the only use of common randomness, when UC security is required[6]. More precisely,

Theorem 2. *A randomized SFE functionality $\mathcal{F}$ has a UC-secure protocol in the $\mathcal{F}_{\mathsf{coin}}$-hybrid model if and only if $\mathcal{F}$ is isomorphic to the SSFE functionality $\mathcal{F}^*$ with output function $\mathcal{F}^*$ such that $\mathcal{F}^*(x, y, r) = (h(x), r)$, where h is a deterministic function.*

Note that a secure protocol for $\mathcal{F}^*(x, y, r)$ above is simple: Alice sends $h(x)$ to Bob, and then they obtain uniformly random coins r from $\mathcal{F}_{\mathsf{coin}}$. Thus, any UC secure protocol for f which uses $\mathcal{F}_{\mathsf{coin}}$ can be replaced by one of the following form: (1) one party sends a function of its input to the other party; (2) both parties access $\mathcal{F}_{\mathsf{coin}}$ to obtain coins r; (3) both parties carry out local computation to produce their outputs.

Given Lemma 1, it is enough to establish our characterization for the special case of *symmetric* SFE functionalities (for which we shall denote the common output by $f(x, y, r)$).

The first step in proving Theorem 2 for SSFE is to show that only one party's input can have influence on the outcome of the other party.

Lemma 4. *If $\mathcal{F}$ is a 2-party randomized SSFE functionality with a UC-secure protocol in the $\mathcal{F}_{\mathsf{coin}}$-hybrid model, then $\mathcal{F}(x, y)$ is distributed as $\mathcal{F}'(x)$ (or $\mathcal{F}'(y)$), where $\mathcal{F}'$ is some randomized function of one input.*

If $\mathcal{F}$ does not have the form $\mathcal{F}'(x)$ or $\mathcal{F}'(y)$, we call it an SSFE functionality with *bidirectional influence.* Using a lemma proven in the full version of this paper [14], we know that if a SSFE $\mathcal{F}$ with bidirectional influence has a UC-secure protocol in the $\mathcal{F}_{\mathsf{coin}}$-hybrid then there exists a semi-honest protocol for OT. However, this is not possible against computationally unbounded adversaries and hence, $\mathcal{F}$ can not have bidirectional influence.

Frontiers of influence. Suppose we are given a protocol π for f in the $\mathcal{F}_{\mathsf{coin}}$-hybrid model, with simulation error ϵ. Without loss of generality, we assume

[6] The full version of this paper [14] contains examples of randomized SSFE for which $\mathcal{F}_{\mathsf{coin}}$ is useful in a more non-trivial way, but for *standalone* security. For one such example, consider the protocol to compute the minimum of the private inputs of the two parties [15,12]: Alice declares whether her input is 0 or 2; next, conditioned on Alice input being 2, Bob declares whether his input is 1 or 3. This protocol is a standalone secure protocol for the deterministic SSFE: $\boxed{\begin{smallmatrix} 0 & 0 \\ 1 & 3 \end{smallmatrix}}$. Now, we randomize the output when Alice input is 1: Based on whether Bob's input is 1 or 3 we use $\mathcal{F}_{\mathsf{coin}}$ to uniformly sample from the set $\{1, 2\}$ or $\{2, 3\}$, respectively. This is a standalone-secure protocol for the randomized SSFE: $\boxed{\begin{smallmatrix} (1,0,0,0) & (1,0,0,0) \\ (0,\frac{1}{2},\frac{1}{2},0) & (0,0,\frac{1}{2},\frac{1}{2}) \end{smallmatrix}}$, where the vectors indicate probability distribution over an output alphabet of size 4.

that the last step of π is to toss a random coin which is included in the output[7]. First, define $\mathcal{O}_v^x$ to be the output distribution of the protocol when executed honestly on (Alice) input x, *starting from partial transcript* v. We use this to define our first frontier:

$$G = \left\{ v \;\middle|\; \begin{array}{c} \forall x', x'' : \mathrm{SD}\left(\mathcal{O}_v^{x'}, \mathcal{O}_v^{x''}\right) < \sqrt{\epsilon} \\ \text{and no ancestor of } v \text{ satisfies the same condition} \end{array} \right\}$$

Intuitively, G represents the point at which Alice's input has first exhausted any "significant" influence on the final output distribution — her input can no longer change the output distribution by more than $\sqrt{\epsilon}$. Next, note that the only way to induce an output distribution in the ideal world is to choose an input x according to some distribution $\mathcal{D}$ and then send x to f, yielding the output distribution $\{f(x)\}_{x \leftarrow \mathcal{D}}$. Let $\mathcal{S}$ be the space of all possible output distributions that can be induced in this way[8]. We use this to define a collection of frontiers, one for each value of x.

$$F_x = \{v \mid \mathrm{SD}(\mathcal{O}_v^x, \mathcal{S}) > \sqrt{\epsilon} \text{ and no ancestor of } v \text{ satisfies the same condition}\}$$

Intuitively F_x represents the first time that randomness has had a "significantly" non-trivial influence on the output when Alice's input is x. Here, the influence of randomness in the protocol is considered non-trivial if the protocol has reached a point such that the conditional output distribution induced by the protocol starting from that point cannot be achieved by Alice in the ideal world.

We now show that in a secure protocol, Alice's input must completely exhaust its influence before the randomness from $\mathcal{F}_{\mathsf{coin}}$ can begin to influence the output distribution.

Lemma 5. *In the above setting, let $F_x < G$ denote the event that the protocol generates a transcript that encounters frontier F_x strictly before encountering frontier G. Then $\Pr[F_x < G|x]$ is negligible for all x.*

Proof (Sketch). Consider a malicious Alice that runs π honestly on input x. Whenever this adversary encounters F_x strictly before G, it reports the resulting partial transcript to the environment. Being in the frontier F_x, this transcript intuitively represents an assertion by the adversary that it can realize an output distribution $\mathcal{O}_v^x$ that is impossible to effect in the ideal world (by continuing hereafter with input x). Being before G, the transcript also indicates an assertion by the adversary that it can still induce two "significantly" different output

[7] To see that this is without loss of generality, define a randomized SSFE f' which on input x, outputs $f(x)$ as well as a random bit. Then define π' to be the protocol which runs π and in the last step uses $\mathcal{F}_{\mathsf{coin}}$ to toss a coin which is included in the output. It is easy to see that if π is a secure protocol for f, then π' is a secure protocol for f', so proving the insecurity of π' establishes the insecurity of π.

[8] Note that $\mathcal{S}$ is the space of convex combinations of $\{f(x) \mid x\}$, where here $f(x)$ denotes the discrete probability distribution itself, represented by a stochastic vector.

distributions (by continuing hereafter with one of the inputs from the condition in the definition of G). The environment can choose to challenge the adversary on any of these choices, and in the real world the adversary can always succeed. However, for any simulator in the ideal world, there must be some challenge for which the simulator must fail. Namely, if the simulator has already sent an input to ideal f at the time it makes its "assertion", then it cannot proceed to induce two significantly different output distributions on command — the output is already fixed. On the other hand, if the simulator has not sent an input to the ideal f, then it cannot proceed to realize an output distribution that is impossible in the ideal world.

Thus this adversary violates the security of f with success proportional to $\Pr[F_x < G|x]$, so we conclude that this probability must be negligible.

Using the previous two lemmas, we can now prove the special case of Theorem 2, restricted to SSFE functionalities:

Lemma 6. *A 2-party randomized SSFE functionality $\mathcal{F}$ has a UC-secure protocol in the $\mathcal{F}_{\mathsf{coin}}$-hybrid model against computationally unbounded adversaries if and only if $\mathcal{F}$ is isomorphic to the SSFE functionality $\mathcal{F}^*$ with output function f^* such that $f^*(x, y, r) = (h(x), r)$, where h is a deterministic function.*

Proof. The complete proof of the lemma is provided in the full version of this paper [14]. Lemma 5 shows that there is a frontier G that separates all of the influence of Alice's input (before G) from all of the influence of $\mathcal{F}_{\mathsf{coin}}$ (after G).

Our analysis relies on the *geometric interpretation of possible output distributions.* As before, let $\mathcal{S}$ denote the space of output distributions that can be realized in the ideal world by randomly choosing an input and sending it to f to obtain a sample of $f(x)$. $\mathcal{S}$ is the convex closure of a finite set of points $f(x)$. Call an input x *fundamental* if $f(x)$ is a corner on the convex hull of $\mathcal{S}$. Without loss of generality, we restrict our attention exclusively to fundamental inputs[9].

Let x be a fundamental input. Since x affects the output of the protocol only negligibly after the transcript reaches G, we have that $f(x)$ is statistically close to a convex combination of $\{\mathcal{O}_v^x \mid v \in G\}$. An overwhelming weight of these distributions $\mathcal{O}_v^x$ are negligibly close (in statistical distance) to the space $\mathcal{S}$. By a geometric argument, since $f(x)$ is a *corner* in the space $\mathcal{S}$, we have that an overwhelming weight of $\mathcal{O}_v^x$ distributions are negligibly close to $f(x)$.

Thus, consider any two inputs x, x' such that $f(x) \not\equiv f(x')$. The statistical distance between these two distributions is a constant Δ. The above argument implies that x and x' must induce distributions over G that have statistical distance negligibly close to 1. In other words, executing the protocol until G unambiguously determines the distribution $f(x)$; after G, x has no more influence on the output. Then it is straight-forward to show that the following simple protocol is also secure for f: Alice sends a description of the distribution $f(x)$ to Bob (say, the lexicographically smallest x^* s.t. the distributions of $f(x)$ and

[9] Any non-fundamental input x is redundant in f and we can remove it to obtain an isomorphic functionality (for details refer to the full version of this paper [14]).

$f(x^*)$ are identical). Both parties use $\mathcal{F}_{\mathsf{coin}}$ to generate random coins r and use them to compute a sample from the distribution $f(x)$. Then it is clear that f has the desired form — the output of this protocol is computed from a deterministic function of x along with independent coins.

On extending to selectable sources. Unlike our results in Section 4, Theorem 2 does **not** generalize to arbitrary selectable sources (instead of just $\mathcal{F}_{\mathsf{coin}}$). To see this, one can easily construct a selectable source f which is not of the form $f(x, y, r) = (h(x), r)$. Then trivially f has a UC-secure protocol using some selectable source (namely, itself), but f is not of the form required by Theorem 2.

Indeed, to prove Theorem 2, we made a crucial distinction between Alice's choice of input influencing the output distribution and $\mathcal{F}_{\mathsf{coin}}$ influencing the output distribution. This distinction is lost if $\mathcal{F}_{\mathsf{coin}}$ is replaced by a functionality in which Alice is allowed to influence the output.

On a common random string (CRS) vs. $\mathcal{F}_{\mathsf{coin}}$. A common random string (CRS) is a source of shared randomness in which all random bits are generated once and for all at the beginning of a protocol interaction, rather than as-needed, as with $\mathcal{F}_{\mathsf{coin}}$. Our proof of Theorem 2 states that the influence of the parties' inputs ends before the influence of the shared randomness begins. Since the influence of a CRS must happen at the start of a protocol, a CRS is useless for SSFEs except those of the form $f(x, y, r) = h(x)$ (no influence from shared randomness) or $f(x, y, r) = h(r)$ (no influence from parties' inputs), for a deterministic h.

6 Randomized Functionalities and Computational Intractability

Our results so far have been presented in the computationally unbounded setting. However, they do extend somewhat to the probabilistic polynomial time (PPT) setting (where all entities, including the adversary and the environment are PPT), and yield interesting connections with computational intractability assumptions. These results are similar in spirit to the connections established in [16], but unlike there, are applicable to randomized functionalities.

Firstly, in the case of deterministic SFE functionalities, we obtain the following unconditional result in the PPT setting

Theorem 3. *For every 2-party deterministic SFE $\mathcal{F}$ and selectable source $\mathcal{G}$, $\mathcal{F}$ has a standalone secure protocol in the $\mathcal{G}$-hybrid model in the PPT setting, if and only if $\mathcal{F}$ has a standalone secure protocol in the plain model in the PPT setting. (for the proof, consult the full version [14]).*

Our other results for the PPT setting are conditional. An important observation in [16] was that, statements of the form "2-party functionality $\mathcal{F}$ has a UC-secure protocol in the $\mathcal{G}$-hybrid world (in the PPT setting)" are either known to be unconditionally true or false, or tend to be *equivalent* to the assumption that one-way functions exist, or the assumption that there is an oblivious transfer

(OT) protocol secure against semi-honest adversaries. [16] study a large class of such statements for deterministic $\mathcal{F}$ and $\mathcal{G}$, and show that for every one of them the corresponding statement falls into one of the four classes listed above. An important problem left open is to understand whether the same pattern holds when considering *randomized* functionalities.

Our results suggest that this may be the case: the only intractability assumptions (other than being known to be unconditionally true or false) that arise among randomized functionalities still seem to be the existence of OWF and the existence of a semi-honest OT protocol. In particular we have the following two results (refer to the full version of this paper [14]):

Theorem 4. *Let $\mathcal{F}$ be any 2-party SFE functionality, possibly randomized. If one-way functions do not exist then $\mathcal{F}$ has a UC-secure protocol in the $\mathcal{F}_{\mathsf{coin}}$-hybrid model in the PPT setting, if and only if $\mathcal{F}$ is a publicly-selectable source.*

Theorem 5. *The following three statements are equivalent:*
1. *There exists a semi-honest OT protocol.*
2. *$\exists$ (possibly randomized) 2-party SSFE $\mathcal{F}$ with bidirectional influence : $\mathcal{F}$ is UC securely-realizable in $\mathcal{F}_{\mathsf{coin}}$-hybrid world.*
3. *$\forall$ (possibly randomized) 2-party SSFE $\mathcal{F}$ with bidirectional influence : $\mathcal{F}$ is UC securely-realizable in $\mathcal{F}_{\mathsf{coin}}$-hybrid world.*

The main task in proving the above is to show that (2) $\Rightarrow$ (1), which shown in the full version. (1) $\Rightarrow$ (3) follows from a result proven in [9,17] on the completeness of $\mathcal{F}_{\mathsf{coin}}$. (3) $\Rightarrow$ (2) is trivial.

7 Conclusion and Future Work

Recently, [16] made a case for "cryptographic complexity theory," trying to understand the qualitative difference between different multiparty functionalities. However, the results there were confined to deterministic functionalities; the universe of randomized functionalities is vastly more complex, and is little understood. Among other things, this work initiates a systematic study of randomized functionalities, by proving the low-complexity nature of certain classes of randomized functionalities. In this work we do not consider randomized functionalities of higher levels of complexity, nor do we seek to classify all kinds of randomized functionalities. Nevertheless, we believe that our proof techniques — both for the computationally unbounded setting and for the PPT setting — will be useful in such a study. We leave this for future work.

References

1. Beaver, D.: Perfect privacy for two-party protocols. In: Feigenbaum, J., Merritt, M. (eds.) Proceedings of DIMACS Workshop on Distributed Computing and Cryptography, vol. 2, pp. 65–77. American Mathematical Society, Providence (1989)
2. Beaver, D.: Precomputing oblivious transfer. In: Coppersmith, D. (ed.) CRYPTO 1995. LNCS, vol. 963, pp. 97–109. Springer, Heidelberg (1995)

3. Blum, M., Feldman, P., Micali, S.: Non-interactive zero-knowledge and its applications (extended abstract). In: STOC, pp. 103–112. ACM, New York (1988)
4. Canetti, R.: Universally composable security: A new paradigm for cryptographic protocols. Electronic Colloquium on Computational Complexity (ECCC) TR01-016 (2001); Previous version "A unified framework for analyzing security of protocols" availabe at the ECCC archive TR01-016, Extended abstract in FOCS 2001
5. Canetti, R., Lindell, Y., Ostrovsky, R., Sahai, A.: Universally composable two-party computation. In: Proc. 34th STOC, pp. 494–503. ACM, New York (2002)
6. Chor, B., Kushilevitz, E.: A zero-one law for boolean privacy (extended abstract). In: STOC, pp. 62–72. ACM, New York (1989)
7. Cleve, R.: Limits on the security of coin flips when half the processors are faulty (extended abstract). In: STOC, pp. 364–369. ACM, New York (1986)
8. Cleve, R., Impagliazzo, R.: Martingales, collective coin flipping and discrete control processes (1993) (manuscript), `http://www.cpsc.ucalgary.ca/~cleve/pubs/martingales.ps`
9. Damgård, I., Nielsen, J.B., Orlandi, C.: On the necessary and sufficient assumptions for UC computation. Cryptology ePrint Archive, Report 2009/247 (2009), `http://eprint.iacr.org/`
10. Impagliazzo, R., Luby, M.: One-way functions are essential for complexity based cryptography. In: Proc. 30th FOCS, pp. 230–235. IEEE, Los Alamitos (1989)
11. Kilian, J.: More general completeness theorems for secure two-party computation. In: Proc. 32th STOC, pp. 316–324. ACM, New York (2000)
12. Künzler, R., Müller-Quade, J., Raub, D.: Secure computability of functions in the it setting with dishonest majority and applications to long-term security. In: Reingold, O. (ed.) TCC 2009. LNCS, vol. 5444, pp. 238–255. Springer, Heidelberg (2009)
13. Kushilevitz, E.: Privacy and communication complexity. In: FOCS, pp. 416–421. IEEE, Los Alamitos (1989)
14. Maji, H.K., Ouppaphan, P., Prabhakaran, M., Rosulek, M.: Exploring the limits of common coins using frontier analysis of protocols. In: Ishai, Y. (ed.) TCC 2011. LNCS, vol. 6597, pp. 486–503. Springer, Heidelberg (2011), `http://eprint.iacr.org/`
15. Maji, H.K., Prabhakaran, M., Rosulek, M.: Complexity of multi-party computation problems: The case of 2-party symmetric secure function evaluation. In: Reingold, O. (ed.) TCC 2009. LNCS, vol. 5444, pp. 256–273. Springer, Heidelberg (2009)
16. Maji, H.K., Prabhakaran, M., Rosulek, M.: Cryptographic complexity classes and computational intractability assumptions. In: Yao, A.C.-C. (ed.) ICS, pp. 266–289. Tsinghua University Press, Beijing (2010)
17. Maji, H.K., Prabhakaran, M., Rosulek, M.: A zero-one law for cryptographic complexity with respect to computational UC security. In: Rabin, T. (ed.) CRYPTO 2010. LNCS, vol. 6223, pp. 595–612. Springer, Heidelberg (2010)
18. Moran, T., Naor, M., Segev, G.: An optimally fair coin toss. In: Reingold, O. (ed.) TCC 2009. LNCS, vol. 5444, pp. 1–18. Springer, Heidelberg (2009)
19. Prabhakaran, M., Rosulek, M.: Cryptographic complexity of multi-party computation problems: Classifications and separations. In: Wagner, D. (ed.) CRYPTO 2008. LNCS, vol. 5157, pp. 262–279. Springer, Heidelberg (2008)
20. Yao, A.C.: Protocols for secure computation. In: Proc. 23rd FOCS, pp. 160–164. IEEE, Los Alamitos (1982)

Limits on the Stretch of Non-adaptive Constructions of Pseudo-Random Generators

Josh Bronson[1], Ali Juma[2], and Periklis A. Papakonstantinou[3,*]

[1] HP TippingPoint
josh.t.bronson@hp.com
[2] University of Toronto
ajuma@cs.toronto.edu
[3] ITCS, Tsinghua University
papakons@tsinghua.edu.cn

Abstract. The standard approach for constructing a large-stretch pseudo-random generator given a one-way permutation or given a smaller-stretch pseudo-random generator involves repeatedly composing the given primitive with itself. In this paper, we consider whether this approach is necessary, that is, whether there are constructions that do not involve composition. More formally, we consider black-box constructions of pseudo-random generators from pseudo-random generators of smaller stretch or from one-way permutations, where the constructions make only non-adaptive queries to the given object. We consider three classes of such constructions, and for each class, we give a black-box impossibility result that demonstrates a contrast between the stretch that can be achieved by adaptive and non-adaptive black-box constructions.

We first consider constructions that make constantly-many non-adaptive queries to a given pseudo-random generator, where the seed length of the construction is at most $O(\log n)$ bits longer than the length n of each oracle query. We show that such constructions cannot achieve stretch that is even a single bit greater than the stretch of the given pseudo-random generator.

We then consider constructions with arbitrarily long seeds, but where oracle queries are collectively chosen in a manner that depends only on a portion of the seed whose length is at most $O(\log n)$ bits longer than the length n of each query. We show that such constructions making constantly-many non-adaptive queries cannot achieve stretch that is $\omega(\log n)$ bits greater than the stretch of the given pseudo-random generator.

Finally, we consider a class of constructions motivated by streaming computation. Specifically, we consider constructions where the computation of each individual output bit depends only on the seed and on the response to a single query to a one-way permutation. We allow the seed to have a public portion that is arbitrarily long but must always be included in the output, and a non-public portion that is at most $O(\log n)$ bits longer than the length n of each oracle query. We show that such

* P.P. is supported in part by the NSF China Grant 61050110452, 60553001, 61073174, 61033001 and the National Basic Research Program of China Grant 2007CB807900, 2007CB807901.

Y. Ishai (Ed.): TCC 2011, LNCS 6597, pp. 504–521, 2011.

constructions whose queries are chosen non-adaptively based only on the non-public portion of the seed cannot achieve linear stretch.

1 Introduction

It is well known that if there exist pseudo-random generators obtaining even one bit of stretch, then for every polynomial $p(n)$, there exist pseudo-random generators obtaining $p(n)$ bits of stretch. The usual approach for constructing a pseudo-random generator of large stretch from a pseudo-random generator of smaller stretch involves composing the smaller-stretch generator with itself repeatedly. Similarly, the usual approach for constructing a pseudo-random generators of large stretch from a one-way permutation involves composing the one-way permutation with itself repeatedly.

In this paper, we consider whether there exist such constructions that do *not* involve composition. To formalize this requirement about composition, we consider constructions that only have oracle access to the given object (a smaller-stretch pseudo-random generator or a one-way permutation) and query this oracle *non-adaptively*. We refer to such constructions as *non-adaptive (oracle) constructions*.

> *Given oracle access to a pseudo-random generator or a one-way permutation is it possible to construct, via non-adaptive oracle queries, a pseudo-random generator of large stretch?*

We give a number of black-box impossibility results for non-adaptive oracle constructions of pseudo-random generators. Some of these arguments are rather technically involved. Roughly speaking, we answer in the negative whether we can obtain, with only a constant number of queries to a pseudo-random generator, a pseudo-random generator of much larger stretch, where *answers to these non-adaptive queries are combined arbitrarily*. The challenge is to deal with this arbitrary computation phase.

Non-adaptive constructions are conceptually related to *streaming cryptography*; that is, computing private-key primitives with a device that uses small space and accesses the seed a small number of times. One of the three non-adaptive settings we consider in this paper is motivated by questions in streaming cryptography.

Our results. Observe that if pseudo-random generators exist, then there exist trivial non-adaptive oracle constructions of large-stretch pseudo-random generators: such constructions can simply ignore their oracle and directly compute a large-stretch pseudo-random generator. Since we are interested in constructions that use their oracle in a non-trivial way, we focus on constructions whose pseudo-randomness is proven using a *black-box reduction* [8] to the the security (pseudo-randomness or one-wayness) of their oracle.

We consider three classes of such constructions, and give bounds on the stretch that can be obtained by each class. For each class, our results demonstrate a contrast between the stretch that can be achieved by adaptive and non-adaptive constructions. We show that, in some sense, whatever was already known regarding algorithms for non-adaptive constructions is the best we can hope for. While we are primarily interested in constructions that are polynomial-time computable, our bounds hold even for computationally-unbounded constructions (where the number of oracle queries is still bounded).

- *Class 1: Constructions with short seeds*
  Suppose we have a pseudo-random generator $f : \{0,1\}^n \to \{0,1\}^{n+s(n)}$ and we wish to obtain a pseudo-random generator with larger stretch, say stretch $2 \cdot s(n)$. We can easily define such a generator $G^f : \{0,1\}^n \to \{0,1\}^{n+2 \cdot s(n)}$ as follows: on input $x \in \{0,1\}^n$, G^f computes $y_0||y_1 = f(x)$ (where $|y_0| = s(n)$ and $|y_1| = n$), and outputs $y_0||f(y_1)$. G^f can be formalized as a fully black-box construction making two *adaptive* oracle queries, each of the same length as G's seed x, to an oracle mapping n bits to $n + s(n)$ bits. This idea can easily be extended to obtain, for every $k \in \mathbb{N}$, a fully black-box construction making k adaptive oracle queries and achieving stretch $k \cdot s(n)$.
  We show that fully black-box constructions making constantly-many queries, each of the same length as their seed length n, *must* make *adaptive* queries even to achieve stretch $s(n) + 1$, that is, even to achieve a one-bit increase in stretch. We show that this also holds for constructions whose seed length is at most $O(\log n)$ bits longer than the length n of each oracle query.

- *Class 2: Constructions with long seeds*
  What about constructions whose seed length is significantly longer than the length of each oracle query? Can we also show that such constructions must make adaptive oracle queries in order to achieve greater stretch than their oracle? In fact, a very simple way for such a construction to make non-adaptive oracle queries, yet achieve greater stretch than its oracle, involves splitting up its seed into two or more portions, and using each portion as an oracle query. For example, if $f : \{0,1\}^n \to \{0,1\}^{n+1}$ is pseudo-random, then the generator $G^f : \{0,1\}^{2n} \to \{0,1\}^{2n+2}$ defined for all $x_1, x_2 \in \{0,1\}^n$ as $G^f(x_1||x_2) = f(x_1)||f(x_2)$ is also pseudo-random. Observe that when this construction is given an input chosen uniformly at random, the oracle queries x_1 and x_2 are chosen independently (and uniformly at random); this property is crucial for the construction's security.
  What about constructions where oracle queries cannot be chosen independently and uniformly at random? Specifically, what if we consider constructions where we place no restriction on the seed length, but insist that oracle queries are collectively chosen in a manner that depends only on a portion of the seed that is not too much longer than the length of each oracle query (making it impossible to simply split up the seed into multiple queries)? While this setting may seem unnatural at first, it *is* possible in this setting to obtain a construction that makes constantly-many non-adaptive oracle queries to a pseudo-random generator and achieves more stretch than its oracle; indeed, even a single query suffices. For example, if $f : \{0,1\}^n \to \{0,1\}^{n+s(n)}$ is pseudo-random, then by the Goldreich-Levin theorem [6] we have that for

all functions $m(n) \in O(\log n)$, the number generator $G^f : \{0,1\}^{n \cdot m(n)+n} \rightarrow \{0,1\}^{n \cdot m(n)+n+s(n)+m(n)}$ defined for all $r_1, r_2, \ldots, r_{m(n)}, x \in \{0,1\}^n$ as

$$G^f\left(r_1 || r_2 || \ldots || r_{m(n)} || x\right) = r_1 || r_2 || \ldots || r_{m(n)} || f(x) || \langle r_1, x \rangle || \langle r_2, x \rangle || \ldots || \langle r_{m(n)}, x \rangle$$

is pseudo-random; the stretch of G^f is $m(n)$ bits greater than the stretch of f. Also observe that the query made by $G^{(\cdot)}$ depends only on a portion of the seed of $G^{(\cdot)}$ whose length is the same as the length of the query (indeed, the query is identical to this portion of the seed). Using this Goldreich-Levin-based approach, it is easy to see that *adaptive* black-box constructions whose input length is much longer than the length n of each oracle query can obtain stretch $k \cdot s(n) + O(\log n)$ by making k queries to an oracle of stretch $s(n)$, even when the portion of the seed that is used to choose oracle queries has length n.

We show that fully black-box constructions $G^{(\cdot)}$ making constantly-many queries of length n to a pseudo-random generator $f : \{0,1\}^n \rightarrow \{0,1\}^{n+s(n)}$, such that only the rightmost $n + O(\log n)$ bits of the seed of $G^{(\cdot)}$ are used to choose oracle queries, *must* make *adaptive* queries in order to achieve stretch $s(n) + \omega(\log n)$. That is, such constructions making *constantly*-many non-adaptive queries cannot achieve greater stretch than the stretch provided by Goldreich-Levin with just a *single* query. This holds no matter how long a seed is used by the construction $G^{(\cdot)}$.

– *Class 3: Goldreich-Levin-like constructions*
The final class of constructions we consider is motivated by the streaming computation of pseudo-random generators. What is the relationship between non-adaptivity and streaming? In what sense could one prove a black-box lower bound that rules out streaming constructions of pseudo-random generator G of linear stretch using a one-way permutation π? A black-box lower-bound stated for a streaming device has to reference the many details of the model. We wish to state a similar thing in a setting that abstracts out a common property of streaming algorithms extended to have oracle access to a one-way permutation. Can a streaming algorithm be adaptive (even when we do not account for the space occupied by the oracle tape), in the sense that a query depends on many bits of previous queries? Given that a random input is incompressible, and the fact that we lack space (so as to store) and passes over the input (so as to recompute), it is plausible to consider non-adaptivity as a clean setting for studying black-box streaming constructions.

We consider a class of constructions where the seed has a public portion that is always included in the output, the choice of each oracle query does not depend on the public portion of the seed, and the computation of each individual output bit depends only on the seed and on the response to a *single* oracle query. We refer to such constructions making non-adaptive oracle queries as *bitwise-nonadaptive* constructions. It is not hard to see that such constructions making polynomially-many *adaptive* queries to a one-way permutation $\pi : \{0,1\}^n \rightarrow \{0,1\}^n$ can achieve arbitrary polynomial stretch; the idea is to repeatedly compose π with itself, outputting a hardcore bit of π on each composition. For example, using the Goldreich-Levin hardcore bit [6], a standard way of constructing a pseudo-random generator G of polynomial stretch $p(n)$ is the following: On input $r, x \in \{0,1\}^n$,

$$G^{\pi}(r||x) = r||\langle r, x\rangle||\langle r, \pi(x)\rangle||\langle r, \pi^2(x)\rangle|| \ldots ||\langle r, \pi^{p(n)+n}(x)\rangle$$

where $\pi^i := \underbrace{\pi \circ \pi \circ \ldots \circ \pi}_{i \text{ times}}$, and $\langle \alpha, \beta\rangle$ denotes the standard inner product of α and β. Observe that the leftmost n bits of the seed of G are public in the sense that they are included in the output. Also observe that each of the remaining output bits of G is computed using only a single output of π along with the input bits of G. Finally, observe that the queries made to π do not depend on the public input bits of G, and the number of non-public input bits is no greater than the length n of each oracle query. It is natural to ask whether the *adaptive* use of π in a construction of this form is necessary. This is particularly interesting if we wish to compute G in a streaming setting where we have small workspace, we are allowed to produce the output bit-by-bit, and we are allowed to re-read the input once per output bit.

We show that fully black-box bitwise-nonadaptive constructions $G^{(\cdot)}$ making queries of length n to a one-way permutation, such that the non-public portion of the seed of $G^{(\cdot)}$ is of length at most $n + O(\log n)$, cannot achieve linear stretch. This holds no matter the length of the public portion of the seed of $G^{(\cdot)}$.

We conclude this paper with some remarks and observations about streaming models for cryptography. Our treatment of streaming models mostly serves the purpose of proposing some new research directions.

Related work. Black-box reductions were formalized by Impagliazzo and Rudich [8], who observed that most proofs of security in cryptography are of this form. Impagliazzo and Rudich also gave the first black-box impossibility results. In their most general form, such results show that for particular security properties P_1 and P_2, it is impossible to give a black-box construction of P_1 from P_2. The same approach can also be applied to particular classes of black-box constructions, such as those making some restricted number of oracle queries or those that query their oracle non-adaptively. A large number of impossibility results have been given using this framework. The results most closely related to the problem we are considering are those of Gennaro *et al* [5], Viola [13], Lu [10], and Miles and Viola [11].

Gennnaro *et al* [5] consider black-box constructions of pseudo-random generators from one-way permutations. They show that such constructions cannot achieve $\omega(\log n)$ bits of stretch per oracle query of length n, even when queries are chosen adaptively. Their result can be extended in a straightforward way to show that for the second class of constructions we consider (and also for a more general class where queries are allowed to depend on the entire seed), for every

$k \in \mathbb{N}$, constructions making k oracle queries to a pseudo-random generator of stretch $s(n)$ cannot achieve stretch $k \cdot s(n) + \omega(\log n)$, even when these queries are chosen adaptively. By contrast, recall that we show that for this class of constructions, for every $k \in \mathbb{N}$, constructions making k *non-adaptive* oracle queries to a pseudo-random generator of stretch $s(n)$ cannot achieve stretch $s(n) + \omega(\log n)$.

Viola [13] considers black-box constructions of pseudo-random generators from one-way *functions* where oracle queries are non-adaptive but chosen in a computationally unbounded way, while the output of the construction is computed from the query responses by an AC^0 (polynomial-size and constant-depth) circuit. He shows that such constructions cannot achieve linear stretch. The class of constructions considered by Viola is, in general, incomparable to the classes we consider. His class is more general in terms of the numbers of queries allowed and the way that queries are chosen: he places no bounds on the number of queries, allows the queries to be chosen arbitrarily based on the seed (while we require queries to be chosen in a computable manner), and places no restrictions on the length of the queries relative to the length of the seed. On the other hand, his class is more restrictive in terms of the computational power allowed after the query responses are received: he only allows AC^0 computation, while we allow unbounded computation.

Lu [10] considers the same class of constructions as Viola, except that Lu allows the output to be computed from the query responses by a subexponential-size constant-depth circuit (rather than an AC^0 circuit). He shows that such constructions cannot achieve linear stretch.

Miles and Viola [11] consider black-box constructions of pseudo-random generators from pseudo-random generators of 1-bit stretch, where the oracle queries are non-adaptive but chosen in a computationally unbounded way, while the output of the construction consists simply of query response bits; that is, these constructions are not allowed to perform any computation on query responses. They show that such constructions cannot achieve linear stretch. Like the constructions considered by Viola [13] and Lu [10], the class of constructions considered by Miles and Viola is, in general, incomparable to the classes we consider: the constructions they consider are more general in the manner in which queries are chosen (they place no restrictions on the length of queries relative to the length of the seed), but much more restrictive in terms of the computational power allowed after query responses are received.

In the positive direction, Haitner *et al* [7] give the first *non-adaptive* black-box construction of a pseudo-random generator from a one-way *function*. Their construction achieves sublinear stretch. They also give a non-adaptive black-box construction achieving linear stretch, but this requires an *exponentially-hard* one-way function. In both of these constructions, the oracle queries are collectively chosen based on a portion of the seed that is significantly longer than the length of each oracle query. By contrast, recall that all of our impossibility results are for constructions where the oracle queries are collectively chosen based on a portion of the seed that is no more than logarithmically-many bits longer than the length of each oracle query.

Organization. Section 2 contains definitions and preliminaries. The impossibility results for constructions with short seeds and long seeds are discussed in Sections 3 and 4 respectively. In Section 5, we state a restriction on the way that constructions choose oracle queries, and under this restriction we extend the results of Sections 3 and 4 to constructions making polynomially-many queries. The impossibility result for Goldreich-Levin-like constructions is found in Section 6. Section 7 contains our remarks on streaming models in cryptography.

2 Preliminaries

Notation. We use "PPT" to denote "probabilistic polynomial time". We denote by $\langle a \rangle_n$ the n-bit binary string representation of $a \in \mathbb{N}$, padded with leading zeros when necessary. If the desired representation length is clear from the context, we write $\langle a \rangle$ instead of $\langle a \rangle_n$. If $a \geq 2^n$, then $\langle a \rangle_n$ denotes the n least significant bits of the binary representation of a. We denote by $x||y$ the concatenation of strings x and y.

2.1 Pseudo-Random Generators and One-Way Functions

A length-increasing function $G : \{0,1\}^{\ell_1(n)} \to \{0,1\}^{\ell_2(n)}$ is a *pseudo-random generator* if for every PPT adversary M, we have

$$\left| \Pr_{x \leftarrow \{0,1\}^{\ell_1(n)}} \left[M\left(G\left(x\right)\right) = 1 \right] - \Pr_{z \leftarrow \{0,1\}^{\ell_2(n)}} \left[M\left(z\right) = 1 \right] \right| \leq 1/n^c$$

for all c and sufficiently large n.

A function $f : \{0,1\}^{\ell_1(n)} \to \{0,1\}^{\ell_2(n)}$ is *one-way* if for every PPT adversary M, we have $\Pr_{x \leftarrow \ell_1(n)} \left[f\left(M\left(f\left(x\right)\right)\right) = f\left(x\right) \right] \leq 1/n^c$ for all c and sufficiently large n.

2.2 Non-adaptive Constructions

Our impossibility results are for constructions that use their oracle in a non-adaptive manner.

Definition 1 (Non-adaptive oracle machine). *Let $M^{(\cdot)}$ be a deterministic oracle Turing machine. We say that $M^{(\cdot)}$ is a* non-adaptive oracle machine *if the oracle queries made by $M^{(\cdot)}$ are determined by only the input to $M^{(\cdot)}$, and, in particular, do not depend on the responses to previous queries.*

We will sometimes need to refer to the *querying function* of a non-adaptive oracle machine.

Definition 2 (Querying function). *Let $\ell_1(n)$, $\ell_2(n)$, and $p(n)$ be polynomials, and let $M^{(\cdot)} : \{0,1\}^{\ell_1(n)} \to \{0,1\}^{\ell_2(n)}$ be a non-adaptive oracle machine that makes $p(n)$ oracle queries, each of length n. The* querying function *of $M^{(\cdot)}$,* *denoted* Q_M*, is the function* $Q_M : \{0,1\}^{\ell_1(n)} \times \{0,1\}^{\log p(n)} \rightarrow \{0,1\}^n$ *such that for all* $x \in \{0,1\}^{\ell_1(n)}$ *and* $0 \leq i < p(n)$*, the* i*-th oracle query made by* $M^{(\cdot)}(x)$ *is* $Q_M(x, \langle i \rangle)$*. When* $p(n) \equiv 1$*, the second argument to* Q_M *is omitted.*

If there exists a polynomial $r(n)$ *such that the queries made by* $M^{(\cdot)}$ *depend only on the rightmost* $r(n)$ *bits of the input of* $M^{(\cdot)}$*, then the* $r(n)$-restricted querying function *of* $M^{(\cdot)}$*, denoted* $Q_M^{r(n)}$*, is the function* $Q_M^{r(n)} : \{0,1\}^{r(n)} \times \{0,1\}^{\log p(n)} \rightarrow \{0,1\}^n$ *such that for all* $v \in \{0,1\}^{\ell_1(n)-r(n)}$*,* $w \in \{0,1\}^{r(n)}$*, and* $0 \leq i < p(n)$*, the* i*-th oracle query made by* $M^{(\cdot)}(v||w)$ *is* $Q_M^{r(n)}(w, \langle i \rangle)$*.*

2.3 Black-Box Reductions

Reingold, Trevisan, and Vadhan [12] give a classification of black-box security reductions. Our impossibility results apply to what Reingold *et al* call *fully-black box* reductions. We avoid defining such reductions in their full generality and instead focus on security reductions for constructions of pseudo-random number generators from pseudo-random generators of smaller stretch.

Definition 3 (Fully black-box reduction [8]). *Let* $G^{(\cdot)} : \{0,1\}^{\ell_1(n)} \rightarrow \{0,1\}^{\ell_2(n)}$ *be a number generator whose construction has access to an oracle for a length-increasing function mapping* $\ell_1'(n)$ *bits to* $\ell_2'(n)$ *bits. There is a* fully black-box *reduction of the pseudo-randomness of* $G^{(\cdot)}$ *to the pseudo-randomness of its oracle if there exists a PPT oracle machine* $M^{(\cdot,\cdot)}$ *such that for every function* $f : \{0,1\}^{\ell_1'(n)} \rightarrow \{0,1\}^{\ell_2'(n)}$ *and every function* $A : \{0,1\}^{\ell_2(n)} \rightarrow \{0,1\}$*, if* A *breaks the pseudo-randomness of* G^f *then* $M^{(f,A)}$ *breaks the pseudo-randomness of* f*.*

Definition 3 can be modified in a straightforward way for constructions of pseudo-random number generators from other primitives, such as from one-way permutations.

An oracle construction whose security is proven using a black-box reduction is called a *black-box construction.*

3 Constructions with Short Seeds

In this section, we consider constructions whose seed length is not more than $O(\log n)$ bits longer than the length n of each oracle query. Recall that such constructions making k adaptive queries to a given pseudo-random generator can achieve stretch that is k times the stretch of the given generator. We show that such constructions making constantly-many *non-adaptive* queries cannot achieve stretch that is *even a single bit* longer than the stretch of the given generator.

Theorem 1. *Let* $k \in \mathbb{N}$*, and let* $\ell_1(n)$ *and* $\ell_2(n)$ *be polynomials such that* $\ell_1(n) \leq n + O(\log n)$ *and* $\ell_2(n) > n$*. Let* $G^{(\cdot)} : \{0,1\}^{\ell_1(n)} \rightarrow \{0,1\}^{\ell_1(n)+(\ell_2(n)-n)+1}$ *be a non-adaptive oracle construction of a number generator, making* k *queries of length* n *to an oracle mapping* n *bits to* $\ell_2(n)$ *bits. Then there is no fully black-box reduction of the pseudo-randomness of* $G^{(\cdot)}$ *to the pseudo-randomness of its oracle.*

The approach we use to prove Theorem 1 does not seem to extend to the case of polynomially-many (or even $\omega(1)$-many) queries. However, a similar approach does work for polynomially-many queries when we place a restriction on the many-oneness of the number generator's querying function. We state this restriction in Section 5.

We give an overview of the proof of Theorem 1 in Section 3.1, and we give the proof details in the full version of this paper.

3.1 Proof Overview for Theorem 1

A simpler case. We first consider the simpler case of constructions making just a single query, where the query made is required to be the same as the construction's input. That is, we consider constructions $G^{(\cdot)} : \{0,1\}^n \to \{0,1\}^{\ell_2(n)+1}$ such that on every input $x \in \{0,1\}^n$, G makes query x to an oracle mapping n bits to $\ell_2(n)$ bits. Fix such a construction $G^{(\cdot)}$. We need to show the existence of functions $f : \{0,1\}^n \to \{0,1\}^{\ell_2(n)}$ and $A : \{0,1\}^{\ell_2(n)} \to \{0,1\}$ such that A breaks the pseudo-randomness of G^f but f is pseudo-random even with respect to adversaries that have oracle access to f and A. Following the approach for proving black-box impossibility results initiated by Impagliazzo and Rudich [8], we actually define a *joint distribution* $(\mathcal{F}, \mathcal{A})$ over pairs of functions, such that with probability one over $(f, A) \leftarrow (\mathcal{F}, \mathcal{A})$, A breaks the pseudo-randomness of G^f but f is pseudo-random even with respect to adversaries that have oracle access to f and A.

Consider how we might define such a joint distribution $(\mathcal{F}, \mathcal{A})$. The most obvious approach is to let $(\mathcal{F}, \mathcal{A})$ be the distribution defined by the following procedure for sampling a tuple $(f, A) \leftarrow (\mathcal{F}, \mathcal{A})$: randomly select f from the (infinite) set of all functions that, for each $n \in \mathbb{N}$, map n bits to $\ell_2(n)$ bits; let A be the function such that for every $z \in \{0,1\}^{\ell_2(n)+1}$, $A(z) = 1$ if and only if there exists an $s \in \{0,1\}^n$ such that $G^f(s) = z$. Following this approach, we have that with probability one over $(f, A) \leftarrow (\mathcal{F}, \mathcal{A})$, A breaks the pseudo-randomness of G^f but f is pseudo-random with respect to adversaries that have oracle access to f alone. However, it is *not* necessarily the case that f is pseudo-random with respect to adversaries that have oracle access to f *and* A. For example, suppose construction G is such that for every $x \in \{0,1\}^{n-1}$ and every $b \in \{0,1\}$, $G^f(x||b) = f(x||b)||b$. In this case, it is easy to use A to break f: on input $y \in \{0,1\}^{\ell_2(n)}$, output 1 if and only if either $A(y||0) = 1$ or $A(y||1) = 1$.

To overcome this problem, we add some "noise" to A. We need to be careful that we add enough noise to A so that it is no longer useful for breaking f, but we do not add so much noise that A no longer breaks G^f. Our basic aproach is to modify A so that instead of only accepting $G^f(s)$ for all $s \in \{0,1\}^n$, A accepts $G^{f_i(s)}$ for all s, all i, and some appropriate collection of functions $\{f_0, f_1, f_2, \ldots\}$ where $f_0 = f$. How should this collection of functions be defined? Since we want to make sure that A still breaks G^f, and since we have that A accepts $G^f(s)$ with probability 1 over $s \leftarrow \{0,1\}^n$, we need to ensure that A accepts randomly chosen strings with probability non-negligibly less than 1. For this, it suffices to ensure that (# of n-bit strings s)*(# of functions f_i) is at most, say, half the number of strings of length $\ell_2(n)+1$. At the same time, to prevent A from helping to break f, we would like it to be the case that, intuitively, A treats strings that are *not* in the image of f on an equal footing with strings that *are* in the image of f. One way to accomplish these objectives, which we follow, is to randomly select a permutation π on $\{0,1\}^{\ell_2(n)}$, define $f(x) = \pi(0^{\ell_2(n)-n}||x)$ for all $x \in \{0,1\}^n$, and define A to accept $G^{\pi(y||\cdot)}(s)$ for every $y \in \{0,1\}^{\ell_2(n)-n}$ and every $s \in \{0,1\}^n$. We formalize this as a joint distribution $(\mathcal{F}, \mathcal{A}, \Pi)$ over tuples (f, A, π) that are sampled in the manner just described.

It is easy to show that with probability one over $(f, A, \pi) \leftarrow (\mathcal{F}, \mathcal{A}, \Pi)$, A does indeed break G^f. It is much more difficult to show that with probability one over $(f, A, \pi) \leftarrow (\mathcal{F}, \mathcal{A}, \Pi)$, f is pseudo-random ever with respect to PPT adversaries that have oracle access to f and A. We argue that it suffices to show that for every PPT oracle machine $D^{(\cdot,\cdot)}$, the probability over $(f, A, \pi) \leftarrow (\mathcal{F}, \mathcal{A}, \Pi)$ and $s \leftarrow \{0,1\}^n$ that $D^{(f,A)}(f(s))$ makes oracle query s to f is negligible. Now, instead of only showing this for every PPT oracle machine $D^{(\cdot,\cdot)}$, we find it more convenient to show this for every computationally unbounded probabilistic oracle machine $D^{(\cdot,\cdot)}$ that makes at most polynomially-many oracle queries. How might we do so? We would like to argue that A does not help D to find s since a computationally unbounded D can try to compute A by itself. More formally, we would like to show that given D, we can build a D' that, given input $f(s)$ and given oracle access only to f, simulates D on input $f(s)$, answers f-queries of D using the given oracle, and "makes up" answers to the A-queries of D in a manner that ensures that the probability that the simulation of D makes query s is very close to the probability that $D^{(f,A)}(f(s))$ makes oracle query s. Of course, D' does not "know" π, so it is not immediately clear how it should answer the A-queries of the simulation of D. If D' simply randomly chooses its own permutation π' and answers A-queries using π' in place of the unknown π, the simulation of D may "notice" this sleight of hand. For example, since D is given $f(s)$ as input, it might (depending on the definition of G) be able to compute the value of $G^f(s)$, and hence make query $G^f(s)$ to A; if this query does not produce response 1, D will "know" that queries are not being responded to properly.

We address this by showing that D' can still compute "most" of A on its own, and that the "rest" of A is not helpful for finding s. Specifically, we split A into two functions, A_1 and A_2, that together can be used to compute A. Function A_1 outputs 1 only on input $G^f(s)$. For every $(\ell_2(n)+1)$-bit string z, $A_2(z) = 1$ if and only if $z \neq G^f(s)$ and $A(z) = 1$. We then argue that querying A_1 provides very little help for finding s. Let X be the set of all strings $x \in \{0,1\}^n$ such that $G^f(x) = G^f(s)$. Roughly speaking, if X is large, then A_1 gives no information about s beyond the fact that $s \in X$. On the other hand, if X is small, then we argue it is unlikely that an adversary making polynomially-many queries to A_1 will receive a non-zero response to any of its queries (in other words, it is unlikely that query $G^f(s)$ will be made). It remains to argue that D' can compute A_2 on its own. We show that if D' randomly selects a permutation π', computes an A_2' based on π' (rather than π), uses this A_2' along with the given A_1 to answer the

A-queries of the simulation of D, and answers the f-queries of the simulation of D based on $\pi'(0^{\ell_2(n)-n}||\cdot)$ (rather than using the given oracle f), then it is unlikely that the simulation of D will make a query that "exposes" the fact that its oracle queries are not being answered by f and A.

The general case. We extend the above argument to constructions $G^{(\cdot)} : \{0,1\}^{\ell_1(n)} \rightarrow \{0,1\}^{\ell_1(n)+(\ell_2(n)-n)+1}$ making constantly-many non-adaptive queries, where the length $\ell_1(n)$ of the construction's input is allowed to be $O(\log n)$ bits longer than the length n of each oracle query. The high-level idea is the same: we define a joint distribution $(\mathcal{F}, \mathcal{A}, \Pi)$ by specifying a procedure for sampling a tuple $(f, A, \pi) \leftarrow (\mathcal{F}, \mathcal{A}, \Pi)$, and the way we sample π and f is (almost) the same as before. But now we change the way A behaves. Our goal is to follow the same style of argument as before. To accomplish this, we would still like it to be the case that when we "split up" A into functions A_1 and A_2, there is still at most one string accepted by A_1 (this helps us ensure that A_1 does not provide too much information about s). Recall that before, when D' was run on an input $f(s)$, the unique string accepted by A_1 was $G^f(s)$. This made sense because in the previous setting, the only input on which $G^{(\cdot)}$ made oracle query s was s itself. But in the current setting, for each $s \in \{0,1\}^n$, there may be many inputs $x \in \{0,1\}^{\ell_1(n)}$ on which $G^{(\cdot)}$ makes oracle query s. We would like to modify the definition of A so that rather than accepting $G^{\pi(y||\cdot)}(x)$ for every $y \in \{0,1\}^{\ell_2(n)-n}$ and *every* $x \in \{0,1\}^{\ell_1(n)}$, A accepts $G^{\pi(y||\cdot)}(x)$ for every $y \in \{0,1\}^{\ell_2(n)-n}$ and x in some subset $\mathit{Good}(n) \subseteq \{0,1\}^{\ell_1(n)}$ such that for every $s \in \{0,1\}^n$, there is at most one $x \in \mathit{Good}(n)$ such that $G^{(\cdot)}$ on input x makes query s. But we cannot do exactly this (and still have that A breaks G^f), since, for example, there might be some string t that $G^{(\cdot)}$ queries no matter what its input is.

Instead, we need to proceed very carefully, partitioning the set of strings t of length n into those that are queried by $G^{(\cdot)}$ for "many" of its inputs $x \in \{0,1\}^{\ell_1(n)}$, and those queried by $G^{(\cdot)}$ for "at most a few" of its inputs $x \in \{0,1\}^{\ell_1(n)}$. We call the former set $Fixed(n)$ and the latter set $NotFixed(n)$. We then define a set $\mathit{Good}(n) \subseteq \{0,1\}^{\ell_1(n)}$ of inputs to $G^{(\cdot)}$ such that for no pair of distinct inputs from $\mathit{Good}(n)$ does $G^{(\cdot)}$ make the same query $t \in NotFixed(n)$. That is, each $t \in NotFixed(n)$ is queried by $G^{(\cdot)}$ for at most one of its inputs $x \in \mathit{Good}(n)$. The challenge, of course, is ensuring that that the set $\mathit{Good}(n)$ defined this way is "large enough".

We define A to accept $G^{\pi(y||\cdot)}(x)$ for every $y \in \{0,1\}^{\ell_2(n)-n}$ and every $x \in \mathit{Good}(n)$. Now we can "split up" A into A_1 and A_2 in a manner similar to what we did before: on input $f(s)$ to D', where $s \in NotFixed(n)$, if there exists a string $x \in \mathit{Good}(n)$ such that $G^{(\cdot)}(x)$ makes query s (note that there can be at most one such string x by definition of $\mathit{Good}(n)$), then A_1 only accepts $G^f(x)$, and if there is no such string x then A_1 does not accept any strings; as before, we define A_2 to accept the remaining strings accepted by A. We then argue as before about the (lack of) usefulness of A_1 and A_2 for helping to find s. Finally, we argue that our definition of $Fixed(n)$ ensures that this set will be of negligible size, and hence it does not hurt to ignore the case $s \in Fixed(n)$ (since this case will occur with negligible probability).

4 Constructions with Long Seeds

In Section 3, we saw that black-box constructions $G^{(\cdot)}$ making constantly-many non-adaptive oracle queries, where the seed length of $G^{(\cdot)}$ is not too much longer than the length of each oracle query, cannot achieve even a single bit more stretch than their oracle. In this section, we consider constructions whose seed length is allowed to be much longer than the length of each oracle query, but where the oracle queries are collectively chosen in a manner that depends only on a portion of the seed whose length is not more than $O(\log n)$ bits longer than the length n of each oracle query. Recall that such constructions making even a single query to a given pseudo-random generator can achieve stretch that is $O(\log n)$ bits longer than the stretch of the given generator [6]. Further, recall that such constructions making k adaptive queries can achieve stretch that is $O(\log n)$ bits longer than k times the stretch of the given generator. We show that such constructions making constantly-many non-adaptive queries cannot achieve stretch that is $\omega(\log n)$ bits longer than the stretch of the given generator.

Theorem 2. *Let $k \in \mathbb{N}$, $c \in \mathbb{R}^+$, and $m(n) \in \omega(\log n)$. Let $\ell_0(n)$, $\ell_1(n)$, and $\ell_2(n)$ be polynomials such that $\ell_1(n) \leq n + c \log n$. Let $G^{(\cdot)} : \{0,1\}^{\ell_0(n)+\ell_1(n)} \to \{0,1\}^{\ell_0(n)+\ell_1(n)+(\ell_2(n)-n)+m(n)}$ be a non-adaptive oracle construction of a number generator that makes k queries of length n to a number generator mapping n bits to $\ell_2(n)$ bits, such that for all $r \in \{0,1\}^{\ell_0(n)}$ and $x \in \{0,1\}^{\ell_1(n)}$, the queries made by $G^{(\cdot)}$ on input $(r||x)$ depend only on x. Then there is no fully black-box reduction of the pseudo-randomness of $G^{(\cdot)}$ to the pseudo-randomness of its oracle.*

As is the case for Theorem 1, the approach we use to prove Theorem 2 does not seem to extend to the case of polynomially-many (or even $\omega(1)$-many) queries. However, a similar approach does work for polynomially-many queries when we place a restriction on the many-oneness of the number generator's querying function. We state this restriction in Section 5.

We give an overview of the proof of Theorem 2 in Section 4.1, and we give the proof details in the full version of this paper.

4.1 Proof Overview for Theorem 2

As in the proof of Theorem 1, it suffices to define a joint distribution $(\mathcal{F}, \mathcal{A})$ over pairs of functions, such that with probability one over $(f, A) \leftarrow (\mathcal{F}, \mathcal{A})$, A breaks the pseudo-randomness of G^f but f is pseudo-random even with respect to adversaries that have oracle access to f and A. Unlike the previous proof, we actually define distributions $\mathcal{F}$ and $\mathcal{A}$ that are independent – in fact, we define $\mathcal{A}$ to be a degenerate distribution that assigns all probability to a fixed function A. We define a set $Good(n) \subseteq \{0,1\}^{\ell_1(n)}$ in a careful manner very similar to the proof of Theorem 1, but taking into account the fact that the queries of $G^{(\cdot)}$ depend only on the rightmost $\ell_1(n)$ bits of its seed. The goal is to ensure that *Good*(n) is sufficiently large and has the property that for every string $x \in Good(n)$, every $r \in \{0,1\}^{\ell_0(n)}$, and every $f \in \mathcal{F}$, A accepts $G^f(r||x)$. Simultaneously, we need to ensure that the total number of strings accepted by A is sufficiently smaller than $2^{\ell_0(n)+\ell_1(n)+(\ell_2(n)-n)+m(n)}$ and that $f \leftarrow \mathcal{F}$ is pseudo-random with probability one even with respect to adversaries that have oracle access to f and A.

If we define $\mathcal{F}$ in a very straightforward way (e.g. as the uniform distribution over all 1-1 functions), the total number of strings that A will need to accept (in order to accept $G^f(r||x)$ for every $f \in \mathcal{F}$, every r, and every $x \in Good(n)$) could be too large. The problem is that when deciding whether to accept a given input, A is existentially quantifying over over a set that is (much) larger than the set of its possible inputs. We need to minimize the number of different $f \in \mathcal{F}$ (while, of course, still ensuring that $f \leftarrow \mathcal{F}$ is pseudo-random with probability one even with respect to adversaries that have oracle access to f and A). At the same time, we need to add some structure to the $f \in \mathcal{F}$ to, intuitively, reduce the amount of new information contained in the responses to the oracle queries made by G^f when run on each $r||x$ where $x \in Good(n)$. The idea is that rather than existentially quantifying over every r, every $x \in Good(n)$, and every $f \in \mathcal{F}$ when deciding whether to accept a particular input z, A will instead existentially quantify over every r, every $x \in Good(n)$, and every possible value for the (small amount of) new information (that is, the information not already determined by x) contained in the responses to oracle queries made by $G^{(\cdot)}$ when run on input $r||x$.

Similarly to the proof of Theorem 1, our procedure for constructing the set *Good*(n) ensures that for every distinct $x, x' \in Good(n)$, each query q made by G, when run on an input whose rightmost bits are x, is either in some small set $Fixed(n)$ or is distinct from every query q' made by G when run on every input whose rightmost bits are x'. This allows us to follow a two-step approach to defining $\mathcal{F}$. We first define a permutation h on $\{0,1\}^n$ that, for each $x \in Good(n)$, maps the queries $q \notin Fixed(n)$ made by G, when run on an input whose rightmost bits are x, to strings that differ in at most a small number of bits, and, in particular, have a common $(m(n)/2)$-bit suffix. Roughly speaking, sampling $f \leftarrow \mathcal{F}$ proceeds as follows. We randomly select a function $f' : \{0,1\}^n \leftarrow \{0,1\}^{\ell_2(n)}$ that is the identity on its first $n - m(n)/2$ input bits, and is 1-1 on its last $m(n)/2$ input bits, mapping them to $\ell_2(n) - n + m(n)/2$ output bits. We then define $f = f' \circ h$. The actual definition of $\mathcal{F}$ that we use in the proof also ensures that for every $q \in Fixed(n)$, the value $f(q)$ is independent of the choice $f \leftarrow \mathcal{F}$ (that is, $f_1(q) = f_2(q)$ for all $f_1, f_2 \in \mathcal{F}$).

Intuitively, this approach ensures that $f \leftarrow \mathcal{F}$ has "just enough" randomness. At the same time, this approach ensures that for every r and every $x \in Good(n)$, the responses to oracle queries made by $G^f(r||x)$ collectively contain at most $\ell_2(n) - n + m(n)/2$ bits of information that depend on the choice $f \leftarrow \mathcal{F}$.

We remark that it is crucial for this proof that $2^{m(n)/2}$ is super-polynomial. It is for this reason that we cannot adapt the current proof in order to obtain a significantly simpler proof of Theorem 1; in Theorem 1, the corresponding value of $m(n)$ (the additional stretch achieved by $G^{(\cdot)}$) is exactly 1.

5 Moving beyond Constantly-Many Queries

In this section we consider extending Theorem 1 and Theorem 2 to the case of polynomially-many queries. We are able to do this for a restricted class of constructions. We begin by defining the restriction we need to place on the querying function of the construction.

Definition 4 (Many-oneness bounded almost everywhere). *Let $\ell(n)$ and $q(n)$ be polynomials, and let $f : \{0,1\}^{\ell(n)} \to \{0,1\}^n$ be a function. f has* many-oneness bounded by $q(n)$ almost everywhere *if for all c and sufficiently large n, there are fewer than $2^n/n^c$ strings $y \in \{0,1\}^n$ such that $|f^{-1}(y)| > q(n)$.*

Theorem 3. *Let $p(n)$, $q(n)$, $\ell_1(n)$, and $\ell_2(n)$ be polynomials such that $\ell_1(n) \leq n + O(\log n)$ and $\ell_2(n) > n$. Let $G^{(\cdot)} : \{0,1\}^{\ell_1(n)} \to \{0,1\}^{\ell_1(n)+(\ell_2(n)-n)+1}$ be a non-adaptive oracle construction of a number generator, making $p(n)$ queries of length n to an oracle mapping n bits to $\ell_2(n)$ bits, such that the querying function of $G^{(\cdot)}$ has many-oneness bounded by $q(n)$ almost everywhere. Then there is no fully black-box reduction of the pseudo-randomness of $G^{(\cdot)}$ to the pseudo-randomness of its oracle.*

Theorem 4. *Let $c \in \mathbb{R}^+$ and $m(n) \in \omega(\log n)$. Let $p(n)$, $q(n)$, $\ell_0(n)$, $\ell_1(n)$, and $\ell_2(n)$ be polynomials such that $\ell_1(n) \leq n + c\log n$. Let $G^{(\cdot)} : \{0,1\}^{\ell_0(n)+\ell_1(n)} \to \{0,1\}^{\ell_0(n)+\ell_1(n)+(\ell_2(n)-n)+m(n)}$ be a non-adaptive oracle construction of a number generator that makes $p(n)$ queries of length n to a number generator mapping n bits to $\ell_2(n)$ bits, such that $G^{(\cdot)}$ has an $\ell_1(n)$-restricted querying function whose many-oneness is bounded by $q(n)$ almost everywhere. Then there is no fully black-box reduction of the pseudo-randomness of $G^{(\cdot)}$ to the pseudo-randomness of its oracle.*

The proofs of Theorem 3 and Theorem 4 follow the same basic structure as the proofs of Theorem 1 and Theorem 2, respectively, but the procedure used to define the set *Good*(n) in each proof is simpler as a result of the restriction on the many-oneness of the querying function. For both Theorem 3 and Theorem 4, the procedure begins by defining $Fixed(n) \subseteq \{0,1\}^n$ to be the set of strings in the image of the querying function Q_G whose many-oneness is *not* bounded by $q(n)$. Then, since the remaining strings in the image of Q_G have bounded many-oneness, it is easy to define a large set $Good(n) \subseteq \{0,1\}^{\ell_1(n)}$ such that for all distinct $x, x' \in Good(n)$ and all $0 \leq i, j < p(n)$, either $Q_G(x, \langle i \rangle) \in Fixed(n)$ or $Q_G(x, \langle i \rangle) \neq Q_G(x', \langle j \rangle)$. The idea is to proceed as follows: initially, every $x \in \{0,1\}^{\ell_1(n)}$ is a candidate for inclusion in *Good*(n); while there are candidates remaining, select an arbitrary candidate x, add it to *Good*(n), and remove from consideration as candidates all x' such that for some $0 \le i, j < p(n)$, we have $Q_G(x, \langle i \rangle) \notin \mathit{Fixed}(n)$ and $Q_G(x, \langle i \rangle) = Q_G(x', \langle j \rangle)$. For every x added to $\mathit{Good}(n)$ by this procedure, at most $p(n)(q(n) - 1)$ are removed from consideration, and hence at the end of this procedure $\mathit{Good}(n)$ has size at least $2^{\ell_1(n)}/(p(n)(q(n) - 1) + 1)$. Further details about these proofs are omitted for the sake of conciseness.

6 Goldreich-Levin-Like Constructions

In this section, we consider constructions where the seed has a public portion that is always included in the output, such that the oracle queries are chosen *non-adaptively* based only on the non-public portion of the seed. We further require that the computation of each individual output bit depends only on the seed and on the response to a single oracle query. We begin by formalizing this class of constructions.

Definition 5 (Bitwise-nonadaptive construction). *Let $\ell_0(n)$, $\ell_1(n)$, and $\ell_2(n)$ be polynomials, and let $G^{(\cdot)} : \{0,1\}^{\ell_0(n)+\ell_1(n)} \to \{0,1\}^{\ell_0(n)+\ell_2(n)}$ be a non-adaptive oracle machine. We say that $G^{(\cdot)}$ is* bitwise-nonadaptive *if there exist uniformly-computable functions*

$$Q_G = \left\{ Q_{G,n} : \{0,1\}^{\ell_1(n)} \times \{0,1\}^{\log \ell_2(n)} \to \{0,1\}^n \right\}$$

and

$$B = \left\{ B_n : \{0,1\}^{\ell_0(n)} \times \{0,1\}^{\ell_1(n)} \times \{0,1\}^n \times \{0,1\}^{\log \ell_2(n)} \to \{0,1\} \right\}$$

such that for all n, all $r \in \{0,1\}^{\ell_0(n)}$, all $x \in \{0,1\}^{\ell_1(n)}$, and all permutations $\pi : \{0,1\}^n \to \{0,1\}^n$, we have $G^\pi(r||x) = r||b_0||b_1|| \dots ||b_{\ell_2(n)-1}$ where $b_i = B_n(r, x, \langle i \rangle, \pi(Q_{G,n}(x, \langle i \rangle)))$ for $0 \le i \le \ell_2(n) - 1$.

Observe that the Goldreich-Levin-based pseudo-random generator $G^\pi(r||x) = r||\pi(x)||\langle r, x \rangle$ is bitwise-nonadaptive.

We show that fully black-box bitwise-nonadaptive constructions $G^{(\cdot)}$ making queries to a one-way permutation, such that the non-public portion of the seed of $G^{(\cdot)}$ is no more that $O(\log n)$ bits longer than the length n of each oracle query, cannot achieve linear stretch.

Theorem 5. *Let $\alpha > 1$, and let $\ell_0(n)$, $\ell_1(n)$, and $\ell_2(n)$ be polynomials such that $\ell_1(n) < n + O(\log n)$ and $\ell_2(n) \ge \alpha \cdot \ell_1(n)$. Let $G^{(\cdot)} : \{0,1\}^{\ell_0(n)+\ell_1(n)} \to \{0,1\}^{\ell_0(n)+\ell_2(n)}$ be a bitwise-nonadaptive number generator that makes queries to a permutation on $\{0,1\}^n$. Then there is no fully black-box reduction of the pseudo-randomness of $G^{(\cdot)}$ to the one-wayness of its oracle.*

To prove Theorem 5, we proceed in a manner similar to the proof of Theorem 2, building up a set $\mathit{Good}'(n)$ whose purpose is similar to the set $\mathit{Good}(n)$ in that proof. The fact that each output bit of G depends only a single oracle query simplifies the construction of $Good'(n)$. Specifically, when constructing $Good'(n)$, we can ignore some of the "more difficult to deal with" queries made by $G^{(\cdot)}$, since we can later define adversary A to also ignore these queries simply by ignoring the corresponding output bits. This is what allows us to handle linearly-many queries in the current setting, even though we could only handle constantly-many queries in the proof of Theorem 2.

Proof details are deferred to the full version of this paper.

7 Some Remarks on Streaming Cryptography

The study of non-adaptivity in Goldreich-Levin-like constructions (Theorem 5) is motivated by questions related to *Streaming Models for Cryptography.* In some sense, impossibility results for non-adaptive black-box-constructions indicate the impossibility of certain type of black-box streaming constructions. We ask whether there is anything positive that can be said in the streaming setting, perhaps using non-black-box techniques. In this section, we put forward the main questions in streaming models for cryptography. Here is the main motivating question:

> *Starting from generic assumptions, is it possible to construct a one-way function or a pseudo-random generator using $O(\log n)$ space and a small $(1, 2, \ldots, \mathsf{constant}, \mathsf{polylog})$ number of passes over the seed?*

Why logarithmic space? Observe that assuming the existence of 2^{n^ϵ}-hard one-way functions (resp. one-way permutations), we can easily construct a one-way function (resp. pseudo-random generator) that uses *poly*-logarithmic space and *reads its input once.* By "2^{n^ϵ}-hard", we mean functions that are hard to invert with probability $\geq 1/2^{n^\epsilon}$ in time $\leq 2^{n^\epsilon}$. Computing such functions in logarithmic space without the ability to recompute (by revisiting the input) seems counter-intuitive. In fact, one can show that unconditionally this cannot be done with any constant number of passes (see the full version of this paper). Are super-constantly-many passes sufficient?

Motivation and related work. Streaming cryptography is motivated both from a theoretical and a practical viewpoint. The practical impact is in settings where on-line or streaming computation of a cryptographic primitive is needed. Theoretical motivation comes from the general theme of computing cryptographic primitives using rudimentary resources. Most relevant to streaming cryptography is the seminal work of Applebaum, Ishai, and Kushilevitz [2,1,4,3], which builds upon the work of Randomizing Polynomials (e.g. [9]), and shows the possibility of *Cryptography in* NC^0: given a "cryptographic function" f, construct a randomized encoding of f, which is a distribution $\{\hat{f}\}$ that (i) preserves the security of f, and (ii) is much simpler to compute than f. This amazing technical achievement brings the combinatorics of cryptographic functions to a simplified setting, and opens the possibility of better understanding cryptographic primitives and non-black-box techniques.

Goals and observations. We wish to be able to state a theorem of the form: if one-way functions exist then one-way functions computable in a streaming manner exist. We believe that this is a difficult thing to show. A potentially more feasible goal would be to show: if 2^{n^ϵ}-hard one-way functions exist then log-space streaming cryptography exists. In fact, by relying on [1,7], one can easily obtain a *non-black-box* construction of a one-way function computable in $O(\log n)$ space with $\log^{O(1)} n$ passes over the input, assuming that both (i) 2^{n^ϵ}-hard one-way functions exist, and (ii) log-space computable one-way functions exist; see the full version of this paper for the details. The latter assumption refers to functions that are just super-polynomially hard, computable with $n^{O(1)}$ many passes. It seems challenging to do the construction relying only on the existence of 2^{n^ϵ}-hard one-way functions. One can take this further to conjecture that it is possible to prove the following statement in some *constructive* way:

> 2^{n^ϵ}*-hard one-way functions exist* $\iff$ $O(\log n)$ *streaming one-way functions exist* $\iff$ *one-way functions computable in* NC^0 *exist*

This is a rather ambitious research direction. In particular, the right-to-left implication is a hardness amplification of some sort. Our intuition is that streaming computation of functions, functions computable by NC^0 circuits of some restricted form (e.g. of bounded treewidth), and 2^{n^ϵ}-hard one-way functions seem to be related.

References

1. Applebaum, B., Ishai, Y., Kushilevitz, E.: Computationally private randomizing polynomials and their applications. Computational Complexity 15(2), 115–162 (2006) (also CCC 2005)
2. Applebaum, B., Ishai, Y., Kushilevitz, E.: Cryptography in NC^0. SIAM J. Comput. 36(4), 845–888 (2006) (also FOCS 2004)
3. Applebaum, B., Ishai, Y., Kushilevitz, E.: Cryptography with constant input locality. In: Menezes, A. (ed.) CRYPTO 2007. LNCS, vol. 4622, pp. 92–110. Springer, Heidelberg (2007)
4. Applebaum, B., Ishai, Y., Kushilevitz, E.: On pseudorandom generators with linear stretch in NC0. Comput. Complexity 17(1), 38{69 (2008); also In: Díaz, J., Jansen, K., Rolim, J.D.P., Zwick, U. (eds.) APPROX 2006 and RANDOM 2006. LNCS, vol. 4110, pp. 260–271. Springer, Heidelberg (2006)
5. Gennaro, R., Gertner, Y., Katz, J., Trevisan, L.: Bounds on the effciency of generic cryptographic constructions. SIAM J. Comput. 35(1), 217–246 (2005)
6. Goldreich, O., Levin, L.A.: A hard-core predicate for all one-way functions. In: STOC 1989, pp. 25–32. ACM, Berlin (1989)
7. Haitner, I., Reingold, O., Vadhan, S.: Efficiency improvements in constructing pseudorandom generators from one-way functions. In: STOC 2010, pp. 437–446. ACM, New York (2010)
8. Impagliazzo, R., Rudich, S.: Limits on the provable consequences of one-way permutations. In: STOC 1989 (1989)
9. Ishai, Y., Kushilevitz, E.: Randomizing polynomials: a new representation with applications to round-efficient secure computation. In: Young, D.C. (ed.) FOCS 2000, pp. 294–304. IEEE Computer Society, Los Alamitos (2000)

10. Lu, C.J.: On the complexity of parallel hardness amplification for one-way functions. In: Halevi, S., Rabin, T. (eds.) TCC 2006. LNCS, vol. 3876, pp. 462–481. Springer, Heidelberg (2006)
11. Miles, E., Viola, E.: On the complexity of increasing the stretch of pseudorandom generators. In: Ishai, Y. (ed.) TCC 2011. LNCS, vol. 6597, pp. 504–521. Springer, Heidelberg (2011)
12. Reingold, O., Trevisan, L., Vadhan, S.: Notions of reducibility between cryptographic primitives. In: Naor, M. (ed.) TCC 2004. LNCS, vol. 2951, pp. 1–20. Springer, Heidelberg (2004)
13. Viola, E.: On constructing parallel pseudorandom generators from one-way functions. In: CCC 2005, pp. 183–197. IEEE Computer Society, Los Alamitos (2005)

On the Complexity of Non-adaptively Increasing the Stretch of Pseudorandom Generators

Eric Miles* and Emanuele Viola**

Northeastern University
{enmiles,viola}@ccs.neu.edu

Abstract. We study the complexity of black-box constructions of linear-stretch pseudorandom generators starting from a 1-bit stretch oracle generator G. We show that there is no construction which makes non-adaptive queries to G and then just outputs bits of the answers. The result extends to constructions that both work in the non-uniform setting and are only black-box in the primitive G (not the proof of correctness), in the sense that any such construction implies NP/poly $\neq$ P/poly. We then argue that not much more can be obtained using our techniques: via a modification of an argument of Reingold, Trevisan, and Vadhan (TCC '04), we prove in the non-uniform setting that there is a construction which only treats the primitive G as black-box, has polynomial stretch, makes non-adaptive queries to the oracle G, and outputs an affine function (i.e., parity or its complement) of the oracle query answers.

1 Introduction

The notion of a pseudorandom generator is fundamental to the study of both cryptography and computational complexity. An efficient algorithm $G : \{0,1\}^n \to \{0,1\}^{n+s}$ is a (cryptographic) pseudorandom generator (PRG) if no efficient adversary can distinguish a random output from a uniformly random string, except with some small advantage. That is, for all efficient adversaries A, we have $|\Pr[A(G(U_n)) = 1] - \Pr[A(U_{n+s}) = 1]| < \epsilon$.

A key parameter for any PRG is its stretch s, the difference between the output and input lengths. Any PRG must have stretch $s \geq 1$ to even satisfy the definition, but in fact such a small amount of stretch is not useful for any cryptographic or derandomization applications of which we are aware. For these, one typically needs the stretch to be larger, e.g. linear ($s = \Omega(n)$). An important and well-known result is that the existence of a PRG with stretch $s = 1$ implies the existence of a PRG with stretch $s = \text{poly}(n)$ for any desired polynomial. We begin by briefly recalling the construction that is typically used to prove this result (due to Goldreich and Micali; see [4] Sect. 3.3.2 for a more thorough treatment).

For a generator $G : \{0,1\}^\ell \to \{0,1\}^{\ell+1}$ and a positive integer k, let $G^k(x)$ denote the $(\ell+1)$-bit string resulting from k iterative applications of G, each

* Supported by NSF grant CCF-0845003.
** Supported by NSF grant CCF-0845003.

Y. Ishai (Ed.): TCC 2011, LNCS 6597, pp. 522–539, 2011.

time using the first ℓ bits of the previous output as input, and using x as input for the first invocation. Then, the "stretch-increasing" construction $H^{(\cdot)} : \{0,1\}^\ell \to \{0,1\}^m$ is defined as

$$H^G(x) := G^1(x)_{\ell+1} \circ G^2(x)_{\ell+1} \circ \cdots \circ G^m(x)_{\ell+1}. \tag{1}$$

That is, H iteratively queries G as described above, and outputs the final bit of each answer.

An aspect of the Goldreich-Micali construction that we would like to stress is that the queries H makes to its oracle are adaptive, in the sense that the ith query can be determined only after the answer to the $(i-1)$th query has been received. The presence of adaptivity in such constructions is of particular importance when considering the existence of cryptographic primitives in "low" complexity classes. The celebrated work of Applebaum et al. [1], in combination with the recent (non-adaptive) construction of Haitner et al. [7], demonstrates the existence of PRGs computable in NC^0 under the assumption that there exist one-way functions computable in NC^1. However, the resulting PRGs have *sub-linear* stretch, and the application of construction (1) would place them outside of NC^0.

Finally, we note that construction (1) only outputs bits of the answers it gets back from the queries, with no additional computation performed.

Informal discussion of our results. In this paper we study the complexity of increasing the stretch of cryptographic PRGs. We work in the setting of black-box constructions, which is explained in detail later. Informally, our first main result says that there can be no linear-stretch construction that makes non-adaptive queries and outputs only bits of its answers. For example, this rules out constructions which make non-adaptive queries and always output the last bit of the query answers, or even the entire query answers. Thus, linear-stretch constructions require either adaptive queries or postprocessing the queries in a more sophisticated way than just projecting. Our proof of this result is similar to one by Gennaro, Gertner, Katz, and Trevisan [3] who prove that constructions of generators with stretch s from one-way permutations must make $\geq \Omega(s/\log(\text{security}))$ queries. But note that our work is incomparable to theirs because we do not bound the number of queries (our constructions make one query per output bit).

Our second main result complements the first by showing that not much more can be obtained with the techniques in this paper. Specifically, we consider a special type of construction which is termed *weakly black-box* by Reingold, Trevisan, and Vadhan in [12], and for which we later advocate the alternative terminology *primitive black-box.* Then we extend an argument also in [12] to prove unconditionally, in the non-uniform setting, the existence of such constructions which have polynomial stretch, make non-adaptive queries, and just compute affine functions (parities or their complement) of the query answers. This complements the previous result because if instead of affine functions one only allows for projections, we show that such a construction implies NP/poly

$\neq$ P/poly. This means that, at least in the non-uniform setting, to extend our negative result to constructions with more postprocessing power requires different techniques from the ones in this paper. Our extension of the argument in [12] combines that argument with the Nisan-Wigderson generator [11].

Black-box constructions and formal statement of our results. To formally state our result, we first define black-box constructions. To explain and motivate the latter, we start by sketching the proof of correctness of the Goldreich-Micali Construction (1). Suppose there is an adversary A that distinguishes $H^G(U_\ell)$ from U_m with advantage greater than $\epsilon \cdot m$. Using a hybrid argument, one can show that there exists a $k \in [m]$ such that A distinguishes the distributions $U_{k-1} \circ \left(H^G(U_\ell)|_{[m-(k-1)]}\right)$ and $U_k \circ \left(H^G(U_\ell)|_{[m-k]}\right)$ with advantage greater than ϵ. Then, we define a probabilistic oracle circuit $C^{(\cdot)}$ as follows: on input $(x, b) \in \{0,1\}^\ell \times \{0,1\}$, $C^{G,A}$ computes $H^G(x)$ using its oracle to G, chooses $y \in \{0,1\}^{k-1}$ uniformly at random, and then outputs $A\left(y \circ b \circ H^G(x)|_{[m-k]}\right)$. Depending on whether (x, b) was chosen from $U_{\ell+1}$ or from $G(U_\ell)$, the input C gives to A will come from one of the two hybrid distributions that A can distinguish between, and so C distinguishes G with advantage greater than ϵ, contradicting G's pseudorandomness.

The above argument is an example of a *black-box reduction*: the argument applies to any (possibly hard to compute) functions G and A, provided that we are given oracle access to them.

Definition 1 (Black-box stretch-increasing construction). *An oracle function $H^{(\cdot)} : \{0,1\}^n \to \{0,1\}^{n+s}$ is a black-box stretch-increasing construction with security reduction size t of a generator with stretch s and error ϵ from any one-bit-stretch oracle generator $G : \{0,1\}^\ell \to \{0,1\}^{\ell+1}$ with error δ if the following holds:*

For every 1-bit stretch generator $G : \{0,1\}^\ell \to \{0,1\}^{\ell+1}$ and every adversary A, if A distinguishes H^G with advantage ϵ, i.e.

$$\left|\Pr[A(H^G(U_n)) = 1] - \Pr[A(U_{n+s}) = 1]\right| \geq \epsilon$$

then there is an oracle circuit $C^{(\cdot)}$ of size t that, when given oracle access to both A and G, distinguishes G with advantage δ, i.e.

$$\left|\Pr[C^{A,G}(G(U_\ell)) = 1] - \Pr[C^{A,G}(U_{\ell+1}) = 1]\right| \geq \delta.$$

Poly-time computable stretch-increasing black-box constructions are useful in constructing efficient PRGs, because if we start with an oracle G that is a poly-time computable PRG with error $\delta(\ell) = 1/\ell^{\omega(1)}$ against circuits of size $s(\ell) = \ell^{\omega(1)}$, and we have $t, n = \text{poly}(\ell)$, then H^G is a poly-time computable PRG with error $\epsilon(n) = 1/n^{\omega(1)}$ against circuits of size $s'(n) = n^{\omega(1)}$. This can be easily seen by noticing that any circuit of size poly(n) that distinguishes H^G with advantage $1/n^{O(1)}$ can be transformed into a circuit of size at most poly(ℓ) that distinguishes G with advantage $1/\ell^{O(1)}$, contradicting G's pseudorandomness.

We can now state our first main result.

Theorem 1. *For all sufficiently large ℓ and for $n \leq 2^{\sqrt{\ell}}$, there is no black-box construction $H^{(\cdot)} : \{0,1\}^n \to \{0,1\}^{n+s}$ of a generator with stretch $s \geq 5n/\log n$ and error $\epsilon \leq 1/4$ from any one-bit stretch generator $G : \{0,1\}^\ell \to \{0,1\}^{\ell+1}$ with error $\delta \geq 2^{-\sqrt{\ell}/30}$ and with security reduction size $t \leq 2^{\sqrt{\ell}/30}$ of the form*

$$H^G(x) := G(q_1(x))_{b_1(x)} \circ \cdots \circ G(q_{n+s}(x))_{b_{n+s}(x)}$$

where $q_i : \{0,1\}^n \to \{0,1\}^\ell$ specifies the i-th query and $b_i : \{0,1\}^n \to [\ell+1]$ specifies the bit of the i-th answer to output.

Note this holds even if the 1-bit stretch generator is hard against circuits (as opposed to uniform algorithms).

An interesting open problem is to understand whether a result like Theorem 1 holds for arbitrary polynomial-time postprocess of the query answers. We do not know how to solve this problem. However, as mentioned before we can show that the techniques in this paper are unlikely to prove a negative result, even if the postprocessing is just computing parities or their complements. For this, we consider a special type of construction, which is termed weakly black-box in [12]. Informally, a reduction is weakly black-box if the construction H treats the primitive G as a black-box, but outputs a generator in the real-world, i.e. the proof of correctness is arbitrary. We suggest the alternative terminology *primitive black-box*, to signify both that only the primitive (and not the adversary) is treated as a black-box, and that this is a "cruder" form of reduction.

We define next primitive constructions, working in the asymptotic setting because the following results are cleaner to state in that setting. We also note that our construction will hold for infinitely many input lengths (as opposed to sufficiently large input), and for conciseness we incorporate this into the definition.

Definition 2 (Primitive black-box stretch-increasing construction). *Let ℓ be a security parameter, and let $n = n(\ell)$ and $s = s(\ell)$ be functions of ℓ. A family of oracle functions $H^{(\cdot)} : \{0,1\}^n \to \{0,1\}^{n+s}$ is a primitive black-box stretch-increasing construction with stretch s from any family of one-bit-stretch generators $G : \{0,1\}^\ell \to \{0,1\}^{\ell+1}$ if the following holds for infinitely many input lengths ℓ:*

For every generator family $G : \{0,1\}^\ell \to \{0,1\}^{\ell+1}$, if there exists a constant c_0 and a circuit family A of size at most n^{c_0} which distinguishes H^G from uniform with advantage at least $1/n^{c_0}$, i.e.

$$\left|\Pr\left[A\left(H^G(U_n)\right) = 1\right] - \Pr\left[A\left(U_{n+s}\right) = 1\right]\right| \geq 1/n^{c_0}$$

then there exists a constant c_1 and a oracle circuit family $C^{(\cdot)}$ of size at most ℓ^{c_1} which distinguishes G from uniform with probability at least $1/\ell^{c_1}$, i.e.

$$\left|\Pr\left[C^G\left(G(U_\ell)\right) = 1\right] - \Pr\left[C^G\left(U_{\ell+1}\right) = 1\right]\right| \geq 1/\ell^{c_1}.$$

Our second main results proves the existence of a non-adaptive primitive black-box stretch-increasing construction of a slightly modified form which computes a polynomial-stretch PRG. The additional power that this construction has is the ability to compute parities or their complements of the query answers, rather than just projections. Recall from Definition 2 that primitive constructions only work for infinitely many input lengths.

Theorem 2. *Let $c > 1$ be any constant. Then, for $n = 17\ell^2$, there exists a primitive black-box stretch-increasing construction $H^{(\cdot)} : \{0,1\}^n \to \{0,1\}^{n^c}$ with stretch $s := n^c - n$ from any family of one bit stretch generators $G : \{0,1\}^\ell \to \{0,1\}^{\ell+1}$. In addition, $H^{(\cdot)}$ is computable by a* poly(n)*-sized circuit family, and has the form*

$$H^G(x) := \langle G(q_1(x)), r_1(x)\rangle \oplus t_1(x) \circ \cdots \circ \langle G(q_{n+s}(x)), r_{n+s}(x)\rangle \oplus t_{n+s}(x)$$

where $q_i : \{0,1\}^n \to \{0,1\}^\ell$ specifies the ith query, $r_i : \{0,1\}^n \to \{0,1\}^{\ell+1}$ specifies the parity function for the ith answer, and $t_i : \{0,1\}^n \to \{0,1\}$ specifies whether to flip the ith bit.

One weakness of the above theorem is that it works in the non-uniform setting. It is an open problem whether something like this can be proved in the uniform one.

To appreciate Theorem 2, we point out that the techniques used for our previous negative result (Theorem 1) "extend" to primitive constructions as well, in the sense that they can be used to show that any such construction implies NP/poly $\not\subseteq$ P/poly. Note that primitive black-box constructions cannot be ruled out without ruling out the existence of pseudorandom generators (if the latter exist, a construction can just ignore the oracle and output a pseudorandom generator).

Theorem 3. *Let $n = n(\ell) \leq 2^{\sqrt{\ell}}$ and $s = s(n) \geq 5n/\log n$. Let $H^{(\cdot)} : \{0,1\}^n \to \{0,1\}^{n+s}$ be a primitive black-box stretch-increasing construction with stretch s from any family of one-bit stretch generators $G : \{0,1\}^\ell \to \{0,1\}^{\ell+1}$. If H has the form*

$$H^G(x) := G(q_1(x))_{b_1(x)} \circ \cdots \circ G(q_{n+s}(x))_{b_{n+s}(x)}$$

and the q_i and b_i are computable by poly(n)*-sized circuits, then NP/poly $\not\subseteq$ P/poly.*

Note that the parameters in Theorem 2 are within the range of parameters considered by Theorem 3 – the only difference is the amount of postprocess.

1.1 More Related Work

The earlier work [13] (which was later extended by [10]) analyzes a type of pseudorandom generator construction that is very similar to ours. The constructions in [13] make non-adaptive queries to an oracle one-way function, and then apply an arbitrary unbounded-fan-in constant-depth circuit (AC^0) to the outputs; [13] shows that such constructions cannot have linear stretch. At first glance this construction is incomparable to Theorem 1, because it starts from a weaker primitive (one-way function instead of one-bit stretch generator) but on the other hand allows for AC^0 postprocessing instead of just projections.

However, it was pointed out to us by Benny Applebaum that a strengthening of Theorem 1 follows from [13] when combined with the works [1] and [7]. Specifically, a version of Theorem 1 holds even if the construction H is allowed to apply an AC^0 circuit to the output of the one-bit stretch oracle PRG G (rather than just taking projections). We now elaborate on this improvement. (We also remark that at the moment this establishes a strengthened negative result only for constructions that start from a uniform hardness assumption, because Theorem 1.1 in [13] is only proved for those).

Let $H^{(\cdot)} : \{0,1\}^n \to \{0,1\}^{n+s}$ be a a black-box construction of a PRG from a one-bit stretch PRG of the form $H^G(x) := C_x(G(q_1(x)), \ldots, G(q_{\mathrm{poly}(n)}(x)))$, where C_x is an AC^0 circuit generated arbitrarily from x and the functions q_i are arbitrary as before. Let $G_{\mathrm{HRV}}^{(\cdot)} : \{0,1\}^\ell \to \{0,1\}^{\ell+1}$ be the black-box construction of a PRG from a OWF given by [7] Theorem 6.1. This construction has the form $G_{\mathrm{HRV}}^f(x) := C'(x, f(x'_1), \ldots, f(x'_t))$ where C' is an NC^1 circuit and the x'_i are disjoint projections of the input x. Then, we can apply the compiler from [1] (cf. Remark 6.7 in that work) to obtain a black-box construction $G_{\mathrm{AIK}}^{(\cdot)} : \{0,1\}^\ell \to \{0,1\}^{\ell+1}$ of a PRG from a OWF of the form $G_{\mathrm{AIK}}^f(x) := C''(x, f(x'_1), \ldots, f(x'_t))$, where now C'' is an NC^0 circuit (and thus is also an AC^0 circuit). (For both G_{HRV} and G_{AIK} the seed length is $\ell = \mathrm{poly}(m)$, where m is the input length of the oracle OWF, though the compiler from [1] does increase the seed length). Finally, by combining H and G_{AIK}, we obtain a black-box construction $H_*^{(\cdot)} : \{0,1\}^n \to \{0,1\}^{n+s}$ of a PRG from a OWF which has the form $H_*^f(x) := C_x'''(f(q_1(x)), \ldots, f(q_{\mathrm{poly}(n)}(x)))$ where C_x''' is an AC^0 circuit. This is a contradiction to Theorem 1.1 of [13] when the oracle $f : \{0,1\}^m \to \{0,1\}^k$ has $\log^{\omega(1)} m < k \le m^{O(1)}$ and the stretch s is greater than $n \cdot \log^{O(1)} m/k = o(n)$.

Finally, we mention that in a concurrent work, Bronson, Juma, and Papakonstantinou [2] also study non-adaptive black-box PRG constructions and obtain results which are incomparable to ours.

1.2 Techniques

We now explain the ideas behind the proof of Theorem 1. For simplicity, we first explain our proof in the case in which the construction always outputs the same bit of the answers, say the first bit (i.e., $b_i(x) = 1$ for every i, in Theorem 1). We start by considering a non-explicit oracle pseudorandom generator $G : \{0,1\}^\ell \to \{0,1\}^{\ell+1}$ that is hard to break even for circuits that have access to G. Such oracles are obtained in an unpublished manuscript of Impagliazzo [9] and in a work by Zimand [15]. (They work in a slightly different setting, however, obtaining PRGs with high probability in the random oracle model, where we instead require an unconditional (but non-explicit) generator that is secure against adversaries which can query it. For completeness we present a streamlined version of their arguments in Sect. 4, and for the moment continue with the description of the proof of Theorem 1). By padding, we can modify our oracle to have the extra property that $G(x)_1 = x_1$ for every x. Now, the point is that the construction doesn't need to query the oracle, since each output bit $G(q_i(x))_{b_i(x)}$ can be replaced with $q_i(x)_1$. So we can consider an adversary A that breaks the construction H by simply checking, given a challenge $z \in \{0,1\}^m$, whether there exists an input to H that produces z. This breaks H as soon as the output length is $\geq |x|+1$. Since H doesn't use G anymore, neither does the adversary A. Hence the ability to access A does not compromise the security of G, contradicting Definition 1.

To obtain the result for primitive constructions, we observe that A can be computed in NP/poly, and hence under the assumption that NP/poly = P/poly we obtain a distinguisher.

Moreover, this simplified argument says nothing about, for example, a construction which outputs the entirety of the answers received from the oracle, which a priori may seem to be a plausible candidate. To generalize our result to constructions that output different bits (i.e. not always the first one), we identify a set of indices $T \subseteq [\ell+1]$ of size $\ell(1-\Theta(1/\log \ell))$, such that for most input strings $x \in \{0,1\}^n$, most of the bits $b_i(x)$ chosen by H fall inside T. We exploit this fact by designing an oracle PRG G that reveals the first $|T|$ bits of its input on the set T; that is, $G(x)|_T = x_1 x_2 \cdots x_{|T|}$ for every input x. We then design an (inefficient) adversary A that distinguishes H^G from uniform by examining, for every $x \in \{0,1\}^n$, only the bits i such that $b_i(x) \in T$, and checking if each bit matches the corresponding bit from the query $q_i(x)$. This turns out to break H as soon as the the output length is $\geq |x| + \Omega(|x|/\log|x|)$ (we do not attempt to optimize this value and content ourselves with anything sublinear). On the other hand, A depends on G just because of the knowledge of the set T, which means that oracle access to A does not compromise the security of G, again contradicting 1.

We now explain the proof of Theorem 2. Here we follow closely an argument of Reingold, Trevisan, and Vadhan in [12]. We proceed by case analysis, depending on the existence or non-existence of one-way functions (OWFs). For our argument, we define OWFs as computable by a family of poly-size circuits and hard to invert by any family of poly-size circuits.

If OWFs exist, we use the result of Håstad et al. [8] that efficiently computable PRGs also exist; the construction then ignores its oracle and simply outputs the PRG, by letting $t_i(x)$ be the ith output bit of the PRG, and setting $r_i = 0$, for every i.

If OWFs do not exist, this means that the oracle cannot be computable by poly-size circuits (since it is assumed to be hard to invert). We can then use Goldreich-Levin [6] to transform the oracle into a Boolean function that is hard to compute by any family of poly-size circuits. Until now this is the argument in [12]. (Actually [12] is more involved because it works even in the uniform setting). Our contribution is to apply at this point the Nisan-Wigderson construction [11] to get a PRG. Since this construction is non-adaptive and has arbitrary polynomial stretch, and the hard function given by Goldreich-Levin is just the inner product of the oracle with a random vector, this has the desired form.

We note that, as is well-known, the proof of correctness of the Nisan-Wigderson construction requires non-uniformity, and this is what prevents this result to apply to the uniform setting.

Organization. In Sect. 2 we prove Theorems 1 and 3, the two negative results. In Sect. 3 we prove Theorem 2, the complementing positive result. Finally, in Sect. 4 we construct the one-bit-stretch oracle generator used in Sect. 2.

2 Black-Box Stretch-Increasing Constructions

In this section we prove Theorems 1 and 3. We use the following definition of an oracle pseudorandom generator.

Definition 3 (Oracle pseudorandom generator). *Let $G : \{0,1\}^n \rightarrow \{0,1\}^{n+s}$ be a function. G is a (T,ϵ)-pseudorandom generator if $s \geq 1$ and for every oracle circuit C of size at most T, we have $\left|\Pr[C^G(G(U_n)) = 1] - [C^G(U_{n+s}) = 1]\right| < \epsilon$. The quantity on the left-hand side of the inequality is referred to as C's advantage in distinguishing G, and s is referred to as G's stretch.*

The key property we require of our one-bit stretch oracle G, stated in the next theorem, is that it reveals a large portion of its input, i.e. most of the output bits are simply copied from the input.

Theorem 4. *Let $\ell, d \in \mathbb{N}$ be sufficiently large with $d \leq \ell/2$. Then, for any subset $T \subseteq [\ell+1]$ with $|T| = \ell - d$ and any oracle A, there exists a generator $G : \{0,1\}^\ell \rightarrow \{0,1\}^{\ell+1}$ such that*

1. *G is a $(2^{d/30}, 2^{-d/30})$-PRG against adversaries with oracle access to A (and G).*
2. *For every input $x \in \{0,1\}^\ell$, $G(x)|_T = x_1 x_2 \cdots x_{\ell-d}$.*

We defer the proof of this theorem to Sect. 4, and instead start by showing how it is used to prove the main theorem. First, we need a simple technical lemma showing that for any stretch-increasing construction of the specified form, we can find a large set of indices inside which most $b_i(x)$ fall for most choices of x.

Lemma 1. *Let $n, d, s, \ell \in \mathbb{N}$ with $d < \ell$. Let $\{b_i : \{0,1\}^n \rightarrow [\ell+1]\}_{i \in [n+s]}$ be a collection of $n+s$ functions. Then, there exists a set $T \subseteq [\ell+1]$ of size $\ell - d$ such that*

$$\Pr_x\left[|\{i : b_i(x) \in T\}| \geq (n+s) \cdot \left(1 - \frac{4(d+1)}{\ell+1}\right)\right] \geq \frac{3}{4}.$$

Proof. Let $S \subseteq [\ell+1]$ denote a random subset of size $d+1$. We have $\Pr_{x,i,S}[b_i(x) \in S] = (d+1)/(\ell+1)$, and so we can fix some S so that $\Pr_{x,i}[b_i(x) \in S] \leq (d+1)/(\ell+1)$. This can be restated as $\mathbb{E}_x[\Pr_i[b_i(x) \in S]] \leq (d+1)/(\ell+1)$, and so by Markov's inequality we have $\Pr_x[\Pr_i[b_i(x) \in S] \geq 4(d+1)/(\ell+1)] \leq 1/4$. Letting $T := [\ell+1] \setminus S$ completes the proof.

We now prove Theorem 1, restated for convenience.

Theorem 1. *For all sufficiently large ℓ and for $n \leq 2^{\sqrt{\ell}}$, there is no black-box construction $H^{(\cdot)} : \{0,1\}^n \to \{0,1\}^{n+s}$ of a generator with stretch $s \geq 5n/\log n$ and error $\epsilon \leq 1/4$ from any one-bit stretch generator $G : \{0,1\}^\ell \to \{0,1\}^{\ell+1}$ with error $\delta \geq 2^{-\sqrt{\ell}/30}$ and with security reduction size $t \leq 2^{\sqrt{\ell}/30}$ of the form*

$$H^G(x) := G(q_1(x))_{b_1(x)} \circ \cdots \circ G(q_{n+s}(x))_{b_{n+s}(x)}$$

where $q_i : \{0,1\}^n \to \{0,1\}^\ell$ specifies the i-th query and $b_i : \{0,1\}^n \to [\ell+1]$ specifies the bit of the i-th answer to output.

Proof. Let $H^{(\cdot)}$ be a construction of the specified form. Fix a parameter $d := \ell/\log n$. Fix $T \subseteq [\ell+1]$ to be the subset of size $\ell - d$ guaranteed by Lemma 1. For each $x \in \{0,1\}^n$, let I_x denote the set $\{i : b_i(x) \in T\} \subseteq [n+s]$. Using $s = 5n/\log n$, the chosen value for d, and the fact that $|I_x|$ is an integer, the bound from Lemma 1 can be restated as $\Pr_x[|I_x| \geq n+1] \geq 3/4$ for sufficiently large n and ℓ. In the remainder of the proof, we refer to x such that $|I_x| \geq n+1$ as *good.*

Let T^{-1} denote a transformation such that $T^{-1}(j) = k$ if j is the kth smallest element of T (this is simply to provide a mapping from G's output bits to the corresponding revealed input bits). The adversary $A : \{0,1\}^{n+s} \to \{0,1\}$ is defined as the function which accepts exactly the set

$$\{z : \exists x \in \{0,1\}^n \text{ such that } x \text{ is good and } \forall i \in I_x, z_i = q_i(x)_{T^{-1}(b_i(x))}\}.$$

Let $G : \{0,1\}^\ell \to \{0,1\}^{\ell+1}$ be the PRG guaranteed by Theorem 4 using these choices of T and A. We claim that A distinguishes $H^G(U_n)$ from U_{n+s} with advantage at least $1/4$. To see this, consider z which is a uniformly chosen output of H^G, i.e. $z = H^G(x)$ for $x \leftarrow U_n$. Because x is good with probability at least $3/4$, and because $H^G(x)_i = q_i(x)_{T^{-1}(b_i(x))}$ for all $i \in I_x$ by item 2 of Theorem 4, we have $\Pr[A(H^G(U_n)) = 1] \geq 3/4$. Conversely, for the case where A's input is chosen from U_{n+s}, we have the following calculation:

$$\begin{aligned}
\Pr_{z \leftarrow U_{n+s}}[A(z) = 1] &= \Pr_z\left[\exists x : \ x \text{ is good} \wedge \forall i \in I_x : z_i = q_i(x)_{T^{-1}(b_i(x))}\right] \\
&\leq \sum_{\substack{x \in \{0,1\}^n \\ x \text{ is good}}} \Pr_z\left[\forall i \in I_x : z_i = q_i(x)_{T^{-1}(b_i(x))}\right] \\
&\leq \sum_{\substack{x \in \{0,1\}^n \\ x \text{ is good}}} 2^{-(n+1)} \\
&\leq \frac{1}{2}.
\end{aligned}$$

(The second inequality follows from the fact that $|I_x| \geq n+1$ for x that are good).

Finally, note that item 1 in Theorem 4 (along with the choice of d and the upper bound on n) implies that there is no oracle circuit C of size at most $2^{\sqrt{\ell}/30}$ such that $C^{A,G}$ distinguishes G with advantage at least $2^{-\sqrt{\ell}/30}$. Therefore, H does not meet the conditions of Definition 1 for the stated parameters.

Next, we show that this theorem can be extended to the primitive black-box setting.

Theorem 3. *Let $n = n(\ell) \leq 2^{\sqrt{\ell}}$ and $s = s(n) \geq 5n/\log n$. Let $H^{(\cdot)} : \{0,1\}^n \to \{0,1\}^{n+s}$ be a primitive black-box stretch-increasing construction with stretch s from any family of one-bit stretch generators $G : \{0,1\}^\ell \to \{0,1\}^{\ell+1}$. If H has the form*

$$H^G(x) := G(q_1(x))_{b_1(x)} \circ \cdots \circ G(q_{n+s}(x))_{b_{n+s}(x)}$$

and the q_i and b_i are computable by $\mathrm{poly}(n)$*-sized circuits, then NP/poly $\not\subseteq$ P/poly.*

Proof. Let H be a primitive black-box stretch-increasing construction of the specified form. Let G and I_x be defined as in Theorem 1 (the oracle A against which G is secure is not relevant here). Because the q_i, b_i functions are computable by $\mathrm{poly}(n)$-size circuits, there is a $\mathrm{poly}(n)$-size circuit family which computes the string $H^G(x)|_{I_x}$ on input x, while making *no* oracle calls to G. As a result, we can define a non-deterministic $\mathrm{poly}(n)$-size circuit family which distinguishes H^G from uniform with advantage $1/4$: on input $z \in \{0,1\}^{n+s}$, the circuit non-deterministically guesses $x \in \{0,1\}^n$, and accepts iff $|I_x| \geq n+1$ and $z|_{I_x} = H^G(x)|_{I_x}$. The proof that this is indeed a distinguisher for H^G is identical to the argument given for Theorem 1.

Now assume for contradiction that NP/poly = P/poly, i.e. that every non-deterministic circuit family can be simulated by a deterministic circuit family with only a polynomial increase in size. Then, there is a $\mathrm{poly}(n)$-size deterministic circuit family which distinguishes H^G from uniform with noticeable advantage. By the definition of a primitive black-box construction, there must also be such a circuit family that distinguishes G, contradicting G's pseudorandomness.

3 A Non-adaptive Primitive Black-Box Construction

In this section we prove Theorem 2, showing that there exists a non-adaptive primitive black-box reduction with slightly more post-processing power (namely the ability to compute inner products) that computes a polynomial-stretch PRG. We remind the reader that this construction produces a PRG on infinitely many input lengths. For the sake of brevity, we state two standard definitions that are used in this section and the next.

Definition 4 (Hard to invert). *Let $f : \{0,1\}^n \to \{0,1\}^m$ be a function. f is (T,ϵ)-hard to invert if for every oracle circuit C of size at most T, we have $\Pr[f(C^f(f(U_n))) = f(U_n)] < \epsilon$.*

Definition 5 (Hard to compute). *Let $f : \{0,1\}^n \to \{0,1\}$ be a Boolean function. f is (T, ϵ)-hard to compute if for every circuit C of size at most T, we have* $\Pr[C(U_n) = f(U_n)] < 1/2 + \epsilon$.

We now state the prior results that we will use, and prove a simple lemma. In what follows, we will sometimes make the assumption that "OWFs do not exist", which means that, for any family of functions $f : \{0,1\}^\ell \to \{0,1\}^{\text{poly}(\ell)}$ that is $(p(\ell), 1/p(\ell))$-hard to invert for all polynomials p and sufficiently large ℓ, every poly(ℓ)-sized circuit family fails to compute f on infinitely many input lengths. This corresponds to one-way functions computable by circuits and hard to invert by circuits.

Theorem 5 ([8]). *Assume that there exists a family of functions $G : \{0,1\}^\ell \to \{0,1\}^{\ell+1}$ that is computable by a* poly(ℓ)*-size circuit family and is $(p(\ell), 1/p(\ell))$-hard to invert for all polynomials p and sufficiently large ℓ. Then, for any constant c, there exists a family of generators $H : \{0,1\}^n \to \{0,1\}^{n^c}$ that is computable by a* poly(n)*-size circuit family and is $(p(n), 1/p(n))$-pseudorandom for all polynomials p and sufficiently large n.*

A version of the following result was proved in [12] for the uniform computation model. This proof, which relies on non-uniformity, is a bit simpler.

Lemma 2. *Assume that OWFs do not exist and let $G : \{0,1\}^\ell \to \{0,1\}^{\ell+1}$ be a generator family. If G is $(p(\ell), 1/p(\ell))$-pseudorandom for all polynomials p and sufficiently large ℓ, then the Boolean function family $f(x,r) := \langle G(x), r\rangle$ is $(p(\ell), 1/p(\ell))$-hard to compute for all polynomials p and infinitely many input lengths.*

Proof. First, we show that if G is $(p(\ell), 1/p(\ell))$-pseudorandom for all polynomials p and sufficiently large ℓ, then it is also $(p(\ell), 1/p(\ell))$-hard to invert for all polynomials p and sufficiently large ℓ. Let C be a poly(ℓ)-size circuit family which, for sufficiently large ℓ, inverts G with probability $= \epsilon$ for some $\epsilon = 1/\text{poly}(\ell)$. Then, define an adversary $A : \{0,1\}^{\ell+1} \to \{0,1\}$ as follows: on input y, A computes $x = C(y)$, uses its oracle to G to check if $G(x) = y$, and outputs 1 iff this holds. We clearly have $\Pr[A(G(U_\ell)) = 1] = \epsilon$. Let $T \subseteq \text{Im}(G)$ be the set of outputs that C inverts, and note that $\sum_{y \in T} \Pr[G(U_\ell) = y] = \epsilon$. For each $y \in T$ we have $\Pr[G(U_\ell) = y] \geq 1/2^\ell$, and so $|T|/2^\ell \leq \epsilon$. Then, since A will only output 1 on inputs that C can invert and since no string outside $\text{Im}(G)$ can be inverted, we have $\Pr[A(U_{\ell+1}) = 1] = |T|/2^{\ell+1} \leq \epsilon/2$, and thus A distinguishes G from uniform with advantage $\geq \epsilon/2 = 1/\text{poly}(\ell)$.

Now, assume for contradiction that there exists a polynomial p and a circuit family C of size $p(\ell)$ which computes f correctly with probability at least $1/2 + 1/p(\ell)$ over the input, for sufficiently large ℓ. Then by the Goldreich-Levin theorem [6], there exists a polynomial p' and a circuit family C' of size $p'(\ell)$ such that $\Pr[C'(U_\ell) = G(U_\ell)] \geq 1/p'(\ell)$, for sufficiently large ℓ. Notice that (the function computed by) C' can only be inverted on strictly less than a $1 - 1/(2p'(\ell))$ fraction of inputs by poly(ℓ)-size circuits, because any circuit which inverts C' on a $1-1/(2p'(\ell))$ fraction of inputs also inverts G on at least a $1/(2p'(\ell))$ fraction of inputs. However, using the standard direct product construction (originally due to Yao [14]; see also [4] Theorem 2.3.2), this implies the existence of a one-way function, contradicting the assumption that OWFs do not exist.

By virtue of the above proof, Theorem 2 actually establishes a primitive black-box stretch-increasing construction which works when the oracle is any hard-to-invert function, and not only the special case of one-bit-stretch PRGs.

In order to apply the Nisan-Wigderson construction, we recall the notion of designs.

Definition 6 (Design). *A collection of sets $S_1, \ldots, S_d \subseteq [n]$ is an (n, d, ℓ, α)-design if*

1. $\forall i : \; |S_i| = \ell.$
2. $\forall i \neq j : \; |S_i \cap S_j| \leq \alpha.$

Lemma 3 ([11]). *For any integers d and ℓ such that $\log d \leq \ell \leq d$, there exists a $\mathrm{poly}(d)$-time constructible collection $S_1, \ldots, S_d$ which is a $(4\ell^2, d, \ell, \log d)$-design.*

We now give the proof of Theorem 2.

Theorem 2. *Let $c > 1$ be any constant. Then, for $n = 17\ell^2$, there exists a primitive black-box stretch-increasing construction $H^{(\cdot)} : \{0,1\}^n \to \{0,1\}^{n^c}$ with stretch $s := n^c - n$ from any family of one bit stretch generators $G : \{0,1\}^\ell \to \{0,1\}^{\ell+1}$. In addition, $H^{(\cdot)}$ is computable by a $\mathrm{poly}(n)$-sized circuit family, and has the form*

$$H^G(x) := \langle G(q_1(x)), r_1(x) \rangle \oplus t_1(x) \;\circ\; \cdots \;\circ\; \langle G(q_{n+s}(x)), r_{n+s}(x) \rangle \oplus t_{n+s}(x)$$

where $q_i : \{0,1\}^n \to \{0,1\}^\ell$ specifies the ith query, $r_i : \{0,1\}^n \to \{0,1\}^{\ell+1}$ specifies the parity function for the ith answer, and $t_i : \{0,1\}^n \to \{0,1\}$ specifies whether to flip the ith bit.

Proof. Assume that OWFs exist, and let $H' : \{0,1\}^n \to \{0,1\}^{n^c}$ be the generator guaranteed by Theorem 5. Then, the construction $H^{(\cdot)}$ is $H^G(z) := H'(z)$ for any oracle G. (Note that this can be achieved in the form stated in the theorem by setting $r_i(z) = 0^{\ell+1}$ for all i and z, and choosing the t_i appropriately to compute each bit of H').

Now assume that OWFs do not exist. Let $G : \{0,1\}^\ell \to \{0,1\}^{\ell+1}$ be any generator family, and define $f : \{0,1\}^{2\ell+1} \to \{0,1\}$ as $f(x, r) := \langle G(x), r \rangle$. Fix a constant $c > 1$, and define $n = 4(2\ell+1)^2$ (which is at most $17\ell^2$ for sufficiently large ℓ). Let $S_1, \ldots, S_{n^c}$ be the $(n, n^c, 2\ell+1, c\log n)$ design guaranteed by Lemma 3. Then, the construction $H^G : \{0,1\}^n \to \{0,1\}^{n^c}$ is defined as

$$H^G(z) := f(z|_{S_1}) \circ \cdots \circ f(z|_{S_{n^c}}).$$

If there exists a polynomial p and a circuit family of size $p(\ell)$ which distinguishes G from uniform with advantage at least $1/p(\ell)$, then the theorem is trivially true. Thus, we can take G to be $(p(\ell), 1/p(\ell))$-pseudorandom for all polynomials p and sufficiently large ℓ. Assume for contradiction that there exists a constant c_0 and a circuit family A of size n^{c_0} that distinguishes $H^G(U_n)$ from U_{n^c} with advantage $1/n^{c_0}$. Using the equivalence of distinguishing and next-bit predicting [14], this implies the existence of an $i \in [n^c]$ and a circuit family $A' : \{0,1\}^{i-1} \to \{0,1\}$ of size $n^{O(c_0)}$ such that $\Pr\left[A'(H^G(U_n)|_{[i-1]}) = H^G(U_n)_i\right] \geq 1/2 + 1/n^{c+c_0}$. Separating out the part of the input indexed by S_i, this can be rewritten as

$$\Pr_{(x,y) \leftarrow (U_{2\ell+1}, U_n)} \left[A'(H^G(z)|_{[i-1]}) = H^G(z)_i\right] \geq 1/2 + 1/n^{c+c_0}, \tag{2}$$

where $z \in \{0,1\}^n$ is defined by $z|_{S_i} = x$ and $z|_{\overline{S_i}} = y|_{\overline{S_i}}$. By an averaging argument, there is a way to fix $y \in \{0,1\}^n$ such that (2) holds; from here on we assume that this y is fixed. For each $j \in [i-1]$, define the function $f_j : \{0,1\}^{2\ell+1} \to \{0,1\}$ as $f_j(x) := f(z)$, where now z is defined by $z|_{S_i \cap S_j} = x_1 x_2 \cdots x_{|S_i \cap S_j|}$ and $z|_{\overline{S_i \cap S_j}} = y|_{\overline{S_i \cap S_j}}$. Note that since $S_i \cap S_j \leq c \log n$ and y is fixed, each f_j is computable by a circuit family of size $\text{poly}(n) = \text{poly}(\ell)$. Finally, define the circuit family $A'' : \{0,1\}^{2\ell+1} \to \{0,1\}$ as $A''(x) := A'(f_1(x), \ldots, f_{i-1}(x))$. It can be easily checked that A'' has size $\text{poly}(\ell)$ and correctly computes f on a random input with probability at least $1/2 + 1/n^{c+c_0}$, contradicting Lemma 2.

4 Constructing the Oracle Generator

In this section we prove Theorem 4 (restated for convenience), which gives the one-bit-stretch oracle generator used in the proofs of our negative results (Theorems 1 and 3).

Theorem 4. *Let $\ell, d \in \mathbb{N}$ be sufficiently large with $d \leq \ell/2$. Then, for any subset $T \subseteq [\ell+1]$ with $|T| = \ell - d$ and any oracle A, there exists a generator $G : \{0,1\}^\ell \to \{0,1\}^{\ell+1}$ such that*

1. G is a $(2^{d/30}, 2^{-d/30})$-PRG against adversaries with oracle access to A (and G).
2. For every input $x \in \{0,1\}^\ell$, $G(x)|_T = x_1 x_2 \cdots x_{\ell-d}$.

On constructing the oracle. A direct proof that a random function $G : \{0,1\}^\ell \to \{0,1\}^{\ell+1}$ is a pseudorandom generator even for circuits that have oracle access to G does not seem immediate to us. The existence of such oracles is shown via an indirect route in an unpublished manuscript of Impagliazzo [9] and – in a slightly different scenario – in a work by Zimand [15]. Both works proceed by considering an oracle one-way function, and then applying standard constructions of generators from one-way functions (for which one can now use [8] or [7]).

We proceed by first considering a hard-to-invert oracle permutation π, and then using the Goldreich-Levin hardcore bit [6] to get one bit of stretch.

This approach will have security exponential in the input length of π, and so we can apply π to the relatively few ($\Theta(\ell/\log \ell)$) bits outside of $|T|$, and then use padding to get a generator G on ℓ bits that reveals most of its input.

We know of two ways to demonstrate the existence of such a permutation π. One is via a theorem in [3] which uses a clever encoding argument to prove that a random permutation is hard to invert with very high probability. They show that if there exists a small circuit which inverts a permutation π on some fraction of inputs, then π can be succinctly encoded when the circuit is given as advice. Then, since only a small number of permutations have succinct encodings, the probability that a random π can be sufficiently inverted by a fixed circuit is small, and a union bound over circuits gives the result.

The second way, and the one that we use here, is an arguably more direct argument showing that any fixed circuit with access to a fixed auxiliary oracle has negligible probability (over the choice of permutation) of sufficiently inverting the permutation. This method is from [9] and [15] (though they consider general length-preserving functions rather than permutations), and hinges on a combinatorial trick which originally appeared in [5]. Briefly, it is shown that for a fixed circuit C, the expected number of subsets of size k that are inverted by C is not too large. Then, Markov's inequality is used to show that the probability that C inverts *any* set of size $m \approx k^2$ is small, since to do so C would have to invert *each* of its $\binom{m}{k}$ subsets of size k (this is the combinatorial trick).

We now turn to the formal proof of Theorem 4. There are two main ingredients; the first is the well-known Goldreich-Levin hard-core bit theorem [6]. It can be checked that the standard proof of this theorem relativizes; we omit the details.

Theorem 6. *Let $f : \{0,1\}^d \to \{0,1\}^m$ be a function, and let A be any oracle. Let C be an oracle circuit of size T such that $\Pr[C^A(f(U_d), U'_d) = \langle U_d, U'_d\rangle] \geq 1/2 + \epsilon$. Then, for d sufficiently large, there exists an oracle circuit B of size at most $\alpha \cdot T \cdot (d/\epsilon)^2$ (where α is a universal constant) such that $\Pr[B^A(f(U_d)) = U_d] \geq \epsilon^3/8d$.*

The second ingredient is the fact that there exist permutations π which are hard to invert even for adversaries that have access to π and to an arbitrary fixed auxiliary oracle.

Theorem 7. *Let $d \in \mathbb{N}$ be sufficiently large. Then for any oracle A, there exists a permutation $\pi : \{0,1\}^d \to \{0,1\}^d$ that is $(2^{d/5}, 2^{-d/5})$-hard to invert against adversaries with oracle access to π and A.*

Before giving the proof, we state and prove two lemmas. The aforementioned combinatorial trick, due to [5], is given by the following lemma.

Lemma 4. *Let U be a finite set, let $\Gamma = \{\phi : U \to \{0,1\}\}$ be a family of predicates on U, and let p_k be an upper bound on the probability that ϕ chosen uniformly from Γ returns true for every element in a subset of size k, i.e.*

$$\forall K \subseteq U, |K| = k : \Pr_{\phi \leftarrow \Gamma}\left[\prod_{x \in K} \phi(x) = 1\right] \leq p_k.$$

Then, for any m such that $k \leq m \leq |U|$, we have

$$\Pr_{\phi \leftarrow \Gamma}\left[\exists M \subseteq U, |M| \geq m : \prod_{x \in M} \phi(x) = 1\right] \leq \frac{\binom{|U|}{k} \cdot p_k}{\binom{m}{k}}.$$

Proof. Let $\phi(X)$ denote $\prod_{x \in X} \phi(x)$. We have $\mathbb{E}\left[|\{K \subseteq U : |K| = k \text{ and } \phi(K) = 1\}|\right] \leq \binom{|U|}{k} \cdot p_k$ by linearity of expectation. Then the lemma follows from double counting, because for any set $M \subseteq U$ of size m, $\phi(M) = 1$ iff $\phi(K) = 1$ for every one of the $\binom{m}{k}$ subsets $K \subseteq M$ of size k.

We now explain why this lemma is helpful. Following [9] and [15], we bound the probability (over the permutation π) that a fixed circuit C of size s inverts a fixed set K of size k; this is done by considering the probability that any k out of the at most ks distinct queries made by C on inputs from K are mapped by π to K; specifically, we bound

$$p_k \leq \binom{ks}{k} \cdot \left(\frac{k}{|U|}\right)^k \approx \frac{s^k}{\binom{|U|}{k}}.$$

The factor of s^k means that we cannot use a union bound over all $\binom{|U|}{k}$ subsets of size k. So we instead use Lemma 4, choosing m so that $\binom{m}{k} \approx s^{2.3k}$, which makes the probability of inverting a set of size m small enough to use a union bound over all circuits.

We also require a bound on the number of oracle circuits of a given size.

Lemma 5. *There are at most $2^{s(3+4\log s)}$ oracle circuits of size s which have access to two oracles π and A.*

Proof. We define the size of a circuit to be the number of wires it has; this is also an upper bound on the number of gates. For each wire in the circuit, we must specify two things:

- which gate it is an output of (or if it is an input wire) and which position it is in for this gate
- which gate it is an input of (or if it is an output wire) and which position it is in for this gate

Note that the positions are relevant for wires incident on oracle gates, as the functions computed by these gates may not be symmetric. Specifying either incident gate for a given wire takes $\log s$ bits (as there are at most s gates), and likewise each position can be specified with $\log s$ bits. Therefore, each of the s wires can be specified with $4 \log s$ bits. Finally, for each gate, we must specify which of the five types it is ($\wedge, \vee, \neg$, π-oracle or A-oracle), which takes three bits.

Proof (Proof of Theorem 7). We will in fact show that a random π has the desired property with probability at least $1 - 2^{-2^{d/4}}$. Fix an oracle A and an oracle circuit C of size s. Fix a subset $K \subseteq \{0,1\}^d$ of size k; we will first bound the probability that C inverts all of K. Let Q_x^π denote the set of at most s distinct queries that $C^{A,\pi}(x)$ makes to π (for some choice of x and π), and let $Q_K^\pi := \bigcup_{x\in K} Q_x^\pi$. We assume without loss of generality that the last query that C makes to π is the string that C outputs (this is justified because any circuit which does not query its output string can be modified into one that does with an increase in size that is so small as to not affect the union bound below).

A necessary condition for C to invert all of K is that $\pi^{-1}(x) \in Q_K^\pi$ for all $x \in K$. Since $|Q_K^\pi| \le ks$, we can bound this by

$$\begin{aligned}\Pr_\pi\left[\forall x \in K : \pi^{-1}(x) \in Q_K^\pi\right] &\le \Pr_\pi\left[\exists X \subseteq Q_K^\pi : \bigcup_{x\in X} \pi(x) = K\right] \\ &\le \binom{ks}{k} \cdot \left(\frac{k}{2^d}\right)\left(\frac{k-1}{2^d-1}\right)\cdots\left(\frac{1}{2^d-k+1}\right) \\ &\le \left(\frac{eks}{2^d}\right)^k .\end{aligned}$$

We now apply Lemma 4 in the obvious way: U is $\{0,1\}^d$, and there is a predicate $\phi_\pi \in \Gamma$ for each permutation π, where $\phi_\pi(x) = 1$ iff $C^{A,\pi}(x) = \pi^{-1}(x)$. By the lemma, the probability that there exists a set M of size $m \ge k$ such that C inverts every element of M is bounded from above by $(e^2 \cdot k \cdot s/m)^k$. Choosing $k = 2^{d/3}, m = 2^{4d/5}$ and $s = 2^{d/5}$, this is bounded by $2^{-2^{d/3}}$ for sufficiently large d. By Lemma 5, there are at most $2^{2^{d/5}\cdot\Theta(d)}$ circuits of size $2^{d/5}$, and so the probability over the choice of π that there *exists* a circuit of size $2^{d/5}$ which inverts a set of size at least $2^{4d/5}$ is at most $2^{-2^{d/3}+2^{d/5}\cdot\Theta(d)} < 2^{-2^{d/4}}$ for sufficiently large d. Therefore, π is $(2^{d/5}, 2^{-d/5})$-hard to invert with probability at least $1 - 2^{-2^{d/4}}$.

We may now give the proof of Theorem 4.

Proof (Proof of Theorem 4). Let the oracle A and the subset T be given. Recall that $|T| = \ell - d$, and let $\pi : \{0,1\}^d \to \{0,1\}^d$ be the permutation guaranteed by Theorem 7 which is $(2^{d/5}, 2^{-d/5})$-hard to invert against adversaries with oracle access to π and A. Then, the generator G treats its input $x \in \{0,1\}^\ell$ as $(x_1, x_2, x_3) \in \{0,1\}^{\ell-2d} \times \{0,1\}^d \times \{0,1\}^d$, and outputs the $(\ell+1)$-bit string defined as follows:

$$G(x)|_{[\ell+1]\setminus T} = \pi(x_3) \circ \langle x_3, x_2\rangle \qquad\qquad G(x)|_T = x_1 \circ x_2 .$$

Now assume for contradiction that there exists an oracle circuit $C : \{0,1\}^{\ell+1} \to \{0,1\}$ of size at most $2^{d/30}$ such that $\Pr[C^{A,G}(G(U_\ell)) = 1] - \Pr[C^{A,G}(U_{\ell+1}) = 1] \ge 2^{-d/30}$ (dropping the absolute value w.l.o.g).. Because the permutation π is the only part of G's output which may be "difficult" to compute, we can take C to have oracles (A, π) instead of (A, G) at the cost of increasing C's size by a factor of poly(d). We construct a probabilistic oracle circuit $IP : \{0,1\}^d \times \{0,1\}^d \to \{0,1\}$ which, on input (x, y), tries to compute $\langle \pi^{-1}(x), y\rangle$. $IP^{A,\pi}(x, y)$ performs the following steps:

1. chooses a random string $z \in \{0,1\}^{\ell-2d}$ and a random bit $b \in \{0,1\}$
2. constructs the $(\ell+1)$-bit string w defined by $w|_{[\ell+1]\setminus T} = x \circ b,\ w|_T = z \circ y$
3. computes $C^{A,\pi}(w)$ and outputs $C^{A,\pi}(w) \oplus 1 \oplus b$

We clearly have $|IP| \leq |C| \cdot \mathrm{poly}(d) \leq 2^{d/30} \cdot \mathrm{poly}(d)$. Consider the behavior of $IP^{A,\pi}$ on a uniformly random input (x,y). It is easy to see that the string w is distributed according to $U_{\ell+1}$. If we condition on the chosen bit b being equal to $\langle \pi^{-1}(x), y\rangle$ (which happens with probability 1/2), then w is distributed according to $G(U_\ell)$. For brevity, let E_{IP} denote the event $IP^{A,\pi}(x,y) = \langle \pi^{-1}(x), y\rangle$, and let E_b denote the event $b = \langle \pi^{-1}(x), y\rangle$. Then,

$$\begin{aligned}
\Pr[E_{IP}] &= \frac{1}{2}\left(\Pr[E_{IP} \mid E_b] + \Pr[E_{IP} \mid \overline{E_b}]\right) \\
&= \frac{1}{2}\left(\Pr[C^{A,\pi}(w) = 1 \mid E_b] + \left(1 - \Pr[C^{A,\pi}(w) = 1 \mid \overline{E_b}]\right)\right) \\
&= 1/2 + \Pr[C^{A,\pi}(w) = 1 \mid E_b] - \Pr[C^{A,\pi}(w) = 1] \\
&= 1/2 + \Pr[C^{A,\pi}(G(U_\ell)) = 1] - \Pr[C^{A,\pi}(U_{\ell+1}) = 1] \\
&\geq 1/2 + 2^{-d/30}.
\end{aligned}$$

The probabilities are over both (x,y) and the internal randomness of IP; by a standard averaging argument, we can fix the internal randomness of IP to get a deterministic circuit which computes $\langle \pi^{-1}(x), y\rangle$ on a random (x,y) with the same success probability. Then for sufficiently large d, Theorem 6 gives an oracle circuit of size at most $2^{d/30} \cdot \mathrm{poly}(d) \cdot O(d^2 \cdot 2^{2d/30}) \leq 2^{d/5}$ that, when given access to A and π, inverts π with probability at least $2^{-3d/30}/8d \geq 2^{-d/5}$ over its input, contradicting the hardness of π.

Acknowledgements. We are very grateful to Benny Applebaum for several useful comments, and especially for pointing out the strengthening of Theorem 1 and allowing us to include a proof in §1.1. We also would like to thank Russell Impagliazzo for sharing [9] with us, and the anonymous TCC referees for helpful feedback.

References

1. Applebaum, B., Ishai, Y., Kushilevitz, E.: Cryptography in NC^0. SIAM J. Comput. 36(4), 845–888 (2006)
2. Bronson, J., Juma, A., Papakonstantinou, P.A.: Limits on the stretch of non-adaptive constructions of pseudo-random generators. In: Ishai, Y. (ed.) TCC 2011. LNCS, vol. 6597, pp. 522–539. Springer, Heidelberg (2011)
3. Gennaro, R., Gertner, Y., Katz, J., Trevisan, L.: Bounds on the efficiency of generic cryptographic constructions. SIAM J. Comput. 35(1), 217–246 (2005)
4. Goldreich, O.: Foundations of Cryptography. Basic Tools, vol. 1. Cambridge University Press, Cambridge (2001)
5. Goldreich, O., Krawczyk, H., Luby, M.: On the existence of pseudorandom generators. SIAM J. Comput. 22(6), 1163–1175 (1993)

6. Goldreich, O., Levin, L.: A hard-core predicate for all one-way functions. In: 21st Annual ACM Symposium on Theory of Computing (STOC), pp. 25–32 (1989)
7. Haitner, I., Reingold, O., Vadhan, S.P.: Efficiency improvements in constructing pseudorandom generators from one-way functions. In: 42nd ACM Symposium on Theory of Computing (STOC), pp. 437–446 (2010)
8. Håstad, J., Impagliazzo, R., Levin, L.A., Luby, M.: A pseudorandom generator from any one-way function. SIAM J. Comput. 28(4), 1364–1396 (1999) (electronic)
9. Impagliazzo, R.: Very strong one-way functions and pseudo-random generators exist relative to a random oracle (1996) (manuscript)
10. Lu, C.-J.: On the complexity of parallel hardness amplification for one-way functions. In: Halevi, S., Rabin, T. (eds.) TCC 2006. LNCS, vol. 3876, pp. 462–481. Springer, Heidelberg (2006)
11. Nisan, N., Wigderson, A.: Hardness vs randomness. J. Computer & Systems Sciences 49(2), 149–167 (1994)
12. Reingold, O., Trevisan, L., Vadhan, S.: Notions of Reducibility between Cryptographic Primitives. In: Naor, M. (ed.) TCC 2004. LNCS, vol. 2951, pp. 1–20. Springer, Heidelberg (2004)
13. Viola, E.: On constructing parallel pseudorandom generators from one-way functions. In: 20th Annual Conference on Computational Complexity (CCC), pp. 183–197. IEEE, Los Alamitos (2005)
14. Yao, A.: Theory and applications of trapdoor functions. In: 23rd Annual Symposium on Foundations of Computer Science (FOCS), pp. 80–91. IEEE, Los Alamitos (1982)
15. Zimand, M.: Efficient privatization of random bits. In: Randomized Algorithms Satellite Workshop of the 23rd International Symposium on Mathematical Foundations of Computer Science (1998), `http://triton.towson.edu/~mzimand/pub/rand-privat.ps`

Concurrent Security and Non-malleability

Rafael Pass

Cornell University

Abstract. The Internet enables concurrent executions of cryptographic protocols. This concurrent setting, however, also brings forth new types of coordinated attacks in which an adversary controls many parties, interleaving the executions of the various protocol instances, and attempts to "maul" messages from one execution to use in another.

In this talk, we will survey some recent methods for achieving concurrent security without relying on any trusted-set up (such as e.g., Common Reference Strings).

Y. Ishai (Ed.): TCC 2011, LNCS 6597, p. 540, 2011.

(Nearly) Round-Optimal Black-Box Constructions of Commitments Secure against Selective Opening Attacks

David Xiao[1,2,⋆]

[1] LIAFA, Université Paris 7, 75205 Paris Cedex 13, France
[2] Université Paris-Sud, 91405 Orsay Cedex, France
dxiao@liafa.fr

Abstract. Selective opening attacks against commitment schemes occur when the commitment scheme is repeated in parallel (or concurrently) and an adversary can choose depending on the commit-phase transcript to see the values and openings to some subset of the committed bits. Commitments are secure under such attacks if one can prove that the remaining, unopened commitments stay secret.

We prove the following black-box constructions and black-box lower bounds for commitments secure against selective opening attacks:

1. For parallel composition, 4 (resp. 5) rounds are necessary and sufficient to build computationally (resp. statistically) binding and computationally hiding commitments. Also, there are no perfectly binding commitments.
2. For parallel composition, $O(1)$-round statistically-hiding commitments are equivalent to $O(1)$-round statistically-binding commitments.
3. For concurrent composition, $\omega(\log n)$ rounds are sufficient to build statistically binding commitments and are necessary even to build computationally binding and computationally hiding commitments, up to $\log \log n$ factors.

Our lower bounds improve upon the parameters obtained by the impossibility results of Bellare *et al.* (EUROCRYPT '09), and are proved in a fundamentally different way, by observing that essentially all known impossibility results for black-box zero-knowledge can also be applied to the case of commitments secure against selective opening attacks.

Keywords: commitments, black-box lower bounds, zero knowledge, selective opening attacks, parallel composition, concurrent composition.

1 Introduction

Commitment schemes have a wide array of applications in cryptography, one of the most notable being the construction of zero knowledge protocols [14, 4].

⋆ Supported by the French ANR Defis program under contract ANR-08-EMER-012.

Y. Ishai (Ed.): TCC 2011, LNCS 6597, pp. 541–558, 2011.

A problem that arises in the use of commitment schemes is whether their hiding property holds when composed in parallel: if some subset of the committed messages are opened, do the remaining unopened messages remain secure? This question arose early in the study of zero knowledge protocols, and is also natural in other cryptographic contexts where commitments are used as building blocks for protocols that might be then used in parallel (*e.g.* secure multi-party computation, etc.).

Although naively one might think because commitments are hiding that no additional information should be leaked by composing them, nevertheless it is unknown how to prove that standard stand-alone commitments (*e.g.* [18]) remain hiding when composed.

More formally, a selective opening attack on a commitment scheme allows a cheating receiver to interact in k parallel (or concurrent) commitments, and then ask the sender to open some subset $I \subseteq [k]$ of the commitments. The question is whether the unopened messages remain hidden in the following sense: is there a simulator strategy for every cheating receiver strategy that outputs a commit-phase transcript, a set $I \subset [k]$, and decommitments to $(b_i)_{i \in I}$ that is indistinguishable from the output of the cheating receiver with an honest sender?

In this paper we show that techniques both for constructions and lower bounds from the study of zero knowledge protocols can be applied to the study of commitments secure against selective opening attacks. We study the minimal round complexity needed to construct such commitments, and give solutions for commitments secure against selective opening attacks that are optimal or nearly optimal up to small factors.

1.1 Our Results

We let PAR denote parallel composition and CC denote concurrent composition. We let CB (resp. SB, PB) denote computational (resp. statistical, perfect) binding and CH (resp. SH) denote computational (resp. statistical) hiding. We give the following constructions:

Theorem 1. *The following hold via fully black-box reductions:*

1. *One-way permutations imply* 4*-round* PAR-CBCH *commitments exist.*
2. *t-round stand-alone* SH *commitments imply* $(t+3)$*-round* PAR-SB *commitments exist.*
3. *t-round stand-alone* SH *commitments imply* $\omega(t \log n)$*-round* CC-SB *commitments exist.*

In particular, Item 2 implies that collision-resistant hash functions (or even just 2-round statistically hiding commitments) suffice to construct 5-round PAR-SB commitments.

Assuming the proof of security for such a commitment scheme is given by a black-box simulator, we prove the following corresponding lower bounds:

Theorem 2 (Informal). *The following hold relative to any oracle:*

1. *There is no 3-round* PAR-CBCH *commitment.*
2. *There is no 4-round* PAR-SB *commitment.*
3. *There is a black-box reduction that uses a $O(1)$-round* PAR-SB *commitment to build a $O(1)$-round statistically hiding commitment.*
4. *There is no $o(\log n / \log\log n)$-round* CC-CBCH *commitment.*

We stress that besides the constraint that the simulator be black-box, these results are otherwise *unconditional*. Namely, Theorem 2 implies that no such commitments exist in the plain model (without oracles), but also implies that such commitments do not exist even in say the random oracle model (or stronger oracle models), where *a priori* one might have hoped to bypass impossibility results in the plain model.

Combining the second item of Theorem 2 with the main theorem of [15], which proves that there is no black-box reduction building a $o(n/\log n)$-round statistically hiding commitment from one-way permutations, we obtain the following corollary:

Corollary 1. *There is no black-box reduction that uses a one-way permutation to build $O(1)$-round* PAR-SB *commitment.*

Wee [23] independently proved via different techniques a theorem similar to Corollary 1 for the very closely related case of trapdoor commitments.

In addition to the above impossibility results, we also prove:

Theorem 3 (Informal). *Relative to any oracle, there exists no* PAR-PB *commitments nor receiver public-coin* PAR-CBCH *commitments.*

1.2 Comparison to Previous Constructions

Notions related to security against selective opening attacks have previously been studied in the literature. Security against selective opening is closely related to chameleon blobs [5, 6], trapdoor commitments [11], and equivocable commitments [2, 9, 8]. Roughly speaking, these notions all allow a simulator that can generate commit-phase transcripts that can be opened in many ways. Indeed, our constructions will be based on the equivocable commitment of [8].

Security against selective opening may be weaker than the notions above, and was directly studied in [10, 3]. Bellare *et al.* [3] give a construction of a scheme that is CC-SB secure, but this construction is non-black-box and requires applying a concurrent zero knowledge proof on a statement regarding the code implementing a one-way permutation. In contrast, all constructions presented in this paper are fully black-box.

Equivalence of statistical hiding and statistical binding. In this work we only study commitments with computational hiding. [3] already noted that stand-alone SH commitments satisfy a notion of PAR-SH security based on indistinguishability (this notion is different from ours). Independent of our work, Zhang *et al.* [24] gave a black-box reduction that uses t-round stand-alone SH commitments and one-way permutations to construct $(t + 3)$-round PAR-SH commitments (under our definition of selective opening security). Their construction is an extension of a recent trapdoor commitment of Pass and Wee [19].

With Item 2 of Theorem 2, this implies that $O(1)$-round statistical hiding and $O(1)$-round statistical binding are *equivalent* via black-box reductions when security against selective opening attacks is required. This contrasts sharply with the stand-alone case, as 2-round statistically binding commitments are equivalent to one-way functions, but no black-box reduction can build $o(n/\log n)$-round statistically hiding commitment from one-way functions [15].

1.3 Comparison to Previous Lower Bounds

Bellare *et al.* [3] proved that non-interactive commitments and perfectly binding commitments secure against selective opening attacks cannot be based on *any* black-box cryptographic assumption. Our lower bounds are stronger than theirs in that we can rule out 3- or 4-round rather than non-interactive commitments, as well as ruling out certain types of commitment with non-zero statistical binding error. However, our proof *technique* is incomparable to theirs.

Ways in which our lower bounds are stronger: first, the lower bounds of [3] assume black-box access to a cryptographic primitive, and therefore do not apply to constructions based on *concrete assumptions* (*e.g.* factoring, discrete log, lattice problems) where one might hope to exploit the specific structure of those problems to achieve security. In contrast, our results immediately rule out all constructions in the plain model.

Second, the lower bounds of [3] prove that non-interactive and perfectly binding commitments secure against selective opening attacks are impossible with respect to a very specific message distribution *that is defined in terms of a random oracle*. One could argue that the message distribution they consider is artificial and would not arise in applications of these commitments. In particular, it may suffice for applications to build commitments that are secure only for particular natural message distributions, such as the uniform distribution or the distributions encountered when using commitments to build zero knowledge proofs for **NP**. [3] does not rule out the existence of commitments that are secure only for these message distributions, while our impossibility results do and in fact apply simultaneously to all message distributions satisfying what we argue are very natural constraints (see Definition 5). In particular, the results of [3] also use the assumptions in Definition 5.

Ways in which our lower bounds are weaker: our results are weaker because they only apply to constructions with black-box simulators, *i.e.* we require that there exists a single simulator that works given black-box access to any cheating receiver. The results of [3] hold even for slightly non-black-box simulation techniques: they only require that for every cheating receiver oracle algorithm $(\mathsf{Rec}')^{(\cdot)}$ that accesses the underlying crypto primitive as a black-box, there exists an efficient oracle algorithm $\mathsf{Sim}^{(\cdot)}$ that accesses the underling crypto primitive as a black box that generates an indistinguishable transcript. However, [3] do *not* rule out techniques such as Barak's simulator [1], because the simulator there includes a PCP encoding of the code of the underlying cryptographic primitive, and thus treats the *crypto primitive itself* in a non-black-box way.

1.4 Our Techniques

Our constructions for parallel composition are essentially the equivocable commitment scheme of [8], while the case for concurrent composition follows in a straight-forward way by combining the commitment of [8] with the preamble from the concurrent zero knowledge proof of [21].

Our lower bounds are proven by observing that most known lower bounds for zero knowledge (*e.g.* [13, 17, 7, 16, 20]) extend naturally to the case of commitment schemes. Lower bounds for zero knowledge show that if a zero knowledge proof for L satisfies certain restrictions (*e.g.* 3 rounds, constant-round public coin [13], etc.), then $L \in \mathbf{BPP}$.

As was observed by [10, 3], plugging a t-round PAR-CBCH commitment into the GMW zero knowledge protocol for **NP** allows the zero knowledge property to be preserved under parallel repetition, thus allowing one to reduce soundness error while preserving zero knowledge and without increasing round complexity. Furthermore, the resulting protocol has $t + 2$ rounds, and has a black-box simulator if the commitment had a black-box simulator. This immediately implies the following:

Proposition 1 ([13], weak impossibility of PAR-CBCH, informal). *In the plain model, there exist no black-box simulator non-interactive or constant-round public-coin PAR-CBCH commitment schemes.*

To see why, suppose there were such a scheme, then by the above discussion one would obtain either a 3-round or constant-round public-coin zero knowledge argument for **NP** with a black-box simulator that remains zero knowledge under parallel repetition. By [13], this implies that $\mathbf{NP} = \mathbf{BPP}$. But this contradicts the existence of a PAR-CBCH commitment scheme, since by the Cook-Levin reduction we can use an algorithm solving **NP** to break any commitment.

Our results improve upon Proposition 1 as they apply to broader categories of commitments (*e.g.* 3-round vs. non-interactive). In addition, Proposition 1 uses the Cook-Levin reduction and therefore does not apply when considering schemes that might use random oracles. In contrast, Theorem 2 does hold relative to any oracle, and in the case of Item 3 of Theorem 2, is *black-box*. This is important for two reasons: first, Proposition 1 does not say whether such constructions are possible in the random oracle model, which is often used to prove the security of schemes for which we cannot prove security in the plain model. Second, if we want to compose our impossibility result with other black-box lower bounds, then our impossibility result had better also be black-box. For example, in order to obtain Corollary 1 we must combine Item 3 of Theorem 2 with the black-box lower bound of Haitner *et al.*. This is only possible if Item 3 of Theorem 2 is a black-box reduction, which would not be true using the approach of the weak impossibility result Proposition 1.

To prove Theorem 2, we construct what we call "equivocal senders": senders that run the commit phase without knowing the bits that must be revealed. We show that the existence of such equivocal senders implies that binding can be broken. We then construct equivocal senders for various kinds of protocols by applying the proof strategy for zero knowledge lower bounds originally outlined by Goldreich and Krawczyk [13]. By arguing directly, we avoid the Cook-Levin step in Proposition 1 and therefore our results hold relative to any oracle.

2 Preliminaries

For a random variable X, we let $x \leftarrow_{\mathrm{R}} X$ denote a sample drawn according to X. We let U_k denote the uniform distribution over $\{0,1\}^k$. For a set S, we let $x \leftarrow_{\mathrm{R}} S$ denote a uniform element of S. Let 2^S denote the set of all subsets of S. All security definitions in this paper are with respect to non-uniform circuits. We say that an event occurs with overwhelming probability if it occurs with probability $1 - n^{-\omega(1)}$, and that it occurs with negligible probability if it occurs with probability $n^{-\omega(1)}$. Two families of random variables $(X_n)_{n\in\mathbb{N}}, (Y_n)_{n\in\mathbb{N}}$ over $\{0,1\}^n$ are computationally indistinguishable, or equivalently $X \approx_c Y$, if for all circuits C of size $\mathrm{poly}(n)$ it holds that $|\Pr[C(X)=1] - \Pr[C(Y)=1]| \leq n^{-\omega(1)}$.

2.1 Commitment Schemes

We formally define commitments for single-bit messages; since we will be concerned with commitments that are composable, multi-bit messages can be handled by just repeating the single-bit protocol in parallel or concurrently.

Definition 1. *A t-round (stand-alone) commitment protocol is a pair of efficient algorithms* Send *and* Rec. *Given a sender input $b \in \{0,1\}$, we define:*

1. *The* commit phase *transcript is $\tau = \langle \mathsf{Send}(b; \omega_{\mathsf{Send}}), \mathsf{Rec}(\omega_{\mathsf{Rec}}) \rangle$ where ω_{Send}, ω_{Rec} are the random coins of the sender and receiver, respectively. Exactly t messages are exchanged in the commit phase t.*
2. *The* decommit phase *transcript consists of* Send *sending (b, open) to* Rec. $\mathsf{Rec}(\tau, b, \mathsf{open}) = 1$ *if* open *is a valid opening, and outputs 0 otherwise.*

Notation and variable definitions: We assume that a commitment scheme is put in a canonical form, where each party alternates speaking. We assume the number of rounds is even and the receiver speaks first. If the number of rounds is $2t$, then we label the sender's messages $\alpha_1, \ldots, \alpha_t$ and the receiver's messages $\beta_1, \ldots, \beta_t$, and we let $\alpha_{[i]} = (\alpha_1, \ldots, \alpha_i)$ and likewise for $\beta_{[i]}$. For a commitment protocol $(\mathsf{Send}, \mathsf{Rec})$, we write that the receiver's i'th response β_i is given by computing $\beta_{[i]} = \mathsf{Rec}(\alpha_{[i-1]}; \omega)$ where $\alpha_{[i-1]}$ are the first $i-1$ sender messages, and ω are the receiver's random coins. We let $\mathsf{Rec}(\bot; \omega) = \beta_1$ denote the first receiver message.

Let k denote the number of parallel or concurrent repetitions of a commitment protocol. Let n denote the security parameter of the protocol. Given a stand-alone commitment $(\mathsf{Send}, \mathsf{Rec})$, let Send^k denote the k-fold repeated sender (context will determine whether we mean parallel or concurrent composition). Let Rec^k denote the k-fold parallel receiver, and let Rec^k_Σ denote the k-fold concurrent receiver with schedule Σ. Underlined variables denote vectors of message bits (*e.g.* $\underline{b} \in \{0,1\}^k$) and plain letters with indices the bit at each coordinate (*e.g.* b_i is the i'th bit of $\underline{b}$).

Definition 2 (Binding). *A commitment scheme* $(\mathsf{Send}, \mathsf{Rec})$ *is computationally (resp. statistically) binding if for all polynomial-time (resp. unbounded) sender strategies* Send', *only with negligible probability can* Send' *interact with an honest* Rec *to generate a commit-phase transcript* τ *and then produce* $\mathsf{open}, \mathsf{open}'$ *such* $\mathsf{Rec}(\tau, 0, \mathsf{open}) = 1$ *and* $\mathsf{Rec}(\tau, 1, \mathsf{open}') = 1$. *A scheme is* perfectly *binding if the above probability of cheating is* 0.

It is straight-forward to prove that all the variants of the binding property are preserved under parallel/concurrent composition.

Hiding under selective opening attacks. We only study the case of computational hiding. In the following, $\mathcal{I} \subseteq 2^{[k]}$ *is a family of subsets of* $[k]$*, which denotes the set of legal subsets of commitments that the receiver is allowed to ask to be opened.*

Definition 3 (Hiding under selective opening: k-fold parallel composition security game). *Sender input:* $\underline{b} \in \{0,1\}^k$. *Let* Rec' *be the (possibly cheating) sender.*

1. $\mathsf{Send}^k, \mathsf{Rec}'$ *run* k *executions of the commit phase in parallel using independent random coins, obtaining* k *commit-phase transcripts* $\tau^k = (\tau_1, \ldots, \tau_k)$.
2. Rec' *chooses a set* $I \leftarrow_R \mathcal{I}$ *and sends it to* Send^k.
3. Send^k *sends* (b_i, ω_i) *for all* $i \in I$*, where* ω_i *is an opening of the* i*'th commitment.*

In Item 2, the honest receiver is defined to pick $I \in \mathcal{I}$ uniformly, while a malicious receiver may pick I adversarially.

Definition 4 (Hiding under selective opening, parallel composition). *Let* $\mathcal{I} \subseteq 2^{[k]}$ *be a family of subsets and* $\underline{\mathcal{B}}$ *be a family of message distributions over* $\{0,1\}^k$ *for all* k. *Let* $(\mathsf{Send}, \mathsf{Rec})$ *be a commitment and* Sim_k *be a simulator. We say that* $(\mathsf{Send}, \mathsf{Rec})$ *is secure against selective opening attacks for* $(\mathcal{I}, \underline{\mathcal{B}})$ *if for all* $k \leq \mathrm{poly}(n)$:

- *Let* $\langle \mathsf{Send}^k(\underline{b}), \mathsf{Rec}' \rangle = (\tau^k, I, \{(b_i, \omega_i)\}_{i \in I})$ *be the complete interaction between* Rec' *and the honest sender, including the commit-phase transcript* τ^k*, the subset* I *of coordinates to be opened and the openings* $(b_i, \omega_i)_{i \in I}$.
- *Let* $(\mathsf{Sim}_k^{\mathsf{Rec}'} \mid \underline{b})$ *denote the following: first,* $\mathsf{Sim}_k^{\mathsf{Rec}'}$ *interacts with* Rec' *(without knowledge of* $\underline{b}$*) and outputs a subset* I *of bits to be opened. Then* Sim_k *is given* $\{b_i\}_{i \in I}$. *Using this,* Sim_k *interacts with* Rec' *some more and outputs a commit-phase transcript* τ^k*, the set* I*, and the openings* $\{(b_i, \omega_i)\}_{i \in I}$.
- *It holds that* $(\mathsf{Sim}_k^{\mathsf{Rec}'} \mid \underline{b}) \approx_c \langle \mathsf{Send}^k(\underline{b}), \mathsf{Rec}' \rangle$ *where* $\underline{b} \leftarrow_R \underline{\mathcal{B}}$.

Definition 5. *We say that* $(\mathcal{I}, \underline{\mathcal{B}})$ *is* non-trivial *if (the uniform distribution over)* $\mathcal{I}, \underline{\mathcal{B}}$ *are efficiently samplable, and (1)* $|\mathcal{I}| = k^{\omega(1)}$ *and (2)* $\Pr_{I \leftarrow_R \mathcal{I}}[H_\infty(\underline{\mathcal{B}}_I) \geq 1/\text{poly}(n)] \geq 1/\text{poly}(n)$.

Here $\underline{\mathcal{B}}_I$ is the joint distribution of bits $\underline{\mathcal{B}}_i$ for $i \in I$. Property 1 says that if the receiver asks for a random set in $\mathcal{I}$ to be opened, then the sender cannot guess the set with noticeable probability. This restriction is natural because in many contexts if the sender can guess the set to be opened then it can cheat in the larger protocol where the commitment is being used (*e.g.* in a zero knowledge proof). Property 2 says that with noticeable probability over the choice of I, there is non-negligible entropy in the bits revealed. This is very natural as otherwise any receiver is trivially simulable since it always sees the same constant bits.

Stronger definitions of hiding. Our definitions are chosen to be as weak as possible in order to make our lower bounds stronger. Nevertheless, our positive results also satisfy a stronger definition of security, where security holds simultaneously for all $\mathcal{I}, \underline{\mathcal{B}}$. For such a notion, we prepend STR to the name of the security property (*e.g.* STR-PAR-SB).

Definition 6 (Security game for k-fold concurrent composition). *Identical to the parallel case, except that the receiver has the power to schedule messages as he wishes, rather than sending them in parallel. In addition, we allow the receiver to pick the set* I *incrementally subject to the constraint that at the end,* $I \in \mathcal{I}$. *For example, the receiver can generate one commit-phase transcript, ask the sender to decommit that instance, then use this information in its interaction to generate the second commit-phase transcript, and so forth.*

Definition 7 (Hiding under selective opening, concurrent composition) *Same as the parallel case, except that the simulator can incrementally ask for the values* $(b_i)_{i \in I}$ *before completing all commit-phase executions, subject to the constraint that at the end* $I \in \mathcal{I}$.

Discussion of definitional choices: One could weaken Definition 6 to require that although all the commit-phase transcripts may be generated concurrently, the openings happen simultaneously. Indeed, this was the definition used in [3]. We do not work with this weakening because it makes the definition not truly concurrent: forcing all the openings to occur simultaneously "synchronizes" the sessions.

3 Constructions

Di Crescenzo and Ostrovsky [8] (see also [9]) showed how to build an *equivocable* commitment scheme. Equivocable means that for every cheating receiver Rec', there exists a simulator that generates a commit-phase transcript that is computationally indistinguishable from a real transcript, but which the simulator can decommit to both 0 and 1. Equivocation seems even stronger than STR-PAR-CBCH security, except that STR-PAR-CBCH explicitly requires security to hold in many parallel sessions. Although it is not clear how to generically convert any stand-alone equivocable commitment to an equivocable commitment that is composable in parallel/concurrently, the particular construction of Di Crescenzo and Ostrovsky can be composed by using a suitable preamble.

The DO construction consists of a preamble, which is a coin-flipping scheme that outputs a random string, followed by running Naor's commitment based on OWF [18] using the random string of the preamble as the receiver's first message. Depending on how the preamble is constructed, we get either a STR-PAR-CBCH, STR-PAR-SB, or STR-CC-SB commitment. Therefore, Theorem 1 follows by Theorem 6 and Theorem 8 about the following protocol.

Protocol 4 ([8, 9, 18]). *Sender's bit: b. Let $G : \{0,1\}^n \to \{0,1\}^{3n}$ be a PRG.*

> ***Preamble:*** *Use a coin-flipping protocol to obtain $\sigma \leftarrow_R \{0,1\}^{3n}$.*
> ***Commit phase:*** *The sender picks random $s \leftarrow_R \{0,1\}^n$ and sends $c = (\sigma \wedge b) \oplus G(s)$ (where we use the notation $(\sigma \wedge b)_i = \sigma_i \wedge b$).*
> ***Decommit phase:*** *The sender sends b, s. Receiver checks that $c = (\sigma \wedge b) \oplus G(s)$.*

We now present three different preambles that when used in the protocol above, provide STR-PAR-CBCH, STR-PAR-SB, and STR-CC-SB security, respectively.

Protocol 5 ([8]). *Preambles for STR-PAR-CBCH or STR-PAR-SB:*

1. *Using the non-interactive stand-alone CH commitment based on one-way permutations (to achieve STR-PAR-CBCH) or a t-round stand-alone SH commitment (to achieve STR-PAR-SB), the receiver sends a commitment to $\alpha \leftarrow_R \{0,1\}^{3n}$.*
2. *The sender replies with $\beta \leftarrow_R \{0,1\}^{3n}$.*
3. *The receiver opens α.*
4. *Output $\sigma = \alpha \oplus \beta$.*

Theorem 6 ([8]). *Protocol 4 with the STR-PAR-CBCH (resp. STR-PAR-SB) version of the preamble of Protocol 5 gives a STR-PAR-CBCH (resp. STR-PAR-SB) commitment.*

Proof (Proof sketch of Theorem 6). We include a proof sketch for the sake of completeness, and refer the reader to [18, 12, 8] for full proofs. The binding properties are easy to verify, given the fact that Naor's commitment scheme is statistically binding.

The following simulator proves security against selective opening attacks for both the computational and statistical binding variants. Consider the k-fold repetition $\mathsf{Send}^k, \mathsf{Rec}^k$ of the protocol. Following the proof of Goldreich and Kahan [12], one can construct a simulator such that, by rewinding the first step of the preamble (*i.e.* Step 1 of Protocol 5), can learn the value of the $\alpha_1, \ldots \alpha_k$ used in each of the k parallel sessions. Care must be taken to ensure this finishes in expected polynomial time, but the same technique as in [12] works in our setting and we refer the reader to that paper for details.

Now for each $i \in [k]$ in the i'th session the simulator can sample $s_0, s_1 \leftarrow_{\mathrm{R}} \{0,1\}^n$ and reply with $\beta_i = G(s_0) \oplus G(s_1) \oplus \alpha_i$. This sets $\sigma_i = G(s_0) \oplus G(s_1)$. Then the simulator sets $c = G(s_0)$. Now the simulator can decommit to both 0 (by sending s_0) and to 1 (by sending s_1). ■

Protocol 7 ([21]). *Preamble for STR-CC-SB:*

1. *The receiver picks $\alpha \leftarrow_R \{0,1\}^{3n}$ and for $\ell = \omega(\log n)$ picks $\alpha^0_{i,j} \leftarrow_R \{0,1\}^{3n}$ for $i, j \in [\ell]$ and sets $\alpha^1_{i,j} = \alpha \oplus \alpha^0_{i,j}$. The receiver commits in parallel to $\alpha, \alpha^0_{i,j}, \alpha^1_{i,j}$ via a t-round statistically hiding commitment.*
2. *For each $j = 1$ to ℓ sequentially, do the following:*
   (a) *The sender sends $q_1, \ldots, q_\ell \leftarrow_R \{0,1\}$.*
   (b) *The receiver opens the commitment to $\alpha^{q_i}_{i,j}$ for all $i \in [\ell]$.*
3. *The sender sends $\beta \leftarrow_R \{0,1\}^{3n}$.*
4. *The receiver opens the commitment to $\alpha, \alpha^0_{i,j}, \alpha^1_{i,j}$ for all $i, j \in [\ell]$.*
5. *The sender checks that indeed $\alpha = \alpha^0_{i,j} \oplus \alpha^1_{i,j}$ for all $i, j \in [\ell]$. If so output $\sigma = \alpha \oplus \beta$, otherwise abort.*

Theorem 8 ([21, 22]). *Protocol 4 using the preamble of Protocol 7 gives a STR-CC-SB commitment.*

Proof. Binding is straightforward. For hiding, observe that this is the preamble of the concurrent zero knowledge proof of Prabhakaran *et al.* [21]. They prove the following:

Theorem 9 (Theorem 5.2 of [21], informal). *There is black-box simulator strategy that, given access to any efficient receiver for Protocol 7 with any concurrent scheduling, outputs with high probability in every session a string α* before *Step* 3 *such that the receiver opens to α in step* 5.

Namely, [21] show that by using an appropriate rewinding schedule, the simulator can obtain the value of α in *all* of the concurrent executions before the sender is supposed to send β, regardless of how the receiver schedules the messages. Once the simulator knows α, one can apply the simulator strategy of [12, 8], as in the proof sketch of Theorem 6. ■

4 Optimality of Constructions

We now define our main tool for proving lower bounds, *equivocal senders*. Intuitively, an equivocal sender must run its commit phase without knowing what it is committing to, so if it can cause the receiver to accept with non-negligible probability, then it must be able to open its commitments in many ways.

4.1 Equivocal Senders

For a pair of algorithms $T = (T_{com}, T_{decom})$, define the following game:

1. $\langle T_{com}, \mathsf{Rec}^k \rangle = (\tau^k, I, \mathsf{state}_{com})$. Here, state_{com} is the internal state of T_{com} to be transmitted to T_{decom}. I is the set Rec^k asks to be opened. Notice T_{com} runs without knowledge of $\underline{b}$, hence T is "equivocal" during the commit phase.
2. $T_{decom}(\underline{b}, \tau^k, I, \mathsf{state}_{com}) = \{(b_i, \mathsf{open}_i)\}_{i \in I}$.

The overall transcript is $(\langle T, \mathsf{Rec}^k \rangle \mid \underline{b}, \mathsf{NoAbort}_T) = (\tau^k, I, \{(b_i, \mathsf{open}_i)\}_{i\in I})$, where $\mathsf{NoAbort}_T$ denotes the event that T does not abort. Say that $(\tau^k, I, \mathsf{state}_{com})$ is δ-*openable* if with probability at least δ over the choice of $\underline{b}$, Rec^k accepts $(\tau^k, I, \{(b_i, \mathsf{open}_i)\}_{i\in I})$, where $\{(b_i, \mathsf{open}_i)\}_{i\in I} = T_{decom}(\underline{b}, \tau^k, I, \mathsf{state}_{com})$..

Definition 8 (Equivocal sender). *We say that* $T = (T_{com}, T_{decom})$ *is a* k-*equivocal sender for* $(\mathsf{Send}, \mathsf{Rec}, \mathsf{Sim}_k)$ *if it holds that*

$$\Pr[(\tau^k, I, \mathsf{state}_{com}) = \langle T_{com}, \mathsf{Rec}^k \rangle \textit{ is } (1 - n^{-\omega(1)})\textit{-openable} \wedge \mathsf{NoAbort}_T] \geq 1/\mathrm{poly}(n)$$

Using equivocal senders to break binding. To prove our impossibility results, we will apply the following theorem, which says that the existence of equivocal senders imply that a commitment is not secure.

Theorem 10. *Fix any non-trivial* $(\mathcal{I}, \underline{\mathcal{B}})$ *and* k*-fold repeated commitment scheme* $(\mathsf{Send}^k, \mathsf{Rec}^k)$ *with a simulator* Sim_k *that proves computational hiding. If this commitment has a* k*-equivocal sender* $T = (T_{com}, T_{decom})$ *for any* $k \leq \mathrm{poly}(n)$*, then this commitment cannot be statistically binding. If furthermore* T *is efficient, then this commitment cannot be computationally binding.*

Proof. The idea is to convert a k-equivocal T sender into a sender Send' that breaks binding in a single execution of the commitment, Send' emulates T internally and chooses one of the k parallel instances to insert its interaction with the real receiver Rec. By the non-triviality of $(\mathcal{I}, \underline{\mathcal{B}})$, with high probability over $I \leftarrow_{\mathrm{R}} \mathcal{I}$ the coordinates in I have significant min-entropy, and in particular some coordinate must have significant min-entropy. Therefore if Send' picks this coordinate, then since T is able to open its commitment with non-trivial probability for $I \leftarrow_{\mathrm{R}} \mathcal{I}$ and $\underline{b} \leftarrow_{\mathrm{R}} \underline{\mathcal{B}}$, it follows that Send' can open its commitment to both 0 and 1 with non-negligible probability.

We now proceed formally by constructing a malicious sender Send' and proving that this sender breaks binding.

Algorithm 11
Malicious sender Send'*, interacting with a single honest receiver* Rec*:*

1. *Pick a random* j*. For each* $j' \neq j$*, sample random coins* $\omega^{(j')}$ *to run an honest receiver.*
2. *Respond to the* i*'th message* β_i *from* Rec *as follows.*
   (a) *If* $i > 1$*, let* $(\alpha^{(1)}_{[i-1]}, \ldots, \alpha^{(k)}_{[i-1]})$ *be* T_{com}*'s response from previous queries.*
   (b) *For* $j' \neq j$*, compute* $\beta^{(j')}_i = \mathsf{Rec}(\alpha^{(j')}_{[i-1]}; \omega^{(j')})$*. Set* $\beta^{(j)}_i = \beta_i$*.*
   (c) *Feed* $(\beta^{(1)}_i, \ldots, \beta^{(k)}_i)$ *to* T_{com} *to obtain response* $(\alpha^{(1)}_{[i]}, \ldots, \alpha^{(k)}_{[i]})$ *(assuming* T_{com} *does not abort).*

   (d) Forward $\alpha_i^{(j)}$ *back to* Rec.
3. *If* T_{com} *does not abort,* Send′ *successfully generates a commit-phase transcript distributed according to* $\langle T_{com}, \mathsf{Rec}^k \rangle$. Send′ *picks a random* $I \leftarrow_R \mathcal{I}$ *to be opened.*
4. *If* $j \notin I$, Send′ *aborts. Otherwise, it independently picks two* $\underline{b}, \underline{b}' \leftarrow_R \mathcal{B}$*, and runs* $T_{decom}(\underline{b}, I)$ *to obtain a decommitment for* $(b_i)_{i \in I}$ *and runs* $T_{decom}(\underline{b}', I)$ *to obtain openings for* $(b_i')_{i \in I}$*. In particular, the malicious sender obtains openings for* b_j *and* b_j'*.*

Analyzing Send′: By hypothesis, T is a $(k, \varepsilon, 1 - n^{-\omega(1)})$-equivocal server for some $\varepsilon = 1/\text{poly}(n)$. This implies that with probability at least ε, $\langle T_{com}, \mathsf{Rec}^k \rangle$ produces an $(1 - n^{-\omega(1)})$-openable $(\tau^k, I, \mathsf{state}_{com})$. Therefore, since the probability of producing an accepting opening for a random $\underline{b}$ at least $(1 - n^{-\omega(1)})$, it holds with probability at least $\varepsilon(1 - n^{-\omega(1)})^2$ that Rec^k accepts both openings $T_{decom}(\underline{b}, \tau^k, I, \mathsf{state}_{com})$ and $T_{decom}(\underline{b}', \tau^k, I, \mathsf{state}_{com})$.

Since $(\mathcal{I}, \mathcal{B})$ is non-trivial, it follows that $\Pr_{\underline{b}, \underline{b}', I}[\forall\, i \in I,\ b_i = b_i'] \leq n^{-\omega(1)}$. Therefore with probability $\varepsilon(1 - n^{-\omega(1)})^2 - n^{-\omega(1)}$, T produces accepting openings for $\underline{b}$ and $\underline{b}'$ and furthermore there exists i such that $\underline{b}_i \neq \underline{b}_i'$. Since the sender picked at random the coordinate j that contains the real interaction, with probability $1/k$ it chooses $j = i$ and therefore with non-negligible probability produces decommitments for both 0 and 1 in an interaction with the real receiver, breaking binding. ■

4.2 Impossibility Results for Parallel Composition

We present the proofs for the case of 3-round PAR-CBCH and 4-round PAR-SB commitments, while the cases in Theorem 3 are deferred to the full version.

We construct equivocal senders using the strategy of Goldreich and Krawczyk [13]. Intuitively, the idea is to construct a sender T whose output distribution is the same as $\mathsf{Sim}_k^{\mathsf{Rec}_h}$. Here, Rec_h is intuitively a cheating receiver that, for each sender message, uses its hash function h to generate a response that looks completely random, and therefore Sim_k gains no advantage by rewinding Rec_h. From this cheating property, we will be able to conclude that T satisfies Definition 8.

Goldreich and Krawczyk [13] observe that we can make the following simplifying assumptions *w.l.o.g.*: **(1)** Sim_k makes exactly $p(n) = \text{poly}(n)$ queries to its receiver black box, **(2)** all queries made by Sim_k are distinct, and **(3)** Sim_k always outputs a transcript τ^k that consists of queries it made to the receiver and the corresponding receiver responses.

The following lemma from [13] says that simply by guessing uniformly at random, one can pick with some noticeable probability the queries/responses that the simulator outputs as its final transcript.

Lemma 1 ([13]). *Fix a black-box simulator* Sim_k *for a protocol with* t *sender messages, and suppose* Sim_k *makes* $p(n)$ *queries. Draw* $u_1, \ldots, u_t \leftarrow_R [p(n)]$*, then with probability* $\geq 1/p(n)^t$*, the final transcript output by* Sim_k *consists of the* $u_1, \ldots, u_t$*'th queries (along with the corresponding receiver responses).*

3-Round Commitments

Theorem 12. *For all non-trivial $(\mathcal{I}, \underline{\mathcal{B}})$ and relative to any oracle, there exists no 3-round PAR-CBCH commitment protocol secure for $(\mathcal{I}, \underline{\mathcal{B}})$.*

Proof. We construct a polynomial-time k-equivocal sender for $(\mathsf{Send}, \mathsf{Rec})$ for $k = n$. By Theorem 10, this contradicts the binding property of the commitment.

Algorithm 13
Equivocal sender $T = (T_{com}, T_{decom})$ for 3-round commitments:

1. *T_{com} picks $u_1, u_2 \leftarrow_R [p(n)]$.*
2. *T_{com} internally runs Sim_k, answering its queries as follows:*
   - *For the u_1, u_2'th queries, if the u_1'th query is a first sender message α_1 and the u_2'th query is a second sender message $\alpha_{[2]}$ that extends α_1, then T_{com} forwards them to the real receiver and forwards the receiver's responses to the simulator. Otherwise, T_{com} aborts.*
   - *For all other queries: if the query is α_1, then T_{com} returns $\mathsf{Rec}^k(\alpha_1; \omega)$ for uniform ω. If the query is $\alpha_{[2]}$ then T returns a random $I \leftarrow_R \mathcal{I}$.*
3. *When Sim_k requests that a subset I of bits be revealed, T_{com} checks to see if I equals the set that the real receiver asked to be opened. If not, T_{com} aborts.*
4. *In the opening phase, T_{decom} receives $\underline{b}$ and feeds $(b_i)_{i \in I}$ to the simulator and obtains $(\tau^k, I, (b_i, \mathsf{open}_i)_{i \in I})$. T_{decom} checks that τ^k and I consists of queries to/from the real receiver, and if not aborts. Otherwise it outputs these openings.*

Analyzing equivocal sender T. It is clear that T runs in polynomial time. Lemma 1 implies that with probability $1/p(n)^2$, Sim_k picks the set to be revealed I using the receiver's responses to the guessed queries u_1, u_2.

Lemma 2. *The probability that Sim_k makes two queries $\alpha_{[2]}, \alpha'_{[2]}$ that are both answered with the same I is negligible*

This claim holds because $|\mathcal{I}| = n^{\omega(1)}$ and Sim_k makes at most $p(n) = \mathrm{poly}(n)$ queries. Lemma 2 implies that when T emulates Sim_k, Sim_k cannot pick I using the real receiver's messages but then find a different commit-phase transcript that leads to the same set I. Therefore the probability that T does not abort and outputs the queries to and responses from the real receiver is at least $1/p(n)^2 - n^{-\omega(1)} \geq 1/\mathrm{poly}(n)$.

Lemma 3. Rec^k *accepts* $(\langle T, \mathsf{Rec}^k \rangle \mid \underline{b}, \mathsf{NoAbort}_T)$ *with overwhelming probability.*

This claim combined with the above assertion that T does not abort with non-negligible probability implies that T satisfies Definition 8.

We now prove Lemma 3 by comparing the output of T to $(\mathsf{Sim}_k^{\mathsf{Rec}_h} \mid \underline{b})$ where Rec_h is defined as follows: h is a $p(n)$-wise independent hash function, it responds to first sender queries α_1 by computing $\beta_1 = \mathsf{Rec}(\alpha_1; h(\alpha_1))$ and to second sender queries $\alpha_{[2]}$ by sampling uniform $I \leftarrow_R \mathcal{I}$ using $h(\alpha_{[2]})$ as random coins.

As observed by [13], $(\langle T, \mathsf{Rec} \rangle \mid \underline{b}, \mathsf{NoAbort}_T) = (\mathsf{Sim}_k^{\mathsf{Rec}_h} \mid \underline{b})$ for a uniform choice of h. Since Rec_h is efficient, by the hiding property this is indistinguishable from $\langle \mathsf{Send}^k(\underline{b}), \mathsf{Rec}_h \rangle$. This in turn is equal to a true interaction $\langle \mathsf{Send}^k(\underline{b}), \mathsf{Rec}^k \rangle$, since by the definition of Rec_h the two receivers Rec_h and Rec^k behave identically when there is no rewinding. Since Rec^k always accepts a real interaction, therefore Rec^k accepts $(\langle T, \mathsf{Rec} \rangle \mid \underline{b}, \mathsf{NoAbort}_T)$ with overwhelming probability. ■

4-Round Commitments

Theorem 14. *For all non-trivial $(\mathcal{I}, \underline{\mathcal{B}})$ and relative to any oracle, there exists no 4-round PAR-SB commitment protocol secure for $(\mathcal{I}, \underline{\mathcal{B}})$.*

Proof (Proof). As before, it suffices to construct a k-equivocal sender for $k = n$.

Algorithm 15
Equivocal sender $T = (T_{com}, T_{decom})$ for 4-round PAR-SB commitments

1. *T_{com} picks $u_1, u_2 \leftarrow_R [p(n)]$.*
2. *T_{com} receives the first message β_1 from the receiver.*
3. *T_{com} internally runs Sim_k, answering its queries as follows:*
   - *For the simulator's u_1, u_2'th queries, if the u_1'th query is a first sender message α_1 and the u_2'th query is a second sender message $\alpha_{[2]}$ that extends α_1, then T_{com} forwards them to the real receiver and forwards the receiver's responses to the simulator. Otherwise, T_{com} aborts.*
   - *For all other queries: if the query is α_1 then T_{com} samples a random $\omega' \leftarrow_R \{\omega \mid \mathsf{Rec}(\perp; \omega) = \beta_1\}$ and returns $\beta_2 = \mathsf{Rec}(\beta_1, \alpha_1; \omega')$ to the simulator. If the query is $\alpha_{[2]}$ then the simulator picks a random $I \leftarrow_R \mathcal{I}$ and returns it to the simulator.*
4. *When Sim_k requests that a subset I of bits be revealed, T_{com} checks to see if I equals the set that the real receiver asked to be opened. If not, T_{com} aborts.*
5. *In the opening phase, T_{decom} receives $\underline{b}$ and feeds $(b_i)_{i \in I}$ to the simulator and obtains $(\tau^k, I, (b_i, \mathsf{open}_i)_{i \in I})$. T_{decom} checks that τ^k and I consists of queries to/from the real receiver, and if not aborts. Otherwise it outputs the openings.*

Analyzing equivocal sender T. T may not run in polynomial time because sampling $\omega' \leftarrow_R \{\omega \mid \beta_1 = \mathsf{Rec}(\perp; \omega)\}$ may be inefficient. This implies the sender breaking binding given by Theorem 10 may be inefficient, which is why we can only handle PAR-SB commitments.

Applying Lemma 1, T does not abort with probability $\geq 1/p(n)^2$. Lemma 2 applies here for the same reason as in the proof of Theorem 12, therefore it holds with probability $1/p(n)^2 - n^{-\omega(1)} \geq 1/\mathrm{poly}(n)$ that T's messages to/from the receiver are exactly those in the output of its emulation of Sim_k.

We claim that Lemma 3 holds in this case as well, which would imply that T satisfies Definition 8. We prove Lemma 3 in this setting by comparing the output of T to $(\mathsf{Sim}_k^{\mathsf{Rec}_h^{\omega_1, \ldots, \omega_s}} \mid \underline{b})$, where we use the cheating receiver strategy $\mathsf{Rec}_h^{\omega_1, \ldots, \omega_s}$ defined by Katz [17]: s will be set below, and the ω_i are random coins for the honest receiver algorithm such that $\mathsf{Rec}(\perp; \omega_i) = \mathsf{Rec}(\perp; \omega_j)$ for all $i, j \in [s]$, and h is a $p(n)$-wise independent hash function with output range $[s]$. The first message of

$\mathsf{Rec}_h^{\omega_1,\ldots,\omega_s}$ is $\beta_1 = \mathsf{Rec}(\perp;\omega_1)$ and given sender message α_1, the second message is $\beta_2 = \mathsf{Rec}(\beta_1,\alpha_1;\omega_{h(\beta_1,\alpha_1)})$. Given sender messages $\alpha_{[2]}$, the set I to be opened is sampled using $\omega_{h(\beta_{[2]},\alpha_{[2]})}$ as random coins.

As observed in [17], for $s = 50p(n)^2/\delta$ it holds that the statistical distance between $(\langle T,\mathsf{Rec}^k\rangle \mid \underline{b},\mathsf{NoAbort}_T)$ and $(\mathsf{Sim}_k^{\mathsf{Rec}_h^{\omega_1,\ldots,\omega_s}} \mid \underline{b})$ is at most δ, where the randomness is over uniform $p(n)$-wise independent h, uniform ω_1 and uniform $\omega_2,\ldots,\omega_s$ conditioned on $\mathsf{Rec}(\perp;\omega_j) = \mathsf{Rec}(\perp;\omega_1)$ for all $j \in [s]$. By the commitment's hiding property this is indistinguishable from $\langle \mathsf{Send}^k(\underline{b}),\mathsf{Rec}_h^{\omega_1,\ldots,\omega_s}\rangle$, which in turn is equal to $\langle \mathsf{Send}^k(\underline{b}),\mathsf{Rec}^k\rangle$ by the definition of $\mathsf{Rec}_h^{\omega_1,\ldots,\omega_s}$. Finally, since Rec^k always accepts a real interaction, therefore it accepts $(\langle T,\mathsf{Rec}^k\rangle \mid \underline{b},\mathsf{NoAbort}_T)$ with probability $1-\delta-n^{-\omega(1)}$.

We can apply the above argument for any $\delta \geq 1/\mathrm{poly}(n)$ to conclude that Rec^k accepts $(\langle T,\mathsf{Rec}^k\rangle \mid \underline{b},\mathsf{NoAbort}_T)$ with probability $1-\delta-n^{-\omega(1)}$ for all $\delta \geq 1/\mathrm{poly}(n)$.

Therefore Rec^k must accept $(\langle T,\mathsf{Rec}^k\rangle \mid \underline{b},\mathsf{NoAbort}_T)$ with probability $1-n^{-\omega(1)}$ and so T satisfies Definition 8. ■

4.3 PAR-SB Commitments Imply (Stand-Alone) SH Commitments

To prove Item 2 of Theorem 2, we show that PAR-SB commitments can be used to generate a gap between real and accessible entropy [16]. Then we apply the transformation of [16] that converts an entropy gap into a statistically hiding commitment.

To simplify the statement of our result, we assume that $\mathcal{I} = 2^{[k]}$ and $\underline{\mathcal{B}} = U_k$; see the full version for the general treatment. We also defer the definitions of real min-entropy and accessible max-entropy and the formal proof of the main technical lemma (Lemma 4) to the full version.

Theorem 16. *For $(\mathcal{I} = 2^{[k]}, \underline{\mathcal{B}} = U_k)$, if there exists $O(1)$-round* (Send, Rec) *that is PAR-SB secure for $(\mathcal{I},\underline{\mathcal{B}})$, then there exists $O(1)$-round statistically hiding commitments.*

Proof. Assume without loss of generality that Rec^k sends all his random coins at the end of the opening phase, and that Rec uses m random coins in a single stand-alone instance.

Lemma 4. Rec^k *has real min-entropy at least $km(1-1/k^{1/3})$ and has context-independent accessible max-entropy $\leq km - k/4$.*

Lemma 4 implies there is an entropy gap, and so the theorem follows by combining it with the black-box construction of statistically hiding commitments from entropy gaps given by Lemmas 6.7, 4.7, and 4.18 of [16]. ■

Proof (Proof of Lemma 4.). The real min-entropy part of the claim follows from the definitions and amplification by parallel repetition (Proposition 3.8 in [16]). For the accessible entropy part, we use the following:

Claim. If there exists efficient A^* sampling context-independent max-entropy $> km - k/4$ for Rec^k, then there exists a k-equivocal sender.

By Theorem 10 this contradicts the binding property of the commitment, and so A^* cannot exist. The proof follows from ideas of [16] and we defer a formal proof to the full version. ∎

4.4 Impossibility Results for Concurrent Composition

Again, we state our theorem for the natural case $\mathcal{I} = 2^{[k]}$ and $\underline{\mathcal{B}} = U_k$, and defer the general statement to the full version.

Theorem 17. *For* $(\mathcal{I} = 2^{[k]}, \underline{\mathcal{B}} = U_k)$*, and relative to any oracle, no* $o(\log n/\log\log n)$*-round commitment is CC-CBCH secure for* $\mathcal{I}, \underline{\mathcal{B}}$*.*

Proof. **Building the equivocal sender:** The message schedule we use is exactly that of [7], which we call Σ, and is defined in the full version. The high-level idea of T, also adapted from [7], is to execute Sim_k and to insert the real receiver's messages into one session j chosen at random, and where T aborts if the simulator tries to rewind queries in session j. Messages for other sessions are computed using the partial transcripts generated by the simulator so far. We defer a more explicit description to the full version.

Analyzing equivocal sender T: [7] prove the following lemma:

Lemma 5 ([7], informal). *It holds with non-negligible probability that there exists a "good session" in the execution of* $\mathsf{Sim}_k^{\mathsf{Rec}^k_\Sigma}$, i.e. *a session where* Sim_k *does not rewind* Rec^k_Σ.

The only place where T may abort is if in its emulation of Sim_k, the simulator tries to rewind the receiver in session j. Therefore, with probability $1/k$, T inserts the real receiver into the good session that is guaranteed to exist by Lemma 5 with non-negligible probability. Furthermore, since the k concurrent simulation is indistinguishable from a real interaction, it follows that Rec^k_Σ accepts $(\langle T, \mathsf{Rec}^k_\Sigma\rangle \mid b, \mathsf{NoAbort}_T)$ with overwhelming probability. ∎

Acknowledgements

The author would like to thank Dennis Hofheinz and Salil Vadhan for helpful conversations.

References

[1] Barak, B.: How to go beyond the black-box simulation barrier. In: Proc. 42nd FOCS, pp. 106–115. IEEE, Los Alamitos (2001)
[2] Beaver, D.: Adaptive zero knowledge and computational equivocation (extended abstract). In: Proc. STOC 1996, pp. 629–638. ACM, New York (1996)

[3] Bellare, M., Hofheinz, D., Yilek, S.: Possibility and impossibility results for encryption and commitment secure under selective opening. In: Joux, A. (ed.) EUROCRYPT 2009. LNCS, vol. 5479, pp. 1–35. Springer, Heidelberg (2009)

[4] Brassard, G., Crépeau, C.: Zero-knowledge simulation of boolean circuits. In: Odlyzko, A.M. (ed.) CRYPTO 1986. LNCS, vol. 263, pp. 223–233. Springer, Heidelberg (1987)

[5] Brassard, G., Chaum, D., Crépeau, C.: Minimum disclosure proofs of knowledge. J. Comput. Syst. Sci. 37(2), 156–189 (1988)

[6] Brassard, G., Crépeau, C., Yung, M.: Everything in NP can be argued in *perfect* zero-knowledge in a *bounded* number of rounds. In: Quisquater, J.-J., Vandewalle, J. (eds.) EUROCRYPT 1989. LNCS, vol. 434, pp. 192–195. Springer, Heidelberg (1990)

[7] Canetti, R., Kilian, J., Petrank, E., Rosen, A.: Black-box concurrent zero-knowledge requires (almost) logarithmically many rounds. SIAM J. Comput. 32(1), 1–47 (2003)

[8] Di Crescenzo, G., Ostrovsky, R.: On concurrent zero-knowledge with pre-processing (Extended abstract). In: Wiener, M. (ed.) CRYPTO 1999. LNCS, vol. 1666, pp. 485–502. Springer, Heidelberg (1999)

[9] Di Crescenzo, G., Ishai, Y., Ostrovsky, R.: Non-interactive and non-malleable commitment. In: Proc. STOC 1998, pp. 141–150. ACM, New York (1998)

[10] Dwork, C., Naor, M., Reingold, O., Stockmeyer, L.: Magic functions. J. ACM 50(6), 852–921 (2003)

[11] Fischlin, M.: Trapdoor Commitment Schemes and Their Applications. Ph.D. Thesis (Doktorarbeit), Department of Mathematics, Goethe-University, Frankfurt, Germany (2001)

[12] Goldreich, O., Kahan, A.: How to construct constant-round zero-knowledge proof systems for NP. Journal of Cryptology 9(3), 167–189 (1996)

[13] Goldreich, O., Krawczyk, H.: On the composition of zero-knowledge proof systems. SIAM J. of Com. 25(1), 169–192 (1996); Preliminary version appeared In: Paterson, M. (ed.) ICALP 1990. LNCS, vol. 443. Springer, Heidelberg (1990)

[14] Goldreich, O., Micali, S., Wigderson, A.: Proofs that yield nothing but their validity or all languages in NP have zero-knowledge proof systems. Journal of the ACM 38(3), 691–729 (1991); Preliminary version in FOCS 1986

[15] Haitner, I., Hoch, J.J., Reingold, O., Segev, G.: Finding collisions in interactive protocols - a tight lower bound on the round complexity of statistically-hiding commitments. In: Proc. FOCS 2007, pp. 669–679 (2007)

[16] Haitner, I., Reingold, O., Vadhan, S., Wee, H.: Inaccessible entropy. In: Proc. STOC 2009, pp. 611–620. ACM, New York (2009)

[17] Katz, J.: Which languages have 4-round zero-knowledge proofs? In: Canetti, R. (ed.) TCC 2008. LNCS, vol. 4948, pp. 73–88. Springer, Heidelberg (2008)

[18] Naor, M.: Bit commitment using pseudorandomness. Journal of Cryptology 4(2), 151–158 (1991); Preliminary version In: Brassard, G. (ed.) CRYPTO 1989. LNCS, vol. 435, pp. 128–136. Springer, Heidelberg (1990)

[19] Pass, R., Wee, H.: Black-box constructions of two-party protocols from one-way functions. In: Reingold, O. (ed.) TCC 2009. LNCS, vol. 5444, pp. 403–418. Springer, Heidelberg (2009)

[20] Pass, R., Tseng, W.-L.D., Wikström, D.: On the composition of public-coin zero-knowledge protocols. In: Halevi, S. (ed.) CRYPTO 2009. LNCS, vol. 5677, pp. 160–176. Springer, Heidelberg (2009)

[21] Prabhakaran, M., Rosen, A., Sahai, A.: Concurrent zero knowledge with logarithmic round-complexity. In: Proc. 43rd FOCS, pp. 366–375. IEEE, Los Alamitos (2002)

[22] Rosen, A.: Concurrent Zero-Knowledge - With Additional Background by Oded Goldreich. Information Security and Cryptography. Springer, Heidelberg (2006)
[23] Wee, H.: On statistically binding trapdoor commitments from one-way functions (2008) (manuscript)
[24] Zhang, Z., Cao, Z., Zhu, H.: Constant-round adaptive zero knowledge proofs for NP (2009) (manuscript)

Limits on the Power of Zero-Knowledge Proofs in Cryptographic Constructions

Zvika Brakerski[1], Jonathan Katz[2], Gil Segev[3], and Arkady Yerukhimovich[2]

[1] Weizmann Institute of Science, Rehovot, Israel
zvika.brakerski@weizmann.ac.il
[2] University of Maryland, College Park, MD, USA
{jkatz,arkady}@cs.umd.edu
[3] Microsoft Research, Mountain View, CA, USA
gil.segev@microsoft.com

Abstract. For over 20 years, black-box impossibility results have been used to argue the infeasibility of constructing certain cryptographic primitives (e.g., key agreement) from others (e.g., one-way functions). A widely recognized limitation of such impossibility results, however, is that they say nothing about the usefulness of (known) *non*black-box techniques. This is unsatisfying, as we would at least like to rule out constructions using the set of techniques we have at our disposal.

With this motivation in mind, we suggest a new framework for black-box constructions that encompasses constructions with a nonblack-box flavor: specifically, those that rely on *zero-knowledge proofs* relative to some oracle. We show that our framework is powerful enough to capture the Naor-Yung/Sahai paradigm for building a (shielding) CCA-secure public-key encryption scheme from a CPA-secure one, something ruled out by prior black-box separation results. On the other hand, we show that several black-box impossibility results still hold even in a setting that allows for zero-knowledge proofs.

1 Introduction

A central goal of theoretical cryptography is to explore relationships between various cryptographic primitives and, in particular, to show constructions of various "high-level" cryptographic objects (encryption schemes, key-agreement protocols, etc). based on "low-level" cryptographic tools (such as one-way functions). This line of research has been very successful in many cases. In other cases, however, constructions of certain primitives from others are unknown: for example, we do not currently know how to construct public-key encryption schemes based on one-way functions. Given this failure, it is natural to wonder whether such constructions are inherently *impossible*. Unfortunately, we cannot rule out *all* such constructions as long as we believe that the object in question exists in the real world: if we believe that RSA encryption (say) is secure, then a valid construction of public-key encryption from any one-way function f consists of simply ignoring f and then outputting the code for the RSA encryption scheme. Yet this is clearly not what is intended.

Y. Ishai (Ed.): TCC 2011, LNCS 6597, pp. 559–578, 2011.

In an effort to capture what is meant by a "natural" construction of one primitive from another, Impagliazzo and Rudich [15] formalized the notion of a *black-box* construction. Informally, a black-box construction of A from B is a construction of A that uses only the input/output characteristics of an implementation of B, but does not rely on any internal details as to how B is implemented. Moreover, A should be "secure" as long as B is "secure" (each in their respective senses). (We refer the reader to the work of Reingold, Trevisan, and Vadhan [19] for a more extensive definitional treatment). Impagliazzo and Rudich show that there does not exist a black-box construction of key agreement from one-way functions, and since their work many more black-box impossibility results have been shown.

A recognized drawback of existing black-box impossibility results is that they say nothing regarding whether these results might be circumvented using *non*black-box techniques. While it is true that most constructions in cryptography are black-box, we have many examples of nonblack-box constructions as well. One striking example is given by the observation that all known constructions of CCA-secure public-key encryption schemes based on trapdoor permutations [18,5,20,16] are, in fact, not black-box. (Interestingly, a partial black-box separation is known [11]). Other nonblack-box constructions include those of [6,4,3,1,9]; we refer the reader to [10] for further discussion and additional examples.

If black-box constructions are supposed to be representative of existing techniques, we should update our definition of what "black-box" means. In this paper, we propose a framework to do exactly this. Specifically, we suggest a model that incorporates a specific class of nonblack-box techniques: those that rely on zero-knowledge proofs. We accomplish this by augmenting the basic, black-box model — in which there is only an oracle $\mathcal{O}$ for some primitive — with a *zero-knowledge (ZK) oracle* that allows parties to prove statements in zero knowledge relative to $\mathcal{O}$. (Technically, a ZK oracle allows zero-knowledge proofs for any language in $\mathcal{NP}^{\mathcal{O}}$. We also find it simpler to work with a *witness-indistinguishability (WI) oracle*, but we show that a WI oracle implies zero-knowledge proofs in the settings we consider. In fact, although we do not define the notion of proofs of *knowledge*, our formulation can also be seen as providing that stronger property). We call any construction using black-box access to $\mathcal{O}$ *and its associated WI oracle* an **augmented black-box** construction. Given primitives A and B, we can then ask whether there exists an *augmented* black-box construction of A from B; an impossibility result demonstrating that no such construction exists rules out a broader class of approaches to constructing one from the other. Of course, as with all impossibility results, such a result says nothing about whether some *other* nonblack-box techniques might apply (and, in fact, the nonblack-box results of, e.g., [3,1] do not fall within our framework); nevertheless, impossibility results are still useful insofar as they show us where we must look if we hope to circumvent them.

Our contributions. We view the primary contribution of this paper as definitional and conceptual. In addition to putting forth the notion of augmented black-box constructions, however, we also show several technical results.

To validate our framework, we show that the Naor-Yung/Sahai [18,20] (shielding) construction of CCA-secure public-key encryption from CPA-secure public-key encryption falls within our framework. (Such a construction is ruled out, in a black-box sense, by the result of Gertner et al. [11]). We note that several other existing nonblack-box constructions also fall within our framework, including those of [6,4,9]. This demonstrates that our framework meaningfully encompasses constructions that lie outside the standard black-box model.

On the negative side, we present two impossibility results for augmented black-box constructions. Generalizing the work of Impagliazzo and Rudich [15], we rule out augmented (fully) black-box constructions of key-agreement protocols with perfect completeness from one-way functions. (We leave the case of protocols without perfect completeness as an open problem). Generalizing results of Haitner et al. [12,13], we rule out augmented (fully) black-box constructions of statistically-hiding commitment schemes with low round complexity or low communication complexity from enhanced trapdoor permutations. Though it may seem "intuitively obvious" to the reader that zero-knowledge proofs cannot help in these settings, the challenge — as in all black-box impossibility proofs — is to prove this intuition. (In fact, under our initial modeling of a random WI proof system there *was* a construction of key agreement from one-way functions).

Outline of the paper. In Section 2 we define the notion of augmented black-box constructions, and in Section 3 we show that our framework encompasses the Naor-Yung/Sahai paradigm for building CCA-secure public-key encryption from CPA-secure schemes. Our main technical results are in the sections that follow. We rule out augmented black-box constructions of key agreement from one-way functions in Section 4, and in Section 5 we prove lower bounds on the round complexity and communication complexity of augmented black-box constructions of statistically-hiding commitments from trapdoor permutations.

2 Augmented Black-Box Constructions

In this section we formally define our notion of *augmented black-box constructions*. Recall that our goal here is to model constructions that use an oracle $\mathcal{O}$ for some primitive as a black box, while also (possibly) using zero-knowledge proofs of $\mathcal{NP}$ statements relative to $\mathcal{O}$. To enable such proofs we introduce an additional set of oracles $(\mathcal{P}, \mathcal{V})$ implementing a "prover" and a "verifier", respectively. We find it easiest to model $(\mathcal{P}, \mathcal{V})$ as a *witness-indistinguishable* (WI) proof system [7], and to prove our impossibility results relative to oracles achieving this notion. In Section 2.2, however, we show that any WI proof system can be used to construct non-interactive zero-knowledge (NIZK) proofs in the common random string model, assuming the existence of one-way functions.

Fix an oracle $\mathcal{O} : \{0,1\}^* \to \{0,1\}^*$. For a language L, we say $L \in \mathcal{NP}^{\mathcal{O}}$ if there exists a polynomial-time oracle machine M running in time polynomial in its first input such that $x \in L$ if and only if there exists a witness w for which $M^{\mathcal{O}}(x, w)$ accepts. (We assume a valid witness w satisfies $|w| = |x|$ without loss of generality). For any $L \in \mathcal{NP}^{\mathcal{O}}$, we let R_L denote an $\mathcal{NP}$-relation associated with L, and we let $L_n \stackrel{\text{def}}{=} L \cap \{0,1\}^n$ and $R_n \stackrel{\text{def}}{=} \{(x,w) \mid (x,w) \in R_L \text{ and } x \in L_n\}$.

We now define what it means for a pair of oracles $(\mathcal{P}, \mathcal{V})$ to be a witness-indistinguishable proof system. (All adversaries are stateful by default).

Definition 1. *Fix an oracle $\mathcal{O}$, a language $L \in \mathcal{NP}^{\mathcal{O}}$, and an $\mathcal{NP}$ relation R_L for L. An oracle $\mathcal{WI} = (\mathcal{P}, \mathcal{V})$ is a* proof system *for R_L if the following hold:*

- **Perfect completeness:** *For any $n \in \mathbb{N}$, $(x,w) \in R_n$, and $r \in \{0,1\}^n$, it holds that $\mathcal{V}_n(x, \mathcal{P}_n(x,w,r)) = 1$.*
- **Perfect soundness:** *For any $x \notin L$ and any π, it holds that $\mathcal{V}_n(x,\pi) = 0$.*

$\mathcal{WI}$ is witness-indistinguishable (WI) *if additionally:*

- **Witness indistinguishability:** *For every probabilistic polynomial-time $\mathcal{A}$, it holds that $|\Pr[\mathsf{ExptWI}_{\mathcal{A}}(n) = 1] - 1/2|$ is negligible, where $\mathsf{ExptWI}_{\mathcal{A}}(n)$ is defined as follows:*

$$\left| \begin{array}{rcl} (x, w_0, w_1) \leftarrow \mathcal{A}^{\mathcal{O},\mathcal{WI}}(1^n); b \leftarrow \{0,1\}; & & \textit{if } (x,w_0),(x,w_1) \in R_n \\ r \leftarrow \{0,1\}^n; \pi \leftarrow \mathcal{P}_n(x, w_b, r) & : & \textit{output 1 iff } b' = b \\ b' = \mathcal{A}^{\mathcal{O},\mathcal{WI}}(1^n, \pi) & & \textit{else, output a random bit} \end{array} \right|$$

When the relation R_L is irrelevant for the discussion at hand, or is clear from the context, we may abuse terminology and call $\mathcal{WI}$ a WI proof system for L. We say that $\mathcal{WI}$ is a WI proof system for $\mathcal{NP}^{\mathcal{O}}$ if it is a WI proof system for the $\mathcal{NP}^{\mathcal{O}}$-complete language CIRCUIT-SAT$^{\mathcal{O}}$ (the set of satisfiable circuits with $\mathcal{O}$ gates) under the natural relation R_L.

We now define our notion of black-box reductions using a base oracle $\mathcal{O}$ and a WI oracle $\mathcal{WI}$ for $\mathcal{NP}^{\mathcal{O}}$. The definitions and terminology are adapted from [19].

Definition 2 (Augmented fully black-box construction). *There is an* augmented fully black-box construction *of primitive Q from primitive P if there exist probabilistic polynomial-time oracle machines G and S such that:*

- *For any $\mathcal{O}, \mathcal{WI}$ such that $\mathcal{O}$ implements P, and $\mathcal{WI}$ is a proof system for $\mathcal{NP}^{\mathcal{O}}$, the algorithm $G^{\mathcal{O},\mathcal{WI}}$ implements Q.*
- *For any $\mathcal{O}, \mathcal{WI}$ and any (possibly inefficient) adversary $\mathcal{A}^{\mathcal{O},\mathcal{WI}}$ that breaks the Q-security of $G^{\mathcal{O},\mathcal{WI}}$, the adversary $S^{\mathcal{A},\mathcal{O},\mathcal{WI}}$ breaks the P-security of $\mathcal{O}$ or the witness indistinguishability of $\mathcal{WI}$.*

Definition 3 (Augmented semi-black-box construction). *There is an* augmented semi-black-box construction *of primitive Q from primitive P if there exists a probabilistic polynomial-time oracle machine G such that:*

- *For any $\mathcal{O}, \mathcal{WI}$ such that $\mathcal{O}$ implements P, and $\mathcal{WI}$ is a proof system for $\mathcal{NP}^{\mathcal{O}}$, the algorithm $G^{\mathcal{O},\mathcal{WI}}$ implements Q.*
- *For any $\mathcal{O}, \mathcal{WI}$ and any probabilistic polynomial-time adversary $\mathcal{A}^{\mathcal{O},\mathcal{WI}}$ that breaks the Q-security of $G^{\mathcal{O},\mathcal{WI}}$, there is a probabilistic polynomial-time S such that $S^{\mathcal{O},\mathcal{WI}}$ breaks the P-security of $\mathcal{O}$ or the witness indistinguishability of $\mathcal{WI}$.*

We remark that our notions of augmented black-box constructions are not transitive: i.e., if there is an augmented black-box construction of Q from P, and an augmented black-box construction of R from Q, this does not imply that there is an augmented black-box construction of R from P. (On the other hand, if either of the given constructions is black-box, that does imply an augmented black-box construction of R from P). The reason is that $\mathcal{WI}$ enables proofs for $\mathcal{NP}^{\mathcal{O}}$ but not $\mathcal{NP}^{\mathcal{O},\mathcal{WI}}$. While it is true that Definition 1 can be meaningfully changed to allow for proofs of $\mathcal{NP}^{\mathcal{O},\mathcal{WI}}$, doing so introduces technical issues (due to circularity) and we were unable to prove our separation results with respect to such a definition. We leave this as an interesting open question.

2.1 Instantiating a WI Proof System

For arbitrary $\mathcal{O}$, we show how to instantiate a WI proof system for $\mathcal{NP}^{\mathcal{O}}$. We begin by describing a distribution such that an oracle sampled according to this distribution is a WI proof system for $\mathcal{NP}^{\mathcal{O}}$ with overwhelming probability (Lemma 2). We then show that this implies that measure 1 of the oracles under this distribution constitute a WI proof system for $\mathcal{NP}^{\mathcal{O}}$ (Lemma 3). Throughout this section, we take L to be CIRCUIT-SAT$^{\mathcal{O}}$.

It is convenient to view the (infinite) oracle $\mathcal{WI}$ as a sequence of oracles $\{\mathcal{WI}_n = (\mathcal{P}_n, \mathcal{V}_n)\}_{n\in\mathbb{N}}$, one for each input length. Consider the distribution over $\mathcal{WI}$ where, for each n, the distribution over $\mathcal{WI}_n$ is defined as follows:

Prover oracle: $\mathcal{P}_n$ is a random function $\mathcal{P}_n : \{0,1\}^{3n} \to \{0,1\}^{7n}$ whose inputs are parsed as tuples of the form $(x, w, r) \in \{0,1\}^n \times \{0,1\}^n \times \{0,1\}^n$. Note that $\mathcal{P}_n$ is defined for all such tuples (x, w, r) of the appropriate length, and not only for those satisfying $(x, w) \in R_L$ (i.e., $\mathcal{P}_n$ does not check whether $(x, w) \in R_L$).

Verifier oracle: The verifier oracle is a function $\mathcal{V}_n : \{0,1\}^{8n} \to \{0,1\}$, whose inputs are parsed as pairs of the form $(x, \pi) \in \{0,1\}^n \times \{0,1\}^{7n}$. The function is defined as:

$$\mathcal{V}_n(x, \pi) = \begin{cases} 1 \text{ if } \exists(w, r) \text{ s.t. } \pi = \mathcal{P}_n(x, w, r) \ \wedge \ (x, w) \in R_L \\ 0 \text{ otherwise} \end{cases}$$

Note that $\mathcal{WI}$ sampled as above is always a proof system. It remains to show that witness indistinguishability holds with overwhelming probability. We begin by proving that, for oracles distributed as above, it is essentially impossible to "spoof" a proof. That is, for n large enough, the only way to generate a proof π such that $\mathcal{V}_n(x, \pi) = 1$ is by querying $\mathcal{P}_n$. This property of the $\mathcal{WI}$ oracle will also be useful later.

Lemma 1. *For an oracle algorithm $\mathcal{A}^{\mathcal{O},\mathcal{WI}}$, let Spoof_n be the event that $\mathcal{A}$ makes a query $\mathcal{V}_n(x, \pi)$ that returns 1, yet π was not output by a previous query $\mathcal{P}_n(x, w, \star)$ with $(x, w) \in R_L$. For any $\mathcal{O}$, any $\mathcal{A}$ making at most q oracle queries, and any n, the probability of Spoof_n is at most $q \cdot 2^{-4n}$ (where the probability is taken over choice of $\mathcal{WI}$ according to the distribution above).*

Proof (sketch). We drop the subscript n for ease of presentation. There are at most 2^{3n} elements in the range of $\mathcal{P}$ and these are distributed uniformly in a space of size 2^{7n}. Since $\mathcal{P}$ is chosen independently of $\mathcal{O}$, queries to $\mathcal{O}$ give no information about the range of $\mathcal{P}$. Each $\mathcal{P}$-query reveals one point in the range of $\mathcal{P}$, but as the other points in the range are chosen independently this does not help to find another element in the range. The probability that any particular query $\mathcal{V}(x, \pi)$ returns 1 if π was not previously output by $\mathcal{P}$ is at most 2^{-4n}. Taking a union bound over all the queries of $\mathcal{A}$ gives the desired result.

Lemma 2. *For any oracle $\mathcal{O}$, every probabilistic polynomial-time oracle machine $\mathcal{A}$, and n large enough:*

$$\left|\Pr\left[\mathsf{ExptWI}_{\mathcal{A}}(n) = 1\right] - \frac{1}{2}\right| \leq 2^{-n/2},$$

where $\mathsf{ExptWI}_{\mathcal{A}}(n)$ is as in Definition 1, and the above probability is also taken over the choice of $\mathcal{WI}$.

Proof. Consider some value of n, and fix the values of $\mathcal{WI}$ other than $\mathcal{WI}_n$. Assume without loss of generality that $\mathcal{A}(1^n)$ outputs values (x, w_0, w_1) with $(x, w_0), (x, w_1) \in R_n$. Then $\mathcal{A}$ is given a proof π and has to identify whether w_0 or w_1 was used to generate it. We observe that for all $k \neq n$ the output of any query to $\mathcal{P}_k$ or $\mathcal{V}_k$ is independent of the bit b. Therefore from this point on we focus on queries to $\mathcal{P}_n$ and $\mathcal{V}_n$. Let q be the total number of oracle queries made by $\mathcal{A}$. We may assume that $\mathcal{A}$ does not query $\mathcal{V}_n$ since it can simulate this oracle by itself to within statistical difference at most 2^{-n} (for n large enough). Indeed, there are three types of queries to $\mathcal{V}_n$:

- The query $\mathcal{V}_n(x, \pi)$. In this case, the output is 1.
- Queries of the form $\mathcal{V}_n(x, \pi')$, where π' was output by a previous query $\mathcal{P}_n(x, w, \star)$ with $(x, w) \in R_n$. Once again, in this case the output is 1. Note that $\mathcal{A}$ can check in polynomial time whether $(x, w) \in R_n$.
- All other queries to $\mathcal{V}_n$. In this case, Lemma 1 shows that the output of all these queries is 0 except with probability at most $q \cdot 2^{-4n}$, which is bounded by 2^{-n} for n sufficiently large.

Given the above, and the fact that $\mathcal{P}_n$ is chosen at random, it follows that $\mathcal{A}$ cannot distinguish which witness was used with probability better than $q \cdot 2^{-n}$, which is bounded by $2^{-n/2}$ for n sufficiently large. The lemma follows.

Lemma 3. *Fix an oracle $\mathcal{O}$. For measure 1 of the oracles $\mathcal{WI}$ under the distribution defined above, $\mathcal{WI}$ is a witness-indistinguishable proof system for L.*

Proof. Completeness and soundness always hold, and so we must only prove witness indistinguishability. To do so we apply a standard argument using the Borel-Cantelli lemma for reversing the order of quantifiers in Lemma 2.

Fix $\mathcal{O}$. For any $n \in \mathbb{N}$ and any probabilistic polynomial-time $\mathcal{A}$, denote by $E_{n,\mathcal{A}}$ the event in which $\mathcal{WI}$ is chosen such that

$$\left|\Pr\left[\mathsf{ExptWI}_{\mathcal{A}}(n) = 1\right] - \frac{1}{2}\right| > 2^{-n/3}.$$

Lemma 2 and an averaging argument imply that for any $\mathcal{A}$ and sufficiently large n the probability of $E_{n,\mathcal{A}}$ is at most $1/n^2$. Then $\sum_n \Pr[E_{n,\mathcal{A}}]$ is finite, and so the Borel-Cantelli lemma implies that the probability over choice of $\mathcal{WI}$ that event $E_{n,\mathcal{A}}$ occurs for infinitely many values of n is zero. Thus, for large enough n and measure 1 of the oracles under the distribution in question we have

$$\left|\Pr\left[\mathsf{ExptWI}_{\mathcal{A}}(n) = 1\right] - \frac{1}{2}\right| \leq 2^{-n/3}.$$

This holds for any specific $\mathcal{A}$, and therefore by removing a set of measure 0 for each of the (countably many) machines $\mathcal{A}$ we obtain that for measure 1 of the oracles $\mathcal{WI}$ it holds that for *all* probabilistic polynomial-time $\mathcal{A}$ the quantity $\left|\Pr\left[\mathsf{ExptWI}_{\mathcal{A}}(n) = 1\right] - \frac{1}{2}\right|$ is negligible.

Before concluding this section we prove a technical result regarding oracles $\mathcal{WI}$ sampled according to the distribution described earlier. We show that if f is one-way relative to $\mathcal{O}$, then for measure 1 of the oracles $\mathcal{WI}$ under the distribution defined above f remains one-way relative to $(\mathcal{O}, \mathcal{WI})$.

Lemma 4. *Let f be a polynomial-time oracle machine such that $f^{\mathcal{O}}$ is one-way relative to $\mathcal{O}$. Then for measure 1 of the oracles $\mathcal{WI}$ under the distribution defined above, $f^{\mathcal{O}}$ is one-way relative to $(\mathcal{O}, \mathcal{WI})$.*

Proof. It suffices to show that for any PPT $\mathcal{A}$ the probability that $\mathcal{A}^{\mathcal{O},\mathcal{WI}}$ inverts $f^{\mathcal{O}}$ is negligible, where the probability is also taken over choice of $\mathcal{WI}$. We can then proceed as in Lemma 3 to obtain the stated result.

Assume toward a contradiction that there exists an algorithm $\mathcal{A}$ and a polynomial $p(n) \geq n$ such that the running time of $\mathcal{A}$ is bounded by $p(n)$ and, for infinitely many n, it holds that $A^{\mathcal{O},\mathcal{WI}}$ inverts $f^{\mathcal{O}}$ with probability at least $1/p(n)$ when $\mathcal{WI}$ is chosen at random. We show how to construct a PPT algorithm $\hat{\mathcal{A}}$ such that $\hat{\mathcal{A}}^{\mathcal{O}}$ inverts $f^{\mathcal{O}}$ with inverse-polynomial probability for infinitely many values of n, a contradiction.

$\hat{\mathcal{A}}(1^n, y)$ runs $\mathcal{A}(1^n, y)$, simulating the $\mathcal{WI}$ oracle for $\mathcal{A}$ as follows. Let $k^* = \log p(n)$. Algorithm $\hat{\mathcal{A}}$ samples $\mathcal{WI}_k = (\mathcal{P}_k, \mathcal{V}_k)$ according to the prescribed distribution for all $k \leq k^*$, and these are used to (perfectly) simulate $\{\mathcal{WI}_k\}_{k \leq k^*}$ to $\mathcal{A}$. Thus, we now only need to deal with the queries of $\mathcal{A}$ to $\mathcal{WI}_k$ for $k > k^*$. When $\mathcal{A}$ queries $\mathcal{P}_k(x, w, r)$, then $\hat{\mathcal{A}}$ returns a random $\pi \in \{0,1\}^{7k}$ as the result. When $\mathcal{A}$ queries $\mathcal{V}_k(x, \pi)$ then $\hat{\mathcal{A}}$ first checks to see whether there was any prior query $\mathcal{P}_k(x, w, \star) = \pi$ with $(x, w) \in R_L$. If not, then $\hat{\mathcal{A}}$ returns 0 in response to this $\mathcal{V}_k$-query. Otherwise, $\hat{\mathcal{A}}$ returns 1.

It follows from Lemma 1 that $\hat{\mathcal{A}}$'s simulation of $\mathcal{A}$ degrades the latter's probability of inversion by at most $1/2p(n)$. This implies that $\hat{\mathcal{A}}^{\mathcal{O}}$ inverts $f^{\mathcal{O}}$ with probability at least $1/2p(n)$ for infinitely many values of n, a contradiction.

2.2 Zero-Knowledge Proofs

We define a notion of zero knowledge, and then discuss appropriate conditions under which zero-knowledge (ZK) proofs can be constructed from WI proofs.

In our context, zero knowledge is most easily expressed in terms of non-interactive zero knowledge in the common random string model.

Definition 4. *Fix an oracle $\mathcal{O}$ and a language $L \in \mathcal{NP}^{\mathcal{O}}$. An oracle $\mathcal{ZK} = (\mathcal{P}, \mathcal{V})$ is a* proof system in the common random string model *for L with relation R_L if there is a polynomial ℓ such that the following hold:*

- **Perfect completeness:** *For all $n \in \mathbb{N}$, all $(x, w) \in R_n$, all $\mathsf{crs} \in \{0,1\}^{\ell(n)}$, and all $r \in \{0,1\}^n$, we have $\mathcal{V}(\mathsf{crs}, x, \mathcal{P}(\mathsf{crs}, x, w, r)) = 1$.*
- **Statistical soundness:** *With all but negligible probability over choice of $\mathsf{crs} \in \{0,1\}^{\ell(n)}$, there do not exist $x \notin L_n$ and π such that $\mathcal{V}(\mathsf{crs}, x, \pi) = 1$.*

$\mathcal{ZK}$ is a non-interactive zero-knowledge (NIZK) proof system *if additionally:*

- **Black-box (adaptive) zero knowledge:** *There exists a* PPT simulator *$\mathcal{S} \stackrel{\text{def}}{=} (\mathcal{S}_1, \mathcal{S}_2)$ such that for all probabilistic polynomial-time $\mathcal{A}$ the following is negligible:*

$$\left| \Pr\left[\begin{array}{c} \mathsf{crs} \leftarrow \{0,1\}^{\ell(n)}; \\ (x,w) \leftarrow \mathcal{A}^{\mathcal{O},\mathcal{ZK}}(\mathsf{crs}); \\ r \leftarrow \{0,1\}^n; \\ \pi \leftarrow \mathcal{P}(\mathsf{crs}, x, w, r) \end{array} : \mathcal{A}^{\mathcal{O},\mathcal{ZK}}(\pi) = 1 \wedge (x,w) \in R_n \right] - \Pr\left[\begin{array}{l} (\mathsf{crs}, s) \leftarrow \mathcal{S}_1^{\mathcal{O},\mathcal{ZK}}(1^n); \\ (x,w) \leftarrow \mathcal{A}^{\mathcal{O},\mathcal{ZK}}(\mathsf{crs}); \\ \pi' \leftarrow \mathcal{S}_2^{\mathcal{A},\mathcal{O},\mathcal{ZK}}(s, x) \end{array} : \mathcal{A}^{\mathcal{O},\mathcal{ZK}}(\pi') = 1 \wedge (x,w) \in R_n \right] \right|.$$

Constructing NIZK proofs from WI proofs. Fix an oracle $\mathcal{O}$, and let $\mathcal{WI} = (\mathcal{P}, \mathcal{V})$ be a WI proof system for $L = \text{CIRCUIT-SAT}^{\mathcal{O}}$. We show that if a one-way function $f^{\mathcal{O}}$ exists relative to $\mathcal{O}, \mathcal{WI}$, then we can construct an NIZK proof system for $\mathcal{NP}^{\mathcal{O}}$.

Assume $f^{\mathcal{O}}$ is one-way relative to $\mathcal{O}, \mathcal{WI}$. Using f, we can construct, in a black-box fashion, a pseudorandom generator $G^{\mathcal{O}} : \{0,1\}^n \to \{0,1\}^{2n}$ (see [14]). Define the following language $L' \in \mathcal{NP}^{\mathcal{O}}$:

$$L' \stackrel{\text{def}}{=} \left\{ (x, \mathsf{crs}) \text{ s.t. } \exists w \in \{0,1\}^n \text{ for which } (x,w) \in R_L \text{ or } \mathsf{crs} = G^{\mathcal{O}}(w) \right\}.$$

A zero-knowledge proof that $x \in L$ can then be constructed [7] by giving a witness-indistinguishable proof that $(x, \mathsf{crs}) \in L'$. In more detail, given a WI proof system $(\mathcal{P}, \mathcal{V})$ for L, consider the following proof system $(\mathcal{P}_{\mathcal{ZK}}, \mathcal{V}_{\mathcal{ZK}})$ for L:

Prover $\mathcal{P}_{\mathcal{ZK}}$: Given crs, x, w with $\mathsf{crs} \in \{0,1\}^{2n}$ and $(x,w) \in R_n$, set $x' = (x, \mathsf{crs})$ and note that $(x', w) \in L'$. Use a Levin reduction to the $\mathcal{NP}^{\mathcal{O}}$-complete language L to obtain $(\hat{x}, \hat{w}) \in L$. Choose $r \leftarrow \{0,1\}^{|\hat{x}|}$ and return the proof $\pi = \mathcal{P}(\hat{x}, \hat{w}, r)$.

Verifier $\mathcal{V}_{\mathcal{ZK}}$: Given crs, x, π, set $x' = (x, \mathsf{crs})$ and use a Levin reduction to the $\mathcal{NP}^{\mathcal{O}}$-complete language L to obtain $\hat{x}$. Then output $\mathcal{V}(\hat{x}, \pi)$.

Theorem 1. *If $(\mathcal{P}, \mathcal{V})$ is a WI proof system for L, then $(\mathcal{P}_{\mathcal{ZK}}, \mathcal{V}_{\mathcal{ZK}})$ is an NIZK proof system for L.*

Proof. Completeness is immediate, and statistical soundness of $(\mathcal{P}_{\mathcal{ZK}}, \mathcal{V}_{\mathcal{ZK}})$ follows from the soundness of $(\mathcal{P}, \mathcal{V})$ and the fact that a uniform $\mathsf{crs} \in \{0,1\}^{2n}$ is in the range of G with only negligible probability.

A simulator $\mathcal{S} = (\mathcal{S}_1, \mathcal{S}_2)$ is given as follows. $\mathcal{S}_1(1^n)$ chooses $w \leftarrow \{0,1\}^n$ computes $\mathsf{crs} = G^{\mathcal{O}}(w)$, and then outputs (crs, w). Given x, simulator $\mathcal{S}_2$ sets $x' = (x, \mathsf{crs})$, applies a Levin reduction to (x', w) to obtain $(\hat{x}, \hat{w}) \in L$, chooses $r \leftarrow \{0,1\}^{|\hat{x}|}$, and outputs $\pi = \mathcal{P}(\hat{x}, \hat{w}, r)$.

The fact that $\mathcal{S}$ provides a good simulation follows from pseudorandomness of G relative to $\mathcal{O}, \mathcal{WI}$, and witness indistinguishability of $\mathcal{WI}$.

3 An Augmented Black-Box Construction

Here we show that the Naor-Yung/Sahai construction of CCA-secure public-key encryption from CPA-secure public-key encryption can be cast as an augmented fully black-box construction. This result is not surprising; the point is to demonstrate that our framework does, indeed, capture constructions that go beyond the usual black-box ones. In particular, the construction is *shielding* in the terminology of [11], something ruled out in that same work in a black-box sense.

Let $\mathcal{O} = (G, E, D)$ be a public-key encryption scheme (with perfect correctness), and let $\mathcal{WI} = (\mathcal{P}, \mathcal{V})$ be a WI proof system for $\mathcal{NP}^{\mathcal{O}}$. Assume $\mathcal{O}$ is CPA-secure relative to $\mathcal{O}, \mathcal{WI}$. As noted in Section 2.2, we can use $\mathcal{WI}$ to construct an NIZK proof system $(\mathcal{P}_{\mathcal{ZK}}, \mathcal{V}_{\mathcal{ZK}})$ for $\mathcal{NP}^{\mathcal{O}}$. (Existence of CPA-secure encryption implies existence of a one-way function). Moreover, we can use the results of Sahai [20] to transform $(\mathcal{P}_{\mathcal{ZK}}, \mathcal{V}_{\mathcal{ZK}})$ into a *simulation-sound* NIZK proof system $\mathsf{ss}\mathcal{ZK} = (\mathcal{P}_{\mathsf{ss}\mathcal{ZK}}, \mathcal{V}_{\mathsf{ss}\mathcal{ZK}})$ for $\mathcal{NP}^{\mathcal{O}}$. (We remark that for $\mathcal{WI}$ sampled according to the distribution described in Section 2.1, the NIZK proof system $(\mathcal{P}_{\mathcal{ZK}}, \mathcal{V}_{\mathcal{ZK}})$ would already satisfy simulation soundness with overwhelming probability. However, here we want a construction starting from *any* WI proof system). For notational convenience, we will treat $\mathsf{ss}\mathcal{ZK}$ as an NIZK proof system for the specific language

$$L \stackrel{\text{def}}{=} \{(c_1, c_2, pk_1, pk_2) \mid \exists m, r_1, r_2 : c_1 = E^{\mathcal{O}}_{pk_1}(m; r_1) \bigwedge c_2 = E^{\mathcal{O}}_{pk_2}(m; r_2)\}.$$

We now describe the construction of a CCA-secure encryption scheme:

KeyGen $\mathcal{G}^{\mathcal{O},\mathsf{ss}\mathcal{ZK}}$: Compute $(pk_1, sk_1) \leftarrow G(1^n)$ and $(pk_2, sk_2) \leftarrow G(1^n)$. Then choose $\mathsf{crs} \leftarrow \{0,1\}^{\ell(n)}$ and set $PK = (pk_1, pk_2, \mathsf{crs})$ and $SK = (sk_1, sk_2)$.

Encryption $\mathcal{E}^{\mathcal{O},\mathsf{ss}\mathcal{ZK}}$: To encrypt plaintext m, choose $r_1, r_2, r \leftarrow \{0,1\}^n$ and then compute the ciphertexts $c_1 = E_{pk_1}(m; r_1)$ and $c_2 = E_{pk_2}(m; r_2)$. Set $x = (c_1, c_2, pk_1, pk_2)$ and $w = (m, r_1, r_2)$ and generate an NIZK proof $\pi = \mathcal{P}_{\mathsf{ss}\mathcal{ZK}}(\mathsf{crs}, x, w, r)$. Output (c_1, c_2, π).

Decryption $\mathcal{D}^{\mathcal{O},\mathsf{ss}\mathcal{ZK}}$: To decrypt (c_1, c_2, π), set $x = (c_1, c_2, pk_1, pk_2)$ and check that $\mathcal{V}_{\mathsf{ss}\mathcal{ZK}}(\mathsf{crs}, x, \pi) = 1$. If not, output $\perp$. Otherwise, output $m = D_{sk_1}(c_1)$.

The following theorem follows from [20, Theorem 4.1].

Theorem 2. *Let $\mathcal{O}$ be an encryption scheme (with perfect correctness) that is CPA-secure relative to $\mathcal{O}, \mathcal{WI}$. Then the above is an augmented fully black-box construction of a CCA-secure encryption scheme from $\mathcal{O}$.*

4 An Impossibility Result for Key Agreement

In this section, we rule out augmented black-box constructions of key agreement with perfect completeness from one-way functions. (We conjecture that the result extends to the case of imperfect completeness, but we were unable to prove this). For the remainder of this section, we only consider 1-bit key-agreement protocols with perfect completeness.

Say (A, B) is a pair of polynomial-time oracle algorithms that is an augmented black-box construction of key agreement from one-way functions. Then:

- For any $\mathcal{O}, \mathcal{WI}$ such that $\mathcal{WI}$ is a proof system for $\mathcal{NP}^{\mathcal{O}}$ and all n, following an execution between $A^{\mathcal{O},\mathcal{WI}}(1^n)$ and $B^{\mathcal{O},\mathcal{WI}}(1^n)$ both parties agree on a common bit $k \in \{0, 1\}$.
- Given (A, B) and E, define the advantage of E by the following experiment:
  1. $A^{\mathcal{O},\mathcal{WI}}(1^n)$ and $B^{\mathcal{O},\mathcal{WI}}(1^n)$ interact, resulting in a shared key k and a transcript T.
  2. E is given T, and outputs a bit k'.

  The advantage of E is $|\Pr[k' = k] - 1/2|$.

  For any $\mathcal{O}$ and $\mathcal{WI}$ such that $\mathcal{O}$ is one-way relative to $(\mathcal{O}, \mathcal{WI})$ and $\mathcal{WI}$ is a WI proof system for $\mathcal{NP}^{\mathcal{O}}$, every unbounded algorithm E making at most polynomially many queries to $\mathcal{O}$ and $\mathcal{WI}$ has negligible advantage.

To prove that no augmented (fully) black-box construction of key agreement from one-way functions exists, fix some (A, B) and consider an execution of (A, B) when $\mathcal{O}$ is chosen at random and $\mathcal{WI}$ is chosen as described in Section 2.1. A random oracle is one-way [15], and Lemma 4 shows that it remains one-way in the presence of $\mathcal{WI}$ chosen from the specified distribution. Moreover, by Lemma 3 we have that $\mathcal{WI}$ is a WI proof system for $\mathcal{NP}^{\mathcal{O}}$. We note that even though these lemmas are stated with respect to polynomial time adversaries, since our proofs relativize, they also hold for computationally unbounded adversaries making at most polynomially many oracle queries. Thus, if (A, B) were an augmented black-box construction of key-agreement from one-way functions, then for any unbounded algorithm E making at most polynomially many oracle queries that has non-negligible advantage, there should exist a polynomial time machine $S^{E,\mathcal{O},\mathcal{WI}}$ that inverts $\mathcal{O}$ or breaks the witness indistinguishability of $\mathcal{WI}$. However, since S makes at most polynomially many queries to $\mathcal{O}, \mathcal{WI}$ (even indirectly through E), such an S does not exist. Therefore, every unbounded algorithm E making at most polynomially many queries to $\mathcal{O}$ and $\mathcal{WI}$ should have negligible advantage. However, we show an explicit E for which this is not the case, thus proving that no augmented (fully) black-box construction of key agreement from one-way functions exists. (In fact, our attack works for *any* oracles $\mathcal{O}, \mathcal{WI}$, not just those chosen according to the distributions stated). E can be made efficient if $\mathcal{P} = \mathcal{NP}$; thus any augmented semi-black-box construction of key agreement from one-way functions would imply $\mathcal{P} \neq \mathcal{NP}$.

4.1 Breaking Key Agreement Relative to a Random Oracle

In this section we provide a warmup for our main proof by ruling out (standard) black-box constructions of key agreement from one-way functions. This proof may also be of independent interest for pedagogical purposes as a simplified version of the proofs in [15,2]. Note, however, that we prove a weaker result: we only rule out constructions of key-agreement protocols with perfect completeness based on one-way functions.

Let (A, B) be a construction of key agreement from one-way functions. Let q_A (resp., q_B) be a polynomial upper bound on the number of queries made by A (resp., B). Consider an attacker E defined as follows. E, given a transcript T of an execution of (A, B) in the presence of a random oracle $\mathcal{O}$, maintains a set $Q(E)$ of query/answer pairs for $\mathcal{O}$, and a multiset of candidate keys K, both initialized to $\emptyset$. Then E runs $2q_B + 1$ iterations of the following attack:

- *Simulation phase:* E finds a view of A consistent with the given transcript and with $Q(E)$. This view contains the randomness r_A used by A, as well as a set of oracle queries/answers $\hat{Q}(A)$ made by A. The set $\hat{Q}(A)$ is chosen to be consistent with any queries/answers in $Q(E)$, but it need not be consistent with the true oracle $\mathcal{O}$.

  Let k denote the key computed by A in the view. Then E adds k to K.
- *Update phase:* E makes all queries in $\hat{Q}(A) \setminus Q(E)$ to the true oracle $\mathcal{O}$, and adds the resulting query/answer pairs to $Q(E)$.

Following the above, E has a multiset K of $2q_B + 1$ possible keys. E outputs the majority value in K.

In each iteration E makes at most q_A queries to $\mathcal{O}$. Thus, E makes $O(q_A \cdot q_B)$ queries overall. We claim that E outputs the key computed by A and B with probability 1. Toward this, we first prove the following:

Claim 1. *Let k denote the actual key computed by A and B in an execution of the protocol. Then in each iteration of the attack, either E adds k to K, or E adds to $Q(E)$ one of the queries made by B in the real execution.*

Proof. Let $Q(B)$ denote the queries made by B in the real execution of the protocol. In a given iteration, there are two possibilities. If $\hat{Q}(A) \cap Q(B) \not\subseteq Q(E)$, then we are done since E makes all queries in $\hat{Q}(A) \setminus Q(E)$ to the true oracle $\mathcal{O}$. If, on the other hand, $\hat{Q}(A) \cap Q(B) \subseteq Q(E)$ then there is an oracle $\tilde{\mathcal{O}}$ that is consistent with the sampled view of A and the view of the real B. That is, there is an execution of the protocol with an oracle $\tilde{\mathcal{O}}$ that yields the observed transcript T, a view for B identical to the view of the real B, and a view for A identical to the view generated by E in the current iteration. Perfect completeness implies that the key k computed by A in this case must match the (actual) key computed by B.

Since B makes at most q_B queries, it follows that there are at most q_B iterations in which E adds an incorrect key to K, and so at least $q_B + 1$ iterations in which E adds the correct key to K. Since E outputs the key that occurs most often, E always outputs the correct key.

4.2 Breaking Key Agreement Relative to $\mathcal{O}, \mathcal{WI}$

Here we prove the main result of this section:

Theorem 3. *There is no augmented fully black-box construction of key agreement with perfect completeness from one-way functions.*

The overall structure of the attack is the same as in the previous section, but there are some key differences. Our attack again proceeds by having E repeatedly find a view of A consistent with a transcript T and the oracle queries $Q(E)$ that E has made thus far. Let $Q(A)$ and $Q(B)$ denote the queries of A and B, respectively, in the actual execution of the protocol, and let $\hat{Q}(A)$ denote the queries of A in the view found by E in some iteration. In the previous section we argued that as long as $\hat{Q}(A) \cap Q(B) \subseteq Q(E)$, the key found by E in the given iteration matches the key computed by the real B. This was because, under that condition, there must exist an oracle $\tilde{\mathcal{O}}$ that is consistent with an execution of the protocol in which party A makes queries $\hat{Q}(A)$, party B makes queries $Q(B)$, and the resulting transcript is T. Here, however, that need not be the case. For example, consider a real execution of the protocol in which B makes a query $\mathcal{V}(x, \pi)$ that returns 1, yet B does not make any corresponding query $\mathcal{P}(x, w, \star) = \pi$ with $(x, w) \in R_L$. If E samples a view of A in which $x \notin L$, then there are no oracles $\tilde{O}, \widetilde{\mathcal{WI}}$ consistent with the sampled view of A and the real view of B, but neither does E necessarily learn any new queries in $Q(B)$.

We deal with the above by modifying the attack and changing the proof. First, we modify the attack by having E sample *extended* views of A, which include a view of A along with additional oracle queries used for "book-keeping". Second, rather than showing that, in every iteration, E either learns the correct key or a query in $Q(B)$, we show that, in every iteration, E either learns the correct key or a query in $Q(AB) \stackrel{\text{def}}{=} Q(A) \cup Q(B)$.

An additional subtlety arises due to the possibility that Spoof_i occurs (cf. Lemma 1) for some i. In our attack we handle this by guaranteeing that $\mathsf{Spoof} = \cup_i \mathsf{Spoof}_i$ occurs with sufficiently small probability, and showing that the attack is successful whenever Spoof does not occur. (Our proof can be significantly simplified if we make the assumption that $A(1^n)$ and $B(1^n)$ only query their oracles on inputs of length n, however we wish to avoid this assumption).

Preliminaries: We view $Q(A), Q(B)$, and $Q(E)$ interchangeably as sets of queries and sets of query/answer pairs. We write, e.g., $[\mathcal{P}(x, w, r) = \pi] \in Q(A)$ to denote that A made the query $\mathcal{P}(x, w, r)$ and received the answer π. As usual, we let L denote the set of satisfiable circuits with $\mathcal{O}$-gates.

We assume any key-agreement construction (A, B) has the following normal form: Before a party queries $\mathcal{P}(x, w, r)$, that party also asks all $\mathcal{O}$-queries necessary to check whether $(x, w) \in R_L$; after receiving the result $\pi = \mathcal{P}(x, w, r)$, that party also asks $\mathcal{V}(x, \pi)$. Any key-agreement protocol can be modified to satisfy this condition with only a polynomial blow-up in the number of queries. We let $q = q(n) \geq n$ denote a polynomial upper bound on the combined running time of A and B (and so in particular a bound on the number of queries they make).

Without loss of generality, assume that for any (circuit) $x \in \{0,1\}^n$ and $w \in \{0,1\}^n$, computation of x on input w queries $\mathcal{O}$ at most n times, each time on input of length at most n.

Extended views of A: In our attack, E will repeatedly sample *extended* views of A which include A's view along with some additional oracle queries/answers. We denote an extended view by $(r_A, \mathcal{O}', \mathcal{WI}')$, where r_A are the random coins of A and $\mathcal{O}', \mathcal{WI}'$ are a set of query/answer pairs that includes all those made by A (using coins r_A and the given transcript). E samples only *consistent* extended views, which we define now.

Definition 5. *Let $Q = (\mathcal{O}', \mathcal{WI}' = (\mathcal{P}', \mathcal{V}'))$ be a set of queries/answers. We say it is* consistent *if*

1. *For every query $[\mathcal{P}'(x, w, r) = \pi] \in \mathcal{WI}'$, oracle $\mathcal{O}'$ contains queries/answers sufficient to determine whether $(x, w) \in R_L$. Moreover, if $(x, w) \in R_L$ then $[\mathcal{V}'(x, \pi) = 1] \in \mathcal{WI}'$, while if $(x, w) \notin R_L$ then $[\mathcal{V}'(x, \pi) = 0] \in \mathcal{WI}'$.*
2. *For every query $[\mathcal{V}'(x, \pi) = 1] \in \mathcal{WI}'$, there exist w, r such that $\mathcal{O}'$ contains queries/answers for which $(x, w) \in R_L$ and $[\mathcal{P}'(x, w, r) = \pi] \in \mathcal{WI}'$.*

Let T be a transcript of an execution between $A(1^n)$ and $B(1^n)$, and let $Q(E)$ be a set of queries/answers. We say the extended view $(r_A, \mathcal{O}', \mathcal{WI}')$ is consistent with T and $Q(E)$ *if $\mathcal{O}', \mathcal{WI}'$ is consistent, and also:*

1. *Every query in $Q(E)$ is in $\mathcal{O}', \mathcal{WI}'$, and is answered the same way.*
2. *$A^{\mathcal{O}', \mathcal{WI}'}(1^n; r_A)$, when fed with incoming messages as in T, would generate outgoing messages consistent with T.*

The attack. Let $t = 4 \log q$. First, E queries $\mathcal{O}(x)$ for all x with $|x| \leq t$; queries $\mathcal{P}(x, w, r)$ for all x, w, r with $|x| = |w| = |r| \leq t$; and queries $V(x, \pi)$ for all x, π with $|x| = |\pi|/7 \leq t$. Denote these queries/answers by $Q^*(E)$. The rest of the attack is similar to that of the previous section. E, given a transcript T of an execution of (A, B), initializes $Q(E) = Q^*(E)$ and $K = \emptyset$, and then runs $2q + 1$ iterations of the following:

- *Simulation phase:* E finds an extended view $(r_A, \mathcal{O}', \mathcal{WI}')$ consistent with T and $Q(E)$, with $\mathcal{O}', \mathcal{WI}'$ of size at most $|Q(E)| + q$. (If no such extended view exists, E aborts). Let k be the key computed by A in this view. E adds k to K.
- *Update phase:* E makes all queries in $(\mathcal{O}' \cup \mathcal{WI}') \setminus Q(E)$ to the true oracles $\mathcal{O}, \mathcal{WI}$. For any queries $[\mathcal{P}'(x, w, r) = \pi]$ just made, E also makes any $\mathcal{O}$ queries needed to determine whether $(x, w) \in R_L$, as well as the query $\mathcal{V}(x, \pi)$. All the resulting query/answer pairs are added to $Q(E)$.

Following the above, E has a multiset K of $2q+1$ possible keys. E outputs the majority value in K.

Analysis. In pre-processing, E makes polynomially many queries. In each iteration of the attack, E makes at most $q + q(q+1) \leq 3q^2$ queries: there are at most q queries in $(\mathcal{O}' \cup \mathcal{WI}') \setminus Q(E)$, and for each such query of the form $[\mathcal{P}'(x, w, r) = \pi]$ we have $|x| \leq q$ and so at most q queries are needed to check whether $(x, w) \in R_L$. Thus, E makes at most $7q^3$ queries after the pre-processing.

For any i, define Spoof_i (cf. Lemma 1) to be the event that there is a query $[\mathcal{V}_i(x, \pi) = 1] \in Q(A) \cup Q(B)$, yet there is no query

$$[\mathcal{P}_i(x, w, \star) = \pi] \in Q(A) \cup Q(B) \cup Q^*(E)$$

with $(x, w) \in R_L$. Let $\mathsf{Spoof} = \bigvee_i \mathsf{Spoof}_i$. We claim that Spoof occurs with probability at most 1/4. Indeed, by construction Spoof_i cannot occur for $i \leq t$, and (by Lemma 1 and a union bound) $\Pr[\bigvee_{i>t} \mathsf{Spoof}_i] \leq 1/8$.

Define Spoof' to be the event that, at some point during the attack, E queries $\mathcal{V}(x, \pi) = 1$ to the real oracle, but there was no previous query $[\mathcal{P}_i(x, w, \star) = \pi]$ made by A, B, or E with $(x, w) \in R_L$. By construction, this can only possibly occur if $|x| > 4 \log q$. Since E makes at most $7q^3$ queries after the pre-processing stage, however, Spoof' occurs with probability at most 1/8.

In the rest of the analysis, we show that as long as neither Spoof nor Spoof' occur, E outputs the key computed by A and B. This suffices, since then E finds the shared key with probability at least 3/4 overall. As in the previous section, then, the following lemma will prove Theorem 3:

Lemma 5. *Let k denote the actual key computed by A and B in an execution of the protocol, and assume neither* Spoof *nor* Spoof' *occur. Then E does not abort, and in each iteration of the attack either E adds k to K, or E adds to $Q(E)$ one of the queries made by A or B in the real execution.*

Proof. Let $Q(AB) \stackrel{\text{def}}{=} Q(A) \cup Q(B)$ denote the queries/answers made/received by A or B in the real execution. We first show that E never aborts. Say $Q(E)$ is consistent at the beginning of some iteration; this is true by construction in the first iteration. Since Spoof did not occur, a consistent, extended view is given by letting $(\mathcal{O}', \mathcal{WI}') = Q(E) \cup Q(AB)$, which is of size at most $|Q(E)| + q$. Moreover, regardless of what consistent, extended view is actually sampled by E, the new set $Q(E)$ defined at the end of the iteration is consistent unless Spoof' occurs.

We now prove the rest of the lemma. Let $(r_A, \mathcal{O}', \mathcal{WI}')$ be the consistent, extended view chosen by E in some iteration. We define three events, and show:

- If one of the events occurs, then, in the update phase of that iteration, E adds to $Q(E)$ some query in $Q(AB)$.
- If none of the events occur then there are oracles $\tilde{\mathcal{O}}, \widetilde{\mathcal{WI}}$ that match (i.e., are not inconsistent with) the extended view of A and the real view of B. (Thus, by perfect completeness, E adds the correct key to K in that iteration).

Before defining the events, we introduce some terminology. Given some set of queries Q, we say Q *fixes* $x \in L$ if either (1) there exists a w and $\mathcal{O}$-queries in Q

such that $(x, w) \in R_L$, or (2) there is a query $[\mathcal{V}(x, \star) = 1] \in Q$. We say Q *fixes* $x \notin L$ if for all w there are $\mathcal{O}$-queries in Q such that, regardless of how any of the $\mathcal{O}$-queries not in Q are answered, it holds that $(x, w) \notin R_L$. We define Q *fixes* $(x, w) \in R_L$ and Q *fixes* $(x, w) \notin R_L$ in the obvious way.

We now define the events of interest:

E_1: $\mathcal{O}', \mathcal{WI}'$ disagrees with $Q(AB)$ on the answer to some $\mathcal{O}$-, $\mathcal{P}$-, or $\mathcal{V}$-query.

E_2: There exists an x such that $Q(AB)$ fixes $x \in L$ but $\mathcal{O}', \mathcal{WI}'$ fixes $x \notin L$, or vice versa.

E_3: A $\mathcal{V}'$-query returning 0 in $\mathcal{WI}'$ is "inconsistent" with the $\mathcal{O}, \mathcal{P}$ queries in $Q(AB)$, or vice versa. Formally, one of the following occurs:
- There is a query $[\mathcal{V}'(x, \pi) = 0] \in \mathcal{WI}'$, but $[\mathcal{P}(x, w, \star) = \pi] \in Q(AB)$ and $Q(AB)$ fixes $(x, w) \in R_L$.
- There is a query $[\mathcal{P}'(x, w, \star) = \pi] \in \mathcal{WI}'$ and $\mathcal{O}'$ fixes $(x, w) \in R_L$, but $[\mathcal{V}(x, \pi) = 0] \in Q(AB)$.

Claim 2. *If any of E_1, E_2, or E_3 occur in the simulation phase of some iteration, then E learns a new query in $Q(AB)$ in the update phase of that iteration.*

Proof. If E_1 occurs, the claim is immediate. ($Q(E)$ contains the answers of the true oracles, and so can never disagree with $Q(AB)$. So any disagreement between $\mathcal{O}', \mathcal{WI}'$ and $Q(AB)$ must be due to some query in $\mathcal{O}', \mathcal{WI}'$ outside of $Q(E)$). If E_2 occurs there are several sub-cases to consider:

1. Say $Q(AB)$ fixes $x \in L$, but $\mathcal{O}', \mathcal{WI}'$ fixes $x \notin L$. The second event implies that for all w oracle $\mathcal{O}'$ fixes $(x, w) \notin R_L$. There are two ways the first event can occur:
   - There exists a w such that $Q(AB)$ fixes $(x, w) \in R_L$. In this case there must be an $\mathcal{O}$-query in $Q(AB)$ that is answered inconsistently with some query in $\mathcal{O}'$, and event E_1 has occurred.
   - There is a query $[\mathcal{V}(x, \pi) = 1] \in Q(AB)$ (for some π). Since Spoof has not occurred, this means that for some w, r there is a query $[\mathcal{P}(x, w, r) = \pi]$ in $Q(AB)$ or $Q^*(E)$. Say $[\mathcal{P}(x, w, r) = \pi] \in Q(AB)$. Then by our normal-form assumption, $Q(AB)$ fixes $(x, w) \in R_L$; this, in turn, implies an $\mathcal{O}$-query in $Q(AB)$ inconsistent with $\mathcal{O}'$ (which, recall, fixed $x \notin L$), and so E_1 has occurred.

     On the other hand, say $[\mathcal{P}(x, w, r) = \pi] \in Q^*(E)$. Then, by definition of $Q^*(E)$, the query $[\mathcal{V}(x, \pi) = 1]$ is also in $Q^*(E)$, and $Q^*(E)$ fixes $(x, w) \in R_L$. But since any queries in $\mathcal{O}'$ must agree with the corresponding $\mathcal{O}$-queries in $Q^*(E)$, this cannot happen.
2. Say $\mathcal{O}', \mathcal{WI}'$ fixes $x \in L$, but $Q(AB)$ fixes $x \notin L$. The second event implies that for all w we have that $Q(AB)$ fixes $(x, w) \notin R_L$. There are two ways the first event can occur:
   - There exists a w for which $\mathcal{O}'$ fixes $(x, w) \in R_L$. In this case there is an $\mathcal{O}$-query in $Q(AB)$ that is answered inconsistently with some query in $\mathcal{O}'$, and event E_1 has occurred.

- There is a query $[\mathcal{V}(x, \pi) = 1] \in \mathcal{WI}'$ for some π. By definition of consistency, there exists w such that $\mathcal{O}'$ fixes $(x, w) \in R_L$. Then there must be an $\mathcal{O}$-query in $Q(AB)$ that is answered inconsistently with $\mathcal{O}'$, and so E_1 has occurred.

Finally, we turn to E_3. Here there are two sub-cases:

1. Say $[\mathcal{V}'(x, \pi) = 0] \in \mathcal{WI}'$, but $[\mathcal{P}(x, w, \star) = \pi] \in Q(AB)$ and furthermore $Q(AB)$ fixes $(x, w) \in R_L$. Because of our normal-form assumption, $[\mathcal{V}(x, \pi) = 1] \in Q(AB)$. Thus there is a $\mathcal{V}$-query in $Q(AB)$ that is answered inconsistently with $\mathcal{WI}'$ and so E_1 has occurred.
2. Say $[\mathcal{P}'(x, w, \star) = \pi] \in \mathcal{WI}'$ and $\mathcal{O}'$ fixes $(x, w) \in R_L$, but we have $[\mathcal{V}(x, \pi) = 0] \in Q(AB)$. By definition of consistency, $[\mathcal{V}(x, \pi) = 1] \in \mathcal{WI}'$. Thus there is a $\mathcal{V}$-query in $Q(AB)$ that is answered inconsistently with $\mathcal{WI}'$, and so E_1 has occurred.

This concludes the proof of Claim 2.

To complete the proof of the lemma, we show that if none of E_1, E_2, or E_3 occur, there exist oracles $\tilde{\mathcal{O}}, \widetilde{\mathcal{WI}}$ (in the support of the distribution from Section 2.1) that match (i.e., do not disagree with) $\mathcal{O}', \mathcal{WI}'$, and $Q(AB)$. This means there is an execution of the protocol with oracles $\tilde{\mathcal{O}}, \widetilde{\mathcal{WI}}$ that yields a view for B identical to the view of the real B, and a view for A identical to the view of A in the extended view sampled by E. Perfect completeness implies that the key k computed by A in that case must match the (actual) key computed by B, as we needed to show.

We construct $\tilde{\mathcal{O}}, \widetilde{\mathcal{WI}}$ as follows. First, answer all queries in $\mathcal{O}', \mathcal{WI}'$, and $Q(AB)$ as answered by those oracles; if E_1 does not occur, this is well-defined as there is no conflict. Answer all other queries in $\tilde{\mathcal{O}}$ arbitrarily. Note that if $\mathcal{O}', \mathcal{WI}', Q(AB)$ fixes $x \in L$ then so does $\tilde{\mathcal{O}}$, and similarly if $\mathcal{O}', \mathcal{WI}', Q(AB)$ fixes $x \notin L$. Note also that with $\tilde{\mathcal{O}}$ fixed, so are $\tilde{L}$ and $\tilde{R}_L$.

For $\tilde{\mathcal{P}}$, proceed as follows. Recall that all $\tilde{\mathcal{P}}_i$ queries for $i \leq t = 4 \log q$ were made by E during pre-processing and so are already fixed. Any other unassigned query $\tilde{\mathcal{P}}(x, w, r)$ with $|x| > t$ is defined as follows:

- If $(x, w) \notin \tilde{R}_L$, the query is answered arbitrarily.
- If $(x, w) \in \tilde{R}_L$, let $\pi^* \in \{0, 1\}^{7|x|}$ be such that $\mathcal{V}(x, \pi^*)$ is not in $\mathcal{WI}'$ or $Q(AB)$. (There must exist such a π^*, by the bound on the number of queries in these sets). Set $\tilde{\mathcal{P}}(x, w, r) = \pi^*$.

With the $\tilde{\mathcal{O}}$ and $\tilde{\mathcal{P}}$ queries fixed, oracle $\tilde{\mathcal{V}}$ is set as in Section 2.1.

We show that $\tilde{\mathcal{O}}, \widetilde{\mathcal{WI}}$ match (i.e., do not disagree with) $\mathcal{O}', \mathcal{WI}'$, and $Q(AB)$. By construction, the only possible conflict can be between $\tilde{\mathcal{V}}$ and some $\mathcal{V}$-query in $\mathcal{WI}'$ or $Q(AB)$. No such conflict is possible:

1. Say $[\mathcal{V}(x, \pi) = 1] \in \mathcal{WI}'$ for some x, π. Then by definition of consistency, there exist w, r such that $\mathcal{O}'$ fixes $(x, w) \in R_L$, and $[\mathcal{P}(x, w, r) = \pi] \in \mathcal{WI}'$. But then $(x, w) \in \tilde{R}_L$ and $\tilde{\mathcal{P}}(x, w, r) = \pi$, and so $\tilde{\mathcal{V}}(x, \pi) = 1$.

2. Say $[\mathcal{V}(x,\pi)=1] \in Q(AB)$ for some x,π. Since Spoof does not occur, there exist w,r such that $\mathcal{O}' \cup Q(AB)$ fixes $(x,w) \in R_L$, and $[\mathcal{P}(x,w,r)=\pi] \in \mathcal{WI}' \cup Q(AB)$. But then $(x,w) \in \tilde{R}_L$ and $\tilde{\mathcal{P}}(x,w,r)=\pi$, and so $\tilde{\mathcal{V}}(x,\pi)=1$.
3. Say $[\mathcal{V}(x,\pi)=0] \in \mathcal{WI}' \cup Q(AB)$ for some x,π. If $x \notin \tilde{L}$ then $\tilde{\mathcal{V}}(x,\pi)=0$ also. If $x \in \tilde{L}$, there is an inconsistency only if there is some w with $\tilde{\mathcal{P}}(x,w,\star)=\pi$ and $(x,w) \in \tilde{R}_L$. Note $\tilde{\mathcal{P}}(x,w,\star)=\pi$ can only occur if $[\mathcal{P}(x,w,\star)=\pi] \in \mathcal{WI}' \cup Q(AB)$, but in that case (since $[\mathcal{V}(x,\pi)=0] \in \mathcal{WI}' \cup Q(AB)$ and E_3 did not occur) either $\mathcal{O}'$ or $Q(AB)$ fix $(x,w) \notin R_L$, and hence $(x,w) \notin \tilde{R}_L$ either.

This completes the proof of Lemma 5.

5 Impossibility for Statistically-Hiding Commitments

We show that the impossibility results of Haitner et al. [12,13] for statistically-hiding commitment schemes can be strengthened to hold even within our new framework. (Our results here do not require perfect completeness).

Theorem 4. *Any augmented fully black-box construction of a statistically-hiding bit-commitment scheme from trapdoor permutations over $\{0,1\}^n$ has an $\Omega(n/\log n)$-round commit stage.*

Theorem 5. *Any augmented fully black-box construction of a statistically-hiding bit-commitment scheme from trapdoor permutations over $\{0,1\}^n$ requires the sender to communicate $\Omega(n)$ bits to the receiver during the commit stage.*

Note that in the above theorems we consider constructions which invoke only trapdoor permutations over n bits, where n is the security parameter. In fact, when considering constructions which may invoke the trapdoor permutations over smaller domains, better upper bounds are known. In particular, it is possible to apply the scheme of Naor et al. [17] using a one-way permutation over n^ϵ bits, which results in a statistically-hiding commitment scheme with an $O(n^\epsilon)$-round commit phase. As already discussed by Haitner et al. [12] this issue is not unique to our setting, but arises in essentially any study of the *efficiency* of cryptographic reductions. The common approach for addressing this issue is by restricting the class of constructions (as in the statements of our theorems); we refer the reader to [12] for a less restrictive approach.

Due to space limitations the proofs of Theorems 4 and 5 are provided in the full version of this work, and here we only give a high-level overview. We consider a set of oracles $\Gamma = (\mathcal{O}, \mathcal{P}, \mathcal{V}, \mathsf{Sam})$, and prove that the following hold with high probability[1]:

1. $\mathcal{O}$ is a collection of trapdoor permutations relative to Γ.
2. $(\mathcal{P}, \mathcal{V})$ is a WI proof system for $\mathcal{NP}^{\mathcal{O}}$ relative to Γ.

[1] We prove our statements with respect to a distribution over oracles. As in Lemma 3, we can also reverse the order of quantifiers and fix a specific oracle.

3. Any statistically-hiding bit-commitment scheme in which the sender and receiver have oracle access to $(\mathcal{O}, \mathcal{P}, \mathcal{V})$ can be broken using Γ. The efficiency and success probability of the attack depend on the round complexity or communication complexity of the commitment scheme.

This suffices because any (augmented) fully black-box construction is also *relativizing* [19].

The oracle Sam is the interactive collision-finding oracle of Haitner et al. [12]. In its most basic and non-interactive form, this oracle takes as input a circuit C, and outputs a random pair of inputs (w, w') such that $C(w) = C(w')$. Relative to such an oracle there are no collision-resistant hash functions [21] or 2-move statistically-hiding commitment schemes [8]. Moreover [21], one-way functions exist relative to Sam. This oracle was generalized by Haitner et al. to an interactive setting: Sam takes as input a circuit C and a "target" value z, and outputs a random input w such that $C(w) = z$. Haitner et al. had to force various restrictions on Sam such that one-way functions continue to exist, yet Sam remains sufficiently powerful to break the binding of (interactive) statistically-hiding commitment schemes.

In our setting, where we also consider a WI oracle $(\mathcal{P}, \mathcal{V})$, the argument that Sam can be used to break statistically-hiding commitment schemes is essentially identical to the corresponding argument of Haitner et al. [12,13]. The challenging part (in which our proof differs from that of Haitner et al.), lies in showing that one-way functions (and, more generally, that trapdoor permutations) still exist relative to Sam, and that $(\mathcal{P}, \mathcal{V})$ is still witness indistinguishable.

The proofs of these properties are more subtle than the corresponding proofs in Section 2. In that section we relied on the fact that any efficient algorithm can issue only a polynomial number of queries to $\mathcal{O}$ and $\mathcal{P}$. Here, however, when considering also the oracle Sam, this is no longer true: although an efficient algorithm with access to Γ may issue only a polynomial number of *direct queries* to $\mathcal{O}$, $\mathcal{P}$, and Sam, the oracles $\mathcal{O}$ and $\mathcal{P}$ may actually be *indirectly queried* an exponential number of times by Sam, and the previous arguments no longer hold.

To circumvent this and several other similar difficulties, we extend the proof of Haitner et al. [12] that manages to distinguish between the amount of "useful information" that is obtained by direct and indirect queries, and uses information-theoretic compression arguments that are oblivious to the (possibly exponential) number of indirect queries. The main difficulty in our setting, when compared to that of [12], is that we need to deal also with the oracles $\mathcal{P}$ and $\mathcal{V}$. Note that $\mathcal{P}$ is simply a random function (and thus can be treated as in [12]), but $\mathcal{V}$ has structure. Technically, proving that $\mathcal{O}$ is one-way is very similar to the corresponding proof in [12], since when compressing the description of $\mathcal{O}$ we are granted unbounded access to $\mathcal{P}$ and $\mathcal{V}$, and this enables us to perfectly simulate their behavior. The main difference is in proving that $(\mathcal{P}, \mathcal{V})$ is a WI proof system, and due to the structure of $\mathcal{V}$ this requires us to refine and adjust the compression arguments from [12] for arguing that $\mathcal{V}$ does not reveal too much "useful information" on $\mathcal{P}$, and can be simulated while compressing $\mathcal{P}$.

Finally, we note that although we prove our impossibility results for non-interactive *witness-indistinguishable* proof systems, our results immediately extend to non-interactive *zero-knowledge* proof systems (the main difference is in allowing the sender and receiver access to a common reference string). This follows from the fact that our impossibility results hold even for commitment schemes in which the hiding property is assumed to hold only with respect to the honest receiver (exactly as in [12,13]). Therefore, in such a case the receiver can choose a common random string that transforms a witness-indistinguishable proof system into a zero-knowledge proof system as in Section 2.2. Specifically, showing that $\mathcal{O}$ is one-way relative to Γ implies the existence of a pseudorandom generator, and therefore the transformation in Section 2.2 can be carried out by having the receiver sample a uniform random string and send it to the sender in the first round.

Acknowledgments

We thank the anonymous referees for their extensive comments, and Dominique Unruh for a thorough proofreading of our results and several useful discussions.

References

1. Applebaum, B., Ishai, Y., Kushilevitz, E.: Computationally private randomizing polynomials and their applications. Computational Complexity 15(2), 115–162 (2006)
2. Barak, B., Mahmoody-Ghidary, M.: Merkle puzzles are optimal — an $o(n^2)$-query attack on any key exchange from a random oracle. In: Halevi, S. (ed.) CRYPTO 2009. LNCS, vol. 5677, pp. 374–390. Springer, Heidelberg (2009)
3. Beaver, D.: Correlated pseudorandomness and the complexity of private computations. In: 28th Annual ACM Symposium on Theory of Computing (STOC), pp. 479–488. ACM Press, New York (1996)
4. Bellare, M., Goldwasser, S.: New paradigms for digital signatures and message authentication based on non-interactive zero knowledge proofs. In: Brassard, G. (ed.) CRYPTO 1989. LNCS, vol. 435, pp. 194–211. Springer, Heidelberg (1990)
5. Dolev, D., Dwork, C., Naor, M.: Nonmalleable cryptography. SIAM Journal on Computing 30(2), 391–437 (2000)
6. Feige, U., Fiat, A., Shamir, A.: Zero-knowledge proofs of identity. Journal of Cryptology 1(2), 77–94 (1988)
7. Feige, U., Lapidot, D., Shamir, A.: Multiple non-interactive zero knowledge proofs under general assumptions. SIAM Journal on Computing 29(1), 1–28 (1999)
8. Fischlin, M.: On the impossibility of constructing non-interactive statistically-secret protocols from any trapdoor one-way function. In: Preneel, B. (ed.) CT-RSA 2002. LNCS, vol. 2271, pp. 79–95. Springer, Heidelberg (2002)
9. Fischlin, M.: Round-optimal composable blind signatures in the common reference string model. In: Dwork, C. (ed.) CRYPTO 2006. LNCS, vol. 4117, pp. 60–77. Springer, Heidelberg (2006)
10. Gennaro, R., Gertner, Y., Katz, J., Trevisan, L.: Bounds on the efficiency of generic cryptographic constructions. SIAM Journal on Computing 35(1), 217–246 (2005)

11. Gertner, Y., Malkin, T.G., Myers, S.: Towards a separation of semantic and CCA security for public key encryption. In: Vadhan, S.P. (ed.) TCC 2007. LNCS, vol. 4392, pp. 434–455. Springer, Heidelberg (2007)
12. Haitner, I., Hoch, J.J., Reingold, O., Segev, G.: Finding collisions in interactive protocols — a tight lower bound on the round complexity of statistically-hiding commitments. In: 48th Annual Symposium on Foundations of Computer Science (FOCS), pp. 669–679. IEEE, Los Alamitos (2007)
13. Haitner, I., Hoch, J.J., Segev, G.: A linear lower bound on the communication complexity of single-server private information retrieval. In: Canetti, R. (ed.) TCC 2008. LNCS, vol. 4948, pp. 445–464. Springer, Heidelberg (2008)
14. Håstad, J., Impagliazzo, R., Levin, L.A., Luby, M.: A pseudorandom generator from any one-way function. SIAM Journal on Computing 28(4), 1364–1396 (1999)
15. Impagliazzo, R., Rudich, S.: Limits on the provable consequences of one-way permutations. In: 21st Annual ACM Symposium on Theory of Computing (STOC), pp. 44–61. ACM Press, New York (1989)
16. Lindell, Y.: A simpler construction of CCA2-secure public-key encryption under general assumptions. In: Biham, E. (ed.) EUROCRYPT 2003. LNCS, vol. 2656, pp. 241–254. Springer, Heidelberg (2003)
17. Naor, M., Ostrovsky, R., Venkatesan, R., Yung, M.: Perfect zero-knowledge arguments for NP using any one-way permutation. Journal of Cryptology 11(2), 87–108 (1998)
18. Naor, M., Yung, M.: Public-key cryptosystems provably secure against chosen ciphertext attacks. In: 22nd Annual ACM Symposium on Theory of Computing (STOC), pp. 427–437. ACM Press, New York (1990)
19. Reingold, O., Trevisan, L., Vadhan, S.P.: Notions of reducibility between cryptographic primitives. In: Naor, M. (ed.) TCC 2004. LNCS, vol. 2951, pp. 1–20. Springer, Heidelberg (2004)
20. Sahai, A.: Non-malleable non-interactive zero knowledge and adaptive chosen-ciphertext security. In: 40th Annual Symposium on Foundations of Computer Science (FOCS), pp. 543–553. IEEE, Los Alamitos (1999)
21. Simon, D.R.: Findings collisions on a one-way street: Can secure hash functions be based on general assumptions? In: Nyberg, K. (ed.) EUROCRYPT 1998. LNCS, vol. 1403, pp. 334–345. Springer, Heidelberg (1998)

Towards Non-Black-Box Lower Bounds in Cryptography

Rafael Pass*, Wei-Lung Dustin Tseng,
and Muthuramakrishnan Venkitasubramaniam

Cornell University,
{rafael,wdtseng,vmuthu}@cs.cornell.edu

Abstract. We consider average-case strengthenings of the traditional assumption that coNP is not contained in AM. Under these assumptions, we rule out generic and potentially *non-black-box* constructions of various cryptographic primitives (e.g., one-way permutations, collision-resistant hash-functions, constant-round statistically hiding commitments, and constant-round black-box zero-knowledge proofs for NP) from one-way functions, assuming the security reductions are *black-box*.

1 Introduction

In the past four decades, many cryptographic tasks have been put under rigorous treatment in an effort to realize these tasks under minimal assumptions. In particular, one-way functions are widely regarded as the most basic cryptographic primitive; their existence is implied by most other cryptographic tasks. Presently, one-way functions are known to imply schemes such as private-key encryption [GM84, GGM86, HILL99], pseudo-random generators [HILL99], statistically-binding commitments [Nao91], statistically-hiding commitments [NOVY98, HR07] and zero-knowledge proofs [GMW91]. At the same time, some other tasks still have no known constructions based on one-way functions (e.g., key agreement schemes or collision-resistant hash functions).

Following the seminal paper by Impagliazzo and Rudich [IR88], many works have addressed this phenomenon by demonstrating *black-box separations*, which rules out constructions of a cryptographic task using the underlying primitive as a *black-box*. For instance, Impagliazzo and Rudich rule out black-box constructions of key-agreement protocols (and thus also trapdoor predicates) from one-way functions; Simon [Sim98] rules out black-box constructions of collision-resistant hash functions from one-way functions. Furthermore, these impossibility results are *unconditional*.

Yet many classical cryptographic constructions (e.g., [FS90, DDN00, GMW91]) are non-black-box. This begs the question: to what extent does black-box separations give us insight into the actual separation of cryptographic primitives?

* Supported in part by a Microsoft New Faculty Fellowship, NSF CAREER Award CCF-0746990, AFOSR Award FA9550-08-1-0197 and BSF Grant 2006317.

Y. Ishai (Ed.): TCC 2011, LNCS 6597, pp. 579–596, 2011.

In this paper, we directly focus on providing lower bounds for *non-black-box* constructions of cryptographic primitives from one-way functions. We emphasize that although we consider non-black-box constructions, we still assume Turing (i.e., black-box) security reductions. For some of our results, we heavily leverage the existing literature on the impossibility of basing cryptography on NP hardness (these works also directly consider a Turing reduction of cryptographic primitives from NP). Perhaps surprisingly, we also make extensive use of known black-box separations. In other words, we demonstrate that some black-box separations *can* be modified to give further insight into the separation of cryptographic primitives.

Before stating our theorems, we first discuss our assumptions. Assumptions are necessary for non-black-box separations assuming black-box reductions; to show that a primitive P cannot be constructed using one-way functions, we must at least assume that a weak notion of so-called somewhere-uninvertable one-way functions exist—i.e. functions that cannot be inverted on *all* input lengths (as opposed to infinitely many lengths as in the traditional definition of one-way functions)[1]. As one of the main contributions of the paper, we introduce general assumptions that we believe are reasonable, and are useful in establishing a variety of non-black-box separations.

1.1 Our Assumptions

Assumption 1. $\mathrm{Dist}^{\mathrm{1sided}}\text{-}\mathsf{coNP} \not\subseteq \mathrm{Heur}_{1/\mathrm{poly}}\mathsf{AM}$ is an average-case extension of the well-studied (and widely believed) classical assumption $\mathsf{coNP} \not\subseteq \mathsf{AM}$. Briefly, $\mathrm{Dist}^{\mathrm{1sided}}\text{-}\mathsf{coNP}$ contains all coNP languages coupled with an efficiently samplable distribution over the no instances of the language. Such a language is considered to be in $\mathrm{Heur}_{1/\mathrm{poly}}\mathsf{AM}$ if there exists an AM (constant-round) protocol that accepts the language, with the relaxation that soundness only needs to hold with high probability over the no instances, as measured by the given distribution. As we prove later, the assumption is equivalent to the existence of an efficiently computable function f that is not *heuristically co-range verifiable*—that is, there does not exist an AM protocol proving that an element is outside the range of f, where soundness holds with high probability for a random instance $f(x)$[2]. Assuming that there exists an efficiently computable function that is not heuristically co-range verifiable seems most reasonable (consider, for instance, proving that an element is not in the range of AES [DR02]). We additionally show that such a function is implied by the existence of pseudorandom generators[3] secure against "promise-$\mathsf{AM} \cap \mathsf{coAM}$".

Assumption 2. Our second assumption is of a different flavor: we assume the existence of one-way functions that are secure against $\mathrm{PPT}^{\mathsf{SAM}_d}$. Here SAM_d refers

[1] If Somewhere-Uninvertable OWFs do not exist, then every cryptographic primitive can be constructed from OWFs, because for every efficiently computable function, there would be a trivial reduction that inverts the function on all input lengths.

[2] See section 3 for a comparison with the literature of "average refutation" [FKO06].

[3] Here we refer to BMY-type pseudo-random generators [BM84, Yao82].

to the depth-d collision finding oracle defined in [Sim98, HHRS07];[4] $\mathrm{PPT}^{\mathsf{SAM}_d}$ refers to the class of probabilistic polynomial time machines with oracle access to SAM_d. This assumption is flexible since we can adjust the parameter d; a larger d implies a stronger assumption (in fact, if $d = n/\log n$, the assumption is simply false since $\mathsf{SAM}_{n/\log n}$ can in fact invert one-way functions [PV10]). In our work, we focus on the case $d = O(1)$ (constant depth), and refer to $\mathsf{SAM}_{O(1)}$ simply as SAM.

Assumption 3. Our final and strongest assumption is $\mathrm{Dist}^{\mathrm{1sided}}\text{-}\mathsf{coNP} \not\subseteq \mathrm{Heur}_{1/\mathrm{poly}}\mathsf{IP}[\mathrm{PPT}^{\mathsf{NP}}]$ (heuristically verified by an interactive protocol where the prover is a probabilistic polynomial time machine with access to a NP oracle). It directly implies assumption 1, and relying on the work of Haitner, Mahmoody-Ghidary and Xiao [HMX10], we show that it implies assumption 2 as well in the case $d = O(1)$. Due to their similarity, $\mathrm{Dist}^{\mathrm{1sided}}\text{-}\mathsf{coNP} \not\subseteq \mathrm{Heur}_{1/\mathrm{poly}}\mathsf{IP}[\mathrm{PPT}^{\mathsf{NP}}]$ inherits many of the justifications as our first assumption in a weaker form (e.g., it is based on the classical assumption $\mathsf{coNP} \not\subseteq \mathsf{IP}[\mathrm{PPT}^{\mathsf{NP}}]$, and is equivalent to the existence of efficient functions whose co-range cannot be verified by $\mathsf{IP}[\mathrm{PPT}^{\mathsf{NP}}]$ protocols). We treat assumption 3 as a unifying (and strongest) assumption that implies all of the results in our work.

Minimizing the assumption. It is natural to ask if the classical assumption $\mathsf{coNP} \not\subseteq \mathsf{AM}$, or perhaps the more standard average-case hardness assumption $\mathrm{Dist}\text{-}\mathsf{coNP} \not\subseteq \mathrm{Heur}_{1/\mathrm{poly}}\mathsf{AM}$, are enough for our theorems (Dist-coNP consists of coNP languages coupled with efficiently samplable distributions that may range over all instances). We argue that it would be unlikely. In order to rule out constructions of cryptographic primitives based on OWFs, we first need to assume the existence of OWFs. But, it is unknown even if hard-on-the-average languages exist assuming only $\mathsf{coNP} \not\subseteq \mathsf{AM}$. Similarly, the stronger assumption $\mathrm{Dist}\text{-}\mathsf{coNP} \not\subseteq \mathrm{Heur}_{1/\mathrm{poly}}\mathsf{AM}$ implies the existence of a hard-on-the-average language, but, as far as we know, does not imply the existence of OWFs (indeed, this is related to the question of whether one-way functions can be based on average-case hardness). Restricting to one-sided distributions (i.e., considering $\mathrm{Dist}^{\mathrm{1sided}}\text{-}\mathsf{coNP}$ instead of Dist-coNP) is the next logical step, and this can be shown to imply a form of one-way functions (see full version).

1.2 Our Results

As mentioned, we are able to prove many separation results by adapting numerous previous works to take advantage of our assumptions. We highlight the main separations here (grouped by their assumptions), and leave the numerous corollaries to the main text.

[4] Given an interactive Turing machine M and a transcript of $\leq d$ rounds, the SAM_d oracle samples uniformly from the set of random tapes on which the M would produce the given transcript.

Based on the work of [Bra83], [AGGM06] and [Pas06], we have

Theorem 1 (Informal). *Assuming* $Dist^{1sided}$-$\mathsf{coNP} \not\subseteq Heur_{1/poly}\mathsf{AM}$, *one-way permutations and constant-round public-coin strongly witness-indistinguishable proofs for all of* NP *cannot be based on one-way functions with a Turing security reduction.*

Based on the work of [Sim98], [HHRS07] and [PV10], we have

Theorem 2 (Informal). *Assuming the existence of one-way functions secure against* $\mathrm{PPT}^{\mathsf{SAM}_{O(1)}}$ *(implied by* $Dist^{1sided}$-$\mathsf{coNP} \not\subseteq Heur_{1/poly}\mathsf{IP}[\mathrm{PPT}^{\mathsf{NP}}]$*), collision-resistant hash functions, constant-round statistically hiding commitments, and constant-round black-box zero-knowledge proofs for all of* NP *cannot be based on one-way functions with a Turing security reduction.*

Remark 1. Based on the work of [HMX10], the results in Theorem 2 can be obtained under the weaker assumption of Theorem 1 if we restrict to security reductions that have constant adaptivity.

In addition to these theorems, we again stress the following philosophical contribution: with the right assumptions, not only are non-black-box separation results possible, many such separations can be based on existing techniques. For example, the black-box separation results of [Sim98], [HHRS07] and [PV10] are essentially "upgraded" to non-black-box separations using our framework.

1.3 Our Techniques

Regarding the first assumption, $\mathrm{Dist}^{\mathrm{1sided}}$-$\mathsf{coNP} \not\subseteq \mathrm{Heur}_{1/\mathrm{poly}}\mathsf{AM}$, our separation results are largely based on previous works in the literature of separating cryptography from NP hardness, specifically ruling out constructions of one-way permutations [Bra83], size-verifiable one-way functions [AGGM06] and public-coin strongly witness-indistinguishable proofs [Pas06]. These works follow a common pattern: they take a (candidate) Turing security reduction of some cryptographic primitive P from NP, transform the reduction into an AM protocol, and conclude that $\mathsf{coNP} \subseteq \mathsf{AM}$, an unlikely consequence. By adapting their techniques, we show that a (candidate) Turing security reduction of the same primitive P from a one-way function can be transformed into an AM protocol that inverts the one-way function, and therefore the AM protocol may verify the co-range of f. This is a contradiction (not surprising since our assumption is an average case generalization of $\mathsf{coNP} \not\subseteq \mathsf{AM}$).

Our second assumption is used in a different fashion. Having justified the assumption that there exist one-way functions secure against $\mathsf{SAM} = \mathsf{SAM}_{O(1)}$, it follows that any cryptographic primitive P whose security can be broken using SAM cannot be based on one-way functions. This is because a Turing security reduction of primitive P from a one-way function f directly gives an algorithm that inverts f by using the SAM oracle, if $\mathsf{SAM}_{O(1)}$ can be used to break the security of primitive P. The $\mathsf{SAM}_{O(1)}$ oracle (as well as its variants) is particularly interesting in this aspect, since it is originally studied in the setting of black-box separations. Therefore, we know from previous works that in a relativized world with the $\mathsf{SAM}_{O(1)}$ oracle, there do not exist collision-resistant hash functions [Sim98], constant-round statistically hiding commitments [HHRS07], and zero-knowledge proofs for all of NP [PV10]. In a similar spirit, other on black-box separations can also be extended also to non-black-box separations; the work then lies in justifying the resulting new assumption.

A note on Turing reductions. In this work, we only consider constructions with Turing security reductions; that is, reductions that use the adversary (supposedly breaking the security of the construction) as a black box. The non-black-box simulation technique of Barak [Bar01] demonstrates how the code of the adversary can be used in security proofs for certain interactive zero-knowledge protocols. Such non-black-box reductions might potentially also be useful in analyzing the security of other cryptographic tasks.

However, as we argue, in the context of basing cryptographic primitives on one another, Turing reductions provide a semantically stronger notion of security than non-black-box reductions. The existence of a Turing reduction from a primitive P to a primitive Q implies that any "physical device"—which might rely on physical phenomena—that breaks the security of primitive Q, can be used to break the security of primitive P. With a non-black-box security reduction, we would instead require an *explicit* description of the code of the attack on primitive Q. Such descriptions might be hard to find: consider, for instance, a "human-aided" computation, where a human is interacting with a computer program in order to break a crypto system;[5] getting an explicit description of the attack would require providing an explicit (and "short") description of the human brain.

2 Preliminaries

We assume familiarity with common complexity classes such as NP, AM, etc., as well as common cryptographic primitives such as one-way functions (OWF), collision-resistant hash-functions (CRH), zero-knowledge proofs (ZK), and witness-indistinguishable proofs (WI).

Let $[n]$ denotes the set $\{1, \ldots, n\}$. Given an interactive protocol (P, V) (a pair of interactive Turing machines), let $\langle P, V \rangle (x)$ denote the output of V (the verifier) at the end of an execution with P (the prover), on common input x. Given a function $f : \{0,1\}^* \to \{0,1\}^*$ and a polynomial $q(n)$, we say g is $q(n)$ **concatenations of** f to mean that for $x_1, \ldots, x_{q(n)} \in \{0,1\}^n$, $g(x_1, \ldots, x_{q(n)}) = (f(x_1), \ldots, f(x_{q(n)}))$ (on other input lengths, g considers part of the input to be padding appropriately).

2.1 Distributional Languages

Definition 3 (Distributional Languages). An **ensemble of distributions** is a collection $D = \{D_1, D_2, \ldots\}$ where D_n is a distribution over $\{0,1\}^n$.

[5] Practical attacks on crypto-systems are often not fully automatized, but do indeed rely on such interactions; see e.g., [AAG $^+$00].

The ensemble is **efficiently samplable** if there exists a probabilistic polynomial-time algorithm S that, on input 1^n, outputs a sample according to D_n. A **distributional language** is a pair (L, D) where L is a standard language and D is an ensemble of distributions.

A well known class of distributional languages is Dist-coNP; it contains the set of distributional languages (L, D) where $L \in \mathsf{coNP}$ and D is efficiently samplable.

2.2 Hardness Amplification of One-Way Functions

The following lemma on hardness amplification of one-way functions is due to Yao [Yao82].

Lemma 4 ([Yao82]). *Let $f : \{0,1\}^* \to \{0,1\}^*$ be an efficiently computable function. Given any polynomial $q(n)$, let g be $q(n)$ concatenations of f. Then there is a PPT oracle machine $\mathcal{A}^{\mathcal{O}}$ such that whenever $\mathcal{O}$ is an oracle that inverts g with non-negligible probability, i.e., there exists some polynomial $p(n)$ such that for some set of n's,*

$$\Pr_{x \leftarrow \{0,1\}^{nq(n)}} \left[\mathcal{O}(g(x)) \in g^{-1}(g(x))\right] \geq 1/p(n)$$

then $\mathcal{A}^{\mathcal{O}}$ inverts f with probability $1 - 1/q(n)$, i.e., for the same set of n's,

$$\Pr_{x \leftarrow \{0,1\}^n} \left[\mathcal{A}^{\mathcal{O}}(f(x)) \in f^{-1}(f(x))\right] \geq 1 - 1/q(n)$$

3 On $\mathrm{Dist}^{\mathrm{1sided}}$-coNP $\not\subseteq$ $\mathrm{Heur}_{1/\mathrm{poly}}$AM

In this section we discuss our first assumption, $\mathrm{Dist}^{\mathrm{1sided}}$-coNP $\not\subseteq \mathrm{Heur}_{1/\mathrm{poly}}\mathsf{AM}$, starting with definitions, followed by its relation to other assumptions, and its implications on basing cryptography on one-way functions.

Definition 5. A distributional language (L, D) is in $\mathrm{Dist}^{\mathrm{1sided}}$-coNP if and only if $L \in \mathsf{coNP}$, D is efficiently samplable, and D only ranges over $\bar{L}$.

Remark 2. In other words, $(L, D) \in \mathrm{Dist}^{\mathrm{1sided}}$-coNP if and only if $(L, D) \in$ Dist-coNP and D only sample instances *not* in L.

Definition 6. A distributional language (L, D) is in $\mathrm{Heur}_{1/\mathrm{poly}}\mathsf{AM}$ if for every polynomial q, there exists an AM (i.e., constant-round public-coin) protocol (P, V) such that:

Completeness: If $x \in L$, $\Pr[\langle P, V\rangle(x) = 1] \geq 2/3$.

Soundness: For every $n \in \mathbb{N}$ and every machine P^*, with probability $1 - 1/q(n)$, an $x \in \{0,1\}^n$ sampled from D_n conditioned on $x \notin L$ satisfies $\Pr[\langle P^*, V\rangle(x) = 1] \leq 1/3$.

Remark 3. As usual, the choice of $2/3$ and $1/3$ is arbitrary and can be amplified to $1 - 2^{-n}$ and 2^{-n}. Intuitively, the soundness condition means that L is almost in AM, except for a fraction of instances in $\bar{L}$ that is sampled with (arbitrarily small) polynomial probability.

Remark 4. In a related work, Feige, Kim and Ofek give positive results in refuting restricted random coSAT instances on average [FKO06]. The main difference between the notion of average refutation and our definition of heuristic verifiability is in where errors are allowed. An average refutation algorithm may not refute a random unsatisfiable instance with small probability, but will never refute a satisfiable instance (i.e., perfect soundness). On a philosophical level, the work of [FKO06] gives a distribution of coSAT instances that may indeed be heuristically verifiable.

The complexity assumption we consider is $\mathrm{Dist}^{\mathrm{1sided}}\text{-}\mathsf{coNP} \not\subseteq \mathrm{Heur}_{1/\mathrm{poly}}\mathsf{AM}$, which is a strengthening of the more standard assumption that $\mathrm{Dist}\text{-}\mathsf{coNP} \not\subseteq \mathrm{Heur}_{1/\mathrm{poly}}\mathsf{AM}$, which in turn is the heuristic analog of $\mathsf{coNP} \not\subseteq \mathsf{AM}$.

Relation to other assumptions. To get a more concrete handle on our assumption, we prove that $\mathrm{Dist}^{\mathrm{1sided}}\text{-}\mathsf{coNP} \not\subseteq \mathrm{Heur}_{1/\mathrm{poly}}\mathsf{AM}$ is equivalent to the existence of an efficiently computable function f that is not heuristically co-range verifiable, i.e., there does not exist an AM protocol proving that an instance is outside the range of f, where soundness holds only with high probability with respect to random instances of $f(x)$. We then present several candidates for such a function (such as AES [DR02] and Learning Parity with Noise [BFKL93]). Using this equivalence, we also show that $\mathrm{Dist}^{\mathrm{1sided}}\text{-}\mathsf{coNP} \not\subseteq \mathrm{Heur}_{1/\mathrm{poly}}\mathsf{AM}$ is implied by the existence of pseudorandom generators secure against $\mathsf{BPP}(\mathsf{Promise}(\mathsf{AM} \cap \mathsf{coAM}))$[6].

3.1 Heuristic co-Range Verifiable Functions

Given a function f, consider the language $\mathsf{Range}_f = \{f(x) \mid x \in \{0,1\}^*\}$.

Definition 7. f is **heuristically co-range verifiable** if for any polynomial p, there exists an AM (i.e., constant-round public-coin) protocol (P, V) such that:

Completeness: For every $y \notin \mathsf{Range}_f$, $\Pr[\langle P, V\rangle(y) = 1] \geq 2/3$.
Soundness: For every $n \in \mathbb{N}$ and every machine P^*, with probability $1 - 1/p(n)$ over $x \leftarrow \{0,1\}^n$, $\Pr[\langle P^*, V\rangle(f(x)) = 1] \leq 1/3$.

Theorem 8. *$Dist^{1sided}$-$\mathsf{coNP} \not\subseteq Heur_{1/poly}\mathsf{AM}$ if and only if there exists an efficiently computable function that is not heuristically co-range verifiable.*

Proof. We show each direction separately.

[6] Traditionally, NW-style [NW94] PRGs against AM have been considered in the literature (see e.g., [MV05]); in contrast, we require a BMY-style [BM84, Yao82] "cryptographic" PRG.

"if" part: Let f be a function that is not heuristically co-range verifiable. By padding the input/output of f, construct another efficiently computable function g that is length preserving (i.e., $|g(x)| = |x|$ for all x). It is easy to see that padding preserves heuristic co-range verifiability, and so g is also not heuristically co-range verifiable. Consider the $\text{Dist}^{\text{1sided}}$-$\mathsf{coNP}$ distributional language (L, D) where $L = \overline{\mathsf{Range}_g}$ and D_n is the distribution that results from computing g on a uniformly random $x \in \{0,1\}^n$. Because g is not heuristically co-range verifiable, $(L, D) \notin \text{Heur}_{1/\text{poly}}\mathsf{AM}$.

"only-if" part: Let (L, D) be a distributional language such that $(L, D) \in \text{Dist}^{\text{1sided}}$-$\mathsf{coNP}$ and $(L, D) \notin \text{Heur}_{1/\text{poly}}\mathsf{AM}$, and let $t(n)$ be a bound on the random bits required to efficiently sample from D_n. Define f on input $x \in \{0,1\}^{t(n)}$ to be the result of sampling from D_n given randomness x (for other input lengths, f may treat part of the input as padding). f is an efficient function since D is efficiently samplable, and f is not heuristically co-range verifiable precisely because $(L, D) \notin \text{Heur}_{1/\text{poly}}\mathsf{AM}$. □

The statement "f is heuristically co-range verifiable" can be viewed as an average-case (heuristic) variant of the statement "$\mathsf{Range}_f \in \mathsf{coAM}$". (Also observe that if f is efficiently computable then $\mathsf{Range}_f \in \mathsf{NP} \subseteq \mathsf{AM}$.) We believe that the existence of such functions is a reasonable average-case generalization of SAT $\notin \mathsf{coAM}$: Just as it seems "unlikely" that there exist AM proofs for proving that a string is outside an arbitrary NP set, it seems "unlikely" that there is a AM proof for proving that a string is outside the range an arbitrary efficiently computable function, even if we only require soundness to hold for a random string in the range of the function.

Candidate functions that are not heuristic co-range verifiable. Although many traditional one-way functions (based for example on the hardness of factoring, RSA, discrete log [Rab80], or lattice-based problems [GG00, AR05]) are co-range verifiable, there are also "natural" one-way functions for which we do not know of co-range verifiability protocols. We here briefly discuss a few functions that are not known to be heuristically co-range verifiable.

Generalized AES: AES is a permutation on 128 bits [DR02]; that is, for a 128-bit seeds, AES_s is a permutation on defined on $\{0,1\}^{128}$. However, due to the algebraic nature of the construction of AES, it can easily be generalized to longer input lengths. Let AES^n denote this generalized version of AES to n-bit inputs. Now, consider the (one-way) function $f(x) = \text{AES}_x^{|x|}(0^{|x|})$. It would seems unlikely that this function is heuristically co-range verifiable.

Random Binary Linear Codes: A random binary linear code is obtained by encoding a message $x \in \{0,1\}^n$ as Ax where A is a random $m \times n$ binary matrix. Given the matrix A and a codeword y, it is easy to find the corresponding message x when $m \geq n$. However, the problem of finding x becomes hard when only a "noisy" codeword is given. The *learning parity with noise* (LPN) problem requires finding a random secret x, given $(A, Ax + e)$ where e is a "short" (binary) error vector. The worst-case variant of the LPN problem (i.e. given a set of equations $Ax = s$ to find x that maximally satisfies the equations) is known to be NP-hard even to approximate [Hås01]. The average-case version of LPN is also believed to be intractable: the $\mathrm{LPN}_{p,m}$ assumption [BFKL93] states that for $p \in (0, \frac{1}{2})$ and polynomial m, there is no PPT algorithm that finds x with more than negligible probability given $(A, Ax + e \bmod 2)$ where A is a random $m \times n$ binary matrix and every component of e is set to 1 independently with probability p. It seems like a reasonable strengthening of the LPN assumption to say that the function $x \mapsto (A, Ax + e \bmod 2)$ is not heuristically co-range verifiable, for some choices of m and p. In other words, there is no AM-proof showing that a binary string y is "far" from Ax for any x, even if soundness only holds for randomly perturbed codewords.

Pseudo-random Generators secure against $\mathsf{BPP}(\mathsf{Promise}(\mathsf{AM} \cap \mathsf{coAM}))$ While not a specific function, we show that this class of PRGs are not heuristically co-range verifiable.

Definition 9. Let U_n denote the distribution of uniform bit-strings of length n. A collection of efficiently computable functions $\mathcal{G} = \{g_n : \{0,1\}^n \to \{0,1\}^{n+1}\}_{n \in \mathbb{N}}$ is a PRG secure against $\mathsf{BPP}(\mathsf{Promise}(\mathsf{AM} \cap \mathsf{coAM}))$ if no PPT adversary with a $\mathsf{Promise}(\mathsf{AM} \cap \mathsf{coAM})$ oracle can distinguish the ensembles $\{g_n(U_n)\}_{n \in \mathbb{N}}$ and $\{U_{n+1}\}_{n \in \mathbb{N}}$ with non-negligible probability in n.

Claim 10. *Let $g : \{0,1\}^n \to \{0,1\}^{n+1}$ be a PRG secure against $\mathsf{BPP}(\mathsf{Promise}(\mathsf{AM} \cap \mathsf{coAM}))$. Then g is not heuristically range verifiable.*

Proof. Assume for contradiction that g is heuristically range verifiable. By the definition of heuristic range verifiability, there is a AM protocol (P, V) such that on input $g(x)$ for a uniformly random $x \in \{0,1\}^n$, V rejects $g(x)$ with probability at least $1 - 1/n$. Let $S = \{x \in \{0,1\}^n \mid \Pr[V \text{ rejects } g(x)] \leq 1/n\}$ (i.e., the set of x where V fails to reject $g(x)$). Then we must have

$$\Pr_{x \leftarrow \{0,1\}^n}[x \in S] \leq 2/n$$

Let $T = \{g(x) \mid x \in S\}$, i.e., the set of inputs where (P, V) has high soundness error. Now consider the promise problem $\Pi = (\Pi_Y, \Pi_N) = (\mathsf{Range}_g - T, \overline{\mathsf{Range}_g})$. Note that Π is trivially in $\mathsf{NP} \subseteq \mathsf{AM}$, and that $\Pi \in \mathsf{coAM}$ by definition of T (via protocol (P, V)). Therefore $\Pi \in \mathsf{AM} \cap \mathsf{coAM}$.

We now describe a polynomial-time distinguisher $\mathcal{D}$ that has oracle access to a decision procedure for the the promise problem Π. On input y, $\mathcal{D}$ simply outputs $\Pi(y)$. To show that $\mathcal{D}$ is a good distinguisher for g, observe that

$$\Pr_{x \leftarrow \{0,1\}^n}[\mathcal{D}(g(x)) = 1] \geq \Pr_x[g(x) \notin T] = \Pr_x[x \notin S] \geq 1 - \frac{2}{n}$$

On the other hand,

$$\Pr_{y \leftarrow \{0,1\}^{n+1}}[\mathcal{D}(y) = 1] \leq \Pr_y[y \notin \mathsf{Range}_g] \leq \frac{1}{2} \quad \square$$

Claim 10 together with forthcoming theorems yields the following trade-off: if certain cryptographic primitives can be based on OWFs, then there does not exist PRGs secure against $\mathsf{BPP}(\mathsf{Promise}(\mathsf{AM} \cap \mathsf{coAM}))$.

3.2 Consequences of $\mathrm{Dist}^{1\mathrm{sided}}$-$\mathsf{coNP} \not\subseteq \mathrm{Heur}_{1/\mathrm{poly}}\mathsf{AM}$

The assumption $\mathrm{Dist}^{1\mathrm{sided}}$-$\mathsf{coNP} \not\subseteq \mathrm{Heur}_{1/\mathrm{poly}}\mathsf{AM}$ implies some impossibility results on basing cryptographic primitives on one-way functions. First, we provide an outline of our proof framework.

Recall that we consider arbitrary non-black-box (and even non explicit) constructions based on one-way functions, but restrict our attention to Turing (black-box) security reductions. This means a primitive P constructed from a one-way function f is accompanied by a PPT oracle reduction $R^{\mathcal{O}}$, such that whenever $\mathcal{O}$ is an oracle that "breaks the security of P, $R^{\mathcal{O}}$ inverts the f with non-negligible probability. We will show that for certain primitives P and respective oracles $\mathcal{O}$ that break the security of P, the reduction $R^{\mathcal{O}}$ can be emulated in an AM protocol, allowing the verifier of the AM protocol to invert the one-way function. Coupled with the Yao's amplification lemma (Lemma 4), the verifier can actually invert f with very high probability, and therefore heuristically verify the co-range of f (by checking for a lack of inverses).

We present the lower-bound result for one-way permutations and Strong WI AM proofs based on OWFs below.

On Basing One-Way Permutations on One-Way Functions. We first formalize the definition of basing one-way permutations (OWP) on one-way functions (OWF) with Turing (black-box) reductions, and show that such a construction is ruled out by the assumption $\mathrm{Dist}^{1\mathrm{sided}}$-$\mathsf{coNP} \not\subseteq \mathrm{Heur}_{1/\mathrm{poly}}\mathsf{AM}$.

Definition 11. We say that **OWPs can be based on OWFs** if:

Construction: There is a mapping that takes the description of any polynomial-time function f (candidate OWF) and outputs the description of a permutation $\phi = \phi_f$ (candidate OWP).

Reduction: For any polynomial-time function f, there is a PPT oracle algorithm R_f such that whenever $\mathcal{O}$ inverts ϕ, i.e., there is a polynomial p such that $\Pr_{x \leftarrow \{0,1\}^n}[\mathcal{O}(\phi(x)) = x] \geq 1/p(n)$, $R_f^{\mathcal{O}}$ inverts f, i.e., there is some polynomial p' such that

$$\Pr_{x \leftarrow \{0,1\}^n}[R_f^{\mathcal{O}}(f(x)) \in f^{-1}(f(x))] \geq 1/p'(n)$$

The following theorem is proved using our framework combined with the work of [Bra83].

Theorem 12. *If OWPs can be based on OWFs, then Dist^{1sided}-$\mathsf{coNP} \subseteq \mathit{Heur}_{1/poly}\mathsf{AM}$ (contradicting our assumption).*

Proof. Suppose that OWPs can be based on OWFs. We will show that every efficiently computable function is heuristically co-range verifiable. Fix any efficient function f and polynomial $q(n)$ (as in the definition of heuristically co-range verifiability), and define g to be $q(n)$ concatenations of f. By assumption, there exists a permutation P_g and an efficient security reduction R_g such that, given an oracle $\mathcal{O}$ that inverts ϕ inverts g, $R_g^{\mathcal{O}}$ inverts g with non-negligible probability.

Using Lemma 4, we can construct a new efficient reduction $\tilde{R}_f$ that, given an oracle $\mathcal{O}$ that inverts ϕ inverts g, $\tilde{R}_f^{\mathcal{O}}$ inverts f with probability $1 - 1/q(n)$.

Next we recall from [Bra83] an AM protocol that allows the verifier to run $\tilde{R}_f$ without access to $\mathcal{O}$. The verifier start by sending the prover a sufficiently long random string to act as the random tape of $\tilde{R}_f$. The prover then runs $\tilde{R}_f$ with the given randomness, solving oracle queries as needed. When $\tilde{R}_f$ terminates, the prover sends the output of $\tilde{R}_f$ as well as any oracle query-answer pairs encountered in the execution of $\tilde{R}_f$ to the verifier. The verifier can check the validity of the oracle query-answer pairs, and the validity of the execution using the given oracle query-answer pairs. On common input y, the verifier accepts if and only if $\tilde{R}_f(y)$ fails to find an inverse.

Completeness: If $y \notin \mathsf{Range}_f$, and if the prover simulates $\tilde{R}_f(y)$ honestly, then the verifier will always accept the simulation, and of course $\tilde{R}_f$ will never find an inverse to y under f. Hence we have completeness probability 1.

Soundness: We may assume that the verifier accepts the execution of $\tilde{R}_f(y)$ provided by the (possibly cheating) prover. In this case, the simulated execution of $\tilde{R}_f(y)$ is identical to a real execution of $\tilde{R}_f^{\mathcal{O}}(y)$ for a "perfect oracle" $\mathcal{O}$ that answers all queries correctly; this is because every oracle has exactly one answer. Therefore:

$$\Pr_{x \leftarrow \{0,1\}^n}[\tilde{R}_f(f(x)) \in f^{-1}(f(x))] > 1 - 1/q(n)$$

By an averaging argument, we have that with probability at least $1 - 3/q(n)$ over a random $x \in \{0,1\}^n$, $y = f(x)$,

$$\Pr[\tilde{R}_f(f(x)) \in f^{-1}(f(x))] > 2/3$$

in which case the verifier would reject.

This concludes that f is heuristically co-range verifiable.

Remark 5. The difficulty of extending Theorem 12 to other cryptographic primitives comes from constructing an AM protocol. For many primitives (e.g., collections of trapdoor one-way functions), an oracle that breaks the security of the primitive suffers from two caveats: some queries have no answers (which cannot be checked by the verifier), and some queries have multiple answers (which allow a cheating prover to adaptively select the answer). These difficulties are well known; see [BT03, AGGM06, HMX10].

Theorem 12 can be extended beyond one-way permutations. For example, it can rule out basing *certified collection of (trapdoor) permutations* on one-way functions [BY96]. In this case, an oracle query consists of a candidate permutation description and a candidate image. The verifier can check whether each description is indeed a valid permutation in the collection (certifiable), and if so expect a unique inverse of the given image. (We may even extend the definition of "certified" to mean certifiable under an AM protocol.)

Another example is to rule out basing *size-verifiable, polynomial-sized pre-image one-way functions* on one-way functions [AGGM06]. In this case, size-verifiable one-way functions allow the verifier to check the pre-image size of any oracle query (in particular the verifier checks whether a pre-image exists). Then, the verifier may ask the prover to provide all polynomially many pre-images to force a unique answer.

On Basing Public-Coin Strongly Witness Indistinguishable Proofs on OWFs. Using the same framework, we rule out the possibility of basing $O(1)$-round public-coin strongly witness-indistinguishable proofs (Strong-WI AM) for languages in NP on OWFs. Below, we provide the result and brief overview of the proof. The complete proof will appear in the full version.

The definition of basing Strong-WI AM proofs on OWFs can be extended similarly to OWPs. Roughly speaking, for any language L, there exists a mapping from the description of any function f to a protocol $(P_f^{\mathsf{sWI}}, V_f^{\mathsf{sWI}})$ and a reduction R such that for any adversary O and pair of ensembles of distributions, $\{D_n^1\}_{n\in\mathbb{N}}$ and $\{D_n^2\}_{n\in N}$, and D_n^1 and D_n^2 are distributions over $L\cap\{0,1\}^n\times\{0,1\}^*$, if O distinguishes proofs of statements using $(P_f^{\mathsf{sWI}}, V_f^{\mathsf{sWI}})$ sampled from the two distributions D_n^1 and D_n^2, then R^O inverts f with non-negligible probability. The main result we obtain using the work of [Pas06] is.

Theorem 13. *If there exists $O(1)$-round Strong-WI AM proof systems with perfect completeness based on OWFs for all* NP*-languages, then* $Dist^{1sided}$-$\mathsf{coNP}\subseteq Heur_{1/poly}\mathsf{AM}$.

On a high-level, [Pas06] shows how to construct a game G^f from any function f using a Strong-WI AM protocol for NP languages based on f such that there exists a reduction from breaking the game to inverting the function f. Additionally, he shows that a worst-case breaking oracle for G^f can be simulated using an AM protocol. We obtain our result using the same game G^f but instead of using any one-way function f, we use the function g obtained from any language $(L, D)\in \mathrm{Dist}^{\mathrm{1sided}}$-$\mathsf{coNP}$ as in the proof for OWP. Since a worst-case breaker can be simulated using an AM protocol, following the proof technique from Theorem 12, it essentially follows that $(L, D)\in \mathrm{Heur}_{1/\mathrm{poly}}\mathsf{AM}$.

4 On One-Way Functions Secure against $\mathrm{PPT}^{\mathsf{SAM}_{O(1)}}$

In this section we explore our second assumption: the existence of one-way functions that cannot be inverted by $\mathrm{PPT}^{\mathsf{SAM}_{O(1)}}$: efficient algorithms that have access to a $\mathsf{SAM}_{O(1)}$ oracle.

4.1 Definition of the SAM Oracle

Let M be a probabilistic interactive Turing machine that runs a d-round protocol. Let $\mathrm{TRANS}_i = (a_1, b_1, \ldots, a_i, b_i)$ be a partial transcript of the messages exchange with $M(1^n)$ in an execution. We use :: to denote appending messages to a transcript. Define $R_{\text{TRANS}_i}(M)$ to be the set of all random tapes τ for which $M_\tau(1^n, a_1, b_1, \ldots, b_{j-1}) = a_j$ for all $j < i$; we say that such a τ is *consistent* with respect to TRANS_i. Without loss of generality, we assume that M sends the first message (i.e., outputs a message on initiation). The oracle $\mathsf{SAM}_{d(n)}$ takes inputs of the form $Q = (M(1^n), \text{TRANS}_i, r)$ where $\text{TRANS}_{i-1} = (a_1, b_1, \ldots, b_{i-1})$ is a partial transcript and $r \in \{0,1\}^*$. On input Q, $\mathsf{SAM}_{d(n)}$ outputs $(\tau', \text{TRANS}_{i-1} :: a_i)$ such that $\tau' \in R_{\text{TRANS}_{i-1}}(M(1^n))$ and $M_{\tau'}(1^n, \text{TRANS}_i) = a_i$,[7] with the following restrictions:

1. If $i > 1$, then $(a_1, b_1, \ldots, a_{i-1})$ was the result of a previous query of the form $(M, (a_1, b_1, \ldots, b_{i-2}), r')$ for some $r' \in \{0,1\}^*$.
2. τ' is uniformly distributed in $R_{\text{TRANS}_{i-1}}(M)$ over the randomness of $\mathsf{SAM}_{d(n)}$, independent of all other queries.
3. $\mathsf{SAM}_{d(n)}$ answers queries only up to a depth $d(n)$, i.e. $i \leq d(n)$.

Otherwise, $\mathsf{SAM}_{d(n)}$ outputs $\bot$. The role of r in the query is to obtain new and independent samples for each r and to allow a verifier to obtain the same sample query by querying on the same r.

Our above description of the $\mathsf{SAM}_{d(n)}$-oracle is a stateful instantiation of the oracle defined in [HHRS07]. Just as in [HHRS07], for our results, we need the oracle to be stateless; [HHRS07] specify how to modify the oracle to achieve this (using "signatures"); we omit the details. When clear from context, we drop the input 1^n to M.

Definition 14. We say that a (one-way) function $f : \{0,1\}^* \to \{0,1\}^*$ is **secure against** (or **hard to invert by**) $\text{PPT}^{\mathsf{SAM}_d}$ if for every oracle PPT machine A there exists a negligible function $\nu(\cdot)$ such that

$$\Pr[x \leftarrow \{0,1\}^n ; y = f(x) : A^{\mathsf{SAM}_d}(y) \in f^{-1}(y)] \leq \nu(n)$$

In this work, we focus on the $\mathsf{SAM}_{O(1)}$ and in the rest of the paper, we refer to this oracle simply by SAM.

Definition 15. We say that a language L is in $\mathsf{BPP}^{\mathsf{SAM}}$ if there exists an oracle PPT machine M such that the following holds:

Completeness: For every $x \in L$, $\Pr[M^{\mathsf{SAM}}(x) = 1] \geq 2/3$
Soundness: For every $x \notin L$, $\Pr[M^{\mathsf{SAM}}(x) = 1] \leq 1/2$

The second assumption that we consider to establish non black-box lower bounds is the existence of one-way functions that are secure against $\text{PPT}^{\mathsf{SAM}}$. We justify our assumption in the next section.

---

[7] It suffices to consider an oracle that merely outputs τ', however, we consider SAM that additionally outputs $\text{TRANS}_{i-1} :: a_i$ for ease of exposition.

4.2 Relation to $\text{Dist}^{\text{1sided}}$-coNP $\not\subseteq$ $\text{Heur}_{1/\text{poly}}\mathsf{IP}[\text{PPT}^{\mathsf{NP}}]$

Definition 16. A distributional language (L, D) is in $\text{Heur}_{1/\text{poly}}\mathsf{IP}[\text{PPT}^{\mathsf{NP}}]$ if for every polynomial q, there exists an interactive protocol (P, V) where $P \in \text{PPT}^{\mathsf{NP}}$ (oracle PPT machine with oracle access to an NP oracle) such that:

Completeness: If $x \in L$, $\Pr[\langle P, V\rangle(x) = 1] \geq 2/3$.

Soundness: For every $n \in \mathbb{N}$ and every machine P^*, with probability $1 - 1/q(n)$, an $x \in \{0,1\}^n$ sampled from D_n conditioned on $x \notin L$ satisfies $\Pr[\langle P^*, V\rangle(x) = 1] \leq 1/3$.

The assumption $\text{Dist}^{\text{1sided}}$-coNP $\not\subseteq$ $\text{Heur}_{1/\text{poly}}\mathsf{IP}[\text{PPT}^{\mathsf{NP}}]$ is a heuristic extension of the worst case assumption coNP $\not\subseteq$ $\mathsf{IP}[\text{PPT}^{\mathsf{NP}}]$, i.e., there are no interactive proofs for coSAT where the prover is efficient with a NP oracle. While coNP $\not\subseteq$ $\mathsf{IP}[\text{PPT}^{\mathsf{NP}}]$ is not as well studied as more standard assumptions like coNP $\not\subseteq$ AM, the search for the aforementioned interactive proof for coSAT has been open since the question was raised by Babai, Fortnow and Lund in 1991 [BFL91]. Next we show that $\text{Dist}^{\text{1sided}}$-coNP $\not\subseteq$ $\text{Heur}_{1/\text{poly}}\mathsf{IP}[\text{PPT}^{\mathsf{NP}}]$ implies the existence of one-way functions secure against $\text{PPT}^{\mathsf{SAM}}$; the bulk of the technical content of the proof is taken from [HMX10].

Lemma 17. *If $\text{Dist}^{\text{1sided}}$-coNP $\not\subseteq$ $\text{Heur}_{1/\text{poly}}\mathsf{IP}[\text{PPT}^{\mathsf{NP}}]$, then there exists a one-way function that is secure against $\text{PPT}^{\mathsf{SAM}}$.*

Proof. We prove the contrapositive. Suppose all efficiently computable functions can be inverted by $\text{PPT}^{\mathsf{SAM}}$. Fix any $(L, D) \in \text{Dist}^{\text{1sided}}$-coNP and any polynomial q as in the definition of $\text{Heur}_{1/\text{poly}}\mathsf{IP}[\text{PPT}^{\mathsf{NP}}]$. We will show that $(L, D) \in \text{Heur}_{1/\text{poly}}\mathsf{IP}[\text{PPT}^{\mathsf{NP}}]$.

Let $t(n)$ be a bound on the randomness required to efficiently sample from D_n, define f on input $x \in \{0,1\}^{t(n)}$ to be the result of sampling from D_n given randomness x, and let $g = g_q$ be $q(n)$ concatenations of f. By assumption, there is a PPT oracle algorithm R such that R^{SAM} inverts g with polynomial probability. By Lemma 4, we can further construct a PPT oracle algorithm $\tilde{R}$ such that $\tilde{R}^{\mathsf{SAM}}$ inverts f with probability $1 - 1/q(n)$.

By the work of Haitner et. al [HMX10], the reduction $\tilde{R}$ can be simulated in an interactive proof (P, V) where the P is an efficient algorithm with access to an NP oracle. Specifically, using Theorem 5.2 of [HMX10][8], with parameter $\delta = 1/q$, (P, V) has two properties:

Completeness: (P, V) has completeness error $1/q(n)$ (the probability that V aborts).

Soundness: For any (possibly cheating) prover P^*, if V does not abort, $\langle P^*, V\rangle(y)$ (the output of V) and the output of $\tilde{R}^{\mathsf{SAM}}(y)$ has statistical difference at most $1/q(n)$.

[8] The theorem number refers to the full version of [HMX10] on ECCC.

We modify the protocol so that V on input y accepts if and only if V does not abort during the simulation of $\tilde{R}$, and that $\tilde{R}$ does not find an inverse of y under f. The resulting protocol shows that $(L, D) \in \text{Heur}_{1/\text{poly}}\mathsf{IP}[\text{PPT}^{\mathsf{NP}}]$:

Completeness: On input $y \in L$, i.e., $y \notin \mathsf{Range}_f$, V only rejects during the simulation of $\tilde{R}$ because $\tilde{R}$ can never find an inverse to y. Therefore V rejects with probability at most $1/q(n)$.

Soundness: Let P^* be an arbitrary machine. On a random input $y \notin L$ distributed according to D_n, i.e., $y = f(x)$ for a random $x \in \{0,1\}^{t(n)}$, $\tilde{R}^{\mathsf{SAM}}(y)$ would find an inverse of y with probability $1 - 1/q(n)$. Therefore, if V does not reject the simulation of $\tilde{R}$ provided by P^*, V would find an inverse of y with probability at least $1 - 2/q(n)$. By an averaging argument, with probably at least $1 - 3/q(n)$ over choosing y from D_n, $\Pr[\langle P^*, V\rangle(y) = 0] \geq 2/3$.

4.3 Consequences of the Existence of One-Way Function Secure w.r.t PPT^{SAM}

Assuming the existence of one-way function secure against PPT^{SAM} we show separation of collision-resistant hash-functions, $O(1)$-round statistically-hiding commitments and $O(1)$-round zero-knowledge proofs for NP from OWFs. On a high-level, for each of these primitives, we show that there exists an adversary that can break the security with oracle access to SAM. Therefore, if these primitives could be based on one-way functions, then we arrive at a contradiction under the assumption.

As with the case of one-way permutations, we consider arbitrary non-black-box (and even non explicit) constructions, but as before restrict attention to Turing (i.e., black-box) security reductions. The definitions of basing CRHs, statistically-hiding commitments and zero-knowledge proofs on one-way functions can be extended analogously from OWP. Below we discuss briefly how the SAM oracle can be used to break each primitive.

Collision-Resistant Hash-Functions: Recall that, the SAM oracle can sample uniform collisions for probabilistic interactive Turing machines. If we consider the efficient Turing machine that computes the CRH function, it follows that SAM can find a collision for a uniform input to the CRH if one exists. Since any length-compressing function with high-probability has collisions for uniformly chosen inputs, SAM breaks any CRH. We remark that it suffices to consider the potentially weaker SAM_1-oracle to break CRHs. As a consequence, we obtain the following theorem.

Theorem 18. *Assuming the existence of one-way functions that are secure against* $\text{PPT}^{\mathsf{SAM}}$*, we have that worst-case CRHs cannot be based on OWFs.*

As a corollary, we also obtain (a potentially weaker statement) that worst-case CRHs cannot be based on OWFs unless $\text{Dist}^{\text{1sided}}\text{-}\mathsf{coNP} \subseteq \text{Heur}_{1/\text{poly}}\mathsf{IP}[\text{PPT}^{\mathsf{NP}}]$.

Statistically-Hiding Commitments: We show that, for every $O(1)$-round statistically hiding commitment based on one-way functions, there exists a cheating sender who with oracle access to SAM violates the binding property of the commitment. Haitner, Hoch, Reingold and Segev [HHRS07] prove that using the stronger SAM^π oracle (that finds collisions for PPT machines that access a random permutation oracle π), there is a cheating committer that can break the binding property of any fully black-box construction of a statistically-hiding commitment scheme based on one-way permutations. It essentially follows using the same proof that without access to any oracle π, SAM can break any statistically-hiding commitment scheme with a PPT committer. As a consequence, we obtain the following theorem.

Theorem 19. *Assuming the existence of one-way functions secure w.r.t.* $\mathrm{PPT}^{\mathsf{SAM}}$*, then there exists no* $O(1)$*-round statistically-hiding bit-commitment scheme based on one-way function.*

As for the case of CRHs, we also have that there exists no $O(1)$-round statistically-hiding bit-commitment scheme unless $\mathrm{Dist}^{1\mathrm{sided}}\text{-}\mathsf{coNP} \subseteq \mathrm{Heur}_{1/\mathrm{poly}}\mathsf{IP}[\mathrm{PPT}^{\mathsf{NP}}]$.

Zero-Knowledge Proofs: Using similar techniques we show how to extend to lower-bound of [PV10] on $O(1)$-round zero-knowledge proofs based on one-way functions. Goldreich-Krawczyk [GK96b] showed that only languages in BPP have constant-round *public-coin* black-box zero-know-ledge protocols. In [PV10], this lower bound was extended to "fully black-box" constructions of black-box zero-knowledge proofs (that could be *private-coin*) based on one-way functions. More precisely, they show that only languages decidable by oracle PPT machines with oracle access to SAM^π (for random permutation π) can have constant-round fully black-box zero-knowledge proofs. On a high-level, they establish this lower-bound, by providing a transformation that takes any private-coin zero-knowledge proof based on OWFs and produces a public-coin zero-knowledge proof in a SAM^π-relativized world and then concluding using the result of Goldreich-Krawczyk for public-coin protocols. Based on the result of [PV10], we obtain the following theorem.

Theorem 20. *Assume the existence of one-way functions that are secure w.r.t.* $\mathrm{PPT}^{\mathsf{SAM}}$*, there does not exist* $O(1)$*-round computational zero-knowledge proofs for all of* NP *based on one-way functions.*

Following the proof of [PV10], we can show that only languages in $\mathrm{PPT}^{\mathsf{SAM}}$ have $O(1)$-round computational zero-knowledge proofs based on one-way functions. We complete the argument by noting that our assumption implies that $\mathsf{NP} \not\subseteq \mathsf{BPP}^{\mathsf{SAM}}$, since otherwise, we can construct an oracle PPT machine that with oracle access to SAM inverts OWFs. We provide the formal proof in the full version.

Finally, we remark that Theorem 20 implies Theorem 19 relying on the result of Goldreich and Kahan [GK96a] and Theorem 19 implies Theorem 18 relying on the result of Damgård, Pedersen and Pfitzmann [DPP98]. Nevertheless, the direct proofs are simpler and as mentioned before, it suffices to assume the weaker SAM_1-oracle for Theorem 18.

Acknowledgements

We are thankful to the anonymous TCC-reviewers for their insightful comments.

References

[AAG+00] Almgren, F., Andersson, G., Granlund, T., Ivansson, L., Ulfberg, S.: How we cracked the code book ciphers (2000) (manuscript), `http://codebook.org/codebook_solution.pdf`

[AGGM06] Akavia, A., Goldreich, O., Goldwasser, S., Moshkovitz, D.: On basing one-way functions on NP-hardness. In: STOC 2006, pp. 701–710 (2006)

[AR05] Aharonov, D., Regev, O.: Lattice problems in NP cap coNP. J. ACM 52(5), 749–765 (2005)

[Bar01] Barak, B.: How to go beyond the black-box simulation barrier. In: FOCS 2001, vol. 0, pp. 106–115 (2001)

[BFKL93] Blum, A., Furst, M., Kearns, M., Lipton, R.: Cryptographic Primitives Based on Hard Learning Problems. In: Stinson, D.R. (ed.) CRYPTO 1993. LNCS, vol. 773, pp. 278–291. Springer, Heidelberg (1994)

[BFL91] Babai, L., Fortnow, L., Lund, C.: Non-deterministic exponential time has two-prover interactive protocols. Computational Complexity 1, 3–40 (1991)

[BM84] Blum, M., Micali, S.: How to generate cryptographically strong sequences of pseudo-random bits. SIAM Journal on Computing 13(4), 850–864 (1984)

[Bra83] Brassard, G.: Relativized cryptography. IEEE Transactions on Information Theory 29(6), 877–893 (1983)

[BT03] Bogdanov, A., Trevisan, L.: On worst-case to average-case reductions for np problems. In: FOCS 2003, pp. 308–317 (2003)

[BY96] Bellare, M., Yung, M.: Certifying permutations: Noninteractive zero-knowledge based on any trapdoor permutation. J. Cryptology 9(3), 149–166 (1996)

[DDN00] Dolev, D., Dwork, C., Naor, M.: Nonmalleable cryptography. SIAM Journal on Computing 30(2), 391–437 (2000)

[DPP98] Damgård, I., Pedersen, T.P., Pfitzmann, B.: Statistical secrecy and multibit commitments. IEEE Transactions on Information Theory 44(3), 1143–1151 (1998)

[DR02] Daemen, J., Rijmen, V.: The Design of Rijndael. Springer, New York, Inc. (2002)

[FKO06] Feige, U., Kim, J.H., Ofek, E.: Witnesses for non-satisfiability of dense random 3cnf formulas. In: FOCS 2006, pp. 497–508 (2006)

[FS90] Feige, U., Shamir, A.: Witness indistinguishable and witness hiding protocols. In: STOC 1990, pp. 416–426 (1990)

[GG00] Goldreich, O., Goldwasser, S.: On the limits of nonapproximability of lattice problems. J. Comput. Syst. Sci. 60(3), 540–563 (2000)

[GGM86] Goldreich, O., Goldwasser, S., Micali, S.: How to construct random functions. Journal of ACM 33(4), 792–807 (1986)

[GK96a] Goldreich, O., Kahan, A.: How to construct constant-round zero-knowledge proof systems for NP. Journal of Cryptology 9(3), 167–190 (1996)

[GK96b] Goldreich, O., Krawczyk, H.: On the composition of zero-knowledge proof systems. SIAM Journal on Computing 25(1), 169–192 (1996)

[GM84] Goldwasser, S., Micali, S.: Probabilistic encryption. J. Comput. Syst. Sci. 28(2), 270–299 (1984)

[GMW91] Goldreich, O., Micali, S., Wigderson, A.: Proofs that yield nothing but their validity for all languages in np have zero-knowledge proof systems. J. ACM 38(3), 691–729 (1991)

[Hås01] Håstad, J.: Some optimal inapproximability results. J. ACM 48(4), 798–859 (2001)

[HHRS07] Haitner, I., Hoch, J.J., Reingold, O., Segev, G.: Finding collisions in interactive protocols - a tight lower bound on the round complexity of statistically-hiding commitments. In: FOCS, pp. 669–679 (2007)

[HILL99] Håstad, J., Impagliazzo, R., Levin, L.A., Luby, M.: A pseudorandom generator from any one-way function. SIAM J. Comput. 28(4), 1364–1396 (1999)

[HMX10] Haitner, I., Mahmoody, M., Xiao, D.: A new sampling protocol and applications to basing cryptographic primitives on the hardness of NP. In: IEEE Conference on Computational Complexity, pp. 76–87 (2010)

[HR07] Haitner, I., Reingold, O.: Statistically-hiding commitment from any one-way function. In: STOC 2007, pp. 1–10 (2007)

[IR88] Impagliazzo, R., Rudich, S.: Limits on the Provable Consequences of One-Way Permutations. In: Goldwasser, S. (ed.) CRYPTO 1988. LNCS, vol. 403, pp. 8–26. Springer, Heidelberg (1990)

[MV05] Miltersen, P.B., Vinodchandran, N.V.: Derandomizing arthur-merlin games using hitting sets. Computational Complexity 14(3), 256–279 (2005)

[Nao91] Naor, M.: Bit commitment using pseudorandomness. J. Cryptology 4(2), 151–158 (1991)

[NOVY98] Naor, M., Ostrovsky, R., Venkatesan, R., Yung, M.: Perfect zero-knowledge arguments for p using any one-way permutation. J. Cryptology 11(2), 87–108 (1998)

[NW94] Nisan, N., Wigderson, A.: Hardness vs randomness. J. Comput. Syst. Sci. 49(2), 149–167 (1994)

[Pas06] Pass, R.: Parallel repetition of zero-knowledge proofs and the possibility of basing cryptography on NP-hardness. In: IEEE Conference on Computational Complexity, pp. 96–110 (2006)

[PV10] Pass, R., Venkitasubramaniam, M.: Private coins versus public coins in zero-knowledge proofs. In: Micciancio, D. (ed.) TCC 2010. LNCS, vol. 5978, pp. 588–605. Springer, Heidelberg (2010)

[Rab80] Rabin, M.O.: Probabilistic algorithm for testing primality. Journal of Number Theory 12(1), 128–138 (1980)

[Sim98] Simon, D.R.: Findings Collisions on a One-Way Street: Can Secure Hash Functions Be Based on General Assumptions? In: Nyberg, K. (ed.) EUROCRYPT 1998. LNCS, vol. 1403, pp. 334–345. Springer, Heidelberg (1998)

[Yao82] Yao, A.C.-C.: Theory and applications of trapdoor functions (extended abstract). In: FOCS 1982, pp. 80–91 (1982)

On Black-Box Separations among Injective One-Way Functions

Takahiro Matsuda* and Kanta Matsuura

The University of Tokyo, Japan
{tmatsuda,kanta}@iis.u-tokyo.ac.jp

Abstract. A one-way permutation (OWP) is one of the most fundamental cryptographic primitives, and can be used as a building block for most of basic symmetric-key cryptographic primitives. However, despite its importance and usefulness, previous black-box separation results have shown that constructing a OWP from another primitive seems hopeless, unless building blocks already achieve "one-way" property and "permutation" property simultaneously. In this paper, in order to clarify more about the constructions of a OWP from other primitives, we study the construction of a OWP from primitives that are very close to a OWP. Concretely, as a negative result, we show that there is no fully black-box construction of a OWP from a length-increasing injective one-way function (OWF), even if the latter is just 1-bit-increasing and achieves strong form of one-wayness which we call *adaptive one-wayness*. As a corollary, we show that there is no fully black-box construction of a OWP from a regular OWF with regularity greater than 1. Since a permutation is length-preserving and injective, and is a regular OWF with regularity 1, our negative result indicates that to construct a OWP from another primitive is quite difficult, even if we use very close primitives to a OWP as building blocks. Moreover, we extend our separation result of a OWP from a length-increasing injective OWF, and show a certain restrictive form of black-box separations among injective OWFs in terms of how much a function stretches its input. This result shows a hierarchy among injective OWFs (including a OWP).

Keywords: black-box separation, injective one-way function, one-way permutation, adaptive one-wayness.

1 Introduction

A one-way permutation (OWP) is one of the most fundamental cryptographic primitives[1]. It has been shown that OWPs are sufficient for constructing most of basic "symmetric-key" primitives, which include, e.g. pseudorandom generators [18,2], pseudorandom functions [4], symmetric-key encryption schemes and

* Takahiro Matsuda is supported by JSPS Research Fellowships for Young Scientists.
[1] Unless otherwise stated explicitly, whenever we say "OWP", we mean a single OWP, not a family of OWPs.

Y. Ishai (Ed.): TCC 2011, LNCS 6597, pp. 597–614, 2011.

message authentication codes. While most of primitives implied by a OWP are later shown to be implied by an ordinary one-way function (OWF) (e.g. a pseudorandom generator from any OWF [7]), it is usual that a primitive built from a OWP is more efficient than the one built from a general OWF. Therefore, a OWP is still quite an important primitive, and constructions of a OWP from other (simpler) primitives (possibly efficiently) is worth studying. However, how to construct a OWP from other primitives has not been studied well, compared to constructions of other primitives that use a OWP as a building block. In this paper, we focus on constructions of OWPs.

There are several negative results on this, in terms of black-box constructions[2]. The combination of the results by Rudich [14] and Kahn et al. [10] shows that there is no black-box construction of a OWP from a OWF. Chang et al. [3] later extend it and show that there is no black-box construction of a OWP from (a family of) trapdoor functions or private information retrieval protocols. These negative results indicate that it is quite difficult to construct a OWP from other primitives. There are also some positive results. However, to the best of our knowledge, the only positive results about the construction of a OWP are the constructions from primitives which already have "one-way" property and "permutation" property simultaneously: Yao [18] shows that the existence of a permutation which is weakly one-way implies that of a (normal) OWP. Goldreich et al. [5] show that if a family of OWPs satisfies some special properties, it can be used to construct a single OWP.

Taking into account these negative and positive results on the constructions of OWPs, a natural question that arises here is: *Which properties of a OWP make it hard to construct a OWP from other primitives?* To clarify this, in this paper, we focus on a special type of OWFs, a length-increasing injective OWF, and tackle the problem of whether we can construct a OWP from it. Recall that a permutation is a function which is both length-preserving and injective. Therefore, a length-increasing injective OWF is one of the primitives that is extremely close to a OWP. Regarding this problem, our answer is negative.

The problem on a OWP versus a length-increasing injective OWF can be generalized into the following form: *Let m, ℓ be integers with $m > \ell \geq 0$. Can we construct an ℓ-bit-increasing injective OWF from an m-bit-increasing injective OWF?* (Note that if $m \leq \ell$, then the answer to the question is trivially yes.) We also tackle this problem and show a partial negative answer.

1.1 Our Contribution

In this paper, we show that even if we use a very close primitive to a OWP, it is impossible to construct a OWP in a black-box manner. More concretely,

[2] Roughly speaking, a construction of a primitive from another is black-box if the constructed primitive does not use the code of the building block primitive, and the reduction algorithm for the security proof does not use the code of an adversary attacking the constructed primitive. In this paper, we only talk about the so-called *fully black-box* constructions (reductions) defined in [13]. See [13] for other types of black-box constructions/reductions.

we show in Section 3 that there is no fully black-box construction[3] of a OWP from a length-increasing injective OWF, even if the latter is just 1-bit-increasing and achieves stronger type of one-wayness which we call *adaptive one-wayness* (see below). We note that this separation result is not implied by the previous results [14,10,3] on a black-box separation of a OWP from other primitives. Actually, one of the results in [3] implies the separation of a OWP from an injective OWF whose range is sparse (e.g. length-tripling functions). However, our impossibility holds as long as the building block injective OWF is length-increasing, regardless of the sparseness of the range. An immediate corollary of the separation result is the non-existence of a fully black-box construction of a OWP from a regular OWF[4] for any regularity greater than 1. Note that a OWP is also a regular OWF with regularity 1. Therefore, our negative results suggest that constructing a OWP is quite difficult (or maybe impossible) unless a building block primitive already achieves both "one-way" property and "permutation" property simultaneously.

Moreover, we extend our above result to show some restricted type of black-box separations among injective OWFs. More precisely, we show in Section 4 that for any integer pair (ℓ, m) satisfying $m > \ell \geq 0$, there is no *range-invariant* fully black-box construction of an ℓ-bit-increasing injective OWF from an m-bit-increasing injective OWF, where a construction of an injective OWF from other primitives is said to be range-invariant if the range of the constructed injective OWF depends only on the construction and is independent of the building block primitives. Note that a OWP is a 0-bit-increasing injective OWF, and any construction of a OWP from other primitives is inherently range-invariant, and thus our first separation result is the special case of the latter one in which $\ell = 0$. Although this range-invariance condition seems a bit heavy when $\ell > 0$, this result shows a hierarchy among injective OWFs (including a OWP), and we think this result is interesting. So far, we are not sure whether we can remove the range-invariance condition from this separation result, and thus we would like to leave it as an open problem.

In order to make our black-box separation results stronger, for both of our separation results, we consider a stronger type of one-wayness, *adaptive one-wayness*, for building block OWFs. Roughly, an injective function is adaptively one-way if it is one-way against adversaries that have access to a "magical" inversion oracle which takes a string (other than a challenge instance that the adversary has to invert) as input and returns the inverse of the value. Our definition of adaptive one-wayness is different from the one introduced by Pandey et al. [11] who considered it in a "tag-based" setting while ours does not consider it. See Section 2 for a formal definition.

---

[3] This is the most restrictive type of black-box constructions formalized in [13]. However, it should be noticed that most cryptographic constructions of a primitive from another primitive is fully black-box.

[4] Roughly, a function is said to be regular if it is length-preserving and each image of the function has the same-sized set of preimages, and the regularity is the size of the set of preimages which map to a same image.

1.2 Technical Overview of Our Separation Results

The combination the results by Rudich [14] and Kahn et al. [10] shows that there is no (so-called ∀∃ semi-)black-box construction of a OWP from a OWF. However, it is known that if we only need to exclude a more restrictive form of black-box constructions, *fully black-box* constructions, of a OWP from a OWF, proving the following statement is sufficient (see also [12]): "*For any oracle probabilistic polynomial time algorithm (PPTA)* P, *if* $\mathsf{P}^{\mathcal{O}}$ *implements a permutation for all random oracles*[5] $\mathcal{O}$, *then there is an oracle PPTA that has access to* $\mathcal{O}$ *and a* PSPACE*-complete oracle and inverts* $\mathsf{P}^{\mathcal{O}}$*.*"[6] Let us call this statement "(A)".

Since we will use the basic proof strategy for the statement (A) in our black-box separation results, we briefly outline the proof, and then explain the problems that arise when proving our results.

Basic Proof Strategy for Separation of OWP and (Ordinary) OWF. To prove the statement (A), we construct a computationally unbounded adversary $\mathcal{A}$ that makes only polynomially many queries to its given random oracle $\mathcal{O}$ and successfully inverts $\mathsf{P}^{\mathcal{O}}$: Given a string $y^* \in \{0,1\}^k$ which $\mathcal{A}$ has to find the preimege under $\mathsf{P}^{\mathcal{O}}$, $\mathcal{A}$ first generates an empty list L that will be used to maintain the "known" query/answer pair of $\mathcal{O}$, and then repeats the following steps 1 to 3 for polynomially many times:

Step 1: find a string x' and an oracle $\widetilde{\mathcal{O}}$ under the condition: $\widetilde{\mathcal{O}}(\alpha) = \mathcal{O}(\alpha)$ for all $\alpha \in L$ and $\mathsf{P}^{\widetilde{\mathcal{O}}}(x') = y^*$.

Step 2: check if $\mathsf{P}^{\mathcal{O}}(x') = y^*$ (note that here we use $\mathcal{O}$, not $\widetilde{\mathcal{O}}$), and terminate with output this x' if this is the case.

Step 3: ask $\mathcal{O}$ all the queries made by $\mathsf{P}^{\widetilde{\mathcal{O}}}(x')$ and update the known query/answer pair list L.

The key observation is that *in each iteration, either (a)* $\mathcal{A}$ *finds a preimage* x^* *such that* $\mathsf{P}^{\mathcal{O}}(x^*) = y^*$ *and terminates at Step 2, or (b) at Step 3* $\mathcal{A}$ *finds (and stores in* L*) at least one query that is also made by* P *during the computation of* $y^* = \mathsf{P}^{\mathcal{O}}(x^*)$ *but has not been contained in the known queries* L. Very roughly, this is because if (a) and (b) are simultaneously false, then we can construct a "hybrid" random oracle $\widehat{\mathcal{O}}$ that behaves like $\mathcal{O}$ on the queries made by $\mathsf{P}^{\mathcal{O}}(x^*)$ and like $\widetilde{\mathcal{O}}$ on those made by $\mathsf{P}^{\widetilde{\mathcal{O}}}(x')$. This hybrid oracle $\widehat{\mathcal{O}}$ has the property that $\mathsf{P}^{\widehat{\mathcal{O}}}(x^*) = \mathsf{P}^{\widehat{\mathcal{O}}}(x') = y^*$ while $x^* \neq x'$, which is a contradiction because $\mathsf{P}^{\widehat{\mathcal{O}}}$ implements a permutation which cannot have a collision (recall that $\mathsf{P}^{\mathcal{O}'}$ implements a permutation for all random oracles $\mathcal{O}'$). Since P makes only polynomially many queries to $\mathcal{O}$, by repeating the above procedure polynomially many times $\mathcal{A}$ eventually finds the preimage x^* and terminates, or we reach the situation where L contains all the queries made by $\mathsf{P}^{\mathcal{O}}(x^*)$. Note that if L contains all

[5] Here, "all random oracles" should be interpreted as "all the possible instances of random oracles".

[6] According to [14, Sect. 9.1], this statement can be shown as a corollary of the result independently discovered in [1,6,16]. For more details, see [14].

the queries made by $\mathsf{P}^{\mathcal{O}}(x^*)$, then the preimage of y^* under the permutation $\mathsf{P}^{\widetilde{\mathcal{O}}}$ (where $\widetilde{\mathcal{O}}$ is the oracle chosen at Step 1) must be x^*, because $\mathcal{O}(\alpha) = \widetilde{\mathcal{O}}(\alpha)$ for all the queries α made by $\mathsf{P}^{\mathcal{O}}(x^*)$, which means that $\mathsf{P}^{\mathcal{O}}(x^*) = \mathsf{P}^{\widetilde{\mathcal{O}}}(x^*) = y^*$. Since Step 1 in each iteration can be simulated by a PPTA with oracle access to a PSPACE-complete oracle[7], $\mathcal{A}$ can actually be a PPTA.

Problems for Separations of OWP and Length-Increasing Injective OWF. For our purpose of (fully) black-box separation of a OWP from a length-increasing injective OWF (say, m-bit-increasing with $m > 0$), we would like to replace the random oracle in the statement (A) with a random instance of oracles $\mathcal{O}$ which is m-bit-increasing and injective (we call it m-bit-increasing injective oracle). We are given an oracle PPTA P such that $\mathsf{P}^{\mathcal{O}'}$ implements a permutation for all m-bit-increasing injective oracles $\mathcal{O}'$. Then, consider the same proof strategy as above, i.e. constructing a PPTA adversary $\mathcal{A}$ that has access to an m-bit-increasing injective oracle $\mathcal{O}$ and PSPACE-complete oracle, and tries to invert a given instance $y^* = \mathsf{P}^{\mathcal{O}}(x^*)$ by the above procedure. However, if we naively do the sampling process of x' and $\widetilde{\mathcal{O}}$ at Step 1, we will meet some problem when arguing the above key observation. More concretely, even though we have $\mathsf{P}^{\widetilde{\mathcal{O}}}(x') = y^*$ and $x^* \neq x'$, it might be the case that the range of $\mathcal{O}$ and $\widetilde{\mathcal{O}}$ have an overlap outside the range corresponding to L, which could prevent the hybrid oracle $\widehat{\mathcal{O}}$ from being injective. If $\widehat{\mathcal{O}}$ is not injective, then it is no longer guaranteed that $\mathsf{P}^{\widehat{\mathcal{O}}}$ is a permutation, and thus having a collision does not cause a contradiction.

Therefore, in order to avoid such a situation, we have to choose the m-bit-increasing injective oracle $\widetilde{\mathcal{O}}$ in Step 1 so that we can always construct an "injective" hybrid oracle $\widehat{\mathcal{O}}$ from $\mathcal{O}$ and $\widetilde{\mathcal{O}}$. Our solution is to choose $\widetilde{\mathcal{O}}$ so that (i) $\widetilde{\mathcal{O}}(\alpha) = \mathcal{O}(\alpha)$ for all $\alpha \in L$ and (ii) for all other inputs from $\{0,1\}^* \backslash L$, the range of $\widetilde{\mathcal{O}}$ and that of $\mathcal{O}$ are disjoint. It will be shown later that picking such $\widetilde{\mathcal{O}}$ is always possible as long as $\mathcal{O}$ is length-increasing.

Another problem that arises here is that it might not be possible to execute Step 1 modified as above by only using a PSPACE-complete oracle and making only polynomially many queries to $\mathcal{O}$, because it seems that we need the entire knowledge about the range of $\mathcal{O}$ in order to pick such $\widetilde{\mathcal{O}}$. However, since $\mathcal{O}$ is a random m-bit-increasing injective oracle, it seems hard to know the range of $\mathcal{O}$ entirely by making only polynomially many queries to $\mathcal{O}$. To solve it, we adopt the so-called "two oracle" separation paradigm introduced by Hsiao and Reyzin [8], which is sufficient for showing the non-existence of fully black-box constructions. More concretely, we introduce another oracle $\mathcal{B}$ (which we call "breaking oracle") that helps $\mathcal{A}$ to do the above procedure of picking x' such that $\mathsf{P}^{\widetilde{\mathcal{O}}}(x') = y^*$ where $\widetilde{\mathcal{O}}$ is chosen as above. To make Step 3 possible, the oracle $\mathcal{B}$ also outputs a query set that is made by $\mathsf{P}^{\widetilde{\mathcal{O}}}(x')$ to $\widetilde{\mathcal{O}}$.

[7] This is because what we need is not the entire oracle $\widetilde{\mathcal{O}}$, but the queries made by $\mathsf{P}^{\widetilde{\mathcal{O}}}(x')$, which is a witness for a certain NP language statement, which can be picked by using a PSPACE-complete oracle.

Now, since we have introduced a new oracle $\mathcal{B}$, we also have to show that an m-bit-increasing injective oracle $\mathcal{O}$ is one-way in the presence of $\mathcal{B}$. We will show that (adaptive) one-wayness of $\mathcal{O}$ in the presence of $\mathcal{B}$ can be reduced to (adaptive) one-wayness of a *random permutation oracle* π against computationally unbounded adversary $\mathcal{S}$ that makes only polynomially many queries.

1.3 Paper Organization

The rest of this paper is organized as follows. In Section 2 we review some basic definitions and facts used for describing our results. In Section 3, we show our main result on a black-box separation of a OWP from a length-increasing injective OWF, and we extend the result for a restricted type of black-box separations among injective OWFs in Section 4.

2 Preliminaries

In this section, we review basic definitions and some fact necessary used for describing our results.

Basic Notations. Throughout this paper, we use the following notations: "$\mathbb{N}$" denotes the set of natural numbers. "$x||y$" denotes a concatenation of x and y. If x is a string, then "$|x|$" denotes the bit length of x. "$x \leftarrow y$" denotes an assignment of y to x. If S is a set, then "$|S|$" denotes its size, and "$x \leftarrow_{\mathtt{R}} S$" denotes that x is chosen uniformly at random from S. If Ψ is a probability distribution, then "$x \leftarrow_{\mathtt{R}} \Psi$" denotes that x is chosen according to Ψ, and "$[\Psi]$" denotes the "support" of Ψ, that is, $[\Psi] = \{x | \Pr_{x' \leftarrow_{\mathtt{R}} \Psi}[x' = x] > 0\}$. "PPTA" denotes *probabilistic polynomial time algorithm.* If $\mathcal{A}$ is a (probabilistic) algorithm, then "$z \leftarrow_{\mathtt{R}} \mathcal{A}(x, y, \dots)$" denotes that $\mathcal{A}$ takes x, y, ... as input and outputs z, and "$\mathcal{A}^{\mathcal{O}}$" denotes that $\mathcal{A}$ has oracle access to an oracle $\mathcal{O}$. If f is a function/algorithm/oracle and S is a (sub)domain of f, then we define $f(S) = \{f(x)|x \in S\}$. We say that an oracle algorithm has query complexity q if the algorithm makes at most q queries to its given oracle. "Perm_n" denotes the set of all permutations over $\{0,1\}^n$. A function $f : \mathbb{N} \rightarrow [0,1]$ is said to be *negligible* in k if $f(k) < 1/p(k)$ for any positive polynomial $p(k)$ and all sufficiently large $k \in \mathbb{N}$.

2.1 One-Way Functions

We say that a function $f : \{0,1\}^* \rightarrow \{0,1\}^*$ is a one-way function (OWF) if there exists a PPTA that for any x computes $f(x)$, and for any PPTA $\mathcal{A}$, the following advantage function $\mathsf{Adv}^{\mathtt{OW}}_{f,\mathcal{A}}(k)$ is negligible in k:

$$\mathsf{Adv}^{\mathtt{OW}}_{f,\mathcal{A}}(k) = \Pr[x^* \leftarrow_{\mathtt{R}} \{0,1\}^k; y^* \leftarrow f(x^*); x' \leftarrow_{\mathtt{R}} \mathcal{A}(1^k, y^*) : f(x') = y^*].$$

If the function f is injective, then we call it an injective OWF. If f is injective and length-preserving (i.e. permutation), we call it a one-way permutation (OWP). If $|f^{-1}(f(x))| = |f^{-1}(f(y))|$ for any $x, y \in \{0,1\}^n$ and any $n \in \mathbb{N}$, we call it a regular OWF, and in particular, if $\alpha(n) = |f^{-1}(f(x))|$, then we call it an α-regular OWF.

Adaptive One-wayness for Injective Functions. In this paper, we will use a stronger type of one-wayness for strengthening our black-box separation results.

We say that an injective function $f : \{0,1\}^* \to \{0,1\}^*$ is adaptive one-way, if there exists a PPTA that for any x computes $f(x)$, and for any PPTA $\mathcal{A}$, the following advantage function $\mathsf{Adv}^{\mathtt{AOW}}_{f,\mathcal{A}}(k)$ is negligible in k.

$$\mathsf{Adv}^{\mathtt{AOW}}_{f,\mathcal{A}}(k) = \Pr[x^* \leftarrow_{\mathtt{R}} \{0,1\}^k; y^* \leftarrow f(x^*); x' \leftarrow_{\mathtt{R}} \mathcal{A}^{f^{-1}_{\neq y^*}(\cdot)}(1^k, y^*) : x' = x^*],$$

where $f^{-1}_{\neq y^*}(\cdot)$ is the *inversion* oracle which takes a string y as input and outputs x such that $f(x) = y$ if such x exists and $y \neq y^*$, or outputs $\perp$ otherwise.

We also say that f is an adaptive one-way function (AOWF). (Whenever we say f is an AOWF, we always mean that f is injective.)

2.2 Basic Fact about Random Permutations

In this paper, we will use a simple fact that a random permutation is adaptively one-way even against a computationally unbounded adversary who is given oracle access to the permutation and its inversion only polynomially many times.

Let $\mathcal{A}$ be an oracle adversary. Consider the following experiment $\mathsf{Expt}^{\mathtt{AOW}}_{\mathrm{RP},\mathcal{A}}(k)$:

$$\mathsf{Expt}^{\mathtt{AOW}}_{\mathrm{RP},\mathcal{A}}(k) : [\pi \leftarrow_{\mathtt{R}} \mathsf{Perm}_k; \alpha^* \leftarrow_{\mathtt{R}} \{0,1\}^k; \beta^* \leftarrow \pi(\alpha^*); \quad \mathtt{Return\ 1\ iff\ } \mathcal{A}^{\pi, \pi^{-1}_{\neq \beta^*}}(1^k, \beta^*) \mathtt{\ returns\ } \alpha^*]$$

(Here, "RP" stands for "random permutation", and note that the experiment includes the choice of the permutation π.) We define the advantage function of an oracle adversary $\mathcal{A}$ by $\mathsf{Adv}^{\mathtt{AOW}}_{\mathrm{RP},\mathcal{A}}(k) = \Pr[\mathsf{Expt}^{\mathtt{AOW}}_{\mathrm{RP},\mathcal{A}}(k) = 1]$. Regarding the above experiment, the following is easy to prove (the proof is omitted due to lack of space).

Lemma 1. *For any (even computationally unbounded) adversary $\mathcal{A}$ with polynomial query complexity, $\mathsf{Adv}^{\mathtt{AOW}}_{\mathrm{RP},\mathcal{A}}(k)$ is negligible in k. Specifically, if $\mathcal{A}$ has query complexity q, then $\mathsf{Adv}^{\mathtt{AOW}}_{\mathrm{RP},\mathcal{A}}(k) \leq \frac{(q+1)}{2^k - q}$.*

3 Black-Box Separation of OWP from Length-Increasing Injective AOWF

In this section, we show that there is no fully black-box construction of a OWP from a length-increasing injective OWF, even if the latter is just 1-bit-increasing and achieves adaptive one-wayness.

We first recall the formal definition of a fully black-box construction [13] of a OWP from an m-bit-increasing injective AOWF.

Definition 1. *Let $m > 0$ be an integer. We say that there exists a fully black-box construction of a OWP from an m-bit-increasing injective AOWF, if there exist oracle PPTAs P and $\mathcal{R}$ such that for all functions f that implement m-bit-increasing injective functions and all algorithms $\mathcal{A}$ (where f and $\mathcal{A}$ are of arbitrary complexity):*

Correctness: P^f *is a permutation.*
Security: *If* $\mathsf{Adv}^{\mathtt{OW}}_{\mathsf{P}^f,\mathcal{A}}(k)$ *is non-negligible, so is* $\mathsf{Adv}^{\mathtt{AOW}}_{f,\mathcal{R}^{f,\mathcal{A}}}(k)$.

Now, we show the following separation between a OWP and a length-increasing injective OWF:

Theorem 1. *For any integer $m > 0$, there is no fully black-box construction of a OWP from an m-bit-increasing injective AOWF.*

To prove Theorem 1, We use a variant of the two oracle separation paradigm [8]:

Lemma 2. *Let $m > 0$ be an integer. Assume that there exists a distribution Ψ of an oracle pair $(\mathcal{O}, \mathcal{B})$ that satisfies the following three conditions:*

(1): $\mathcal{O}$ implements an m-bit-increasing injective function for all $(\mathcal{O}, \mathcal{B}) \in [\Psi]$.
(2): For any oracle PPTA $\mathcal{A}$, $\mathbf{E}_{(\mathcal{O},\mathcal{B})\leftarrow_R\Psi}[\mathsf{Adv}^{\mathtt{AOW}}_{\mathcal{O},\mathcal{A}^{\mathcal{O},\mathcal{B}}}(k)]$ is negligible.
(3): For any oracle PPTA P, if $\mathsf{P}^{\mathcal{O}'}$ implements a permutation for all $(\mathcal{O}', \mathcal{B}') \in [\Psi]$, then there is an oracle PPTA $\mathcal{A}$ s. t. $\mathbf{E}_{(\mathcal{O},\mathcal{B})\leftarrow_R\Psi}[\mathsf{Adv}^{\mathtt{OW}}_{\mathsf{P}^{\mathcal{O}},\mathcal{A}^{\mathcal{O},\mathcal{B}}}(k)] = 1$.

Then, there is no fully black-box construction of a OWP from an m-bit-increasing injective AOWF.

In order to use Lemma 2, we define the distribution Ψ of an oracle pair $(\mathcal{O}, \mathcal{B})$ in Section 3.1. Then we show that Ψ satisfies the conditions (1) and (2) in Section 3.2, and Ψ satisfies the condition (3) in Section 3.3. Finally in Section 3.4 we show the formal proof of Theorem 1.

3.1 Definitions of Oracles and Their Distribution

m-Bit-Increasing Injective Oracle $\mathcal{O}$. Let $m > 0$ be an integer. We say that an oracle $\mathcal{O}$ is an *m-bit-increasing injective oracle* if (1) for every $n \in \mathbb{N}$, $\mathcal{O}$ is of the form $\mathcal{O} : \{0,1\}^n \to \{0,1\}^{n+m}$, and (2) $\mathcal{O}$ is injective. Let $\mathbb{O}_m$ be the set of all m-bit-increasing injective oracles.

"Breaking" Oracle $\mathcal{B}$. Before describing the definition of our breaking oracle, we introduce the following notation.

Definition 2. *Let P be an oracle algorithm, $\mathcal{O} \in \mathbb{O}_m$ be an m-bit-increasing injective oracle, and x be a string. A* query set *with respect to $(\mathsf{P}, \mathcal{O}, x)$, denoted by $\mathsf{QS}_{\mathsf{P}^{\mathcal{O}}(x)}$, is a set of all the queries to $\mathcal{O}$ made by $\mathsf{P}^{\mathcal{O}}(x)$, i.e., $\mathsf{QS}_{\mathsf{P}^{\mathcal{O}}(x)} = \{ \alpha \mid \mathsf{P}^{\mathcal{O}}(x)$ makes a query α to $\mathcal{O} \}$.*

By definition, if the running time of P is bounded by τ_{P}, then $|\mathsf{QS}_{\mathsf{P}^{\mathcal{O}}(x)}| \le \tau_{\mathsf{P}}$.

Now, we describe the definition of the breaking oracle $\mathcal{B}$ formally. Our breaking oracle $\mathcal{B}$ is associated with an m-bit-increasing injective oracle $\mathcal{O} \in \mathbb{O}_m$. Since $\mathcal{B}$ generates a slightly different m-bit-increasing injective oracle $\widetilde{\mathcal{O}}$ based on $\mathcal{O}$ during its procedure, in order not to mix up them we refer to the oracle $\mathcal{O}$ with which $\mathcal{B}$ is associated as the "original oracle", and the oracle $\widetilde{\mathcal{O}}$ that is generated in the procedure of $\mathcal{B}$ as a "modified oracle".

$\mathcal{B}$ takes the following three objects as input:

1. a description of an oracle algorithm P that is a candidate of a OWP.
2. a set of strings $L \subseteq \{0,1\}^*$
3. a string y

$\mathcal{B}$ then does the following:

Step 1 (Correctness check): Check if $\mathsf{P}^{\mathcal{O}'}$ implements a permutation over $\{0,1\}^{|y|}$ for all the possible instances of m-bit-increasing injective oracles $\mathcal{O}' \in \mathbb{O}_m$. If the check fails, output $\perp$ and stop.

Step 2 (Generating a modified oracle $\widetilde{\mathcal{O}}$): For each $n \in \mathbb{N}$, pick uniformly an m-bit-increasing injective function $g'_n : \{0,1\}^n \to (\{0,1\}^{n+m} \backslash \mathcal{O}(\{0,1\}^n))$. Here, note that the size of the set $\{0,1\}^{n+m} \backslash \mathcal{O}(\{0,1\}^n)$ is $2^{n+m} - 2^n \geq 2^n$ (because $m > 0$), and thus picking such an injective function g'_n is always possible. Then, a "modified" oracle $\widetilde{\mathcal{O}}$ is defined as:

$$\widetilde{\mathcal{O}}(\alpha) = \begin{cases} \mathcal{O}(\alpha) & \text{if } \alpha \in L \\ g'_{|\alpha|}(\alpha) & \text{otherwise} \end{cases}$$

Note that $\widetilde{\mathcal{O}} \in \mathbb{O}_m$: clearly $\widetilde{\mathcal{O}}$ is m-bit-increasing, and is injective for each of the subdomains L and $\{0,1\}^* \backslash L$; the set $\widetilde{\mathcal{O}}(L) = \mathcal{O}(L)$ and the set $\widetilde{\mathcal{O}}(\{0,1\}^* \backslash L) = \{g'_{|\alpha|}(\alpha) | \alpha \in \{0,1\}^* \backslash L\}$ are always disjoint, and thus there is no pair $(\alpha, \alpha') \in L \times (\{0,1\}^* \backslash L)$ such that $\widetilde{\mathcal{O}}(\alpha) = \widetilde{\mathcal{O}}(\alpha')$.

Step 3 (Finding a preimage and a query set wrt. $\widetilde{\mathcal{O}}$): Find x such that $\mathsf{P}^{\widetilde{\mathcal{O}}}(x) = y$. (Note that $x \in \{0,1\}^{|y|}$ is unique since it is guaranteed by Correctness check that $\mathsf{P}^{\widetilde{\mathcal{O}}}$ implements a permutation over $\{0,1\}^{|y|}$.) Output x and the corresponding query set $\mathsf{QS}_{\mathsf{P}^{\widetilde{\mathcal{O}}}(x)}$.

The above completes the description of $\mathcal{B}$. Although $\mathcal{B}$ is probabilistic (see Step 2), in order to make $\mathcal{B}$ behave as a deterministic function, we assume that the randomness $\mathcal{B}$ uses to pick functions $\{g'_n(\cdot)\}_{n \in \mathbb{N}}$ is fixed for each input to $\mathcal{B}$, and make $\mathcal{B}$ choose the same functions $\{g'_n(\cdot)\}_{n \in \mathbb{N}}$ for the same input (P, L, y).)

Let τ_{P} be the maximum running time of P, where the maximum is over all oracles $\mathcal{O} \in \mathbb{O}_m$ and all inputs of length $|y|$. Similarly to [15] and [17], we count each $\mathcal{B}$-query as $|L| + \tau_{\mathsf{P}}$ queries, rather than naively counting it as a single query, and make $\mathcal{B}$ output the response after these steps have passed from the point $\mathcal{B}$ receives the input. This is to prevent an adversary from making a very "large" $\mathcal{B}$-query that may give too much information about $\mathcal{O}$ to the adversary.

We call a $\mathcal{B}$-query *valid* if the correctness check passes, and *invalid* otherwise.

What should be noticed about the modified m-bit-increasing injective oracle $\widetilde{\mathcal{O}}$ is: For any $L \subseteq \{0,1\}^*$,

– $\widetilde{\mathcal{O}}(\alpha) = \mathcal{O}(\alpha)$ for all $\alpha \in L$
– the set $\mathcal{O}(\{0,1\}^*)$ and the set $\widetilde{\mathcal{O}}(\{0,1\}^* \backslash L)$ are always disjoint

Moreover, the following property of $\mathcal{B}$ will be used later for breaking any candidate of OWP that is constructed from $\mathcal{O} \in \mathbb{O}_m$.

Lemma 3. *Let* P *be a PPTA such that* $\mathsf{P}^{\mathcal{O}'}$ *implements a permutation for all* $\mathcal{O}' \in \mathbb{O}_m$*. For any string* x*, any* $L \subseteq \{0,1\}^*$*, and any* $\mathcal{O} \in \mathbb{O}_m$*, if* $\mathcal{B}$ *is associated with* $\mathcal{O}$ *and* $\mathsf{QS}_{\mathsf{P}^{\mathcal{O}}(x)} \subseteq L$*, then* $\mathcal{B}(\mathsf{P}, L, \mathsf{P}^{\mathcal{O}}(x))$ *returns* x *and* $\mathsf{QS}_{\mathsf{P}^{\mathcal{O}}(x)}$*.*

Proof. Fix a string x, a set $L \subseteq \{0,1\}^*$ such that $\mathsf{QS}_{\mathsf{P}^{\mathcal{O}}(x)} \subseteq L$, and $\mathcal{O} \in \mathbb{O}_m$. Let $y = \mathsf{P}^{\mathcal{O}}(x)$. By the given condition, P will pass the correctness check. Let $\widetilde{\mathcal{O}}$ be the modified m-bit-increasing injective oracle generated in the step 2 of $\mathcal{B}$. By the definition of $\mathcal{B}$ and the given condition $\mathsf{QS}_{\mathsf{P}^{\mathcal{O}}(x)} \subseteq L$, it holds that $\widetilde{\mathcal{O}}(\alpha) = \mathcal{O}(\alpha)$ for all $\alpha \in \mathsf{QS}_{\mathsf{P}^{\mathcal{O}}(x)}$. This means that $\mathsf{P}^{\widetilde{\mathcal{O}}}(x) = \mathsf{P}^{\mathcal{O}}(x) = y$ and $\mathsf{QS}_{\mathsf{P}^{\widetilde{\mathcal{O}}}(x)} = \mathsf{QS}_{\mathsf{P}^{\mathcal{O}}(x)}$. Since $\mathsf{P}^{\widetilde{\mathcal{O}}}(\cdot)$ is a permutation, the preimage of y under $\mathsf{P}^{\widetilde{\mathcal{O}}}(\cdot)$ is unique and must be x, which is found and output in Step 3 of $\mathcal{B}$ together with the query set $\mathsf{QS}_{\mathsf{P}^{\widetilde{\mathcal{O}}}(x)} = \mathsf{QS}_{\mathsf{P}^{\mathcal{O}}(x)}$. This completes the proof of Lemma 3. □

Distribution Ψ *of Oracles* $(\mathcal{O}, \mathcal{B})$. We define "how to pick oracles $(\mathcal{O}, \mathcal{B})$ according to the distribution Ψ" as follows: Pick an m-bit-increasing injective oracle $\mathcal{O} \in \mathbb{O}_m$ uniformly at random, and then pick the breaking oracle $\mathcal{B}$ associated with $\mathcal{O}$ ($\mathcal{B}$'s internal randomness is fixed in this step).

3.2 Adaptive One-Wayness of $\mathcal{O}$ in the Presence of $\mathcal{B}$

From the definition of Ψ in the previous subsection, it is clear that for any $(\mathcal{O}, \mathcal{B}) \in [\Psi]$, $\mathcal{O}$ implements an m-bit-increasing injective function. The rest of this subsection is devoted to proving the following.

Lemma 4. *For any oracle PPTA* $\mathcal{A}$*,* $\mathbf{E}_{(\mathcal{O},\mathcal{B}) \leftarrow_{\mathrm{R}} \Psi}[\mathsf{Adv}^{\mathsf{AOW}}_{\mathcal{O},\mathcal{A}^{\mathcal{O},\mathcal{B}}}(k)]$ *is negligible.*

Proof. Fix an arbitrary PPTA adversary $\mathcal{A}$, and let $\tau_{\mathcal{A}} = \tau_{\mathcal{A}}(k)$ be $\mathcal{A}$'s maximum running time (when run with input 1^k). Since $\mathcal{A}$ is a PPTA, $\tau_{\mathcal{A}}$ is polynomial.

The expectation (over the choice of $(\mathcal{O}, \mathcal{B})$) of the advantage of the adversary $\mathcal{A}$ attacking adaptive one-wayness of $\mathcal{O}$ can be written as:

$$\mathop{\mathbf{E}}_{(\mathcal{O},\mathcal{B}) \leftarrow_{\mathrm{R}} \Psi} \left[\mathsf{Adv}^{\mathsf{AOW}}_{\mathcal{O},\mathcal{A}^{\mathcal{O},\mathcal{B}}}(k) \right]$$

$$= \Pr[(\mathcal{O}, \mathcal{B}) \leftarrow_{\mathrm{R}} \Psi; x^* \leftarrow_{\mathrm{R}} \{0,1\}^k; y^* \leftarrow \mathcal{O}(x^*); x' \leftarrow_{\mathrm{R}} \mathcal{A}^{\mathcal{O},\mathcal{O}^{-1}_{\neq y^*},\mathcal{B}}(y^*) : x' = x^*]$$

For notational convenience, we denote by $\widetilde{\mathsf{Expt}}^{\mathsf{AOW}}_{\mathbb{O}_m,\mathcal{A}}(k)$ the experiment

$$\widetilde{\mathsf{Expt}}^{\mathsf{AOW}}_{\mathbb{O}_m,\mathcal{A}}(k) : [(\mathcal{O}, \mathcal{B}) \leftarrow_{\mathrm{R}} \Psi; x^* \leftarrow_{\mathrm{R}} \{0,1\}^k; y^* \leftarrow \mathcal{O}(x^*); x' \leftarrow_{\mathrm{R}} \mathcal{A}^{\mathcal{O},\mathcal{O}^{-1}_{\neq y^*},\mathcal{B}}(y^*)],$$

and we write $\widetilde{\mathsf{Adv}}^{\mathsf{AOW}}_{\mathbb{O}_m,\mathcal{A}}(k) = \mathbf{E}_{(\mathcal{O},\mathcal{B}) \leftarrow_{\mathrm{R}} \Psi} \left[\mathsf{Adv}^{\mathsf{AOW}}_{\mathcal{O},\mathcal{A}^{\mathcal{O},\mathcal{B}}}(k) \right]$.

Assume towards a contradiction that $\widetilde{\mathsf{Adv}}^{\mathtt{AOW}}_{\mathbb{O}_m,\mathcal{A}}(k)$ is not negligible. Then, we show that we can construct another *computationally unbounded* adversary $\mathcal{S}$ that has query complexity at most $\tau_{\mathcal{A}}$ (and thus has polynomial query complexity) and has non-negligible advantage in the experiment $\mathsf{Expt}^{\mathtt{AOW}}_{\mathrm{RP},\mathcal{S}}(k)$, which will contradict Lemma 1. $\mathcal{S}$ is given 1^k and an image $\beta^* = \pi(\alpha^*)$ for randomly chosen $\alpha^* \in \{0,1\}^k$ and $\pi \in \mathsf{Perm}_k$, is given oracle access to π and $\pi^{-1}_{\neq\beta^*}$, and has to find α^*. The description of $\mathcal{S}^{\pi,\pi^{-1}_{\neq\beta^*}}(1^k, \beta^*)$ is as follows.

Description of $\mathcal{S}^{\pi,\pi^{-1}_{\neq\beta^*}}(1^k, \beta^* = \pi(\alpha^*))$: (Below, recall that $\mathcal{S}$ is computationally unbounded and thus can do very powerful things such as picking an injective function uniformly, etc.) Firstly, $\mathcal{S}$ picks its own m-bit-increasing injective oracle $\mathcal{O} \in \mathbb{O}_m$ uniformly at random[8]. Next, $\mathcal{S}$ defines a slightly modified m-bit-increasing injective oracle $\mathcal{O}_\pi$ into which $\mathcal{S}$'s oracle π is "embedded" as follows:

$$\mathcal{O}_\pi(\alpha) = \begin{cases} \mathcal{O}(\pi(\alpha)) & \text{if } |\alpha| = k \\ \mathcal{O}(\alpha) & \text{otherwise} \end{cases}$$

Note that this modification does not change the range of the m-bit-increasing injective oracle, i.e. $\mathcal{O}_\pi(\{0,1\}^*) = \mathcal{O}(\{0,1\}^*)$.

Then, $\mathcal{S}$ sets $y^* \leftarrow \mathcal{O}(\beta^*) = \mathcal{O}(\pi(\alpha^*)) = \mathcal{O}_\pi(\alpha^*)$, and runs $\mathcal{A}$ with input $(1^k, y^*)$. Hereafter, $\mathcal{S}$ starts simulating $\widetilde{\mathsf{Expt}}^{\mathtt{AOW}}_{\mathbb{O}_m,\mathcal{A}}$ for $\mathcal{A}$ in which the m-bit-increasing injective oracle is $\mathcal{O}_\pi$. We note that since π is not entirely known to $\mathcal{S}$, $\mathcal{O}_\pi$ for input length k (and the corresponding inversion) is not entirely known to $\mathcal{S}$. However, $\mathcal{S}$ knows all the knowledge about $\mathcal{O}$ (recall that $\mathcal{O}$ is picked by $\mathcal{S}$) and $\mathcal{S}$ can access to π and $\pi^{-1}_{\neq\beta^*}$, and thus $\mathcal{S}$ can perfectly simulate the responses of $\mathcal{O}$-queries and the inversion (i.e. $\mathcal{O}^{-1}_{\neq y^*}$-)queries from $\mathcal{A}$.

When $\mathcal{A}$ makes a $\mathcal{B}$-query (P, L, y), if (P, L, y) has been queried before, the same answer is used as a response. Otherwise, $\mathcal{S}$ responds as follows.

1. Firstly, $\mathcal{S}$ does Correctness check of P as is done in Step 1 in $\mathcal{B}$, by its computationally unbounded power (e.g. exhaustively checking if $\mathsf{P}^{\mathcal{O}'}$ implements a permutation over $\{0,1\}^{|y|}$ for all $\mathcal{O}' \in \mathbb{O}_m$)[9]. If the check fails, $\mathcal{S}$ returns $\bot$ to $\mathcal{A}$ after $|L| + \tau_{\mathsf{P}}$ steps from the point $\mathcal{A}$ made the $\mathcal{B}$-query, where τ_{P} is the maximum running time of P.

   Next, for each $\alpha \in L \cap \{0,1\}^k$, $\mathcal{S}$ issues α to π and obtains the corresponding value $\beta = \pi(\alpha)$. Then, for each response $\beta \in \pi(L \cap \{0,1\}^k)$ obtained in the above procedure, $\mathcal{S}$ computes $\gamma \leftarrow \mathcal{O}(\beta)$.
2. $\mathcal{S}$ generates the "modified" m-bit-increasing injective oracle $\widetilde{\mathcal{O}}_\pi$ (corresponding to $\mathcal{O}_\pi$) as is done in Step 2 of $\mathcal{B}$[10]. As mentioned earlier, $\mathcal{S}$ does not have

[8] Here, actually it is enough to pick $\mathcal{O}$ for input length up to $\tau_{\mathcal{A}}$, because $\mathcal{A}$ cannot issue an $\mathcal{O}$-query (and an $\mathcal{O}^{-1}_{\neq y^*}$-query) of length more than $\tau_{\mathcal{A}}$.

[9] Here, it is enough to check the correctness of P with oracles $\mathcal{O}$ that is defined up to input length τ_{P}, because P cannot issue an $\mathcal{O}$-query of length longer than τ_{P}.

[10] With similar reasons to what we mentioned in the previous footnotes, the new oracle $\widetilde{\mathcal{O}}_\pi$ is only needed to be defined for input length up to $\tau_{\mathcal{A}}$.

all the knowledge about the m-bit-increasing injective oracle $\mathcal{O}_\pi$ that $\mathcal{S}$ is using for simulating $\widetilde{\mathsf{Expt}}^{\mathtt{AOW}}_{\mathbb{O}_m,\mathcal{A}}$ for $\mathcal{A}$, because π is not entirely known to $\mathcal{S}$. However, for the strings $\alpha \in L \cap \{0,1\}^k$, the corresponding evaluations $\gamma = \mathcal{O}_\pi(\alpha) = \mathcal{O}(\pi(\alpha))$ are already calculated in the above step, and to generate $\widetilde{\mathcal{O}}_\pi$, no further knowledge about $\pi(\cdot)$ is needed. Note also that $\mathcal{O}$ itself was generated by $\mathcal{S}$, and the range of $\mathcal{O}_\pi$ is identical to that of $\mathcal{O}$ (i.e. $\mathcal{O}_\pi(\{0,1\}^*) = \mathcal{O}(\{0,1\}^*)$), which means that $\mathcal{S}$ can appropriately pick an injective function $g'_n : \{0,1\}^n \rightarrow (\{0,1\}^{n+m} \backslash \mathcal{O}_\pi(\{0,1\}^n))$ for each input length $n \in \mathbb{N}$.
3. $\mathcal{S}$ finds x such that $\mathsf{P}^{\widetilde{\mathcal{O}}_\pi}(x) = y$ and the corresponding query set $\mathsf{QS}_{\mathsf{P}^{\widetilde{\mathcal{O}}_\pi}(x)}$ by using $\mathcal{S}$'s computationally unbounded power and the entire knowledge about $\widetilde{\mathcal{O}}_\pi$. Finally, $\mathcal{S}$ returns x and $\mathsf{QS}_{\mathsf{P}^{\widetilde{\mathcal{O}}}(x)}$ to $\mathcal{A}$ after $|L| + \tau_\mathsf{P}$ steps from the point $\mathcal{A}$ made this $\mathcal{B}$-query.

When $\mathcal{A}$ terminates with an output, $\mathcal{S}$ outputs what $\mathcal{A}$ outputs as a candidate of the preimage of β^* under π and terminates.

The above completes the description of $\mathcal{S}$. Let us confirm the query complexity of $\mathcal{S}$. $\mathcal{S}$ issues at most one query to π (resp. $\pi^{-1}_{\neq \beta^*}$) for simulating a response to an $\mathcal{O}$-query (resp. $\mathcal{O}^{-1}_{\neq y^*}$-query), and at most $|L|$ queries to π for simulating a response to a $\mathcal{B}$-query. Recall that in the original experiment $\widetilde{\mathsf{Expt}}^{\mathtt{AOW}}_{\mathbb{O}_m,\mathcal{A}}$, each $\mathcal{B}$-query (P, L, y) is responded after $|L| + \tau_\mathsf{P}$ steps from the point $\mathcal{B}$ is invoked, where τ_P is the maximum running time of P (for $|y|$-bit input). These imply that the number of $\mathcal{S}$'s queries never exceeds $\mathcal{A}$'s running time. Since the running time of $\mathcal{A}$ is at most $\tau_\mathcal{A}$, $\mathcal{S}$'s query complexity is at most $\tau_\mathcal{A}$.

Moreover, as we have explained in the description of $\mathcal{S}$, $\mathcal{S}$ perfectly simulates the experiment $\widetilde{\mathsf{Expt}}^{\mathtt{AOW}}_{\mathbb{O}_m,\mathcal{A}}(k)$ for $\mathcal{A}$ so that the m-bit-increasing injective oracle is $\mathcal{O}_\pi$. Under this situation, if $\mathcal{A}$ succeeds in outputting a preimage $\mathcal{O}^{-1}_\pi(y^*) = \pi^{-1}(\mathcal{O}^{-1}(y^*))$, since $\beta^* = \mathcal{O}^{-1}(y^*)$, $\mathcal{S}$ also succeeds in outputting the preimage $\alpha^* = \pi^{-1}(\beta^*)$. Therefore, we have $\mathsf{Adv}^{\mathtt{AOW}}_{\mathrm{RP},\mathcal{S}}(k) = \widetilde{\mathsf{Adv}}^{\mathtt{AOW}}_{\mathbb{O}_m,\mathcal{A}}(k)$, which means that if $\widetilde{\mathsf{Adv}}^{\mathtt{AOW}}_{\mathbb{O}_m,\mathcal{A}}(k) = \mathbf{E}_{(\mathcal{O},\mathcal{B}) \leftarrow_\mathrm{R} \Psi} \left[\mathsf{Adv}^{\mathtt{AOW}}_{\mathcal{O},\mathcal{A}^{\mathcal{O},\mathcal{B}}}(k) \right]$ is non-negligible, so is $\mathsf{Adv}^{\mathtt{AOW}}_{\mathrm{RP},\mathcal{S}}(k)$. Since $\mathcal{S}$ has only polynomial query complexity (at most polynomial $\tau_\mathcal{A}$), this contradicts Lemma 1, and thus $\mathbf{E}_{(\mathcal{O},\mathcal{B}) \leftarrow_\mathrm{R} \Psi} \left[\mathsf{Adv}^{\mathtt{AOW}}_{\mathcal{O},\mathcal{A}^{\mathcal{O},\mathcal{B}}}(k) \right]$ must be negligible. This completes the proof of Lemma 4. □

3.3 Breaking Any Candidate of One-Way Permutation with $\mathcal{B}$

In this subsection, we show that for any candidate of a OWP that is constructed using an m-bit-increasing injective oracle $\mathcal{O} \in \mathbb{O}_m$, there is a *perfect* inverter that has access to $\mathcal{O}$ and $\mathcal{B}$, where $(\mathcal{O}, \mathcal{B})$ are chosen according to Ψ defined in Section 3.1.

Lemma 5. *For any oracle PPTA* P*, if* $\mathsf{P}^{\mathcal{O}'}$ *implements a permutation for all* $(\mathcal{O}', \mathcal{B}') \in [\Psi]$*, then there is an oracle PPTA* $\mathcal{A}$ *such that* $\mathbf{E}_{(\mathcal{O},\mathcal{B}) \leftarrow_\mathrm{R} \Psi}[\mathsf{Adv}^{\mathtt{OW}}_{\mathsf{P}^{\mathcal{O}},\mathcal{A}^{\mathcal{O},\mathcal{B}}}(k)] = 1$.

Proof. Fix an oracle PPTA P such that $\mathsf{P}^{\mathcal{O}'}$ implements a permutation for all $\mathcal{O}' \in \mathbb{O}_m$. Let $\tau_{\mathsf{P}} = \tau_{\mathsf{P}}(k)$ be the maximum running time of P on input k-bit strings. Since P is a PPTA, τ_{P} is a polynomial.

The expectation (over the choice of $(\mathcal{O}, \mathcal{B})$) of the advantage of an oracle PPTA adversary $\mathcal{A}$ attacking the one-wayness of $\mathsf{P}^{\mathcal{O}}$ can be written as:

$$\mathop{\mathbf{E}}_{(\mathcal{O},\mathcal{B})\leftarrow_{\mathtt{R}}\Psi}\left[\ \mathsf{Adv}^{\mathtt{OW}}_{\mathsf{P}^{\mathcal{O}},\mathcal{A}^{\mathcal{O},\mathcal{B}}}(k)\ \right]$$
$$= \Pr\left[(\mathcal{O},\mathcal{B}) \leftarrow_{\mathtt{R}} \Psi; x^* \leftarrow_{\mathtt{R}} \{0,1\}^k; y^* \leftarrow \mathsf{P}^{\mathcal{O}}(x^*); x' \leftarrow_{\mathtt{R}} \mathcal{A}^{\mathcal{O},\mathcal{B}}(y^*) : x' = x^*\right]$$

For notational convenience, we denote by $\widetilde{\mathsf{Expt}}^{\mathtt{OW}}_{\mathsf{P},\mathcal{A}}(k)$ the experiment

$$\widetilde{\mathsf{Expt}}^{\mathtt{OW}}_{\mathsf{P},\mathcal{A}}(k) : [(\mathcal{O},\mathcal{B}) \leftarrow_{\mathtt{R}} \Psi; x^* \leftarrow_{\mathtt{R}} \{0,1\}^k; y^* \leftarrow \mathsf{P}^{\mathcal{O}}(x^*); x' \leftarrow_{\mathtt{R}} \mathcal{A}^{\mathcal{O},\mathcal{B}}(y^*)]$$

($\widetilde{\mathsf{Expt}}^{\mathtt{OW}}_{\mathsf{P},\mathcal{A}}(k)$ includes the sampling of oracles $(\mathcal{O}, \mathcal{B})$ according to Ψ), and we write $\widetilde{\mathsf{Adv}}^{\mathtt{OW}}_{\mathsf{P},\mathcal{A}}(k) = \mathbf{E}_{(\mathcal{O},\mathcal{B})\leftarrow_{\mathtt{R}}\Psi}\left[\ \mathsf{Adv}^{\mathtt{OW}}_{\mathsf{P}^{\mathcal{O}},\mathcal{A}^{\mathcal{O},\mathcal{B}}}(k)\ \right]$.

We show that there is an oracle PPTA adversary $\mathcal{A}$ such that $\widetilde{\mathsf{Adv}}^{\mathtt{OW}}_{\mathsf{P},\mathcal{A}}(k) = 1$. That is, $\mathcal{A}$ is given 1^k and $y^* \in \{0,1\}^k$, has access to two oracles $(\mathcal{O}, \mathcal{B})$, and will always find the preimage $x^* \in \{0,1\}^k$ of y^* under the permutation $\mathsf{P}^{\mathcal{O}}$. The description of $\mathcal{A}^{\mathcal{O},\mathcal{B}}(1^k, y^*)$ is as follows:

Description of $\mathcal{A}^{\mathcal{O},\mathcal{B}}(1^k, y^*)$: Firstly, $\mathcal{A}$ generates an empty list $L_1 = \emptyset$. Then for $1 \le i \le \tau_{\mathsf{P}} + 1$, $\mathcal{A}$ does the following.

Iterations for $1 \le i \le \tau_{\mathsf{P}} + 1$: $\mathcal{A}$ issues a $\mathcal{B}$-query (P, L_i, y^*). Let x_i and $\mathsf{QS}_{\mathsf{P}^{\tilde{\mathcal{O}}_i}(x_i)}$ be the response from $\mathcal{B}$, where $\widetilde{\mathcal{O}}_i$ is the modified m-bit-increasing injective oracle generated in the step 2 of $\mathcal{B}$ in the i-th iteration, x_i is a string such that $\mathsf{P}^{\tilde{\mathcal{O}}_i}(x_i) = y^*$, and $\mathsf{QS}_{\mathsf{P}^{\tilde{\mathcal{O}}_i}(x_i)}$ is the corresponding query set. (Since $\mathsf{P}^{\mathcal{O}'}$ implements a permutation for all oracles $\mathcal{O}' \in \mathbb{O}_m$, $\mathcal{A}$'s $\mathcal{B}$-query is always valid.) Then, $\mathcal{A}$ computes $y_i \leftarrow \mathsf{P}^{\mathcal{O}}(x_i)$. If $y_i = y^*$, $\mathcal{A}$ terminates with output x_i as the preimage of y^* under the permutation $\mathsf{P}^{\mathcal{O}}$. Otherwise (i.e. $y_i \ne y^*$), $\mathcal{A}$ updates the list by $L_{i+1} \leftarrow L_i \cup \mathsf{QS}_{\mathsf{P}^{\tilde{\mathcal{O}}_i}(x_i)}$, and goes to the next iteration.

If $\mathcal{A}$ does not terminate after $\tau_{\mathsf{P}} + 1$ iterations, $\mathcal{A}$ simply gives up and aborts.

The above completes the description of $\mathcal{A}$. We first confirm the query complexity of $\mathcal{A}$. Clearly, the query complexity becomes maximum if $\mathcal{A}$ performs all $\tau_{\mathsf{P}} + 1$ iterations without discovering the preimage. Thus we consider this case. In the i-th iteration, $\mathcal{A}$ makes one $\mathcal{B}$-query (P, L_i, y^*) which is counted as $|L_i| + \tau_{\mathsf{P}}$ queries, and executes $\mathsf{P}^{\mathcal{O}}$ once during which at most τ_{P} queries are made to $\mathcal{O}$. Therefore, in the i-th iteration, the query complexity can increase by at most $|L_i| + 2\tau_{\mathsf{P}}$. Moreover, recall that in each iteration the list L_i is updated to $L_{i+1} \leftarrow L_i \cup \mathsf{QS}_{\mathsf{P}^{\tilde{\mathcal{O}}_i}(x_i)}$. Since $|L_1| = 0$ and it is always the case that $|\mathsf{QS}_{\mathsf{P}^{\tilde{\mathcal{O}}_i}(x_i)}| \le \tau_{\mathsf{P}}$, we have $|L_i| \le (i-1)\tau_{\mathsf{P}} \le \tau_{\mathsf{P}}^2$ for $1 \le i \le \tau_{\mathsf{P}} + 1$. This implies that in the i-th iteration, the query complexity can increase by at most $|L_i| + 2\tau_{\mathsf{P}} \le \tau_{\mathsf{P}}^2 + 2\tau_{\mathsf{P}} = \tau_{\mathsf{P}}(\tau_{\mathsf{P}} + 2)$. Thus, after $\tau_{\mathsf{P}} + 1$ iterations, $\mathcal{A}$'s total query complexity is at most $\tau_{\mathsf{P}}(\tau_{\mathsf{P}}+1)(\tau_{\mathsf{P}}+2)$, which is polynomial in k. Therefore, the number of steps incurred by the use of oracles $(\mathcal{O}, \mathcal{B})$ is at most polynomial, which means $\mathcal{A}$ works in polynomial time.

Next, we show that $\mathcal{A}$ can always find the preimage x^* of y^* under $\mathsf{P}^{\mathcal{O}}$. We show the following key claim, which states that in each iteration $\mathcal{A}$ either finds the preimage x^* or finds at least one "undiscovered" $\mathcal{O}$-query $\alpha \in \mathsf{QS}_{\mathsf{P}^{\mathcal{O}}(x^*)} \backslash L_i$, i.e. a query to $\mathcal{O}$ that is made by P during a computation of $\mathsf{P}^{\mathcal{O}}(x^*)$ but is not contained in L_i.

Claim 1. For every $i \in \{1, \ldots, \tau_{\mathsf{P}}\}$, at least one of the following two is true:

(a) $x_i = x^*$
(b) the set $\mathsf{QS}_{\mathsf{P}^{\widetilde{\mathcal{O}}_i}(x_i)} \backslash L_i$ contains at least one $\alpha \in \mathsf{QS}_{\mathsf{P}^{\mathcal{O}}(x^*)} \backslash L_i$

Proof of Claim 1. Assume towards a contradiction that we have both $\overline{\text{(a)}}$ $x_i \neq x^*$ and $\overline{\text{(b)}}$ no $\alpha \in \mathsf{QS}_{\mathsf{P}^{\mathcal{O}}(x^*)} \backslash L_i$ is contained in the set $\mathsf{QS}_{\mathsf{P}^{\widetilde{\mathcal{O}}_i}(x_i)} \backslash L_i$. The latter means that the set $\mathsf{QS}_{\mathsf{P}^{\mathcal{O}}(x^*)} \backslash L_i$ and the set $\mathsf{QS}_{\mathsf{P}^{\widetilde{\mathcal{O}}_i}(x_i)} \backslash L_i$ are disjoint.

Consider the following "hybrid" oracle $\widehat{\mathcal{O}}$ defined by:

$$\widehat{\mathcal{O}}(\alpha) = \begin{cases} \widetilde{\mathcal{O}}_i(\alpha) & \text{if } \alpha \in (\mathsf{QS}_{\mathsf{P}^{\widetilde{\mathcal{O}}_i}(x_i)} \backslash L_i) \\ \mathcal{O}(\alpha) & \text{otherwise} \end{cases}$$

We argue $\widehat{\mathcal{O}} \in \mathbb{O}_m$, i.e., $\widehat{\mathcal{O}}$ is also a possible instance of an m-bit-increasing injective oracle. For notational convenience, we write $A = (\mathsf{QS}_{\mathsf{P}^{\widetilde{\mathcal{O}}_i}(x_i)} \backslash L_i)$. Firstly, it is clear that $\widehat{\mathcal{O}}$ is m-bit-increasing, and is injective for each of the subdomains A and $\{0,1\}^* \backslash A$, because so are $\mathcal{O}$ and $\widetilde{\mathcal{O}}_i$. Secondly, recall that the sets $\widetilde{\mathcal{O}}_i(\{0,1\}^* \backslash L_i)$ and $\mathcal{O}(\{0,1\}^*)$ are always disjoint (see the explanation after the description of $\mathcal{B}$ in Section 3.1). Since we trivially have $A \subseteq (\{0,1\}^* \backslash L_i)$ and $(\{0,1\}^* \backslash A) \subseteq \{0,1\}^*$, the set $\widehat{\mathcal{O}}(A) = \widetilde{\mathcal{O}}_i(A)$ and the set $\widehat{\mathcal{O}}(\{0,1\}^* \backslash A) = \mathcal{O}(\{0,1\}^* \backslash A)$ are also disjoint, which means that there is no pair $(\alpha, \alpha') \in A \times (\{0,1\}^* \backslash A)$ such that $\widehat{\mathcal{O}}(\alpha) = \widehat{\mathcal{O}}(\alpha')$. These imply $\widehat{\mathcal{O}} \in \mathbb{O}_m$.

Therefore, $\mathsf{P}^{\widehat{\mathcal{O}}}$ also implements a permutation.

Next, we confirm the property of this hybrid oracle $\widehat{\mathcal{O}}$. The condition $\overline{\text{(b)}}$ and the definitions of $\mathcal{O}$, $\widetilde{\mathcal{O}}_i$, and $\widehat{\mathcal{O}}$ imply the following relations among these oracles:

(1): $\widehat{\mathcal{O}}(\alpha) = \mathcal{O}(\alpha) = \widetilde{\mathcal{O}}_i(\alpha)$ for all $\alpha \in L_i$
(2): $\widehat{\mathcal{O}}(\alpha) = \mathcal{O}(\alpha)$ for all $\alpha \in \mathsf{QS}_{\mathsf{P}^{\mathcal{O}}(x^*)} \backslash L_i$
(3): $\widehat{\mathcal{O}}(\alpha) = \widetilde{\mathcal{O}}_i(\alpha)$ for all $\alpha \in \mathsf{QS}_{\mathsf{P}^{\widetilde{\mathcal{O}}_i}(x_i)} \backslash L_i$

where (2) is due to the condition $\overline{(b)}$, i.e. the set $\mathsf{QS}_{\mathsf{P}^{\mathcal{O}}(x^*)} \backslash L_i$ and the set $\mathsf{QS}_{\mathsf{P}^{\widetilde{\mathcal{O}}_i}(x_i)} \backslash L_i$ are disjoint. On the one hand, (1) and (2) imply $\widehat{\mathcal{O}}(\alpha) = \mathcal{O}(\alpha)$ for all $\alpha \in \mathsf{QS}_{\mathsf{P}^{\mathcal{O}}(x^*)}$, which in turn implies $\mathsf{P}^{\widehat{\mathcal{O}}}(x^*) = \mathsf{P}^{\mathcal{O}}(x^*) = y^*$. On the other hand, (1) and (3) imply $\widehat{\mathcal{O}}(\alpha) = \widetilde{\mathcal{O}}_i(\alpha)$ for all $\alpha \in \mathsf{QS}_{\mathsf{P}^{\widetilde{\mathcal{O}}_i}(x_i)}$, which in turn implies $\mathsf{P}^{\widehat{\mathcal{O}}}(x_i) = \mathsf{P}^{\widetilde{\mathcal{O}}_i}(x_i) = y^*$.

In summary, we have $\mathsf{P}^{\widehat{\mathcal{O}}}(x^*) = \mathsf{P}^{\widehat{\mathcal{O}}}(x_i) = y^*$, while we also have $x^* \neq x_i$ by the condition $\overline{\text{(a)}}$. This is a contradiction because $\mathsf{P}^{\widehat{\mathcal{O}}}$ is a permutation, which cannot have a collision. Therefore, our hypothesis must be false, and (a) or (b) must be true. This completes the proof of Claim 1. □

Due to Claim 1, in the i-th iteration (for $1 \leq i \leq \tau_{\mathsf{P}}$) $\mathcal{A}$ finds the preimage x^* and stops, or the set $\mathsf{QS}_{\mathsf{P}^{\tilde{\mathcal{O}}_i}(x_i)} \backslash L_i$ contains at least one α which is contained in $\mathsf{QS}_{\mathsf{P}^{\mathcal{O}}(x^*)}$ but is not contained in L_i. In the latter case, since the list L_i is updated as $L_{i+1} \leftarrow L_i \cup \mathsf{QS}_{\mathsf{P}^{\tilde{\mathcal{O}}_i}(x_i)}$, the size of the set of "undiscovered queries" $\mathsf{QS}_{\mathsf{P}^{\mathcal{O}}(x^*)} \backslash L_i$ decreases at least by one in each iteration. Recall that $|\mathsf{QS}_{\mathsf{P}^{\mathcal{O}}(x^*)}| \leq \tau_{\mathsf{P}}$. Therefore, when $\mathcal{A}$ is executed, $\mathcal{A}$ finds the preimage x^* and stops within τ_{P} iterations, or after τ_{P} iterations we have $\mathsf{QS}_{\mathsf{P}^{\mathcal{O}}(x^*)} \subseteq L_{\tau_{\mathsf{P}}+1}$. In the latter case, by Lemma 3, the $(\tau_{\mathsf{P}}+1)$-th $\mathcal{B}$-query $(\mathsf{P}, L_{\tau_{\mathsf{P}}+1}, y^*)$ results in x^* (and $\mathsf{QS}_{\mathsf{P}^{\mathcal{O}}(x^*)}$).

These imply that $\mathcal{A}$ always outputs x^* within $\tau_{\mathsf{P}}+1$ iterations, which means that we have $\widetilde{\mathsf{Adv}}^{\mathtt{OW}}_{\mathsf{P},\mathcal{A}} = \mathbf{E}_{(\mathcal{O},\mathcal{B}) \leftarrow_{\mathtt{R}} \Psi}[\mathsf{Adv}^{\mathtt{OW}}_{\mathsf{P}^{\mathcal{O}},\mathcal{A}^{\mathcal{O},\mathcal{B}}}(k)] = 1$. This completes the proof of Lemma 5. □

3.4 Putting Everything Together

Here, we show the formal proof of Theorem 1.

Proof of Theorem 1. Let $m > 0$ be an integer. We use the distribution Ψ of oracles $(\mathcal{O}, \mathcal{B})$ defined in Section 3.1 for Lemma 2. Then, by definition any $\mathcal{O}$ where $(\mathcal{O}, \mathcal{B}) \in [\Psi]$ is m-bit-increasing and injective, and thus the assumption (1) in Lemma 2 is satisfied. Moreover, the assumptions (2) and (3) are satisfied due to Lemma 4 in Section 3.2 and Lemma 5 in Section 3.3, respectively. Since the distribution Ψ satisfies all the assumptions in Lemma 2, it follows that there is no fully black-box construction of a OWP from an m-bit-increasing injective AOWF. This statement holds for any integer $m > 0$. This completes the proof of Theorem 1. □

An immediate corollary of Theorem 1 is the following:

Corollary 1. *For any integer $m > 0$, there is no fully black-box construction of a OWP from a 2^m-regular OWF.*

Proof. Let $m > 0$. Suppose $f : \{0,1\}^n \rightarrow \{0,1\}^{n+m}$ is an m-bit-increasing injective OWF. From f, construct a new function $g : \{0,1\}^{n+m} \rightarrow \{0,1\}^{n+m}$ by $g(x\|x') = f(x)$ where $|x| = n$ and $|x'| = m$ (i.e, g ignores the last m-bits of the input). Then, this function g is a 2^m-regular OWF (for security parameter 1^n) as long as g is an m-bit-increasing injective OWF (for security parameter 1^n). Trivially, the construction of g from f and the security proof are black-box, and thus for any integer $m > 0$, there is a fully black-box construction of a 2^m-regular OWF from an m-bit-increasing injective OWF. Since a fully-black-box construction of a primitive from another primitive is a transitive relation, we have the claimed result. □

Remark. In Theorem 1, the "stretch" m of the building block (injective) AOWFs has been treated as a constant. However, our negative results can be extended to the case in which $m = m(k)$ is a function of the input-length k (i.e. security parameter), as long as $m(k) > 0$ for all $k > 0$. m in Corollary 1 can be a function of the security parameter as well.

4 Restricted Type of Black-Box Separations among Injective OWFs

In this section, we show that Theorem 1 can be extended to show the impossibility of a restricted type of black-box constructions, which we call *range-invariant* fully black-box constructions, of an injective OWF from another injective OWF.

Definition 3. *Let $\ell, m \geq 0$ be integers. We say there exists a* range-invariant *fully black-box construction of an ℓ-bit-increasing injective OWF from an m-bit-increasing injective AOWF, if there exist oracle PPTAs G and $\mathcal{R}$ such that:*

(Correctness). *For all m-bit-increasing injective functions f (of arbitrary complexity), G^f is an ℓ-bit-increasing injective function and* has the same range[11].

(Security). *For all m-bit-increasing injective functions f and all algorithms $\mathcal{A}$ (where f and $\mathcal{A}$ are of arbitrary complexity), if $\mathsf{Adv}^{\mathtt{OW}}_{\mathsf{G}^f,\mathcal{A}}(k)$ is non-negligible, so is $\mathsf{Adv}^{\mathtt{AOW}}_{f,\mathcal{R}^{f,\mathcal{A}}}(k)$.*

In other words, the range of the constructed ℓ-bit-increasing injective function G^f depends solely on G, and independent of the building block f.

Now, we state our black-box separation result among injective OWFs.

Theorem 2. *For any integer pair (ℓ, m) satisfying $m > \ell \geq 0$, there is no range-invariant fully black-box construction of an ℓ-bit-increasing injective OWF from an m-bit-increasing injective AOWF.*

Note that a permutation is a 0-bit-increasing injective function. Moreover, any construction of a permutation from other primitives has the same range, namely, a set of all strings, and thus is inherently range-invariant. Therefore, Theorem 1 is the special case of Theorem 2 in which $\ell = 0$.

Since Theorem 2 can be proved very similarly to Theorem 1, below we only show the outline of the proof, highlighting the different points from the proof of Theorem 1 we need to care.

As in the case of Theorem 1, in order to show Theorem 2, we can use the generalized version of Lemma 2 (which can be proved similarly to Lemma 2):

Lemma 6. *Let ℓ and m be integers satisfying $m > \ell \geq 0$. Assume that there exists a distribution Ψ of an oracle pair $(\mathcal{O}, \mathcal{B})$ that satisfies the following three conditions:*

[11] That is, G^f and $\mathsf{G}^{f'}$ have the same range for any f, f' implementing m-bit-increasing injective functions.

(1) and (2): Same as Lemma 2.
(3): For any oracle PPTA G*, if* $\mathsf{G}^{\mathcal{O}'}$ *implements an* ℓ*-bit-increasing injective function and* has the same range *for all* $(\mathcal{O}', \mathcal{B}') \in [\Psi]$*, then there exists an oracle PPTA* $\mathcal{A}$ *such that* $\mathbf{E}_{(\mathcal{O},\mathcal{B}) \leftarrow_R \Psi}[\mathsf{Adv}^{\mathtt{OW}}_{\mathsf{G}^{\mathcal{O}}, \mathcal{A}^{\mathcal{O},\mathcal{B}}}(k)] = 1$.

Then, there is no range-invariant fully black-box construction of an ℓ*-bit-increasing injective OWF from an* m*-bit-increasing injective AOWF.*

We remark that the range-invariance condition in Theorem 2 is due to the additional condition on the range of $\mathsf{G}^{\mathcal{O}}$ in (3) in the above lemma. The reason why we need this additional condition on the range of G is to ensure that if a string y is in the range of $\mathsf{G}^{\mathcal{O}}$ for some $\mathcal{O} \in \mathbb{O}_m$, then y is also in the range of $\mathsf{G}^{\mathcal{O}'}$ for any $\mathcal{O}' \in \mathbb{O}_m$, and thus the preimage of y under $\mathsf{G}^{\mathcal{O}'}$ always exists.

To use Lemma 6 to prove Theorem 2, it remains to show the definition of an oracle pair $(\mathcal{O}, \mathcal{B})$ and their distibution Ψ, and the conditions (1) to (3) of Lemma 6 (i.e. the statements corresponding to Lemmas 4 and 5). For $\mathcal{O}$, we again use an m-bit-increasing injective oracle. The breaking oracle $\mathcal{B}$ needs to be modified slightly: in Step 1, $\mathcal{B}$ checks if a given description of an algorithm G implements an ℓ-bit-increasing injective function and has the same range for all m-bit-increasing injective oracles, and also checks if a given string y belongs to the range of $\mathsf{G}^{\mathcal{O}}$. If G and y pass the check, then it is guaranteed that there always exist x satisfying $y^* = \mathsf{G}^{\mathcal{O}}(x^*) = \mathsf{G}^{\widetilde{\mathcal{O}}}(x)$ where $\widetilde{\mathcal{O}}$ is the modified oracle generated in the step 2 of $\mathcal{B}$ and x is a string found in the step 3 of $\mathcal{B}$. (Without the checks, it is possible that such x does not exist, which we want to avoid.) The distribution Ψ is also naturally defined.

The statement corresponding to Lemma 4, which roughly states that a random m-bit-increasing injective function is adaptively one-way even against adversaries with access to $\mathcal{B}$ that is modified as above, can be similarly proved as Lemma 4.

To show the statement corresponding to Lemma 5, which roughly states that no ℓ-bit-increasing injective function $\mathsf{G}^{\mathcal{O}}$ with the "range-invariance" can be one-way if $\mathcal{B}$ is available, we need to rely on the additional assumption on the range of G, especially when showing the statement analogous to Claim 1. Recall that in order to prove Claim 1 by contradiction we need a property that under several different oracles, namely the original oracle $\mathcal{O}$, the modified oracle $\widetilde{\mathcal{O}}_i$, and the "hybrid oracle" $\widehat{\mathcal{O}}$, P always implements a permutation and causes a situation in which $y^* = \mathsf{P}^{\widehat{\mathcal{O}}}(x^*) = \mathsf{P}^{\widehat{\mathcal{O}}}(x_i)$ and $x^* \neq x_i$. In order for the same strategy to work, the challenge instance y^* needs to belong to the range of the ℓ-bit-increasing injective functions $\mathsf{G}^{\mathcal{O}}$, $\mathsf{G}^{\widetilde{\mathcal{O}}_i}$, and $\mathsf{G}^{\widehat{\mathcal{O}}}$. Due to the assumption we have made, however, it is guaranteed that y^* always belong to the range of these ℓ-bit-increasing injective functions, and we can cause a situation in which $y^* = \mathsf{G}^{\widehat{\mathcal{O}}}(x^*) = \mathsf{G}^{\widehat{\mathcal{O}}}(x_i)$ and $x^* \neq x_i$.

Acknowledgement

The authors would like to thank Jacob Schuldt for his helpful discussion. The authors also would like to thank anonymous reviewers of TCC 2011 for their invaluable comments.

References

1. Blum, M., Impabliazzo, R.: Generic oracles and oracle classes (extended abstract). In: FOCS 1987, pp. 118–126 (1987)
2. Blum, M., Micali, S.: How to generate cryptographically strong sequences of pseudo-random bits. SIAM J. Computing 13(4), 850–864 (1984)
3. Chang, Y.-C., Hsiao, C.-Y., Lu, C.-J.: The impossibility of basing one-way permutations on central cryptographic primitives. J. of Cryptology 19(1), 97–114 (2006)
4. Goldreich, O., Goldwasser, S., Micali, S.: How to construct random functions. J. ACM 33(4), 792–807 (1986)
5. Goldreich, O., Levin, L.A., Nisan, N.: On constructing 1-1 one-way functions. Electronic Colloquium on Computational Complexity (ECCC) 2(29), 1–1 (1995)
6. Hartmanis, J., Hemachandra, L.A.: One-way functions and the nonisomorphism of NP-complete sets. Theor. Comput. Sci. 81(1), 155–183 (1991)
7. Håstad, J., Impagliazzo, R., Levin, L., Luby, M.: Construction of a pseudorandom generator from any one-way function. SIAM J. Computing 28(4), 1364–1396 (1999)
8. Hsiao, C.-Y., Reyzin, L.: Finding collisions on a public road, or do secure hash functions need secret coins? In: Franklin, M. (ed.) CRYPTO 2004. LNCS, vol. 3152, pp. 92–105. Springer, Heidelberg (2004)
9. Impagliazzo, R., Rudich, S.: Limits on the provable consequences of one-way permutations. In: STOC 1989, pp. 44–61 (1989)
10. Kahn, J., Saks, M., Smyth, C.: A dual version of Reimer's inequality and a proof of Rudich's conjecture. In: CoCo 2000, pp. 98–103 (2000)
11. Pandey, O., Pass, R., Vaikuntanathan, V.: Adaptive one-way functions and applications. In: Wagner, D. (ed.) CRYPTO 2008. LNCS, vol. 5157, pp. 57–74. Springer, Heidelberg (2008)
12. Reingold, O.: On black-box separations in cryptography (2006), One of Tutorials in TCC 2006 (2006), A slide file, `http://research.ihost.com/tcc06/slides/Reingold-tutorial.ppt`
13. Reingold, O., Trevisan, L., Vadhan, S.: Notions of reducibility between cryptographic primitives. In: Naor, M. (ed.) TCC 2004. LNCS, vol. 2951, pp. 1–20. Springer, Heidelberg (2004)
14. Rudich, S.: Limits on the provable consequences of one-way functions. PhD thesis, University of California at Berkeley (1988)
15. Simon, D.R.: Findings collisions on a one-way street: Can secure hash functions be based on general assumptions? In: Nyberg, K. (ed.) EUROCRYPT 1998. LNCS, vol. 1403, pp. 334–345. Springer, Heidelberg (1998)
16. Tardos, G.: Query complexity, or why is it difficult to separate $NP^A \cup coNP^A$ from P^A by random oracles A? Combinatorica 9(4), 385–392 (1989)
17. Vahlis, Y.: Two is a crowd? a black-box separation of one-wayness and security under correlated inputs. In: Micciancio, D. (ed.) TCC 2010. LNCS, vol. 5978, pp. 165–182. Springer, Heidelberg (2010)
18. Yao, A.C.: Theory and application of trapdoor functions. In: FOCS 1982, pp. 80–91 (1982)

Impossibility of Blind Signatures from One-Way Permutations

Jonathan Katz[1], Dominique Schröder[2,⋆], and Arkady Yerukhimovich[1]

[1] University of Maryland, USA
{jkatz,arkady}@cs.umd.edu
[2] Darmstadt University of Technology, Germany
schroeder@me.com

Abstract. A seminal result in cryptography is that signature schemes can be constructed (in a black-box fashion) from any one-way function. The minimal assumptions needed to construct *blind* signature schemes, however, have remained unclear. Here, we rule out black-box constructions of blind signature schemes from one-way functions. In fact, we rule out constructions even from a random permutation oracle, and our results hold even for blind signature schemes for 1-bit messages that achieve security only against honest-but-curious behavior.

1 Introduction

Blind signature schemes, introduced by Chaum [10], allow a signer to interactively issue signatures for a user in such a way that, roughly, the signer learns nothing about the message being signed (*blindness*) while the user cannot compute any additional signatures without the help of the signer (*unforgeability*). Classical applications of blind signatures include e-cash, where a bank blindly signs coins withdrawn by users, and e-voting, where an authority blindly signs public keys that voters later use to cast their votes.

Several constructions of blind signature schemes are known in either the random oracle model [23,1,6,7,3] or the standard model [19,8,22,12,15,16,20,13,2]. The *minimal* assumptions needed to construct blind signatures, however, are unclear. On the positive side, there exist constructions of blind signatures based on (doubly) enhanced trapdoor permutations [19,12,16]. Interestingly, these constructions are all *nonblack-box* even in the honest-but-curious setting, relying as they do on either generic secure two-party computation or non-interactive zero-knowledge proofs. (More recently, protocols for secure two-party computation making only black-box use of enhanced trapdoor permutations have been shown [18]; these could be used in conjunction with [16] to give a black-box construction of blind signatures from certified enhanced trapdoor permutations). On the other hand, for standard signatures we know that one-way functions suffice [21,25], and there is no reason *a priori* to believe that blind signatures cannot be constructed from one-way functions also.

⋆ This work was done while visiting the University of Maryland.

Y. Ishai (Ed.): TCC 2011, LNCS 6597, pp. 615–629, 2011.

Previous work of Camenisch, Neven, and Shelat [9] (see also [13]) shows that any *unique* blind signature scheme implies oblivious transfer. Combined with known results showing that oblivious transfer cannot be constructed in a black-box fashion from one-way permutations, this at first may appear to rule out black-box constructions of blind signatures from one-way permutations. The uniqueness requirement, however, is quite strong: Fiore and Schröder show that unique signatures (even without the blindness requirement) cannot be constructed in a black-box fashion even from trapdoor permutations [11]. More importantly, uniqueness is not a standard desideratum for blind signatures and the result of Camenisch et al. implies nothing for blind signatures without the uniqueness property. In another line of work, Fischlin and Schröder [14] show that *three-round* blind signature schemes with signature-derivation checks cannot be constructed in a black-box way from any non-interactive problem. Their result, however, says nothing about protocols with more rounds, or for schemes that do not have signature-derivation checks. We refer to the reader to [26] for a comprehensive survey of the above results.

As our main result, we show:

Theorem 1 (Main theorem). *There is no black-box construction of blind signature schemes from one-way functions.*

Our result imposes no restrictions on the blind signature scheme, and applies even to schemes with imperfect completeness. Our result is actually more general than the above theorem indicates; it also applies to constructions based on one-way permutations or random oracles, and even rules out constructions of blind signature schemes for 1-bit messages that achieve security only against honest-but-curious behavior.

The proof of our impossibility result requires a careful combination of prior techniques in the area of black-box separations. At a high level, our basic framework is similar to the one used by Barak and Mahmoody-Ghidary in studying black-box constructions of (standard) signature schemes from one-way functions [4]. Our setting introduces several additional difficulties, however, not least of which is that we must deal with the case of *interactive* protocols. Also, Barak and Mahmoody-Ghidary prove limits on the efficiency of constructions, whereas we are interested in proving impossibility. To deal with these complications, we also rely on techniques used in analyzing constructions of key-agreement protocols from one-way functions [17,5]. A more detailed overview of our proof is given in Section 2.

Black-box separations. In cryptography, constructions are usually proven secure by reduction to the security of some "low-level" primitive. Most known constructions are *black-box*, in that they treat the underlying primitive as an oracle and do not use any internal structure of the primitive; see [24] for extensive discussion and formal definitions. Impagliazzo and Rudich [17] initiated work showing impossibility of black-box constructions; in their paper they showed impossibility of constructing key-exchange protocols in a black-box manner from one-way functions. It is important to bear in mind that several nonblack-box constructions are known; nevertheless, black-box impossibility are useful insofar as they rule out a particular approach to a problem. Nonblack-box constructions also tend to be orders of magnitude less efficient than black-box constructions.

Organization. We provide on overview of our proof in Section 2. In Section 3 we present definitions of blind signatures, and we prove our main result in Section 4. In Section 5 we discuss extensions of our result to handle schemes with imperfect completeness, and to rule out constructions from one-way permutations.

2 Overview of Our Techniques

We consider interactive signature-issue protocols between a *signer* and a *user*. The input of the signer is a private key *sk*, and the user's input is a public key *pk* and a message m; at the end of this protocol the user outputs a signature σ on the message m. The algorithms run by both the signer and the user are given black-box access to a one-way function (OWF); we allow the parties to be computationally unbounded, but require that they only query the one-way function a polynomial number of times. For our impossibility result, we assume that both parties follow the protocol and are just honest-but-curious (i.e., semi-honest). This assumption only strengthens our result.

In the setting of blind signatures, security demands that:

Unforgeability. The user should not be able to output two valid signatures after interacting with the signer once. (More generally, the user should be unable to output $k+1$ valid signatures on distinct messages after interacting with the signer k times).

Blindness. If the user executes the signature-issue protocol twice, once using a message m_0 and once using a message m_1, then the signer should be unable to tell in which order these executions were run. This should hold even if the signer is given both of the resulting signatures.

We show that if we wish to satisfy both conditions above then OWFs are not sufficient. To illustrate the main idea why this is true, consider the setting where both the user and signer are given access to a random oracle. Let Q denote the oracle queries made by the signer in generating its public and private keys. Now consider two protocol executions in which the user first obtains a signature on the message m_0 and then obtains a signature on the message m_1. Correctness intuitively requires that in each interaction the user learns sufficiently many of the queries in Q in order to be abe to derive a valid signature. Unforgeability requires that the user does not learn "too many" of the queries in Q in each interaction; in particular, the user should not learn enough queries in the first interaction to derive a valid signature on m_1. Finally, blindness implies that, from the point of view of the signer, the queries the user learns in the first interaction should be distributed identically to the queries the user learns in the second interaction. We show that all these requirements are in conflict.

More formally, we rely on results of [17,5] showing that for any two-party protocol there is an algorithm Find that takes as input a transcript of an execution of the protocol and outputs, with high probability, a set that contains every oracle query that was asked by both parties ("intersection queries"). Noting that the signer can run this algorithm, the blindness requirement thus implies that the set obtained by running Find on the signature-issue protocol for m_0 must contain a set of intersection queries that are sufficient to derive a signature on the message m_1. (Else the signer knows that the first execution could not possibly have been for m_1). We use this to construct a forger, which is a more efficient version of the one given in [4]. Our forger runs a single protocol execution honestly to obtain a signature on m_0, and then runs Find to learn all the intersection queries. By what we have just said, this set will contain enough information to allow the forger to also compute a valid signature on m_1.

From a technical point of view, our proof technique can be viewed as following the general framework proposed by Barak and Mahmoody-Ghidary [4], who show a forger for any (standard) signature scheme constructed from one-way functions in a black-box fashion. Our work differs from theirs in the following respects:

- The obvious difference is that we consider an *interactive* signing protocol, whereas in [4] the signing algorithm was non-interactive. Moreover, Barak and Mahmoody-Ghidary assume that signing is deterministic. This assumption is without loss of generality for standard signatures, but is more subtle in the case of blind signatures.
- While we use the same "usefulness" property as in [4], our proof that usefulness holds is very different from the analogous proof in their work: they assume a large message and argue that usefulness occurs for some pair of messages with high probability, whereas in our case we rely on blindness and show (roughly) that usefulness holds for any two messages with all but negligible probability. This allows us to simplify the attack and obtain a forger that makes only polynomially many oracle queries regardless of how many oracle queries the construction uses. (In the work of Barak and Mahmoody-Ghidary the number of queries made by the forger depends exponentially on the number of queries made by the construction).

3 Definitions

3.1 Blind Signatures

The notation $A^{\mathcal{O}}(x)$ refers to an algorithm A that on input x gets black-box access to an oracle $\mathcal{O}$. By $(a, b) \leftarrow \langle \mathcal{X}(x), \mathcal{Y}(y) \rangle$ we denote interactive execution of algorithms $\mathcal{X}$ and $\mathcal{Y}$, where x (resp., y) is the private input of $\mathcal{X}$ (resp., $\mathcal{Y}$), and a (resp., b) is the private output of $\mathcal{X}$ (resp., $\mathcal{Y}$). We write $\mathcal{Y}^{\langle \mathcal{X}(x), \cdot \rangle}(y)$ if $\mathcal{Y}$ can invoke a *single* execution of the protocol with $\mathcal{X}$. Accordingly, $\mathcal{X}^{\langle \cdot, \mathcal{Y}(y_0) \rangle, \langle \cdot, \mathcal{Y}(y_1) \rangle}(x)$ denotes that $\mathcal{X}$ can invoke *one execution each* with $\mathcal{Y}(y_0)$ and $\mathcal{Y}(y_1)$.

We define blind signatures for 1-bit messages; since we are proving impossibility, this only makes our results stronger.

Definition 1 (Oracle blind signature scheme). *An oracle blind signature scheme is a tuple of polynomial-time algorithms* $\mathsf{BS} = (\mathsf{Gen}^{(\cdot)}, \mathcal{S}^{(\cdot)}, \mathcal{U}^{(\cdot)}, \mathsf{Vrfy}^{(\cdot)})$, *where for any* $\lambda \in \mathbb{N}$ *and any oracle* $\mathcal{O} : \{0,1\}^\lambda \to \{0,1\}^\lambda$ *we have:*

- $\mathsf{Gen}^{\mathcal{O}}(1^\lambda)$ *generates a key pair* (sk, pk).
- *The joint execution of* $\mathcal{S}^{\mathcal{O}}(sk)$ *and* $\mathcal{U}^{\mathcal{O}}(pk, m)$, *where* $m \in \{0,1\}$, *generates an output* σ *for the user and no output for the signer. We write this as* $(\perp, \sigma) \leftarrow \langle \mathcal{S}^{\mathcal{O}}(sk), \mathcal{U}^{\mathcal{O}}(pk, m)\rangle$.
- *Algorithm* $\mathsf{Vrfy}^{\mathcal{O}}(pk, m, \sigma)$ *outputs a bit* b.

We assume[1] perfect completeness*: i.e., for any* $\lambda \in \mathbb{N}$ *and* $\mathcal{O} : \{0,1\}^\lambda \to \{0,1\}^\lambda$, *any* $(sk, pk) \leftarrow \mathsf{Gen}^{\mathcal{O}}(1^\lambda)$, *any* $m \in \{0,1\}$, *and any signature* σ *output by* $\mathcal{U}^{\mathcal{O}}$ *in the joint execution of* $\mathcal{S}^{\mathcal{O}}(sk)$ *and* $\mathcal{U}^{\mathcal{O}}(pk, m)$, *it holds that* $\mathsf{Vrfy}^{\mathcal{O}}(pk, m, \sigma) = 1$.

3.2 Security of Blind Signatures

Blind signatures must satisfy two properties: unforgeability and blindness. For unforgeability we require that a user who runs a single execution of the signature-issuing protocol should be unable to forge a valid signature on two messages. For blindness we require that in two executions of the protocol, in which the user obtains signatures on both possible messages, the signer should be unable to determine which message was signed in which execution. In both cases, we assume semi-honest behavior. Our definitions of security are weaker than those usually considered; since we show impossibility, this only strengthens our results.

In the definitions that follow we consider an execution of an oracle blind signature scheme BS relative to a random oracle $\mathcal{O}$. Since a random oracle is one-way with overwhelming probability, any construction of blind signatures from one-way functions must give an oracle blind signature scheme satisfying these definitions. We remark that our definitions consider unbounded adversaries who make polynomially many queries to $\mathcal{O}$; however, we could have stated our definitions in terms of polynomial-time adversaries given access to an **NP** oracle.

Definition 2 (Unforgeability). *Oracle blind signature scheme* $\mathsf{BS} = (\mathsf{Gen}, \mathcal{S}, \mathcal{U}, \mathsf{Vrfy})$ *is* unforgeable *if for any semi-honest algorithm* $\mathcal{U}^*$ *that makes at most* $\mathsf{poly}(\lambda)$ *queries to* $\mathcal{O}$, *the probability that experiment* $\mathsf{Forge}^{\mathsf{BS}}_{\mathcal{U}^*}(\lambda)$ *evaluates to* 1 *is negligible (in* λ*), where*

Experiment $\mathsf{Forge}^{\mathsf{BS}}_{\mathcal{U}^*}(\lambda)$:
Oracle $\mathcal{O} : \{0,1\}^\lambda \to \{0,1\}^\lambda$ *is chosen at random*
$(sk, pk) \leftarrow \mathsf{Gen}^{\mathcal{O}}(1^\lambda)$
$(\sigma_0, \sigma_1) \leftarrow \mathcal{U}^{*\langle \mathcal{S}(sk), \cdot\rangle, \mathcal{O}}(pk)$ *(where* $\mathcal{U}^*$ *runs an honest execution of* $\mathcal{U}(pk, 0)$ *with* $\mathcal{S}$ *and then outputs signatures of its choice)*
Return 1 *iff* $\mathsf{Vrfy}^{\mathcal{O}}(pk, 0, \sigma_0) = 1$ *and* $\mathsf{Vrfy}^{\mathcal{O}}(pk, 1, \sigma_1) = 1$.

Definition 3 (Blindness). *Oracle blind signature scheme* $\mathsf{BS} = (\mathsf{Gen}, \mathcal{S}, \mathcal{U}, \mathsf{Vrfy})$ *satisfies* blindness *if for any semi-honest algorithm* $\mathcal{S}^*$ *that makes at most* $\mathsf{poly}(\lambda)$ *queries to* $\mathcal{O}$, *the probability that experiment* $\mathsf{Unblind}^{\mathsf{BS}}_{\mathcal{S}^*}(\lambda)$ *evaluates to* 1 *is negligibly close to* 1/2, *where*

[1] We relax this requirement in Section 5.1.

Experiment $\mathsf{Unblind}_{\mathcal{S}^*}^{\mathsf{BS}}(\lambda)$
Oracle $\mathcal{O} : \{0,1\}^\lambda \to \{0,1\}^\lambda$ *is chosen at random*
$r \leftarrow \{0,1\}^\lambda$; $b \leftarrow \{0,1\}$
$(sk, pk) \leftarrow \mathsf{Gen}^{\mathcal{O}}(1^\lambda; r)$
$st \leftarrow \mathcal{S}^{*\langle\cdot,\mathcal{U}(pk,b)\rangle,\langle\cdot,\mathcal{U}(pk,\bar{b})\rangle,\mathcal{O}}(sk, pk, r)$ *(where* $\mathcal{S}^*$ *runs an honest execution of the protocol with each instance of* $\mathcal{U}$*)*
Let σ_b, σ_{1-b} *denote the local outputs of each instance of* $\mathcal{U}$
$b' \leftarrow \mathcal{S}^{*\mathcal{O}}(st, \sigma_0, \sigma_1)$
Return 1 *iff* $b = b'$.

Throughout this work we make the simplifying assumption that the signing algorithm $\mathcal{S}$ is deterministic. This is without loss of generality when we consider the above definitions of security, as a blind signature scheme with randomized signer $\mathcal{S}$ can always be converted to a scheme with deterministic signer $\mathcal{S}'$ by (1) including a key for a pairwise-independent hash function as part of the signer's private key; (2) having the user send a random nonce as its first message in the signing protocol; and then (3) having $\mathcal{S}'$ apply the hash function to the user's first message to generate random coins that it then uses to run $\mathcal{S}$.

4 Attacking Black-Box Constructions of Blind Signatures

In this section we show that there is no black-box construction of blind signatures from one-way functions. To this end, we show that any oracle blind signature scheme $\mathsf{BS}^{(\cdot)}$ fails to satisfy either blindness or unforgeability when instantiated with a random oracle $\mathcal{O} : \{0,1\}^\lambda \to \{0,1\}^\lambda$.

4.1 Preliminaries

We begin by reviewing a lemma from previous work [17,5] that we utilize in our proof. Informally, it states that for any two-party protocol Π where each party has access to a random oracle there exists an algorithm that, upon observing the transcript of an interaction, finds with high probability all the intersection queries (queries to the oracle that have been asked by both parties).

Lemma 1 ([5]). *Let* Π *be a two-party (randomized) protocol where each party asks at most* q *queries to an oracle. Then for every* $\delta \in (0,1)$*, there is an algorithm* Find_δ *that makes* $\mathcal{O}((q/\delta)^2)$ *oracle queries, such that when* Find_δ *is given the transcript of an execution of the protocol between the parties in the presence of a random oracle, the queries made by* Find_δ *contain all the intersection queries of the two parties with probability at least* $1-\delta$*. (The probability is taken over the coins of* Find_δ *and the parties, as well as choice of the random oracle).*

We apply this in our setting in the following way. Corresponding to any oracle blind signature scheme $\mathsf{BS}^{(\cdot)}$, define the following two-party protocol Π between a signer $\mathcal{S}$ and a user $\mathcal{U}$:

1. $\mathcal{S}$ runs $(sk, pk) \leftarrow \mathsf{Gen}^{\mathcal{O}}(1^\lambda)$ and sends pk to $\mathcal{U}$.
2. $\mathcal{U}$ and $\mathcal{S}$ then run the signature-issuing protocol on the message 1, at the end of which $\mathcal{U}$ obtains a signature σ_1.
3. $\mathcal{U}$ runs $\mathsf{Vrfy}^{\mathcal{O}}(pk, 1, \sigma_1)$.

For the remainder of Section 4, fix some δ and define Find_δ (as per Lemma 1) relative to the above protocol Π. Say the above protocol is run in the presence of a random oracle $\mathcal{O}$. If we let $Q(\mathcal{S}_\Pi)$ and $Q(\mathcal{U}_\Pi)$ denote the $\mathcal{O}$-queries made by each party during an execution of the above protocol that resulted in transcript trans, then Lemma 1 guarantees that, with high probability,

$$Q(\mathcal{S}_\Pi) \cap Q(\mathcal{U}_\Pi) \subseteq \mathsf{Find}_\delta^{\mathcal{O}}(\mathsf{trans}).$$

4.2 From Blindness to Usefulness

In this section we study the question of what blindness implies with regard to the set of queries $\mathcal{I}$ output by the Find algorithm. The main observation is that due to blindness the set $\mathcal{I}$ (that contains all intersection queries with high probability) must be somehow "independent" of the actual message being signed. Recall that in the blindness game the semi-honest signer interacts with two honest user instances in a random order. The task for the attacker is to guess which instance used which message. Now, consider two protocol executions and suppose that the set of intersection queries depended on the message being used. Then just by looking at those queries it would be possible to determine the order of the messages.

To formalize this intuition, we first define some notation. Consider an execution of the blindness experiment. We write $Q(\mathsf{Gen})$ to represent the set of $\mathcal{O}$-queries made during key generation. In the interaction between $\mathcal{S}$ and $\mathcal{U}(pk, 0)$, let $Q(\mathcal{S}_0)$ denote the $\mathcal{O}$-queries made by $\mathcal{S}$; let trans_0 denote the resulting transcript; let σ_0 be the signature that $\mathcal{U}$ outputs; and let $Q(\mathsf{Vrfy}_0)$ be the set of $\mathcal{O}$-queries made by the verification algorithm $\mathsf{Vrfy}^{\mathcal{O}}(pk, 0, \sigma_0)$. Define $Q(\mathcal{S}_1)$, trans_1, and $Q(\mathsf{Vrfy}_1)$ analogously for the interaction between $\mathcal{S}$ and $\mathcal{U}(pk, 1)$. (Note that by perfect completeness and the assumption of semi-honest behavior by $\mathcal{S}$, both user instances always obtain a valid signature on their message).

Consider a (semi-honest) signer $\mathcal{S}^*$ in the blindness game. Say the adversary runs Find using trans_1. It follows from Lemma 1 and the definition of Π in the previous section that, with high probability,

$$Q(\mathsf{Vrfy}_1) \cap (Q(\mathsf{Gen}) \cup Q(\mathcal{S}_1)) \subseteq \mathsf{Find}(\mathsf{trans}_1). \tag{1}$$

$\mathcal{S}^*$ can check whether Equation (1) holds by computing $\mathsf{Vrfy}^{\mathcal{O}}(pk, 1, \sigma_1)$ itself. But then blindness implies that Equation (1) must hold with high probability even when Find is run on the "wrong" interaction; i.e.,

$$Q(\mathsf{Vrfy}_1) \cap (Q(\mathsf{Gen}) \cup Q(\mathcal{S}_0)) \subseteq \mathsf{Find}(\mathsf{trans}_0).$$

In the language of [4], this means that the message '0' is "useful" for the message '1' with high probability.

We now give the formal proof.

Lemma 2. *Let* BS *be an oracle blind signature scheme satisfying blindness. Consider an execution of the blindness experiment (cf. Definition 3), and let* $Q(\mathsf{Gen})$, $Q(\mathcal{S}_b)$, trans_b, *and* $Q(\mathsf{Vrfy}_b)$ *be as defined above. Then with probability at least* $1 - \delta - \mathsf{negl}(\lambda)$ *over the random coins of the experiment it holds that*

$$Q(\mathsf{Vrfy}_1) \cap (Q(\mathsf{Gen}) \cup Q(\mathcal{S}_0)) \subseteq \mathsf{Find}_\delta(\mathsf{trans}_0).$$

Proof. We first observe that with probability at least $1 - \delta$ we have

$$Q(\mathsf{Vrfy}_1) \cap (Q(\mathsf{Gen}) \cup Q(\mathcal{S}_1)) \subseteq \mathsf{Find}_\delta(\mathsf{trans}_1).$$

This follows immediately from Lemma 1 and our definition of protocol Π in the previous section.

Consider now the following adversary $\mathcal{S}^*$:

1. $\mathcal{S}^*$ runs the honest key-generation algorithm to obtain (sk, pk). It records the $\mathcal{O}$-queries $Q(\mathsf{Gen})$ made during this step.
2. $\mathcal{S}^*$ then runs the honest signing protocol with the first user instance. Let trans denote the transcript of this execution, and let $Q(\mathcal{S})$ denote the $\mathcal{O}$-queries made during this step.
3. $\mathcal{S}^*$ then runs the honest signing protocol with the second user instance.
4. $\mathcal{S}^*$ is given signatures σ_0, σ_1 on the messages 0 and 1, respectively. (By perfect completeness, both user instances always obtain valid signatures). $\mathcal{S}^*$ verifies σ_1 and records the $\mathcal{O}$-queries $Q(\mathsf{Vrfy}_1)$ made in doing so.
5. Finally, $\mathcal{S}^*$ outputs 1 iff $Q(\mathsf{Vrfy}_1) \cap (Q(\mathsf{Gen}) \cup Q(\mathcal{S})) \subseteq \mathsf{Find}_\delta(\mathsf{trans})$.

If $b = 1$, and so the first user instance represents an interaction with $\mathcal{U}(pk, 1)$, then $\mathsf{trans} = \mathsf{trans}_1$ and $Q(\mathcal{S}) = Q(\mathcal{S}_1)$ and so $\mathcal{S}^*$ outputs 1 with probability at least $1-\delta$. The blindness property thus implies that $\mathcal{S}^*$ outputs 1 with probability at least $1 - \delta - \mathsf{negl}(\lambda)$ when $b = 0$ (and the first user instance represents an interaction with $\mathcal{U}(pk, 0)$). This concludes the proof.

4.3 Forging a Signature

Before presenting our forger, we begin by discussing the ideas behind our attack. The main observation is that due to the blindness of the signature scheme the intersection queries between the signer and user are somehow "independent" of the message. This was formalized in Lemma 2, where we showed that (with high probability)

$$Q(\mathsf{Vrfy}_1) \cap (Q(\mathsf{Gen}) \cup Q(\mathcal{S}_0)) \subseteq \mathsf{Find}(\mathsf{trans}_0).$$

Intuitively, this means that all the "important" queries needed to verify a signature on the message '1' must already be contained in the set of queries that are found when signing and verifying the message '0'. Thus, in the language of Barak and Mahmoody-Ghidary [4], we have shown that 0 is "useful" for 1 with high probability. As in that paper, we use this property to show an attack.

The above condition seems to suggest that the set of intersection queries for '0' is sufficient to generate a signature on '1'. However, this is not quite true. The problem is that there may be queries that the user makes with high probability when generating and verifying a signature for 1 that are not in the set $\mathsf{Find}(\mathsf{trans}_0)$; this could cause technical problems because our forger must get the answers to these queries right when constructing a forged signature. For a concrete example, consider a blind signature scheme where the user, on input a message b, always queries $y = \mathcal{O}(b)$ and includes y as part of the signature; verification checks whether $\mathcal{O}(b) = y$ (among other things). In such a case the query $\mathcal{O}(1)$ may not be in the set $\mathsf{Find}(\mathsf{trans}_0)$.

As in [4], we handle this issue by introducing a phase in which the forger makes any "heavy" queries that are made by the user with high probability. If the forger knows the correct answers to all these high-probability queries then it is very unlikely that it will incorrectly answer some query asked during the verification of the forged signature.

Given this intuition we now present the details of the attack. The main structure of the attack is based on [4] with necessary changes to adapt the proof to our setting. In particular, our attack makes only polynomially many oracle queries (regardless of the number of queries the scheme itself makes).

Theorem 2. *Let* BS *be an oracle blind signature scheme satisfying blindness. Then there exists an adversary* $\mathcal{U}^*$ *for which* $\mathsf{Forge}_{\mathcal{U}^*}^{\mathsf{BS}^{\mathcal{O}}}(\lambda)$ *(cf. Definition 2) is not negligible.*

Proof. Consider the following adversary $\mathcal{U}^*$:

Setup. The input of $\mathcal{U}^*$ is a public key pk.

Step 1: Requesting a signature. $\mathcal{U}^*$ runs the honest signing protocol (using message '0') with the signer, eventually obtaining a valid signature σ_0. Let trans_0 be the transcript (i.e., the messages exchanged) for this execution. $\mathcal{U}^*$ verifies the received signature and records the oracle queries $Q(\mathsf{Vrfy}_0)$ made. $\mathcal{U}^*$ then computes $\mathsf{Find}_\delta(\mathsf{trans}_0)$ with $\delta = 1/10$.

Denote by T_0 the complete transcript of the entire experiment run so far; i.e., T_0 contains the entire views of both the signer and $\mathcal{U}^*$. Note that $\mathcal{U}^*$ has only partial knowledge about T_0.

Step 2: Learning query/answer pairs. Let L_0 be the information that $\mathcal{U}^*$ has about T_0 and the oracle $\mathcal{O}$ following Step 1. Let q be an upper bound on the total number of queries asked when running each of the algorithms in BS once. Let $\epsilon = \delta/q$ and $M = q/\epsilon\delta = 100q^2$. For $i = 1, \ldots, M$ do the following:

1. Let $\mathbf{D}_{i-1}$ be the distribution of T_0, the transcript of the first step, conditioned on L_{i-1}.
2. Denote by $Q(L_{i-1})$ the oracle queries that appear in L_{i-1}. If a query $x \in \{0,1\}^\lambda \backslash Q(L_{i-1})$ appears with probability at least ϵ in $\mathbf{D}_{i-1}$, then $\mathcal{U}^*$ queries $\mathcal{O}(x)$ and adds the query/answer pair to L_{i-1} to obtain L_i. (If there is more than one such x, then $\mathcal{U}^*$ adds the lexicographically first one).

Step 3: Sampling a possible transcript. $\mathcal{U}^*$ samples a random transcript $\widetilde{T}_0$ according to the distribution $\mathbf{D}_M$. Observe that $\widetilde{T}_0$ also defines a secret key $\widetilde{sk}$ that may be distinct from the real secret key sk. Moreover, $\widetilde{T}_0$ may include some new mappings that were not defined in L_M. We let $\widetilde{\mathcal{O}}$ be the following oracle: If a query x appears in $\widetilde{T}_0$ then $\widetilde{\mathcal{O}}(x)$ returns the value contained in $\widetilde{T}_0$; otherwise, $\widetilde{\mathcal{O}}(x) = \mathcal{O}(x)$.

Step 4: Forging. $\mathcal{U}^*$ runs the interactive signing protocol for the message '1' locally, playing the role of both the signer and the user, using $\widetilde{sk}$ and $\widetilde{\mathcal{O}}$; that is, it computes $\sigma_1 \leftarrow \left\langle \mathcal{S}^{\widetilde{\mathcal{O}}}(\widetilde{sk}), \mathcal{U}^{\widetilde{\mathcal{O}}}(pk, 1)\right\rangle$. For technical reasons, we also have $\mathcal{U}^*$ verify σ_1 (using $\mathcal{O}$). Finally, $\mathcal{U}^*$ outputs the two signatures σ_0, σ_1.

Analysis. It is easy to see that $\mathcal{U}^*$ makes polynomially many queries to $\mathcal{O}$. Since $\mathcal{U}^*$ runs the honest user protocol in its execution with the signer (in step 1), σ_0 is always a valid signature on '0'. In the remainder of the proof, we show that σ_1 is a valid signature on the message '1' with probability at least $4/5 - \delta - \mathsf{negl}(\lambda)$.

In the following we show that, with high probability, verification of σ_1 never asks a query on which oracles $\widetilde{\mathcal{O}}$ and $\mathcal{O}$ disagree. Assuming this to be the case, it follows (by the perfect completeness of the signature scheme) that σ_1 is a valid signature on '1' with respect to the true oracle $\mathcal{O}$.

Lemma 3. *Let $Q(\mathsf{Vrfy}_1)$ denote the set of oracle queries made when $\mathcal{U}^*$ verifies the signature σ_1. Let $\widetilde{Q}(\mathsf{Gen})$ and $\widetilde{Q}(\mathcal{S}_0)$ denote the set of oracle queries made by the key-generation and signing algorithms, respectively, in the sampled transcript $\widetilde{T}_0$. Then with probability at least $\frac{4}{5} - \delta - \mathsf{negl}(\lambda)$ it holds that*

$$Q(\mathsf{Vrfy}_1) \cap \left(\widetilde{Q}(\mathsf{Gen}) \cup \widetilde{Q}(\mathcal{S}_0)\right) \subseteq \mathsf{Find}_\delta(\mathsf{trans}_0).$$

Lemma 3 implies Theorem 2. To see this, note that $\mathsf{Vrfy}^{\widetilde{\mathcal{O}}}(pk, 1, \sigma_1) = 1$ by perfect completeness of the signature scheme. But the only queries on which $\widetilde{\mathcal{O}}$ and $\mathcal{O}$ can possibly differ are queries in $\left(\widetilde{Q}(\mathsf{Gen}) \cup \widetilde{Q}(\mathcal{S}_0)\right) \setminus \mathsf{Find}_\delta(\mathsf{trans}_0)$. If verification makes no such queries, then

$$\mathsf{Vrfy}^{\mathcal{O}}(pk, 1, \sigma_1) = \mathsf{Vrfy}^{\widetilde{\mathcal{O}}}(pk, 1, \sigma_1) = 1.$$

Let E denote the event considered in Lemma 3. The proof of Lemma 3 follows the proof in [4]: we define a sequence of hybrid distributions, and analyze the probability of E in each of them. The biggest difference between the proof in [4] and the proof here is when we analyze the probability that E happens in the final hybrid distribution.

Definition of hybrid distributions. We define four hybrid distributions $\mathbf{H}^0$, $\mathbf{H}^1$, $\mathbf{H}^2$, and $\mathbf{H}^3$ as follows:

$\mathbf{H}^0$. The first hybrid is the distribution $(\widetilde{T}_0, T_1)$, where $\widetilde{T}_0$ is the transcript sampled by $\mathcal{U}^*$ in Step 3, and T_1 is the transcript of Step 4 (i.e., generation and verification of σ_1). Note that $\widetilde{T}_0$ includes the queries of the key-generation algorithm.

$\mathbf{H}^1$. The second hybrid is defined identically to $\mathbf{H}^0$, except that we use $\widetilde{\mathcal{O}}$ to verify σ_1. (In $\mathbf{H}^0$, oracle $\mathcal{O}$ was used when verifying σ_1).

$\mathbf{H}^2$. The third hybrid has the same distribution as $\mathbf{H}^1$, except that we change the definition of $\widetilde{\mathcal{O}}$ as follows. Recall that L_M is the set of $\mathcal{O}$ query/answer pairs that $\mathcal{U}^*$ knows after the learning queries step (Step 2). We define $\widetilde{\mathcal{O}}$ to answer any query contained in L_M with the answer stored there and all other queries with a randomly chosen value. This modification results in an oracle $\widetilde{\mathcal{O}}$ that agrees with $\mathcal{O}$ on all the queries $\mathcal{U}^*$ has queried to $\mathcal{O}$ until the end of Step 2; all other queries are answered completely at random.

$\mathbf{H}^3$. The distribution of the last hybrid is the same as $\mathbf{H}^2$ except that $\widetilde{T}_0$ is replaced with T_0. Thus the output of this hybrid is (T_0, T_1) which describes the experiment where we first compute $(sk, pk) \leftarrow \mathsf{Gen}$; then run $\sigma_0 \leftarrow \langle \mathcal{S}^{\mathcal{O}}(sk), \mathcal{U}^{\mathcal{O}}(pk, 0)\rangle$ and $\sigma_1 \leftarrow \langle \mathcal{S}^{\mathcal{O}}(sk), \mathcal{U}^{\mathcal{O}}(pk, 1)\rangle$; and finally verify both signatures. Note that all algorithms here use the "real" oracle $\mathcal{O}$ and thus verification succeeds for both signatures.

The distributions considered in each hybrid are taken over random choice of the oracle and random coins of the key-generation algorithm, the signer, and the adversary. We prove Lemma 3 by showing that (1) event E occurs with high probability in $\mathbf{H}^3$ and (2) the probability that event E occurs in $\mathbf{H}^0$ is not much smaller than its probability in $\mathbf{H}^3$.

We first show that E occurs with high probability in $\mathbf{H}^3$. The following is an immediate consequence of Lemma 2.

Claim. $\Pr_{\mathbf{H}^3}[E] \geq 1 - \delta - \mathsf{negl}(\lambda)$.

We next show that the probability of E remains unchanged when we move from $\mathbf{H}^3$ to $\mathbf{H}^2$.

Claim. $\mathbf{H}^2 \equiv \mathbf{H}^3$. Thus, $\Pr_{\mathbf{H}^2}[E] = \Pr_{\mathbf{H}^3}[E]$.

Proof. The proof here is the same as in [4]. We can view $\mathbf{H}^3$ as being sampled as follows: first, fix L_M; then choose the transcript T_0 at random from $\mathbf{D}_M$. This, however, is exactly the same distribution as $\mathbf{H}^2$ where L_M is fixed and we then choose $\widetilde{T}_0$ from $\mathbf{D}_M$.

For the next claim, we need the following definition.

Definition 4 (Statistical distance). *If X, Y are two random variables taking values in a finite set A, then* $\mathsf{SD}(X, Y) = 1/2 \cdot \sum_{a \in A} |\Pr[X = a] - \Pr[Y = a]|$.

We now show that $\mathbf{H}^1$ and $\mathbf{H}^2$ are "close".

Claim. $\mathsf{SD}(\mathbf{H}^1, \mathbf{H}^2) \leq \frac{1}{5}$. Thus, $\Pr_{\mathbf{H}^1}[E] \geq \Pr_{\mathbf{H}^2}[E] - \frac{1}{5}$.

Proof. Let $Q(T_0)$ be the queries contained in the transcript T_0. Let B be the event that $\mathcal{U}^*$ ever asks a query in $Q(T_0) \setminus Q(L_M)$. It is clear that $\mathbf{H}^1 = \mathbf{H}^2$ as long as event B does not occur in either of them, since in both distributions any queries outside of $Q(T_0)$ are answered randomly. This implies that $\Pr_{\mathbf{H}^1}[B] =$

$\Pr_{\mathbf{H}^2}[B]$, and $\mathsf{SD}(\mathbf{H}^1, \mathbf{H}^2) \leq \Pr_{\mathbf{H}^2}[B]$. We now show that $\Pr_{\mathbf{H}^2}[B] \leq \frac{1}{5}$. (In the following, all probabilities are in $\mathbf{H}^2$).

Recall that in Step 2 of the attack, we set $\epsilon = \delta/q$ and $\mathcal{U}^*$ learns at most $M = 100q^2$ query/answer pairs from $\mathcal{O}$. Let $\mathbf{D}_i$ be the distribution of T_0 sampled in this step by $\mathcal{U}^*$ given the set L_i of known query/answer pairs. Let C be the event that there are more than M queries that become likely during the attack. That is, C is the event that there exists a query $x \notin Q(L_M)$ such that x is asked in $\mathbf{D}_M$ with probability at least ϵ. Below, we show that $\Pr[C] \leq \delta = \frac{1}{10}$ and $\Pr[B \mid \neg C] \leq \delta = \frac{1}{10}$. This completes the proof, since then

$$\begin{aligned}\Pr[B] &= \Pr[C] \cdot \Pr[B \mid C] + \Pr[\neg C] \cdot \Pr[B \mid \neg C] \\ &\leq \Pr[C] + \Pr[B \mid \neg C] \leq 2\delta = \frac{1}{5}.\end{aligned}$$

The following two claims complete the proof that $\mathbf{H}^1$ and $\mathbf{H}^2$ are close.

Claim. Let C be the event defined in the proof of the previous claim. Then $\Pr_{\mathbf{H}^2}[C] \leq \delta$.

Proof. All probabilities here are in $\mathbf{H}^2$. Consider an arbitrary query x and let hit_x be the event that x is queried to $\mathcal{O}$ by the signer and then by the user when generating the signature on '0'. Let $q_x = \Pr[\mathsf{hit}_x]$. Finally, let $A_x(i)$ be the event that x is asked in the ith iteration of Step 2; let $p_x(i) = \Pr[A_x(i)]$; and let $p_x = \Pr[\cup_i A_x(i)]$. Note that $\sum_x q_x \leq q$ since q is an upper bound on the total number of queries asked when running each algorithm of the blind signature scheme. Furthermore, $q_x \geq \epsilon p_x$ because

$$q_x = \Pr[\mathsf{hit}_x] \geq \sum_i \Pr[\mathsf{hit}_x \mid A_x(i)] \cdot \Pr[A_x(i)],$$

and $\mathcal{U}^*$ adds a query to its list only if the probability that this query is asked is at least ϵ. Thus, $\Pr[\mathsf{hit}_x \mid A_x(i)] \geq \epsilon$ and so $q_x \geq \epsilon \sum_i \Pr[A_x(i)] = \epsilon p_x$.

Assume for the sake of contradiction that $\Pr[C] > \delta$. Since C is the event that M queries are learned in Step 2, this implies that the expected number of queries asked, $\sum_x p_x$, is larger than δM. But this would imply

$$\delta M < \sum_x p_x \leq \sum_x q_x/\epsilon \leq q/\epsilon,$$

contradicting the fact that $M = q/\delta\epsilon$.

Claim. Let B and C be as defined earlier. Then $\Pr_{\mathbf{H}^2}[B \mid \neg C] \leq \delta$.

Proof. Recall that in Step 4 $\mathcal{U}^*$ relies only on the mappings stored in L_M, and all queries from $Q(T_0) \backslash Q(L_M)$ are answered at random. But then $\mathbf{H}^2$ is independent of T_0 conditioned on L_M (whereas L_M has the distribution $\mathbf{D}_M$). This means that we can imagine defining $\mathbf{H}^2$ by choosing L_M first, then running $\mathcal{U}^*$ (using L_M) to sample $\mathbf{H}^2$, and then choosing T_0 conditioned on L_M and $\mathbf{H}^2$. Recall that event C is determined by L_M, and assume that L_M is such that event

$\neg C$ occurs. This implies that every query asked by $\mathcal{U}^*$ that is not in $Q(L_M)$ must appear in $\mathbf{D}_M$ with probability less than ϵ. Since $\mathcal{U}^*$ asks at most q queries in Step 4, the probability that $Q(T_0)\backslash Q(L_M)$ contains one of these queries is at most $\epsilon q = \delta$.

Finally, we show that E occurs with the same probability in $\mathbf{H}^0$ and $\mathbf{H}^1$.

Claim. $\Pr_{\mathbf{H}^0}[E] = \Pr_{\mathbf{H}^1}[E]$.

Proof. This claim follows easily if both hybrid distributions $\mathbf{H}^0$ and $\mathbf{H}^1$ use the same oracle $\mathcal{O}$ and if they are sampled using the same random coins for key generation and the adversary (note that the randomness of the adversary fully determines the randomness used to run the honest user algorithm during the signature-issue protocol). But then it follows that event E occurs in $\mathbf{H}^0$ if and only if it also occurs in $\mathbf{H}^1$.

This completes the proof of Lemma 3, and thus the proof of Theorem 2.

5 Extensions

In this section we briefly discuss how to extend our impossibility result to the case of blind signature schemes with imperfect completeness, and to constructions from one-way permutations.

5.1 Imperfect Completeness

Let $\mathsf{BS}^{(\cdot)}$ be an oracle blind signature scheme for which correctness holds with all but negligible probability; i.e., for any $\mathcal{O}$ and any $m \in \{0,1\}$, we have

$$\Pr\left[\begin{array}{c}(sk, pk) \leftarrow \mathsf{Gen}^{\mathcal{O}}(1^\lambda);\\ (\bot, \sigma) \leftarrow \langle \mathcal{S}^{\mathcal{O}}(sk), \mathcal{U}^{\mathcal{O}}(pk, m)\rangle\end{array} : \mathsf{Vrfy}^{\mathcal{O}}(pk, m, \sigma) = 1\right] \geq 1 - \mathsf{negl}(\lambda).$$

Our results from the previous section can be easily extended to such schemes. The proof of Lemma 2 is largely identical, with the only modification being to explicitly consider what happens if either of the signatures computed by the two user instances are invalid.

The forgery attack also proceeds just as in the previous section. Since the probability that one of the signatures is invalid is negligible, this only affects the forgery probability by a negligible amount.

5.2 One-Way Permutations

We now discuss how to extend our impossibility result to also rule out constructions from one-way permutations. As noted in [5], the Find algorithm can be modified to work in the random permutation model with a polynomial blow-up in the number of queries. It follows that an analogue of Lemma 2 holds when $\mathcal{O}$ is chosen as a random permutation. (Again, a random permutation oracle is one-way with all but negligible probability). For the forgery attack we modify the proof of Theorem 2 as in [4]. We omit the details here.

Acknowledgments

The authors thank Marc Fischlin, Dov Gordon, and Mohammad Mahmoody-Ghidary for helpful discussions. We also thank the anonymous reviewers for valuable comments. Dominique Schröder is supported by the Emmy Noether Program Fi 940/2-1 of the German Research Foundation (DFG) and by the German Academic Exchange Service (DAAD).

References

1. Abe, M.: A secure three-move blind signature scheme for polynomially many signatures. In: Pfitzmann, B. (ed.) EUROCRYPT 2001. LNCS, vol. 2045, pp. 136–151. Springer, Heidelberg (2001)
2. Abe, M., Fuchsbauer, G., Groth, J., Haralambiev, K., Ohkubo, M.: Structure-preserving signatures and commitments to group elements. In: Rabin, T. (ed.) CRYPTO 2010. LNCS, vol. 6223, pp. 209–236. Springer, Heidelberg (2010)
3. Abe, M., Ohkubo, M.: A framework for universally composable non-committing blind signatures. In: Matsui, M. (ed.) ASIACRYPT 2009. LNCS, vol. 5912, pp. 435–450. Springer, Heidelberg (2009)
4. Barak, B., Mahmoody-Ghidary, M.: Lower bounds on signatures from symmetric primitives. In: 48th Annual Symposium on Foundations of Computer Science (FOCS), pp. 680–688. IEEE, Los Alamitos (2007)
5. Barak, B., Mahmoody-Ghidary, M.: Merkle puzzles are optimal — an $o(n^2)$-query attack on any key exchange from a random oracle. In: Halevi, S. (ed.) CRYPTO 2009. LNCS, vol. 5677, pp. 374–390. Springer, Heidelberg (2009)
6. Bellare, M., Namprempre, C., Pointcheval, D., Semanko, M.: The one-more-RSA-inversion problems and the security of Chaum's blind signature scheme. Journal of Cryptology 16(3), 185–215 (2003)
7. Boldyreva, A.: Threshold signatures, multisignatures and blind signatures based on the gap-diffie-hellman-group signature scheme. In: Desmedt, Y.G. (ed.) PKC 2003. LNCS, vol. 2567, pp. 31–46. Springer, Heidelberg (2003)
8. Camenisch, J.L., Koprowski, M., Warinschi, B.: Efficient blind signatures without random oracles. In: Blundo, C., Cimato, S. (eds.) SCN 2004. LNCS, vol. 3352, pp. 134–148. Springer, Heidelberg (2005)
9. Camenisch, J.L., Neven, G., Shelat, A.: Simulatable adaptive oblivious transfer. In: Naor, M. (ed.) EUROCRYPT 2007. LNCS, vol. 4515, pp. 573–590. Springer, Heidelberg (2007)
10. Chaum, D.: Blind signatures for untraceable payments. In: Advances in Cryptology — Crypto 1982, pp. 199–203. Plenum Press, New York (1983)
11. Fiore, D., Schröder, D.: Uniqueness is a different story: Impossibility of verifiable random functions from trapdoor permutations. Cryptology ePrint Archive, Report 2010/648 (2010), `http://eprint.iacr.org/2010/648`
12. Fischlin, M.: Round-optimal composable blind signatures in the common reference string model. In: Dwork, C. (ed.) CRYPTO 2006. LNCS, vol. 4117, pp. 60–77. Springer, Heidelberg (2006)
13. Fischlin, M., Schröder, D.: Security of blind signatures under aborts. In: Jarecki, S., Tsudik, G. (eds.) PKC 2009. LNCS, vol. 5443, pp. 297–316. Springer, Heidelberg (2009)

14. Fischlin, M., Schröder, D.: On the impossibility of three-move blind signature schemes. In: Gilbert, H. (ed.) EUROCRYPT 2010. LNCS, vol. 6110, pp. 197–215. Springer, Heidelberg (2010)
15. Hazay, C., Katz, J., Koo, C.-Y., Lindell, Y.: Concurrently-secure blind signatures without random oracles or setup assumptions. In: Vadhan, S.P. (ed.) TCC 2007. LNCS, vol. 4392, pp. 323–341. Springer, Heidelberg (2007)
16. Horvitz, O., Katz, J.: Universally-composable two-party computation in two rounds. In: Menezes, A. (ed.) CRYPTO 2007. LNCS, vol. 4622, pp. 111–129. Springer, Heidelberg (2007)
17. Impagliazzo, R., Rudich, S.: Limits on the provable consequences of one-way permutations. In: 21st Annual ACM Symposium on Theory of Computing (STOC), pp. 44–61. ACM Press, New York (1989)
18. Ishai, Y., Kushilevitz, E., Lindell, Y., Petrank, E.: Black-box constructions for secure computation. In: 38th Annual ACM Symposium on Theory of Computing (STOC), pp. 99–108. ACM Press, New York (2006)
19. Juels, A., Luby, M., Ostrovsky, R.: Security of blind digital signatures. In: Kaliski Jr., B.S. (ed.) CRYPTO 1997. LNCS, vol. 1294, pp. 150–164. Springer, Heidelberg (1997)
20. Kiayias, A., Zhou, H.-S.: Equivocal blind signatures and adaptive UC-security. In: Canetti, R. (ed.) TCC 2008. LNCS, vol. 4948, pp. 340–355. Springer, Heidelberg (2008)
21. Naor, M., Yung, M.: Universal one-way hash functions and their cryptographic applications. In: 21st Annual ACM Symposium on Theory of Computing (STOC), pp. 33–43. ACM Press, New York (1989)
22. Okamoto, T.: Efficient blind and partially blind signatures without random oracles. In: Halevi, S., Rabin, T. (eds.) TCC 2006. LNCS, vol. 3876, pp. 80–99. Springer, Heidelberg (2006)
23. Pointcheval, D., Stern, J.: Security arguments for digital signatures and blind signatures. Journal of Cryptology 13(3), 361–396 (2000)
24. Reingold, O., Trevisan, L., Vadhan, S.P.: Notions of reducibility between cryptographic primitives. In: Naor, M. (ed.) TCC 2004. LNCS, vol. 2951, pp. 1–20. Springer, Heidelberg (2004)
25. Rompel, J.: One-way functions are necessary and sufficient for secure signatures. In: 22nd Annual ACM Symposium on Theory of Computing (STOC), pp. 387–394. ACM Press, New York (1990)
26. Schröder, D.: On the Complexity of Blind Signatures. PhD thesis, Darmstadt University of Technology (2010)

Author Index

GPSR Compliance

The European Union's (EU) General Product Safety Regulation (GPSR) is a set of rules that requires consumer products to be safe and our obligations to ensure this.

If you have any concerns about our products, you can contact us on ProductSafety@springernature.com

In case Publisher is established outside the EU, the EU authorized representative is:

Springer Nature Customer Service Center GmbH
Europaplatz 3
69115 Heidelberg, Germany

Batch number: 10838023

Printed by Printforce, the Netherlands